Acid Deposition at High Elevation Sites

NATO ASI Series

Advanced Science Institutes Series

A Series presenting the results of activities sponsored by the NATO Science Committee, which aims at the dissemination of advanced scientific and technological knowledge, with a view to strengthening links between scientific communities.

The Series is published by an international board of publishers in conjunction with the NATO Scientific Affairs Division

A	Life Sciences	Plenum Publishing Corporation
B	Physics	London and New York
C	Mathematical and Physical Sciences	Kluwer Academic Publishers Dordrecht, Boston and London
D	Behavioural and Social Sciences	
E	Applied Sciences	
F	Computer and Systems Sciences	Springer-Verlag
G	Ecological Sciences	Berlin, Heidelberg, New York, London,
H	Cell Biology	Paris and Tokyo

Acid Deposition at High Elevation Sites

edited by

M. H. Unsworth

and

D. Fowler

Institute of Terrestrial Ecology,
Edinburgh, Scotland

Kluwer Academic Publishers

Dordrecht / Boston / London

Published in cooperation with NATO Scientific Affairs Division

Proceedings of the NATO Advanced Research Workshop on
Acid Deposition Processes at High Elevation Sites
Edinburgh, Scotland
8–13 September 1986

Library of Congress Cataloging in Publication Data

NATO Advanced Research Workshop on Acid Deposition Processes at High
  Elevation Sites (1986 : Edinburgh, Scotland)
    Acid deposition processes at high elevation sites : proceedings of
  the NATO Advanced Research Workshop on Acid Deposition Processes at
  High Elevation Sites, Edinburgh, Scotland, 8-13 September 1986 /
  edited by M.H. Unsworth and D. Fowler.
      p.   cm. -- (NATO ASI series. Series C, Mathematical and
  physical sciences ; 252)
    Includes index.
    ISBN 978-94-010-7883-2   ISBN 978-94-009-3079-7 (eBook)
    DOI 10.1007/978-94-009-3079-7
    1. Acid deposition--Congresses. 2. Acid precipitation
  (Meteorology)--Congresses. 3. Atmospheric chemistry--Congresses.
  I. Unsworth, M. H. II. Fowler, D., 1950-   . III. North Atlantic
  Treaty Organization. IV. Title. V. Series: NATO ASI series.
  Series C, Mathematical and physical sciences ; no. 252.
  TD196.A25N38 1986
  363.7'386--dc19   88-25219
                                                           CIP

ISBN 978-94-010-7883-2

Published by Kluwer Academic Publishers,
P.O. Box 17, 3300 AA Dordrecht, The Netherlands.

Kluwer Academic Publishers incorporates the publishing programmes of
D. Reidel, Martinus Nijhoff, Dr W. Junk, and MTP Press.

Sold and distributed in the U.S.A. and Canada
by Kluwer Academic Publishers,
101 Philip Drive, Norwell, MA 02061, U.S.A.

In all other countries, sold and distributed
by Kluwer Academic Publishers Group,
P.O. Box 322, 3300 AH Dordrecht, The Netherlands.

This book contains the proceedings of a NATO Advanced Research Workshop held within the programme of activities of the NATO Special Programme on Global Transport Mechanisms in the Geo-Sciences running from 1983 to 1988 as part of the activities of the NATO Science Committee.

Other books previously published as a result of the activities of the Special Programme are

BUAT-MENARD, P. (Ed.) – *The Role of Air-Sea Exchange in Geochemical Cycling* (C185) 1986

CAZENAVE, A. (Ed.) – *Earth Rotation: Solved and Unsolved Problems* (C187) 1986

WILLEBRAND, J. and ANDERSON, D. L. T. (Eds.) – *Large-Scale Transport Processes in Oceans and Atmosphere* (C190) 1986

NICOLIS, C. and NICOLIS, G. (Eds.) – *Irreversible Phenomena and Dynamical Systems Analysis in Geosciences* (C192) 1986

PARSONS, I. (Ed.) – *Origins of Igneous Layering* (C196) 1987

LOPER, E. (Ed.) – *Structure and Dynamics of Partially Solidified Systems* (E125) 1987

VAUGHAN, R. A. (Ed.) – *Remote Sensing Applications in Meteorology and Climatology* (C201) 1987

BERGER, W. H. and LABEYRIE, L. D. (Eds.) – *Abrupt Climatic Change – Evidence and Implications* (C216) 1987

VISCONTI, G. and GARCIA, R. (Eds.) – *Transport Processes in the Middle Atmosphere* (C213) 1987

SIMMERS, I. (Ed.) – *Estimation of Natural Recharge of Groundwater* (C222) 1987

HELGESON, H. C. (Ed.) – *Chemical Transport in Metasomatic Processes* (C218) 1987

CUSTODIO, E., GURGUI, A. and LOBO FERREIRA, J. P. (Eds.) – *Groundwater Flow and Quality Modelling* (C224) 1987

ISAKSEN, I.S.A. (Ed.) – *Tropospheric Ozone* (C227) 1988

SCHLESINGER, M. E. (Ed.) – *Physically-Based Modelling and Simulation of Climate and Climatic Change* 2 vols. (C243) 1988

TABLE OF CONTENTS

Part III

CLOUD AND RAIN CHEMISTRY: MONITORING STUDIES

Part IV

DRY, WET AND OCCULT DEPOSITION

Part V

PROCESSES AT THE SURFACE

There is no shortage of general books on the subject of acid rain, or of symposium proceedings reviewing work ranging from atmospheric chemistry and deposition processes to freshwater acidification and effects on vegetation. In contrast, the collection of papers from this Workshop is focussed on a much smaller subject, the processes of acid deposition at high altitude sites.

Interest in deposition at high elevation sites comes largely from observed vertical gradients in the degree of forest damage at sites in the Federal Republic of Germany and the eastern United States. These gradients show that damage to Norway spruce and fir increases with altitude at sites in Bavaria and the Black Forest, and that Red spruce are declining at high elevation sites in the Appalachian Mountains. With the large scale of scientific interest in forest decline, many research groups, during the last five years, have been examining atmospheric chemistry, deposition processes, and effects on vegetation and soils at upland sites. In particular there have been many recent studies of cloud and precipitation chemistry, which show much larger concentrations of all ions in cloud water than in rain or snow. These studies have also shown that processes of wet and dry deposition and also the chemistry of the air at hill tops are modified strongly by orographic effects. The special conditions at some field sites have allowed reaction rates for important oxidation processes to be determined in the field: some of this work is described in this volume. The modification to processes of deposition and therefore to patterns of wet and dry deposition on hills is also considered in some detail.

This meeting was necessarily multidisciplinary, with a mixture of physicists, chemists and biologists, but, unlike many acid deposition meetings, the links between the groups were strong as a result of the common interest in the high elevation sites and their special properties.

Modelling studies offer an effective method for extrapolating from detailed mechanistic studies of chemistry, physics and biological processes to large scale budgets for reactive trace substances in the atmosphere. Several modelling approaches are therefore described in this volume, although these have been restricted to local (less than 20 km) and regional (less than 500 km) scales. The biological effects of deposited acidity and related ions were not included at this meeting, though important biologically relevant contributions to the meeting were made on surface effects and the behaviour of droplets on leaf surfaces, and on rates of accumulation of particles on vegetation.

Despite the considerable interest during the last five years in this subject, there is no clear link between deposition of any single pollutant and the widespread decline in health of forests in central Europe and at the high elevation sites in North America. The work described in this volume shows important directions in which future research should be focussed, in particular concerning the properties of the very polluted cloud water experienced at high elevation sites.

Much of the work reported here is based on recent field measurements.

Many individuals helped with the meeting and with refereeing and modifying the contributions to this volume. In particular we would like to thank Carol Morris who played a major part in organising the meeting and who has played a most important role in the preparation of this publication. Other members of ITE Edinburgh who helped with this meeting included Ian Leith, Alan Crossley and Neil Cape.

The meeting was sponsored by the NATO Scientific Affairs Division, the UK Department of the Environment, the British Petroleum Company plc, and the Central Electricity Generating Board. Their support in making possible this timely meeting is gratefully acknowledged.

M.H. Unsworth and D. Fowler

* Present address: Department of Physiology and Environmental Science
University of Nottingham
School of Agriculture
Sutton Bonington
Loughborough LE12 5RD
UK.

PARTICIPANTS IN THE NATO ADVANCED RESEARCH WORKSHOP: Acid Deposition
Processes at High Elevation Sites
Edinburgh, Scotland - 8-13 September 1986

M Baer
Institute fur Meteorologie
 und Klimaforschung
Kernforschungszentrum Karlsruhe
Universitat Karlsruhe
Postfach 3640
D-7500 Karlsruhe 1
West Germany

M Bailey
An Foras Forbartha
St Martins House
Waterloo Road
Dublin 4
Republic of Ireland

F C Barrett
Warren Spring Laboratory
Gunnels Wood Road
Stevenage
Herts SG1 2BX
UK

M Bredemeier
Forschungszentrum Waldökosysteme/
 Waldsterben
University of Goettingen
Büsgenweg 2
D-3400 Goettingen
West Germany

J N Cape
Institute of Terrestrial Ecology
Bush Estate
Penicuik
Midlothian
Scotland
UK

A C Chamberlain
1 Fifth Road
Newbury
RG14 6DN
UK

A S Chandler
University of Manchester
Institute of Science
 and Technology
Manchester M60 1QD
UK

R J Charlson
Department of Atmospheric
 Sciences
University of Washington
Seattle
Washington 98195
USA

N Chaumerliac
Laboratoire Associé de
 Meteorologie Physique
Université de Clermont II
63170 Aubière
France

T W Choularton
Department of Physics
UMIST
Sackville Street
Manchester M60 1QD
UK

P Clark
Central Electicity Research
 Laboratories
Leatherhead
Surrey
UK

A Crossley
Institute of Terrestrial Ecology
Bush Estate
Penicuik
Midlothian
Scotland
UK

P H Daum
Environmental Chemistry Division
Department of Applied Science
Brookhaven National Laboratory
Upton, New York 11973
USA

T D Davies
School of Environmental Sciences
University of East Anglia
Norwich NR4 7TJ
UK

V Delmas
Laboratoire de Glaciologie
 et Geophysique de
 l'Environment
B P 96 38402
St Martin d'Hères Cedex
France

R G Derwent
Environmental Sciences Division
AERE Harwell
Oxon OX11 ORA
UK

G J Dollard
Environmental and Medical
 Sciences Division
AERE Harwell
Didcot
Oxfordshire
OX11 ORA

A E J Eggleton
Environmental Sciences Division
AERE Harwell
Oxon OX11 ORA
UK

D Fowler
Institute of Terrestrial Ecology
Bush Estate
Penicuik
Midlothian
Scotland
UK

S Fuzzi
Istituto FISBAT – C.N.R.
Via de'Castagnoli 1
40126 Bologna
Italy

M W Gallagher
The University of Manchester
Institute of Science
 and Technology
P O Box 88
Manchester M60 1QG
UK

M J Gay
The University of Manchester
Institute of Science
 and Technology
P O Box 88
Manchester M60 1QG
UK

G P Gervat
Central Electricity Research
 Laboratories
Kelvin Avenue
Leatherhead
Surrey
UK

L Granat
Department of Meteorology
University of Stockholm
Arrhenius Laboratory
S-106 91 Stockholm
Sweden

J Godt
Gesamthochschule Kassel
University of Hessen
Dept Landscapeecology/
  Soil Science
Henschelstr 2
P O Box 101380 Hessen
West Germany

G Gravenhorst
Alfred Wegewer Institute for
  Polar and Marine Research
Handelhasen 12
2850 Bremmerhaven
West Germany

W C Graustein
Department of Geology
  and Geophysics
Yale University
Kline Geology Laboratory
New Haven
Connecticut 06511
USA

B B Hicks
NOAA/ARL Atmos Turbulence &
  Diffusion Division
P O Box 2456
Oak Ridge
TN 37831
USA

T A Hill
University of Manchester
Institute of Science
  and Technology
P O Box 88
Manchester
UK

A M Hough
Modelling and Assessment Group
Environmental and Medical
  Sciences Division
Harwell Laboratory
Didcot, Oxfordshire
UK

D J Jacob
Dept of Earth and Planetary
  Sciences
Division of Applied Sciences
Harvard University
Cambridge
Massachusetts 02138
USA

A Jones
The University of Manchester
Institute of Science
  and Technology
P O Box 88
Manchester M60 1QG
UK

B M R Jones
Environmental Sciences Division
AERE Harwell
Oxon OX11 ORA
UK

J Kadlecek
ASRC-SUNYA, ES-324
1400 Washington Avenue
Albany
New York 12222
USA

A S Kallend
Central Electricity Research
  Laboratories
Kelvin Avenue
Leatherhead
Surrey
UK

Bo Richter-Larsen
Research Station RISO
Department of Energetics
(Afd. f. Energitenknik)
Postbox 49
4000 Roskilde
Denmark

A Liberti
Consiglio Nazionale
 delle Richerche
Istituto Sull'Inquinamenta
Atmosferico
Via Salria Km. 29,300 - CP 10
Rome
Italy

S E Lindberg
Environmental Sciences Division
Oak Ridge National Laboratory
Oak Ridge
Tennessee 37831
USA

G M Lovett
Institute of Ecosystem Studies
The New York Botanical Garden
Mary Flagler Cary Arboretum
Box AB
Millbrook NY 12545
USA

R K A M Mallant
Netherlands Energy
 Research Foundation ECN
P O Box 1
1755 ZG Petten
The Netherlands

A R Marsh
Central Electricity Reserach
 Laboratories
Leatherhead
Surrey
UK

F X Meixner
Fraunhofer-Institute for
 Atmospheric Environmental
 Research
Kreuzeckbahnstrasse 19
D-8100 Garmisch-Partenkirchen
West Germany

A P Morse
The University of Manchester
Institute of Science
 and Technology
P O Box 88
Manchester M60 1QG
UK

H Mueller
Fraunhofer-Institute for
 Atmospheric Environmental
 Research
Kreuzeckbahnstrasse 19
D-8100 Garmisch-Partenkirchen
West Germany

K Munzert
Fraunhofer-Institute for
 Atmospheric Environmental
 Research
Kreuzeckbahnstrasse 19
D-8100 Garmisch-Partenkirchen
West Germany

M G Nestlen
Fraunhofer-Institute for
 Atmospheric Environmental
 Research
Kreuzeckbahnstr. 19
D-8100 Garmisch-Partenkirchen
West Germany

J A Ogren
Department of Meteorology
University of Stockholm
S-106 91 Stockholm
Sweden

S A Penkett
University of East Anglia
Norwich NR4 7TJ
UK

H Puxbaum
Institute fur Analytische Chemie
Technische Universitat Wien
Getreidemarkt 9/151
A-1060 Wien
Austria

Institute of Terrestrial Ecology
Bangor Research Station
Penhros Road
Bangor
Wales
UK

U Teichmann Lehrstuhl für
 Bioklimatologie und Angewandte
 Meteorologie der Universität
 München
Amalienstrasse 52/111
D-8000 München 40
West Germany

M Radojevic
The University of Manchester
Institute of Science
 and Technology
P O Box 88
Manchester M60 1QG
UK

B J Tyler
The University of Manchester
Institute of Science
 and Technology
P O Box 88
Manchester M60 1QG
UK

F Ronseaux
Laboratoire de Glaciologie
P O Box 68 F-38402
Saint Martin d'Heres
France

C J Walcek
National Center for Atmospheric
 Research
P O Box 3000
Boulder
Colorado 80307-3000
USA

R S Schemenauer
Atmospheric Environment Service
Downsview
Ontario M3H 5T4
Canada

K C Weathers
Institute of Ecosystem Studies
The New York Botanical Garden
Mary Flagler Cary Arboretum
Box AB
Millbrook
New York 12545
USA

G Schmitt
Institute of Meteorology
 and Geophysics
University of Frankfurt
Feldbergstr. 47
6000 Frankfurt.Main
West Germany

K J Weston
Department of Meteorology
University of Edinburgh
King's Buildings
Mayfield Road
Edinburgh
Scotland
UK

P H Schuepp
Dept of Renewable Resources
Macdonald College of
 McGill University
Ste. Anne-de-Bellevue
Quebec   H9X 1C0
Canada

ATMOSPHERIC CHEMISTRY AT ELEVATED SITES - A DISCUSSION OF THE
PROCESSES INVOLVED AND THE LIMITS OF CURRENT UNDERSTANDING

A.M. Hough
Environmental and Medical Sciences Division
Harwell Laboratory
Didcot
Oxfordshire
England

ABSTRACT.   A discussion of the chemical processes which lead to the
formation of atmospheric acidity is presented.   These include gas and
aqueous phase chemistry, heterogeneous equilibria, and scavenging by
aerosol particles and cloud droplets.   Sulphur oxides, nitrogen oxides
and hydrocarbons all play important roles in these processes, and the
ratio between the concentrations of nitrogen oxides and non-methane
hydrocarbons is important in determining the photochemical behaviour
of an airmass, and its oxidising potential.   The mechanisms for dry,
wet and occult deposition are presented, together with a discussion of
the contribution made to dry deposition rates by the nature of the
surface and how these rates vary over the diurnal cycle.   The
heterogeneous processes which contribute to the wet deposition of
acidity, such as cloud entrainment, droplet scavenging, rainout and
washout, are less well understood than those for dry deposition.   For
long-term (yearly) deposition fields, then model results and
observations can be in good agreement, but on a daily basis, the
inhomogeneities in both the atmospheric conditions and the terrain,
which are associated with many elevated sites, render most large-scale
models inadequate and dictate that many measurements are
non-representative.   Progress in our understanding of the chemistry
of elevated sites will be made by concerted application of models and
experimental observations aimed at improving our knowledge of aqueous
phase atmospheric chemistry, scavenging processes, entrainment into
clouds and the deposition of small particles and droplets.

1.   INTRODUCTION

In recent years, there has been widespread and increasing concern into
the possible ecological impact of atmospheric acidity and
photochemical oxidants.   In Europe this concern has been focussed
especially on the problems of lake acidification in Scandinavia and

1

M. H. Unsworth and D. Fowler (eds.), Acid Deposition at High Elevation Sites, 1–47.
© 1988 by the UKAEA.

2

forest dieback in the Federal Republic of Germany. Elevated sites
are thought to be particularly at risk for a number of reasons
including poorly buffered soils, the high incidence of hillcloud and
topographical features which expose vegetation to a high deposition of
windborne pollutants. Various aspects of the problem have been
reviewed by Bielke and Elshout, 1982; Guderian, 1983; United Kingdom
Review Group on Acid Rain, 1983; Calvert, 1985; Legge and Krupa,
1986.

In Western Europe some fifty percent of the land area has an
altitude of over one thousand feet above sea level and over five
percent is higher than three thousand feet. In North America the

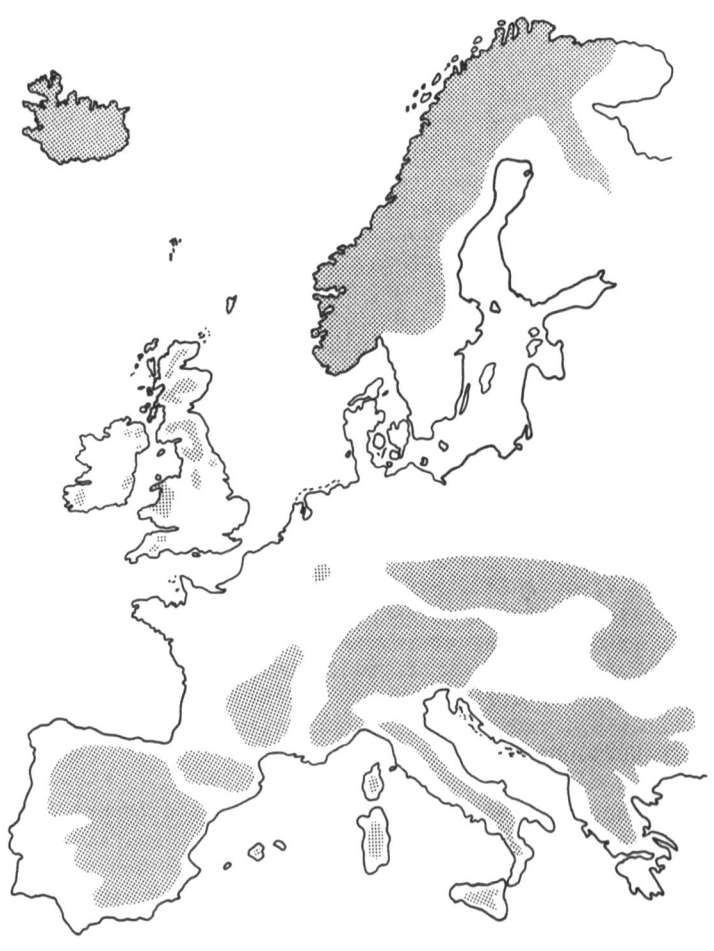

Figure 1.   Areas of western-Europe containing elevated sites.

respective figures are some sixty percent and twenty-five percent respectively. However, height alone is inadequate to define an elevated site in the current context. Some sites in Britain which would genuinely fall into this category are at heights where this would not be the case in, for instance, North America. A suitable definition of an elevated site, with respect to the chemistry which occurs, might be one at which occluding cloud exists for a significant fraction of the time, or which is elevated above the surrounding area and has a terrain which is markedly inhomogeneous. The areas in Western Europe where such sites occur are illustrated in Figure 1.

The chemistry which occurs at elevated sites will depend on the chemical composition of the airmasses arriving there, which are, in turn, determined by the trajectories to these sites. The meteorology which governs such trajectories is beyond the scope of the present review. The chemical compounds of relevance to acid deposition are listed in Table 1.

The oxidation of sulphur dioxide and nitrogen dioxide to give sulphuric and nitric acids respectively, is controlled by the wide range of photochemical processes which occur in the atmosphere. The organic acids are produced as by-products of the photochemical oxidation of hydrocarbons (including natural hydrocarbons, such as isoprene and the terpenes), and are also emitted directly by natural sources (vegetation and soils). The deposition of acidic species cannot therefore be divorced from considerations of photochemical oxidants, especially as approximately one third of deposited acidity

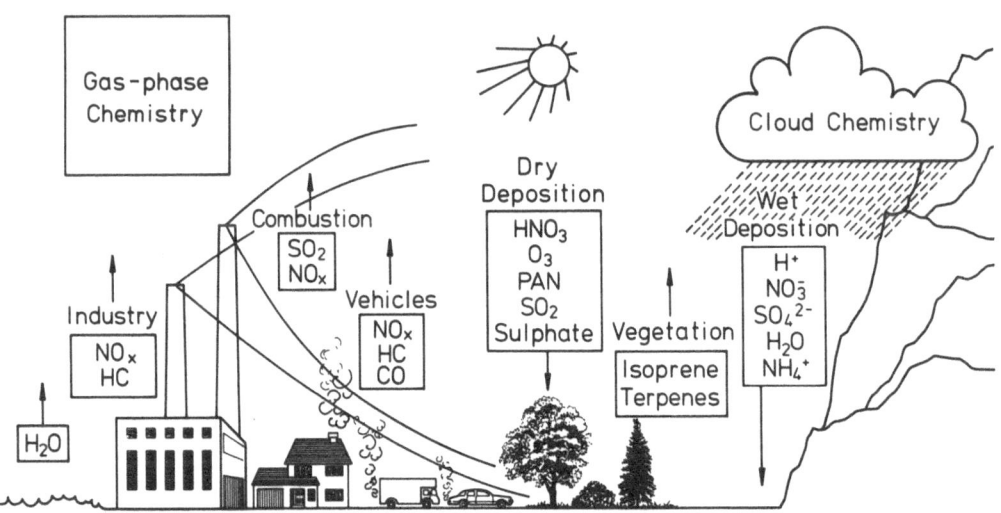

Figure 2. The chemical processes which contribute to acid deposition.

TABLE I   Chemical species of importance to atmospheric acidity

Species	Source
Acidic species	
SO_2	emitted from the stationary combustion of fossil fuels.
H_2SO_4	produced by atmospheric oxidation of SO_2.
NO_2, NO	emitted from the combustion of fossil fuels, both in stationary plant and in motor vehicles. NO is oxidised to NO_2 in the atmosphere.
HNO_3	produced by the atmospheric oxidation of NO_2.
HCl	emitted form the stationary combustion of fossil fuels, and the incineration of PVC wastes.
RCOOH	produced by the photochemical oxidation of hydrocarbons, and emitted directly by natural sources.
Basic species	
NH_3	emitted from biological processes.
$CO_3{}^{2-}$	airborne dust collected from the soils of limestone regions

in Europe is due to nitrogenous species (United Kingdom Review Group
on Acid Rain, 1983; Cape et al., 1984).   The overall pattern of the
atmospheric chemistry which is involved is shown in Figure 2.

When considering the formation of atmospheric acidity, account
should also be taken of the potential which exists for its
neutralisation.   The principal agent for this process is ammonia,
which is emitted from biological processes, notably the degradation of
animal excrement.   Carbonate dusts, collected by the wind, can also
contribute to the neutralisation of atmospheric acidity (Loye-Pilot,
1986).

In the past decade, considerable effort has been expended into
quantifying the sources, lifecycles and fates of the various compounds
involved in acid deposition.   Even so, considerable uncertainties
still exist in emission inventories, particularly those for the
hydrocarbons, in which case the compilation also has to encompass the
problem of speciation as different hydrocarbons have different
degradation mechanisms and produce different amounts of photochemical
oxidants.   The lifecycles of sulphurous and nitrogenous species have
been discussed, for instance, by Moller, 1984; Derwent, 1982;
Derwent, 1986a and within Europe, emission inventories have been
detailed in, for instance, references (Amble, 1981; Sembe, and Amble,
1981).   Similar consideration has also been given to ozone
(Altshuller, 1986; Logan, 1985) and to ammonia (Derwent, 1986b).   It
is important to remember that our estimates and calculations of the
sources and sinks for the elements involved in the atmospheric
processes should be in agreement, after allowing for any long term
changes in atmospheric composition.   Studies of the lifecycle of a
particular component of acidity may help us to understand the relative
importance of the various production and removal processes which are
in operation, as well as ensuring that our calculations are, at least,
internally consistent.

The present account is not intended to be a comprehensive review
of either atmospheric chemistry or of deposition processes.   Such
accounts may be found elsewhere (Finlayson-Pitts and Pitts, 1986;
Seinfeld, 1986).   Its aim is to present a broad picture of the wide
range of chemical processes which occur in the atmosphere, together
with their uncertainties and the way in which they interact with each
other to produce the deposition of acidity at elevated sites.   In
this way the emphasis is placed on the interdependence of all these
various processes and the fact that they should all be considered if
we are to gain an increased knowledge and understanding in this field.

2.   FUNDAMENTALS AND UNCERTAINTIES IN THE CHEMISTRY

Atmospheric chemistry occurs in three distinct phases; gas phase,
liquid (aqueous) phase and in aerosol particles which may be a
combination of solid and concentrated aqueous phases.   In addition
there are heterogeneous equilibria which result in the mass transfer
of chemical species between the gas phase and the liquid and aerosol

6

phases.　These processes have been reviewed in detail in a number of places, e.g. (Finlayson-Pitts and Pitts, 1986;　Seinfeld, 1986; Peterson and Seinfeld, 1980), and their inter-relation is illustarted by Figure 3.　Because each of these systems has its own pattern of chemical behaviour, it is convenient to consider each of them separately.

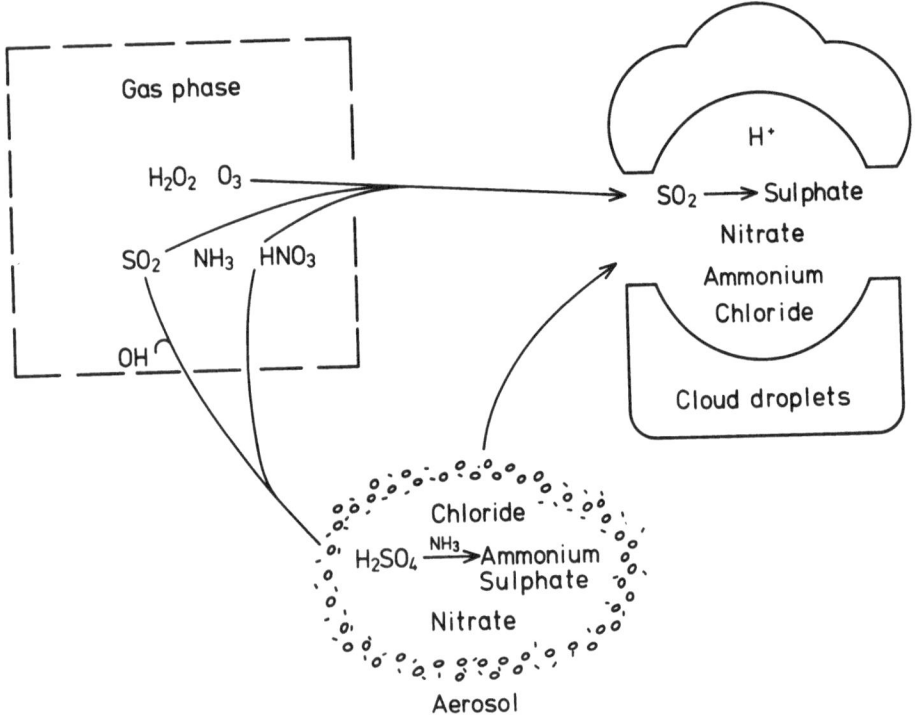

Figure 3.　The interception between the gas phase, aerosol particles and cloud droplets.

2.1　Gas-phase chemistry

The large amount of effort extended into the study of the homogeneous gas phase chemistry of the atmosphere has resulted in this being an area which is now relatively well understood, although there are still a number of areas of uncertainty as discussed below.　Laboratory measurements have produced rate constants for most of the important reactions which occur, and the use of computational models has enabled the determination of the relative importance of such reactions. However, the validation of such studies by comparison with observed concentrations is hindered by uncertainties in the emissions and the atmospheric transport and by a lack of suitable detailed measurements.　The rates of the various reactions have been critically reviewed and tabulated in a number of places (Atkinson and Lloyd, 1984; Baulch, 1980; Demore, 1985; Kerr and Calvert, 1984).

At the heart of the gas-phase chemistry is the pseudo-steady state which exists between nitric oxide (NO), nitrogen dioxide (NO_2) and ozone (O_3).

$$NO_2 \xrightarrow{h\nu} NO + O(^3P) \qquad (R1)$$

$$O(^3P) + O_2 \rightarrow O_3 \qquad (R2)$$

$$O_3 + NO \rightarrow NO_2 + O_2 \qquad (R3)$$

For any given set of conditions (temperature and light intensity) these three compounds would exist in equilibrium if it were not for the presence of competing processes. However, ozone is also photolysed, a reaction which allows the formation of hydroxyl radicals. The hydroxyl radicals perform the gas phase oxidation of sulphur dioxide and nitrogen dioxide, as well as initiating the degradation of hydrocarbons, (all of which are represented here by RH), and carbon monoxide.

$$O_3 \xrightarrow{h\nu} O(^1D) + O_2 \qquad (R4)$$

$$O(^1D) + H_2O \rightarrow 2\ OH \qquad (R5)$$

$$NO_2 + OH \rightarrow HNO_3 \qquad (R6)$$

$$SO_2 + OH \rightarrow HSO_3 \qquad (R7a)$$

$$HSO_3 \xrightarrow{O_2,\ H_2O} H_2SO_4 + HO_2 \qquad (R7b)$$

$$RH + OH \longrightarrow R\cdot + H_2O \qquad (R8a)$$

$$R\cdot + O_2 \longrightarrow RO_2 \qquad (R8b)$$

$$CO + OH \longrightarrow CO_2 + HO_2 \qquad (R9)$$

Reactions (R4)-(R9) appear to remove ozone from the system. However in the presence of nitrogen oxides, the subsequent reactions of the RO_2 radicals have the effect of producing ozone and other photochemical oxidants such as peroxyacyl nitrate (PAN) and hydrogen peroxide. Each hydrocarbon has its own individual reaction mechanism, and the following scheme can only give an outline of the important processes.

$$RO_2 + NO \rightarrow RO + NO_2 \qquad (R10)$$

$$RO + O_2 \rightarrow HCHO, RCHO, RR'CO; + HO_2 \qquad (R11)$$

$$RCHO + OH \xrightarrow{O_2} RCOO_2 + HO_2 \qquad (R12)$$

$$RCOO_2 + NO_2 \rightleftharpoons PANs \qquad (R13)$$

$$RCOO_2 + NO \rightarrow R. + NO_2 + CO_2 \qquad (R14)$$

$$HCHO + OH \xrightarrow{O_2} H_2O + HO_2 + CO \qquad (R15)$$

$$HCHO \xrightarrow{hv} H_2 + CO \qquad (R16a)$$

$$HCHO \xrightarrow{hv, O_2} CO + 2HO_2 \qquad (R16b)$$

The degradation has the overall effect of oxidising NO to NO_2, and producing the hydroperoxyl radical (HO_2). The NO_2 will react via reaction (R1) to produce more ozone, and the ketones (RR'CO) also react to produce further NO_2 and HO_2. The hydroperoxyl radical also has several reactions of importance.

$$HO_2 + NO \rightarrow NO_2 + OH \qquad (R17)$$

$$HO_2 + HO_2 \rightarrow H_2O_2 + O_2 \qquad (R18)$$

$$HO_2 + RO_2 \rightarrow ROOH + O_2 \qquad (R19)$$

$$HO_2 + O_3 \rightarrow 2O_2 + OH \qquad (R20)$$

Reaction (R17) is the dominant reaction of HO_2 in the presence of NO and has the overall effect of producing more ozone and recycling hydroxyl radicals to degrade more hydrocarbons. It also effectively precludes the formation of significant amounts of hydrogen peroxide via equation (R18). Ozone and hydrogen peroxide are both important photochemical oxidants in the liquid phase, which emphasises the importance of the gas phase photochemical mechanism. Organic acids, such as formic acid and acetic acid, can be produced by the degradation of alkenes and natural product hydrocarbons following their reaction with ozone. The overall ratio between non methane hydrocarbons (NHMC) and nitrogen oxides is very important in determining the chemical behaviour of an airmass. With low concentrations of nitrogen oxides, there is very little formation of ozone, and reactions (R18)-(R20) dominate the chemistry of HO_2 and RO_2, producing peroxides, removing ozone from the system, and slowing the degradation of hydrocarbons. If NO and NO_2 are present in excess, then they act as sinks for the hydroxyl and hydroperoxy radicals, again limiting the hydrocarbon reactions and the formation of ozone and hydrogen peroxide. For the production of photochemical oxidants in large quantities, the concentrations of nitrogen oxides need to lie between these two extremes.

The above mechanisms are dependant on light for the formation of acidic species (apart from the ozone-alkene reactions). However, the formation of nitric acid can still proceed in the absence of light.

$$NO_2 + O_3 \rightarrow NO_3 + O_2 \qquad (R21)$$

$$NO_3 + NO_2 \rightleftarrows N_2O_5 \tag{R22}$$

$$N_2O_5 + H_2O \rightarrow 2HNO_3 \tag{R23}$$

This pathway is unimportant in daylight because of the rapid photolysis of NO_3, but may be an important night-time source of nitric acid in an airmass with high concentrations of NO_2 and sufficient ozone remaining from the previous day, e.g. Atkinson, 1986. Modelling studies have been used to condense our knowledge of these various processes and examine their interaction with each other, as well as the effects of reducing man-made emissions. Some typical results are shown by Hov, 1983; Dodge, 1984; Leone and Seinfeld, 1984; Hough and Derwent, 1987

It should be clear from the above, that the presence of hydrocarbons and light is essential for the atmospheric oxidation of nitrogen dioxide and sulphur dioxide to occur (sulphur dioxide also requires the presence of nitrogen oxides). Although this will usually be the case for low level emissions, which are dominated by vehicle exhausts, it will not be the case for power station plumes emitted into a stable atmosphere. Because mixing with the surrounding air may be very slow, such plumes will be deficient in hydrocarbons, so that oxidation proceeds very slowly indeed. Such plumes can remain intact for over one thousand kilometres (Fisher and Callander, 1984).

The oxidation of sulphur dioxide to sulphuric acid is a relatively slow process, and although the rate is controlled by the concentration of OH, which varies both diurnally and with the time of year, an average of about 1 percent per hour is thought to be typical, and is used in many statistical models (e.g. Smith, 1981; Eliassen and Saltbones, 1983). The oxidation of nitrogen dioxide is rather more rapid, the corresponding figure being 10 percent per hour. Sulphur dioxide therefore has a much greater potential for long range transport to regions remote from man made emissions than does nitrogen dioxide. However, this does not preclude the deposition of nitrogeneous species at such sites, due in part to the formation of atmospheric aerosol. In the absence of rain systems, aerosol particles have great potential for long range transport, due to their very slow deposition to the ground. As discussed in section 3, compounds such as ozone, nitrogen dioxide, sulphur dioxide and especially nitric acid suffer a much more rapid depletion through this mechanism. However, this does not preclude the long range transport of such species, a fact which is particularly important in the current context, as most elevated sites are relatively remote from emission areas.

As mentioned above, there are still uncertainties involved in the gas-phase chemistry, these being in reaction rates, their products and in the significance or otherwise of various processes. The reactions of ozone with alkenes constitute one set of reactions for which the products and their fates are uncertain. Other sets concern the reactions of nitrous acid and the formation of organic peroxides and adducts with NO and NO_2. In addition, photolysis rates are strongly

influenced by season, time, latitude and altitude, due to the
atmospheric adsorption of sunlight. The variation of some photolysis
rates with an altitude and solar zenith angle have been tabulated in
Demerjian (1980). The inherant uncertainties in all reaction rate
constants also contributes to our lack of understanding.

From the aspect of measurements, in-situ concentrations of
hydroxyl and hydroperoxyl radicals have the potential to make a
significant contribution to our understanding of the atmospheric
chemistry at specific locations and times. Some typical
concentrations of various compounds are given in Table II, and a range
of lifetimes are tabulated in Table III.

2.2 Atmospheric Aerosol

Aerosol particles are formed from a variety of sources, including
combustion, evaporation of water from sea spray, and photochemical
pollution. In the current context our attention is directed towards
the third of these categories, which is the cause of the reduction in
visibility which occurs during stable anticyclonic conditions.

Although molecular gas-phase sulphuric acid is formed by the
gas-phase oxidation of sulphur dioxide, it does not remain in this
form for long. Thermodynamically, sulphuric acid molecules have a
tendency to combine with water molecules and this strongly exoergic
process produces small particles of concentrated sulphuric acid. The
gas-phase sulphuric acid molecules may also be adsorbed onto the
aerosols formed by combustion products, or into sea-salt aerosol. In
the latter case, hydrogen chloride may be released to the atmosphere.
The structure, composition and particle sizes of such aerosol
particles have been discussed and measured in a number of places, see
for instance Finlayson-Pitts and Pitts, 1986; Whitby and Sverdrup,
1980; Tanner, 1982; Harrison and Pio, 1983a; Jacob, 1986. In general,
however, particles resulting from photochemical pollution have
diameteres of the order of 1 μm. Aerosols may play an important role
in atmospheric chemistry by providing a surface to catalyse gas-phase
processes, for instance, the conversion of N_2O_5 to nitric acid.

Various other gaseous compounds will also be adsorbed by the
aerosol particles, at a rate governed by the collision frequency and
the chance that the molecule will be adsorbed into the particle after
they collide (the accommodation coefficient). The theory of aerosol
scavenging has been considered in detail by Fuchs and Sutugin (1982)
who give equation (1) for the flux F_i of compound i to a single
droplet:

$$F_i = \frac{4}{3} \pi 1 \left[\frac{8kT}{m_i \pi} \right]^{1/2} \cdot \left[\frac{a}{1 + (0.7 + (4(1-\alpha_i/3\alpha_i))(1/a)} \right] n_i \quad (1)$$

TABLE II. Typical non-urban concentrations of some gas-phase species.

Species	Concentration		
SO_2	2-5 ppb	'clean' British air mean England	(Fisher and Callander, 1984)
	7 ppb		(Dollard and Jones, this volume)
	50-200 ppb	British plumes over north sea	(Fisher and Callander, 1984)
	50-100 ppb	Maximum observed, Pennines, England	(Dollard and Jones, this volume)
SO_4^{2-} aerosol	2 ppb	annual mean British value	(Harrison and Pio, 1983; Willison et al., 1985)
	12.5 ppb	daily British maximum	(Harrison and Pio, 1983)
	1-14 ppb	British plumes over north sea	(Fisher and Callander, 1984)
	1.2-5.8 ppb	California	(Waldman et al., 1985)
NO_x	42 ppb)	Maximum and minimum monthly means in southern England	(Sandalls and Leonard, 1986)
	5 ppb)		
	2 ppb	'clean' British air	(Fisher and Callander, 1984)
NO_3^- aerosol	2 ppb	annual mean British rural value	(Harrison and Pio, 1983)
		British plumes over north sea	(Fisher and Callander, 1984)
	1.6-7 ppb	California	(Waldmann et al., 1985)
O_3	24 ppb	1985 annual mean, southern England	(Sandalls and Chadwick, 1985)
	114 ppb	1984 maximum, southern England	(Sandalls and Gibson, 1986)
NH_3	3.6 ppb	annual mean British rural value	(Harrison and Pio, 1983)
H_2O_2	0.5 ppb	typical British summer value	(Dollard and Jones, this volume)
OH	3×10^6 molecules cm^{-3}	midday, midsummer, Origen	(Hard et al., 1986)

TABLE III Lifetimes in the gas-phase

Species	Lifetime
A) From reaction with OH (2×10^6 molecules cm^{-3})	
SO_2	150 hours
NO_2	10 hours
CO	26 days
CH_4	2 years
n-Butane	54 hours
But-1-ene	4 hours
Toluene	21 hours
B) From dry deposition to dry grass in daylight, from a mixing height of 1000 metres	
SO_2	35 hours
NO_2	45 hours
HNO_3	10 hours
O_3	35 hours
PAN	90 hours

where l is the mean free path in the gas phase, m_i the molecular mass, k the Botlzmann constant, a the droplet radius, α_i the accommodation coefficient and n_i the gas-phase concentration of the species concerned. Although compounds such as nitric acid have accommodation coefficients with values close to unity, the values for radicals such as OH and HO_2 are likely to be rather lower than this (Chameides and Davis, 1982).

Although the bulk solubility of a gas in aqueous solution can be described by Henry's Law equilibria as discussed below in section 2.3, the solubility of gases in the water associated with aqueous aerosol particles will be altered from the bulk laboratory values by the Kelvin effect, which describes the change in vapour pressure over a curved surface relative to a plane surface. However, this effect is small for most droplets of interest, being 0.2% and 2% for droplets of diameter 1 μm and 0.1 μm respectively although it increases to 19% for droplets with a radius of 0.01 μm (Peterson and Seinfeld, 1980).

The solubilities (and the ionic equilibria) are also affected by the high concentrations of solute present in the aqueous phase. In any ionic solution which is not infinitely dilute, the ionic equilibria (of an ion, i) are written in terms of activities, a_i, rather than the concentrations c_i, where the two are related by the activity coefficient γ_i

$$a_i = \gamma_i \, c_i \qquad (2)$$

The ionic strength of the solution, I, is defined by equation (3), where z_i is the charge on ion i.

$$I = \tfrac{1}{2} \; \sum_i z^2_i \, c_i \qquad (3)$$

A number of theories have been devised to link the value of γ_i to the ionic strength of the solution, but these are usually valid only for solutions with ionic strengths less than $10^{-2} - 10^{-1}$, whereas the solutions in aerosol particles will often be somewhat more concentrated than this. The expressions resulting from these theories have been considered by Hamer (1968). The heterogeneous processes involved in the scavenging have been reviewed by Peterson and Seinfeld (1980) and the thermodynamics of multicomponent electrolytic aerosols by Saxena and Peterson (1981). The condensation of water has been reviewed by Mozurkewich (1986).

The principal species scavenged by aerosol particles will be those which are highly soluble, such as nitric acid and ammonia, although free radicals such as OH and HO_2 and a range of organic compounds will also be involved. Gas phase N_2O_5 is also converted to nitric acid at the surface of aqueous aerosol (Tuazon, 1983; Richards, 1983) and it is likely that a similar conversion also occurs for NO_3. The ammonia neutralises a corresponding proportion of the acidity, and allows the formation of a range of salts in the solid phase. A list of some of the more important of these salts and of the ions present in aerosol particles is given in Table IV.

TABLE IV Species present in the aerosol phase

Ions (etc)	Identified solid phases
H_2SO_4	$(NH_4)_2SO_4$
HSO_4^-	NH_4HSO_4
SO_4^{2-}	$(NH_4)_2SO_4 \cdot NH_4HSO_4$
NO_3^-	NH_4NO_3
Cl^-	$(NH_4)_2SO_4 \cdot 2NH_4NO_3$
H^+	NH_4Cl
Na^+	
K^+	Na_2SO_4
Ca^{2+}	$NaNO_3$
Mg^{2+}	$Na_2SO_4 \cdot NaNOa_3 \cdot H_2O$
various heavy metals (eg. Fe, Mn, Cu, Pb)	$Na_2SO_4 \cdot (NH_4)_2SO_4 \cdot 4H_2O$
	$CaSO_4 \cdot (NH_4)_2SO_4 \cdot H_2O$
H_2O	Other mixed metal ammonium sulphates

However, the aqueous phase will also contain small amounts of the full
range of pollutants present in the gas phase. It therefore acts as a
site for the oxidation of sulphur dioxide to sulphuric acid, and for a
range of aqueous reactions. In general these will be the same as
those occurring in the cloud droplets (see section 2.3) although their
equilibria and rates will be altered by the very much higher
concentrations which are present in the aerosol. Activities in such
concentrated solutions are not well described by present theories, and
by their very nature, aerosol processes are difficult to study in
laboratory investigations although this has not precluded studies of
rates, e.g. Huntzicker, 1980; McMurry, 1983 and equilibria with the
gas phase e.g. Stelson and Seinfeld, 1982a,b. Experimental data for
concentrated solutions can also be used, and these have enabled
theoretical studies to be made of the of the sulphate/nitrate/
ammonium/water aerosol system (Jacob, 1986; Bassett and Seinfeld,
1983).

Atmospheric aerosol formed from photochemical pollution tends to
be concentrated into particles within the diameter range 0.1 - 10 μm.
Measurements of aerosol particles has shown a bimodal size
distribution, with peaks at around 1 μm and 5-8 μm eg. (Harrison and
Pio, 1983b). As the smaller particles tend to be associated with
ammonium salts, and the larger particles with sodium salts, it is
thought that the latter result from reactions with sea salt aerosol -
primarily the absorption of nitric acid and loss of hydrogen chloride,
e.g. Harrison and Pio, 1983a.

In the absence of rain, aerosol particles tend to accumulate in
the lower troposphere and can be transported over large distances.
For these reasons they have the potential to make a significant
contribution to deposited acidity at elevated sites remote from source
regions, or at those sites where there is orographic enhancement of
the rainfall. Much more study is necessary if we are to gain an
adequate quantitative understanding of aerosol formation and of the
way in which aerosols interact with gas phase chemistry through their
equilibria, their role as sinks and also through the area which they
provide for surface catalysed reactions.

2.3 Aqueous-phase Chemistry

The chemistry of cloud droplets and raindrops is much more uncertain
than is that of the gas phase, although the reactions involved have
been reviewed in a number of places (Graedel and Weschler, 1981;
Hoffman and Calvert, 1985). Many of the measurements of rate and
equilibrium constants which have been made were not designed
specifically for atmospheric studies, and are therefore contained in
more general compilations, e.g. (Farhartaziz and Ross, 1977; Ross and
Neta, 1979; Bielski, 1985). The compilations which are available
specifically for the chemistry of the atmosphere are those associated
with a range of modelling studies (Chameides and Davis, 1982;
Hoffman and Calvert, 1985; Jacob and Hoffmann, 1983).

Cloud droplets condense on atmospheric aerosol particles, and grow
by further condensation and by the coalescence of small droplets to

produce larger ones. The initial chemical composition of the
droplets is therefore controlled by the composition of the aerosol,
and studies have shown that the uptake of aerosol by cloud can be up
to 100 percent. Compounds which are soluble in water will exist in
equilibrium between the gas and aqueous phases, with their
distribution being described by the appropriate partition (Henry's
Law) coefficient, as discussed by Brimblecombe and Dawson, 1984.
However, such equilibria are not instantaneous, with the uptake of
molecules from the gas-phase to solution being governed, as for
aerosol particles, by the collision rate between the molecules and the
droplets, and by the accommodation coefficient for the compound
concerned. Some of the important gas-liquid equilibria and
scavenging processeds are shown in Table V.

Although Henry's Law constants have been measured for many of the
compounds of interest, there are others for which data is lacking.
In addition, the temperature dependance of the Henry's Law constants
is important due to the change in solubility which occurs for some
gases over the range of temperature which occur in the atmosphere.
Where these have not been measured they can be calculated from
standard thermodynamic data, as indeed can the Henry's Law constants
themselves. It should also be remembered, that Henry's Law is,
strictly, only applicable to ideally dilute solutions. In using it
for cloud droplets we are assuming that the other uncertainties
involved will outweigh the error which this introduces. It is also
important to ascertain whether the Henry's Law measurement refers to
the total solubility of the gas concerned, including the products of
its aqueous equilibria, or whether it refers only to the molecules of
the gas themselves. Such misinterpretation can result in a large
under or over estimate of the amount of material in solution.

Within solution a wide range of chemical processes occur,
encompassing both ionic equilibria and reactions. However, both the
rates of the reactions and the relative importance of many of these
processes are much less well determined than is the case for the gas
phase. This is complicated by the fact that the equilibrium and rate
constants refer not to ionic concentrations, but instead to activities
of the various ions as discussed in section 2.2. This highlights the
need for a comprehensive aqueous-phase chemistry. Each solvated ion
will contribute to the ionic strength of the solution, thereby
altering the activities, rates and equilibria for all other ions.
The pH (controlled in many cases by solvation of ammonia) can also
effect the uptake of various gaseous species, such as sulphur dioxide
and the concentration of the species involved in their acid-base
equilibria.

Unlike those in aerosols, the solutions found in cloud droplets
are often sufficiently dilute for the activity coefficients to be
described by simple formulae, such as the Debye-Huckel limiting law;

$$\log \gamma \pm = -0.509 |z^+ z^-| I^{\frac{1}{2}} \qquad (4)$$

TABLE V   Gas phase-aqueous phase equilibria and scavenging processes

Process	Henry's Law constant (moles 1^{-1} atm^{-1}) (298K)
SO_2 (g) = 'G_2SO_3'	1.24
HNO_3 (g) = HNO_3 (aq)	1.98×10^5
HNO_2 (g) = HNO_2 (aq)	47.6
HCl (g) = HCl (aq)	20.3
HCHO (g) = HCHO (aq)	7.04×10^3
NH_3 (g) = NH_3 (aq)	57
H_2O_2 (g) = H_2O_2 (aq)	7.1×10^4
O_3 (g) = O_3 (aq)	1.14×10^{-2}
NO_2 (g) = NO_2 (aq)	7.0×10^{-3}
CO_2 (g) = H_2CO_3 (aq)	3.4×10^{-2}
O_2 (g) = O_2 (aq)	1.26×10^{-3}

OH (g) – OH (aq)

HO_2 – HO_2 (aq)

N_2O_5 (g) – $2HNO_3$ (aq)

NO_3 (g) – HNO_3 (aq) + OH (aq)

HO_2NO_2 (g) – HO_2 (aq) + NO_2 (q)

The solubility of hydrogen chloride is much higher than suggested above, due to its ionization in solution.

Unless otherwise shown, Henry's Law constants are taken from Seigneur, C. and Saxena, P. (1984).

18

or one of its more complex modifications (see Hamer, 1968).

The most important atmospheric reactions in aqueous solution are those for the oxidation of sulphur (IV) to sulphur (VI), which is effectively the production of sulphuric acid from dissolved sulphur dioxide. The five most important proven mechanisms by which this process is believed to occur are listed, with their rates, in Table VI. Their relative importance will clearly depend on the composition of the airmass in question. However, each has a rather different character and needs separate consideration. The rates for the oxidation of aqueous sulphur dioxide by hydrogen peroxide and by ozone are shown in Figure 4.

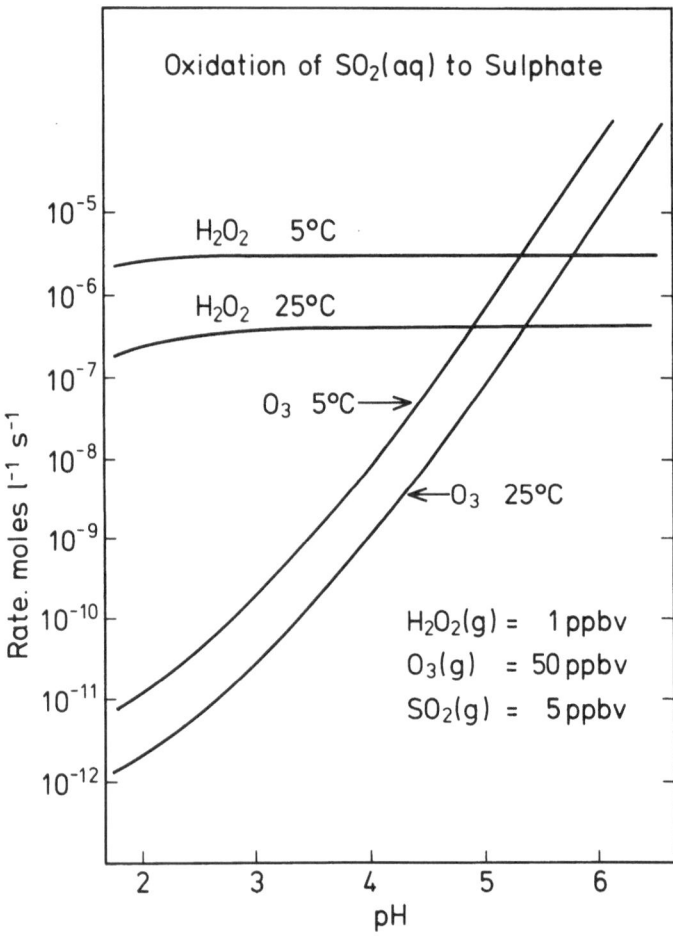

Figure 4.   The rate of production of sulphate in cloud droplets, as a function of pH, due to the oxidation of aqueous sulphur dioxide by hydrogen peroxide and ozone.

TABLE VI. Aqueous phase processes for the oxidation of sulphur
dioxide

Process	Rate expression (moles l^{-1} s^{-1}
$S(IV) + H_2O_2 \longrightarrow S(VI)$	8.9×10^4 $[SO_2.H_2O]$ $[H_2O_2]/0.1+[H^+])$
$S(IV) + O_3 \longrightarrow S(VI)$	$[O_3]$ $(5.9 \times 10^2$ $[SO_2.H_2O]$ +
	3.1×10^5 $[HSO_3-]$ +
	2.2×10^9 $[SO_3^{2-}])$
$S(IV) + O_2 \xrightarrow{\text{Fe, Mn}} S(VI)$	complex pH dependent mechanism.
$S(IV) + N(III) \longrightarrow (S(VI)+\frac{1}{2}N_2O$	$142[H^+]^{0.5}([SO_2.H_2O] + [HSO_3^-])$.
	$([HNO_2] + [NO_2^-])$
$S(IV) + OH \xrightarrow{O_2} S(VI) + SO_4^- +$	9.5×10^9 $[OH.][HSO_3^-]$
	5.5×10^9 $[OH.][SO_3^{2-}]$

Strictly all the concentrations shown should be replaced by the
corresponding activities.

The reaction with hydrogen peroxide is potentially the most rapid of the five reactions (hydrogen peroxide is much more soluble than ozone), but the oxidation is usually limited by the combined total amount of hydrogen peroxide which is present in the airmass and which can be produced in solution. (For solutions with pH<3, the rate of production in solution may exceed that in the gas-phase (McElroy, 1986a), but in a newly condensed cloud system, the gas-phase concentration already present will dominate). The oxidation of Sulphur (IV) is first order in the concentrations of both hydrogen peroxide and of hydrated aqueous sulphur dioxide. (Sulphurous acid, H_2SO_3, does not actually exist, although it is often convenient to suppose that it does). The rate is therefore unaffected by acidity, except at very low pH values (less than 2) when the term in the denominator becomes significant. The rate of production of gaseous hydrogen peroxide is proportional to the square of the concentration of HO_2, so that low NO concentrations are necessary if appreciable amounts are to be formed. Moreover, since HO_2 is produced by a photochemical pathway, there should be a strong seasonal variation in the concentration of H_2O_2. However, this does not preclude the occurrence of this reaction at night, as hydrogen peroxide may remain from the previous day, particularly in a dry airmass which has been trapped aloft.

Although the reaction with ozone is not limited by the amount of the gas which is present, it is limited by the solubility of ozone, and by the fact that as sulphuric acid is produced, so the pH falls, and since hydrated sulphur dioxide is only a weak acid, this lowers the concentrations of HSO_3^- (aq) and SO_3^{2-} (aq) and hence also lowers the rate of the reaction. However, in the presence of gaseous ammonia, the sulphuric acid produced will be neutralised by the uptake of ammonia and the formation of ammonium ions, enabling further sulphur dioxide to be oxidised. Neutralisation may also be affected by the presence of calcium carbonates. Although it has a strong pH dependance, this mechanism can be important at all times of the day and year, since most airmasses contain several tens of parts per billion (10^9) of ozone, sufficient to oxidise appreciable amounts of sulphur dioxide.

The aqueous concentration of the hydroxyl radical is controlled by the balance between its scavenging rate from the gas-phase, and the numerous reactions which it undergoes in solution. The mechanism of its reaction with sulphur (IV) is rather more complex than that shown in Table VI, but has been discussed in detail in a number of places, for instance, reference (McElroy, 1986b). Because of the photochemical source of OH radicals, and their short lifetime (about one second in very clean dry air) the reaction can only be important during daylight hours, and should have a strong seasonal dependance.

The fourth mechanism is that for oxidation by oxygen in the presence of iron or magnanese ions. Unlike the previous routes, this mechanism does not rely on photochemical oxidants and is equally applicable throughout seasonal and diurnal cycles. The mechanism and kinetics both embody a significant amount of uncertainty (see Jacob and Hoffmann, 1983 for the full kinetic expressions). Iron is

involved in the Fe(III) state, which is reduced to Fe(II) during the
course of the reaction. For catalysis to occur this Fe(II) must be
rapidly re-oxidised to Fe(III). It should be noted that the
solubility of Fe(III) has a strong pH dependence which can affect its
catalytic properties. Manganese is thought to participate as Mn(II)
but the nature of other magnanese species involved remains unknown.
The mechanism itself does not rely on photochemical processes, and
could therefore operate over a wide range of conditions, provided that
sufficient concentrations of metal ions are present. However, the
recyling of the oxidation states of the ions may rely on
photochemically produced species. This is an area which requires
further investigation.

The oxidation of sulphur (IV) by nitrogen (III) is not rapid
enough to be important.

All of the above mechanisms can be interfered with, because of
the potential for the complexation of sulphur-containing ions in
solution. Complexes can form both with formaldehyde and also with the
wide range of metal ions which are present, including iron, manganese
and copper.

Formation of nitrate in solution is slow, the rate being
typically a factor of 10^4 lower than for sulphate. This is governed
by a combination of the kinetics involved, the low solubility of
nitrogen dioxide, and the relatively low concentrations of gas phase
nitrous acid. Nitrogen dioxide in solution undergoes
disproportionation to a mixture of nitrate and nitrite, whilst nitrous
acid is oxidised in mechanisms which are analogous to those for
aqueous sulphur dioxide.

Organic acids may be formed in solution, principally by hydrogen
abstraction from hydrolysed aldehydes by hydroxyl radicals. The
largest production of acidity via this mechanism is likely to be from
formaldehyde which is almost completely hydrolysed and therefore, much
more soluble than the other aldehydes, which exist largely in their
unhydroysed forms. However, the formic acid which is produced in
solution and scavenged from the gas phase can in turn be oxidised in
solution, again by reaction with hydroxyl radicals, to produce carbon
dioxide.

The other reactions which occur in solution are principally
radical exchange processes involving chloride and carbonate ions,
ozone and hydrogen peroxide, together with hydroxyl and hydroperoxyl
radicals. The carbonate ions occur due to the dissolution of carbon
dioxide.

It is important to realise that the chemistry of a cloud system
is far from simple, especially that chemistry which occurs during and
following condensation in a previously cloud-free airmass. Under
these conditions the gas phase has the potential to contain high
concentrations of acidic species and radicals which may be rapidly
scavenged by the growing water droplets. The complex nature of the
interactions which occur, and the way in which the gas phase
concentrations of, for instance, nitric acid, hydrogen perioxide and
ammonia can become depleted has been illustrated by the results of
modelling studies (Hough, 1987). Two typical sets of calculations

are shown in Figure 5.   These illustrate the uptake of gaseous
species during cloud formation and the subsequent changes in
concentration which occur during the oxidation of aqueous sulphur
dioxide.   The model included 44 gas phase variables, and an aqueous
chemistry including all the species and equilibria referred to above.
 The results illustrated can therefore only illustrate a small part of
the work.   During the first 250 seconds condensation occurs in a dry
airmass, following which a constant cloud-water content is
maintained.   The airmass illustrated in Figure 5(a) gave low initial
pH values in the cloudwater, but also contained an appreciable
concentration of hydrogen peroxide, which lead to the production of

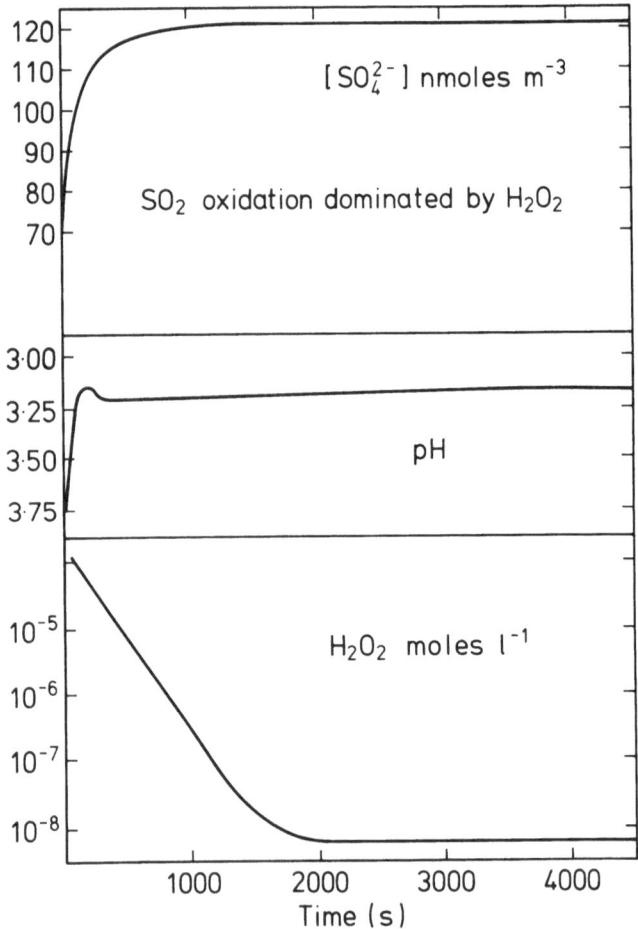

Figure 5a.   The results of some cloud chemistry model calculations.
              Droplet condensation occurs during the first 250 seconds
              of the simulation, after which the liquid water content
              remains constant.
                   Airmass with appreciable acidic aerosol and hydrogen
              perioxide concentrations.

significant amounts of sulphate within five minutes.   Figure 5(b)
shows the results for an airmass with a lower concentration of
hydrogen peroxide, but also a low initial cloud water acidity.   This
high initial pH allowed sulphate formation via the ozone mechanism.
However, as the pH fell, the rate for this process decreased and the
radical mechanisms became dominant, although by no means as rapid as
the oxidation by hydrogen peroxide in 5(a) or by ozone earlier in

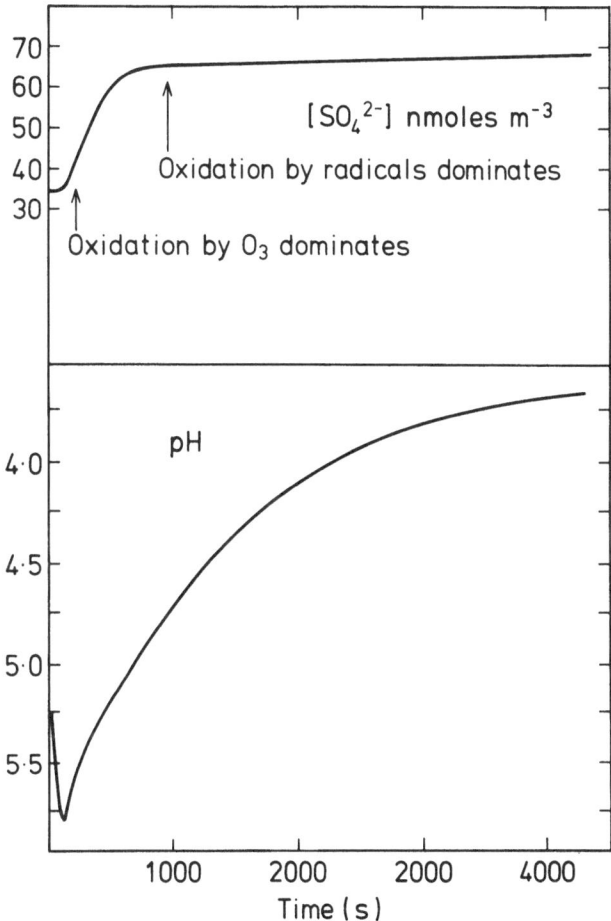

Figure 5b.   The results of some cloud chemistry model calculations.
              Droplet condensation occurs during the first 250 seconds
              of the simulation, after which the liquid water content
              remains constant.
                    Airmass with low acidic aerosol and hydrogen
              peroxide concentrations.

5(b). In both cases the formation and scavenging of gas-phase nitric
acid also made an important contribution throughout the model time
period.

Although there are many occasions when the products of the
droplet-phase chemistry will be deposited on the earth in
precipitation, there are also many occasions when the cloud
evaporates. If this is the case, then the chemical species in
solution are either returned to the gas phase or form aerosol
particles, depending on their volatility, their tendency to self-
nucleate and their solubility. Indeed, during the lifetime of a
cloud, turbulent air-motion will ensure that many aerosol
particles/droplets undergo many cycles of condensation and
evaporation.

3. DEPOSITION PROCESSES

Deposition of atmospheric material to the earths' surface occurs via
three principal mechanisms, namely dry deposition of gases and aerosol
particles, precipitation and occult deposition. Although the first
two of these processes are of importance at all sites (apart from
desert regions where precipitation is absent), occult deposition is of
special significance at elevated or high altitude sites where cloud
may extend to the surface of the earth for a significant amount of the
time. The outline details of each of these processes are considered
below.

3.1 Dry deposition

The physics of dry deposition is relatively well understood and has
been reviewed in a number of places, e.g. Sehmel (1980), Garland
(1983) and Chamberlain (1986). In addition, more specific reviews
have dealt with the dry deposition of, for instance, sulphur dioxide,
Garland (1977) and ozone, Colbeck and Harrison (1985).

The rate of dry deposition is controlled by the resistance which
is offered to the transfer of material from the gas phase. Such
resistance is usually divided into three categories:

r_a: The aerodynamic resistance, depending on air turbulence.

r_b: Surface boundary layer resistance, dependent on the
       molecular diffusivity and the surface roughness. This
       term is usually small compared to r_a for natural surfaces,
       but becomes important in urban areas.

r_c: Canopy or surface resistance, dependent on the affinity of
       the surface for the gas or aerosol concerned.

These processes are illustrated in Figure 6. In general, r_c
will be the dominant term although for very reactive gases, such as

nitric acid, it will be very small or even absent. For deposition to vegetation it is affected by stomatal diffusion, so that the resistance increases at night when the stomata are closed. For any given gas, the nature of the surface can markedly alter the deposition velocity. When calculating the impact of dry deposition on a given area, it is therefore important to consider the land use carefully.

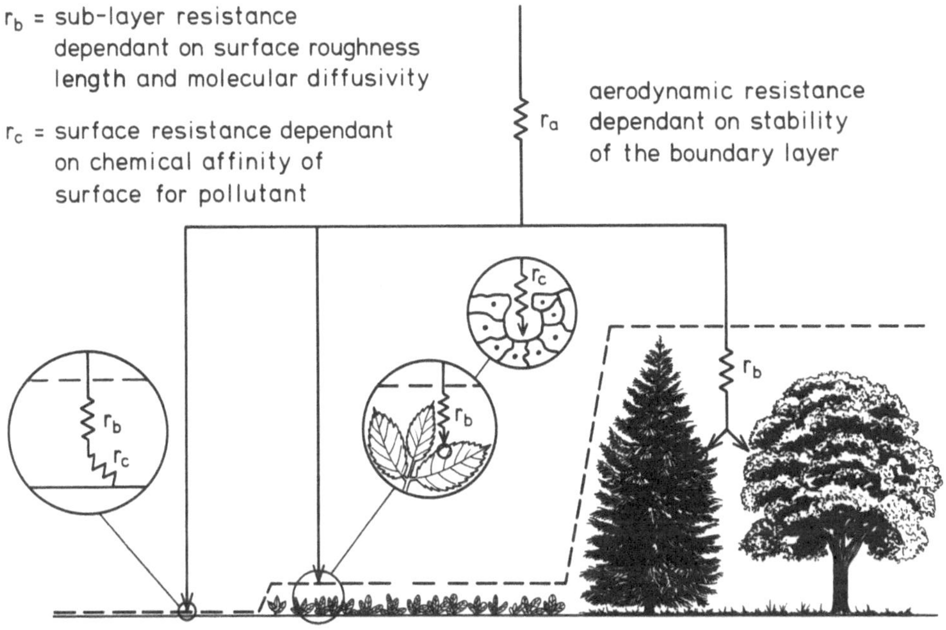

r_b = sub-layer resistance dependant on surface roughness length and molecular diffusivity

r_c = surface resistance dependant on chemical affinity of surface for pollutant

r_a aerodynamic resistance dependant on stability of the boundary layer

Figure 6. The resistance offered to dry deposition by the atmosphere, the surface boundary layer and the nature of the surface.

The reciprocal of the total resistance, r, is the deposition velocity, Vg defined in terms of the constant vertical flux, f, to a surface, and the concentration, x, at a reference height, z above that surface.

$$f = Vg.x_z \tag{5}$$

When considering the depletion of a well mixed layer of height h, the rate of change of the concentration, is given by:

$$dx/dt = -x.Vgh \tag{6}$$

Dry deposition velocities have been measured for a range of gases depositing onto many different surfaces under a variety of conditions. Values have been tabulated in the references referred to above, and some of these are reproduced in Table VII. There are several points which are worth noting. For highly reactive gases, such as nitric acid, the surface resistance will often be zero for deposition to vegetation during the daytime when the stomata are open. For sulphur dioxide, absorption is likely to be by dissolution in moisture following stomatal uptake, so that although deposition to vegetation is therefore favoured, the resistance of a plain surface, such as bare soil, will depend heavily on whether or not the surface is wet. For ozone, which is largely insoluble, experimental evidence shows that deposition velocities are markedly lower on wet surfaces, such as water or snow, than they are for dry surfaces such as brick, (Colbeck and Harrison, 1985). Consideration should also be given as to whether synergic effects exist, such that the presence of one gas slows or hastens the uptake of another. Such questions are at present unanswered.

Evidence for the diurnal variation in dry deposition velocities to vegetation has been presented by a number of workers, e.g. for ozone, Colbeck and Harrison, 1985; Droppo, 1985; Garland and Derwent, 1979, and for sulphur dioxide, Garland and Derwent, 1979. Model calculations have shown that the diurnal variation in the case of ozone can have a pronounced effect on the depletion of ozone from a shallow nocturnal boundary layer, as would be expected (Hough, 1986). The daytime maxima were, however, unaltered.

In considering deposition to vegetation it is also important to take account of the total exposed surface area of the plants or trees, which will be much higher than for a plain surface of the same shape. The part which this factor plays in the deposition of aerosol particles is considered in Jones and Heinemann (1985). In the case of chemical species for which the deposition velocity is limited by the surface resistance, then for a forest or group of trees, the deposition velocity is a factor of between 2 and 20 higher than for grass, depending on the type of tree being considered. In the case of more reactive species for which the surface resistance is zero, the deposition rate will be limited by the aerodynamic resistance, and no such rate enhancement will occur. Clearly, as the effective surface area increases then the role of aerodynamic resistance will become more important for all species.

When considering the dry deposition of aerosol particles, then a wide range of behaviour is demonstrated depending on the diameter of the particles. Taking the case of small particles, then in the limiting case these will behave in exactly the same manner as gas molecules. However, their rate of diffusion to the surface is much lower than for gases, making r_b the dominant resistance term, and dry deposition rather slower than for a gas. Whereas gas molecules have diffusion coefficients of the order of 0.1 cm^2 s^{-1}, that for a 0.1 μm diameter particle is 7 x 10^{-6} cm^2 s^{-1} (Chamberlain, 1986).

For large particles, with diameters in excess of ~ 5 μm, deposition occurs predominantly by impaction, and sedimentation, so that as the particle size increases, the air resistance becomes

TABLE VII   Some dry deposition velocities

Species	Velocity (cm s^{-1})	Surface	
SO$_2$	0.8 (day), 0.4 (night)	Average over eastern U.S.A.	(Walcek et al., 1986)
	1.2	Calcereous	(Garland, 1977)
	0.8	Short grass (mean value)	(Garland, 1977)
	1.2 (day), 0.3 (night)	Wheat	(Fowler and Unsworth, 1979)
	0.5 (day), 0.1 (night)	Dry Forest(Scots pine)	(Fowler and Cape, 1983)
	0.3 (day), 0.3 (night)	Wet Forest(Scots pine)	
NO$_3$	0.1 - 0.6	Short grass	(Delany and Davies, 1983)
HNO$_3$	2.5 (day), 1.8 (night)	Average over eastern U.S.A.	(Walcek et al., 1986)
	2.5 (day)	Short grass	(Huebert and Robert, 1985)
	13.5 (day)	Wheat field, turbulent air	(Dollard et al.)
O$_3$	0.49 (mean value)	grass	(Droppo, 1985)
	0.1 (night) 1.0 (day max)	grass	(Droppo, 1985)
	0.53 (mean value)	grass	(Colbeck and Harrison, 1985)
	0.05	water	(Galbally and Roy, 1980)
HCl	13.0 (day)	Wheat field, turbulent air	
PAN	0.25	grass, soil	(Garland and Penkett, 1976)
Sulphate	0.22	long-term mean-vegetation	(Wesely et al.)
	0.25 (day) 0.1 (night)	Average over eastern U.S.A.	(Walcek et al., 1986)

Aerosol
1 μm diameter 0.01	)		
2 μm diameter 0.03	)	grass	(Jones and Heinemann, 1985)
3 μm diameter 0.06	)		
4 μm diameter 0.14	)	average atmospheric	
5 μm diameter 0.44	)	stability	

can be 2-40 times higher for deposition to trees

The values of 13.5 and 13 cm s^{-1} for HNO$_3$ and HCl were obtained over a
high surface area wheat crop under turbulent conditions.   They
indicate how much variability is involved in dry deposition, depending
on the conditions.

negligible, and the behaviour tends towards that associated with macroscopic objects. It is possible to define an efficiency of impaction, E, for a particle driven in a wind of velocity u towards a solid object, as the ratio of the number of particles which impact on the object, divided by the number which would have passed through the volume it occupies if it was not there. Because the streamlines of flow will tend to bend round the object, E is less than unity. If we define the effective deposition velocity, Vg, relative to the surface area perpendicular to the wind, then Vg is given by:

$$Vg = p.u.E.$$

(7)

where p is the fraction of the particles impacting on the object which actually stick to its surface and do not bounce off. For wet particles p will be unity, but may be rather lower for dry particles.

E depends on the Stokes number for the system, which is defined as the ratio of the stopping distance of the particle to the characteristic length of the object. (The characteristic length is the diameter of an equivalent cylinder). If S is the stopping distance in a wind of velocity u, L is the characteristic length, V_s the terminal velocity of the particle and g the acceleration due to gravity, then:

$$\text{Stokes number} - S/L = uV_s/gL = mu/3\,\pi\eta dL$$

(8)

where m is the particle mass, d the particle diameter, and η is the viscosity of air (more generally the fluid medium). As a result, the dry deposition velocity is proportional to the square of the windspeed.

As discussed in section 2.3, that aerosol which is due to photochemical pollution and which contributes to atmospheric acidity, has a typical particle diameter which falls between those discussed above. As a result, deposition from both diffusion and sedimentation are low. The result is that deposition velocities for such particles may be as low as 10^{-3} cm s^{-1}, and are determined by the surface roughness and the friction velocity. The above behaviour has been discussed by, for instance, (Sehmel, 1980) who has presented values of Vg for a wide range of particle sizes and atmospheric conditions. A typical pattern of behaviour is shown in Figure 7. However, because of the sensitivity of the deposition velocity to the particle diameter and the nature of the surface, it is difficult to calculate accurately the deposition rate for aerosol and the contribution which it makes to the acidic input at one specific site. Further uncertainties are introduced if we wish to consider an area over which there is a significant variation in roughness length, such as the complex terrain associated with elevated sites.

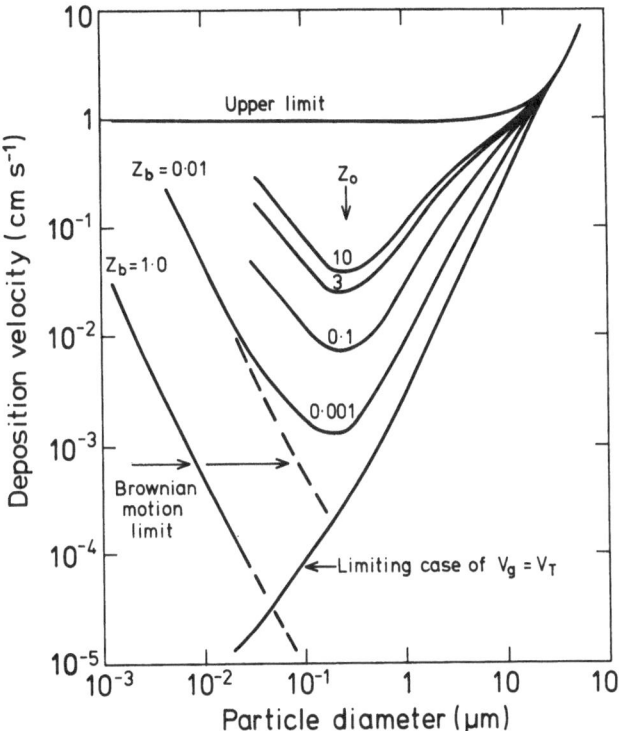

Figure 7.   The variation of the dry deposition velocity for aerosol
             particles with particle diameter. The left hand side
             shows the limit of brownian diffusion and the right hand
             side the limit for sedimentation. For medium sized
             particles (~ 1μm) where deposition is slow, the nature of
             the surface is very important.

     The principal areas of uncertainty in calculating dry deposition
at present are therefore: the diurnal variation in deposition rates
of gases to vegetation, the nature of the reaction between the gas and
the surface, the synergic effects which could occur between different
gases, the deposition rates for aerosol particles, and the effect of
land use and surface roughness, particularly when these vary over the
region of interest.

30

Care should also be exercised when using measured deposition velocities in models, in that it is important to determine precisely what has been measured. Because deposition velocities for a given pollutant vary by over an order of magnitude, depending on the receptor, time of day and meteorological conditions, it is important to choose appropriate values. At elevated sites, the physical shape of the terrain also contributes to the uncertainty which is involved. It is very difficult to predict wind patterns in complex terrain. The status, data requirements and needs of deposition models have been discussed by Hosker (1986).

3.2 Wet deposition

Wet deposition processes are closely interwoven with the meteorology which causes rainfall, a subject which is beyond the scope of the present account. However, there are a distinct number of stages which contribute to the overall deposition process, each of which needs to be considered if we are to obtain a realistic understanding of the deposition which occurs. The overall pattern is illustrated in Figure 8.

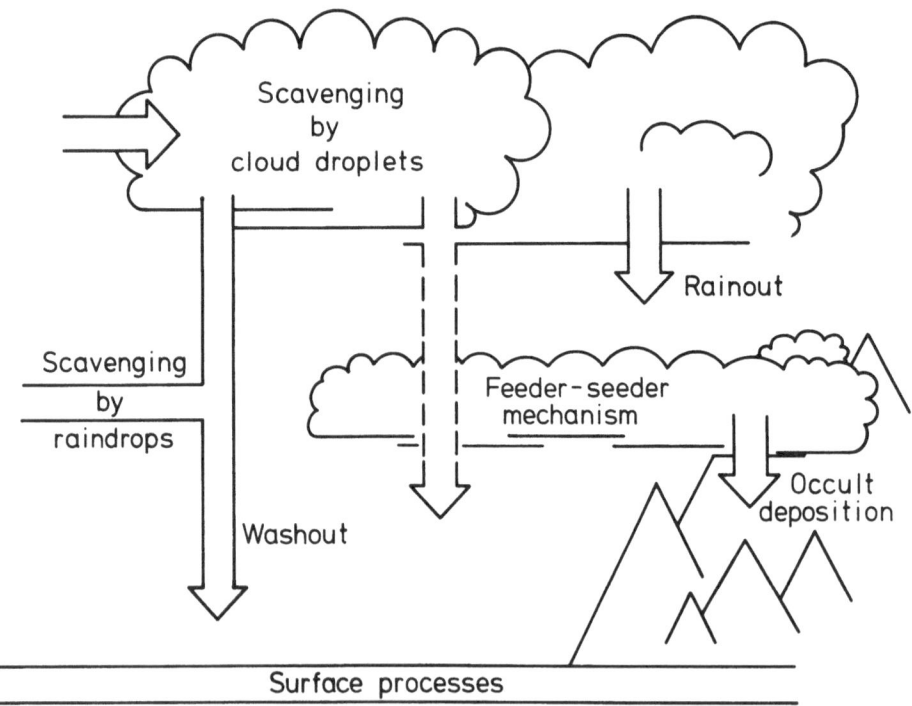

Figure 8.   The physical processes which contribute to wet deposition.

Rainfall is initiated either by the aggregation of cloud droplets and the condensation of water to their surface, or by the riming and coalescence of small ice particles formed in clouds (the Bergeron process). Whenever liquid water is present, the scavenging of aerosol particles and gaseous molecules can occur, so that precipitation results in the rainout of the chemical material which was present in the cloud droplets. The raindrops will, in general, consist of more dilute solutions than the cloud droplets due to the additional condensation which has occurred. Further chemical processes will occur in the falling raindrops. However, as the raindrops fall, they will continue to scavenge aerosol particles from the surrounding air, and to exchange gases with the airmass. This washout process will alter the composition of the droplets and make a significant contribution to the deposition pattern. This will be particularly true if the airmass concerned has not been exposed to cloud droplet scavenging processes in its recent history. A number of models have been used in attempts to describe these processes, details of some of these may be found in Hales (1982); Giorgi and Chameides (1985); Kumar (1986); Gradel and Goldberg (1983); Durham (1981) and Fisher (1982).

Falling raindrops may also enter a second lower layer of cloud, which is itself not precipitating. In this case, the raindrops will scavenge droplets from the second cloud layer. This is the so-called 'feeder-seeder' mechanism (Carruthers and Choularton, 1983) which will result in the wet deposition of material which may have been emitted from a different source to that in the higher cloud, and which would not have been deposited in its absence. This mechanism is particularly important in hilly regions, where a polluted airmass may condense to form a low layer of orographic cloud.

Further complications occur due to the formation of ice in clouds, and in precipitation occurring as snow or hail rather than as rain. Understanding of the chemistry of these systems is far from complete.

Two parameters have been used in attempts to quantify wet deposition, particularly in the long term time scale, namely the washout ratio, W_r, and a wet deposition velocity, V_w, analogous to the dry deposition velocity (see section 3.1). The washout ratio is given by

$$W_r = c/x \qquad (9)$$

where c and x are the pollutant concentrations in precipitation and air respectively, both measured at the same height above the ground. If J is the rainfall rate (or its equivalent for snow and hail) then the wet flux to the ground is Jc, and hence the wet deposition velocity is given by:

$$V_w = Jc/x = W_r J \qquad (10)$$

and the depletion of the gaseous concentration of a compound in an airmass during a rainfall event by:

$$\frac{dx}{dt} = - x \, W_r \, J/h \qquad (11)$$

The effective rate coefficient in equation (11), i.e. $W_r \, J/h$, is sometimes referred to as the scavenging rate (γ), or wet loss rate, and is used, with a fixed value, in many modelling studies, eg. Eliassen and Saltbones (1983). For a uniform aerosol distribution, then it can be shown that W_r may be calculated theoretically from the expression:

$$W_r = \frac{0.5 \, E_r h}{Rm}$$

where E_r is the particle-raindrop collision efficiency and Rm the radius of the raindrop. For a height of 500 m, a droplet radius of 0.5 mm, a rainfall rate of 1 mm hr^{-1} and a collision efficiency of 0.2, then $Wr \sim 1 \times 10^5$ and $V_w \sim 2.8$ cm s^{-1} (Hosker, 1986). This wet deposition velocity is very much higher than the corresponding dry value.

The washout of gases is much less important than for particulate matter. If Henry's Law is applicable and equilibrium is reached during the fall time of the raindrop, then the washout coefficient is simply the reciprocal of the Henry's law constant, so that if H is the Henry's law constant,

$$W_r = {}^1/H \qquad (12)$$

$$V_w = {}^J/H \qquad (13)$$

For sulphur dioxide, $V_2 = 0.8$ cm s^{-1} at a rainfall rate of 1 mm hr^{-1} (Hosker, 1986). This is clearly a much lower rate than that for sulphate aerosol as given above.

The problem with using the washout ratio, is that it assumes the presence of a uniform concentration of material throughout the depth of the airmass. In practice, vertical gradients in concentration will occur, making the situation rather less easy to parameterise.

Very little is known at present, about the spatial and temporal variability of precipitation chemistry and, in particular, how the chemistry varies over short distances. This is caused by both a lack of suitable measurements and by our lack of understanding of the fundamental processes involved. In addition to the uncertainties in source strengths and dispersion, which impinge on all deposition routes, wet deposition is affected by a number of other processes. These include the occurrence of convective storms, which can cause widescale vertical mixing and chemical transport from high in the free troposphere, and the amount of gas and liquid phase chemistry which

has occurred prior to precipitation. Cloud microphysics and the
competition with other deposition processes also contribute to the
uncertainty. For long term calculations, the treatment of rainfall as
a stochastic process can produce results which are in good agreement
with observation, provided that the significance of regions which are
typically dry or wet is taken into consideration; e.g. Smith (1981).
For elevated sites, the enhanced orographic rainfall and terrain
induced turbulence will both contribute to the uncertainties
involved. In attempting to model or predict wet deposition, the
uncertainties involved in modelling the airflow which occurs in
frontal systems places severe limitations on the usefulness of the
calculations.

3.3 Occult deposition

Of the three categories of deposition considered, occult deposition is
the one process which occurs predominantly at elevated sites which are
significantly higher than the surrounding terrain and therefore either
extend into the cloud layer, or cause the formation of orographic
cloud. The presence of such cloud, close to the earth's surface is
useful, in that it allows us to perform measurements on cloudwater,
and to examine the chemistry of cloud formation before, during and
after condensation has occurred, by the use of ground based
instruments. This is both convenient, and also significantly less
expensive than equipping an aircraft for making in-cloud measurements.
   The chemistry which occurs during the deposition process is the
same as that which occurs in other clouds (both aqueous and gaseous
phase chemistry), but the concentrations are not diluted as is the
case in raindrops. The radii of cloud droplets tend to be in the
range 5-20 μm, depending on the airmass, and the deposition mechanism
will be similar to that for the correspondingly sized aerosol
particles. This is therefore dominated by impaction as discussed in
section 3.1. The impact of such droplets to a grass surface has been
measured by Dollard and Unsworth (1983). The fact that the cloud
droplets impact to the surface without dilution is significant, as it
allows solutions with a low pH to interact with the vegetation. As
an example, in a limited number of measurements, median pH values of
2.86 and 2.96 were measured in cloudwater near Los Angeles during 1982
and 1983 with a corresponding value in rainfall of 4.6 (Waldman,
1985). The lowest value measured in cloudwater was 2.06. Dollard
and coworkers have measured the occult deposition of nitrate, sulphate
and acidity to an upland region in Northern England (Dollard, 1983).
Concentrations were around 10^{-3} M in cloudwater, resulting in the
deposition of several kilograms of sulphate and nitrate per hectare
per year.
   Occult deposition therefore has the potential to provide a
significant part of the atmospheric input to elevated regions as well
as providing solutions which contain high concentrations of
pollutants. More detailed measurements are required in order to
assess accurately the actual deposition which does occur via this
pathway, and the potential impact which this can have on the
environment.

4   DEPOSITION FIELDS FOR ELEVATED SITES: MEASUREMENTS AND MODELS

Deposition fields may be obtained either from a series of measurements
made throughout the region of interest, or from the use of
computational models.   Ideally, these two methods should be used in
conjunction with each other.   The measurements provide us with
the real values of the deposition and show the extent to which the
model may be valid, whereas the model can help us to understand the
experimental measurements and can provide results on a far larger

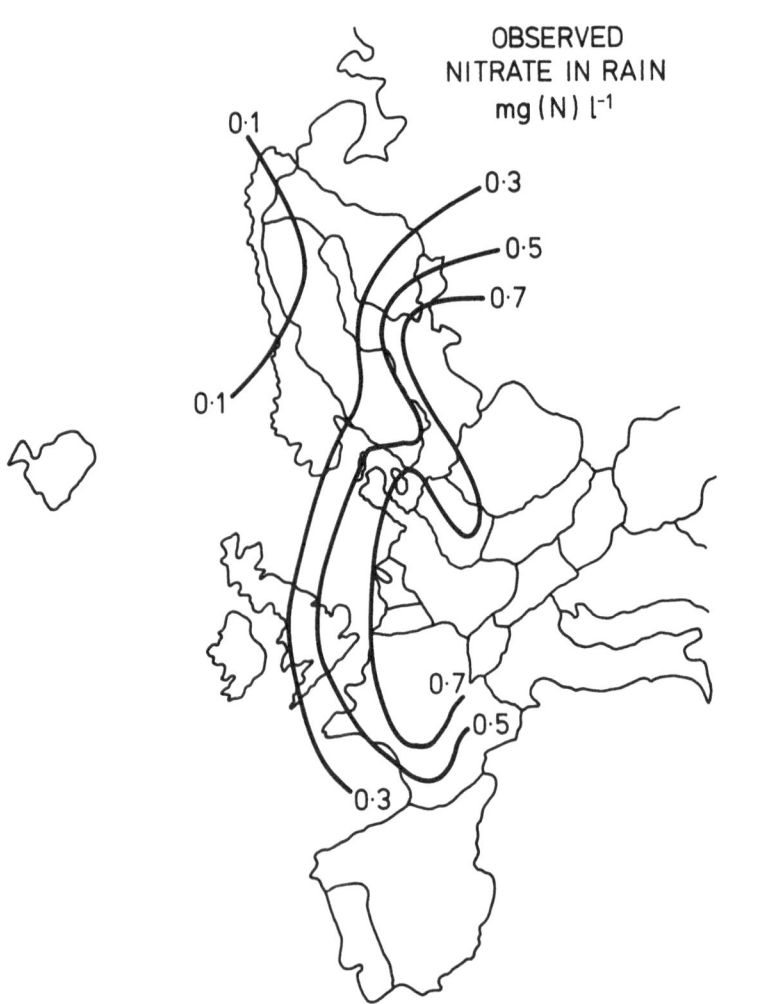

Figure 9a.   The deposition of nitrate in rainwater over western
             Europe – observed.

scale than is possible in a field programme.   As the interest in the
present paper is in principles rather than a review of specific
examples, attention will be focussed on some of the work applied to
Western Europe rather than trying to give a comprehensive account for
both Europe and North America.

Before considering details specific to elevated sites, it is
worth considering the adequacy of our knowledge of deposition amounts
and pollutant concentrations more generally.   In Western-Europe the
extent of our knowledge varies widely between countries depending on
the scale and nature of the national monitoring networks.   Gas phase
concentrations near to the ground are relatively easy to obtain,

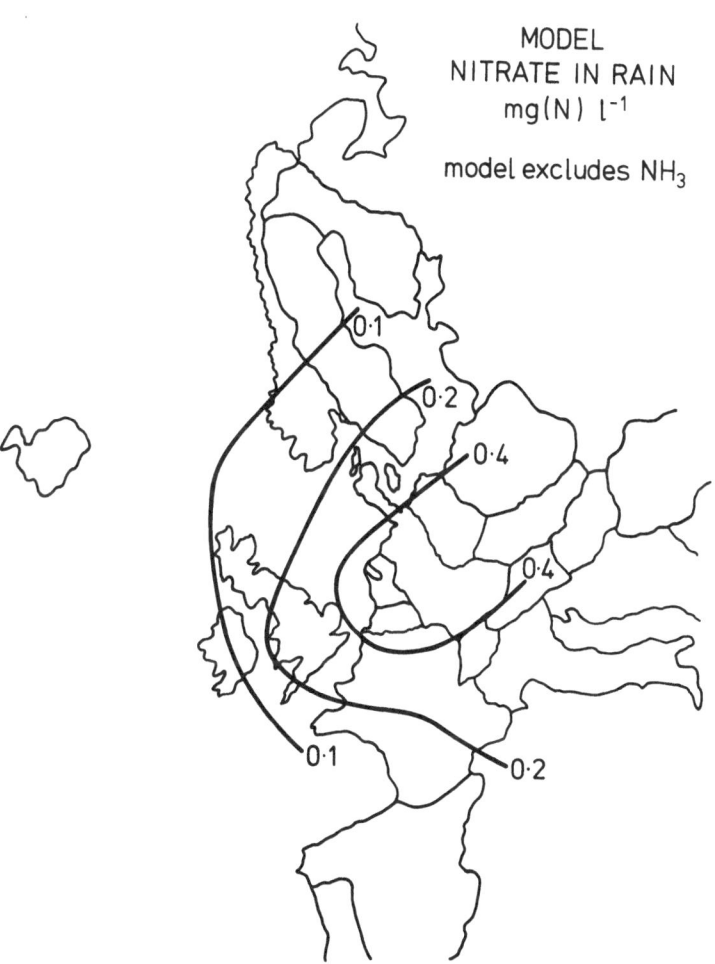

Figure 9b.   The deposition of nitrate in rainwater over western
              Europe - model calculations excluding ammonia.

36

especially if passive monitoring devices are used to obtain weekly (or similar) averages. Even so, such measurements are only made on a routine basis for a small number of sites, although the sulphur dioxide monitoring network in the Netherlands is one carefully thought-out exception. For deposition measurements, the results obtained at the 64 stations operating under the 'European Monitoring and Evaluation Programme' (EMEP) enable deposition maps for sulphate and nitrate to be drawn for Western Europe. Figure 9 (taken from Derwent and Noopod (1986)) shows such a deposition map (plotted in terms of the concentration of nitrate in rainwater), together with the corresponding map drawn from model calculations, as referred to below. Similar maps for the deposition of sulphate have been presented elsewhere (Elieassen and Saltbones, 1983).

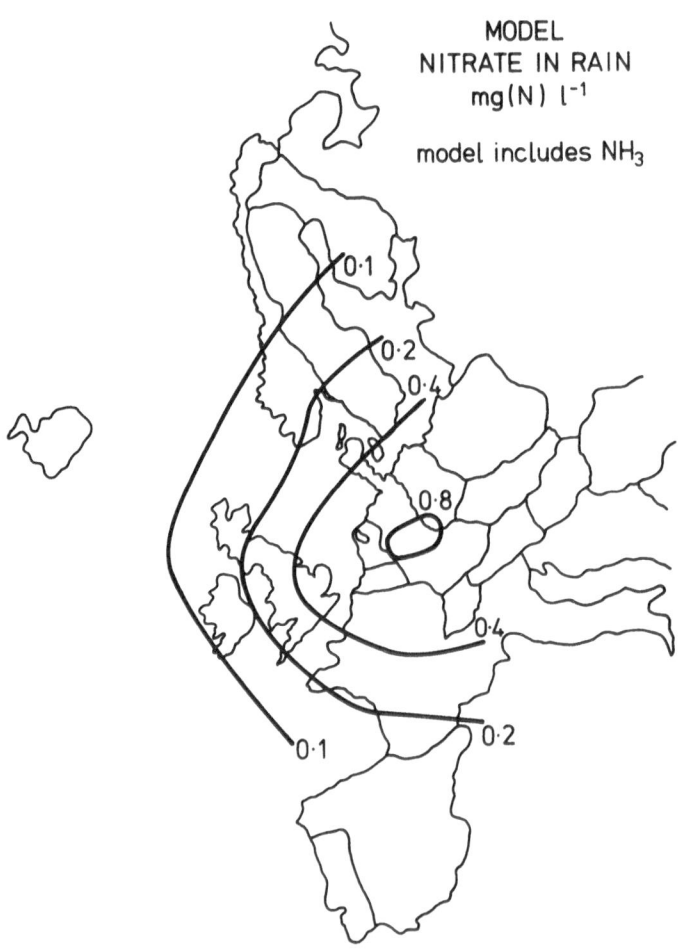

Figure 9c. The deposition of nitrate in rainwater over western Europe - model calculations including ammonia.

Figure 9 shows that agreement between models and observations can be quite good, even when a simple chemical scheme is used. It is worth noting, however, that this agreement was only obtained after the effect of ammonia was included in the calculations. The model used was a simplified version of that used to calculate sulphur deposition over Europe, as considered in detail in Eliassen and Saltbons, (1983) and Eliassen (1978).

Some models are only adequate to obtain long-term average deposition patterns. The results in Figure 9 were obtained by running straight line trajectories based on the statistical analysis of the windfields at each of the EMEP monitoring sites. Such an approach is clearly inadequate to consider a single event. This is illustrated by the EMEP model referred to above, which uses 96 hour back trajectories to 720 sites calculated every 6 hours from measured meteorological data. Although the long-term average values for deposition which are calculated are in good agreement with those which are observed, the agreement on any given day is far less satisfactory. Other statistical models also give good agreement for long term average sulphur deposition, e.g. Fisher (1978). The assumptions embodied in such models, such as the smoothing of rainfall data and the constant boundary layer height, have been discussed in a number of places (Eliassen and Saltbones, 1983 and Fisher, 1983).

In order to calculate daily deposition patterns more sophisticated models are needed. Within Europe such a model is now being implemented under the auspices of the PHOXA (Photochemical Oxidant and Acid Deposition) modelling project, using an Eulerian model to calculate the concentrations of photochemical oxidants and the deposition of acidic species over Western-Europe. A number of large grid models are also being applied in North America with similar height resolution models being applied to areas with dimensions of a few hundred kilometres, e.g. Liu (1984) and Carmichael (1986). The processes involved in constructing an acid deposition model have been discussed by Hough and Eggleton (1986).

When deposition at high altitude sites is considered, then both the experimental programme and the model calculations suffer from the same limitation - a lack of spatial resolution. The deposition fields which may be deduced from Figure 9 both exhibit slow and smoothly varying changes for wet deposition, whereas in practice, there will be large and rapid changes in areas with complex terrain - usually high altitude sites. Evidence for this is provided by examination of rainfall maps as well as by theoretical considerations. In lowland regions with a relatively homogeneous terrain and surface roughness, the measurements made at one site may be typical of those over quite a wide area, especially if the site has been carefully selected. However, in the complex terrain which is usually associated with high altitude sites this is not the case. Rainfall may vary markedly over distances of a few kilometres, and the effect of the terrain on the wind pattern can cause changes in dry deposition on similar scales. To resolve this variation in deposition would require a measurements programme with a large number of closely and strategically spaced collectors. Although such a programme would be possible for a limited time and area, it is impracticable on a continental or national scale. Although the patterns in Figure 9 may be quite a good representation for lowland sites, they are not reliable for high altitudes.

In model studies, the same problem of lack of spatial resolution emerges. Regional Eulerian (grid) models have a typical grid size of 40 x 40 km (e.g. the PHOXA project). This is inadequate for the resolution of the local variability which occurs in an area with complex terrain. It is also inadequate for calculating dry deposition in homogeneous terrain if there is a significant variation in local scale landuse patterns. With the use of langrangian (trajectory) models, a small error in the path of the trajectory could position it in an area with a markedly different topography from that which it actually encountered. Such errors are well within the uncertainties associated with such calculations. The entrainment of polluted airmasses into already existing cloud systems is another area of uncertainty, and one for which data does not exist on a scale suitable for use in regional or mesoscale models. The inability of meteorological forecasting models to cope with the local changes in mountainous areas is further evidence for this lack of ability.

Even if a model with sufficient resolution could be constructed, two major probels would remain. Firstly, input data would be required on the same scale as the calculations. This is a problem which is somewhat analogous to the measurement difficulties discussed above. Secondly, there is the difficulty associated with modelling dispersion in complex terrain (Pasquill and Smith, 1983). The vortices and eddies which occur in such areas are beyond the capabilities of any modelling study which could realistically be applied.

The effect of complex terrain can, however, be incorporated into models in a simple form. Rainfall is known with a relatively high resolution, and observations of rainfall, either for a real episode or for a long-term average, can be used to scale the wet deposition in a suitable manner throughout the model domain. The incorporation of landuse data to produce spatically variant dry deposition velocities can be used to improve the adequacy of the treatment of dry removal.

When analysing the results of computational models it is important to remember the basis behind their construction, and not to draw conclusions which the model is not capable of actually supporting. In this context it is important to perform uncertainty analyses on models to examine the stability of the results with respect to the chosen input data and parameters (see, for instance, the treatment used in Derwent, 1986b).

5   FUTURE DEVELOPMENTS

Our future understanding of high altitude chemistry will be improved by progress in three areas: Laboratory measurements, field measurements and model development together with the associated theoretical understanding which is involved. A list of some of the areas of particular importance is given in Table VIII. The emphasis in this list is placed on chemical aspects of our understanding, and the items are deliberately defined in broad terms. Their translation into specific research projects would be a task for a specialist in the individual fields concerned.

Laboratory measurements are needed to increase our knowledge and understanding of the processes involved in the chemistry involved in atmospheric processes, especially those occurring in cloud water droplets and aerosol particles and the heterogeneous equilibria which exist between these systems and the gas-phase. The degradation mechanisms of hydrocarbons are also important. Such measurements should assist with the development and construction of sound models, and should ideally be used to address those areas which are of particular importance to model construction.

Field measurements progress is concerned with the use of existing technology to make measurements in new locations and of species which were previously unmeasured, together with the development of the technology to make measurements which are not possible at present. In doing this it is important that the methods used should be tested under facile conditions before being exposed to a more demanding environment. Instrumentation which is to be used at high altitude sites or in aircraft should operate in a reliable and verifiable manner in the laboratory and local sites before it is used in the more extreme conditions. There is little to be gained in making unreliable measurements, and producing data which are unreliable or inaccurate.

In the same way as the methods chosen need to be reliable, so the species which they are used to measure and the locations where the measurements are to be made need to be chosen with care. It is clearly not possible to measure everything everywhere, and although there are distinct limitations placed on our measured deposition fields and atmospheric concentrations by the number of measurements which it is possible to make, a carefully chosen set of measurements will contribute much more to our understanding than will a sample taken at random. We need to ask ourselves precisely what it is that we are trying to investigate. Measurements are much more useful if a number of species can be measured together than if they are measured in isolation. Such observations enable us to obtain information on the overall chemical balance of the atmosphere at a given time and place. Having obtained the measurements it is then important that they are interpreted adequately and thoroughly. It is in this area that the interaction between model and experiment is important.

Concerted sets of measurements are useful in that they can be compared with the results of suitable models. Model verification and validation relies on a change of suitable measurements being available for the conditions and situation which the model was used to study. There is, indeed, a need for much closer cooperation and dialogue between those who construct (or decide on the construction of) models and those who plan and implement field measurements. Observations

TABLE VIII   Chemical aspects of importance to our increased
understanding of the chemistry of elevated sites.

---

a)   Increased understanding of the scavenging of gaseous species by
     aerosol particles, cloud droplets and precipitation.

b)   Increased knowledge of the chemical reactions which occur in
     cloud droplets, including their rates and temperature dependence.

c)   Increased understanding of the physical chemistry of strong ionic
     solutions.

d)   Consideration of the significance and chemical consequences of a
     dry air-mass entrained into a cloud, especially during occult
     deposition processes.

e)   Further study of the dry deposition of photochemical and sulphate
     aerosol particles to a variety of surfaces under a range of
     conditions.

---

are needed by models for both initialisation and the setting of parameter values. Moreover, although models may produce interesting results, if there are no observations against which this output can be compared then that model is not nearly as valuable as if such measurements did exist. This is true even if a great deal of care time and effort has been used in the model construction. The solution of this problem lies jointly in the hands of modellers and experimentalists.

The interaction of models and observations is, however, not a one-way process. A well constructed model may be used to identify further measurements which it would be instructive to make in order to improve our understanding and knowledge, and to discriminate between several possible sets of measurements of which only a limited number are actually possible. Cooperation can result in benefits for both experimentalist and theoretician.

If we believe in the importance of atmospheric chemistry and deposition processes, then the currently available database is inadequate to realistically assess the impact of pollutants on elevated sites. In order to redress this deficiency careful measurements are needed, both in the laboratory and in suitable chosen locations, run in conjunction with relevantly constructed models. Such programmes have the potential to make a significant contribution to our knowledge and understanding.

ACKNOWLEDGEMENTS

Discussions with Dr Geoff Dollard are gratefully acknowledged. This work was supported by funding from the United Kingdom Department of the Environment.

REFERENCES

Altshuller, A.P. 1986. The role of nitrogen oxides in nonurban ozone formation in the planetary boundary layer over N. America, W. Europe and adjacent areas of ocean. Atmospheric Environment, **20**, 245-268.

Amble, E. 1981. Estimation of the spatial distribution of the population and the SO_2 emissions in Europe and Turkey, within the EMEP grid. NILU Teknisk Rapport NR 8/81. (Norwegian Institute for Air Research, Lillestrom, Norway).

Atkinson, R. and Lloyd, A.C. 1984. Evaluation of kinetic and mechanistic data for modelling of photochemical smog. J. Phys. Chem. Ref. Data, **13**, 315-444.

Atkinson, R., Winer, A.M. and Pitts, J.N. 1986. Estimation of night-time N_2O_5 concentrations from ambient NO_2 and NO_3 radical concentrations and the role of N_2O_5 in night-time chemistry. Atmospheric Environment, **20**, 331-339.

Bassett, M. and Seinfeld, J.H. 1983. Atmospheric equilibrium model of sulphate and nitrate aerosols. Atmospheric Environment, **17**, 2237-2252.

Baulch, D.L., Cox, R.A., Hampson, R.F., Kerr, J.A., Troe, J. and Watson, R.J. 1980. Evaluated kinetic and photochemical data for atmospheric chemistry. J. Phys. Chem. Ref. Data 9, 295–471. Ibid (1982) 'Supplement 1' J. Physc. Chem. Ref. Data 11, 327–496. Ibid (1984). 'Supplement 2'. J. Phys. Chem. Ref. Data 13, 1259–1375.

Bielke, S. and Elshout, A.J. (eds) 1982 Acid deposition, Proceedings of CEC Workshop., Berlin (Commission of the European Communities).

Bielski, B.H.J., Cabelli, D.E., Arudi, R.L. and Ross, A.B. 1985. Reactivity of HO_2/O_2^- radicals in aqueous solution. J. Phys. Chem. Ref. Data, 14 , 1041–1100.

Brimblecombe, P. and Dawson, G.A. 1984. Wet removal of highly soluble gases. J. Atmos. Chem., 2 , 95–107.

Buijsman, E., Maas, J.F.M. and Asman, W.A.H. 1985. Ammonia emissions in Europe. IMOU report R-85-2. (Institute for Meteorology and Oceanography, State University, Utrecht).

Calvert, J.G., Lazrus, A., Kok, G.L., Heikes, B.G., Walega, J.G., Lind, J. and Cantrell, C.A. 1985. Chemical mehcanisms of acid generation in the troposphere. Nature 317 , 27–35.

Cape, J.N., Fowler, D., Kinnaird, J.W., Paterson, I.S., Leith, I.D. and Nicholson, I.A. 1984. Chemical composition of rainfall and wet deposition over northern Britain. Atmospheric Environment, 18 , 1921–1932.

Carmichael, G.R., Peters, L.K. and Kitada, T. 1986. A second generation model for regional-scale transport/chemistry/deposition. Atmospheric Environment, 20 , 173–188.

Carruthers, D.J. and Choularton, T.W. 1983. A model of the feeder-seeder mechanism of orographic rain including stratification and wind-drift effects. Quart. J.R. Met. Soc., 109 , 575–588.

Carruthers, D.J. and Choularton, T.W. 1984. Acid deposition in rain over hills. Atmospheric Environment 18 , 1905–1908.

Chamberlain, A.C. 1986. Deposition of gases and particles on vegetation and soils. In: Air Pollutants and their Effects on the Terrestrial Ecosystems, eds. Legge and Krupa, 189–210, Wiley, New York.

Chameides, W.L. and Davis, D.D. 1982. The free radical chemistry of cloud droplets and its impact upon the composition of rain. J. Geophys. Res., 87 , 4863–4877.

Colbeck, I. and Harrison, R.M. 1985. Dry deposition of ozone: some measurements of deposition velocity and of vertical profiles to 100 metres. Atmospheric Environment, 19 , 1807–1818.

Delaney, A.C. and Davies, T.D. 1983. Dry deposition of NO_x to grass grass in rural East Anglia. Atmospheric Environment, 17 , 1391–1394.

Demerjian, K.L., Sehere, K.L. and Peterson, J.T. 1980. Theoretical estimates of actinic (spherically integrated) flux and photolytic rate constants of atmospheric species in the lower troposphere. Adv. Env. Sci. Tech. 10 , 369–495.

Demore, W.B., Golden, D.M., Hampson, R.F., Howard, C.J., Kurylo, M.J. Margitan, J.J., Molina, M.J., Ravishankara, A.R. and Watson, R.T. 1985. Chemical kinetics and photochemical data for use in stratospheric modelling, Evaluation No. 7. JPL 85-37. (JPL, California Institute of Technology, Pasadena, California).

Derwent, R.G. 1982. The sources and fates of atmospheric sulphur compounds in the United Kingdom. Harwell Report R-10567. (HMSO, London)

Derwent, R.G. 1986a. The nitrogen budget for the United Kingdom and North-West Europe. ETSU Report R.37 (United Kingdom Department of Energy).

Derwent, R.G. 1986b. Atmospheric ozone and its precursors. ETSU Report R.38 (United Kingdom Department of Energy).

Derwent, R.G. and Nopod, K. 1986. Long range transport and deposition of acidic nitrogen species in North-West Europe. Nature **324** , 356-358.

Dodge, M.C. 1984. Combined effects of organic reactivity and NHMC/ NO_x ratio on photochemical oxidant formation - a modelling study. Atmospheric Environment **18** , 1657-1665.

Dollard, G.J., Atkins, D.H.F., Healy, C. and Davies, T.J. Concentrations and dry deposition velocities of nitric acid. Nature **326** , 481-483.

Dollard, J.G. and Unsworth, M.H. 1983. Field measurements of turbulent fluxes of wind-driven fog drops to a grass surface. Atmospheric Environment **17** , 775-780.

Dollard, G.J., Unsworth, M.H. and Harvey, M.J. 1983. Pollutant transfer in upland regions by occult precipitation. Nature **302** , 241-243.

Droppo, J.G. 1985. Concurrent measurements of ozone dry deposition using eddy correlation and profile flux methods. J. Geophys. Res. **90** , 2111-2118.

Durham, J.L., Overton, J.H. and Aneja, V.P. 1981. Influence of gaseous nitric acid on sulphate production and acidity in rain. Atmospheric Environment **15** , 1059-1068.

Eliassen, A. 1978. The OECD study of long range transport of air pollutants: long range transport modelling. Atmospheric Environment **12** , 479-487.

Eliassen, A. and Saltbones, J. 1983. Modelling of long-range transport of sulphur over Europe: a two-year model run and some model experiments. Atmospheric Environment **17** , 1457-1473.

Farhartazis and Ross, A.B. 1977. Selected specific rates of reactions of transients from water in aqueous solution, III. Hydroxyl radical and perhydroxyl radical and their radical ions. NSRDS-NBS 59, (National Bureau of Standards, Washington).

Finlayson-Pitts, B.J. and Pitts, J.N. 1986. Atmospheric Chemistry. Fundamentals and experimental techniques. (Wiley-Interscience, New York).

Fisher, B.E.A. 1978. The calculation of long term sulphur deposition in Europe. Atmospheric Environment **12** , 489-501.

Fisher, B.E.A. 1982. The transport and removal of sulphur dioxide in a rain system. Atmospheric Environment **16** , 775-783.

44

Fisher, B.E.A. 1983. A review of the processes and models of long-range oxidant model and application to the north eastern United States. Atmospheric Environment **18** , 1145-1161.

Fisher, B.E.A. and Callander, B.A. 1984. Mass balances of sulphur and nitrogen oxides over Great Britain. Atmospheric Environment **18** , 1751-1757.

Fowler, D. and Cape, N. 1983. Dry deposition of SO_2 onto a Scots Pine Forest. in Precipitation scavenging dry deposition and resuspension, ed. Puppacher, H.R., Seminon R.G. and Slinn, W.G.N., (Elsevier, New York).

Fowler, D. and Unsworth, M.H. 1979. Turbulent transfer of sulphur dioxide to a wheat crop. Quart. J.R. Met. Soc. **10** , 767-783.

Fuchs, N.A. and Sutugin, A.G. 1971. High-dispersed aerosols. in International Reviews of Aerosol Physics and Chemistry, 2, ed. Hidy, G.M. and Brock, J.R. (Pergamon, New York).

Galbally, I.E. and Roy, C.R. 1980. Destruction of ozone at the Earth's surface. Quart. J.R. Met. Soc. **106** , 599-620.

Garland, J.A. 1977 The dry deposition of sulphur dioxide to land and water surfaces. Proc. R. Soc. Lond., A354, 245-268.

Garland, J.A. 1983. Principles of dry deposition: Application to acidic species of ozone. VDI-Berichte Nr. 500.

Garland, J.A. and Derwent, R.G. 1979. Destruction at the ground and the diurnal cycle of concentration of ozone and other gases. Quart. J.R. Met. Soc. **105** , 169-183.

Garland, J.A. and Penkett, S.A. 1976. Absorption of peroxyacetyl nitrate and ozone by natural surfaces. Atmospheric Environment **10** , 1127-1131.

Giorgi, F. and Chameides, W.L. 1985. The rainout parameterisation in a photochemical model. J. Geophys. Res. **90** , 7872-7880.

Graedel, T.E. and Goldberg, K.I. 1983. Kinetic studies of raindrop chemistry. 1. Inorganic and organic processes. J. Geophys. Res. **88** , 10865-10882.

Graedal, T.E. and Weschler, C.J. 1981. Chemistry within aqueous atmospheric aerosols and raindrops. Rev. Geophys. Space, Phys. **19** , 595-539.

Guderian, R. (ed) 1983. Air Pollution by Photochemical Oxidants. Ecological Studies 52 (Springer-Verlag, Berlin).

Hales, J.M. 1986. Mechanistic analysis of precipitation scavenging using a one-dimensional, time-variant model. Atmospheric Environment **16** , 1775-1783.

Hamer, W.J. 1968. Theoretical mean activity coefficients of strong electrolytes in aqueous solutions from 0 to 00°C. National Bureau of Science publication - NSRDS-NBS24 (Department of Commerce, Washington D.C.).

Hard, T.M., Chan, C.Y., Mehrabzadeh, A.A., Pan, W.H. and O'Brien, R.J. 1986. Diurnal cycle of tropospheric OH. Nature **322** , 617-620.

Harrison, R.M. and Pio, C.A. 1982a. Major ion composition and chemical associations of inorganic atmospheric aerosols. Environ. Sci. Technol. **17** , 169-174.

Harrison, R.M. and Pio, C.A. 1983b. Size-differentiated composition of inorganic atmospheric aerosols of both marine and polluted continental origin. Atmospheric Environment **17** , 1733-1738.

Hoffmann, M.R. and Calvert, J.G. 1985. Chemical transformation modules for eulerian acid deposition models, volume 2. The aqueous-phase chemistry. Acid Deposition Modelling Project, National Centre for Atmospheric Research, P.O. Box 3000, Boulder, Colorado. (prepared for U.S.E.P.A., Research Triangle Park, NC 27711).

Hosker, R.P. 1986. Practical application of air pollutant deposition models - current status, data requirements and research needs. in Legge and Krupa (1986), p. 505-567.

Hough, A.M. 1986. The significance of physical and chemical processes in a photochemical oxidant model. Harwell Report R-12294, (1987) (HMSO London).

Hough, A.M. 1987. A computer modelling study of the chemistry occurring during cloud formation over hills. Atmospheric Environment 21 , 1073-1095.

Hough, A.M. and Derwent, R.G. 1987. The impact of motor vehicle control technologies on future photochemical ozone formation in the United Kingdom. Envir. Pollut. 44 , 109-118.

Hough, A.M. and Eggleton, A.E.J. 1986. Acid deposition modelling: a discussion of the processes involved and the options for future development. Sci. Prog. Oxf. 70 , 353-379.

Huebert, B.J. and Robert, C.H. 1985. The dry deposition of nitric acid to grass. J. Geophys. Res. 90 , 2085-2090.

Huntzicker, J.J., Cary, R.A. and Ling, C-S 1980. The neutralisation of sulphuric acid aerosol by ammonia. Environ. Sci. Technol. 14 , 819-824.

Jacob, D.J. and Hoffmann, M.R. 1983. A dynamic model for the production of H^+, NO_3^- and SO_4^{2-} in urban fog. J. Geophys. Res 88 , 6611-6621.

Jacob, D.J., Waldman, J.M., Munger, J.W. and Hoffmann, M.R. 1986). The SH_2SO_4 - HNO_3 - NH_3 system at high humidities and in fogs 2. Comparison of field data with thermodynamic calculations. J. Geophys. Res. 91 , 1089-1096.

Jones, R. and Heinemann, K. 1985. Studies on the dry deposition of aerosol particles on vegetation and plane surfaces. J. Aerosol Sci. 16 , 463-471.

Kerr, J.A. and Calvert, J.G. 1984. Chemical transformation modules for Eulerian acid deposition models. Volume 1: The gas-phase chemistry. Acid Deposition Modelling Project, National Centre for Atmospheric Research, P.O. Box 3000, Boulder, Colorado. (prepared for U.S.E.P.A., Research Triangle Park, NC 27711).

Kumar, S. 1986. Reactive scavenging of pollutants by rain: a modelling approach. Atmospheric Environment 20 , 1015-1024.

Legge, A.H. and Krupa, S.V. (eds.) 1986. Air pollutants and their effects on the terrestrial ecosystem. (Wiley, New York).

Leone, J.A. and Seinfeld, J.H. 1984. Analysis of the characteristics of complex reaction mechanisms: application to photochemical smog chemistry. Environ. Sci. Technol. 18 , 280-287.

Liu M-K., Morris, R.E. and Killus, J.P. 1984. Development of a regional oxidant model and application to the north eastern United States. Atmospheric Environment 18 , 1145-1161.

46

Logan, J.A. 1985. Tropospheric ozone: seasonal behaviour, trends, and anthropogenic influence. J. Geophys. Res. **90** , 10463–10482.

Loye-Pilot, M.D., Martin, J.M. and Morelli, J. 1986. Influence of Saharan dust, on the rain acidity and atmospheric input to the Mediterranean. Nature **321** , 427–429.

McElroy, W.J. 1986a. Sources of hydrogen peroxide in cloud-water. Atmospheric Environment **20** , 427–438.

McElroy, W.J. 1986b. The aqueous oxidation of SO_2 by OH radicals. Atmospheric Environment **20** , 323–330.

McMurry, P.H., Takano, H. and Anderson, G.R. 1983. Study of the ammonia (gas) - sulphuric acid (aerosol) reaction rate. Environ. Sci. Technol. **17** , 347–352.

Martens, C.S., Wesolowsky, J.J., Harris, R.C. and Kaifer, R. 1973. Chlorine loss from Puerto-Rican and S. Francisco bay area maritime aerosol. J. Geophys. Res. **78** , 8778–8972.

Martin, A. 1980. Sulphur in rain and deposited from air and rain over Great Britain and Ireland. Envir. Pollut. B1, 177–193.

Moller, D. 1984. Estimation of the global man-made sulphur emission. Atmospheric Environment **18** , 29–39.

Mozurkewich, M. 1986. Aerosol growth and the condensation coefficient for water: a review. Aerosol Science and Technology, 223–236.

Pasquill, F. and Smith, F.B. 1983. Atmospheric Diffusion, 3rd Ed. (Ellis Horwood, Chichester, England). See pages 283 et seq.

Peterson, T.W. and Seinfeld, J.H. 1980. Heterogeneous condensation and chemical reaction in droplets - application to the heterogeneous atmospheric oxidation of SO_2. Adv. Env. Sci. Tech. **10** , 125–180.

Richards, L.W. 1983. Comments on the oxidation of NO_2 to nitrate - day and night. Atmospheric Environment **17** , 397–402.

Ross, A.B. and Neta, P. 1979. Rate constants for radicals of inorganic radicals in aqueous solution. NSRDS-NBS65, (National Bureau of Standards, Washington).

Sandalls, F.J. and Chadwick, M. 1985. Ambient ozone measurements at Harwell, January-December 1985. Harwell Report R-12173, (HMSO, London).

Sandalls, F.J. and Gibson, A.R. 1986. Ambient ozone measurements at Harwell, January-December 1985. Harwell Report R-12173, (HMSO, London).

Sandalls, F.J. and Leonard, A.W. 1986. Measurements of nitric oxide and nitrogen oxide in ambient air at Harwell. Harwell Report R-12160, (HMSO, London).

Saxena, P. and Peterson, T.W. 1981. Thermodynamics of multicomponent electrolytic aerosols. J. Colloid and Interface Sci. **79** , 496–510.

Sehmel, G.A. 1980. Particle and gas dry deposition: a Review. Atmospheric Environment **14** , 983–1011.

Seigneur, C. and Saxena, P. 1984. A study of atmospheric acid formation in different environments. Atmospheric Environment **18** , 2109–2124.

Seinfeld, J.H. 1986. Atmospheric Chemistry and Physics of Air Pollution. (Wiley, New York).

Semb, A. and Amble, E. 1981. Emission of nitrogen oxides from fossil fuel combustion in Europe. NILU Teknisk Rapport NR 13/81. (Norwegian Institute for Air Research, Lillestrom, Norway).

Smith, F.B. 1981. The significance of wet and dry synoptic regions on long-range transport of pollution and its deposition. Atmospheric Environment **15** , 863-873.

Stelson, A.W. and Seinfeld, J.H. 1982a. Relative humidity and temperature dependence of the ammonium nitrate dissociation constant. Atmospheric Environment **161** , 983-992.

Stelson, A.W. and Seinfeld, J.H. 1982b. Relative humidity and pH dependence of the vapour pressure of ammonium nitrate - nitric acid solutions at 25°C. Atmospheric Environment **16** , 993-1000.

Tanner, R.L. 1982. An ambient experimental study of phase equilibrium in the atmospheric system: aerosol H^+, NH_4^+, SO_4^{2-}, NO_3^- - NH_3 (g), HNO_3 (g). Atmospheric Environment **16** , 2935-2942.

Tuazon, E.C., Atkinson, R., Plum, C.N., Winer, A.M. and Pitts, J.N. 1983. The reaction of gas phase N_2O_5 with water vapour. Geophys. Res. Letts. **10** , 953-956.

United Kingdom Review Group on Acid Rain 1983. Acid Deposition in the United Kingdom. (Warren Spring Laboratory, Stevenage, UK).

Walcek, C.J., Brost, R.A., Chang, J.S. and Wesely, M.L. 1986. SO_2 sulphate and HNO_3 deposition velocities computed using reginal use and meteorological data. Atmospheric Environment **20** , 949-964.

Waldman, J.M., Munger, J.W., Jacob, D.J. and Hoffmann, M.R. 1985. Chemical characterisation of stratus cloudwater and its role as a vector for pollutant deposition in a Los Angeles pine forest. Tellus, 37B, 91-108.

Whitby, K.T. and Sverdrup, G.M. 1980. California aerosols: their physical and chemical characteristics. Adv. Env. Sci. Tech. **10** , 477-538.

Willison, M.J., Clarke, A.G. and Zeki, E.M. 1985. Seasonal variation in atmospheric aerosol concentration and composition at urban and rural sites in northern England. Atmospheric Environment **19** , 1081-1089.

METEOROLOGICAL AND CHEMICAL FACTORS INFLUENCING CLOUDWATER COMPOSITION IN A NON-PRECIPITATING, LIQUID-WATER UPDRAFT

C.J. Walcek
National Centre for Atmospheric Research
P O Box 3000
Boulder CO 80307
USA

ABSTRACT. An aqueous chemical model has been combined with an entraining updraft model to predict cloudwater composition above cloud base. Meteorological factors such as liquid water content, entrainment, pressure and temperature were allowed to vary as they might in a cloudy environment. Chemical reactions producing sulphuric acid, formic acid and hydrogen peroxide (H_2O_2) are extremely sensitive to the microphysical, dynamical, and radiative properties within a cloud. Several sulphate production pathways are sensitive to cloudwater pH, which is largely determined by sulphate aerosol concentration and cloud liquid water content. Hydrogen peroxide production may occur in both the gas and aqueous phase, and is sensitive to the amount of light available to initiate photochemical reactions. Near cloud base, H_2O_2 may be rapidly destroyed by SO_2 in polluted areas. At higher cloud levels, where light intensities are higher and SO_2 concentrations are usually lower, H_2O_2 may be produced at rates of up to 0.5 ppb h^{-1}. Vertical or horizontal variations in meteorological factors alone will produce significant changes in cloudwater composition. Entrainment of cleaner mid-tropospheric air, pressure reductions, and increases in liquid water content lead to a decrease in solute concentrations in cloudwater. In contrast, reductions in temperature and aqueous-phase acid generation cause solute concentrations to increase in cloudwater. At high altitude sites, the composition of cloudwater will vary significantly with altitude. Near cloud base, concentrations of sulphate and nitrate in cloudwater will probably be highest, although cloud liquid water contents and droplet sizes will be lower. At higher elevations above cloud base, liquid water contents and drop sizes are larger, while solute concentrations may be lower. These model estimates suggest that several chemical, cloud-dynamical and cloud-microphysical parameters must be well characterised in order to predict cloudwater composition.

M. H. Unsworth and D. Fowler (eds.), Acid Deposition at High Elevation Sites, 49–72.
© *1988 by Kluwer Academic Publishers.*

1. INTRODUCTION

Since mountaintop locations are frequently engulfed in cloud, a
significant amount of acidic deposition at these sites can occur via
cloudwater impaction on material and biological surfaces. In order to
quantify the level of acidity in cloudwater at these locations, it is
useful to develop an understanding of the chemical and meteorological
factors which influence cloudwater composition as air rises within a
cloud updraft.
    Several previous researchers have developed models of cloudwater
composition (e.g. Chameides, 1984; Graedel and Goldberg, 1983; Jacob
and Hoffmann, 1983) capable of describing the chemical composition of
cloudy, rainy or foggy environments. These models assume that a cloud
or raining environment is a meteorologically "closed" reaction chamber,
where variables such as liquid water content, temperature or pressure
are held constant. In addition, entrainment of noncloudy environmental
air into the reaction system is ignored. Numerous observations have
demonstrated that the environment within a cloud is very dynamic, with
each of these parameters varying considerably over the vertical and
horizontal extent of a given cloud. To alleviate these deficiences,
Walcek and Taylor (1986) developed a useful model for computing
cloudwater composition in liquid water clouds by assuming that cloud
base air is diluted with entrained mid-tropospheric air as parcels of
air rise above cloud base. Cloudwater composition is thus computed
using a more realistic meteorological characterisation of the cloudy
environment.
    In the following discussion, the methods and results of Walcek and
Taylor (1986) are further analysed. Sensivity studies of a cloud
chemical model are also presented which identify several important
meteorological parameters which will influence cloudwater composition.
A chemical model of a cloud updraft is then applied to conditions which
may be representative of non-precipitating, liquid water clouds present
at high elevation sites.

2. CLOUD UPDRAFT METEOROLOGY

2.1 Entrainment - microphysical and dynamical implications

Numerous observations (e.g. Paluch, 1979; Hill and Choularton, 1985) of
both stratiform and cumuliform clouds have shown that at a given level
in a cloud, there are significant variations in several microphysical
and dynamical cloud properties over distances less than 100-200 m.
These observations also show that parameters such as temperature,
liquid water content, and updraft velocity are strongly correlated with
each other. Regions of higher temperature are associated with greater
liquid water content and higher updraft velocity. This correlation can
be explained through a simple entrainment mechanism: air rising from
cloud base encounters discrete entrainment "events". As dry
environmental air is mixed with this rising moist air, there is an
immediate reduction in updraft speed, since essentially stagnant air is

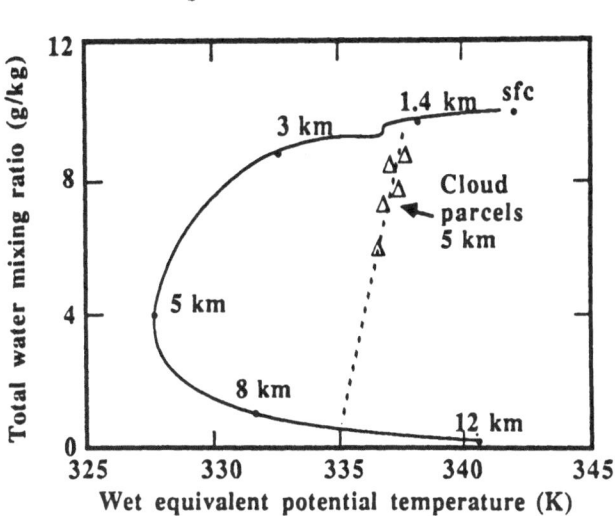

Figure 1. Schematic plot of wet equivalent potential temperature vs. total water content for a particular sounding in which clouds develop. In-cloud points are shown as triangles. Solid curve represents environment around cloud.

content will be reduced, through both dilution and evaporation. Any evaporation which occurs will reduce the parcel's temperature, and thus its bouyancy and vertical acceleration will also be reduced. Different degrees of entrainment will result in discrete mixtures of air with a unique thermodynamic, dynamical and microphysical makeup.

It is possible to quantify this entrainment process through the use of environmental tracers. Paluch (1979) utilised two tracers, saturated equivalent potential temperature and total water mixing ratio, to determine the source of air in a cloudy parcel. Since each of these parameters is roughly conserved during vertical displacements, mixtures of air from cloud base and higher levels would produce air with linear combinations of the properties of the air which make up the cloudy parcel. Figure 1 shows a schematic of a sounding through a cloud as well as a sounding of these two parameters in the environment surrounding the cloud. Based on diagrams such as these, Hill and Choularton (1985) were able to identify source regions of air within clouds.

It is well recognized that horizontally averaged measurements of liquid water content in clouds are significantly lower than one might expect if parcels of air were to rise adiabatically from cloud base and retain their condensed water load. Warner (1970) has shown that the ratio of the observed liquid water content to the adiabatic liquid water content is near unity near cloud base, but within a kilometer above cloud base, liquid water contents are only about 20% of the adiabatic amount. This suggests that entrainment events are more likely to occur the further an air parcel rises above cloud base. Undilute parcels of air have been observed at relatively high elevations above cloud base, although they occupy only a small fraction of the total cloud area. In a non-precipitating cloud, if the ratio of liquid water content to adiabatic liquid water content is known for a given parcel, it is possible to quantify the degree of entrainment within that parcel. In the following discussion, a technique is presented for determining the relationship between entrainment, temperature, and liquid water content. Once these parameters are known, it is possible to estimate the chemical composition of cloudwater at a given level above cloudbase.

2.2 Diagnostic model of entrainment

The adiabatic liquid water mixing ratio, q_{lad} ($kg_{water}kg_{air}^{-1}$), at a given level above cloud base in a non-precipitating cloud can be determined if the temperature and pressure of the cloud base are known using the following relationship

$$q_{lad} = q_s(T_{cb}, P_{cb}) - q_s(T_{ad}, P_c) \qquad (1)$$

where $q_s(T_{cb}, P_{cb})$ is the saturated water vapour mixing ratio at cloud base temperature (T_{cb}) and pressure (P_{cb}), and $q_s(T_{ad}, P_c)$ is the saturated water vapour mixing ratio at a given cloud pressure level (P_c) if a parcel of air were to rise adiabatically from cloud base at its moist adiabatic temperature (T_{ad}). The saturation vapour pressure for water is given by the following relationship

$$q_s(T, P) = \frac{e_s(T) \varepsilon}{[P - e_s(T)]} \qquad (2)$$

where ε is the ratio of the molecular weight of water to air (= 0.622), P is the pressure, and $e_s(T)$ is the saturation vapour pressure of water. A reasonable approximation for this saturation vapour pressure over water as a function of temperature is

$$e_s(T) = 6.11 \exp\left(19.83 - \frac{5417.4}{T}\right) \quad (mb) \qquad (3)$$

The adiabatic temperature can be determined if the cloudbase temperature and pressure are known using the following relationship

$$T_{ad} = T_{cb} - \frac{\Delta P R T}{P C_p} \frac{\left[1 + \dfrac{L_v q_s(T, P)}{RT}\right]}{\left[1 + \dfrac{L_v^2 q_s(T, P)\varepsilon}{RT^2 C_p}\right]} \qquad (4)$$

where ΔP is the difference between cloudbase pressure and the pressure at the level where T_{ad} is being computed ($P_{cb} - P_c$), R is the ideal gas

constant for air (287 J deg^{-1}kg^{-1}), L_v is the latent heat of
vapourisation for water (2.5 x 10^6 J kg^{-1}), C_p is the heat capacity of
air at constant pressure (1004 J deg^{-1}kg^{-1}). Values for the pressure
(P) and temperature (T) used in this expression should ideally be the
average over the depth ΔP along a moist adiabat. Under most conditions
encountered in clouds, the following forms yield reasonable results: P
= P_{cb} – $\Delta P/2$; T = T_{cb} – 0.065 $\Delta P/2$. Thus, once cloud base pressure and
temperature are determined, the adiabatic cloudwater content can be
determined at levels above cloud base using Eqs. (1-4).

The actual liquid water mixing ratio in a cloudy parcel (q_{lc}) is
observed to be a fraction α of this adiabatic water content

$$q_{lc} = \alpha q_{lad} \tag{5}$$

In non-precipitating clouds, this reduction can be attributed to
entrainment of mid-tropospheric air into each cloudy parcel. Through
conservation of total water mass (vapour + condensed), the amount of
water in a mixed cloudy parcel will equal the proportional amount of
water in the air rising from cloud base and the air entrained from some
level above cloud base. By assuming that the cloudy parcel is
saturated at the cloud temperature, the total water in the mixed parcel
will be given by

$$q_s(T_{cb}, P_{cb}) (1-f) + q_e f = q_s(T_c, P_c) + q_{lc} \tag{6}$$

where f is the fraction of air in the cloudy parcel which was entrained
into the parcel, q_e is the water vapour mixing ratio of the air
entrained into the parcel, and T_c, P_c are the temperature and pressure
of the mixed cloudy parcel. Total energy will also be conserved after
air from outside the rising cloudy parcel is mixed. Adiabatically
heated air from above the parcel level can mix and evaporate liquid
water, which will cool the mixture. This energy conservation can be
approximated as

$$C_p(T_c - T_i) = L_v \Delta q_l \tag{7}$$

with T_i being the initial temperature of the mixture of air before any
evaporation has occurred:

$$T_i = T_{ad}(1-f) + fT_e\left(\frac{P_c}{P_e}\right)^{R/C_p} \tag{8}$$

where P_e and T_e refer to the pressure and temperature of the entrained
air at the level where it originated in the environment outside the
cloud. In this expression, the minor variations in heat capacity due
to vapour or liquid water present in the air has been neglected. The
change in liquid water mixing ratio (Δq_l) in Eq. 7 due to evaporation
is

$$\Delta q_l = q_{lad}(1-f) - q_{lc} \tag{9}$$

If one specifies the fraction α of adiabatic water present in a cloud parcel, then equations (5-9) represent an implicit relationship between the final cloud parcel temperature and fraction of cloudy air entrained into the parcel, which can be solved through iterative techniques. Figure 2 shows an example of how liquid water content and temperature

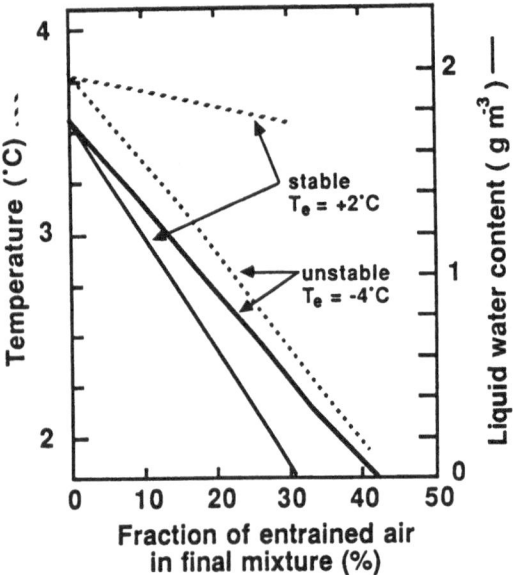

Figure 2.   Liquid water content and temperature vs. amount of air entrained into final cloudy mixture. T_e refers to the temperature of air entrained into cloud. Conditions of air in cloud and air entrained into cloud are described in the text.

vary with different degrees of entrained fraction in the mixed cloudy parcel. The conditions for this simulation are as follows: pressure at cloud level - 100 mb lower than cloud base pressure; cloud base at 900 mb, 9°C. Air entrained into the cloud as assumed to come from the 700 mb level with a water vapour mixing ratio of 2 g kg^{-1}. With no air entrained into the parcel (f = 0) adiabatic temperature and liquid water contents are present. As more air is entrained into the parcel, the liquid water content and temperature of the mixed parcel are reduced. Under stable conditions (T_e = +2°C) the mixed parcel temperature does not vary significantly with amount of entrainment. In contrast, under more unstable conditions (T_e = -4°C), the parcel temperature is fairly sensitive to the amount of entrainment.

3. AQUEOUS CHEMISTRY

Cloudwater composition is determined by the composition of air entering into the cloud from either cloud base or entrained into the cloud from levels above cloud base. Soluble aerosols provide condensation nuclei for cloud drops, and gases dissolve and equilibrate with the cloudwater. Irreversible chemical reactions oxidise SO_2 or other chemical species into more acidic forms. A model of aqueous chemistry which can predict the composition of cloudwater has been developed which incorporates the features outline above. Each of these chemical processes will be discussed in further detail in the following sections.

3.1 Nucleation Effects

A significant portion of the containment load within cloudwater is incorporated into drops during their formation. Measurements of fogwater made immediately after condensation (Munger et al., 1983) showed relatively high concentrations of a number of ions including NH_4^+, $SO_4^=$, H^+, NO_3^-, Cl^-, Na^+, F^+, Ca^{2+} and Mg^{2+}. These ions most likely represent the soluble portions of the nucleating aerosol. In this chemical model, soluble aerosols are assumed to be a source of sulphates, carbonates, chlorides and trace metals in cloudwater. Sulphate is assumed to be apportioned between sulphuric acid (H_2SO_4) and a neutralised ammonium sulphate ($(NH_4)_2SO_4$). The sulphuric acid portion will strongly influence cloudwater acidity. Carbonate-laden soil dust can strongly neutralise a solution as calcium or magnesium ions (anions assumed to be associated with the $CO_3^=$ ion) replace H^+ as the dominant cation. Sea salts (assumed to be sodium or potassium chloride) will influence the ionic stength of cloudwater, and may also affect aqueous-phase free radical chemical reactions converting dissolved SO_2 to sulphuric acid. Other trace metals such as Fe^{3+} and Mn^{2+} can act as catalysts in the oxidation of SO_2. If each of these aerosol concentrations is specified in the air above and below cloudbase, then concentrations of the corresponding ions in cloudwater can be derived using the following relationship

$$[i] = E \frac{[\mu_b(1-f) + \mu_e f] \, P_c(273.15)}{LTM_w 10^3 (1013.25)} \tag{10}$$

where μ_b and μ_e are the loading of aerosol ($\mu g \, m^{-3}$ at 1 atmosphere pressure, 273K) in the region below cloud base and region of entrainment into the cloudy parcel, E is the aerosol scavenging efficiency, M_w is the molecular weight of the aerosol, L is the cloud liquid water content ($= 10^5 P_c q_{1c}/(RT_c) \, g \, m^{-3}$), and f is the fraction of air entrained into the cloudy parcel from above cloudbase, derived in section 2.2. The bracketed quantity [i] refers to concentration of ion in cloudwater (moles $litre_{water}^{-1}$).

3.2 Trace Gases

Numerous trace gases will dissolve in cloudwater to form ionic or acidic species. An equilibrium is quickly established between the gas and aqueous phases in a cloud if droplet sizes are small and chemical reaction rates are relatively slow. Equilibrium relations used to define concentrations of trace gases in cloud water are listed in Table 1. All equilibrium constants have the following temperature dependence

$$K(T) = K(T_0) \exp \left[\frac{\Delta E_a}{R} \left(\frac{1}{T} - \frac{1}{T_0} \right) \right] \tag{11}$$

where ΔE_a is the enthalpy change for the equilibrium reactions involved $K(T_0)$ is the equilibrium constant at a reference temperature, T_0. Concentrations of dissolved ions will be influenced by the ionic strength of the cloudwater. An activity coefficient correction will account for these non-ideal effects in cloudwater solutions, although this effect is usually minor. Partial pressures of trace gases in cloudy parcels can be estimated by assuming that these pollutants are mixed into the parcel from above and below cloud base

$$P^\circ = [\mu_{gb}(1-f) + \mu_{ge}f] \frac{P_c}{1013.25} \tag{12}$$

where P° refers to the total partial pressure (gas + dissolved concentration expressed in atmospheres) of trace gas in the cloudy parcel, μ_{gb} and μ_{ge} are the volume mixing ratio of trace gas in the below- and above-cloudbase regions, f is the fraction of air entrained into the cloud mixture, and P_c is the pressure (mb) of the cloudy parcel. By assuming a mass balance and equilibrium between the gas and aqueous phase, the concentration of trace gas in cloud water can be computed as follows:

$$[G] = \frac{H'P^\circ}{(1 + LRTH'/10^6)} \tag{13}$$

where [G] is the molar concentration of all dissolved forms of the trace gas of interest, R is the universal gas constant (0.082 atm 1 mole^{-1}K^{-1}) and H' is the "effective" Henry's Law equilibrium constant for the gas, defined as the ratio of the equilibrium concentrations of all hydrated and ionic forms of a trace gas to the partial pressure in the gas phase (molar atm^{-1}). For example, SO_2 has three dissolved forms in solution: and aquated form (SO_2H_2O), and two dissociated forms (HSO_3^- and $SO_3^=$). The "effective" Henry's Law constant for SO_2 is therefore a function of the cloudwater hydrogen ion concentration [H$^+$]:

$$H'_{SO2} = H_s \left[1 + \frac{K_{1s}}{[H^+]} \left(1 + \frac{K_{2s}}{[H^+]} \right) \right] \tag{14}$$

TABLE 1. Aqueous-phase Equilibria in Cloudwater

Equilibrium Expression		Equilibrium Constant @ 278 K		$\Delta E_a/R$ (K)	Reference
(E1) SO_2 (g)	$\Leftrightarrow SO_2 \cdot H_2O$	H_s = 2.58×10^0	(M atm^{-1})	3120	NBS
(E2) $SO_2 \cdot H_2O$	$\Leftrightarrow HSO_3^- + H^+$	K_{1s} = 2.12×10^{-2}	(M)	1964	M
(E3) HSO_3^-	$\Leftrightarrow SO_3^- + H^+$	K_{2s} = 9.07×10^{-8}	(M)	1432	M
(E4) HSO_4^-	$\Leftrightarrow SO_4^- + H^+$	K_{s6} = 1.96×10^{-2}	(M)	2714	SM
(E5) HNO_3(g)	$\Leftrightarrow NO_3^- + H^+$	H_{n3} = 2.09×10^7	(M^2 atm^{-1})	8700	SW
(E6) NH_3(g)	$\Leftrightarrow NH_4OH$	H_a = 1.54×10^3	(M atm^{-1})	4085	NBS
(E7) NH_4OH	$\Leftrightarrow OH^- + NH_4^+$	K_{1a} = 1.60×10^{-5}	(M)	-477	SM
(E8) HCl(g)	$\Leftrightarrow Cl^- + H^+$	H_{cl} = 1.90×10^7	(M^2 atm^{-1})	9901	CB
(E9) $HCOOH$(g)	$\Leftrightarrow HCOOH$ (aq)	H_o = 1.45×10^4	(M atm^{-1})	5700	W
(E10) $HCOOH$(aq)	$\Leftrightarrow HCOO^- + H^+$	K_{1o} = 1.71×10^{-4}	(M)	†	W
(E11) CO_2 (g)	$\Leftrightarrow CO_2 \cdot H_2O$	H_c = 5.56×10^{-2}	(M atm^{-1})	2423	NBS
(E12) $CO_2 \cdot H_2O$	$\Leftrightarrow HCO_3^- + H^+$	K_{c1} = 4.74×10^{-7}	(M)	-913	NBS
(E13) HCO_3^-	$\Leftrightarrow CO_3^- + H^+$	K_{c2} = 2.68×10^{-11}	(M)	-1760	SM
(E14) H_2O	$\Leftrightarrow OH^- + H^+$	K_h = 1.87×10^{-15}	(M^2)	-6706	SM
(E15) $HCHO$(g)	$\Leftrightarrow HCH(OH)_2$	H_f = 3.26×10^4	(M atm^{-1})	6470	LB
(E16) $HCHO$(aq)	$\Leftrightarrow HCH(OH)_2$	K_f = 1.82×10^3	(M M^{-1})	4017	LH
(E17) O_3(g)	$\Leftrightarrow O_3$(aq)	H_{o3} = 2.12×10^{-2}	(M atm^{-1})	2560	NBS
(E18) H_2O_2(g)	$\Leftrightarrow H_2O_2$(aq)	H_p = 4.71×10^5	(M atm^{-1})	6600	NBS
(E19) $CH_3(CO)OOH$(g)	$\Leftrightarrow CH_3(CO)OOH$(aq)	H_{pa} = 2.07×10^3	(M atm^{-1})	6171	LK
(E20) CH_3OOH(g)	$\Leftrightarrow CH_3OOH$(aq)	H_{mh} = 8.53×10^2	(M atm^{-1})	5607	LK
(E21) HO(g)	$\Leftrightarrow HO$(aq)	H_{ho} = 7.55×10^1	(M atm^{-1})	†	Sp
(E22) HO_2(g)	$\Leftrightarrow HO_2$(aq)	H_{h2} = 1.99×10^3	(M atm^{-1})	†	S
(E23) HO_2(aq)	$\Leftrightarrow O_2^- + H^+$	K_{1h2} = 2.05×10^{-5}	(M)	†	B
(E24) NO_3(g)	$\Leftrightarrow NO_3$(aq)	H_{nr} = 3×10^{-2}	(M atm^{-1})	†	BB
(E25) Cl_2^-	$\Leftrightarrow Cl^- + Cl$	K_{cl} = 5×10^{-6}	(M)	†	Ja

† Temperature dependence uncertain

References: B - *Bielski* [1978]   NBS- *National Bureau of Standards* [1965]
BB - *Berdnikov and Bazhin* [1970]   S   - *Schwartz* [1984]
CB - *Clegg and Brimblecombe* [1985] SM - *Smith and Martell* [1976]
LB - *Lebury and Blair* [1925]   Sp - *Schwartz*, personal communication
LH - *LeHenaff* [1966]   SW - *Schwartz and White* [1981]
LK - *Lind and Kok* [1986]   W - *Weast* [1982]
M - *Mauhs* [1982]

where H_s is the traditionally defined Henry's Law equilibrium constant for SO_2, and K_{1s} and K_{2s} are the first and second dissociation constant for SO_2.

In cloud or rainwater, the following electroneutrality relationship is established, equating the concentrations of positive and negative charge

$$[H^+] + [NH_4^+] + 2([Ca^{2+}] + [Mg^{2+}]) + [Na^+] + [K^+] +$$
$$= 2([SO_4^=] + [CO_3^=] + [SO_3^=]) + [NO_3^-] + [Cl^-] + \quad (15)$$
$$+ [HCOO^-] + [HSO_3^-] + [HCO_3^-] + [OH^-] + [HSO_4^-]$$

Each of these ions can be expressed as a function of the initial aerosol or trace gas concentration in the cloudy parcel, the equilibrium constraints imposed in Table 1, or the hydrogen ion concentration. When these expressions are incorporated into Eq. 15, an implicit relationship results for the cloudwater $[H^+]$, which can be solved iteratively.

3.3 Kinetic Chemistry

Superimposed on the above described equilibria in cloudwater are a series of irreversible reactions which chemically convert several dissolved species. Rapid acidification of cloudwater can occur as SO_2 and formaldehyde are oxidised into sulphuric and organic acid. Table 2 summarizes a list of potentially important aqueous-phase reactions which either directly oxidise SO_2, or produce other oxidants which may be important in chemical conversion processes in clouds. Many of these reactions involve short-lived free radical species.

Most of the reactions in Table 2 have been distilled from several previous compilations of aqueous-phase reactions (Chameides, 1984; Graedel and Goldberg, 1983; Jacob, 1986). In polluted areas, probably the most important chemical reaction in cloudwater involves the oxidation of SO_2 by hydrogen peroxide (R1). Oxidation of dissolved SO_2 by ozone (R2) will be an important reaction in regions where cloudwater pH's are higher than about 5. The oxidation of SO_2 by dissolved free radicals (R3-R13) may also be important, although uncertainties exist regarding the kinetics, mechanisms and products of this oxidation process (McElroy, 1986). Recently, organic peroxides methylhydrogen peroxide and peroxyacetic acid (MHP, PAA) have also been identified in this oxidation process (R14 and R15). Trace metals (R16) are a known catalyst of SO_2 oxidation in aqueous solutions. SO_2 can also complex with dissolved formaldehyde to produce hydroxymethanesulphonic acid, although this rate is relatively slow if cloudwater pHs are less than about 5. Other SO_2 oxidation pathways have been identified, although their rates relative to the above reactions are significantly slower.

The remaining reactions in Table 2 involve free radical reactions which are required in order to compute concentrations of several free radicals in cloudwater. Dissolved HO_2 and chloride radicals dissociate and rapidly equilibriate between different ionic forms in solution (E23 and E25). For these species, reactions are written in terms of the total concentration of radical in solution ($Cl^R = Cl + Cl_2^-$; $HO_2^R =$

TABLE 2. Chemical Mechanism: Aqueous-Phase Reactions

Reactions	Rate Constants[a]	Reference

Sulfur Chemistry

Reactions					Rate Constants[a]	Reference
(R1)	SO_2H_2O	+ H_2O_2	$\Rightarrow H_2SO_4$	+ H_2O	$k_1 = 3.3 \times 10^4 (0.1 + [H^+])^{-1}$	MD
(R2)	S(IV)	+ O_3	$\Rightarrow H_2SO_4$	+ prod	$k_2 = 1.55 \times 10^5 + 79.3/[H^+]$ pH > 2.7	JH
					$k_2 = 1.9 \times 10^4 [H^+]^{-0.5}$ pH < 2.7	JH
(R3)	HSO_3^-	+ HO	$\Rightarrow SO_3^-$	+ H_2O	$k_3 = 9.5 \times 10^9$	AB
(R4)	SO_3^-	+ HO	$\Rightarrow SO_3^-$	+ OH^-	$k_4 = 5.5 \times 10^9$	AB
(R5)	SO_3^-	+ HO_2^R	$\Rightarrow$ HO	+ SO_4^-	$k_5 = 10^7 (1 + K_{1h2}/[H^+])^{-1}$	GG
(R6)	SO_3^-	+ O_2	$\Rightarrow SO_5^-$		$k_6 = 1.5 \times 10^9$	HN
(R7)[‡]	SO_5^-	+ HSO_3^-	$\Rightarrow SO_4^-$	+ SO_4^-		
			$\Rightarrow SO_3^-$	+ HSO_5^-	$k_7 = 3 \times 10^6$	HN
			$\Rightarrow$ HO	+ $2SO_4^-$		
(R8)	SO_5^-	+ SO_5^-	$\Rightarrow 2SO_4^-$		$k_8 = 2 \times 10^8$	J
(R9)	SO_5^-	+ HO_2^R	$\Rightarrow HSO_5^-$		$k_9 = 10^7 (1 + [H^+]/K_{1h2})^{-1}$	J
(R10)	SO_4^-	+ H_2O	$\Rightarrow$ HO	+ SO_4^-	$k_{10} = 10^3$	H
(R11)	SO_4^-	+ H_2O_2	$\Rightarrow HO_2^R$	+ H^+ + SO_4^-	$k_{11} = 1.2 \times 10^7$	MN
(R12)	SO_4^-	+ HSO_3^-	$\Rightarrow SO_3^-$	+ H^+ + SO_4^-	$k_{12} = 10^9$	MN
(R13)	HSO_5^-		$\Rightarrow H_2SO_4$	+ prod	$k_{13} = 2.5 \times 10^{-3}$	†
(R14)	HSO_3^-	+ PAA	$\Rightarrow H_2SO_4$	+ prod	$k_{14} = 1.39 \times 10^7([H^+] + 1.65 \times 10^{-5})$	L
(R15)	HSO_3^-	+ MHP	$\Rightarrow H_2SO_4$	+ prod	$k_{15} = 7.0 \times 10^6[H^+]$	L
(R16)	S(IV) (Fe^{3+}, Mn^{2+})		$\Rightarrow H_2SO_4$	+ prod	k_{16} = see JH reference	JH
(R17)	HSO_3^-	+ HCHO	$\Rightarrow$ HMSA		$k_{17} = 7.9 \times 10^2$	BH
(R18)	SO_3^-	+ HCHO	$\Rightarrow$ HMSA		$k_{18} = 2.5 \times 10^7$	BH

Oxygen-Hydrogen Chemistry

Reactions					Rate Constants[a]	Reference
(R19)	HO_2^R	+ O_3	$\Rightarrow$ HO	+ $2O_2$	$k_{19} = 1.5 \times 10^9(1 + [H^+]/K_{1h2})^{-1}$	S83
(R20)	HO	+ $CH_2(OH)_2$	$\Rightarrow HO_2^R$	+ HCOOH	$k_{20} = 2 \times 10^9$	AN
(R21)	HO	+ $HCOO^-$	$\Rightarrow HO_2^R$	+ H_2O + CO_2	$k_{21} = 3 \times 10^9$	N
(R22)	HO	+ HCOOH	$\Rightarrow HO_2^R$	+ H_2O + CO_2	$k_{22} = 1.5 \times 10^8$	N
(R23)	HO	+ H_2O_2	$\Rightarrow HO_2^R$	+ H_2O	$k_{23} = 2.7 \times 10^7$	Ch
(R24)	HO_2^R	+ HO_2^R	$\Rightarrow H_2O_2$	+ O_2	$k_{24} = \dfrac{[H^+](10^8 K_{1o} + 8.6 \times 10^5[H^+])}{([H^+] + K_{1h2})^2}$	B
(R25)	H_2O_2	(λ < 390nm)	$\Rightarrow$ 2HO		$k_{25} = 1.74 \times 10^{-6}$	*
(R26)	HO	+ HO_2^R	$\Rightarrow H_2O$	+ prod	$k_{26} = \dfrac{7 \times 10^9[H^+] + 10^{10}K_{1o}}{[H^+] + K_{1h2}}$	S68

TABLE 2 (cont.). Chemical Mechanism: Aqueous-Phase Reactions

Reactions					Rate Constants @ 298 K	Reference

Carbonate Chemistry

(R27)	HCO_3^-	+ HO	$\Rightarrow$	CO_3^-	+ H_2O	$k_{27} = 4.9 \times 10^7$	FR
(R28)	CO_3^-	+ H_2O_2	$\Rightarrow$	HO_2^R	+ HCO_3^-	$k_{28} = 8 \times 10^5$	Be
(R29)	CO_3^-	+ HO_2^R	$\Rightarrow$	HCO_3^-	+ O_2	$k_{29} = 4 \times 10^8 (1 + [H^+]/K_{1h2})^{-1}$	Be

Chlorine Chemistry

(R30)	Cl^-	+ SO_4^-	$\Rightarrow$	Cl^R	+ SO_4^-	$k_{30} = 2 \times 10^8$	RN
(R31)	Cl^-	+ HO	$\Rightarrow$	$ClOH^-$		$k_{31} = 4.3 \times 10^9$	Ja
(R32)	$ClOH^-$		$\Rightarrow$	HO	+ Cl^-	$k_{32} = 6.1 \times 10^9$	Ja
(R33)	$ClOH^-$		$\Rightarrow$	Cl^R	+ $2H_2O$	$k_{33} = 2.1 \times 10^{10} [H^+]$	Ja
(R34)	Cl^R	+ H_2O	$\Rightarrow$	$ClOH^-$	+ H^+	$k_{34} = 1.3 \times 10^3 (1 + K_{cl}/[Cl^-])^-$	Ja
(R35)	Cl^R	+ HSO_3^-	$\Rightarrow$	SO_3^-	+ Cl^-	$k_{35} = 7 \times 10^7 (1 + K_{cl}/[Cl^-])^{-1}$	J
(R36)	Cl^R	+ H_2O_2	$\Rightarrow$	HO_2^R	+ $2Cl^-$ + H^+	$k_{36} = 1.4 \times 10^5 (1 + K_{cl}/[Cl^-])^{-1}$	HaN
(R37)	Cl^R	+ HO_2^R	$\Rightarrow$	$2Cl^-$	+ H^+ + O_2	$k_{37} = \dfrac{(4.5 + K_{1h2}/[H^+]) \, 10^9}{(1 + K_{cl}/[Cl^-])(1 + K_{1h2}/[H^+])}$	RN

Nirtate Radical Chemistry

(R38)	NO_3	+ Cl^-	$\Rightarrow$	Cl^R	+ NO_3^-	$k_{38} = 10^8$	RN
(R39)	NO_3	+ HSO_3^-	$\Rightarrow$	SO_3^-	+ NO_3^- + H^+	$k_{39} = 2 \times 10^9$	C86
(R40)	NO_3	+ H_2O_2	$\Rightarrow$	HO_2^R	+ NO_3^- + H^+	$k_{40} = 10^6$	C86

[R] Refers to a free radical species composed of more than one chemical form $HO_2^R \equiv HO_2 + O_2^-$, $Cl^R \equiv Cl + Cl_2^-$, prod refers to inert reactant products such as CO_2, H^+, OH^-, O_2, MHP = methylhydrogen peroxide, PAA = peroxyacetic acid, HMSA = hydroxymethylsulfonic acid, $S(IV) = SO_2H_2O + HSO_3^- + SO_3^-$.

* Scaled with respect to gas-phase photolysis of H_2O_2. Solar angle of $40°$ for photolysis reactions.

† Estimated

‡ Products of this reaction are uncertain, 3 possible pathways have been suggested by Jacob (1986)

§ Temperature dependence for many of these reactions are uncertain, when information available T = 278.15 K is used

References:
AB - *Adams and Boag* [1964]
AN - *Anbar and Neta* [1967]
B - *Bielski* [1978]
Be - *Behar et al.* [1970]
BH - *Boyce and Hoffmann* [1984]
C86 - *Chameides* [1986]
Ch - *Christensen et al.* [1982]
FR - *Farhataziz and Ross* [1977]
GG - *Graedel and Goldberg* [1983]
H - *Hayon et al.* [1972]
HaN - *Hagesawa and Neta* [1978]

HN - *Huie and Neta* [1984]
J - *Jacob* [1986]
Ja - *Jason et al.* [1973]
JH - *Jacob and Hoffmann* [1984]
L - *Lind et al.* [1986]
MD - *Martin and Damschen* [1981]
MN - *Maruthamuthu and Neta* [1978]
N - *Nenadovic et al.* [1972]
RN - *Ross and Neta* [1979]
S68 - *Sehested et al.* [1968]
S83 - *Sehested et al.* [1983]

$HO_2(aq) + O_2^-$). This treatment allows for a simpler handling of the
mass transfer limitations of these reactions, and also allows the
reader to directly see the pH dependence of many of these reaction rate
constants. Radical reactions in cloudwater are responsible for the
oxidation of both SO_2 and formaldehyde (R20). Significant amounts of
formic acid will result from the later oxidation. The oxygen-hydrogen
chemistry reactions (R19-R26) represent the major reactions which
produce and destroy HO and HO_2 in cloudwater. These reactions can
produce hydrogen peroxide, which can then oxidise SO_2. Carbonate
chemical reactions (R27-R29) produce and destroy radical CO_3^-, which
may be an important sink for free-radicals in cloudwater (R29).
Chlorine chemistry (R30-R37) may play an important role in the
oxidation of SO_2 by free radicals, if reaction pathways producing SO_4^-
radical prove to be important (McElroy, 1986). Nitric acid can be
produced (R38-R40) in solution if significant amounts of NO_3 radical
are present in the gas phase.

The reaction list in Table 2 should not be considered
comprehensive, although many of the important reactions identified to
date have been included. Many reactions involving dissolved NO_x and
higher molecular weight organic species have been neglected. These
reactions are usually slow due to the low solubility of the reactants,
and it was found that their inclusion does not significantly affect
cloudwater chemistry.

3.4 Sensitivity to Meteorological Factors

As discussed earlier, several meteorological factors will vary
significantly within clouds. In this section, the sensitivity of the
previously described cloud chemical model to meteorological variations
will be discussed. Individual factors such as entrainment,
temperature, light intensity, liquid water content and drop radius will
be allowed to range over values which may be encountered in a cloudy
environment. The effect of these variations on critical chemical
concentrations or reaction rates will be explored. In the following
section, an updraft model which allows several of these parameters to
vary simultaneously will be presented.

In polluted cloudy areas, the oxidation of dissolved SO_2 by
hydrogen peroxide or ozone will be an important reaction responsible
for the acidification of cloudwater. Figure 3 shows sulphate
production rates via these two reactions as a function of cloud water
content. There is a distinct limit to the sulphate production rate by
H_2O_2 as the liquid water content of the cloud increases. With liquid
water content exceeding about 1 g m^{-3}, most H_2O_2 is partitioned into
the aqueous phase, and further increases in liquid water content act to
dilute the aqueous concentration, and thus the sulphate production rate
no longer increases. In contrast, ozone is relatively insoluble and
partitions primarily into the gas phase. Increases in liquid water
content will lead to a proportional increase in the amount of sulphate
produced via this reaction since there is more water present in which
the reaction can occur.

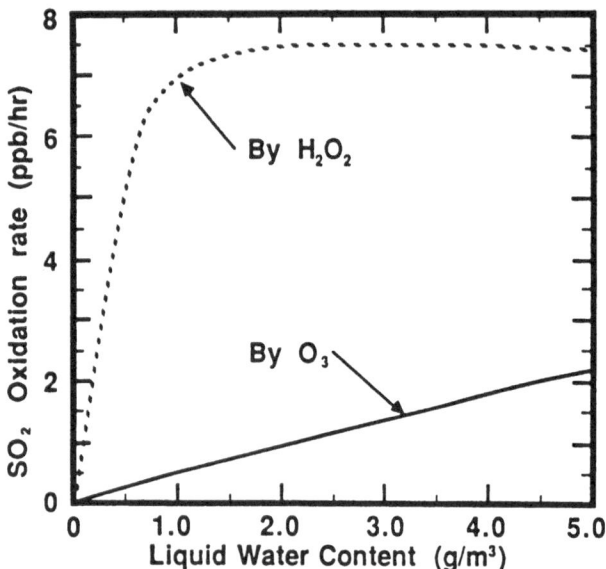

Figure 3.   Aqueous sulphate production rate by O_3 and H_2O_2 vs. cloud liquid water content.   T = 15 C, $[H_2O_2]$ = 1 ppb, $[O_3]$ = 50 ppb, $[SO_2]$ = 3 ppb, P = 0.85 atm.

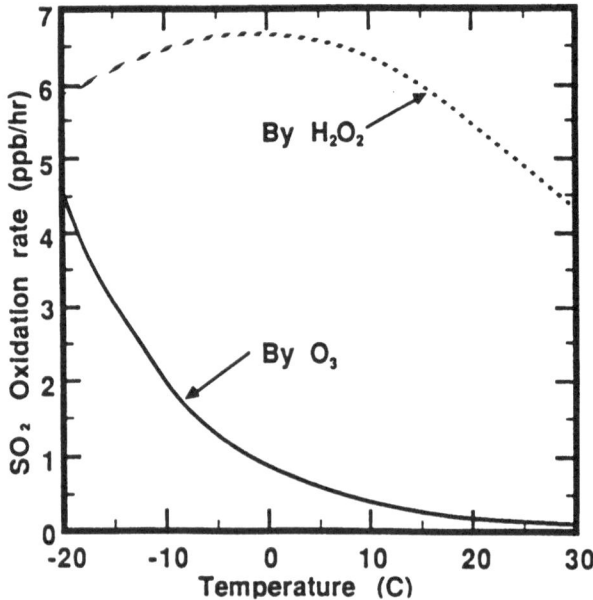

Figure 4.   Aqueous sulphate production rate by O_3 and H_2O_2 vs. cloud temperature L = 0.5 g m^{-3}, $[H_2O_2]$ = 1 ppb, $[O_3]$ = 50 ppb, $[SO_2]$ = 3 ppb, P = 0.85 atm.

The sensitivity of sulphate production to cloud temperature is shown in Figure 4. For H_2O_2 reactions, the decreased solubility at higher temperatures is nearly compensated for by the increased reaction rate with SO_2, and thus the overall SO_2 oxidation rate by H_2O_2 is relatively insensitive to cloud temperature. In contrast, SO_2 oxidation by ozone is more sensitive to cloud temperature since the solubility of the reactants greatly decreases at higher temperatures.

Photochemical reactions occurring in both the gas and aqueous phases are responsible for the formation of numerous oxidants such as H_2O_2 and O_3. In the vicinity of a cloud, photolysis rates will be significantly altered. Above a cloud, the high albedo will produce effective photolysis rates which may be double the rate in a cloud-free environment. Within the upper regions of a cloud, the highly scattering environment can increase the intensity of sunlight significantly. In contrast, below a cloud the solar intensity can be reduced well below a clear-air value. Figure 5 shows the sensitivity

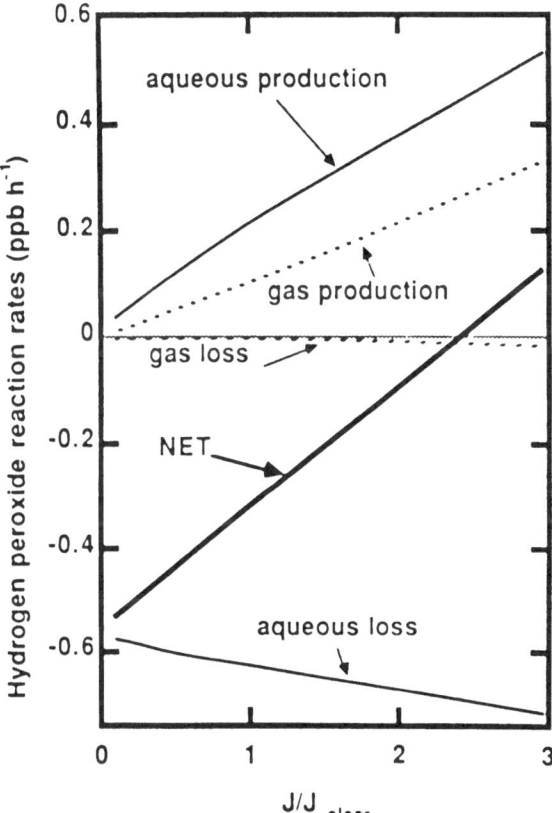

Figure 5. H_2O_2 production and loss rate vs. chemical photolysis rates predicted using aqueous radical chemical model. J/J_{clear} is the ratio of the in-cloud photolysis rate to clear-air photolysis rate.

of H_2O_2 production to variations in the photolysis rates which may be encountered within different regions of a cloud. H_2O_2 is destroyed in clouds rapidly through its reaction with SO_2. Under high photolysis rate conditions, the aqueous and gas production may exceed this destruction rate, implying that some cloudy regions may be important net source areas for H_2O_2 in the troposphere.

The size distribution of liquid water drops can also influence the rate of mass transfer of various chemically reactive species into the aqueous phase. In general, mass transfer to larger drops is less efficient than mass transfer to smaller drops due to a lower surface area to volume ratio. For sulphur oxidation by H_2O_2, O_3, or organic oxidants, the aqueous phase chemical reaction rates are not significantly limited by the rate of mass transfer of the reactants into the cloud droplets. In contrast, for gas-phase radicals which react rapidly in solution, the overall reaction rate may be limited by the mass transfer process, especially if the accommodation coefficient (ratio of the number of molecules which deliquesce into the cloud drop to the number which impinge on the drop surface) for the reactants is less than about 10^{-2}. Both H_2O_2 and organic acids (HCOOH) are produced by reactions involving dissolved HO and HO_2. Figure 6 shows how variations in cloud drop radius can affect H_2O_2 and HCOOH production rates in a cloud. In general, these aqueous reactions become slower in larger drops due to the less efficient mass transfer of free radicals into the cloudwater.

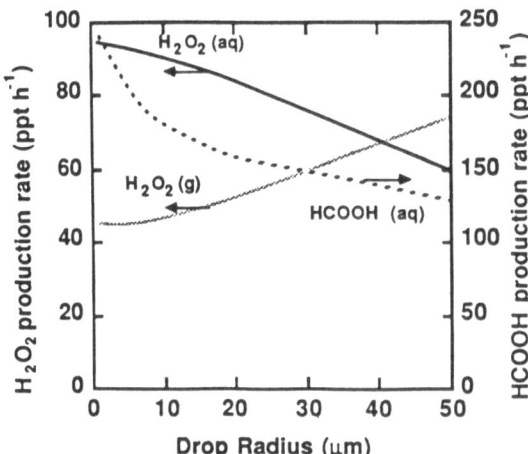

Figure 6. H_2O_2 and formic acid production rate in cloud vs. cloud drop radius computed using aqueous radical chemical model.

Entrainment will not only evaporate liquid water in a rising parcel, but may also introduce air of different composition into the cloudy air. Thus pollutant concentrations will be strongly influenced by the degree of entrainment which a rising parcel is exposed to.

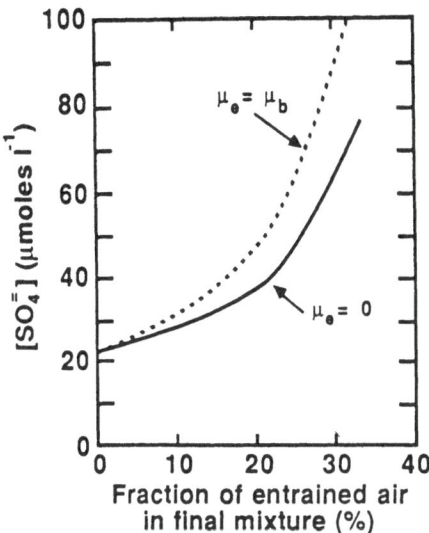

Figure 7. Sulphate concentration in cloudwater vs. fraction of air entrained into adiabatic cloudy parcel. Conditions as in Figure 2. μ_b and μ_e are the concentration of sulphate ($\mu g\ m^{-3}$) entering into cloud base and entrained into cloud.

Figure 7 shows the concentration of sulphate in cloudwater as predicted using Eq. 10 for the unstable conditions of Figure 2. The sulphate nucleation scavenging efficiency E was assumed to be 100% when computing sulphate concentrations for Figure 7. Sulphate concentrations are inversely proportional to liquid water content, and therefore become quite large as more air is entrained into the rising cloudy parcel, evaporating liquid water. The two curves in this figure show the most likely theoretical limits of the final sulphate concentration in cloudwater depending on the composition of the air entrained into the cloud updraft. For the curve with the highest concentrations, the air entrained into the cloudy parcel was assumed to have the same sulphate aerosol loading as the air entering at cloud base. If the air entrained into the cloud has no sulphate, then concentrations of sulphate in the cloudwater are significantly lower, although cloudwater sulphate concentrations still increase with increasing entrainment.

4. CLOUD UPDRAFT CHEMISTRY

All of the above described meteorological factors will vary simultaneously as a given parcel of air rises above cloudbase. In this section, a simple entraining parcel model of cloud updraft chemistry is described which accounts for this meteorological variability. Results are presented which predict the composition of cloudwater at various

elevations above cloudbase under several different pollutant
conditions.

4.1 Updraft Model

A parcel of air was assumed to rise through a cloudy environment with a
steady updraft speed of 2 m s^{-1}. As the parcel rises, meteorological
parameters were assumed to vary according to the conditions shown in
Figure 8. These parameters were predicted by assuming that sufficient

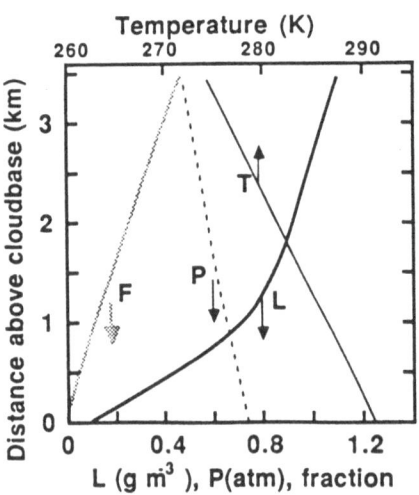

Figure 8.   Liquid water content (L), temperature (T), pressure (P) and
entrained fraction (F) variations with height in a typical
cloud updraft used in updraft model analysis.

air was entrained into each level of the cloud to match the observed
liquid water contents of Warner (1970). Entrainment of environmental
air into the parcel was prescribed to occur at a rate which matches the
predicted entrainment fractions at each level shown in Figure 8. This
method of computing cloudwater composition obviously over-simplifies
the complex dynamics of upward motion and mixing in cloud updrafts.
This technique is employed here to perform a preliminary sensitivity
analysis of the meteorological and chemical factors that influence
cloudwater composition.

4.2 Results

Figure 9a shows the pH (= $-\log_{10}[H^+]$) as a function above cloudbase as
predicted using this rising parcel model for the following mixture of
pollutants at cloud base: 10 ppb SO_2; 10 ppt HNO_3; 340 ppm CO_2; 50 ppt
HCOOH (formic acid); 1 ppb H_2O_2; 1 ppb methylhydrogen peroxide (MHP); 1
ppb peroxyacetic acid (PAA); 50 ppb O_3; 5 μg m^{-3} sulphate (50%
ammonium, 50% sulphuric acid); 10 ng m$^{-3}Fe^{3+}$; 5 ng m$^{-3}Mn^{2+}$. Air

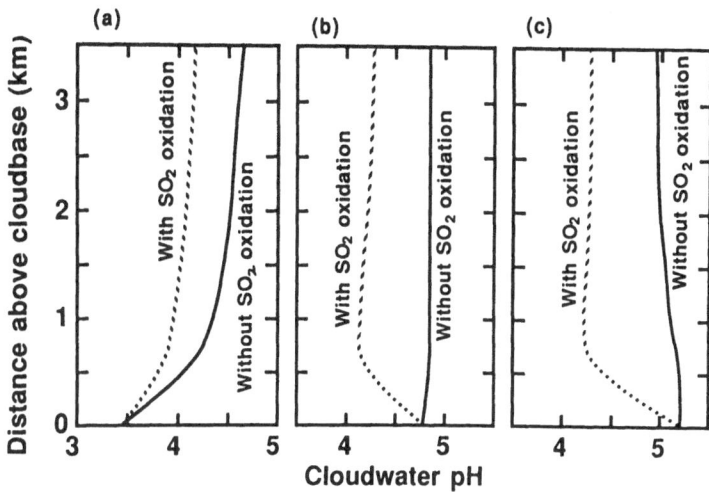

Figure 9. pH vs. height in a typical cloud updraft. Meteorology is given in Figure 8. Figure 9a is a parcel with 5 µg m^{-3} sulphate; 9b is without sulphate load; 9c is with sulphate aerosol, NH$_3$ gas, and carbonate aerosol present. Other chemical conditions are described in text.

entrained into the cloud was assumed to have no sulphate, trace metals, SO$_2$, HNO$_3$ or HCOOH, although concentrations of H$_2$O$_2$, O$_3$, MHP, PAA and CO$_2$ were assumed to be the same throughout the cloud layer. Cloudwater pH is found to rapidly increase above cloud base. This rise in pH with height can be attributed primarily to the dilution of the acidic condensation aerosol by higher liquid water content. Even with the production of sulphuric acid in the parcel, cloudwater is predicted to become less acidic with elevation. This prediction agrees qualitatively with the observations of Leaitch et al. (1983). In Figure 9b, the conditions are the same as in Figure 9a, except it was assumed that there was no sulphate in the air below cloud base. In this case therefore, pH is somewhat higher, and rapid acidification of cloudwater occurs due to the production of sulphuric acid above cloud base. Without this oxidation, cloudwater pH is determined by the partial pressures of SO$_2$, HCOOH and HNO$_3$. The solubility and dissociation of each of these gases increases significantly with

decreasing temperature. The temperature effect alone would
significantly acidify the cloudwater above cloudbase were it not for
the reduction in pressure, entrainment of cleaner air, and liquid water
content increase, which all act to reduce trace gas concentrations in
cloudwater. As a result, without SO_2 oxidation, pH remains roughly
constant with elevation above cloudbase. In Figure 9c, conditions are
the same as Figure 9a, except 1 µg m^{-3} of carbonate-containing aerosol
and 1 ppb NH_3 gas have been added to the air brought into the cloud
from cloudbase. Again, the cloudwater rapidly acidifies due to the
production of sulphuric acid. Even without any aqueous chemistry
occurring, pH is predicted to decrease with height above cloudbase,
since the dilution of these neutralising agent concentrations will lead
to an increase in acidity depending on their concentrations relative to
acidic gas and aerosol concentrations.

In an attempt to quantify importance of the numerous chemical and
meteorological factors which influence cloudwater composition, one
effect at a time will be analysed. Using the conditions which produced
Figure 9a above, Figure 10 shows the artificial effect of allowing only

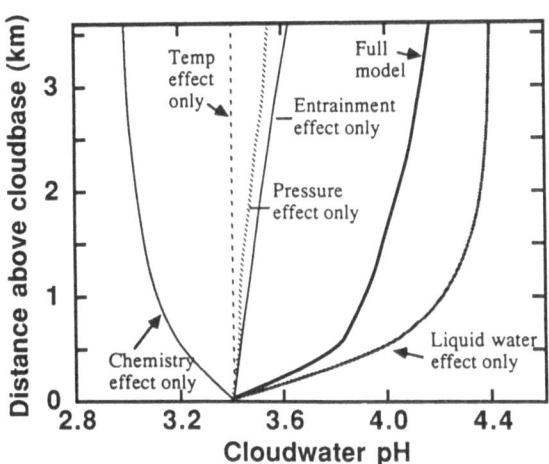

Figure 10. Cloudwater pH vs. distance above cloudbase for conditions
of Figure 9a. Dashed surves represent artificial effect of
allowing only one physiochemical effect to vary with height
as air parcel rises in cloud while all other factors are
maintained at their cloudbase value.

one parameter to vary at a time. All other meteorological parameters
are held at their cloudbase value. In order to "stop" the chemistry,
the concentration of all SO_2 oxidants are set to zero. While it is
physically unrealistic to vary only one parameter at a time, it does
lend some insight to the importance of each variable in determining the
pH distribution in a cloud. One can see that the dominant effect which
acidifies a cloud is the chemical oxidation of SO_2. In contrast,
several other effects act to dilute concentrations of acids in

cloudwater. In order of importance, it appears that increasing liquid water content, entrainment of cleaner air, and pressure reductions will each cause cloudwater solute concentrations to decrease. For the chemical conditions shown here, pH actually decreases with height above cloudbase, despite the efficient production of sulphuric acid. As shown in Figures 9b and 9c, this trend of increasing pH above cloud base may not always be the case, especially if significant amounts of neutralising agents and oxidants are present in the air entering into the cloud from cloudbase.

5. CONCLUSIONS

An aqueous chemical model has been combined with an entraining updraft model to predict cloudwater composition above cloud base. The aqueous composition of cloudwater is predicted to be highly variable over the vertical and horizontal extent of a given cloud. Meteorological factors such as liquid water content, entrainment, pressure and temperature will strongly influence the composition of cloudwater. Chemical reactions producing sulphuric acid, formic acid and hydrogen peroxide (H_2O_2) are extremely sensitive to the microphysical, dynamical, and radiative properties within a cloud. In addition, concentrations of acidic or basic gases and aerosols which are drawn into a cloud from below will influence cloudwater composition. Several sulphate production pathways are extremely sensitive to cloudwater pH, which is largely determined by sulphate aerosol concentration and cloud liquid water content. Hydrogen peroxide production may occur in both the gas and aqueous phase, and is sensitive to the amount of light available to initiate photochemical reactions. Near cloud base, H_2O_2 may be rapidly destroyed by SO_2 in polluted areas. At higher cloud levels, light intensities are higher and SO_2 concentrations are usually lower, causing cloudy areas to be potentially important sources for H_2O_2 in the troposphere. Entrainment of cleaner mid-tropospheric air, pressure reductions, and increases in liquid water content lead to a decrease in solute concentrations above cloud base. In contrast, reductions in temperature and aqueous-phase acid generation cause solute concentrations to increase in cloudwater. At high altitude sites, the composition of cloudwater will probably be highest, although cloud liquid water contents and droplet sizes will be lower. At higher elevations above cloud base, liquid water contents and drop sizes are larger, while solute concentrations may be lower. These model estimates suggest that several chemical, cloud-dynamical and cloud-microphysical parameters must be well characterised in order to predict cloudwater composition.

ACKNOWLEDGEMENTS

Although the research described in this article has been funded as part of the National Acid Precipitation Assessment Programme by the US Environmental Protection Agency under interagency agreement DW930144-1

to the National Centre for Atmospheric Research, the results have not
been subject to the Agency's peer and policy review and therefore do
not necessarily reflect the views of the Agency and no official
endorsement should be inferred.

REFERENCES

Adams, G.E. and Boag, J.W. 1964. 'Spectroscopic studies of the OH
    radical'. Proc. Chem Soc. 112 .
Anbar, M. and Neta, P. 1967. 'A compilation of specific bimolecular
    rate constants for the reactions of hydrated electrons, hydrogen
    atoms and hydroxyl radicals with inorganic and organic compounds
    in aqueous solution'. Int. J. Appl. Rad. Isotopes 18 , 493-523.
Behar, D., Czapski, G. and Duchovny, I. 1970. 'Carbonate radical flash
    photolysis and pulse radiolosis of aqueous carbonate solutions'.
    J. Physc. Chem. 74 , 2206-2210.
Berdnikov, V.M. and Bazhin, N.M. 1970. 'Oxidation-reduction potentials
    of certain inorganic radicals in aqueous solutions'. Russian J.
    Phys. Chem. 44 , 395-398.
Bielski,, B.H.J. 1978. 'Re-evaluation of the spectral and linetic
    properties of HO_2 and O_2-free radicals'. Photochem. Photobiol.
    28 , 645-649.
Boyce, S.D. and Hoffmann, M.R. 1984. 'Kinetics and mechanism of the
    formation of hydroxymethanesulphonic acid at low pH'. J. Phys.
    Chem. 88 , 4740-4746.
Chameides, W.L. 1984. 'The photochemistry of a remote marine
    stratiform cloud'. J. Geophys. Res. 89 , 4739-4755.
Chameides, W.L. 1986. 'Possible role of NO_3 in the night-time
    chemistry of a cloud'. J. Geophys. Res. 91 , 5331-5337.
Christensen, H., Sehested, K. and Corfitzen, H. 1982. Reactions of
    hydroxyl radical with hydrogen peroxide at ambient and elevated
    temperatures'. J. Phys. Chem. 86 , 1588-1590.
Clegg, S.L. and Brimblecombe, P. 1985. 'Potential degassing of
    hydrogen chloride from acidified sodium chloride droplets'.
    Atmos. Environ. 19 , 465-470.
Frahataziz, and Ross, A.B. 1977. 'Selected specific rates of reactions
    of transients from water in aqueous solution. III. Hydroxy
    radical and perhydroxyl radical and their radical ions. Rept. #
    NSRDS-NBS-59, National Bureau of Standards, Washington DC, USA.
Graedel, T.E. and Goldberg, K.I. 1983. 'Kinetic studies of raindrop
    chemistry. 1: Inorganic and organic processes'. J. Geophys.
    Res. 88 , 10865-10882.
Hagesawa, K. and Neta, P. 1978. 'Rate constants and mechanisms of
    reaction for Cl_2^- radicals'. J. Phys. Chem. 82 , 854-857.
Hayon, E., Treinin, A. and Wilf, J. 1972. 'Electronic spectra,
    photochemistry and autoxidation mechanism of the
    sulphite-bisulphite pyrosulphite systems: The SO_2^-, SO_3^-, SO_4^-
    and SO_5^- radicals'. J. Am. Chem. Soc. 94 , 47-57.

Hill, T.A. and Choularton, T.W. 1985. 'An airborne study of the microphysical structure of cumulus clouds'. Quart. J. R. Met. Soc. **111** , 517-544.

Huie, R.E. and Neta, P. 1984. 'Chemical behaviour of SO_3^- and SO_5^- radicals in aqueous solutions'. J. Phys. Chem. **88** , 5665-5669.

Jacob,, D.J. 1986. 'The chemistry of OH in remote clouds and its role in the production of formic acid and peroxymonosulphate'. J. Geophys. Res. **91** , in press.

Jacob, D. and Hoffmann, M.R. 1983. 'A dynamic model for the production of H^+, NO_3^- and $SO_4^=$ in urban fog'. J. Geophys. Res. **88** , 6611-6621.

Jayson, G.G., Parsons, B.J. and Swallow, A.J. 1973. 'Some simple, highly reactive, inorganic chlorine derivatives in aqueous solution'. J. Chem. Soc. Faraday Trans. **69** , 1597-1607.

Leaitch, W.R., Strapp, R.J.W., Wiebe, A. and Isaac, G.A. 1983. 'Measurements of scavenging and transformation of aerosol inside cumulus. Precipitation Scavenging, Dry Deposition and Resuspension. Vol. 1 Elsevier, 53-69.

Lebury, W. and Blair, E.W. 1925. 'The partial formaldehyde vapour pressure of aqueous solutions of formaldehyde II'. J. Chem. Soc., 2832-2839.

Le Henaff, P. 1966. 'Methodes d'etude et propreites des hydrates, hemiacetals et hemithioacetals derives des aldehydes et des cetones'. Bull. Soc. Chim. France, 4687-4700.

Lind, J.A., Kok, G.L. 1986. 'Henry's law determinations for hydrogen peroxide, methylhydroperoxide and peroxyacetic acid '. J. Geophys. Res. **91** , 7889-7895.

Lind, J., Lazrus, A.L. and Kok, G.L. 1986. 'Aqueous-phase oxidation of S(IV) by hydrogen peroxide, methylhydroperoxide and peroxyacetic acid'. J. Geophys. Res. **91** , in press.

Maahs, H.G. 1982. 'Sulphur dioxide/water equilibria between 0° and 50°C. An examination of data at low concentrations'. In: Heterogeneous Atmospheric Chemistry, Schreyer, D.R. ed. Amer. Geophys Union, Washington DC, 187-195.

Martin, L.R. and Damschen, D.E. 1981. 'Aqueous oxidation of sulphur dioxide by hydrogen peroxide at low pH'. Atmos. Environ. **15** , 1615-1621.

Maruthamuthu, P. and Neta, P. 1978. 'Spectra, acid-base equilibria, and reactions with inorganic compounds'. J. Phys. Chem. **82** , 710-713.

McElroy, W.J. 1986. 'The aqueous oxidation of SO_2 by OH radicals'. Atmos. Environ. **20** , 323-330.

Munger, J.W., Jacob, D.J., Waldman, J.M. and Hoffmann, M.R. 1983. 'Fogwater chemistry in an urban atmosphere'. J. Geophys. Res. **88** , 5109-5121.

National Bureau of Standards, 1965. 'Selected values of chemical thermodynamic properties'. NBS Tech note # 270-1, 124 pp.

Nenadovic, M.T., Draganic, Z.D. and Kidric, B. 1972. 'Radiolosis of formic acid-oxygen solution of pD 1.3-13 and the yields of primary product in σ-radiolosis of heavy water'. Proc. Tihany. Symp. Radiat. Che. **3** , 1269-1280.

72

Paluch, I.R. 1979. 'The entrainment mechanism in Colorado cumuli'. J. Atmos. Sci. **36** , 2467-2478.

Ross, A.B. and Neta, P. 1979. 'Rate constants for reactions of inorganic radicals in aqueous solution'. Rept. # NSRDS-NBS-65, US Department of Commerce, Washington DC, USA.

Schwartz, S.E. 1984. 'Gas and aqueous chemistry of HO_2 in liquid water clouds'. J. Geophys. Res. **89** , 11589-11598.

Schwartz, S.E. and White, W.H. 1981. 'Solubility equilibria of the nitrogen oxides and oxyacids in dilute aqueous solution'. Adv. Env. Sci. Eng. **4** , 1-45.

Sehested, K., Rasmussen, O.L. and Fricke, H. 1968. 'Rate constants of OH with HO_2, O_2^- and $H_2O_2^+$ from hydrogen peroxide formation in pulse-irradiated oxygenated water'. J. Phys. Chem. **72** , 626-631.

Sehested, K., Holeman, J. and Hart, E.J. 1983. 'Rate constants and products of the reactions of e_{aq}^-, O_2^- and H with ozone in aqueous solutions'. J. Phys. Chem. **87** , 1951-1954.

Smith, R.M. and Martell, A.E. 1976. Critical Solubility Constants. Vol. 4: Inorganic Complexes. Plenum Press, New York. 257 pp.

Walcek, C.J. and Taylor, G.R. 1986. 'A theoretical method for computing vertical distributions of acidity and sulphate production within cumulus clouds'. J. Atmos. Sci. **43** , 339-355.

Warner, J. 1970. 'On steady-state one-dimensional models of cumulus convection'. J. Atmos. Sci. **27** , 1035-1040.

Weast, R.C. (ed.) 1978. Handbook of Chemistry and Physics, 59th edition. Chemical Rubber Company, Cleveland Ohio.

PHOTOCHEMICAL PRODUCTION OF CARBOXYLIC ACIDS IN A REMOTE CONTINENTAL ATMOSPHERE

Daniel J. Jacob and Steven C. Wofsy
Department of Earth and Planetary Sciences
  and Division of Applied Sciences
Harvard University
Cambridge
Massachusetts 02138

ABSTRACT. Model calculations are conducted to investigate the production of carboxylic acids from photochemical decomposition of isoprene, one of the main natural hydrocarbons emitted from vegetation. Both gas-phase and aqueous-phase chemical reaction pathways are examined. A simple dynamical model is proposed to simulate the boundary layer of the Amazon rain forest, and model predictions are compared to measurements made in that region in July 1985. It is found that formic acid, methacrylic acid, and pyruvic acid can be produced in significant quantities by gas-phase decomposition of isoprene. In the Amazon basin, this source may yield concentrations of these acids in the order of 1 ppb, 0.1 ppb, and 0.02 ppb, respectively. Production of formic acid in cloud by aqueous-phase oxidation of CH_2O does not greatly increase the formic acid concentration predicted from the gas-phase mechanism; cloud droplets with pH > 4 are actually expected to constitute net sinks for formic acid. No significant production of acetic acid is expected from the photochemical decomposition of isoprene. Comparisons of model predictions with field data indicates that isoprene could be a major source of formic acid and pyruvic acid observed in the gas phase and in rainwater; however, acetic acid must originate from another source.

1. INTRODUCTION

Carboxylic acids have been recognised as major contributors to the acidity of precipitation in remote atmospheres (Keene et al., 1983), but little is known of their sources. Keene and Galloway (1986) observed that concentrations of HCOOH and CH_3COOH in rainwater of Central Virginia are much higher in May–September than in October–March, and noted a sharp drop in October coinciding with the senescence of the vegetation. They further noted that the $HCOOH/CH_3COOH$ concentration ratios in rainwater collected at a number

73

M. H. Unsworth and D. Fowler (eds.), Acid Deposition at High Elevation Sites, 73–92.
© 1988 by Kluwer Academic Publishers.

of rural and remote continental sites were remarkably similar. They
inferred that emissions from vegetation are the main atmospheric source
of these two acids. Organic acids are emitted directly by vegetation
(Graedel et al., 1986), but they can also be produced in the atmosphere
by photochemical decomposition of natural hydrocarbons released from
vegetation, in particular isoprene. The relative contributions of
these two types of sources is uncertain. The object of this paper is
to examine the importance of isoprene as a possible atmospheric source
of carboxylic acids. Both gas-phase and aqueous-phase chemical
reaction pathways will be investigated. A simple dynamical model will
be used to simulate concentrations observed during a recent field study
in the Amazon rain forest.

Formic acid is generally the most abundant organic acid found in
rainwater (Keene et al., 1983; Keene and Galloway, 1986). It is
released directly by vegetation, both naturally and during biomass
burning (Graedel et al., 1986; Talbot et al., 1987). It is also
produced in the atmosphere by reaction of CH_2O with HO_2 (Su et al.,
1979), but this reaction is very slow at atmospheric concentrations
(Jacob, 1986). A more important atmospheric source appears to be the
reaction of olefins with O_3 (Atkinson and Lloyd, 1984). In
particular, HCOOH has been observed as a product of the irradiation of
isoprene-NO_x mixtures (Arnts and Gay, 1979). Formation of HCOOH in
this system is thought to proceed by ozonation of the C=C bonds
producing the Criegee biradical $\cdot CH_2OO\cdot$, followed by reaction of
$\cdot CH_2OO\cdot$ with H_2O to give HCOOH (Hatakeyama et al., 1981). The
atmospheric chemistry of Criegee biradicals has been reviewed by
Atkinson and Lloyd (1984).

Some recent papers have argued that aqueous-phase oxidation of
formaldehyde in clouds may represent a major global source of formic
acid (Chameides, 1984; Adewuyi et al., 1984). Formaldehyde scavenged
by cloud droplets hydrolyzes to $H_2C(OH)_2$, which is then rapidly
oxidized by OH(aq) to HCOOH. Formate is in turn rapidly oxidized by
OH(aq), so that the fate of formic acid in a cloud is strongly
dependent on cloudwater pH (Jacob, 1986). Concentrations of HCOOH
predicted solely from this aqueous-phase mechanism are consistent with
measurements in rainwater at remote marine sites (Jacob, 1986; Keene
and Galloway, 1986).

However, as pointed out by Keene and Galloway (1986), the above
aqueous-phase model studies do not account for the much larger HCOOH
concentrations often observed at continental sites. Possibly this
difference could be due to direct emissions of HCOOH from vegetation;
however, it is of interest to explore whether the source could be
explained from the atmospheric chemistry of isoprene. Isoprene is the
main natural hydrocarbon emitted by deciduous trees (Lamb et al.,
1985), and its reaction with O_3 may be a substantial source of HCOOH.
Further, the photochemical decomposition of isoprene yields large
amounts of CH_2O, which may then be oxidized to HCOOH in cloud. The
photochemical decomposition of isoprene may produce other carboxylic
acids by hydrolysis of higher Criegee biradicals, and may produce
acetic acid in cloud by aqueous-phase oxidation of acetaldehyde. The
importances of these pathways will be evaluated.

Our discussion will be based on simulations of boundary layer
chemistry in the Amazon rain forest. Extensive data on atmospheric

concentrations have recently been collected in this region as part of the Amazon Boundary Layer Experiment (ABLE-2A). Isoprene was found to be by far the main non-methane hydrocarbon, with daytime concentrations typically in the range 1-5 ppb (Rasmussen and Khalil, 1987; Zimmerman et al., 1987). Average gas-phase concentrations of formic and acetic acids at ground level were 1.6 and 2.2 ppb, respectively (Andreae et al., 1987a). Data on NO_x and O_3 concentrations, and their fluxes at the ground (Kaplan et al., 1987) were collected. These data provide means to calibrate the model and compare its predictions to actual measurements.

We will begin by a brief description of the model, in particular the gas-phase and aqueous-phase chemistry determining the production and removal of carboxylic acids. We will then present simulations of isoprene chemistry in the Amazon basin, with emphasis on the behaviour of carboxylic acids. The effect of cloud formation on acid production will be investigated. Predicted acid concentrations will be compared to observations.

2. MODEL DESCRIPTION

2.1. Gas-phase chemistry

The gas-phase chemistry of the remote troposphere is simulated with a standard H_xO_y - N_xO_y - CH_4 - CO mechanism (Logan et al., 1981). The photochemistry of isoprene has been added, with the mechanism proposed by Lloyd et al. (1983) and the rate constants of Lurmann et al. (1986). The rate constant for the isoprene + OH reaction has been set to 2.5×10^{11} exp(409/T), following the recommendation by Atkinson (1986). Some necessary alterations have been made to the Lloyd et al. mechanism, in particular the reactions of RO_2 radicals with HO_2 are included and the production of organic acids is explicitly considered. Isoprene reacts with OH, O_3, and NO_3 to produce methylvinylketone and methacrolein. Criegee biradicals are produced following addition of O_3 to the C=C double bonds of isoprene, methylvinylketone, and methacrolein, and subsequent cleavage of the C-O and O-O bonds (Lloyd et al., 1983):

$CH_2CHC(CH_3)CH_2 + O_3 \rightarrow$ 0.5 CH_2O + 0.2 $CH_2CHC(O)CH_3$
   (isoprene)   (methylvinylketone)

$\qquad$ + 0.3 $CH_2C(CH_3)CHO$ + 0.2 $\cdot CH_2OO\cdot$ + 0.06 HO_2
   $\qquad$ (methacrolein)

$\qquad$ + 0.2 $CH_2CHC(\overset{\cdot}{O}\overset{\cdot}{O})CH_3$ + 0.3 $CH_2C(CH_3)\cdot CHOO\cdot$
   $\qquad$ (MVKO)   (MAOO)

$\hfill$ (RG1)

$CH_2CHC(O)CH_3 + O_3 \rightarrow$ 0.5 CH_2O + 0.2 $\cdot CH_2OO\cdot$ + 0.21 HO_2
   (methylvinylketone)

$\qquad$ + 0.2 $CH_3C(O)\cdot CHOO\cdot$ + 0.15 CH_3CHO
   $\qquad$ (MCRG)

$$+ 0.5 \ CH_3C(O)CHO + 0.15 \ CH_3C(O)O_2$$

$$\text{(RG2)}$$

$CH_2C(CH_3)CHO + O_3 \rightarrow 0.5 \ CH_2O + 0.2 \ \cdot CH_2OO\cdot + 0.21 \ HO_2 + 0.15 \ CH_3O_2$
(methacrolein)

$$+ 0.5 \ CH_3C(O)CHO + 0.2 \ CH_3\overset{\bullet}{C}(O\overset{\bullet}{O})CHO \qquad \text{(RG3)}$$
$$\text{(MGLYO)}$$

The thermalized Criegee biradicals $\cdot CH_2OO\cdot$, MVKO, MAOO, MCRG and MGLYO react rapidly with NO, NO_2, SO_2, aldehydes, and H_2O. Except under very polluted conditions, reaction with H_2O is the dominant sink. The hydrolysis of $\cdot CH_2OO\cdot$ produces HCOOH (Hatakeyama et al., 1981). It is likely that MCRG and MAOO are hydrolyzed by a similar mechanism to form pyruvic acid and methacrylic acid, respectively:

$$\cdot CH_2OO\cdot + H \longrightarrow HCOOH + H_2O \qquad \text{(RG4)}$$

$$MCRG + H_2O \longrightarrow H_3CC(O)COOH + H_2O \qquad \text{(RG5)}$$
$$\text{(pyruvic acid)}$$

$$MAOO + H_2O \longrightarrow H_2CC(CH_3)COOH + H_2O \qquad \text{(RG6)}$$
$$\text{(methacrylic acid)}$$

A rate constant $k = 3.4 \times 10^{-18}$ cm^3 sec^{-1} has been recommended for (RG4) (Atkinson and Lloyd, 1984), and the same rate constant will be assumed for (RG5) and (RG6). In the case of MGLYO and MVKO, the radical carbon is fully substituted and reaction with H_2O (if it occurs) would not produce a carboxylic acid. We assume here that MGLYO and MVKO decompose to unreactive products.

Therefore, the gas-phase oxidation of isoprene may produce formic pyruvic acid, and methacrylic acid. One notes that acetic acid is not an expected product. Formic acid is mostly removed by dry deposition and washout. It does not photolyze (Calvert and Pitts, 1966), and its reaction with OH is very slow ($k = 4.5 \times 10^{13}$ cm^3 sec^{-1}; Wine et al., 1985). On the other hand, pyruvic acid photolyzes in the atmosphere on a time scale of the order of a day (Grosjean, 1984). Pyruvic acid absorbs radiation up to 370 nm with a near-unity quantum yield for photodissociation to CO_2 (Yamamoto and Black, 1985). Methacrylic acid does not appear to photolyze at tropospheric wavelengths (Rosenfeld and Weiner, 1983), but it probably reacts rapidly with OH by addition to the C=C bond. It is assumed here that the reaction of OH with methacrylic acid proceeds with the same rate constant as the reaction of OH with 2-methyl propene ($k = 6 \times 10^{-11}$ cm^3 sec^{-1}; Atkinson, 1986).

2.2 Aqueous-phase chemistry

The mechanism used is that presented by Jacob (1986). The aqueous-phase chemical reactions of sulfur, NO_x, chloride, and carbonate originally present in the Jacob (1986) mechanism have not been included since they do not significantly affect the chemistry of interest here.   Formic acid is rapidly produced in cloud droplets during the day following the hydrolysis of $CH_2O(aq)$ and oxidation of $H_2C(OH)_2$ by $OH(aq)$:

$$CH_2O(g) \;===\; CH_2O(aq) \tag{H1}$$

$$CH_2O(aq) + H_2O \;===\; H_2C(OH)_2 \tag{RA1}$$

$$H_2C(OH)_2 + OH \longrightarrow HC(OH)_2 \longrightarrow HCOOH + HO_2 \tag{RA2}$$

Production of HCOOH by this pathway is very fast because of the high hydration constant of CH_2O ($K_{A1} = 1.4x10^3$) and because (RA2) is very rapid ($k_{A2} = 2x10^9$ M^{-1} sec^{-1}).   In the aqueous phase, HCOOH is rapidly oxidized by $OH(aq)$:

$$HCOOH \;====\; HCOO^- + H^+ \tag{RA3}$$

$$HCOOH + OH \longrightarrow CO_2 + H_2O + HO_2 \tag{RA4}$$

$$HCOO^- + OH \longrightarrow CO_2 + H_2O + O_2^- \tag{RA5}$$

with $k_{A4} = 2x10^8$ M^{-1} sec^{-1}, $k_{A5} = 2.5x10^9$ M^{-1} sec^{-1}.   Removal of $HCOO^-$ by (RA5) is fast, therefore a cloud is not an efficient source of HCOOH if the HCOOH produced remains in the aqueous phase (Jacob, 1986). However, HCOOH produced in the aqueous phase can be stabilized by volatilizing to the gas phase.   The volatilization of HCOOH depends on the degree of $HCOOH(aq)/HCOO^-$ dissociation in the droplet ($pK_{A3} = 3.75$) and thus is strongly pH dependent.   Figure 1 shows the

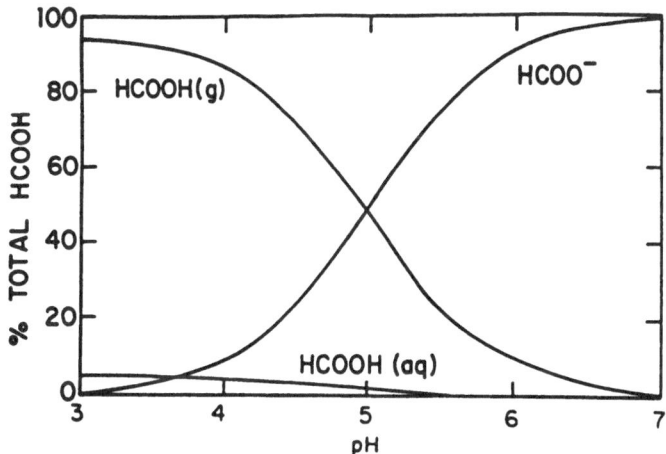

Figure 1.   Equilibrium speciation of HCOOH in cloud (liquid water content 0.5 g m^{-3}, T = 291K).

Table 1. Aqueous-phase mechanism: acetaldehyde, peroxyacetic acid, acetic acid.

a) **Henry's Law constants**

		K_H M atm^{-1}	ΔH kcal mole^{-1}	Reference
(H2)	CH_3CHO	13.	-11.	Snider and Dawson (1985)
(H3)	$CH_3C(O)OOH$	4.4(2)	-12.3	Lind and Kok (1986)
(H4)	CH_3COOH	8.8(3)	-12.8	CRC (1986)
(H5)	$CH_3C(O)OO$	12.	-12.3	footnote a

b) **Aqueous-phase equilibrium constants**

		K M atm^{-1}	ΔH kcal mol^{-1}	Reference
(RA6)	$CH_3COOH = CH_3COO^- + H^+$	1.7(-5)	-0.1	Sillen and Martell (1964)
(RA7)	$CH_3CHO \xrightarrow{\ H_2O\ } CH_3CH(OH)_2$	1.2^b	-5.	Bell (1966)

c) **Aqueous-phase reaction rate constants**

		k sec^{-1},M^{-1}sec^{-1}	E_a kcal mole^{-1}	Reference
(RA8)	$CH_3CHO + OH \xrightarrow{\ O_2\ } CH_3C(O)OO + H_2O$	5.(8)	3.*	Merz and Waters (1949)
(RA9)	$CH_3C(O)OO + HO_2 \longrightarrow CH_3C(O)OOH + O_2$	4.3(5)	6.*	$\frac{1}{2}kHO_2+HO_2$
(RA10)	$CH_3C(O)OO + O_2^- \xrightarrow{\ H_2O\ } CH_3C(O)OOH + OH^- + O_2$	5.(7)	3.2*	$\frac{1}{2}kHO_2+O_2^-$
(RA11)	$CH_3C(O)OOH \xrightarrow{\ H_2O\ } CH_3COOH + H_2O_2$	< 5.(-6)^b		Koubek and Edwards (1963)
(RA12)	$CH_3CH(OH)_2 + OH \longrightarrow CH_3COOH + H_2O + HO_2$	5.(8)	3.*	Merz and Waters (1949)
(RA13)	$CH_3COOH + OH \longrightarrow \ldots$	2.(7)	3.7*	Fahrataziz and Ross (1977)
(RA14)	$CH_3COO^- + OH \longrightarrow \ldots$	7.(7)	3.*	ibid.
(RA15)	$CH_3C(O)OOH + OH \longrightarrow \ldots$	2.(7)	3.7*	estimated

(a) calculated from K_{H3} scaled to the ratio of Henry's Law constants of HO_2 and H_2O_2 (Jacob, 1986).
(b) This upper limit value is used in the simulation.

* These activation energies are estimated following the method of Jacob (1986).
Read 4.4(2) as 4.4×10^2

speciation of HCOOH at 291K, as a function of pH.   It is clear that
the efficiency of a droplet as a source of HCOOH increases with the
droplet acidity.   This effect will be quantitatively demonstrated in
the model simulations.

An issue of interest is whether CH_3COOH may be produced in the
aqueous phase from acetaldehyde or peroxyacetic acid.   Both of these
species are major products of the gas-phase photochemical decomposition
of isoprene (Jacob and Wofsy, 1987).   A mechanism for aqueous-phase
production and loss of CH_3COOH was added to the model of Jacob (1986),
and is shown in Table 1.   Acetic acid may be produced in the same way
as formic acid following hydration of CH_3CHO.   In addition, it may be
produced by hydrolysis of peroxyacetic acid, however this process is
quite slow at the pH values found in cloudwater.   Peroxyacetic acid
may be either scavenged from the gas phase or produced within the
aqueous-phase from the reaction of CH_3CHO(aq) with OH.   Note that the
intermediate CH_3CHO(aq) does not react with NO or NO_2 because of the
low solubility of these species.

2.3 Dynamical model

We attempt to simulate undisturbed conditions in the planetary boundary
layer (PBL) over the Amazon rain forest during the dry season (Gregory
et al., 1987).   The depth of the mixed layer was found during ABLE-2A
to respond quickly to the radiation balance.   At night, the inversion
was based a few tens of meters above the canopy top; at sunrise, the
mixed layer deepened rapidly, and grew at a rate of 5-10 cm sec^{-1}
during the morning hours.   On a typical day, the inversion stabilized
at noon at an altitude of about 1500 m, and remained at that level
until late afternoon.   The mixed layer then decayed very rapidly.   A
simple two-box model extending from canopy top to 1500 m (Figure 2) is
used here to describe the boundary layer dynamics.   At night, an
inversion based 50 m above the canopy top is assumed, separating layer
1 (mixed layer) from layer 2 (remnant PBL).   Eddy diffusion exchange
between layers 1 and 2 is allowed, with an eddy diffusion coefficient
of $2x10^3$ cm^2 sec^{-1}.   This exchange is necessary to account for the
observed persistence of O_3 near the canopy top at night (Kaplan et al.,
1987).   In the morning, air from layer 2 is entrained into layer 2 as
the mixed layer grows, until layer 2 disappears at noon.   Layer 2 is
reconstituted at sunset with the concentrations of layer 1.   A typical
diurnal temperature profile for the mixed layer (D.R. Fitzjarrald,
personal communication, 1986) is adopted and is shown in Figure 2.
The temperature of layer 2 is assumed constant at 296K.

Other model conditions are listed in Table 2.   Kaplan et al.
(1987) have found that the soil of the Amazon forest is a strong source
of NO; their reported value for the emission flux is used here, where
it is assumed (as observed by Kaplan et al.) that NO has been entirely
converted to NO_2 by the time it reaches the top of the canopy (at
night) or has reached a steady state with respect to NO_2 (in the day).
Isoprene emissions are assumed to proceed from sunrise to sunset only,
and to be unsensitive to changes in the intensity of solar radiation
during the day (Sanadze and Kalandadze, 1965).   An exponential

Table 2. Model conditions.

Radiation field: 0° latitude (equator), clear sky, surface albedo = 0.18, solar declination = 20°; vertical columns of ozone, water vapor, and aerosol from Logan et al. (1981).

Temperature: See text and Figure 2.

Water vapor: 16 g/kg air

Condensation nuclei: NH_4^+, H^+, NO_3^-, SO_4^{2-} (see text)

Cloud physics: Liquid water content: 0.5 g m^{-3}
Droplet radius: 10 μm
Sticking coefficient: 0.1

Species with fixed concentrations: CH_4: 1700 ppb (Logan et al., 1981)
C_2H_6: 2 ppb ibid.
CO: 140 ppb (S.C. Wofsy, unpublished data, 1986).

Emission fluxes: NO: 5.2×10^{10} molecules cm^{-1}s^{-1} (Kaplan et al., 1987)
Isoprene: $5 \times 10^{11}e^{0.2(T-298)}$ molecules cm^{-2}s^{-1}
(Lamb et al., 1985; see text)

Deposition velocities: V_d (cm s^{-1}):

Species	V_d	Species	V_d
O_3	2	HCOOH	0.5
Isoprene	2	pyruvic acid	0.5
HNO_3	3	methacrylic acid	0.5
NO_3	3	peroxyacetic acid	0.5
CH_2O	0.5	H_2O_2	0.5
		CH_3CHO	0.5

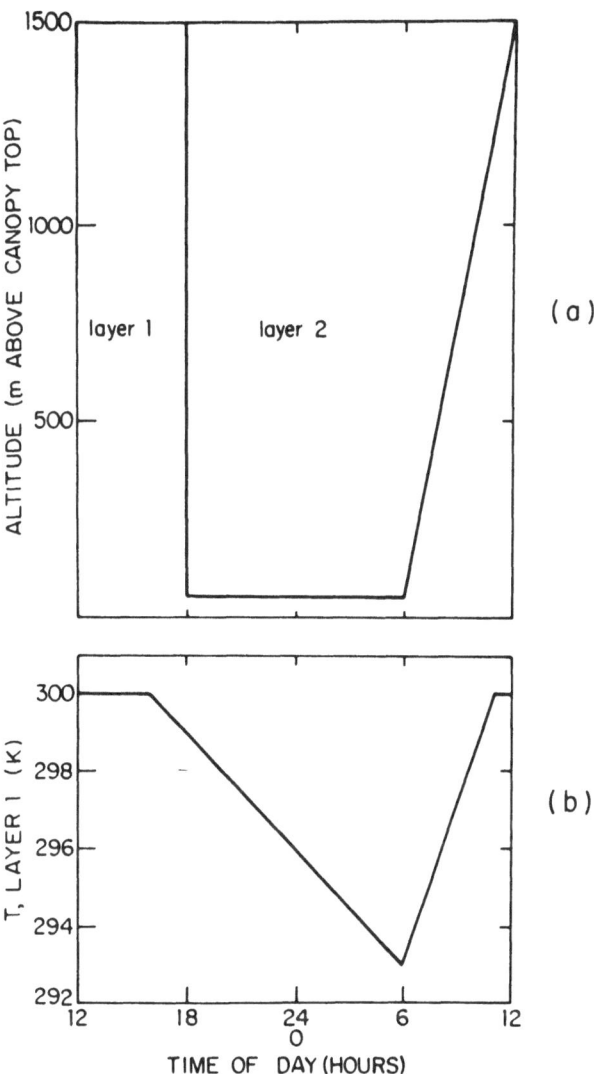

Figure 2   (a) Height of the mixed layer as a function of time of day.
            (b) Diurnal cycle of temperature in the mixed layer (layer
            1).

dependence of the isoprene emission rate on temperature is assumed from
Lamb et al. (1985, 1986), and the emission rate at 298K is selected as
an adjustable parameter to best fit the data.   We thus find a flux of
5×10^{11} molecules cm^{-2} sec^{-1} at 298K, which is in the range reported by
Lamb et al. (1985) for forests in the United States.   The sensitivity
of model predictions to the isoprene emission rate will be discussed.

A deposition velocity for O_3 of 2 cm sec^{-1} was taken from Kaplan et al. (1987). The vertical gradient of O_3 was found by Kaplan et al. to be mostly confined to the lower ten meters above the ground, even at night, indicating little aerodynamic resistance to deposition at the top of the canopy. The nighttime isoprene concentration profiles measured in ABLE-2A (Rasmussen and Khalil, 1987) show evidence of a rapid sink for isoprene at the ground. The nature of this nighttime sink is uncertain; in particular, reaction of isoprene with NO_3 should be slow in view of the low nighttime NO and O_3 concentrations measured near the ground by Kaplan et al. (1987). We elect to simulate the nighttime isoprene sink as a deposition velocity at the top of the canopy, which is the lower boundary of our model. Based on the observed nighttime gradients within the canopy and an eddy diffusion coefficient of $2x10^3$ cm^2 sec^{-1}, we estimate a deposition velocity for isoprene of 2 cm sec^{-1}, similar to that of O_3.

Model simulations in the absence of cloud were iterated over 10 successive diurnal cycles, starting from reasonable initial conditions. After 10 days of simulation most of the species were in steady state, i.e. the same diurnal pattern was repeated from day to day. Separate simulations of cloud chemistry were conducted by cooling noontime boundary layer air down to a temperature at which a liquid water content of 0.5 g m^{-3} was achieved (Table 2). The cloud droplets were assumed to reach instantaneously a size of 10 μm radius, and the sticking coefficient for all species was assumed to be 0.1. The model results are not very sensitive to these assumptions, as discussed by Jacob (1986). An accurate treatment of gas-droplet transfer was used in the cloud chemistry model; this treatment includes consideration of both gas-phase and aqueous-phase gradients of concentrations near the gas-droplet interface. For further details the reader is referred to Jacob (1986).

3. RESULTS

3.1. Gas-phase production of carboxylic acids.

We show in Figure 3 some daily patterns of concentrations obtained after iterating the gas-phase model over 10 successive diurnal cycles. The concentrations and diurnal variations of NO, O_3, and isoprene reproduce observations fairly well. Detailed comparisons of model predictions with observations are presented in a separate paper (Jacob and Wofsy, 1987). Ozone is produced in the boundary layer during the day, reflecting the influence of isoprene photochemistry and high NO emissions from soils; a maximum concentration of 26 ppb is predicted in late afternoon, which is consistent with observations. At night, O_3 is depleted near the ground, but remains at over 20 ppb in the remnant PBL. Maxima in isoprene concentrations are predicted in the early morning and late afternoon, when the source is on and the main sink (reaction with OH) is slow. At night, isoprene is depleted from layer 1 by deposition, but remains at a high concentration aloft because the nighttime chemical sinks (reactions with O_3 and NO_3) are slow.

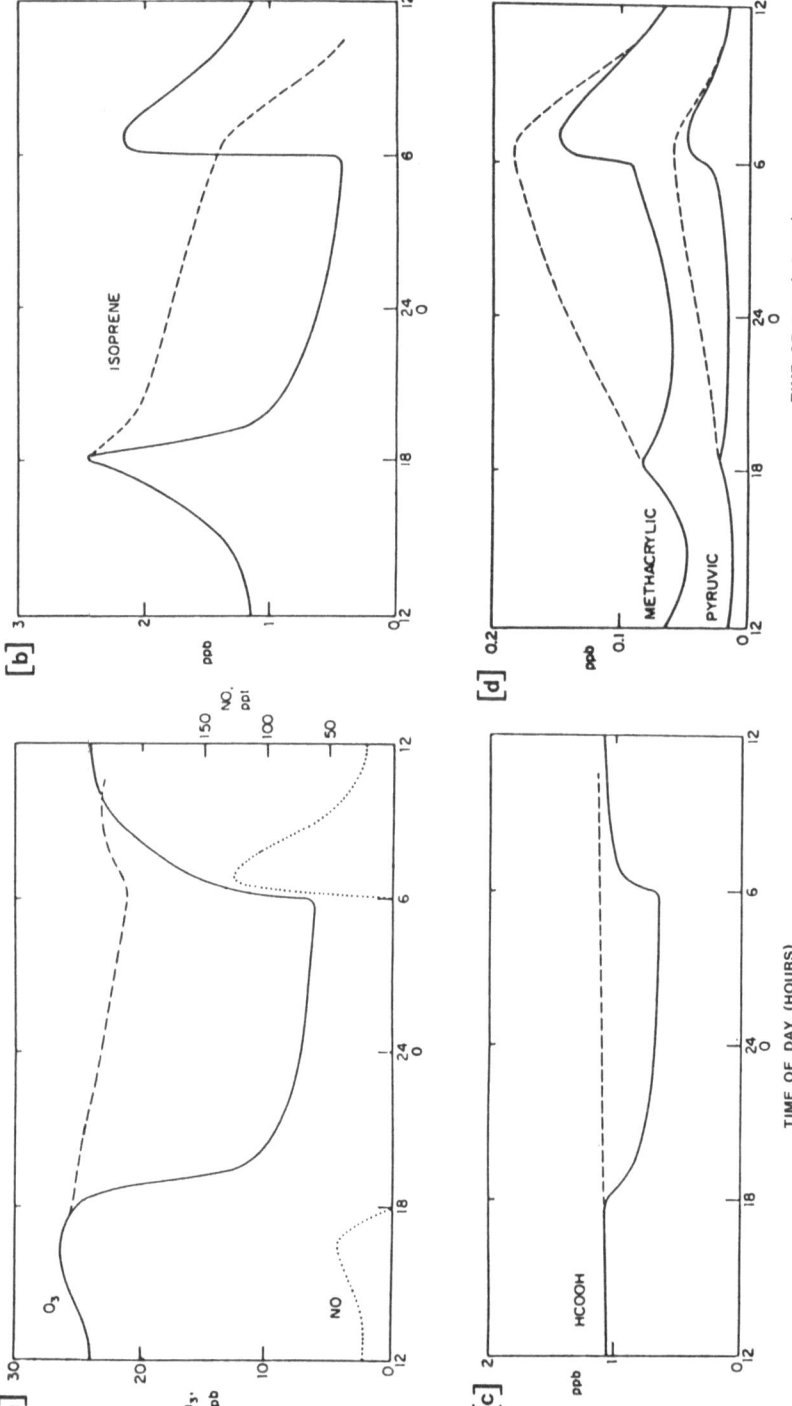

Figure 3. Diurnal variations in the concentrations of (a) ozone, (b) isoprene, (c) formic acid, and (d) methacrylic and pyruvic acids, for layer 1 (solid line) and layer 2 (dashed line), with the conditions of Table 2 and no cloud present. Dotted line is the concentration of NO in layer 1.

84

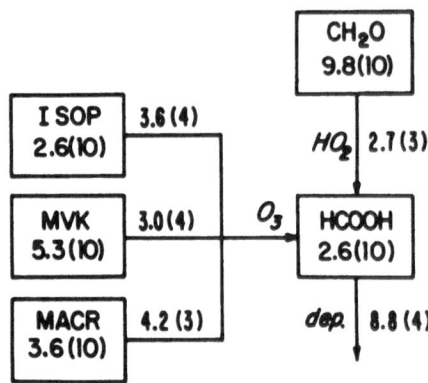

Figure 4   Formic acid production and loss rates at noon, with the
            model conditions of Table 2 and no cloud present.
            Concentrations are in units of molecules cm^{-3}, and
            transformation rates are in units of molecules cm^{-3} sec^{-1}.

The isoprene source yields a formic acid concentration of about
1 ppb (Figures 3 and 4).   Predicted concentrations of methacrylic acid
and pyruvic acid are lower, of the order of 0.1 and 0.02 ppb,
respectively.   Concentrations of methacrylic acid and pyruvic acid are
highest at night because of rapid daytime sinks from reaction with OH
and photolysis, respectively.   Formic acid does not show such a
diurnal variation because of its long chemical lifetime; the main
feature in the diurnal concentration pattern of HCOOH is the nighttime
removal in layer 1 by deposition.   It should be noted that the
predicted concentrations of HCOOH scale almost linearly to the
deposition velocity, a quantity difficult to estimate.   The deposition
velocity assumed here (0.5 cm sec^{-1}) is probably uncertain to about a
factor of 5, and a similar uncertainty would apply to predicted HCOOH
concentrations.

Andreae et al. (1987a) have reported measurements of gas-phase
HCOOH within the forest canopy during ABLE-2A.   The HCOOH
concentrations observed by Andreae et al. were approximately constant
during the daylight hours, and dropped to low values at night, a result
in harmony with model predictions.   Daytime concentrations were in the
range 1-3ppb, in fair agreement with simulated values.   Therefore, it
appears that isoprene could be a major source of HCOOH in continental
atmospheres.

According to our mechanism, photochemical decomposition of
isoprene should produce methacrylic acid and pyruvic acid in addition
to HCOOH.   Both methacrylic acid and pyruvic acid have fairly rapid
photochemical sinks, which would tend to control their concentrations.
We are not aware of any measurements of methacrylic acid in the
atmosphere, but measurements of pyruvic acid were made by Andreae et
al. (1987b) during ABLE-2A.   Pyruvic acid was present ubiquitously in

gas-phase, aerosol, and precipitation samples. The formic-to-pyruvic ratios in precipitation were in the range 13 to 62, with a mean value of 40. Gas-phase concentrations of pyruvic acid within the canopy ranged from 90 to 400 ppt, with a mean value of 180 ppt. These results are roughly consistent with an isoprene source for pyruvic acid as simulated by the model. Observed gas-phase concentrations are higher than predicted, but the significance of this discrepancy is difficult to assess in view of the uncertainties on the rate of (RG5) and the yield of MCRG from (RG2). Pyruvic acid is unlikely to be

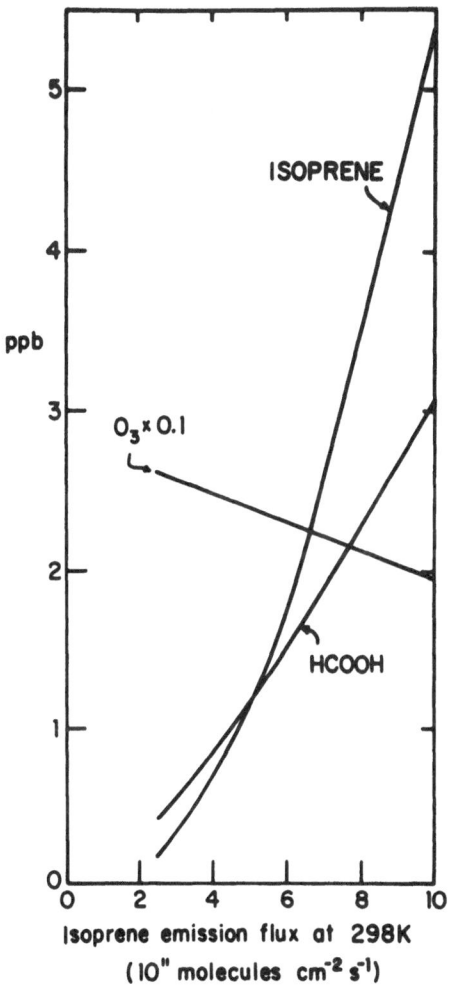

Figure 5   Noontime isoprene, HCOOH, and O_3 concentrations, as a function of the isoprene source strength. The isoprene source E is given as $E = A \, e^{0.2(T-298)}$ molecules $cm^{-2} \, s^{-1}$, where the preexponential constant (isoprene flux at 298K) is the adjustable parameter.

released from vegetation because of its high acidity constant (pK 2.4), and appears to have no significant atmospheric sources other than the oxidation of isoprene. Grosjean (1984) has reported the photochemical production of pyruvic acid from o-cresol, but this source should be significant only in polluted urban atmospheres.

Isoprene fluxes from forests have been observed to vary over a wide range, although the exponential dependence on temperature as given by Table 2 appears to be universal (Lamb et al., 1985). The sensitivity of the HCOOH concentration to the isoprene source strength was explored by a series of simulations where various values of the preexponential constant (value of the isoprene flux at 298K) were considered (Figure 5). The dependence of the isoprene concentration on the isoprene emission flux is strongly non-linear because OH, which provides the dominant isoprene sink, is depleted by isoprene and its decomposition products (methylvinylketone, methacrolein, aldehydes, organic peroxides). The concentration of formic acid also shows a strong dependence on the isoprene emission flux.

3.2 The role of cloud chemistry

We now consider a cloud forming under noontime conditions and with initial gas-phase concentrations obtained from the 10-day standard gas-phase simulation discussed above. Cloud formation enhances radiation in the upper region of the cloud (because of scattering from

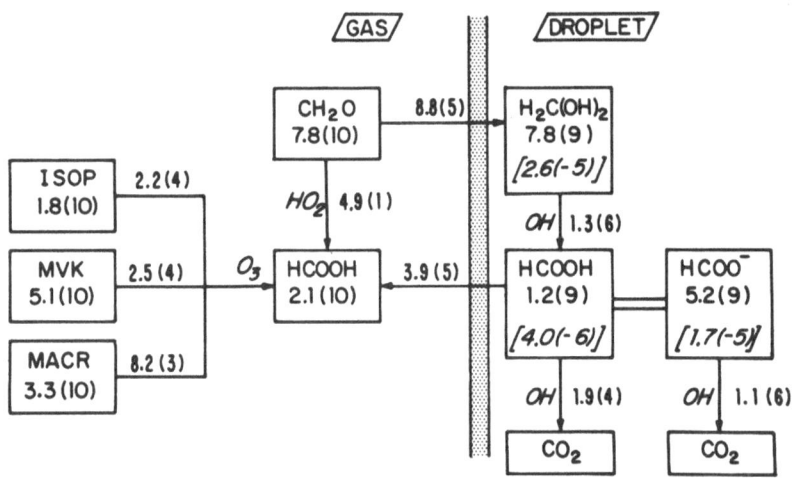

Figure 6   Production and loss rates of HCOOH one hour after cloud formation, with the conditions of Table 2 and a noontime radiation field. Concentrations are in units of molecules cm^{-3}, and transformation rates are in units of molecules cm^{-3} sec^{-1}. Aqueous-phase concentrations (M) are given in brackets. Gas-droplet transfer rates are given as net fluxes.

below), and weakens radiation in the lower region of the cloud (because of increased optical depth). An intermediate level is considered here where cloud formation causes no net change in the radiation field. The cloud is assumed to be decoupled from the ground, i.e. emission and deposition fluxes are set to zero upon cloud formation. The condensation nuclei are assumed to be (H^+, NH_4^+, NO_3^-, SO_4^{2-}) mixtures, initially at equilibrium with $HNO_3(g)$ (specified from the gas-phase model) and $NH_3(g)$. Various initial nuclei acidities and $NH_3(g)$ concentrations are used in the different simulations, in order to investigate a range of cloudwater pH regimes and study the sensitivityof the aqueous-phase chemistry to droplet pH.

Figure 6 shows the production and removal of HCOOH in cloud one hour after cloud formation. In this simulation the condensation nuclei were assumed to be acid-base neutral, and the initial $NH_3(g)$ concentration was set to zero. The pH of the droplets one hour after cloud formation is 4.39, and is still decreasing slowly because of continuing production of HCOOH which is a major component of the droplet acidity. Comparison of Figures 4 and 6 indicates that the aqueous-phase source of HCOOH in cloud is about 20 times faster than the gas-phase source; however, the aqueous-phase sink is also much faster than the gas-phase sink. Let $HCOOH_T$ represent total HCOOH in cloud (HCOOH(g) + HCOOH(aq) + $HCOO^-$); we see from Figures 4 and 6 that the $HCOOH_T$ concentration one hour after cloud formation is close to the initial HCOOH(g) concentration.

The concentration of HCOOH in cloud is strongly dependent on cloudwater pH, mainly because the partitioning of HCOOH between the gas and aqueous phases has a major effect on its removal rate. Figure 7 shows the concentrations of CH_2O and HCOOH species one hour after cloud formation for a range of cloudwater pH values. At low pH, most of the HCOOH produced in the aqueous phase volatilizes to the gas phase, where it has a long lifetime against oxidation by OH(g). As the pH increases, however, an increasing fraction of the HCOOH produced remains in the aqueous phase as $HCOO^-$, and is oxidised by the reaction $HCOO^-$ + OH. As a result, the $HCOOH_T$ concentration decreases with increasing cloudwater pH; at high pH, the cloud is a net sink for HCOOH. A maximum in the aqueous-phase concentration of $HCOO^-$ is found at pH 5.5. Keene and Galloway (1986) have reported average $HCOO^-$ concentrations in rainwater over central Brazil of 17 μeq 1^{-1}, with an average pH of 4.63; our simulations predict a cloudwater concentration of 23 μeq 1^{-1} for that pH. This good agreement indicates that the aqueous-phase mechanism for HCOOH production could explain the observed concentrations in precipitation at continental sites.

Compared to marine clouds (Chameides, 1984; Jacob, 1986), the relative effect of aqueous-phase chemistry on HCOOH concentrations in continental clouds is much less dramatic. Even under acidic conditions, where net production is maximum, the $HCOOH_T$ concentration one hour after cloud formation is 1.2 ppb, as compared to the initial concentration of 1 ppb. Although CH_2O concentrations are much higher in continental clouds than in marine clouds, aqueous-phase production of HCOOH is not correspondingly faster because the reaction $H_2C(OH)_2$ + OH is the dominant sink for OH(aq) at high CH_2O concentrations;

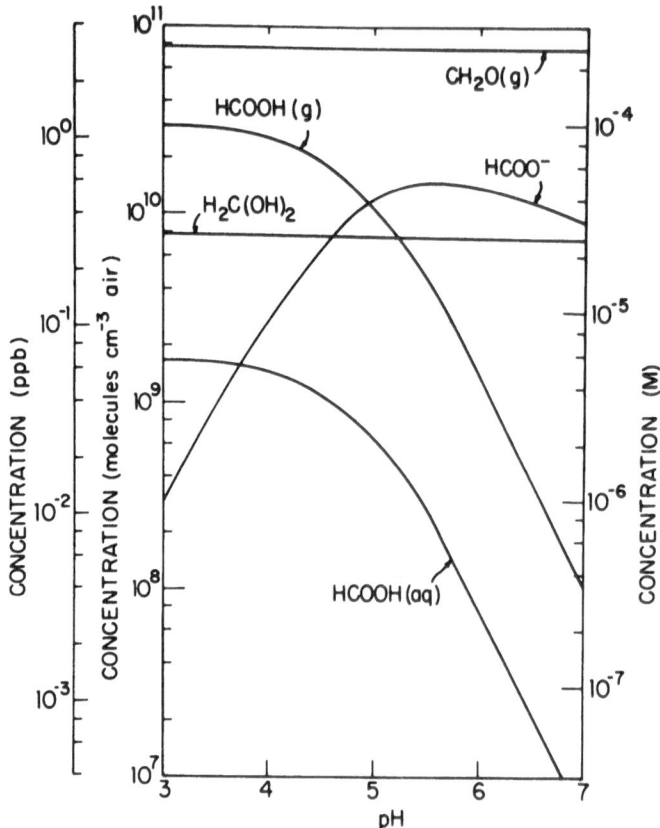

Figure 7   Concentrations of CH$_2$O and HCOOH species, one hour after
            cloud formation, as a function of cloudwater pH.   Model
            conditions are those of Table 2 with a noontime radiation
            field.   Gas-phase concentrations are in units of molecules
            cm^{-3} air and ppb (left-hand scales), and aqueous-phase
            concentrations are in units of M (right-hand scale).

therefore, increasing the CH$_2$O concentration leads to a corresponding
decrease in the OH(aq) concentration, with little change in the rate of
HCOOH production.   The OH(aq) concentrations predicted in the present
simulations range fromm 1.0x10^{-13} M at pH 3 to 3.3x10^{-13} M at pH 7, and
are much lower than the concentrations predicted in a remote marine
cloud under similar radiation conditions (from 3.8x10^{-13} M at pH 3 to
1.5x10^{-12} M at pH 7;   Jacob, 1986).

      Figure 8 shows the concentration of acetic acid one hour after
cloud formation, as a function of cloudwater pH.   Production from
aceltaldehyde oxidation and peroxyacetic acid hydrolysis is very slow.
 Aqueous-phase oxidation of CH$_3$CHO is much slower than for CH$_2$O becaus
the hydration constant for CH$_3$CHO is three orders of magnitude small′

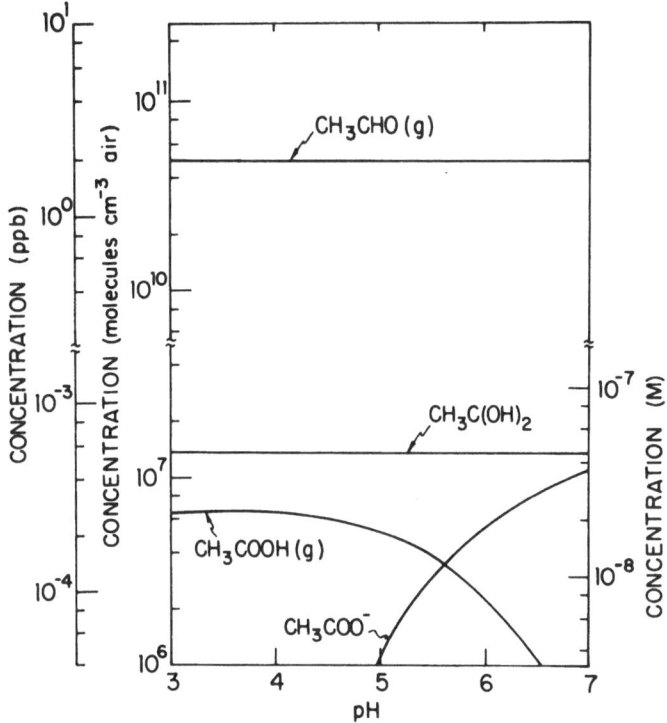

Figure 8   Concentrations of CH3CHO and CH3COOH species, one hour after
            cloud formation, as a function of cloudwater pH.   Model
            conditions are those of Table 2 with a noontime radiation
            field.   Gas-phase concentrations are in units of molecules
            cm^{-3} air and ppb (left-hand scales), and aqueous-phase
            concentrations are in units of M (right-hand scale).

than that of CH_2O.   The predicted concentrations of CH_3COO^- can be
compared to the average concentration of 3.7 μeq l^{-1} measured by Keene
and Galloway (1986) in rainwater over central Brazil.   Clearly,
photochemical decomposition of isoprene appears to be a negligible
source of CH_3COOH.   Other sources of acetic acid, probably direct
emissions from vegetation, must be present to produce the
concentrations observed in the gas phase by Andreae et al. (1987a) and
in rainwater by Keene and Galloway (1986).

4.   CONCLUSIONS

The production of carboxylic acids from atmospheric oxidation of
isoprene in a remote continental atmosphere has been studied with a

detailed chemical model. The model was applied to a simulation of atmospheric chemistry in the boundary layer of the Amazon basin, and predictions were compared to field measurements made during ABLE-2A. The model reproduces fairly well the observed diurnal pattern in the concentrations of isoprene, O_3, and NO; therefore, it should provide a fairly realistic representation of the boundary layer chemistry. The reader is referred to Jacob and Wofsy (1987) for a detailed discussion of the photochemistry of biogenic emissions over the Amazon Basin.

According to our chemical mechanism, the gas-phase photochemical decomposition of isoprene produces formic acid, methacrylic acid, and pyruvic acid. Concentrations of these acids predicted from the isoprene source in the Amazon basin are of the order of 1 ppb, 0.1 ppb, and 0.02 ppb, respectively. Production of formic acid in cloud by aqueous-phase oxidation of CH_2O does not greatly increase the formic acid concentration predicted from the gas-phase mechanism; cloud droplets with pH > 4 are actually expected to constitute net sinks for formic acid. No significant production of acetic acid is expected from the photochemical decomposition of isoprene, either in the gas phase or in the aqueous phase. Comparison of model predictions with field data indicates that isoprene could be a major source of formic acid and pyruvic acid observed in the gas phase and in rainwater; however, acetic acid must originate from another source.

ACKNOWLEDGEMENTS

This research was supported by funds from the National Aeronautics and Space Administration (grant NASA NAG1-55), the National Science Foundation (grant NSF-ATM 84-13153), and the Coordinating Research Council (grant CRC-CAPA-22-83).

REFERENCES

Adewuyi, Y.G., Cho, S.-Y, Tsay, R.-P & Carmichael, G.R. 1984. Importance of formaldehyde in cloud chemistry, Atmos. Environ. 18, 2413-2420.

Andreae, M.O., Talbot, R.W., Andreae, R.T.& R.C. Harris. 1987a. Formic and acetic acids over the Central Amazon region, Brazil. J. Geophys. Res. (in press).

Andreae, M.O., Talbot, R.W. & Li, S.M. 1987b. Atmospheric measurements of pyruvic and formic acid. J. Geophys. Res. 92 , 6635-6641.

Arnts, R.R. & Gay, B., Jr. 1979. Photochemistry of some naturally emitted hydrocarbons. U.S. Environmental Protection Agency, Research Triangle Park, NC, rpt. EPA-600/3-79-081.

Atkinson, R. 1986. Kinetics and mechanisms of the gas-phase reactions of the hydroxyl radical with organic compounds under atmospheric conditions. Chem. Rev., 86, 69-201.

Atkinson, R. & Lloyd, A.L. 1984. Evaluation of kinetic and mechanism data for modeling of photochemical smog. J. Phys. Chem. Ref. Data, 13, 315-444.

Bell, R.P. 1966. The reversible hydration of carbonyl compounds. Adv. Phys. Org. Chem., 4, V. Gold, ed., 1-29.

Bothe, E. & Schulte-Frohlinde, D. 1980. Reaction of dihydroxymethyl radical with molecular oxygen in aqueous solution. Z. Naturforsch., 35, 1035-1039.

Calvert, J.G. & Pitts, J.N. 1966. Photochemistry, Wiley, New York.

Chameides, W.L. 1984. The photochemistry of a remote marine stratiform cloud, J. Geophys. Res., 89, 4739-4755.

Chemical Rubber Company. 1986. Handbook of Chemistry and Physics, 66th ed., R.C. Weast, ed., Cleveland, Ohio.

Fahrataziz, & Ross, A.B. 1977. Selected specific rates of reactions of transients from water in aqueous solution. III. Hydroxyl radical and perhydroxyl radical and their radical ions, NSRDS-NBS, 59, U.S. Dept. of Commercie, Washington, D.C.

Graedel, T.E., Hawkins, D.T. & Claxton, L.D. 1986. Atmospheric chemical compounds: sources, occurrence and bioassay. Academic Press.

Gregory, G.L., Browell, E.V. & Gahan, L.S. 1987. Boundary layer ozone: an airborne survey across the Amazon Basin, J. Geophys. Res. (in press).

Grosjean, D. 1983. Atmospheric reactions of pyruvic acid. Atmos. Environ., 17, 2379-2382.

Grosjean, D. 1984. Atmospheric reactions of orthocresol: gas phase and aerosol products. Atmos. Environ., 18, 1641-1652.

Hatakeyama, S., Bandow, H., Okuda, M. & Akimoto, H. 1981. Reactions of .CH$_2$OO and CH$_2$(^{1}A$_1$) with H$_2$O in the gas phase. J. Phys. Chem., 85, 2249-2254.

Jacob, D.J. 1986. The chemistry of OH in remote clouds and its role in the production of formic acid and peroxymonosulfate. J. Geophys. Res., 91, 9807-9826.

Jacob, D.J. & Wofsy, S.C. 1987. Photochemistry of biogenic emissions over the Amazon forest, J. Geophys. Res. (in press).

Kaplan, W.A., Wofsy, S.C., Keller, M. & da Cost, J.M.N. 1987. Emission of NO and deposition of O$_3$ in a tropical forest system. J. Geophys. Res. (in press).

Keene, W.C., Galloway, J.N. & Holden, H.D. Jr. 1983. Measurements of weak organic acidity in precipitation from remote areas of the world. J. Geophys. Res., 88, 5122-5130.

Keene, W.C. & Galloway, J.N. 1987. Considerations regarding sources for formic and acetic acids in the troposphere. J. Geophys. Res., 91, 14466-14474.

Koubek, E. & Edwards, J.O. 1963. The aqueous photochemistry of peroxychloroacetic acid. Inorg. Chem., 28, 2157-2160.

Lamb, B., Westberg, H. & Allwine, G. 1985. Biogenic hydrocarbon emissions from deciduous and coniferous trees in the United States. J. Geophys. Res., 90, 2380-2390.

Lamb, B. 1986. Isoprene emission fluxes determined by an atmospheric tracer technique. Atmos. Environ., 20, 1-8.

Lind, J.A. & Kok, G.L. 1986. Hendry's Law determinations for aqueous solutions of hydrogcen peroxide, methylhydroperoxide, and peroxyacetic acid. J. Geophys. Res., 91, 7889-7895.

Lloyd, A.C., Atkinson, R., Lurmann, F.W. & Nitta, B. 1983. Modeling potential ozone impacts from natural hydrocarbons - I. Development and testing of a chemical mechanism for the NO_x-air photooxidations of isoprene and α-pinene under ambient conditions. Atmos. Environ. 17, 1931-1950.

Logan, J.A., Prather, M.J., Wofsy, S.C.& McElroy, M.B. 1981. Tropospheric chemistry: a global perspective. J. Geophys. Res., 86, 7210-7254.

Lurmann, F.W., Lloyd, A.C. & Atkinson, R. 1986. A chemical mechanism for use in long-range transport/acid deposition computer modeling. J. Geophys. Res., 91, 1905-1936.

Merz, J.H. & Waters, W.A. 1949. Some oxidations involving the free hydroxyl radical. J. Chem. Soc., S15-S25.

Rasmussen, R.A. & Khalil, M.A.K. 1987. Isoprene over the Amazon Basin: distributions, fluxes, lifetimes, and OH concentrations. J. Geophys. Res. (in press).

Rosenfeld, R.N. & Weiner, B.R. 1983. Photofragmentation of acrylic acid and methacrylic acid in the gas phase. J. Am. Chem. Soc., 105, 6233-6236.

Sanadze, G.A. & Kalandadze, A.N. 1965. Light and temperature curves of the evolution of C_5H_8. Fiziologiya Rastenii, 13, 458-461.

Sillen, G.L. & Martell, A.E. 1964. Stability constants of metal-ion complexes. Spec. Publ., 17, Chem. Soc. London.

Snider, J.R. & Dawson, G.A. 1985. Tropospheric light alcohols, carbonyls, and acetonitrile: concentrations in the southwestern United States and Henry's Law data. J. Geophys. Res., 90, 3979-3805.

Su, F., Calvert, J.G. & J.H. Shaw. 1979. Mechanism of the photooxidation of gaseous formaldehyde. J. Phys.Chem. 83, 3185-3191.

Talbot, R.W., Stein, K.M., Harris, R.C. & Cofer, W.R. III. 1987. Atmospheric geochemistry of formic and acetic acids. J. Geophys. Res. (in press),

Wine, P.H., Atsalos, R.J. & Mauldin, R.L. III. 1985. Kinetic and mechanistic study of the OH + HCOOH reaction. J. Phys. Chem., 89, 2620.

Yamamoto, S. & Black, R.A. 1985. The photolysis and thermal decomposition of pyruvic acid in the gas phase. Can. J. Chem., 63, 549-554.

Zimmerman, P.R., Greenberg, J.P. and Westberg, C.E. 1987. Measurements of atmospheric hydrocarbons and biogenic emission fluxes in the Amazon boundary layer. J. Geophys. Res. (in press).

SULPHUR SCAVENGING IN A MESOSCALE MODEL WITH QUASI-SPECTRAL MICROPHYSICS

N. Chaumerliac
L.A.M.P./O.P.G.C.
LA/C.N.R.S.n°267, BP 45
63170 Aubière
FRANCE

ABSTRACT. A three-dimensional mesoscale numerical model (Nickerson et al., 1986) with quasi-spectral microphysics has been employed for sulphur scavenging by different cloud systems. Combining meteorological predictions and pollutant scavenging parameterizations, two-dimensional sensitivity tests have been performed for continental and maritime clouds over an idealized bell-shaped mountain. Model results indicate that nucleation scavenging is the most efficient in-cloud removal mechanism. Furthermore, according to cloud droplet spectra in continental versus maritime cases, nucleation scavenging, in-solution oxidation of SO_2 by ozone and hydrogen peroxide and dynamical capture of sulphate particles by hydrometeors contribute differently to wet sulphate deposition. This reflects the complexity of the non-linear interactions between winds, microphysics and physicochemistry.

1. INTRODUCTION

Wet deposition is the result of complex non-linear interactions between dynamics, microphysics and physicochemistry. To get better understanding of those interactive processes, it seems quite appropriate to study wet removal of aerosols and gases within the framework of a meteorological model.

Our purpose here is to demonstrate the accuracy of a quasi-spectral paramaterization for liquid water and aerosols so as to describe nucleation scavenging and SO_2 dissolution into cloud droplets. A microphysical scheme, allowing for an improved explicit representation of the condensation/evaporation processes has been incorporated into a three-dimensional mesoscale model (Nickerson et al., 1986) for simulating precipitation. Two-dimensional sensitivity tests are next performed to evaluate the relative contributions of each scavenging process in typical continental and maritime clouds.

M. H. Unsworth and D. Fowler (eds.), Acid Deposition at High Elevation Sites, 93–107.

2. MODEL PRESENTATION

We shall focus here on the microphysical and physicochemical aspects of the model only. For further details, the reader is referred to Nickerson et al., (1986) and Chaumerliac et al., (1986).

The model includes the following warm rain microphysical processes: nucleation, condensation, evaporation, accretion, autoconversion, self-collection and sedimentation. According to the activation spectrum:

$$N = CS^k \qquad (1)$$

where S is the supersaturation, C and k are empirical constants (C = 3500, k = 0.9 in the continental case; C = 100, k = 0.7 in the maritime case), cloud condensation nuclei are activated to form cloud droplets.

Predictions are made of both liquid (cloud and rain) water mixing ratio (q) and total number concentration N, assuming a log-normal drop (cloud droplet and raindrop) distribution. Thus, droplet concentration in the diameter range D to D + dD is given by:

$$dN = \frac{N}{2\pi \sigma D} \exp \left(- \frac{1}{2\sigma^2} \ln^2 \frac{D}{D_0}\right) dD \qquad (2)$$

where and D_0 are distribution parameters.

Integration of (1) over the entire spectrum of droplets (total mass $\rho_w \pi D^3/6$) yields the following expression for the liquid water mixing ratio:

$$q = N/\rho_a (\pi/6 \ D_0^3 \ \rho_w) \exp (9/2 \ \sigma^2)$$

where ρ_a and ρ_w are the densities of air and liquid water respectively.

To close the system of equations, we assume σ is constant and compute D_0. For continental clouds, is 0.157, for maritime clouds σ is 0.277 and for the raindrops distribution σ is 0.547.

Extensive use of the work of Berry and Reinhardt (1974) has been made in developing parameterizations for autoconversion, accretion and self collection processes appropriate to the log-normal distribution.

As for the hydrometeors, aerosol particles are assumed to be log-normally distributed, the dispersion parameter being set to ln2 (Dana and Hales, 1976). One discriminates between three categories of particles: those free in the air and others attached respectively to cloud droplets or raindrops. Only sulphate particles are considered here. An overview of the physicochemical processes is given in Figure 1. Sulphate volume concentration in cloud and rainwater can be increased through conversion of dissolved SO_2. Separate prognostic equations are written for each sulphate category for both number and volume concentration.

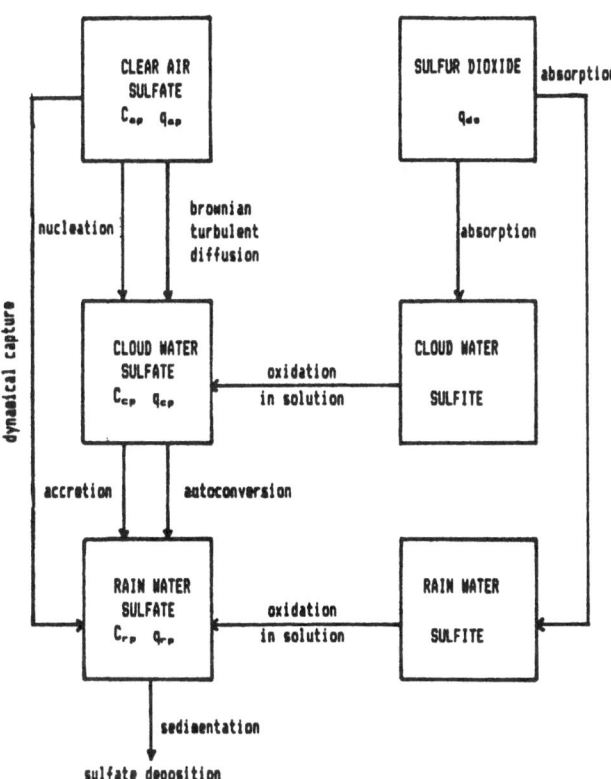

Figure 1. Synopsis of the physicochemical processes in the mesoscale
model.

Aerosol scavenging by hydrometers occurs through two different
mechanisms.   First, aerosols are scavenged through direct capture by
raindrops: the associated capture rate is calculated after Dana and
Hales (1976).   Secondly, indirect scavenging can take place through a
two-stage process (Pruppacher and Klett, 1978): in the first stage,
most aerosols are incorporated into cloud water by nucleation and by
Brownian and turbulent diffusion.   In the second stage, this polluted
cloud water is accreted by falling raindrops or cloud droplets to form
larger raindrops. The parameterizations developed for Brownian and
turbulent diffusion rates are from Dingle and Lee (1973).

Considering now gaseous pollutants, prediction is made of the SO_2
mixing ratio.   Sulphur dioxide is absorbed by cloud droplets and
raindrops before being oxidized in solution. Only major SO_2 oxidants
in aqueous phase, ozone and hydrogen peroxide are considered here.

Parameterizations of SO_2 dissolution and its subsequent oxidation
are derived from Hegg et al. (1984 b). O_3 and H_2O_2 are respectively

50 ppb and 1 ppb initially in the gas phase and the rates of oxidation of S (IV)are based on Maahs (1983) for O_3 and on Martin (1983) for H_2O_2.

3. MODEL RESULTS

3.1 Initial conditions

In order to simultaneously test the microphysical and physicochemical parameterizations, a series of two-dimensional sensitivity tests were carried out. This choice has been made because two-dimensional simulations are less time consuming than 3-D ones and because such idealized tests are well documented as regards the dynamics (Klemp and Lilly, 1978; Mahrer and Pielke, 1978). Comparative tests between dry, continental and maritime conditions have been performed. Due to complex interactions between meteorological and physicochemical processes, such tests are especially useful for identifying the relative contributions of each mechanism.

The model is run over an idealized bell-shaped mountain, 1 km high and 25 km in halfwidth. A two-layer atmosphere is considered, with a lower layer of constant lapse rate up to 8 km and an isothermal layer aloft. The initial horizontal wind speed is uniformly 20 m s^{-1} and the relative humidity is 80% below 3 km.

In addition, the SO_2 and sulphate variables are initialized with exponential profiles (Hegg et al., 1984b):

$$q_{ds} (z) = 4.3 \ 10^{-9} \exp (-z/2000)$$
$$q_{ap} (z) = 4. \ 10^{-12} \exp (-z/3500)$$
$$C_{ap} (z) = 10^9 \exp (-z/3500)$$

where z is the altitude, all units being in MKS.

Our selected case is typical of background pollution in Western Europe (Georgii, 1978). Tests to evaluate interactions between microphysical and physicochemical processes together with comparisons between continental and maritime cases are performed over 6-hour simulation periods. In the comparison between continental and maritime clouds, one discriminates between their dynamical features, their spectra, and the relative efficiency with which they can scavenge aerosols and gases.

3.2 Meteorological fields

First, the dynamics of mountain waves are investigated under three cases: dry case (no moisture) as opposed to continental and maritime clouds. Figure 2-a displays the results of a dry simulation, while Figures 2-b and 2-c respectively refer to simulations for continental and maritime cases. When comparing Figure 2-a and Figures 2-b and 2-c, one can observe that moist waves are weaker in amplitude than drywaves (Durran and Klemp, 1983). Between Figures 2-b and 2-c, noticeable differences appear in the dynamical fields, due to differences in the latent heat release effects between the continental and maritime cases.

potential temperature (K)

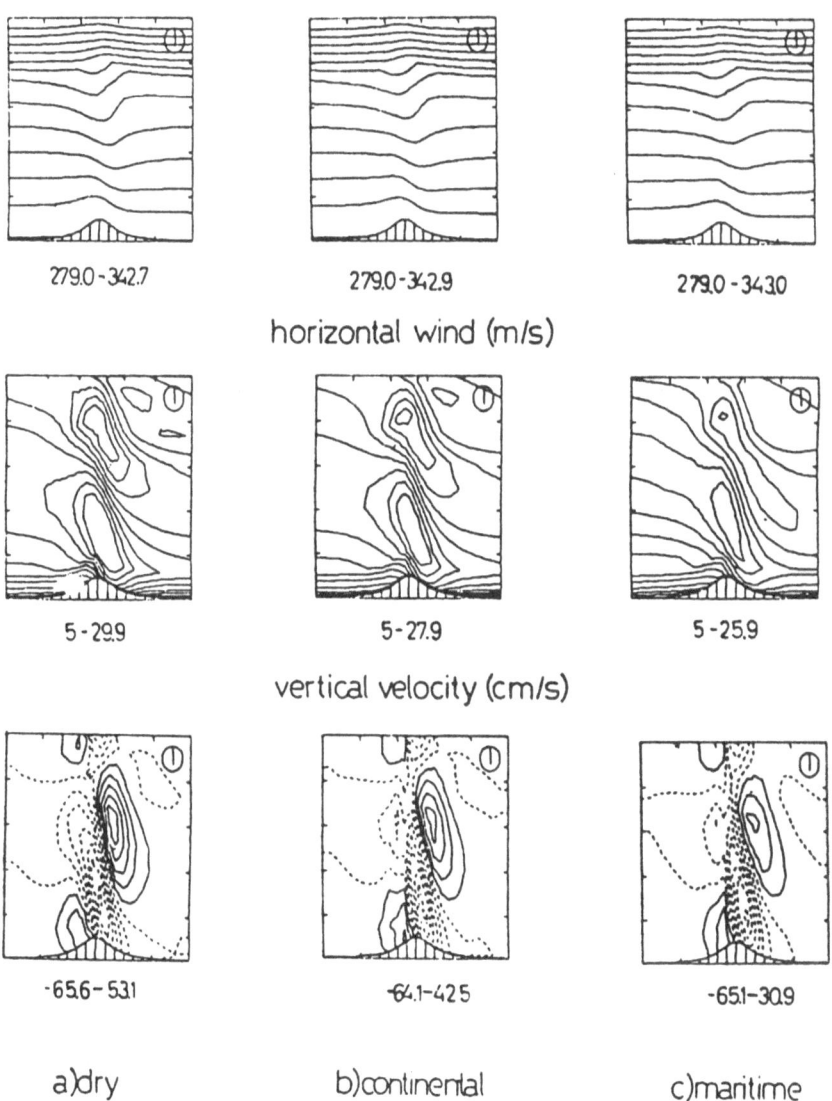

Figure 2.   Potential temperature, horizontal and vertical velocities
for a) dry, b) continental and c) maritime cases in the
presence of orography. Extreme values of the isocontours
are indicated at the bottom of each figure.

98

The preceding meteorological study of continental and maritime cases has given prominence to some characteristic differences between the two types of clouds which should have some repercussions on their scavenging ability.

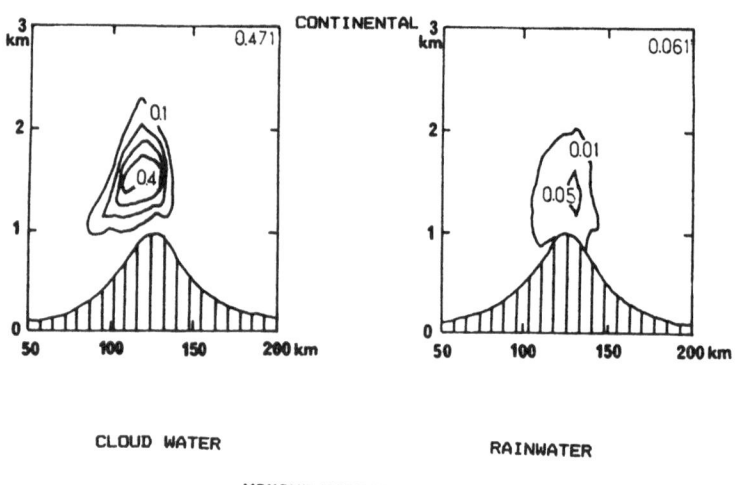

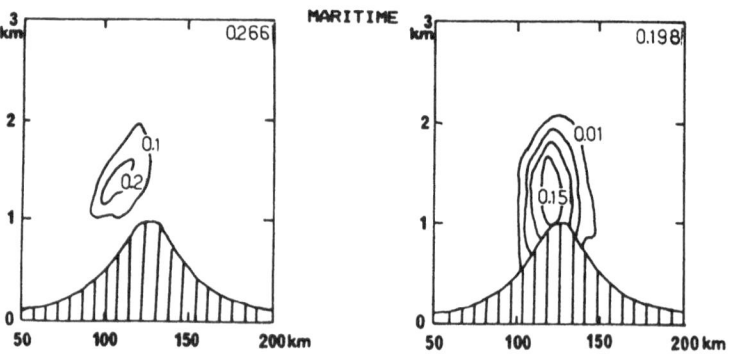

Figure 3. Vertical cross-sections of a) cloud water mixing ratio and b) rain water mixing ratio after 6-hour model time in continental and maritime cases. The maximum values of q_{cw} and q_{rw} are reported in the right corner of each figure.

3.3 Aerosol scavenging results

Nucleation is widely recognized as the most efficient process among all scavenging mechanisms (Flossmann et al., 1985; Jensen and Charlson, 1984; Hegg et al., 1984a). In order to evaluate the effects of nucleation scavenging, we proceed to two sensitivity tests. In the first run, all the scavenging processes are considered, while in the second one, the nucleation scavenging term is set equal to zero in the cloudwater sulphate equation. These two cases (with and without nucleation) are considered for both maritime and continental clouds with the same initial conditions for the insterstitial aerosols.

Results are presented in Figures 4 and 5 in the form of vertical cross-sections for concentrations of aerosols attached to cloud droplets (Figure 4) and aerosols removed by rain (Figure 5) after a 6-hour run. A large difference in the number of aerosols collected by cloud water is observed between the run accounting only for the dynamicalcapture of aerosols (Figure 4-a) and the run including all the scavenging processes (Figure 4-b). There are one to two orders of magnitude difference between the results of the simulations with and without nucleation in continental and maritime cases respectively. The continental cloud is more efficient for in-cloud scavenging (Figure 4) and the contribution of scavenging by Brownian and turbulent diffusions (Figure 4-a) is not completely negligible for small cloud droplets with diameter of about 10 μm which are apt to collect aerosols of radii less than 0.1 μm. This has already been suggested by Garland (1978). On the contrary, maritime clouds are much more efficient for removing aerosols (Figure 5), even when the nucleation term has been omitted (Figure 5-a). This is essentially due to the predominance of autoconversion in the initiation of the rain process in maritime conditions. As already underlined, it can be inferred that sulphate wet deposition will be more effective in the case of maritime cloud. This will be discussed later on.

3.4 Gas scavenging results

Before studying sulphate deposition, SO_2 dissolution and oxidation by H_2O_2 and O_3, and its subsequent transformation to sulphates must be considered.

Results of sensitivity tests are first presented in order to study the relative impact of O_3 and H_2O_2 oxidation on the conversion of SO_2 into sulphates. Vertical cross-sections of SO_2 mixing ratio after oxidation by ozone (Figure 6-a) and after oxidation by both O_3 and H_2O_2 (Figure 6-b) have been superimposed on vertical cross-sections, for which only advection effects have been considered (dashed isocontours in Figure 6). Clear SO_2 depletion coinciding with the presence of cloud (shaded zones) is observed in Figure 6 when oxidation occurs. The reduction in SO_2 mixing ratio is even greater when considering simultaneous oxidation by O_3 and H_2O_2 (Figure 6-b) than when

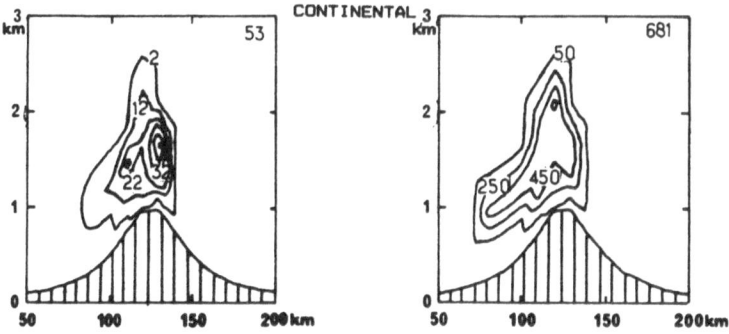

IN—CLOUD AEROSOL CONCENTRATION (N/CM³)

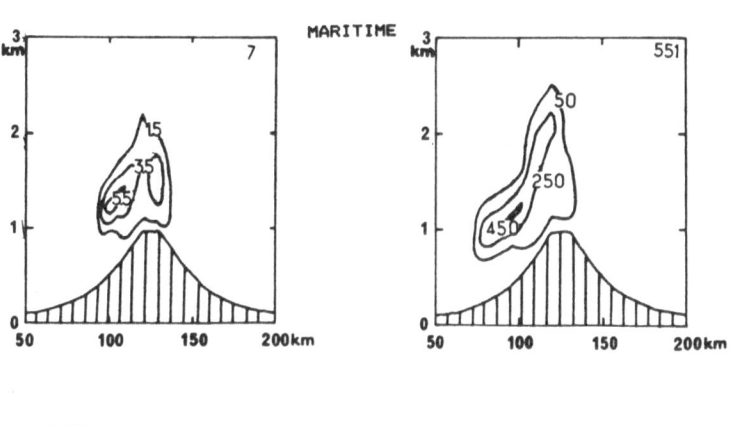

WITHOUT NUCLEATION   WITH NUCLEATION

Figure 4. Vertical cross-sections of in-cloud aerosol concentration
           C_{cp} (in number cm^{-3}) in continental and maritime clouds, a)
           without nucleation, and b) with nucleation.   The maximum
           value of C_{cp} is reported in the right corner of each
           figures.

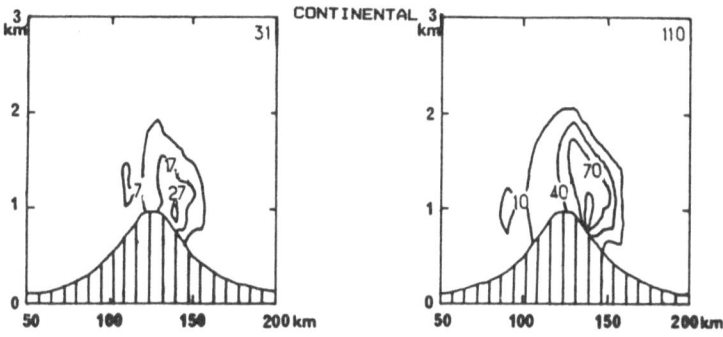

REMOVED AEROSOL CONCENTRATION (N/CM³)

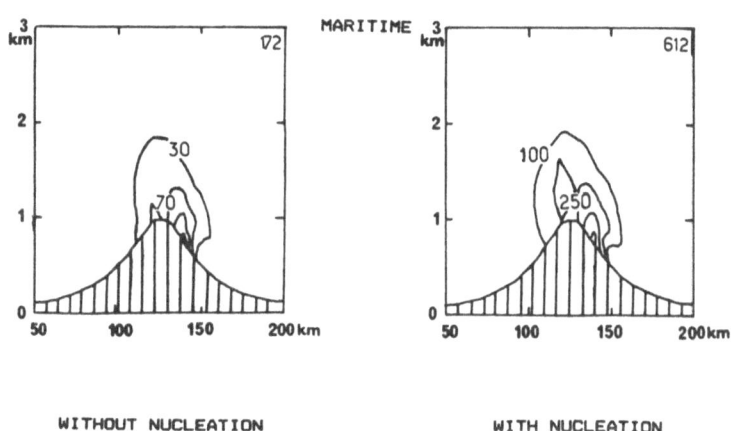

WITHOUT NUCLEATION   WITH NUCLEATION

Figure 5. Id. as Figure 4 but removed aerosol concentration C_{rp}.

considering oxidation by O_3 only (Figure 6-a).   If we now focus
(Figure 7) on the time evolution of SO_2 mixing ratio, the oxidation by
H_2O_2 appears to occur quite rapidly (after about half an hour model
time) after which slower oxidation by O_3 takes place for several
hours.   Oxidation by H_2O_2 is quick and intense, implying a rapid
depletion of aqueous H_2O_2 and a decrease of the pH.   The pH of the
environment was initially set to 5: it falls down to a value less than
4.   When only oxidation by O_3 occurs, the pH is not very much affected
and differences between curves 2 and 3 in Figure 7 can be explained by
the fact that O_3 contribution to sulphate formation is more efficient

102

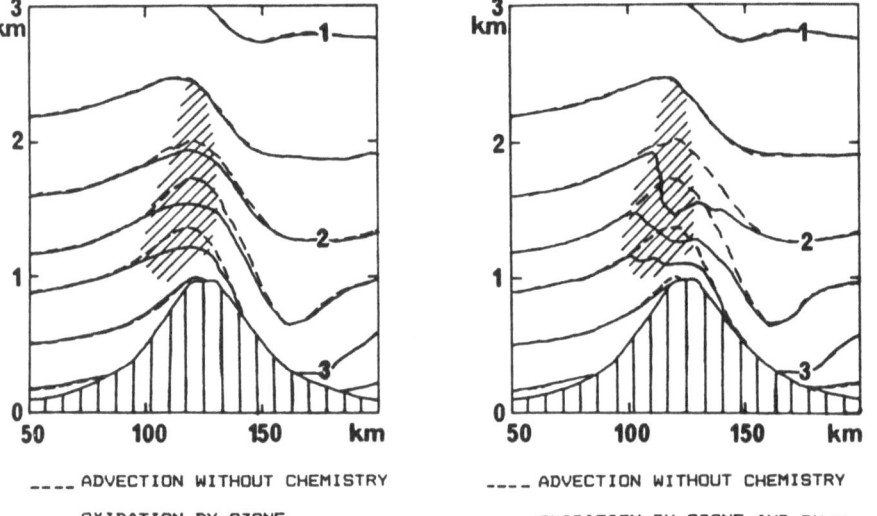

SO$_2$ MIXING RATIO (µG/KG)

---- ADVECTION WITHOUT CHEMISTRY   ---- ADVECTION WITHOUT CHEMISTRY

——— OXIDATION BY OZONE   ——— OXIDATION BY OZONE AND BY H$_2$O$_2$

Figure 6.   Vertical cross-sections of the SO$_2$ mixing ratio in maritime
            clouds after ½ h: a) without gas chemistry (dashed lines),
            b) with oxidation by H$_2$O$_2$ and ozone (solid lines). Shaded
            zones show the presence of cloudwater.   The vertical and
            horizontal scales are indicated.

when the pH is greater than 4 (Hoffman and Jacob, 1984; Seigneur and
Saxena, 1984).   Surprisingly, we did not find any striking difference
between maritime and continental clouds as regards the SO$_2$ mixing
ratios after oxidation in the aqueous phase.   Presumably, some
compensating effect arises between the dynamical and physicochemical
processes leading to such a similarity.   In order to get the same
dynamical conditions, we remove the latent heat release effects so as
to switch off dynamical/microphysical interactions in the model.
Thus, the dynamical conditions of Figure 3-a are artificially imposed
on the continental and maritime clouds.   Figure 8 displays the
vertical cross-sections of the SO$_2$ mixing ratio for both clouds.   The
continental cloud is more efficient than the maritime one for SO$_2$
scavenging.   Clearly a rapid SO$_2$ disappearance occurs as the cloud
begins to form in the zone where condensation processes act upwind of
the mountain top.   The differences observed in the SO$_2$ fields can be
attributed to smaller continental droplets with longer residence times
than maritime cloud ones.

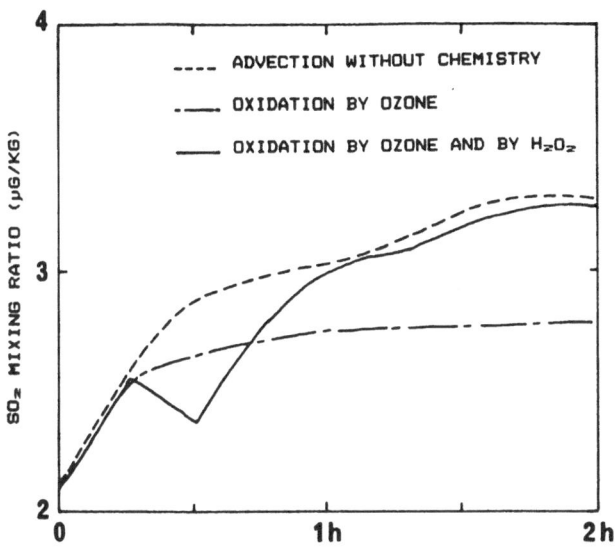

Figure 7. Temporal evolution of SO$_2$ mixing ratio over 2 h period for
three sensitivity tests:
a) without gas chemistry
b) with oxidation by ozone
c) with oxidation by ozone and H$_2$O$_2$

     To explain the changes in behaviour of the SO$_2$ mixing ratio in
Figure 8, it is necessary to recall what happens when dynamics and
microphysics are allowed to fully interact.   As seen previously, the
effect of moisture considerably smoothes out the horizontal and
vertical wind fields.   Consequently, the clouds computed within the
dynamical context of Figure 3-a are much more vigorous (higher cloud
water contents) than with the dynamics in Figures 3-b or 3-c.   This
can partly explain the differences observed in SO$_2$ oxidation rates in
Figures 6 and 8.   Hence, continental clouds appear more effective for
dissolving gases than maritime clouds.   This can be attributed to
smaller continental droplets with longer residence times than maritime
cloud ones (Hong and Carmichael, 1983;   Chameides, 1984).
     Therefore, our results suggest that some compensation mechanisms
exist between dynamical, microphysical and physiocochemical processes,
involved in wet pollutant removal.

3.5 Sulphate wet deposition

Deposition is the ultimate stage in a chain of very complex
interactions between pollutants, winds, clouds and rain.   It
represents a concrete and measurable parameter useful in comparative
studies.   In Figure 9, comparative tests are synthetized in the form
of wet sulphate deposition (in μm), accumulated over a 6-hour period.

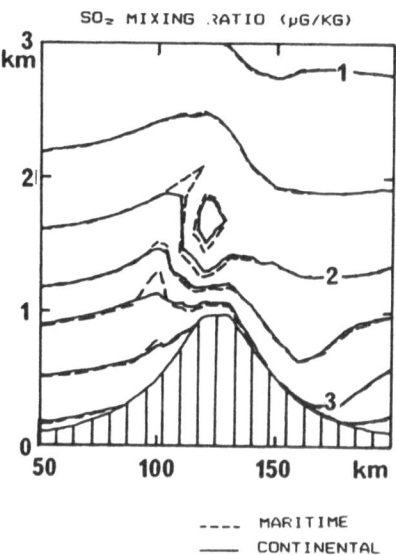

SO₂ MIXING RATIO (μG/KG)

Figure 8. Vertical cross-sections of SO$_2$ mixing ratio for maritime (dashed lines) and continental (solid lines) clouds after $\frac{1}{2}$ h with oxidation by H$_2$O$_2$ and ozone. Latent heat release effects have been removed.

The amount of deposited sulphate is about five times larger in the maritime case than in the continental one.

The maximum of sulphate deposition is not located at the same position for both clouds. There is a slight shift, probably associated to the precipitation drift. Continental small cloud droplets are more sensitive to advection, leading to longer growth times before effective production of rain by coalescence.

4. CONCLUSIONS

A quasi-spectral parameterization for liquid water (both cloud water and rainwater) has been included in the framework of a mesoscale model to quantify the interactive processes at work in pollutant wet removal.

First, sensitivity tests have been performed to establish the following hierarchy among physicochemical processes. Nucleation scavenging has been found to be the most efficient in-cloud scavenging process: its computation requires knowledge of an explicit nucleation rate and is a function of supersaturation and cloud spectrum. During the cloud formation a part of total wet deposition is due to SO$_2$ scavenging, which itself is dependent upon the liquid water content and

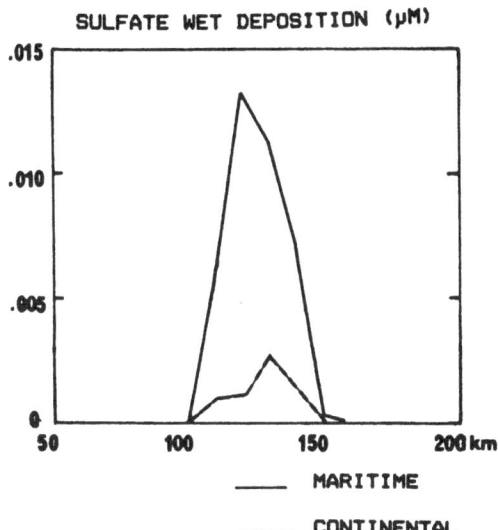

Figure 9.   Sulphate wet deposition on the mountain top accumulated over
6-hour for continental versus maritime cases.

the droplet sizes.   It should be emphasized that the spectral
information given by our microphysical scheme is well suited to the
treatment of SO_2 dissolution and oxidation in liquid phase, due to
detailed representation of the condensation/evaporation processes.   In
addition, Brownian and turbulent diffusions may significantly
contribute to in-cloud scavenging, in the case of small continental
droplets capturing particles with radii less than 0.1 μm.   Through its
quasi-spectral treatment of both hydrometers and aerosols, the model
has the ability to duplicate such effects.

The second point to be emphasized is the methodology adopted for
studying typical continental and maritime clouds through idealized
tests.   In the mesoscale, simultaneous collection of meteorological and
physiocochemical data is difficult to carry out experimentally.   The
results of this study suggest that a comprehensive numerical model
could be effective in isolating the underlying interactions of winds,
microphysics and physicochemistry and should be able to discriminate
between their relative effects.   This can be of use in defining
experimental strategies or in interpreting data.

ACKNOWLEDGEMENTS

This work was supported by funds from Electricité de France and the
authors gratefully acknowledge their support.   The computer

simulations were performed with the support and assistance of C.C.V.R., Palaiseu, France, project no. 3736, the E.C.M.W.F., Reading, England and also Météorologie Nationale, Paris.

REFERENCES

Berry, E.X. and R.L. Reinhardt. 1974. An analysis of cloud drop growth by collection, J. Atmos. Sci. **31**, 1814-2135.

Chameides, W.L. 1984. The photochemistry of a remote marine stratiform cloud, J. Geophys. Res. **89**, 4739-4755.

Chaumerliac, N., E.C. Nickerson, and R. Rosset. 1986. A 3D mesoscale model as a potential tool for the evaluation of sulphate particles scavenging. Annales Geophysicae **4**, B, 3, 345-352.

Dana, J. and J.M. Hales. 1976. Statistical aspects of the washout of polydisperse aerosols, Atmos. Environ. **10**, 45-50.

Dingle, A.N. and Y. Lee. 1973. An analysis of in-cloud scavenging, J. Appl. Meteor. **12**, 1295-1302.

Durran, D.R. and J.B. Klemp. 1983. A compressible model for the simulation of moist mountain waves, Mon. Wea. Rev. **111**, 2341-2361.

Flossman, A.I., W.D. Hall and H.R. Pruppacher. 1985. A theoretical study of the wet removal of atmospheric pollutants. Part I: The redistribution of aerosol particles captured through nucleation and impaction scavenging by growing cloud drops, J. Atmos. Sci. **42**, 583-606.

Garland, J.A. 1978. Dry and wet removal of sulphur from the atmosphere, Atmos. Environ. **12**, 349-362.

Georgii, H.W. 1978. Large-scale spatial and temporal distribution of sulphur compounds. Atmos. Environ. **12**, 681-690.

Hegg, D.A., P.V. Hobbs and L.F. Radke. 1984a. Measurements of the scavenging of sulphate and nitrate in clouds, Atmos. Environ. **18**, 1939-1946.

Hoffman, M.R. and D.J. Jacob. 1984. Kinetics and mechanisms of the catalytic oxidation of dissolved sulphur dioxide in aqueous solution: an application to nighttime fog water chemistry, SO_2, NO, and NO_2 oxidation mechanisms, Atmospheric Consideration, J.G. Calvert, Ed. Butterworth, 254pp.

Hong, H.S. and G.R. Carmichael. 1983. An investigation of sulphate production in clouds using a flow-through chemical reactor model approach, J. Geophys. Res. **88**, 10733-10743.

Mahrer, Y. and R.A. Pielke. 1978. A test of an upstream spline interpolation technique for the advective terms in a numerical memoscale model, Mon. Wea. Rev. **106**, 818-830.

Martin, L.R. 1978. Kinetics studies of sulfite oxidation in aqueous solution, Acid Precipitation, J.G. Calvert, Ed., Ann Arbor Science.

Nickerson, E.C., E. Richard, R. Rosset and D.R. SMith. 1986. The numerical simulation of clouds, rain and airflow over the Vosges and Black Forest montains; a meso-β model with parameterized microphysics, Mon. Wea. Rev. **114**, 398-414.

Pruppacher, H.R. and J.D. Klett. 1978. Microphysics of clouds and precipitation, 714 pp, D. Reidel, Boston, Mass.

Seigneur, C. and P. Saxena. 1984. A study of atmospheric acid formation in different environments, Atmos. Environ. **18,** 2109–2124.

PHYSICAL INFLUENCES OF ALTITUDE ON THE CHEMICAL PROPERTIES OF CLOUDS AND OF WATER DEPOSITED FROM THE TROPOSPHERE

R.J. Charlson, C.H. Twohy and P.K. Quinn
University of Washington, FX-10
Seattle, WA 98195
USA

ABSTRACT. The physical structure of the atmosphere dictates that its temperature usually decreases as altitude increases, often at a rate approaching the adiabatic limit of ca. -1C per 100 m. This requires consideration of the temperature dependencies of both chemical and physical processes of importance to cloud and deposition chemistry. It also requires recognition of the effects of the natural thermal layering of the atmosphere on the vertical distribution of trace substances and on systematic differences of long distance transport and scavenging with altitude. Among the fundamental temperature dependencies are the vapor pressure of water (and its role in determining the liquid water content of clouds and precipitation amount), the presence or absence of the ice phase, and deposition of supercooled water as occult precipitation by riming. Additional temperature dependencies of the chemical equilibrium constants and Henry's Law coefficients should cause marked altitude dependencies of pH. The sign of this depends on the species in control of the equilibrium. Altitude dependencies of air composition are not as readily described with physical and chemical theory, but can often be described effectively by a two layer system of the planetary boundary layer (PBL) and the free troposphere. Gases and fine particles (sub µm) are seen to behave similarly with coarse particles being limited mainly to the PBL. Long distance transport is likely to be much more efficient at high altitudes, but only for the substances present there. Finally, the influences of orography and climate are involved as important determinants of the local workings of the hydrologic cycle. Data will be presented demonstrating the central features of these altitude dependencies.

1. SCOPE

It is convenient to limit ourselves at the outset to a discussion of a class of processes, the altitude dependence of which is due largely to the dependence of temperature on altitude. We focus on nucleation scavenging and its effects on the chemical composition of and deposition by rain, graupel or hail and snow. Other processes such as dry

M. H. Unsworth and D. Fowler (eds.), Acid Deposition at High Elevation Sites, 109–124.
© *1988 by Kluwer Academic Publishers.*

deposition are also important, but do not fit as easily into this set of interdependencies with altitude and are therefore omitted for convenience.

Rain is produced by three different processes:

(1) Coalescence of liquid droplets into raindrops which are large enough to fall,

(2) growth of ice particles of precipitable size by deposition of water vapor in the presence of supercooled liquid droplets since air saturated with respect to liquid water is supersaturated with respect to ice. The ice phase (e.g. as snowflakes) then melts into raindrops in the warmer air below the melting level, and

(3) growth of graupel particles at T<0C due to collision of the ice particles with supercooled water droplets, followed by melting of the ice phase.

Snow is produced only by the second process, with rimed snowflakes being produced by the third process. Snowflakes may aggregate into larger snowflakes; however, this is not expected to alter chemical composition. Graupel or hail pellets are initially formed by the second process but grow mainly by the third process.

These three processes have been segregated because each is associated with unique chemical consequences. In addition, altitude often dictates which of these processes occurs. This paper will review the altitude dependencies which lead to a control of the processes forming precipitation and the chemical deposition associated with it.

2. VERTICAL STRUCTURE OF THE ATMOSPHERE AND ITS ROLE IN DETERMINING THE PHYSICAL STATE OF THE CLOUD-CHEMICAL SYSTEM

The specific heat at constant pressure for air, C_p, and the earth's gravitational constant, g, determine the dry adiabatic lapse rate in the troposphere, $dT/dZ = - g/C_p \cong -10C/km$, where T is temperature and Z is altitude. The dry atmosphere cannot sustain lapse rates of much larger magnitude because of convective instability but can sustain smaller values due, for example, to inversions. When heat release from the condensation of water is included, the moist adiabatic lapse rate is smaller in magnitude and dependent upon the amount of water in the air. Lapse rates ranging between -4 and -7C/km are commonly observed within clouds, with the more negative values being associated with drier air. In all cases, however, air that is being mixed in the vertical direction becomes cooler as it is carried upward. This simple fact, along with the height of typical clouds, dictates that temperature changes of -10 to -30C or more occur from the bottom to the top of clouds. Temperature differences of this magnitude are large enough to cause important shifts in chemical equilibria and reaction rates, in particular those processes involving the exchange of trace gases with liquid droplets, as will be seen below. These temperature differences also dictate that liquid water can freeze; however, its propensity for supercooling results in most condensed water in the atmosphere being in the liquid phase. Even in mixed clouds (i.e. those containing ice) the hydrometeors are composed primarily of the liquid phase.

Besides determining the phase and temperature of water in clouds, the vertical temperature profile of the troposphere is the dominant factor governing mixing of the atmosphere in that dimension. This mixing, in turn, governs the dependence of the chemical composition of air upon altitude. Frequently, the lowest layers of the atmosphere (the planetary boundary layer, PBL) perhaps up to 3 km altitude are slightly stable, that is with equivalent potential temperature increasing slightly with altitude. This results in short-lived surface ($Z = 0$) generated trace substances being concentrated in the PBL. Water, too, is trapped in the PBL, due to both the constraint on vertical mixing and to low temperatures aloft causing it to be removed from the air. Because of the altitude dependence of water content and the location of trace substances which are scavenged by condensed water, the primary atmospheric flux of these substances involves the PBL.

3. SCAVENGING AND DEPOSITION

The process of nucleation scavenging, that is, the incorporation of solute and insoluble substances into liquid cloud droplets in the nucleation process itself, is singularly important to hydrometeor composition and may, in some instances, be described quantitatively. Junge (1963) proposed an equation relating solute molarity, [X], to the atmospheric concentration of aerosol substance, $X(g/m^3)$:

$$[X] = \frac{\varepsilon X}{ML}$$

where M is the molecular weight of X and L is liquid water content in litres/m^3. The scavenging efficiency, ε, is assumed to be close to unity for nucleation scavenging. Hence, we are interested in defining, where possible, the altitude and temperature dependence of L. There is a growing consensus that nucleation scavenging is the key process which incorporates sub-μm aerosol mass into cloud droplets (Sievering et al., 1984; Charlson et al., 1983).

Deposition, F_X, can also be defined simply as the product of aqueous phase concentration and precipitation amount:

$$F_X = [X]V$$

where V is the amount of precipitation, e.g. in litres/m^2 year. [X] here is the volume weighted mean concentration. However, the lack of a simple equation relating L and V precludes further simplification.

4. PHYSICAL EFFECTS OF THE ALTITUDE-TEMPERATURE DEPENDENCE

A simple and general quantitative statement cannot be made for the altitude dependence of deposition fluxes of key species. The reasons for this lie in the non-linearity of the dependence of saturation vapor

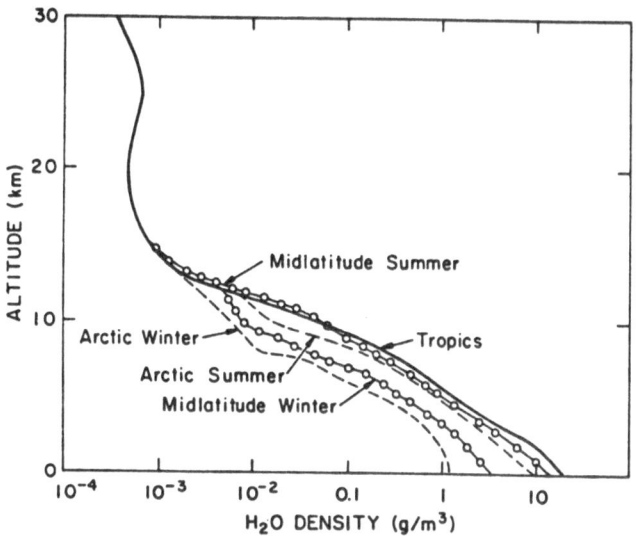

Figure 1.   Total water content profiles for the atmosphere for
             different latitudes and seasons after Ramanathan (1977) and
             McClatchey et al. (1972).

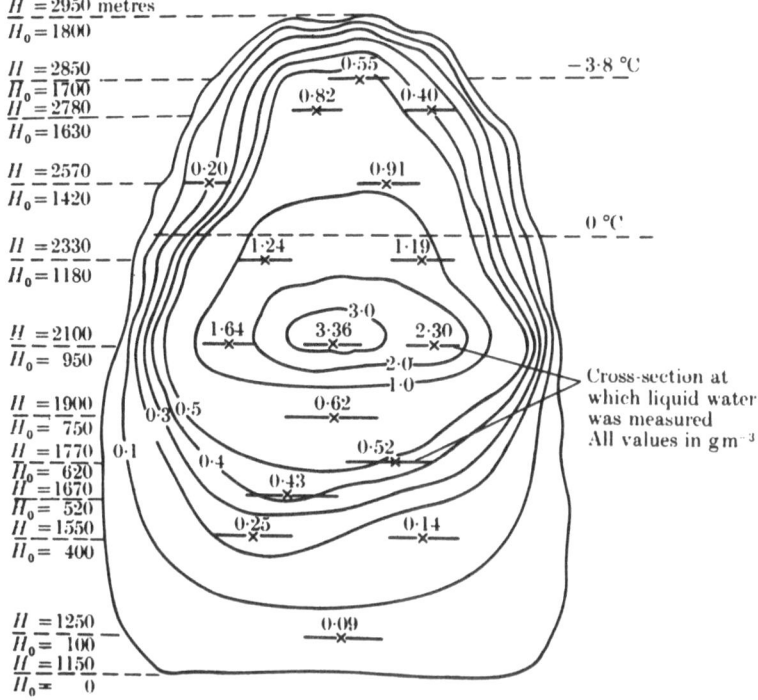

Figure 2.   Liquid water content of cumulus cloud as reported by Zaitsev
             (1950), reproduced from Mason (1971).

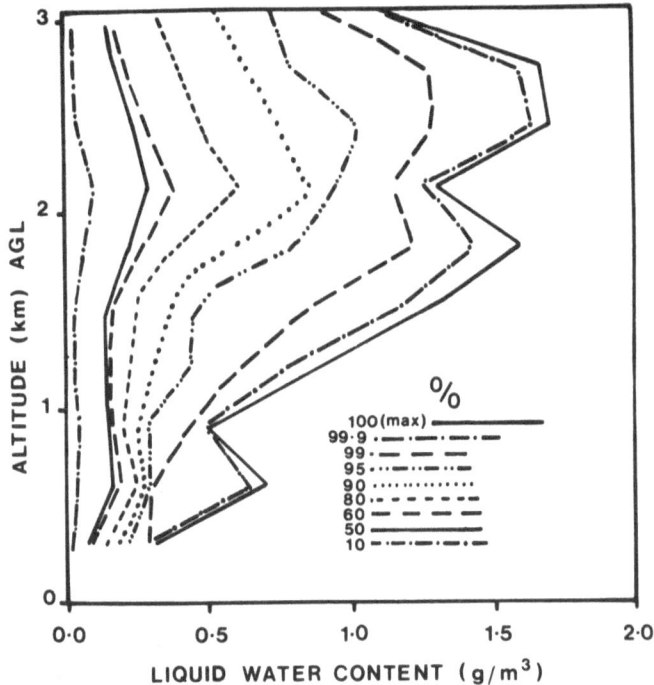

Figure 3.   Altitude variation of supercooled liquid water content (L)
            for all cloud types as reported by Jeck (1983).   At each
            level the nth percentile curves indicate the value of L that
            was exceeded in n% of the distances flown in supercooled
            clouds within the altitude interval immediately below.

pressure of water with temperature and with still more complex
dependencies of L and scavenging efficiencies as a function of
altitude.   The generally decreasing temperature and total water content
with altitude (Figure 1) in the atmosphere, along with the vapour
pressure dependence on temperature, yield clouds which have a complex
dependence on temperature, yield clouds which have a complex dependence
of liquid water content as a function of altitude.   Individual clouds
no doubt have eddies of higher and lower L, as can be seen in Figure 2
for a cumulus cloud.   However, on the average for supercooled clouds at
mid-latitudes, it appears that there is a maximum in L at around 2-3 km
altitude (Figure 3).   The decrease in L above the maximum is due in
part to the decreasing availability of water and the entrainment of dry
air near cloud top (Paluch, 1979; Jensen, 1985).   Comparison of Figures
2 and 3 with Figure 1 illustrates that most of the water in the
troposphere is in the gas phase instead of the condensed phases.   The
variation of L with Z in Figure 3 often is seen to involve more than a
factor of two in L over the height of normal clouds (say one to a few
km).   Hence, all other factors being constant, the concentration in

cloud water of substances scavenged per Junge's (1963) equation for
nucleation scavenging will vary with altitude by more than a factor of
two, just due to variations in L.   The sense of this dependence is that
concentration increases as L decreases, e.g. due to an altitude
increase.   Miller's (1981) data showing decreasing pH with altitude on
the island of Hawaii can be explained at least in part as a result of
the nucleation scavenging of non-sea-salt (NSS) sulfate aerosol by
decreasing amounts of water as altitude increases.

However, this dependence of L is not the sole factor that is
involved in determining the altitude dependence of cloud composition and
deposition fluxes.   The latter is influenced by the rainfall amount,
which is also a function of altitude.   Figure 4 depicts the
precipitation amount as a function of height on a particularly tall,
isolated mountain, Mauna Loa.   There is a decrease in precipitation
amount with altitude of ca. a factor of ten from sea level to the

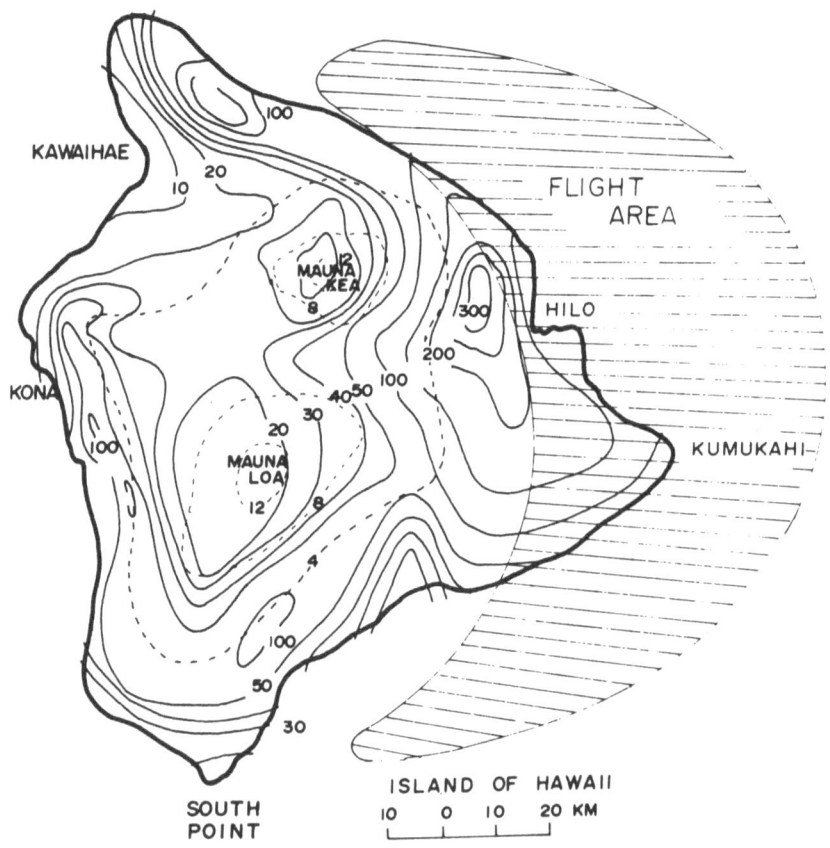

Figure 4.   Solid lines represent median annual rainfall on the island
of Hawaii (1937-1957) as reported by Takahashi (1977).
Dotted lines are approximate elevation contours with Mauna
Loa being ca. 3.3 km high.

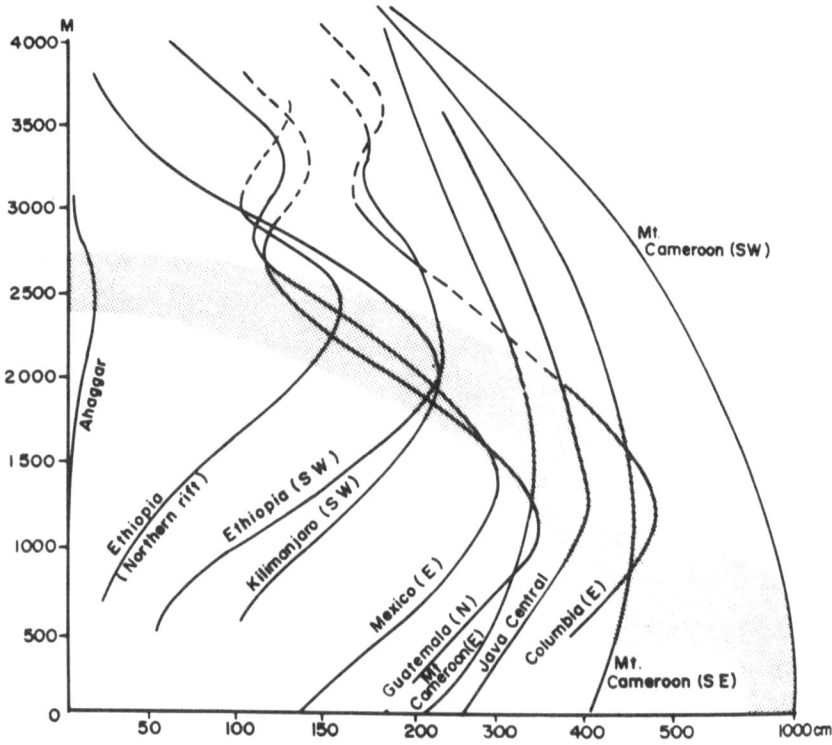

Figure 5.   Mean annual precipitation amount (horizontal axis, cm) as a
            function of altitude (m) for several tropic locations.
            Similar profiles are expected in mid-latitudes. From Barry
            (1981), reproduced from Lauer (1975).

observatory at 3300 m.   Thus we might expect the deposition flux to be
compensated all or in part since as V decreases with greater altitude,
[X] increases.   Other locations also exhibit variation of precipitation
amount as a function of altitude that tend to mirror the altitude
dependence of L.   Figure 5 includes several tropical locations,
although similar L profiles ought to be expected for the mid-latitudes.
    This picture is confounded further by the increasing importance of
the ice phase with increasing Z, and with the decreased rainwater
concentrations to be expected due to the growth of solid hydrometeors by
deposition rather than by coalescence or riming of the products of
nucleation (Scott, 1981).   Table 1 compares compositions of unpolluted
maritime rainwater and snow (glacial ice).   The concentration of
nucleation scavenged material such as NSS SO_4^- in air with low
concentrations of SO_2 is a factor of 5 lower in snow than in rain.   To
investigate this comparison further, an experiment is needed in a single
homogeneous air mass.

Table 1. Compositions of unpolluted maritime rainwater and snow.

Phase	Site/Date	$\frac{NO_3^-(\mu mol)}{\ell}$	$\frac{NSS\ SO_4^-(\mu mol)}{\ell}$	References
Rain	Lake Ozette, WA 1985.	2.66 ± 1.16	3.20 ± 1.43	Vong, 1985.
	Cheeka Peak, WA 1985.	5.45 ± 1.03	4.48 ± 0.708	Vong, 1985
	Amazon Basin	2.1	4.4	Stallard and Edmond, 1981.
	Global Estimate		22	Ryaboshapko, 1983.
Snow	Camp Century, 1976-1977 (NO_3^-) 1967-1971 (SO_4^-)	1.9	1.17	Herron, 1982 Weiss et al., 1975
	Dye 3, Greenland 1979-1980	2.21	0.91*	Herron, 1982
	C-16, Antarctica 1971-1977	0.63	0.42	Herron, 1982
	Q-13, Antarctica 1973-1977	0.66	0.59	Herron, 1982

*Not corrected for sea salt sulfate

5. CHEMICAL EFFECTS OF THE ALTITUDE-TEMPERATURE DEPENDENCE

Returning to the simple picture of the role of L in controlling the concentration of solutes that are incorporated in cloud droplets by nucleation scavenging, it is necessary to include the exchange of gases with droplets and the temperature dependence of these equilibria in order to provide a more complete and realistic model.   Table 2 provides information on the temperature dependence for the equilibria of the key inorganic components of cloud and rainwater: CO_2, SO_2, NH_3 and $H_2O(1)$. Here we see that all of the equilibrium constants are temperature dependent and that the Henry's law expression and K_w temperature dependencies are the strongest.   However, the overall effect of temperature on the equilibria in cloud and rainwater requires a model of the whole system (Taylor et al., 1983).   Taking the system CO_2, SO_2, NH_3, H_2SO_4 and $H_2O(1)$ as the simplest model of real clouds and rain, we see that the influence of temperature indeed is real and of important magnitude.

Figure 6 is a pair of Sillén diagrams (Charlson and Rodhe, 1982) for 278 and 298K for a case where the ionic balance and pH are

Table 2. Temperature dependencies for equilibria of CO_2, SO_2, NH_3 and H_2O

Equilibrium	Equilibrium Constant	Value at 5°C	Value at 25°C	References
$H_2O = H^+ \; OH^-$	$K_w = [H^+][OH^-]$	1.82×10^{-15} mol^2 litre^{-2}	1.01×10^{-14} mol^2 litre^{-2}	Harned and Owen, 1958.
$SO_2(g)+H_2O = SO_2 \cdot H_2O$	$K_{HS} = \dfrac{P_{SO_2}}{[SO_2 \cdot H_2O]}$	0.379 atm M^{-1}	0.752 atm M^{-1}	Johnstone and Leppla, 1934
$SO_2 \cdot H_2O = H^+ + HSO_3^-$	$K_{1S} = \dfrac{[H^+][HSO_3^-]}{[SO_2 \cdot H_2O]}$	2.6×10^{-2} mol litre^{-1}	1.51×10^{-2} mol litre^{-1}	Perrin,1982
$HSO_3^- = H^+ + SO_3^-$	$K_{2S} = \dfrac{[H^+][SO_3^-]}{[HSO_3^-]}$	8.88×10^{-8} mol litre^{-1}	6.61×10^{-8} mol litre^{-1}	Yui, 1940; Perrin,1982
$NH_3(g)+H_2O = NH_3 \cdot H_2O$	$K_{HN} = \dfrac{P_{NH_3}}{[NH_3 \cdot H_2O]}$	7.11×10^{-3} atm M^{-1}	1.62×10^{-2} atm M^{-1}	Freiburg, 1974 Liljestrand & Morgan, 1981
$NH_3 \cdot H_2O = NH_4^+ + OH^-$	$K_{bN} = \dfrac{[NH_4^+][OH^-]}{[NH_3 \cdot H_2O]}$	1.48×10^{-5} mol litre^{-1}	1.77×10^{-5} mol litre^{-1}	Perrin, 1982
$CO_2(g)+H_2O = CO_2 \cdot H_2O$	$K_{HC} = \dfrac{P_{CO_2}}{[CO_2 \cdot H2O]}$	16.6 atm M^{-1}	27.3 atm M^{-1}	Robinson & Stokes, 1959
$CO_2 \cdot H_2O = H^+ + HCO_3^-$	$K_{1C} = \dfrac{[H^+][HCO_3^-]}{[CO_2 \cdot H2O]}$	2.94×10^{-7} mol litre^{-1}	5.01×10^{-7} mol litre^{-1}	Robinson & Stokes, 1959
$HCO_3^- = H^+ + CO_3^-$	$K_{2C} = \dfrac{[H^+][CO_3^-]}{[HCO_3^-]}$	2.74×10^{-11} mol litre^{-1}	5.01×10^{-11} mol litre^{-1}	Robinson & Stokes, 1959

118

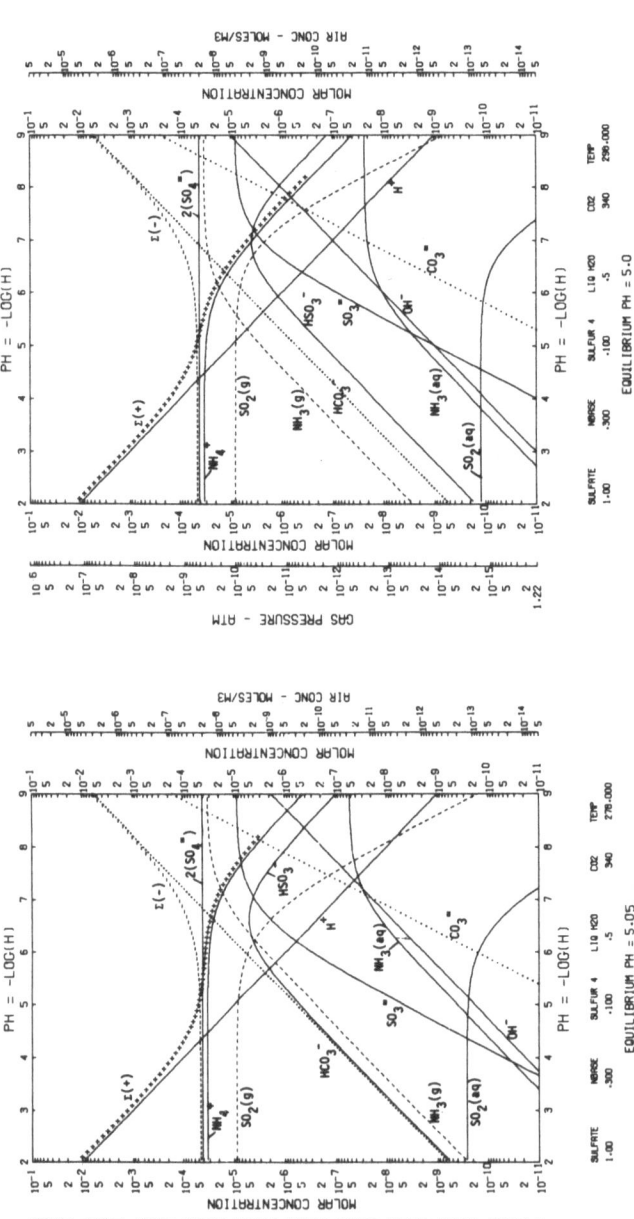

Figure 6. In this and the following two figures the ionic and gaseous species concentration are plotted as functions of pH for the input conditions given below. The independent variables are pH and the total amounts of input substances, while the individual species concentrations are the dependent variables. The singular equilibrium pH for a set of input conditions is given by the pH at which charge balance is satisfied, ie where $\Sigma[+] = \Sigma[-]$. Further explanation of this graphical technique can be found in Vong and Charlson (1985).
Master variable diagrams for 278 and 298K where pH is determined by strong acid concentration. Note temperature independence. Input conditions: sulfate aerosol $- 1 \mu g/m^3$, total ammonia $- 0.3 \mu g/m^3$, oxidation state IV sulfur $- 0.1$ ppbv, liquid water content $- 0.5 g/m^3$, carbon dioxide $- 340$ ppmv.

controlled by the strong acid H_2SO_4. Here, very little temperature effect on pH is noted, as expected due to the control by fully ionized species. Figure 7 shows a situation where SO_2 dissociation is important to the charge balance and, as a result, pH increases with increasing T. Figure 8 is an example of dominance by NH_3 (aq) dissociation yielding a <u>decrease</u> of pH with increasing T.

SO_2 and NH_3 are likely to be the dominant labile acid-base species which can be partitioned variably between the gas and droplet phases. Formic acid may also be important but is not included here for simplicity. Nitric acid is also potentially labile, i.e., it should return to the gas phase when a cloud droplet evaporates; however, while in a cloud, virtually all HNO_3 is in solution as NO_3^- ion. For purposes of these model resuls, NO_3^- would simply add to the total strong acid anion concentration as represented by $[SO_4^=]$.

Because of the importance of SO_2 and NH_3, especially in polluted continental air, it is useful to examine the results of Figures 6-8 more thoroughly. First, the dependence of the equilibrium constants on temperatures and their inclusion in the simple model result in a complex set of interdependent effects. Simple results for increased temperature include the expected increase of pH with SO_2 dominance and decrease of pH with NH_3 dominance. The cause of the former is largely the shifting of SO_2 from the liquid to the gas phase. A decrease in $[HSO_3^-]$ results which, in turn, leads to a higher pH (Figure 7). However, this same effect of T on $[HSO_3^-]$ should result in changes in the reaction rate of HSO_3^- by decreasing the $[HSO_3^-]$ from 6×10^{-5} to 3×10^{-5} M. Whether the rate constant for a given reaction increases enough to compensate for this decrease in concentration is an open question.

The location of the "knee" of the $[NH_4^+]$ - pH dependence is a strong function of T as can be seen in Figure 8. It is this knee that governs the titration curves for this atmospheric system. At the lower temperature most of the NH_3 is in the NH_4^+ form at all pH values below about 6. As a result, the pH of the system is extremely sensitive to additions (or subtractions) of strong acid (here, H_2SO_4). At the higher temperature, the knee has receded by about one pH unit, and the system becomes less sensitive to strong acid. At the lower temperature a mere doubling of the input H_2SO_4 causes pH to drop from 6.15 to around 4.0, while at the higher temperature it only drops from 5.2 to 4.2. This effect is greatest near the end-point of the titration where the amounts of base and acid are equal. Figure 9 summarizes these dependencies in the form of titration curves for the two temperatures.

Liquid water content also mediates this sensitivity. The higher the L, the lower the sensitivity. Comparison of Figures 6 and 8 reveals that the cause of this is the greater dilution of all solutes, and the increasing importance of $[HCO_3^-]$ as L increases.

6. DISCUSSION

The change in temperature with altitude results in numerous atmospheric effects. These include the physical effect of temperature on liquid

120

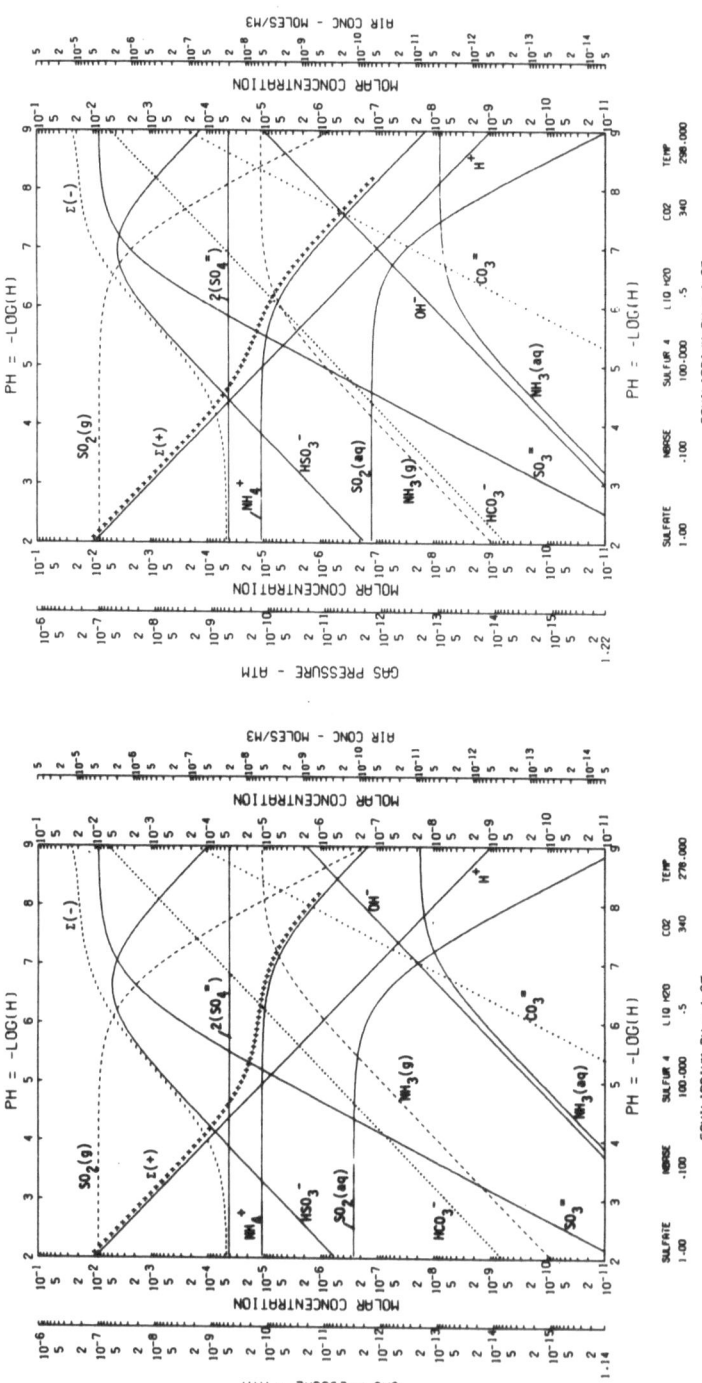

Figure 7. Master variable diagrams for 278 and 298K for a case where the acid-base balance is dominated by sulfur in oxidation state IV. Note that increasing T results in increasing pH, all input conditions being held constant. Input conditions: sulfate aerosol – 1 µg/m³, total ammonia – 0.1 µg/m³, oxidation state IV sulfur – 100 ppbv, liquid water content – 0.5 g/m³, carbon dioxide – 340 ppmv.

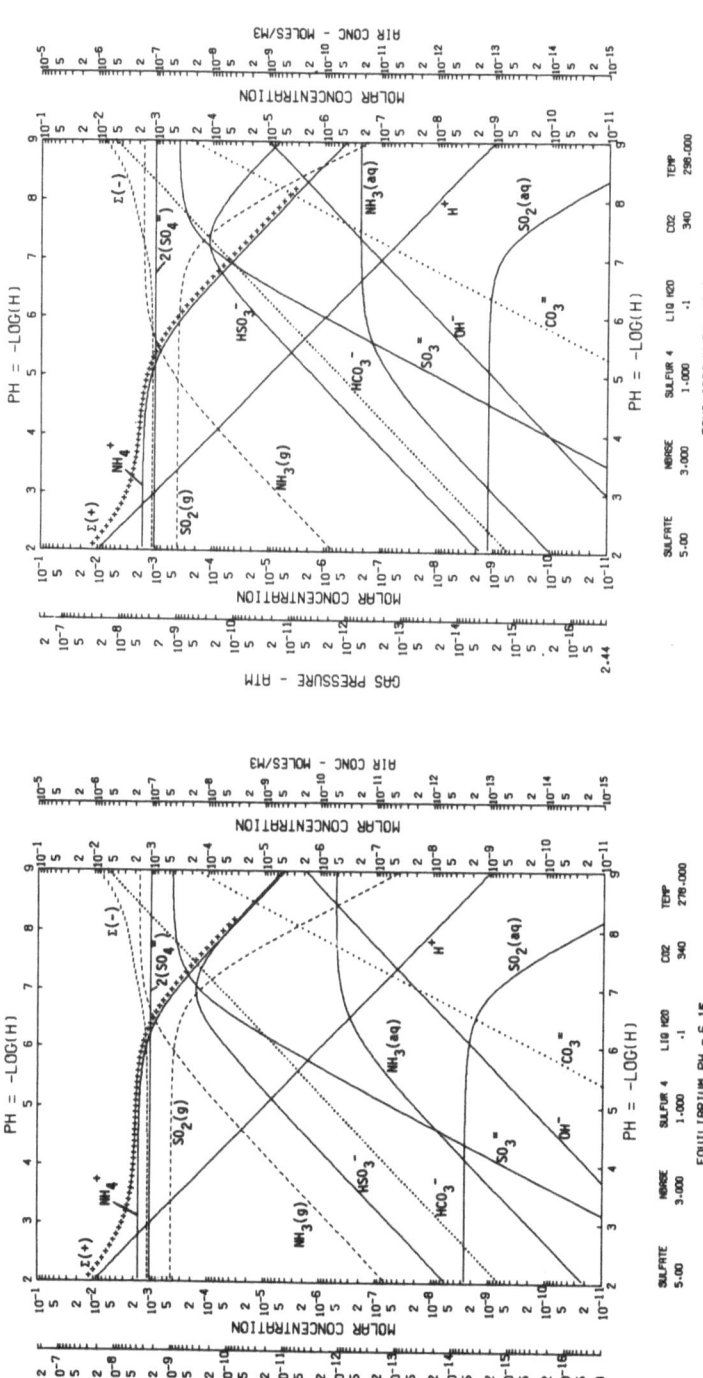

Figure 8. Master variable diagrams for 278 and 298K for a case where the acid–base balance is dominated by the weak base, ammonia. Note that increasing T causes a decrease in pH, all input conditions being held constant. Input conditions: sulfate aerosol – 5 µg/m³, total ammonia – 3 µg/m³, oxidation state IV sulfur – 1 ppbv, liquid water content – 0.1 g/m³, carbon dioxide – 340 ppmv.

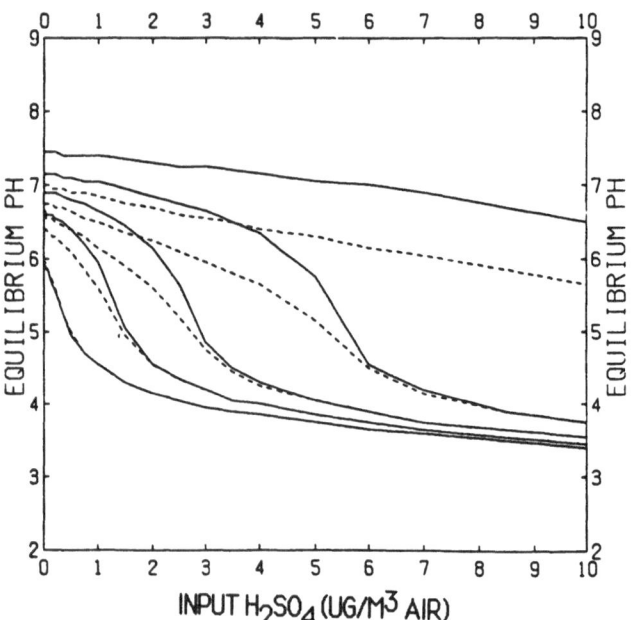

PH – SENSITIVITY TO SO4 AND NH4
TOTAL NH4= 0.1 0.5 1.0 2.0 5.0 UG/M3 AIR

Figure 9.   Cloud water pH as a function of input H_2SO_4 aerosol for L =
0.5 g/m^3, 0.2 ppb SO_2 and no HNO_3.   The dashed curves are
for 298K, the solid ones for 278K.   The parameter is the
amount of total ammonia ranging from 0.1 (bottom) to 5 (top)
µg/m^3.

water content and deposition amount and temperature influences on
chemical equilibria.   The most important effects of lower temperatures
at higher altitudes would appear to be:
   a) decreased L with corresponding increase in the concentration of
      solutes from the dissolution of aerosol particles acting as CCN
   b) increased sensitivity of the cloud and rainwater pH to input
      acid due to decreased L and
   c) increased sensitivity of pH to input acid due to shifts in the
      chemical equilibria, particularly that of the ammonia species.
Thus, the addition of a given amount of acid, as modelled here by H_2SO_4
aerosol, would cause a much larger depression in pH in the high
altitude, low T, low L situation than it would at sea level with higher
T and higher L.

REFERENCES

Charlson, R.J. and H. Rodhe. 1982. Factors controlling the acidity of
    natural rainwater, Nature **295**, 683–685.

Charlson, R.J., R. Vong and D.A. Hegg. 1983. The sources of sulfate in precipitation (2), sensitivities to chemical variables. J. Geophys. Res. **88**, 1375-1377.

Freiberg, J.E. 1974. Effects of relative humidity and temperature on iron catalyzed oxidation of SO_2 in atmospheric aerosols. Environ. Sci. Technol. **8**, 731-734.

Harned and Owen. 1958. The Physical Chemistry of Electrolytic Solutions. New York: Van Nostrand Reinhold.

Herron, M.M. 1982. Impurity sources of F^-, Cl^-, NO_3^- and SO_4^- in Greenland and Antarctic precipitation. J. Geophys. Res. **87**, 3051-3060.

Jeck, R.K. 1983. A new data base of supercooled cloud variables for altitudes up to 10,000 feet AGL and the implications for low altitude aircraft icing, Report No. DOT/FAA/CT-83/21 (NRL Report 8738), U.S. Dept. of Transportation, Federal Aviation Administration Technical Center, New Jersey.

Jensen, J.B. 1985. Turbulent mixing, droplet spectral evolution and dynamics of warm cumulus clouds, Ph.D. dissertation, University of Washington, Seattle, WA.

Johnstone, H.F. and P.W. Leppla. 1934. The solubility of sulfur dioxide at low pressures, J. Am. Chem. Soc. **56**, 2233-2238.

Junge, C.E. 1963. Air Chemistry and Radioactivity. Academic Press, New York.

Lauer, W. 1975. Klimatische Grundzuge der Hohenstufung tropischer Gebuge, in Tagungsbericht und Wissenschaft lische Abhandlungen, 40 Deutscher Geographentag, F. Stein, Innsbruck, 76-90, 1975. (Reproduced from Barry, R.G., Mountain Weather and Climate, p. 186, Methuen and Co., London, 1981).

Liljistrand, H.M. and J.J. Morgan. 1981. Spatial variations of acid precipitation in Southern California. Environ. Sci. Technol. **15**, 333-338.

McClatchey, R.A., R.W. Fenn, J.E.A. Selby, F.E. Volz and J.S. Garing. 1977. Optical properties of the atmosphere, Environ. Res. Pap. No. 411, Air Force Cambridge Res. Lab. [NTIS No. 0497], 1972. (Reproduced from Ramanathan, V., Interactions between ice-albedo, lapse-rate and cloud-top feedbacks: An analysis of the nonlinear response of a GCM climate model, J. Atmos. Sci., **34**, 1885-1897.

Miller, J.M. and A.M. Yoshinaga. 1981. The pH of Hawaiian precipitation: A preliminary report, Geophys. Res. Lett. **8**, 779-782.

Paluch, I.R. 1979. The Entrainment Mechanisms in Colorado Cumuli. J. Atm. Sci. **36**, 2467-2478.

Perrin, D.D. 1982. Ionization Constants of Inorganic Acids and Bases in Aqueous Solutions. New York: Pergamon Press.

Robinson, R.A. and R.H. Stokes. 1959. Electrolytic Solutions. Butterworths, London.

Ryboshapko, A.G. 1983. The atmospheric sulfur cycle. In: The Global Biogeochemical Sulfur Cycle (M.V. Ivanov and G.R. Freney, eds.). New York: Wiley, 203-296.

Scott, B.C. 1981. Sulfate washout ratios in winter storms, J. Appl. Met. **20**, 619-625.

Sievering, H., C.C. Van Valin, E.W. Barrett and R.F. Pueschel. 1984. Cloud scavenging of aerosol sulfur: Two case studies. Atmos. Env. 18, 2685-2690.

Stallard, R.F. and J.M. Edmund. 1981. Geochemistry of the Amazon. 1: Precipitation chemistry and the marine contribution to the dissolved load at the time of peak discharge. J. Geophys. Res. 86, 9844-9858.

Takahashi, T. 1977. A study of Hawaiian warm rain showers based on aircraft observation. J. Atmos. Sci. 34, 1775.

Taylor, G.S., M.B. Baker and R.J. Charlson. 1982. Heterogeneous interactions of the C, N, and S cycles in the atmosphere: the role of aerosols and clouds. In: Interactions of the Biogeochemical Cycles, edited by B. Bolin and R. Cook, Wiley, London.

Vong, R.J. 1985. Simultaneous observations of rainwater and aerosol chemistry at a remote mid-latitude site, Ph.D. dissertation, University of Washington.

Vong, R.J. and R.J. Charlson. 1986. The equilibrium pH of a raindrop: A computer-based solution for a six-component system. J. Chem. Ed. 62, 141-143.

Weiss, H., K. Bertine, M., M. Koidew and E.D. Goldberg. The chemical composition of a Greenland Glacier. Geochim. Cosmochim. Acta. 39, 1-10.

Yui, T. 1940. On the electrolytic dissociation constant of sulfurous acid. Tokyo Inst. Phys. Chem. Res. Bull. 19, 1229-1236.

Zaitsev, V.A. 1950. Vodnost i raspredelenie kapel v kuchevykh oblakakh (Liquid water content and distribution of drops in cumulus clouds), Trans. Trudy Glav. Geofiz. Observ., 19, p. 122. (Reproduced from Mason, B.J., The Physics of Clouds, 2nd ed., p. 120, Clarendon Press, Oxford, 1971).

MEASUREMENTS OF THE SHORT-TERM VARIABILITY OF AQUEOUS-PHASE MASS CONCENTRATIONS IN CLOUD DROPLETS

J.A. Ogren, J. Heintzenberg, A. Zuber, H.-C. Hansson
Department of Meteorology
University of Stockholm
S-106 91 Stockholm
Sweden

K.J. Noone*, D.S. Covert** and R.J. Charlson***
* Department of Civil Engineering
** Department of Environmental Health
*** Department of Atmospheric Sciences
University of Washington
Seattle
Washington 98195
USA

ABSTRACT.   In-situ sampling of clouds with a droplet to aerosol converter allows determination of several important chemical and physical properties of cloud droplets over short spatial and temporal scales.   During a measurement programme in July and August 1986 on the Swedish peak Areskutan (1250 m asl), the number concentration, water content, and aerosol light scattering coefficient of evaporated cloud droplets were determined in-situ.   These measurements allow estimation of the mass concentration of non-volatile materials dissolved or suspended in the cloud droplets, the average size of the sampled cloud droplets, and the average size of the aerosol particles that result from evaporation of the sampled cloud droplets.   The results illustrate the inhomogeneous character of clouds, with substantial variations of droplet number concentration, liquid water content, and residual aerosol light scattering coefficient over distances of less than 1 km. The observed, smaller variations in derived properties such as aqueous-phase mass concentration and mean droplet radius may be due to entrainment of subsaturated air causing complete evaporation of some of the droplets.

1. INTRODUCTION

Chemical processes within cloud droplets are most commonly studied by collecting a sample of water from the cloud and analysing its chemical composition.   This approach has a number of drawbacks, among the most

125

M. H. Unsworth and D. Fowler (eds.), Acid Deposition at High Elevation Sites, 125–137.
© 1988 by Kluwer Academic Publishers.

serious of which is the averaging which occurs during sample collection. One type of averaging is over droplet size, where the collected sample contains contributions from droplets of widely varying sizes. In many cases, the collection efficiency as a function of droplet size is unknown or, at best, partially characterised. Another type of averaging is over time and space, as all the droplet collectors require a finite amount of time to obtain sufficient sample for chemical analysis.

The effect of these kinds of averaging is to prevent the study of chemical and physical processes that depend on droplet properties that are destroyed during the averaging. For example, chemical reactions that depend on the aqueous concentration of reactants in cloud droplets cannot be studied properly from averaged samples if any variabilities in reactant concentrations are removed by the averaging process. In an earlier study, Noone et al. (1987a) observed that aqueous mass concentrations in maritime stratus clouds increased with increasing droplet size, and deduced that the chemical composition of the droplets was probably also a function of droplet size.

In the present study, measurements similar to those of Noone et al. (1987a) have been performed in maritime stratocumulus clouds. These measurements allow investigation of the variability of aqueous-phase mass concentrations in cloud droplets over short temporal and spatial scales.

2. MEASUREMENT TECHNIQUES

Site characteristics. Sampling was performed during July and August, 1986 at an elevation of 1250 m asl on the Swedish peak Areskutan (63°26'N, 13°6'E). The site is 400 m above the timberline and roughly 800 m above the surrounding terrain. Maritime stratocumulus clouds are common at the site, which is about 180 km from the Norwegian Sea. The chemical composition of bulk cloudwater is highly variable, depending on the recent history of the air mass sampled. Cloudwater in air masses from the Norwegian Sea can give sulphate concentrations as low as 1 μeq/l, while air that has recently passed over continental Europe can have sulphate concentrations that are a factor of 1000 higher (Ogren and Rodhe, 1986).

Droplet to aerosol converter. The method used for sampling cloud droplets is a counterflow virtual impactor (CVI) that separates the droplets from the surrounding air and evaporates them in a flow of warm, dry, particle-free air (Ogren et al., 1985, Noone et al., 1987a,b). A cross-sectional view of the CVI is shown in Figure 1. Warm, dry, particle-free air flows through the annular region of two concentric tubes to the tip of the impactor. The wall of the inner tube at the tip is constructed of porous stainless steel, which allows the dry air to flow into the inner tube. A fraction of the air entering the inner tube is sucked back into the sampler, while the rest blows out the tip. A consequence of this flow scheme is that somewhere along the length of the porous tube there is a stagnation plane where

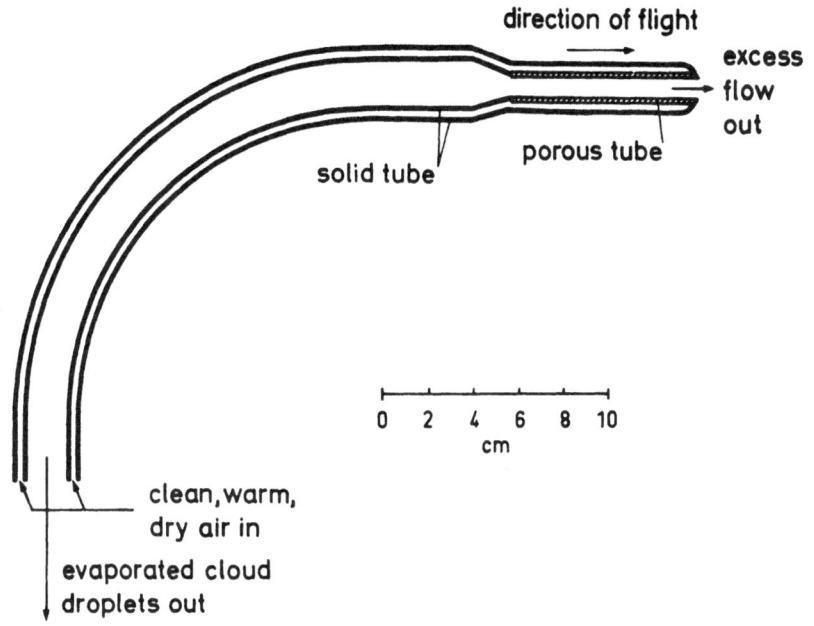

direction of flight

excess flow out

porous tube

solid tube

0 2 4 6 8 10
cm

clean, warm, dry air in

evaporated cloud droplets out

Figure 1   Internal geometry of counterflow virtual impactor.

there is no net axial flow. Tipward of this plane the air is flowing
towards the tip, while inward of this plane the air is flowing back
into the sampler. The distance from the stagnation plane to the tip
can be varied by adjusting the air flow rates to the tip and back into
the sampler.

Cloud droplets approaching the CVI can either be deflected around
the inlet or swept into the inner tube. Gas molecules and
submicrometer aerosol particles follow the air streamlines around the
tip of the impactor, while larger particles (or droplets) diverge from
the streamlines due to their higher inertia. Those droplets that enter
the inner tube are either blown back out the tip or sucked into the
sample flow, depending on whether they have sufficient inertia to pass
through the region off tipwards-flowing air. The parameter controlling
droplet collection is the stop distance of the droplets, which is the
distance a droplet with a given initial velocity travels in motionless
air before coming to rest (Fuchs, 1964). Values of the stop distance
for conditions applicable to the collection system used on Areskutan
are listed in Table 1. The minimum radius of droplets that can be
sampled is determined by the radius of the outer tube of the sampler (1
cm); droplets with stop distances less than about 1 cm will follow the
air streamlines around the tip of the probe. The length of the porous
inner tube in the CVI (8 cm) determines the maximum size of droplets
which can be rejected. A comparison of the dimensions of the CVI with
the values in Tables 1 indicates that the lower size limit of droplets

that are sampled can be varied from ca. 4 μm to ca. 15 μm. The results of calibrations performed by Noone et al. (1987b) show that the sampler performs in accordance with these theoretical predictions.

Large droplets that enter the CVI are removed by impaction on a 90° bend in the tubing located 75 cm downstream of the tip. This bend has a radius of curvature of 2.5 cm, and the sample air velocity is typically 3 m s^{-1}. A first-order analysis based on the Stokes number of the droplets (Fuchs, 1964) shows that droplets with radii larger than ca. 50 μm will be removed at this bend.

The flow rate of air swept out by the sampler is equal to the cross-sectional area of the inlet multiplied by the velocity of the air flowing past the tip; on Areskutan this amounted to ca. 130 lpm. Those droplets with sufficient intertia to enter the sampler become embedded in a flow of ca. 10 lpm. As a result, droplet number concentrations, liquid water content, and the concentrations of material dissolved or suspended in the cloud droplets are enhanced inside the sampler by a factor of ca. 13 relative to ambient air. The results presented below have been corrected for this preconcentration effect, and thus refer to ambient conditions.

The sampled cloud droplets evaporate quite quickly in the warm, dry air inside the CVI. As a first-order estimate, the time required to evaporate a droplet has been calculated by neglecting the effects of ventilation, curvature, and solute concentration, and by assuming that the temperature of the surface film of the droplet is constant. The results of this calculation for typical conditions encountered during sampling (sample air temperature 30°C, ambient air temperature 5°C) are included in Table 1. Although a rigorous calculation would have to include the effects of evaporation on the stop distance, the first-order estimates shown in Table 1 indicate that the maximum droplet radius that can be sampled is between 50 and 100 μm.

Table 1 Stop distance of cloud droplets with an initial velocity of . 100 m s^{-1}, along with their evaporation time in dry air.

Droplet radius (μm)	Stop distance (cm)	Evaporation time (s)
1	0.1	0.001
2	0.3	0.006
5	1.3	0.037
10	3.9	0.15
15	7.3	0.33
20	11	0.58
50	45	3.6
100	124	15

Ambient gases and submicrometer aerosol particles are rejected with almost 100% effectiveness. A measure of this effectiveness is the rejection ratio, defined as the ratio of the concentration of a species in ambient, cloud-free air to the concentration within the CVI. Measurements of the rejection ratio for condensation nuclei resulted in values of 10^2 to 10^5, depending on the wind speed and the distance from the sampler inlet to the stagnation plane. For the worst case combination of a rejection ratio of 10^2, a cloud droplet number concentration of 50 cm^{-3}, and an ambient (within cloud) condensation nucleus concentration of 10^3 cm^{-3}, contamination by submicrometer aerosol particles would result in a surplus of 10 cm^{-3}, amounting to a 17% error in the measured droplet number concentration. As the data presented below exemplify, the actual contamination was much less, as the baseline of the droplet number concentration trace was under 1 cm^{-3}.

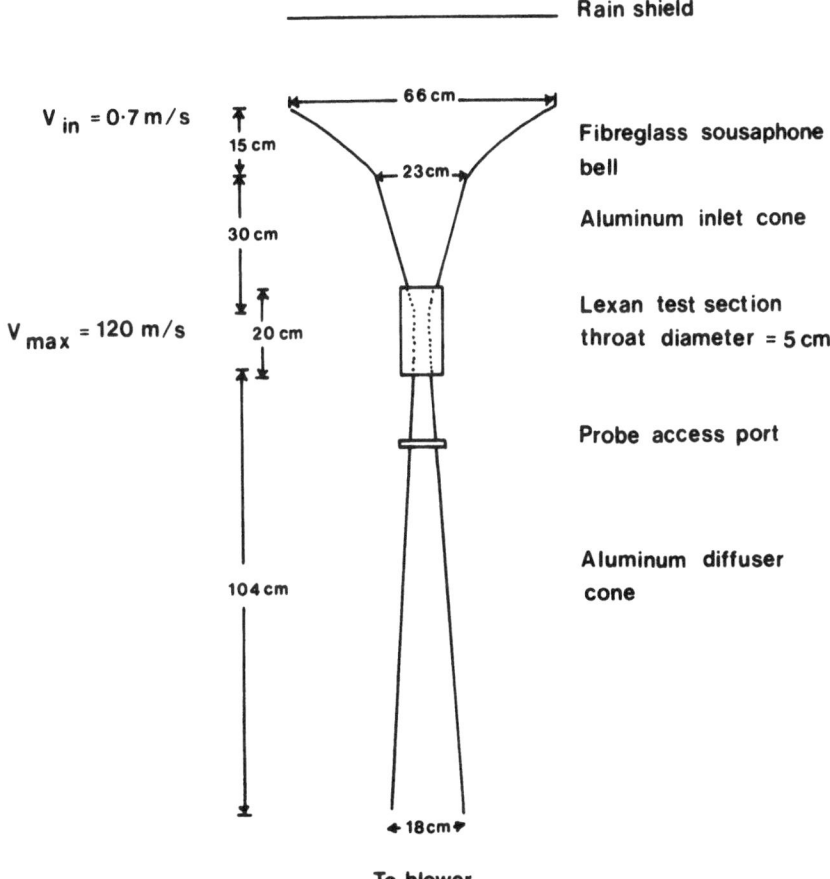

Figure 2   Cross-sectional view of vertical wind tunnel.

Vertical wind tunnel. The first application of the CVI described above was an airborne study (Ogren et al., 1985), where an airplane provided the velocity needed to separate cloud droplets from the surrounding air. For the present work, a vertical wind tunnel (Fig. 2) was developed to accelerate the droplets to ca. 115 m s^{-1} prior to sampling (Noone et al., 1987b). The inlet section of the wind tunnel was constructed from the horn of a damaged sousaphone, leading to the nickname "cloudaphone" for the complete sampler. Air passing through the horn-shaped section was accelerated from 0.7 to 6 m s^{-1} over a distance of 15 cm. A conical section of length 30 cm provided the remaining acceleration to the maximum speed of 115 m s^{-1} in the 5 cm diameter test section, corresponding to a flow rate of 14 m^3 min^{-1}. A 7° conical deceleration section of length 100 cm recovered much of the energy from the accelerated air. To avoid choking off the airflow, the tip of the CVI was located 5 cm downstream of the test secton.

Calibration of the sampling efficiency of the vertical wind tunnel as a function of droplet size and wind speed has not been performed to date. Field experience with the system indicates that serious losses occur with high wind speeds, as very few droplets were detected at wind speeds in excess of ca. 15 m s^{-1}. An inlet with a similar geometry was evaluated by Zebel (1978) and tested by Armbruster and Zebel (1985). If their results can be scaled up to the geometry used on Areskutan, then the sampling efficiency of the vertical wind tunnel for wind speeds of 10 m s^{-1} and 20 m s^{-1} should be 96% and 92%, respectively, for 10 μm radius droplets; the corresponding values for 50 μm radius droplets are 50% and 33%. As these values would seem to be too high, based on experience in the field, a wind tunnel calibration is clearly called for. The lack of calibration data for the vertical wind tunnel do not affect the conclusions of this study, as wind speeds were fairly constant over the periods of time when fast variations in the measured parameters were observed.

In-situ analysers. The aerosol resulting from the evaporation of sampled cloud droplets was analysed for water vapour concentration, particle number concentration, and particulate light scattering coefficient. Cloud liquid water content (LWC) was directly calculated from the water vapour concentration, since the only water in the system was that resulting from the evaporation of cloud droplets. Likewise, the cloud droplet number concentration was directly calculated from the aerosol particle number concentration inside the sampler, assuming that the droplets did not coagulate or fragment in the sampler.

Water vapour was determined with both a cooled-mirror, dew point hygrometer (General Eastern model 1100) and a differential absorption, Lyman alpha hygrometer (Zuber and Witt, 1986). Both instruments were connected into the flow system at the point where the sample leaves the vertical wind tunnel in order to minimize the uptake of water vapour on the sample tubing. As a result of the inability of the dew point hygrometer to cool the mirror more than ca. 30°C below ambient temperature, this instrument was incapable of determining cloud liquid water contents below ca. 100 mg m^{-3}.

The Lyman alpha hygrometer measured the differential absorption of Lyman alpha radiation (121.6 nm) between dry air and a mixture of dry air and sample air from the CVI. For the sampling configuration used on Areskutan, an absolute accuracy of 10% was achieved for ambient cloud liquid water contents of 100 mg m^{-3} or more; accuracy was limited by reductions in the dynamic range of the instrument that resulted from a decrease of the lamp's output during the experiment. The detection limit of the instrument was limited by zero drift during the sampling day, which typically corresponded to an ambient liquid water content of 10-20 mg m^{-3}.

Aerosol number concentrations in the sample flow were determined by a continuous flow, condensation nucleus counter (CNC, TSI model 3020), which has a detection limit of 0.01 cm^{-3}. Aerosol size distribution measurements were obtained with a Knollenberg LAS-X-HS optical particle counter, which has a measurement range from 0.03 to 0.5 μm radius. The light scattering coefficient of the aerosol particles (δ_{sp}) was determined at a wavelength of 500 nm using an integrating nephelometer similar to that described by Heintzenberg and Bäcklin (1984). This instrument is calibrated by filling the sensing volume with filtered air and measuring the resulting scattering signal along with the temperature and pressure. The detection limit is ca. 10^{-7} m^{-1} for an averaging time of 10 s. The nephelometer contains a built-in 16 channel analog data acquisition system, which was used to collect data and status information from the CNC, LWC hygrometers, and system flowmeters. Digital and analog data were transferred from the nephelometer over a serial link to a microcomputer (Stride model 440), which recorded the results on a hard disk and provided a real-time display of the incoming data. Depending on the particular experiment being performed, data were recorded with a time resolution of 2 s to 60 s.

3. RESULTS

Cloud sampling was performed on a total of 19 days during July and August, 1986. Only a small fraction of the data have been processed to date, which means that the results presented here can only provide an example of the data that were obtained, their short-term variations, and the inferences that can be drawn from the data. Figure 3 illustrates the variations in liquid water content, number concentration, and particulate light scattering coefficient of evaporated cloud droplets that were observed during a 45 minute period on 22 July. The CVI was run with an internal stop distance of 10 mm, resulting in a radius for 50% collection efficiency of ca. 6 μm. The wind during this period was from the southeast at 8 m s^{-1} and the temperature was 8°C. The day was characterised by the frequent passage of cumulus clouds, and visual observations recorded that the site was right at cloud top for the last 7 minutes of the period. The visual observations are confirmed by the LWC trace, which shows that only during this time were LWC values of zero measured.

132

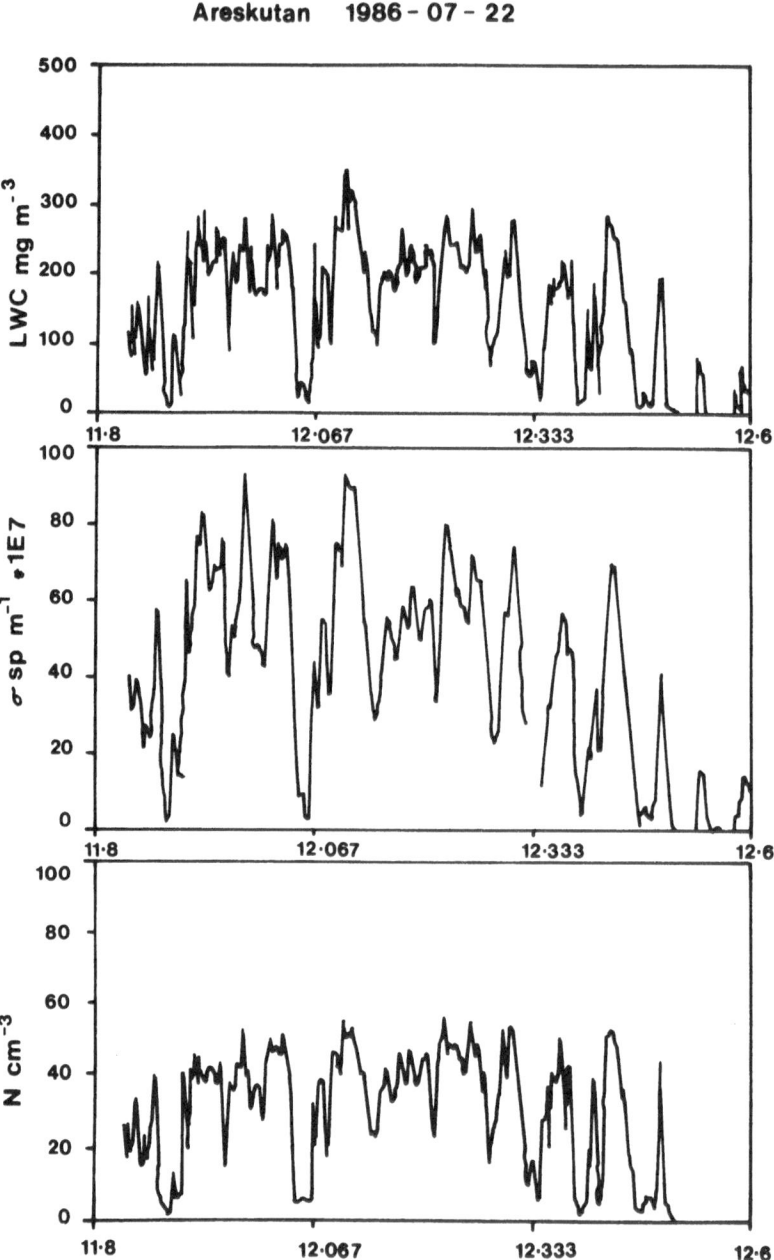

Areskutan   1986 – 07 – 22

Figure 3   Time series of water content (LWC), number concentration
(N), and light scattering coefficient (δ_{sp}) of evaporated
cloud droplets measured on 22 July 1986.

Measured droplet parameters are seen to vary over nearly two orders of magnitude during the sampling period. This behaviour has been reported previosuly for measurements of LWC in cumulus clouds (Warner, 1969), and can be attributed to turbulent mixing of cloudy and cloud-free air. Cloud droplet number concentrations show variations similar to the LWC, as does the light scattering coefficient of the evaporated cloud droplets. At the measured wind speed of 8 m s^{-1}, the regular dips in the LWC trace occurring every 4 minutes or so correspond to a distance of 2 km.

A plot of the ambient aerosol light scattering coefficient, measured with a separate nephelometer sampling from an air stream without cloud droplets, is shown in Figure 4. The passage of the cloud results in a reduction of the submicrometer aerosol light scattering coefficient by a factor of ca. 5, indicating that about 80% of the mass of these particles was incorporated into the cloud droplets. Adding together the measured values for δ_{sp} in the ambient air and the evaporated cloud droplets, it is seen that roughly 75% of the pre-cloud value of δ_{sp} is accounted for by the measurements in the cloud; the remainder is presumably associated with droplets in the 1-6 µm radius interval excluded from the two nephelometers.

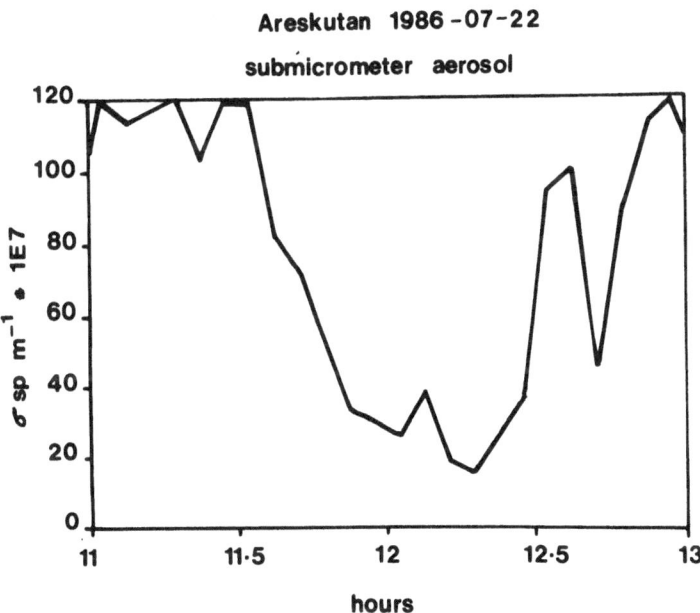

Figure 4   Time series of light scattering coefficient (δ_{sp}) of ambient, submicrometer aerosol particles measured on 22 July 1986 (5 minute median values).

By examining the ratios of the three parameters shown in Figure 3, it is possible to calculate the aqueous mass concentration of nonvolatile material dissolved or suspended in the cloud droplets, the volume mean size of the cloud droplets, and the volume mean size of the particles remaining after evaporation of the cloud droplets. An estimate of the mass of non-volatile material dissolved or suspended in the cloud droplets, can be obtained from the measured values of δ_{sp}. Aerosol size distribution measurements of the evaporated cloud droplets frequently showed a volume mean radius in the range 0.10-0.15 μm, similar to the value of 0.16 μm in background air reported by Whitby (1978). If the aerosol size distributions are similar, then the efficiency for light scattering of the evaporated cloud droplets should also be similar to values measured in clean air. Studies of the ratio of $_{sp}$ to the mass concentration of submicrometer particles have shown that a value of 3.1 ± 0.2 m^2 g^{-1} can be used in both rural and urban air (Waggoner et al., 1981). In this preliminary report, it is assumed that this ratio can also be applied to the light scattering coefficient of the particles resulting from evaporation of cloud droplets in order to estimate their mass concentration in air. Dividing the mass concentration estimated in this fashion by the LWC of the cloud droplets provides an estimate of the aqueous-phase mass concentration of non-volatile material in the cloud droplets.

A similar calculation yields the average volume of the cloud droplets by dividing the LWC by the droplet number concentration. A cube root transformation then results in the volume mean radius of the cloud droplets. Likewise, the mass concentration of the aerosol particles remaining after the droplets evaporate, divided by the number concentration of these particles, yields the average mass of the aerosol particles. Assuming that the density of the particles is 1.5, a cube root transformation provides the mass mean radius of the aerosol particles.

Plots of the three parameters for the same time period as in Figure 3 are shown in Figure 5. The times in Figure 3 when the values dropped near to zero are seen as periods of very strong variability in Figure 5. This variability is an artifact of the analysis resulting from taking the ratio of two numbers that are very close to zero (and thus quite uncertain). On the other hand, variations in the aqueous mass concentration by a factor of two are seen to occur over time scales of less than one minute during periods when the LWC does not approach zero (e.g. at 12.2 hours). Such short variations do not have much effect on the time-averaged bulk composition that would be determined with bulk cloudwater collectors, but may have a larger effect on chemical reactions that are concentration dependent.

Droplet and residual aerosol particle mean sizes are seen to show large variations during periods when LWC was near zero, but much smaller variations at other times. In particular, calculated sizes during the time interval between 12.1 and 12.3 hours were remarkably stable, even though the LWC varied by over a factor of three. This behaviour would indicate that changes in LWC are associated with the total removal of some of the droplets, rather than a decrease in the average size of the droplets. Entrainment of subsaturated air from

Areskutan   1986 - 07 - 22

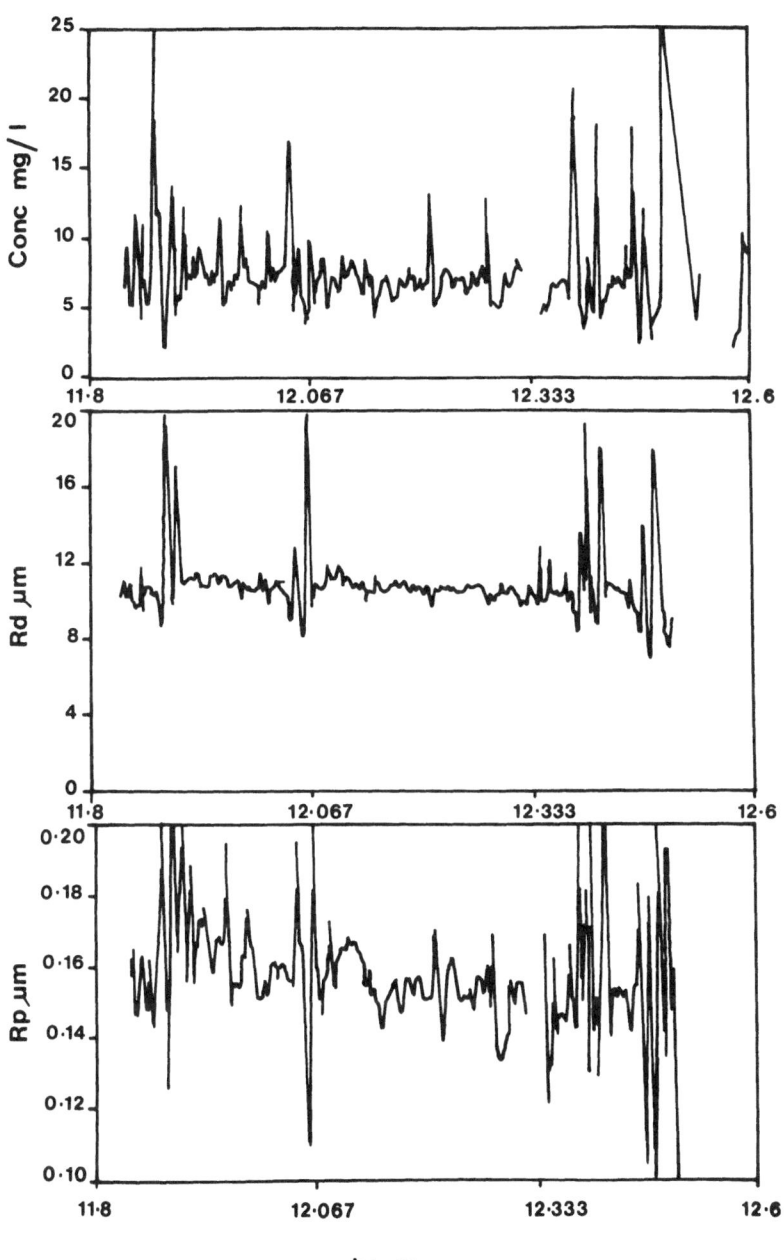

hours

Figure 5   Time series of calculated aqueous-phase mass concentration
(Conc), mean droplet radius (Rd), and mean residual
particle radius (Rp) for the measurements shown in Fig. 3.

outside the cloud could produce such a situation (Baker et al., 1984; Choularton et al, 1986). In this context, it is important to note that the time required to evaporate a 10 μm radius droplet is of order 1 s at a relative humidity of 80%, and the time resolution of the aerosol detectors is in the range 1-10 seconds. Thus, the time resolution of the instrumentation precludes study of the transition region where the droplets are evaporating.

4. CONCLUSIONS

Basic properties of the cumulus cloud described above, namely liquid water content, droplet number concentration, and the light scattering coefficient of the evaporated droplets, are seen to vary by over an order of magnitude on spatial scales of less than 1 km. Much smaller variations are seen in the calculated aqueous-phase mass concentration, volume mean droplet radius, and mass mean residual aerosol particle radius. This behaviour is that which would be observed if parcels of subsaturated air were being entrained into the cloud, causing the complete evaporation of some of the droplets in some regions of the cloud.

REFERENCES

Armbruster, L., and Zebel, G. 1985. Theoretical and experimental studies for determining the aerosol sampling efficiency of annular slot probes. J. Aerosol Sci. 4 , 335-341.

Baker, M.B., Breidenthal, R.E., Choularton, T. and Latham, J. 1984. The effects of turbulent mixing in clouds. J. Atmos. Sci. 41 , 399-304.

Choularton, T.W., Consterdine, I.E., Gardiner, B.A., Gay, M.J., Hill, M.K., Latham, J. and Stromberg, I.M. 1986. Field studies of the optical and microphysical characteristics of clouds enveloping Great Dun Fell. J. R. Met. Soc. 112 , 131-148.

Fuchs, N.A. 1964. The Mechanics of Aerosols. Pergamon Press, Oxford.

Heintzenberg, J. and Bäcklin, L. 1983. A high sensitivity integrating nephelometer for airborne air pollution studies. Atmos. Environ. 17 , 433-436.

Noone, K.J., Charlson, R.J., Covert, D.S., Ogren, J.A. and Heintzenberg, J. 1987a. Cloud droplets: solute concentration is size dependent. Submitted to J. Geophys. Res.

Noone, K.J., Charlson, R.J., Covert, D.S., Ogren, J.A. and Heintzenberg, J. 1987b. Design and calibration of a counterflow virtual impactor for sampling of atmospheric fog and cloud droplets. Aerosol Sci. Tech., in press.

Ogren, J.A., Heintzenberg, J. and Charlson, R.J. 1985. In-situ sampling of clouds with a droplet to aerosol converter. Geophys. Res. Lett. 12 , 121-124.

Ogren, J.A. and Rodhe, H. 1986. Measurements of the chemical composition of cloudwater at a clean air site in central Scandinavia. Tellus 38B, 190-196.

Waggoner, A.P. et al. 1981. Optical characteristics of atmospheric aerosols. Atmos. Environ. **15** , 1891-1909.

Warner, J. 1969. The microstructure of cumulus clouds. Part I. General features of the droplet spectrum. J. Atmos. Sci. **26** , 1049-1059.

Whitby, K.T. 1978. Physical characteristics of sulphur aerosols. Atmos. Enviorn. **12** , 135-159.

Zebel, G. 1978. Some problems in the sampling of aerosols. In Recent Developments in Aerosol Science (ed. D.T. Shaw), pp. 167-185. Wiley, New York.

Zuber, A. and Witt, G. 1987. An optical hygrometer using differential absorption of the hydrogen Lyman alpha emission. Appl. Optics **26** , 3083-3089.

PROCESSES DETERMINING CLOUDWATER COMPOSITION: INFERENCES FROM FIELD
MEASUREMENTS

Peter H Daum
Environmental Chemistry Division
Department of Applied Science
Brookhaven National Laboratory
Upton, New York 11973
USA

ABSTRACT. This paper examines field measurements bearing upon
potential mechanisms responsible for the composition of cloudwater.
Mechanisms considered include: 1) nucleation scavenging of aerosol; 2)
scavenging of soluble gases; and 3) reactive scavenging of less soluble
gases. The efficiency of aerosol scavenging by nucleation has been
examined by use of a variety of physical and chemical techniques. The
data are consistent with essentially complete (> 90%) scavenging of
both particles mass and sulphate by cloudwater. Similarly, comparisons
of interstitial HNO_3 and cloudwater nitrate concentrations are
consistent with essentially complete scavenging of nitric acid by cloud
droplets. Studies of reactive scavenging have focussed on the aqueous
phase reactions of SO_2 with H_2O_2 and O_3. Measurements of O_3 and SO_2
concentrations in and around clouds suggests that the O_3/SO_2 reaction
is slow with respect to cloud lifetimes, consistent with calculations
using laboratory derived rate parameters and typical atmospheric
conditions. In contrast, the H_2O_2/SO_2 reaction appears to be rapid for
representative atmospheric conditions. Cloudwater H_2O_2 concentrations
observed in our measurements ranged from less than 0.1 µM to 100 µM;
equivalent gas phase concentrations, evaluated from aqueous phase
concentrations, cloud liquid water content, and H_2O_2 solubility
typically ranged from less than 0.05 and 1 ppb during the summer, and
less than 0.05 ppb during a winter study. A strong negative
correlation between cloudwater H_2O_2 and interstitial SO_2 concentrations
has been found in the data consistent with the rapid and quantitative
in-cloud reaction of SO_2 with H_2O_2 in which either these species acts
as the limiting reagent. Low level stratus clouds sampled during a
winter study generally exhibited high concentrations of interstitial
SO_2 and very low concentrations of cloudwater H_2O_2, suggesting at most
a small contribution from aqueous transformation of SO_2 by this
oxidant. No direct evidence for the in-cloud oxidation of either NO or
NO_2 to HNO_3 has been found in field data.

M. H. Unsworth and D. Fowler (eds.), Acid Deposition at High Elevation Sites, 139–153.
© 1988 by Kluwer Academic Publishers.

1. INTRODUCTION

The composition of precipitation represents the end result of a series of complex atmospheric processes. Since clouds are the immediate precursors to precipitation, mechanisms incorporating substances into cloudwater are of much interest. Of key importance is the extent to which cloud composition represents simple dissolution of all or some fraction of the soluble gases and aerosols present in pre-cloud air, or whether chemical reactions occurring within clouds contribute significantly to the composition of the cloudwater. In this regard, we note that clouds represent an ideal medium for atmospheric aqueous-phase chemical reaction in view of their high liquid water content, high dispersion (which enhances mass transport) and the generally high thermodynamic driving force resulting from the dissolution of soluble materials in water.

Materials are incorporated into cloudwater principally by three processes. Clouds form by condensation of water vapour on activated aerosol particles in response to a supersaturation in water vapour concentration. During droplet formation and growth, soluble ionic materials in the aerosol are dissolved in the cloudwater. This process, known as nucleation scavenging, is thought on theoretical grounds to be highly efficient for incorporation of both aerosol mass and sulphate. Concurrent with cloud droplet formation, gases will be scavenged to the extent dictated by their physical solubility. Such considerations suggest that highly soluble gases such as HNO_3 will be incorporated almost entirely into the cloudwater, while less soluble gases will be distributed between cloudwater and cloud interstitial air to an extent determined by their Henry's Law coefficient. In some cases (e.g. SO_2), Henry's Law solubility is enhanced by the occurrence of rapidly achieved aqueous phase equilibria. Gaseous substances can also be incorporated into cloudwater by reactive scavenging. In this mechanism sparingly soluble gases react in cloud via gas and/or aqueous phase pathways to form soluble species that are dissolved in the cloudwater.

Study of these processes in natural clouds (Hegg and Hobbs, 1981, 1982, 1983, 1984, 1986; Daum et al., 1983; 1984a, 1984b, 1986; Leaitch et al., 1983, 1986; Castillo et al., 1984; Radke, 1983; Lazrus et al., 1983; Castillo and Jiusto, 1984; Clark et al., 1984; Sievering et al., 1984, ten Brink et al., 1987), is being actively pursued. Estimates of nucleation scavenging efficiences obtained by a variety of methods give results indicating that this process occurs at efficiencies ranging from 10 – 100%. However, the results of many of these studies are ambiguous because of large measurement uncertainties, or because the methods were not capable of distinguishing materials incorporated into cloudwater via nucleation scavenging from that resulting from other processes. Studies of soluble gas scavenging have been concerned almost exclusively with HNO_3. Although data are few (Daum et al., 1983, 1984a, 1984b, 1986; Hegg and Hobbs, 1984, 1986), and in many cases subject to considerable uncertainty, they are consistent with essentially complete incorporation of HNO_3 into cloudwater.

Studies of reactive scavenging have focussed on the aqueous phase oxidation of SO_2 by H_2O_2 and by O_3 to produce dissolved H_2SO_4, and conversion of NO_2 to HNO_3 by means of a gas phase mechanism involving NO_3 and/or N_2O_5 followed by uptake of NO_3 or N_2O_5 by the cloudwater. Significant evidence has been accumulated (Daum et al., 1983, 1984b; Kadlecek et al., 1983; Römer et al., 1985; Kelly et al., 1985) indicating that the SO_2/H_2O_2 reaction is important in warm clouds during seasons of the year when solar intensity is high. Direct evidence for the occurrence of either the O_3/SO_2 pathway to cloudwater sulphate or any rapid pathway for incorporation of NO_2 into cloudwater has not been identified from field measurements.

We present here an overview of the results of field measurements we have made over the last several years in the eastern US and Canada that have been directed at characterizing scavenging mechanisms for incorporation of sulphur and nitrogen compounds into cloudwater. In the course of these measurements we have examined the interphase distribution of these species between cloudwater and interstitial cloud air, compared concentrations by key species in cloudwater to concentrations in associated cloud-free air, and have measured the concentrations of suspect oxidants and determined their relationship to concentrations of sulphur and nitrogen acids and acid precursors. The results of these field measurements support the supposition that nucleation scavenging of aerosol and scavenging of gaseous HNO_3 by cloudwater are efficient processes, and are strongly suggestive that reactive scavenging of SO_2 by H_2O_2 is an important mechanism for incorporation of SO_2 into cloudwater as sulphuric acid.

2. EXPERIMENTAL METHODS

Observations were performed using aircraft at a number of locations in the eastern US and in south-central Ontario, Canada during various field studies conducted between 1982 and 1985. Instrumentation and procedures utilized in these programmes are described in detail in other publications (Daum et al., 1983, 1984a, 1984b, 1986; Kelly et al., 1985; ten Brink et al., 1987; Isaac and Daum, 1987). Measurements typically included continuous in-situ monitoring of NO, NO_x, O_3, SO_2, aerosol sulphur, b_{scat}, and in some cases aerosol and cloud droplet number concentrations and size distributions. Filter techniques were used to obtain integrated samples of SO_2, HNO_3, and aerosol. The aerosol samples were analysed for strong acid, ammonium, sulphate and nitrate. Collected cloudwater samples were analysed for major ionic species (H^+, NH_4^+, NA^+, NO_3^-, $SO_4^=$ and Cl^-) and for H_2O_2. Field measurements discussed here are limited to those in non-precipitating clouds. The cloud type was predominantly stratus and strato-cumulus. Sampling altitudes ranged from 500 m to 3000 m MSL, but were typically 1000 - 2000 m MSL.

3. RESULTS AND DISCUSSION

Nucleation scavenging. The efficiency of nucleation scavenging for the incorporation of aerosol and sulphate has been examined using a variety of methods. These methods include: 1) comparing the composition of pre-cloud air to that of interstitial cloud air by use of chemical and/or physical techniques; 2) comparing interstitial and cloudwater concentrations; 3) comparing concentrations of species in subsaturated regions of clouds to interstial concentrations in saturated regions. Examples are now given illustrating the various techniques that have been employed and showing that the data are generally consistent with large fractional incorporation of aerosol mass and sulphate into cloudwater.

We first consider a case where comparison of both chemical and physical parameters of below and in-cloud air were used to infer nucleation scavenging efficiencies for both particles and $SO_4^=$. Data used in this illustration were collected during a study conducted in the vicinity of North Bay Ontario, Canada during the winter of 1984 (Isaac and Daum, 1987) and are discussed in detail in Daum et al., 1986. Figure 1 shows a vertical profile of aerosol ($0.2 > d < 3$ μm) and cloud droplet ($1 > d < 30$ μm) number concentrations through a stratus cloud layer. Note that the aerosol number concentrations decrease initially with height from ca. 300 cm^{-3} at cloud base to ca. 20 cm^{-3} approximately 150 m above cloud base and thereafter remain constant at ca. 20 cm^{-3} to cloud top. In contrast, the cloud droplet concentration increased with height above cloud base to ca. 300 cm^{-3} 150 m above cloud base, and thereafter remained constant to cloup top. It may be seen that the sum of the interstitial aerosol and cloud droplet number concentrations was nearly constant from below-cloud base through the cloud layer, implying that the cloud formed in air with a particle concentration that can be represented by the particle concentration of the below-cloud air, and furthermore, that a large fraction (ca. 90%) of the aerosol particles was scavenged by nucleation and growth to cloud droplets. To the extent that sulphate mass is proportional to particle number concentrations in this size range, the data also imply that nucleation scavenging of sulphate was also efficient.

Data from this case study also allow direct calculation of the nucleation scavenging efficiency for $SO_4^=$ using below-cloud and interstitial $SO_4^=$ concentrations determined by filter methods. Two such comparisons were possible. In the first comparison, the below-cloud sulphate concentration was 0.69 ppb and the interstitial sulphate concentration was 0.08 ppb giving a scavenging efficiency of 88%. In the second instance the below-cloud and interstitial sulphate concentrations were 0.81 and 0.09 ppb respectively, giving a scavenging efficiency of 89%. These results are essentially identical to the scavenging efficiencies for this case calculated above from consideration of particle number concentrations.

We have also determined scavenging efficiencies by comparing interstitial and cloudwater sulphate concentrations (with cloudwater sulphate concentrations expressed as equivalent gas phase concentrations). This procedure is subject to some ambiguity since it does not distinguish the amount of cloudwater sulphate derived from any reactive scavenging of SO_2 from that resulting from nucleation scavenging of aerosol. The technique is also subject to some

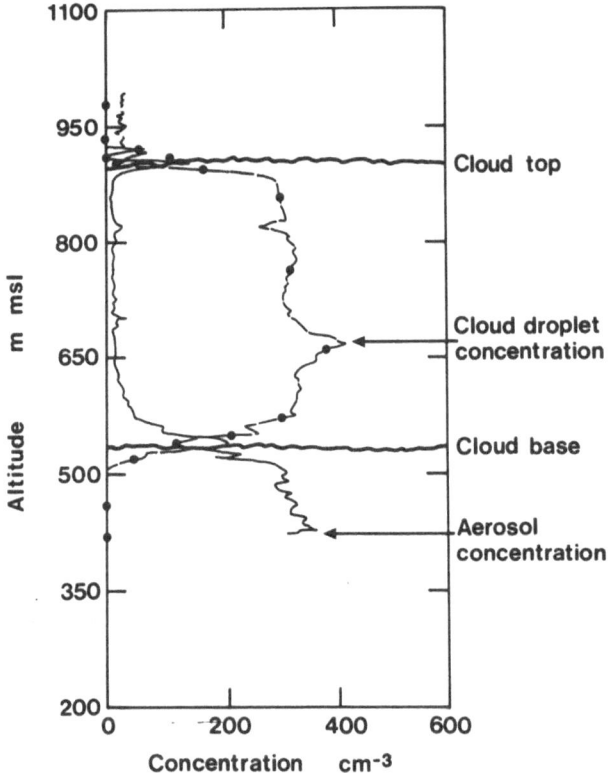

Figure 1 Vertical profile of aerosol ($0.2 < d < 3$ μm) and _____ cloud
droplet ($3 > d < 30$ μm) number concentrations — . — through a
stratiform cloud layer in southeastern Ontario, Canada,
February 20, 1984.

uncertainty due to cloud inhomogeneities (i.e. subsaturated regions
within clouds), since integrative filter techniques requiring
collection times on the order of 30 minutes are required to determine
interstitial sulphate concentrations. On the other hand, interstitial
sulphate concentrations were frequently below the detection limit of

144

the method, allowing only lower bound for the scavenging efficiency to be calculated. Nonetheless, estimates of sulphate scavenging efficiencies using this technique (Daum et al., 1984b), are generally consistent with high fractional scavenging of sulphate by clouds.

The final method for determining scavenging efficiencies that will be discussed is comparison of both continuously measured sulphate concentration (using a modified flame photometric detector, Garber et al., 1983) and light scattering coefficient, b_{scat} in cloud interstitial air with these quantities in adjacent subsaturated regions of clouds. An example of such measurements is given in Figure 2, which

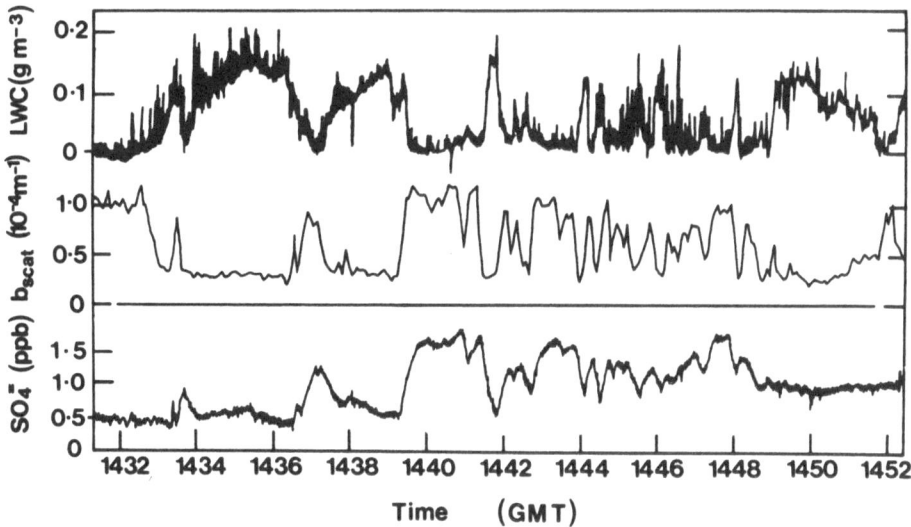

Figure 2 Plots of sulphate, b_{scat}, and cloud liquid water content in a broken stratus cloud layer, Birmingham, Alabama, US, November 17, 1982.

shows time plots of cloud liquid water content, b_{scat}, and $SO_4^=$ concentration for a broken stratus deck sampled in the vicinity of Birmingham, Alabama, USA on November 17, 1982. Pertinent features of this figure are that both the sulphate concentration and b_{scat} are anti-correlated with cloud liquid water content, indicative of nucleation scavenging of sulphate aerosol and its surrogate b_{scat} in regions of the cloud layer where cloud liquid water is present. To the extent that air in the unsaturated regions of the layer have a common origin and chemical composition with the saturated regions (for a detailed discussion concerning this point see ten Brink et al., 1987), scavenging efficiencies for sulphate can be calculated by comparison of concentrations in saturated regions to those in unsaturated regions. In the great majority of instances where this technique has been employed, measured values of interstitial cloud b_{scat} after appropriate baseline subtraction, have been close to the detection limit of the instrument, indicating high fractional incorporation (> 90%) of light scattering aerosol into cloudwater. Direct measurements of interstitial aerosol sulphate with the flame photometric detector are consistent with the b_{scat} measurements although somewhat lower bounds must be placed on the scavenging efficiency (> 75%) in accordance with the detection limits and frequency response of the instrument. Nonetheless, the results are consistent with a substantial, and perhaps complete decrease in the concentration of aerosol sulphate in the cloud interstitial air.

Scavenging of soluble gases. The species of principal concern here has been nitric acid. Measurements have focussed on measuring interstitial concentrations in relation to cloudwater nitrate concentrations. Two methods have been used; collection and analysis of filter samples (LOD < 0.1 ppb), and continuous measurement with a sensitivity enhanced dual channel O_3 chemiluminescent instrument (Tanner et al., 1983), in which nitric acid is determined by difference (LOD = 0.4 ppb). For neither of these methods has HNO_3 ever been found at concentrations exceeding the limit of detection (Daum et al., 1984b). This result is consistent with the expectation that HNO_3 is essentially irreversibly scavenged by cloudwater and that this scavenging takes place on a time scale of seconds (Levine and Schwartz, 1982).

Since it appears that scavenging of HNO_3 is rapid and quantitative, the question arises as to the importance of this process relative to other potential processes, in determining cloudwater acidity and nitrate concentrations. Although we do not have sufficient data to address this question in detail, data from a recent study (Isaac and Daum, 1987) suggest that scavenging of gaseous HNO_3 was a significant (and perhaps dominant) contributor to cloudwater NO_3^- in low level stratus clouds in winter. During this study, conducted in the vicinity of North Bay, Ontario during the winter of 1984, below cloud HNO_3 concentrations were frequently substantial (0.5 - 2 ppb) and were nearly the same as gas phase equivalent NO_3^- concentrations as calculated from the collected cloudwater samples.

Reactive scavenging. Studies of reactive scavenging have been directed to examination of the aqueous phase oxidation of SO_2 and NO_x

(NO and NO$_2$) to their corresponding acids. Our approach has been to measure the concentrations of these species and related oxidants in cloudwater, interstitial cloud air, and associated non-cloud air to determine whether relationships between the concentrations of various species indicate the occurrence of in-cloud chemical reactions and to establish whether these substances behave in-situ according to predictions based on laboratory studies. In-situ cloud data are also used in combination with laboratory kinetic data to evaluate in-cloud reaction rates. These are then compared with field observations for consistency.

Studies of the reactive scavenging of SO$_2$ have concentrated on the aqueous-phase transformation of SO$_2$ by either O$_3$ or H$_2$O$_2$. The kinetics of the O$_3$/S(IV) reaction are second order, with a pH-dependent rate constant that decreases with decreasing pH. The pH dependence of the rate constant, coupled with the decreasing solubility of S(IV) with decreasing pH, causes the overall reaction rate to decrease strongly with decreasing pH. Because of this strong pH dependence, the reaction tends to be self-limiting since in produces H$^+$ as it proceeds. The S(IV)-H$_2$O$_2$ reaction is acid catalyzed with the logarithm of the second-order rate constant increasing linearly with decreasing pH over the pH range 2-6, with a slope of -1. The increase of the rate constant with decreasing pH essentially compensates for the decreased solubility of SO$_2$ with decreasing pH giving an overall reaction rate that is nearly independent of pH. Because of this pH independence, the reaction is not self-limiting and can proceed at a high rate even at low pH. (For a more complete discussion of S(IV)-H$_2$O$_2$ and O$_3$ kinetics and their potential importance in cloud chemistry see Penkett et al., 1979, and Schwartz, 1984, and references therein).

Both gas phase O$_3$ and aqueous phase H$_2$O$_2$ have been measured during our field studies of clouds. Interstitial O$_3$ concentrations were found to be typical of those measured in clear-air associated with clouds, even in the presence of significant concentratons (> 10 ppb) of SO$_2$. In broken clouds (ten Brink et al., 1987) O$_3$ concentrations appeared to be unaffected by the presence or absence of liquid water. The observations are consistent with the low solubility and aqueous phase reactivity of O$_3$ found in laboratory studies and suggest that O$_3$ reacts with SO$_2$ in cloud at a rate much slower than the rate at which clouds form and decay in the atmosphere and mix with the surrounding air. It should be noted however, that direct evidence for reaction of O$_3$ with SO$_2$ would be difficult to obtain from field measurements since O$_3$ is usually present in large excess and concentration variations on the order of 1-2 ppb caused by reaction with SO$_2$ would be hard to detect.

The concentrations of H$_2$O$_2$ we have found in cloudwater ranged broadly from less than 0.1 µM to over 100 µM; the median value was less than 10 µM (Kelly et al., 1985). The highest concentrations were measured in summer; winter concentrations measured in a field programme conducted in North Bay, Ontario in 1984, were typically less than 0.5 µM (Isaac and Daum, 1986). Equivalent gas phase concentrations, evaluated from aqueous phase peroxide concentrations, cloud liquid water content, and the Henry's Law coeficient for H$_2$O$_2$ were typically < 0.01 to 1 ppb gas phase during and summer, and < 0.01 to

0.05 ppb during the winter, suggesting that there is a strong seasonal dependence of the rate of H_2O_2 production.

Although sufficient O_3 is usually present to oxidize large quantities of SO_2, evaluation of the oxidation rates using laboratory derived rate parameters, and assuming conditions typically encountered during our field studies (Daum et al., 1983; Kelly et al., 1985) shows that the rate of oxidation by this process is not sufficiently fast to contribute significantly to production of cloudwater sulphate or acidity unless cloud lifetimes are very long (i.e. 100 hours), or unless air is repeatedly cycled through clouds. In contrast the H_2O_2 reaction is found to be rapid in producing $SO_4^=$ and H^+ and in depleting interstitial SO_2, and under typical conditions is sufficiently fast to approach completion on a timescale shorter that the lifetime of a typical statiform cloud (hours). In most cases the extent of the reaction is limited by the availability of either H_2O_2 or SO_2.

An important consideration in evaluating the rates of reaction of S(IV) by either O_3 or H_2O_2 is the extent to which laboratory derived rate parameters are applicable to evaluation of field data. In the case of the S(IV)-H_2O_2 reaction the applicability has been recently tested in our laboratory (Lee et al., 1986) by examining the chemical kinetics of this reaction in authentic precipitation samples. Kinetic studies were carried out on more than 300 precipitation samples collected over a two year period. The reaction was initiated by adding a small quantity of either H_2O_2 or S(IV) to a portion of the sample. Although values of the effective second order rate constant determined in these samples exhibit considerable scatter, they fall close to the rate constant for this reaction in "pure water". Furthermore, much of the scatter can be attributed to the uncertainty in measuring pH. These results support the use of laboratory derived rate parameters in evaluating the importance of this reaction in clouds. Similar studies of the S(IV)-O_3 reaction have not yet been carried out, and hence the possibility cannot be precluded that this reaction is faster in clouds or precipitation than in pure water via an undefined catalytic process.

The results of these calculations and laboratory studies are consistent with the observation in field measurements of a strong anti-correlation between cloudwater H_2O_2 and interstitial SO_2 concentrations; i.e. when H_2O_2 concentrations were high interstitial SO_2 concentrations were low, and vice-versa (Daum et al., 1984b). This is illustrated in Figure 3 where values of cloudwater H_2O_2 (expressed in gas phase equivalent units) are plotted as a function of the simultaneously measured interstitial SO_2 concentration. This pattern is consistent with the rapid and quantitative reaction of SO_2 with H_2O_2 in which either the SO_2 or the H_2O_2 acts as the limiting reagent. The few points which do not lie along the axes suggest situations in which clouds were newly formed and the reaction was incomplete and/or the presence of subsaturated regions within clouds.

It is interesting to extract winter data from Figure 3 and to display this separately. Figure 4 shows the data collected in a study conducted during January and February of 1984. Although the trend of mutual exclusiveness illustrated by Figure 3 is preserved in the plot

148

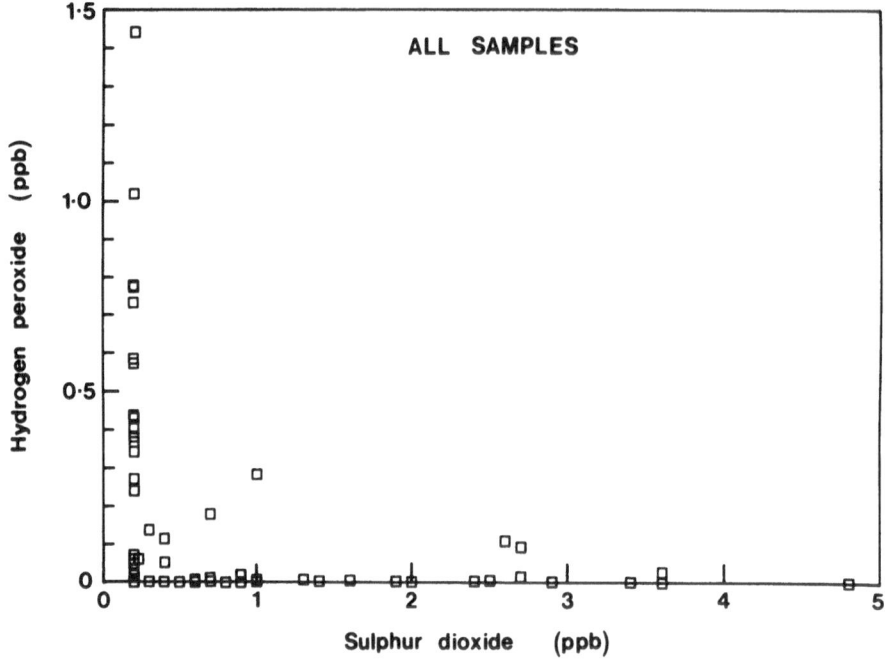

Figure 3   Cloudwater hydrogen peroxide (expressed in gas phase
           equivalent units) vs. interstitial SO_2 in non-precipitating
           clouds. Data collected at various times and locations in the
           eastern US and Canada, 80 data points are shown in the
           figure. Limit of detection for SO_2 was 0.1 - 0.2 ppb.

for samples collected during the winter study, it should be noted that
few samples exhibited excess H_2O_2; apparently because of a limited
quantity of oxidant. Consequently, the winter data imply that the
H_2O_2/SO_2 reaction is of lesser importance than in summer due to lower
H_2O_2 concentrations. This is consistent with expectation based on a
photochemical source of H_2O_2 since solar intensity is greatly reduced
in winter in comparison to in summer at the latitude of these
measurements. Similar relationships for O_3 and SO_2 were not observed
even in the presence of high concentrations of SO_2 (20 - 40 ppb) and O_3

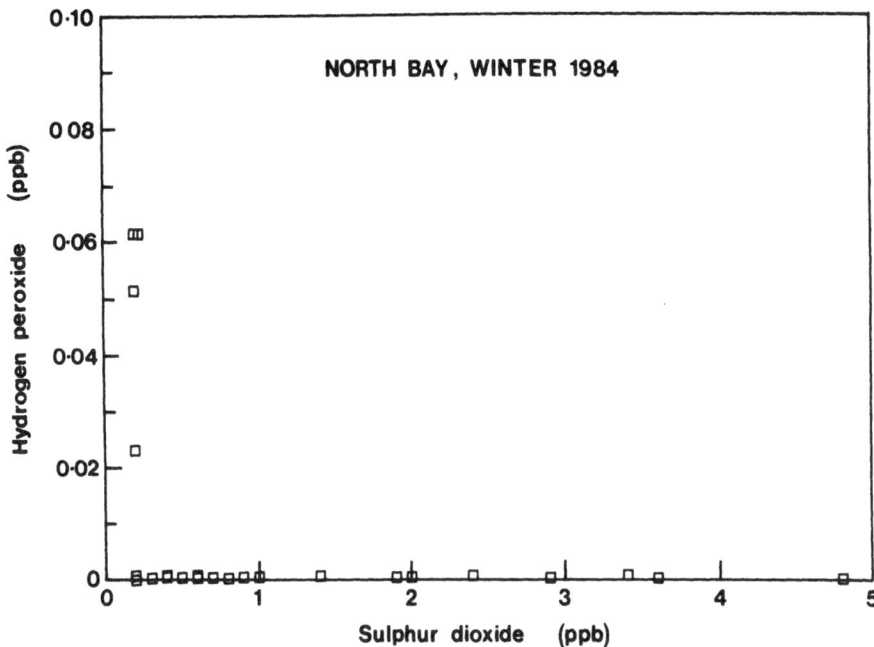

Figure 4   Cloudwater hydrogen peroxide vs. interstitial SO_2 for samples
collected in non-precipitating clouds, North Bay Ontario,
Canada, January and February 1984, 29 data points are shown
in the figure. Limit of detection for SO_2 was 0.1 - 0.2 ppb.

(100 ppb), consistent with our perception that this reaction is much
slower. However, despite the relative slowness of the $S(IV)-O_3$
reaction, it may nonetheless be the dominant in-cloud process in the
winter, since H_2O_2 concentrations seem low and cloud lifetimes tend to
be long.

Examination of the NO_2 concentration data does not reveal a
similar signal indicating the occurrence of an in-cloud reaction
converting this species to cloudwater nitrate. Interstitial
concentrations were typical of those found in clear-air surrounding
clouds, (NO_x, 0.5 to 10 ppb) and concentrations did not appear to be
affected by the presence of liquid water in broken clouds. This

behaviour is consistent with the low solubility and reactivity of this species observed in laboratory studies (Lee and Schwartz, 1981a, 1981b). The lack of any apparent signal for the occurrence of a reaction in the field data indicates that if such a reaction occurs, it must have a $1/e$ time that is substantially greater than the lifetime of a typical cloud.

4. CONCLUSIONS

Principal conclusions concerning the process contributing to cloudwater composition that have been derived thus far from field measurements are as follows:

1. Nucleation scavenging by clouds is an efficient process leading to generally high fractional (> 90%) incorporation of both aerosol and sulphate during cloud droplet formation and growth.
2. Scavenging of HNO_3 is an efficient process consistent with the high solubility of this species in water. Scavenging of HNO_3 may be an important means for incorporation of HNO_3 into cloudwater, especially in the winter in northeastern North America because of the high concentrations of HNO_3.
3. Reactive scavenging of SO_2 by H_2O_2 is an important atmospheric process in situations where both species are available for in-cloud reaction. The amount of cloudwater H_2SO_4 produced by this reaction appears to be limited by either SO_2 or H_2O_2, whichever reagent is present in a lesser amount. Based on a limited number of measurements to date, H_2O_2 available for this reaction apears to be less in winter than in summer.
4. The rate of the SO_2/O_3 reaction appears to be less than that of the SO_2/H_2O_2 reaction based on calculations for representative concentrations; however the abundance of ozone suggests that this reaction may be significant, especially in persistent clouds.
5. Field data give no definitive signal indicating the occurrence of an in-cloud reaction converting NO to HNO_3.

ACKNOWLEDGEMENT

The author gratefully acknowledges support provided by the Electric Power Research Institute under grant RP2023-1 and to the US Department of Energy under the PRECP programme. The work was performed under auspices of the United States Department of Energy under Contract No. DE-AC02-76CH00016.

REFERENCES

Castillo, R.A. and Jiusto, J.E. 1984. A preliminary study of the chemical modification of cloud condensation nuclei in stratiform clouds at Whiteface Mountain, New York. Atmos. Environ. **18** , 1933-1937.

Castillo, R.A., Juisto, J.E. and McLaren, E. 1983. The pH and ionic composition of stratiform cloudwater. Atmos. Environ. 17 , 1497-1505.

Clark, P.A., Fletcher, I.S., Kallend, A.S., McElroy, W.J., Marsh, A.R.W. and Webb, A.H. 1984. Observations of cloud chemistry during long range transport of power plant plumes. Atmos. Environ. 18 , 1849-1858.

Daum, P.H., Schwartz, S.E. and Newman, L. 1983. Studies of the gas- and aqueous-phase composition of stratiform clouds. In Precipitation Scavenging, Dry Deposition andn Resuspension, Vol 1 (edited by Pruppacher, H.R., Semonin, R.G. and Slinn, W.G.N.), pp. 31-52, Elsevier, New York.

Daum, P.H., Schwartz, S.E. and Newman, L. 1984a. Acidic and related constituents in liquid water stratiform clouds. J. Geophys. Res. 89 , 1447-1458.

Daum, P.H., Kelly, T.J., Schwartz, S.E. and Newman, L. 1984b. Measurements of the chemical composition of stratiform clouds. Atmos. Environ. 12 , 2671-2684.

Daum, P.H., Kelly, T.J., Strapp, J.W., Leaitch, W.R., Joe, P., Schemenauer, R.S., Isaac, G.A., Anlauf, K.G. and Wiebe, H.A. 1987. Chemistry and physics of a winter stratus cloud layer: A case study. J. Geophys. Res., 92 , 8426-8436.

Garber, R.W., Daum, P.H., Doering, R.F., D'Ottavio, T. and Tanner, R.L. 1983. Determination of ambient aerosol and gaseous sulphur using a continuous FPD, III. Design and characterization of a monitor for airborne applications. Atmos. Environ. 17 , 1381-1385.

Hegg, D.A. and Hobbs, P.V. 1981. Cloudwater chemistry and production of sulphate in clouds. Atmos. Environ. 15 , 1957-1604.

Hegg, D.A. and Hobbs, P.V. 1982. Measurements of sulphate production in natural clouds. Atmos. Environ. 16 , 2663-2668.

Hegg, D.A. and Hobbs, P.V. 1983. Preliminary measurements of the scavenging of sulphate and nitrate in clouds. In Precipitation Scavenging, Dry Deposition and Resuspension, Vol 1 (edited by Pruppacher, H.R., Semonin, R.G. and Slinn, W.G.N.), pp. 79-89, Elsevier, New York.

Hegg, D.A. and Hobbs, P.V. 1984. Measurements of the scavenging of sulphate and nitrate in clouds. Atmos. Environ. 18 , 1939-1946.

Isaac, G.A. and Daum, P.H. 1987. A winter study of air, cloud and precipitation chemistry in Ontario, Canada. Atmos. Environ. 21 , 1587-1600.

Kadlecek, J., McLaren, S., Camarota, N., Mohnen, V. and Wilson, J. 1983. Cloud water chemistry at Whiteface Mountain. In Precipitation Scavenging, Dry Deposition and Resuspension, Vol 1 (edited by Pruppacher, H.R., Semonin, R.G. and Slinn, W.G.N.), pp. 31-52, Elsevier, New York.

Kelly, T.J., Daum, P.H. and Schwartz, S.E. 1985. Peroxide measurements in cloudwater and rain. J. Geophys. Res. 90 , 7861-7871.

Lazrus, A.L., Haagenson, P.L., Kok, G.L., Huebert, B.J., Kreitzberg, C.W., Likens, G.E., Mohnen, V.A., Wilson, W.E. and Winchester, J.W. 1983a. Acidity in air and water in a case of warm frontal precipitation. Atmos. Environ. 17 , 581-591.

Leaitch, W.R., Strapp, J.W., Wiebe, H.A. and Isaac, G.A. 1983. Measurements of scavenging and transformation of aerosol inside cumulus. In Precipitation Scavenging, Dry Deposition and Resuspension, Vol 1 (edited by Pruppacher, H.R., Semonin, R.G. and Slinn, W.G.N.), pp. 53-69, Elsevier, New York,

Leaitch, W.R., Strapp, J.W., Wiebe, H.A., Anlauf, K.G. and Isaac, G.A. 1986. Chemical and microphysical studies on nonprecipitating summer cloud in Ontario, Canada. J. Geophys. Res., submitted.

Lee, Y-N and Schwartz, S.E. 1981a. Reaction kinetics of nitrogen dioxide with liquid water at low partial pressure. J. Phys. Chem. 85, 840-848.

Lee, Y-N and Schwartz, S.W. 1981b. Evaluation of the rate of uptake of nitrogen dioxide by atmospheric and surface liquid water. J. Geophys. Res. 86, 11971-11983.

Lee, Y-N, Shen, J., Klotz, P.J., Schwartz, S.E. and Newman, L. 1986. Kinetics of the hydrogen peroxide-sulphur (IV) reaction in rainwater collected at a northeastern US site. J. Geophys. Res., 91, 13264-13274.

Levine, S.Z. and Schwartz, S.E. 1982. In-cloud and below-cloud scavenging of nitric acid vapour. Atmos. Environ. 16, 1725-1734.

Penkett, S.A., Jones, B.M.R., Brice, K.A., Eggleton, J. 1979. The importance of atmospheric ozone and hydrogen peroxide in oxidising sulphur dioxide in cloud and rainwater. Atmos. Environ. 13, 123-137.

Radke, L.F. 1983. Preliminary measurements of the size distribution of the cloud interstitial aerosol. In Precipitation Scavenging, Dry Deposition and Resuspension, Vol. 1 (edited by Pruppacher, H.R., Semonin, R.G. and Slinn, W.G.N.), pp. 71-78, Elsevier, New York.

Romer, F.G., Viljeer, J.W., Van den Beld, L., Slangewal, H.J., Veldkamp, A.A. and Reijnders, H.F.R. 1983. Preliminary measurements from an aircraft into the chemical composition of clouds. In Acid Deposition (edited by Beilke, S. and Elshout, A.J.), pp. 195-203, D. Reidel, Dordrecht.

Sievering, H., Van Valin, C.C., Barrett, E.W. and Pueschel, R.F. 1984. Cloud scavenging of aerosol sulphur: Two case studies. Atmos. Environ. 18, 2685-2690.

Schwartz, S.E. 1984. Gas-aqueous reactions of sulphur and nitrogen oxides in liquid-water clouds. In SO2, NO and NO2 Oxidation Mechanisms: Atmospheric Considerations (edited by J.G. Calvert), Butterworth, Boston, pp. 173-208.

Schwartz, S.E., Daum, P.H., Hjelmfelt, M.R. and Newman, L. 1983. Cloudwater acidity measurements and formation mechanisms - experimental design. In Precipitation Scavenging, Dry Deposition and Resuspension (edited by Prupacher, H.R., Semonin, R.G. and Slinn, W.G.N.), Elsevier, New York, pp. 15-30.

Tanner, R.L., Daum, P.H. and Kelly, T.J. 1983. New instrumentation for airborne acid rain research. Int. J. Envir. Anal. Chem. 13, 323-335.

ten Brink, H.M., Schwartz, S.E. and Daum, P.H. 1987. Efficient scavenging of aerosol sulphate by liquid water clouds. _Atmos. Environ._ **21** , 2035-2052.

CLOUD MICROPHYSICAL PROCESSES RELEVANT TO CLOUD CHEMISTRY

T.W. Choularton and T.A. Hill
Department of Physics
UMIST
Sackville Street
Manchester
M60 1QD
UK

ABSTRACT. In this paper a review is presented of recent field studies
of cloud microphysics in a hill cap cloud and in growing cumulus
clouds. In each case the implications of the results for cloud
chemical processes are discussed. In particular it is suggested that
the entrainment of dry air into the cloud from the free troposphere may
reduce the lifetime of individual droplets to values considerably below
that of the cloud and may introduce extra oxidant which may
substantially increase the sulphate production. A new model of cloud
microphysics and chemistry in a growing cumulus cloud is presented to
quantify the magnitude of this effect.

1. INTRODUCTION

During this conference a number of papers will be presented describing
investigations of cloud chemistry in the hill cap cloud which envelopes
the UMIST field station on Great Dun Fell (GDF) in Cumbria, England.
In these papers the close interaction of the physics and chemistry is
emphasised. In this paper the main cloud microphysical processes
which operate in the cap cloud and also in other cloud types which are
relevant to cloud chemistry will be discussed. We will discuss only
liquid phase processes, although a considerable fraction of
precipitation particularly in high latitudes forms via the ice phase.
The complexities of the ice phase would require a separate paper.
    Data will be presented from recent field investigations of the
cloud microphysics and dynamics at GDF and also from the Co-operative
Convective Precipitation Experiment (CCOPE) conducted in the USA during
the summer of 1981. The microphysical data described in these
sections has been published in detail elsewhere, see Choularton et al
(1986) and Hill and Choularton (1985), and so only brief summaries are
presented. In this paper, however, the implications of the results
for cloud chemical processes are explored. UMIST participated in the
CCOPE experiment and we have played a major role in the analysis of

155

M. H. Unsworth and D. Fowler (eds.), Acid Deposition at High Elevation Sites, 155–174.
© *1988 by Kluwer Academic Publishers.*

156

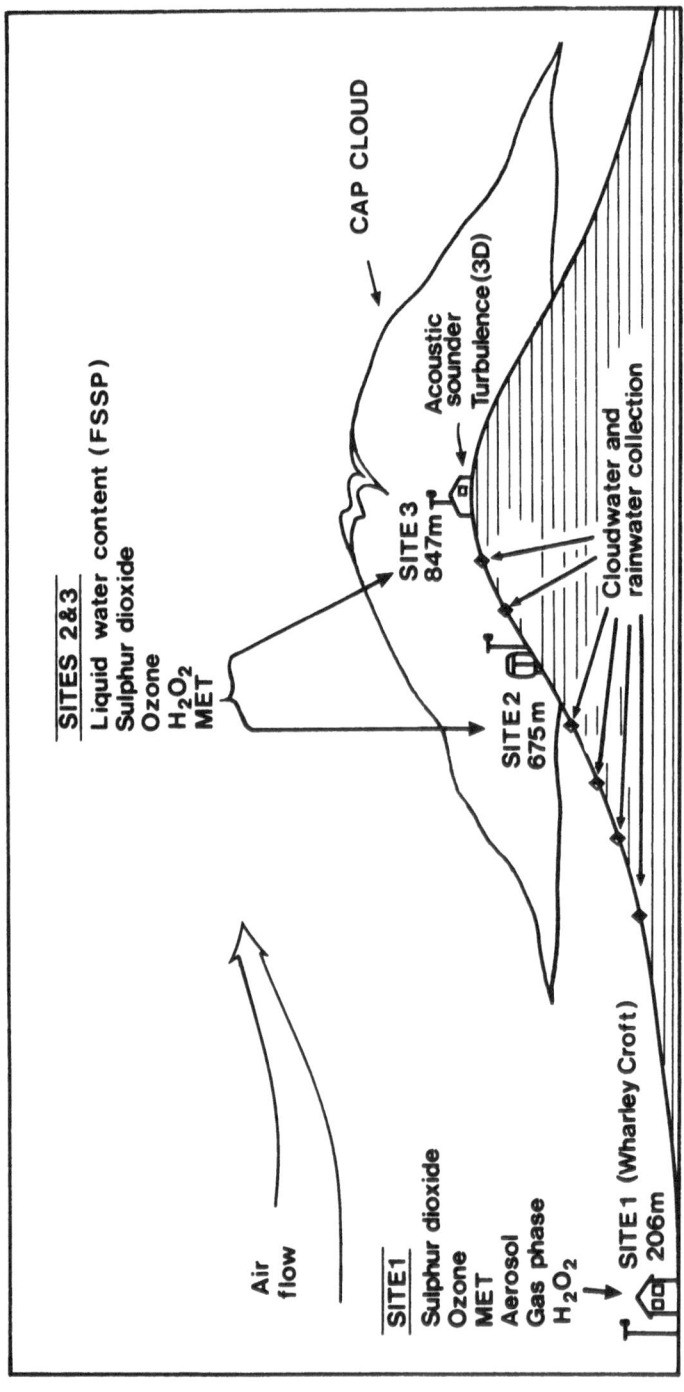

Figure 1. Schematic diagram of the equipment at Great Dun Fell.

microphysical data gathered during aircraft penetrations of cumulus
clouds, Hill and Choularton 1985, 1986.   In section 2 results from 2
recent field studies on the GDF cap cloud designed to investigate the
effects of radiative cooling, dry air entrainment and cloud base
inhomogeneities on its microphysical structure will be presented, full
details of this work may be found in Choularton et al (1986).   In
section 3 data describing the major influences of dry air entrainment
on cumulus cloud microphysics and dynamics will be described.   This
work is described in more detail by Hill and Choularton (1985).   In
section 4 a new model will be described which quantifies the
significance of some of these results to the oxidation of SO_2 in a
cumulus cloud and in section 5 the relevance of these results to cloud
chemical processes will be explored.

2.   CLOUD PHYSICS STUDIES AT GDF

Figure 1 shows the deployment of the major microphysical measuring
equipment on Great Dun Fell.   In this paper we will concentrate on
measurements made at the summit site where measurements of the cloud
droplet size distribution and liquid water content were made using a
Knollenberg Forward Scattering Spectrometer Probe (FSSP) and the
vertical structure of the atmosphere, immediately above the hill top,
was examined using a monostatic accoustic sounder.   Further details of
the experimental procedures and equipment may be found in other
papers in these proceedings and in Choularton et al (1986).

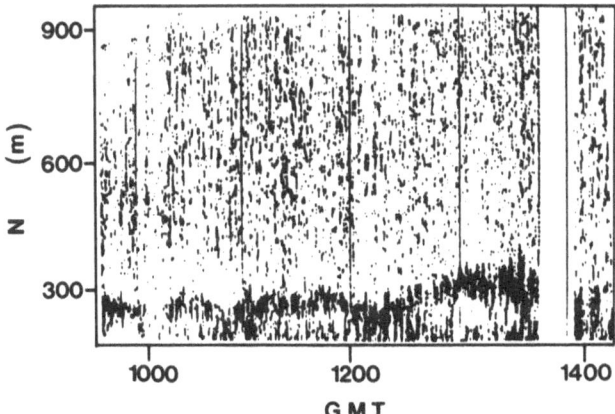

Figure 2.   Acoustic sounder record for 6 April 1982.

2.1 Results

The data presented below were gathered on 6 April 1982.   A light SW
airflow covered the area and winds on the summit of GDF were about
$6m\ s^{-1}$.   Figure 2 shows the acoustic sounder trace from the summit of

GDF.   A region of strong echo is present between 200m and 300m above the mountain summit.   This corresponds to a sharp temperature inversion at the top of the hill cap cloud with drier and warmer free tropospheric air above.   It can be seen that the interface is highly irregular due to blobs of the warmer drier air being entrained into the cap cloud, sometimes penetrating to ground level.   An example of this is labelled 'L' in Figure 2.   Figure 3a shows plots of the droplet number concentration, N, the cloud liquid water content, L and the droplet mean diameter d for the period 1035Z to 1105Z surrounding the entrainment event 'L'.   During this period the adiabatic liquid water content, derived from the observed height of cloud base, fell from 0.5 g m^{-3} to 0.4g m^{-3} and so the observed liquid water content was generally sub-diabatic due to the entrainment of dry air from above the cloud top.

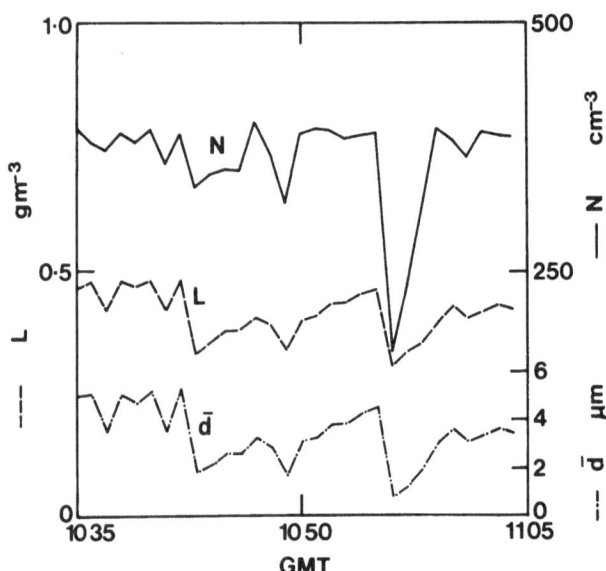

Figure 3a.   Measured time variations at the GDF summit of liquid water content, L, droplet concentration, N, the mean diameter d on 6 April 1982.   1-minute averages.

The relatively variable entrainment rate, illustrated in figure 2, results in relatively small scale variations in the liquid water content.   In this particular cloud the reductions in the liquid water content occur as a result of a reduction in the mean droplet diameter, due to the partial evaporation of droplets and a reduction in the total droplet number concentration as some droplets are completely evaporated and converted back to cloud condensation nuclei.

Figure 3b shows the droplet spectrum observed (averaged over the period 101400Z to 101600Z) and compared to the droplet spectrum computed assuming completely adiabatic cloud growth to the hill summit.   The adiabatic droplet spectrum was calculated using the model

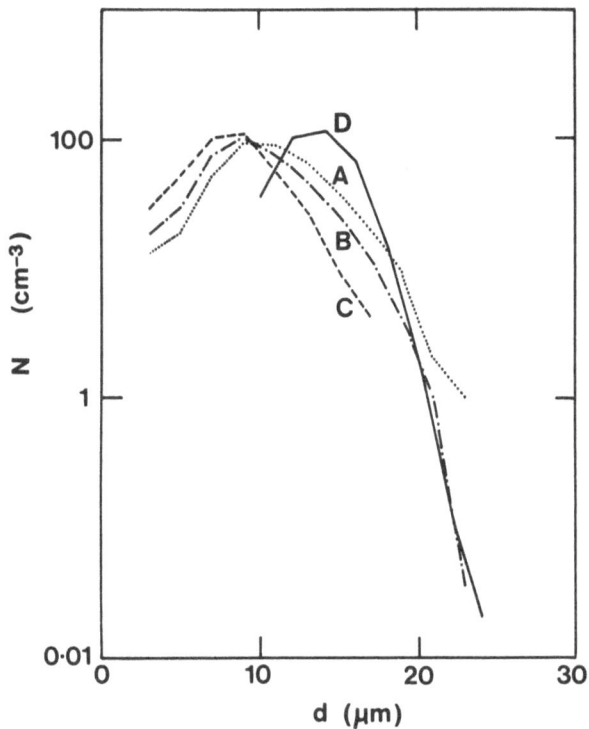

Figure 3b.   Maximum (A), average (B), minimum (C) and adiabatic (D)
             droplet spectra at the GDF summit for the 2-minute period
             commencing 101400 GMT on 6 April 1982. (Curve A is for the
             regions with the 10% highest liquid water content regions,
             curve C the 10% lowest liquid water content regions and
             curve B the remainder).

described by Carruthers and Choularton (1986). It can be seen that the
observed spectrum is considerably broader than the adiabatic
distribution with the mode droplet radius reduced.   These effects are
due to the varying amounts of partial evaporation experienced by
individual droplets.

2.2   Implications for Cloud Chemistry

Calvert et al (1985) have shown that the concentrations of one of the
principle photo-oxidants, hydrogen peroxide are considerably higher in
the free troposphere than in the boundary layer.   In the boundary
layer the concentrations of primary pollutants, eg. sulphur dioxide,
can be considerably in excess of the oxidant concentrations and hence
the entrainment process described above can play a major role in the
cloud chemistry by introducing extra oxidant.   This theme is developed
by other papers in this conference.

The total evaporation of some droplets may reduce the time
available for chemical reactions to occur, to considerably less than
the lifetime of the cloud (or the transit time of an air parcel through
the cloud). This is discussed further below. However, in the case
of a hill cap cloud, where the entrainment process is dominated by
small scale Kelvin-Helmholtz billows the effect of dry air entrainment
is mostly to partially evaporate droplets rather than totally remove
them (see Baker et al 1982) and so this is of secondary importance.

Another major effect of entrainment into a cloud is to produce a
wide range of droplet sizes and droplet ages at all points in the cloud
and, therefore, on many occasions it will be necessary to consider the
deta led microphysics to adequately model the chemical evolution.

3. CUMULUS CLOUDS.

The data presented in this section were gathered during several passes
of the University of Wyoming King Air aircraft through developing
cumulus clouds on 12 July 1981. This was part of the CCOPE
experiment. A much more comprehensive description of the experiment,
the data and their interpretation may be found in Hill and Choularton
(1985). However, the data presented is typical of results found in
growing cumulus clouds approaching precipitating size.

3.1 Results

Figures 4(a)-(d) show the liquid water content (obtained from
integrated FSSP data sampled at 10Hz) (L); droplet number
concentration sampled at 10Hz (N); the dry-bulb temperature measured
using a reverse flow thermometer (T); and the vertical wind (W)
obtained on a single passthrough a growing cumulus cloud. The
observed value of L is typically 2 g m^{-3} with maximum values of around
3.3 g m^{-3} compared to an adiabatic value calculated from the observed
height of cloud base of 4.2 g m^{-3} indicating that entrainment has
affected all parts of the cloud. (Comparison of the FSSP derived
liquid water contents with those from a Johnson-Williams device also
carried by the aircraft suggest an uncertainty of around 10%. The
adiabatic value has an uncertainty of about 0.3g m^{-3}). The regions
marked A, B, C represent fairly homogenous regions of cloud with
different mixing histories.

Figure 5a shows the variation of the droplet number sampled by the
FSSP at 50Hz, this is not a concentration as on the aircraft the FSSP
sample volume cannot be defined for frequencies higher than 10Hz.
Figure 5b shows the variation of C_N/C_P where C_N is the
coefficient of variability in the droplet number against distance
across the cloud, $C_N=(N^2+N^2)/^{1/2}/N$. In the limit in which the
fluctuations follow a Poisson distribution about the mean, C_N becomes
C_P. Thus statistically significant fluctuations show as significant
departures from $C_N/Cp=1$. C_N was calculated using a running mean
of ten points corresponding to 0.2s. When the droplet number is
examined it can be seen that with the exception of the region labelled

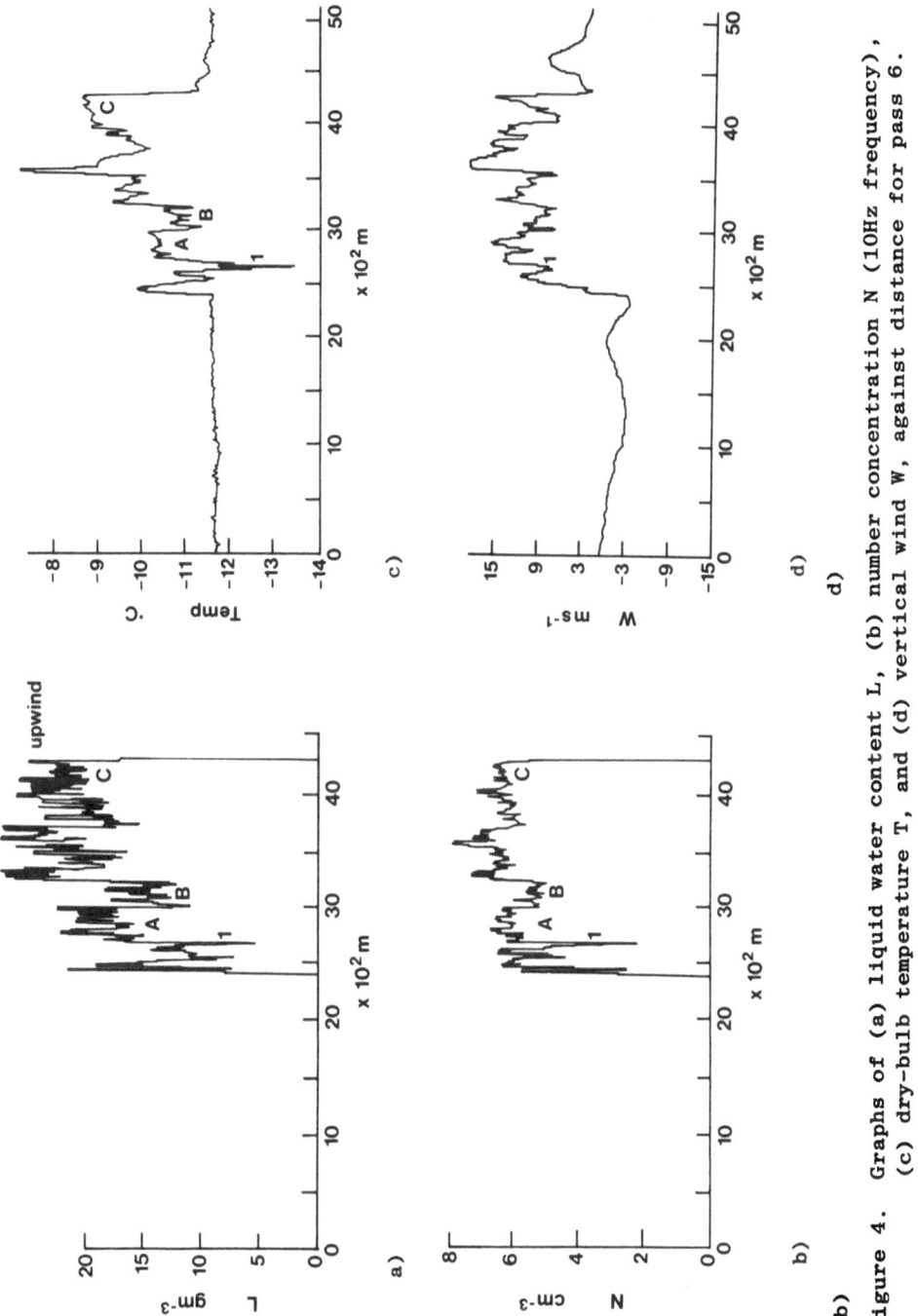

Figure 4. Graphs of (a) liquid water content L, (b) number concentration N (10Hz frequency), (c) dry-bulb temperature T, and (d) vertical wind W, against distance for pass 6.

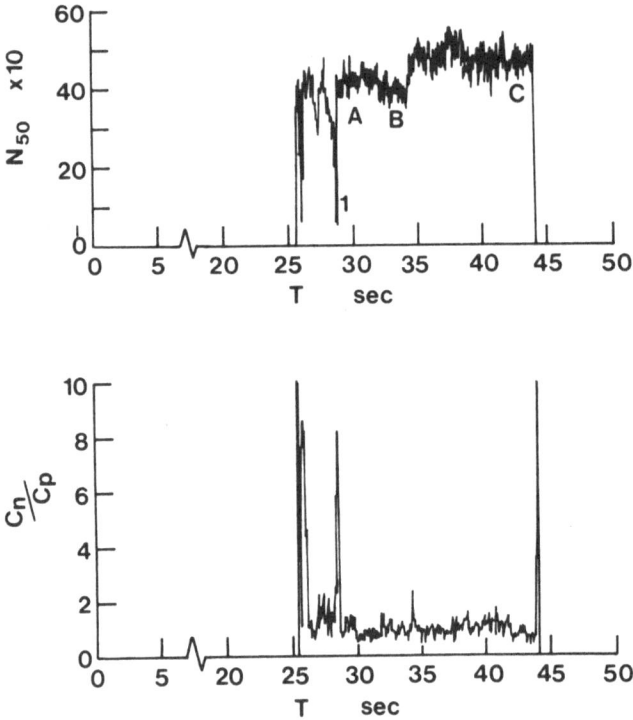

Figure 5.   The droplet number count from the FSSP measured at 50Hz.
(b) The ratio of the coefficient of variation C_N to that
expected from sample error C_p against time.   Regions of
large fluctuation labelled 1.

'1' within which substantial structure is evident, the bulk of the
cloud is close to being homogenous.   Narrow regions of this kind were
frequently found in passes through ascending turrets.   Figures 6 and 7
show two more examples.   Again the regions of large fluctuation are
labelled '1'.

These regions characteristically have lower temperatures, low
liquid water content and rapid, small scale fluctuations in both L and
N.   In order to clarify the effects of mixing on the droplet spectral
shape, the droplet spectra in the three mixing regions identified above
can be compared with the average spectra for the regions close by.
These spectra are averages over 1.7, 0.8 and 2s respectively for the
three examples.   Figures 8(a)-(c) show the results.   All three
regions possess temperatures that are low enough to make it clear that
significant evaporation must have taken place.   In all cases there is
a large reduction in the number of droplets in the main peak of the
spectrum and not a shift in the peak radius to smaller sizes.

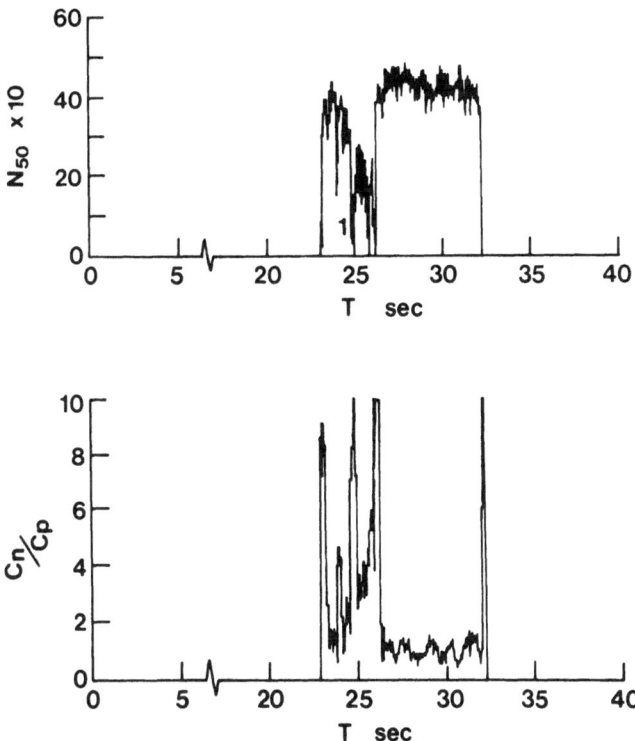

Figure 6. As figure 5 but for pass 4.

These observations indicate that these regions are the product of recent dry air entrainment events and that these occur on a small scale compared with the total cloud width, producing evaporation locally rather than across a large portion of the cloud. The effect of this evaporation is to totally remove a fraction of the droplets in the spectrum, irrespective of size (within the range present) rather than partially evaporating a larger fraction of the droplets.

The different pattern of evaporation in the cumulus cloud may be explained by the larger scale of entrainment than in the hill cap cloud. As discussed by Baker et al this results in the time constant for turbulent mixing of the entrained air in the cumulus cloud being long compared to the droplet evaporation time and hence almost complete evaporation of the droplets occurs in a small fraction of the cloud. In the GDF cap cloud partial evaporation of droplets occurs through a larger fraction of the cloud.

3.2 Implications for Cloud Chemistry

Cumulus clouds are strongly affected by entrainment of free tropospheric air from outside their boundaries and this process may play a crucial role by introducing extra oxidant into the cloud to

164

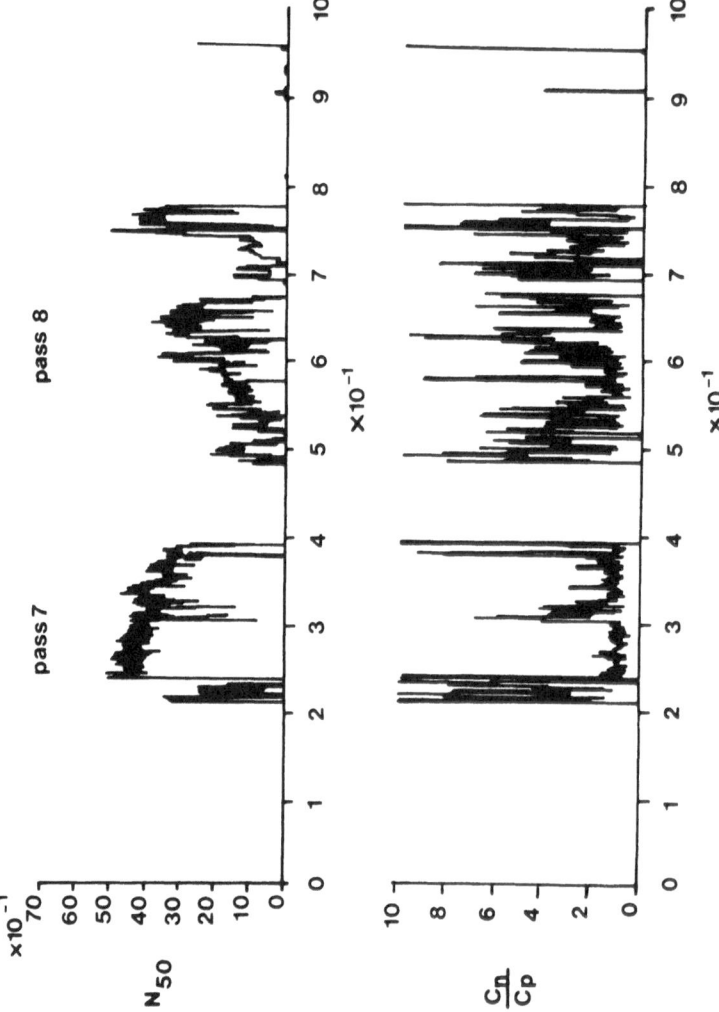

Figure 7. As Figure 6 but for passes 7 and 8.

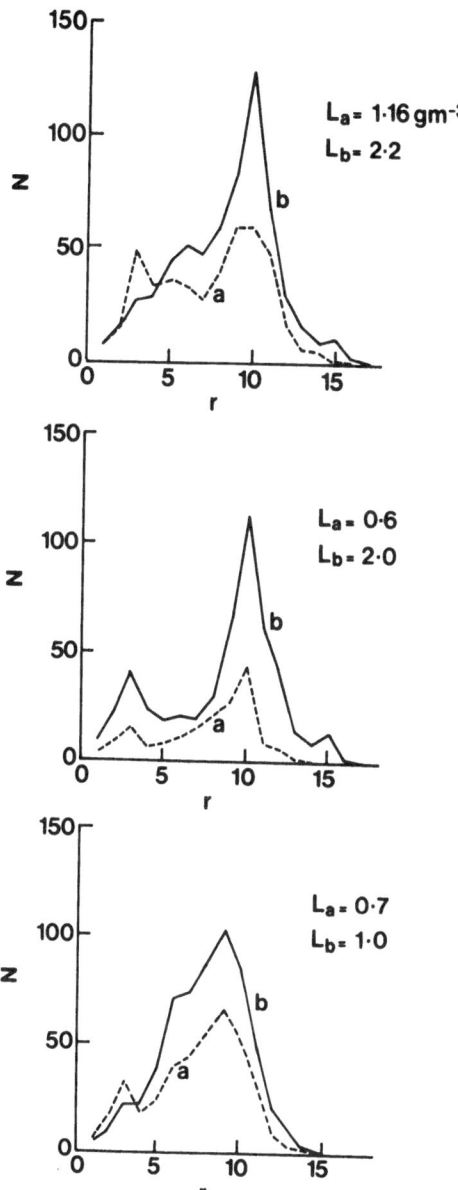

Figure 8.   Comparison of droplet spectral shape in regions
            strongly affected by dry air entrainment labelled
            1 in figures 5, 6 and 7 with immediately
            neighbouring cloud, labelled a and b respectively
            (1) for pass 6, (2) for pass 4 (3) for pass 7.
            Vertical axis, N cm^{-3}; horizontal axis, radius,
            μm.

promote otherwise strongly oxidant limited reactions.   The other major effect is the reduction in mean droplet lifetime resulting from the total evaporation of a fraction of the droplets following dry air entrainment.   The typical lifetime of an individual cumulus cloud is typically 20 to 40 minutes, as suggested by pilot observations during CCOPE, and from the size of the droplets in the cloud it is estimated that the lifetime of an individual droplet will be reduced to about 10 minutes.   This is sufficient to substantially reduce the contribution to the sulphate concentration in the cloud droplets due to the aqueous phase oxidation of sulphur dioxide by ozone in acid cloud water.

4.   CUMULUS CLOUD MODEL

In this section we describe models of the effects of entrainment of extra oxidant on the chemistry of a growing cumulus turret.   Model 1 excludes the effects of detailed microphysics, however, for adiabatic growth the results are compared to model 2 in which the chemical evolution of individual drop categories are followed.

4.1 Chemistry Models

The chemical equations employed in these models are identical to those described in (Hill, Choularton and Penkett 1986) and are set out below:

$$H_2O \rightleftharpoons H^+ + OH^- \quad K_1 = 10^{-14}\exp -6176. \ (1/T-1/298) \quad (1)$$

$$SO_{2(g)} + H_2O \rightleftharpoons SO_2.H_2O \quad K_2 = 1.23\exp 3120.(1/T-1/298) \quad (2)$$

$$CO_{2(g)} + H_2O \rightleftharpoons CO_2.H_2O \quad K_3=3.11\times10^{-2}\exp 2423.(1/T-1/298) \quad (3)$$

$$SO_2H_2O \rightleftharpoons H^+ + HSO_3^- \quad K_4 = 1.7\times10^{-2}\exp 2090.(1/T-1/298) \quad (4)$$

$$CO_2H_2O \rightleftharpoons H^+ + HCO_3^- \quad K_5 = 4.3\times10^{-7}\exp 913(1/T-1/298) \quad (5)$$

$$NH_{3(g)} + H_2O \rightleftharpoons NH_3.H_2O \quad K_6 = 1.7\times10^{-5}\exp -4325.(1/T-1/298) \quad (6)$$

$$NH_3.H_2O \rightleftharpoons NH_4^+ + OH^- \quad K_7 = 58.0\exp 4035(1/T-1/298.) \quad (7)$$

$$H_2O_{2(g)} + H_2O \rightleftharpoons H_2O_2.H_2O \quad K_8= 9.7\times10^4\exp 6600. \ (1/T-1/298.) \quad (8)$$

$$O_{3(g)} + H_2O \rightleftharpoons O_3.H_2O \quad K_9 = 1.15\times10^{-2}\exp 2560.(1/T-1.298.) \quad (9)$$

The rate at which gases are absorbed or desorbed by droplets is taken to be either infinite – assuming spontaneous equilibrium or finite using the formulae developed by Schwartz 1983.

Hence, assuming equilibrium is instantaneously established between gas and liquid phases, the partial pressure of each gas, P_g at temperature T is given by

$$P_g = N(t)/\left[\frac{KL(t)}{10^3} + \frac{1.01 \times 10^5}{8.3T}\right] - (10)$$

where $N(t)$ is the total number of moles in 1 m^3 of air both in the gas phase and in solution and K is the effective solubility of the gas. For SO_2 we have

$$\frac{\Delta N(t)}{\Delta(t)} = \frac{1}{P} \frac{\Delta P}{\Delta t} . N(t) - \begin{array}{l} K_{H_2O_2}[HSO_3^-][H_2O_2] \\ + KO_3 [HSO_3^-][O_3] \end{array} - (11)$$

P is the pressure change during interval t.

The equations are closed by the neutrality condition

$$[H^+] + [NH_4^+] = [OH^-] + 2[SO_4^=] + [HSO_3^-] + [HCO_3^-]$$

and solved for $[H^+]$ after interval Δt.

The rate constants for the aqueous phase oxidation of sulphur dioxide by hydrogen peroxide and ozone are taken from Martin and Damschen (1981) and Maahs (1983) respectively.

4.2 Cloud Models

Two cloud models are used. Model 1 is a conventional entrainment model in which air is entrained into a rising spherical thermal at a rate proportional to its change in height. The equations used for this model are identical to those in Hill and Choularton 1986 with the exception that the droplet growth equations are excluded. The effect of this on the results will be discussed later. The entrainment parameter takes the form

$$\frac{\Delta M}{M} = \frac{\mu U}{R} \Delta t \qquad -(12)$$

where M is the mass of 1 m^3 of cloud, or the number of molecules of a gas in 1 m^3 of cloud and is in the range 10^{-3} to 10^{-4} s^{-1}. μ is a constant = 0.6 derived from experiment and R is the radius of the thermal.

Consequently for a gas, equation 12 is provided an extra term

$$-N(t) \frac{\mu U}{R} + N_{trop} \frac{\mu U}{R} \qquad (13)$$

where N_{trop} is the concentration of a gas in the surrounding atmosphere.

In model 2 we use the basic 'adiabatic' cloud model described in (Hil and Choularton 1986) and calculate the concentration variation in up t nine droplet size categories explicitly.

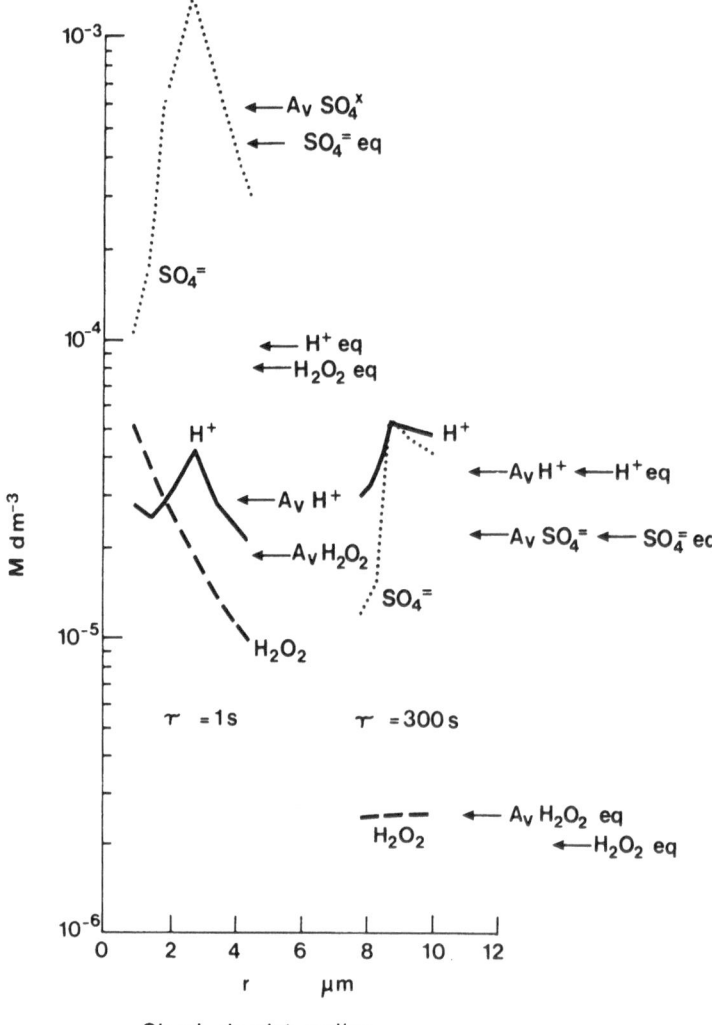

Figure 9. Concentration of various ionic species versus
droplet radius. Also indicated are the average
values and the values assuming instantaneous
equilibrium.

4.3 Results

Adiabatic cloud

The entrainment parameter was set to zero in model 1 and the results compared with those of model 2 with no entrainment events as the cloud was allowed to rise at 1 ms^{-1} by 300m.   The liquid water content rose from zero to .64 gm^{-3}.   It is assumed in model 1 that all gases establish equilibrium instantaneously whilst in model 2 NH_3, H_2O_2 and SO_2 are allowed to be absorbed at a finite rate.   It was found that O_3 always reached equilibrium well within 0.02 s, the time step used throughout.

The SO_2 concentration entering cloud base was 5ppbv, H_2O_2 was 0.2ppbv, NH_3 was 0 and the sulphate aerosol loading was 0.6µg m^{-3}.

Figure 9 shows the concentrations of various ions as a function of droplet size.   Note that 7 droplet size categories are shown instead of 9 because the supersaturation above cloud base was insufficient to activate all CCN.   Since each droplet category has a differing ionic content the droplets have not been re-bind into 1 micron intervals. Two sets of graphs are shown, one pertaining to 1 second after the

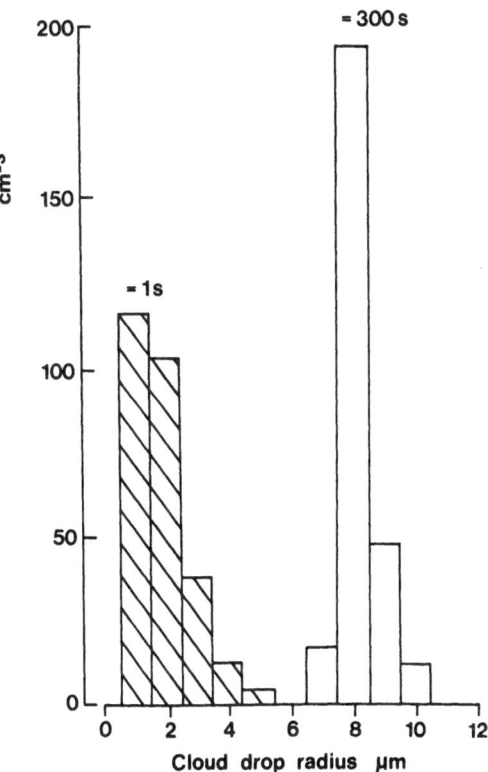

Figure 10.   Droplet distributions after 1s and 300s.   Concentration per micron radius interval.   Water contents are 0.01 and 0.64 gm^{-3}.

chemistry was initiated and one 300 seconds later. The droplet
distributions at these two times may be seen in figure 10 (after
re-binning).

Although the change in concentration of the ions is strongly
affected by dilution, several features are immediately apparent.
Firstly, the hydrogen ion concentrations increase during this period
and this indicates that the oxidation is overcoming the dilution
effect.   Secondly, the H_2O_2 concentrations vary considerably
initially, showingthat equilibrium is far from established at this
time.   In fact, the smallest droplets are about 60% saturated whilst
the largest are only 10%.   After 300s all droplets are very close to
saturation with respect to H_2O_2 and closer examination of the data
shows that it was reached after about 60 s.   It can also be seen that
there is a peak concentration in both sulphate and hydrogen ion within
each collection of droplets.   This is because the largest CCN observed
in the atmosphere are predominantly sodium chloride and since the
largest droplets were produced from the largest CCN (in this instance)
the concentration of ammonium sulphate is lower than for some of the
smaller droplets.   When the droplets are first formed there is a net
outgassing of ammonia because it is assumed absent from the free
atmosphere.   The mid-range droplets outgas to a greater extent because
the concentration is highest in these and hence it is these which
become the most acidic.   It was found throughout this run that the
droplets were never in equilibrium with respect to ammonia, the smaller
droplets never reaching more than 60% of the equilibrium value whilst
the largest droplets remaining supersaturated at 100-200%.   This is
partly due to the high solubility of NH_3 but also because the hydrogen
ion concentration is constantly changing and it takes time for ammonia
to respond.

Shown also in figure 9 are the bulk concentrations predicted by
model 1 and the average concentrations predicted by model 2.   After 1
second the H^+ and H_2O_2 average equilibrium concentrations are much
higher.   However the values after 300s show that this is a transient
affect and that the amount of oxidation is about the same in both
cases.

Model 1 including entrainment.

The model was run for 500s whilst varying the chemical input but
keeping the entrainment rate constant and the updraft fixed at 1 m s^{-1}
in order to simulate a small growing cumulus cloud.   The SO_4^{2-}
concentration both in the air rising through cloud base and in the free
atmosphere was kept constant at 0.6 μgm^{-3}.   This was assumed to be
neutralised by NH_4^+ ions.   The SO_2 and H_2O_2 concentrations in the
cloud base were 5ppbv and 0.2ppbv respectively.   In the remaining
atmosphere the concentration of SO_2 and NH_3 were kept at 0ppbv whilst
that of H_2O_2 was varied.

The concentration of O_3 throughout the atmosphere was assumed to
be 30ppbv.   These values are typical of those observed at GDF and
elsewhere (Calvert et al 1985).

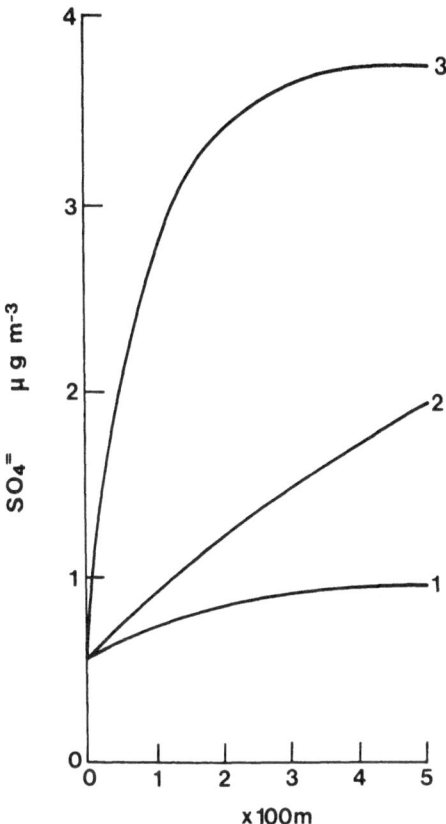

Figure 11a.   Concentration of sulphate in cloud water versus height
above cloud base.   Model 1.

Figure 11a shows the mass of SO_4^{2-} in $1m^3$ of cloud as the cloud
rose 500m.   During this time the liquid water content rose
monotonically to $0.31gm^{-3}$ compared with the adiabatic value of 1.06
gm^{-3}.   Curve 1 shows the case in which H_2O_2 is entirely absent from
the air into which the cloud is penetrating.   It was found that H_2O_2
was almost completelyexhausted by the end of the run (see figure 11b)
and that O_3 did not contribute significantly to SO_4^{2-} production.
Also the pH of the water rapidly fell to about 4 at which value the
oxidation rate of the O_3 reaction is nearly two orders of magnitude
smaller than for the H_2O_2.
    If 1.0ppbv H_2O_2 is present in the entrained air (curve 2) then
twice as much sulphate is produced after 500s and figure 11b shows that
there is over 10 times as much H_2O_2 in the cloud water than in the
previous case indicating that the oxidant is being replenished
significantly.   Once again O_3 is not contributing to the sulphate
production because the pH is too low.

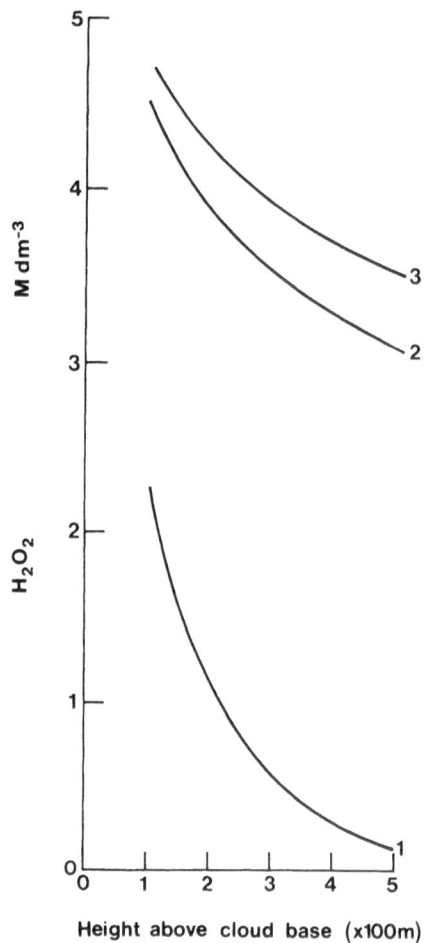

Height above cloud base (x100m)

Figure 11b.   Concentration of hydrogen peroxide in cloud water versus
height above cloudbase. Model 1.

In order to investigate whether O_3 could ever contribute
significantly, a case is included in which there is an additional 4ppbv
of NH_3 in the cloud base air.   The pH now remains over 5 for much of
the time dropping to 4.88 after 500s.   Curve 3 figure 11a shows that
there is now a dramatic increase in the amount of sulphate produced
during the first 200s and this is because O_3 is now oxidising SO_2 at 10
times the H_2O_2 rate.   As the pH falls so does the O_3 oxidation rate,
until by the end of the run H_2O_2 is once again faster.   This
illustrates an interesting point, namely that SO_2 would be oxidised
much more rapidly in clouds if the pH could be maintained in the 5.5-6
range.   However, the production of acidity by these rections produces
a negative feedback effect which reduces the effectiveness of the O_3
oxidation process.   The ambient level of ammonia near the ground in

the Eastern United States of America was observed to be about 2ppbv
(Harward et al 1982) and since this is largely biological in origin, it
is to be expected that higher in the atmosphere the level will be much
lower suggesting that the value used here is very much an upper limit.

5. DISCUSSION

In this paper results have been presented which suggest that the mixing
between a cloud and its environment can have a marked effect on its
chemical evolution. The entrainment of environmental air can add
extra oxidant when the chemistry would otherwise be strongly oxidant
limited. The entrainment of dry air and the subsequent evaporation of
droplets can considerably shorten the average lifetime of droplets
within the cloud thus limiting the chemical reactions which may occur
and also generates a broad spectrum of droplet sizes at all levels in
the cloud.
    In many cloud types mixing between a cloud and its environment can
have other important effects. The evaporative cooling of parts of the
cloud following entrainment can generate downdraughts which alter the
dynamic structure of the cloud and distribute entrained oxidant through
the depth of the cloud. In addition, as discussed by Hill and
Choularton (1985) the effects of dry air entrainment can promote the
growth of a small fraction of the droplets so that they become large
enough for coalescence to be initiated and precipitation formed faster
than would occur in a purely adiabatic cloud.

REFERENCES

Baker, M.B., Carruthers, D.J., Caughey, S.J., Choularton, T.W., Conway,
    B.J., Fullarton, G., Gay, M.J., Mill, C.S., Smith, M.H. &
    Stromberg, I.M. 1982. Field Studies of the Effects of Entrainment
    upon the Structure of Clouds at Great Dun Fell. Quart. J.R. Met.
    Soc. 108 , 899-916.
Calvert, J.G., Lazrus, A., Kok, G.L., Heikes, B.G., Walega, J.G., Lind,
    J. & Cantrell, C.A. 1985. Chemical Mechanisms of Acid Generation
    in the Troposphere. Nature 317 , 27-35.
Carruthers, D.J. & Choularton, T.W. 1982. Airflow over Hills of
    Moderate Slope. Quart. J.R. Met. Soc. 108 , 603-624.
Carruthers, D.J. & Choularton, T.W. 1986. The Microstructure of Hill
    Cap Clouds. Ibid. 112 , 113-129.
Choularton, T.W., Consterdine, I.E., Gardiner, B.A., Gay, M.J., Hill,
    M.K., Latham, J. & Stromberg, I.M. 1986. Field Studies of the
    Optical and Microphysical Characteristics of Clouds Enveloping
    Great Dun Fell. Quart. J.R. Met. Soc. 112 , 131-148.
Harward, C.N., McClenny, W.A., Hoell, J.M., Williams, J.A. & Williams,
    B.S. 1982. Ambient ammonia measurements in coastal south-eastern
    Virginia. Atmos. Env. 16 , 2497-2500.
Hill, T.A. & Choularton, T.W. 1985. An Airborne Study of the
    Microphysical Structure of Cumulus Clouds. Quart. J.R. Met. Soc.
    111 , 517-544.

Hill, T.A. & Choularton, T.W. 1986. A model of the development of the droplet spectrum in a growing cumulus turret. Ibid. 112 , 531–544.

Hill, T.A., Choularton, T.W. & Penkett, S.A. 1986. A Model of Sulphate Production in a Cap Cloud and Subsequent Turbulent Deposition onto the Hill Surface. Atmos. Env. 20 , 1763–1771.

Maahs, H.G. 1983. Kinetics and Mechanisms of the Oxidation of S(IV) by Ozone in Aqueous Solution with Particular Reference to Sulphur Dioxide Conversion in Nonurban Tropospheric Clouds. J. Geophys. Res. 88 , 10721–10732.

Martin, L.R. & Damschen, D.E. 1981. Aqueous Phase Oxidation of Sulphur Dioxide by Hydrogen Peroxide at Low pH. Atmos. Env. 15 , 1615–1621.

Schwartz, S.E. 1983. Gas- and aqueous-phase chemistry of OH_2 in liquid-water clouds. Meeting of an the American Chemical Society, Division of Environmental Chemistry, Washington D.C. September 1983.

MODELLING WET DEPOSITION ONTO ELEVATED TOPOGRAPHY

T.A. Hill, A. Jones and T.W. Choularton
Department of Pure & Applied Physics
University of Manchester
Institute of Science and Technology
Manchester
UK

ABSTRACT. Models predicting deposition rates of chemical species onto the surface of hills by washout and turbulent deposition have been developed. The models include aqueous phase sulphur dioxide chemistry which is allowed to proceed within an evolving dynamical-microphysical framework. Individual droplet chemistry may be examined and allowance made for the finite time required for equilibrium to be achieved between droplets and soluble gases. Comparisons are made with comprehensive field data obtained at the UMIST Great Dun Fell site. Predicted conversion rates for sulphur dioxide to sulphate are found to be in good agreement with field measurements. The concentration of sulphate in collected rainwater is also found to be consistent with that observed.

1. INTRODUCTION

Experiments are currently being undertaken at the Great Dun Fell mountain site in Cumbria, U.K. to investigate the formation of acidic pollutants in the cloud system associated with the hill and their subsequent deposition to the ground either by rain or by turbulent deposition - the latter also referred to as 'occult' deposition (Dollard et al., 1983). In order to interpret the results of these experiments models have been developed which take into account previously gained knowledge of the physics of the system and which also have incorporated into them the chemistry most important to the situation. In this paper we discuss these models and their predictions. At present we have concentrated on the oxidation of SO_2 by H_2O_2 and O_3. These oxidants have been identified as the most important in the aqueous phase (Penkett et al., 1979; Middleton et al., 1980) for clouds remote from pollution sources. It is not the purpose of this paper to describe the models in detail. More comprehensive descriptions may be found in the papers indicated in the text.

175

M. H. Unsworth and D. Fowler (eds.), Acid Deposition at High Elevation Sites, 175–188.
© 1988 by Kluwer Academic Publishers.

2. CAP CLOUD MODEL AND TURBULENT DEPOSITION

One of the experiments being performed at the Great Dun Fell site is
the determination of a field rate constant for the oxidation process.
This involves the collection of cloud water from several different
points on the surface of the hill and a determination of its
composition. However, before a comparison can be made between the
samples their age and history must be determined.

In order to examine the time dependency of the oxidation process
we require a representative transit time from cloud base to a measuring
site on the hill surface. We therefore firstly require an airflow
model of the hill. The wind velocities need to be predicted close to
the hill surface with some accuracy as well as throughout the cloud
depth because all measurements are made within 10 metres of the
ground. We use the model of Carruthers and Choularton (1982) which was
developed specifically for Great Dun Fell and which has received
experimental verification at this site.

We solve Helmholtz's equation for inviscid flow

$$\nabla^2 \zeta + \mu_i^2 \zeta = 0$$

where μ_i is the Scorer parameter N_i/U_i in layer i, and U_i is the
geostrophic wind, ζ is the streamline displacement, and use $f(x)$ as the
lower boundary condition where

$$f(x) = \frac{H}{(1 + x^2/L^2)}$$

The buoyancy frequency N_i is defined

$$N_i = \frac{\left(\frac{g}{\Theta_i} \frac{d\Theta_i}{dz} \right)^{\frac{1}{2}}}{U_i}$$

where Θ_i is the mean potential temperature of the layer and g the
acceleration due to gravity.

The atmosphere may be divided up into several layers and this is
important because the stability of the air above the turbulent boundary
layer can have a large effect on the flow of air in this lowest layer
(the layer in which the cloud frequently resides). Close to the hill
surface the Reynolds stress and viscous terms cannot be left out of the
momentum equation and therefore in the region just above the ground the
wind velocities are calculated by the method described in Jackson and
Hunt (1975). It is not proposed to discuss this model in detail here.

Given the wind field over the hill surface, the next step is to
incorporate the cap cloud. In this model the cloud either begins when
the rising air has cooled to saturation or we can specify cloud base as
occurring at some height based on observation. The depth of the cloud
is estimated from soundings in the vicinity of the hill. The density

changes due to release of latent heat during the condensation process are insufficient to alter the dynamics for a hill of this size.

The position of any streamline and the wind velocity along any point of it can be calculated and used to find the liquid water content and droplet size distribution variation over the hill using the standard droplet growth equations. Hence we can produce droplet size distributions and droplet composition concentrations for a given position on the hill.

The concentration of chemical species may be allowed to vary within the collection of droplets according to the equations set out in section 2, and gases exchanged between the droplets and the surrounding air.

All measurements at the moment have to be made on the ground and therefore the model must be able to predict concentrations within 10 m of the surface. The boundary layer in which the cap cloud resides is turbulent so that droplets tend to mix vertically during transit. The approximate distance through which the droplet will move in time Δt is given by $U*\Delta t$ where U_* is the characteristic velocity of the turbulent motion. Given the approximate transit time of cloudy air from cloud base to the point of measurement, also Δt, we can arrive at a vertical distance such that droplets throughout this depth will have an equal chance of being intercepted at the measuring site.

Figure 1 shows schematically one such region of cloud that this argument defines, stretching back upstream from a particular point on the hill towards cloud base. An average cloud parcel trajectory can now be calculated and hence a vertical wind profile for use in determining droplet growth.

Knowing the composition of the cloud which is in contact with the hill allows us to estimate the turbulent deposition rate of a chemical species to the hill surface of known roughness length at a given position, A. We solve

$$K(z)\frac{dC(z)}{dz} + V_t C(z) = V_d(z)C(z) = \text{constant}$$

where $C(z)$ is the droplet concentration and $V_d(z)$ their deposition velocity. $K(z)$ is the eddy diffusivity $= kzU*$ and the droplet terminal velocity $V_t << U_*$ and U_* is defined by

$$U(z) = \frac{U*\ln(Z/Z_0)}{k}$$

solving for $V_d(z)$ at a point A, assuming the droplet flux is independent of height near the ground, gives

$$V_d(z) = k \frac{U*_A}{\ln(z/z_0)}$$

In the absence of detailed experimental data to the contrary we assume that the deposition is limited by turbulent diffusion. The

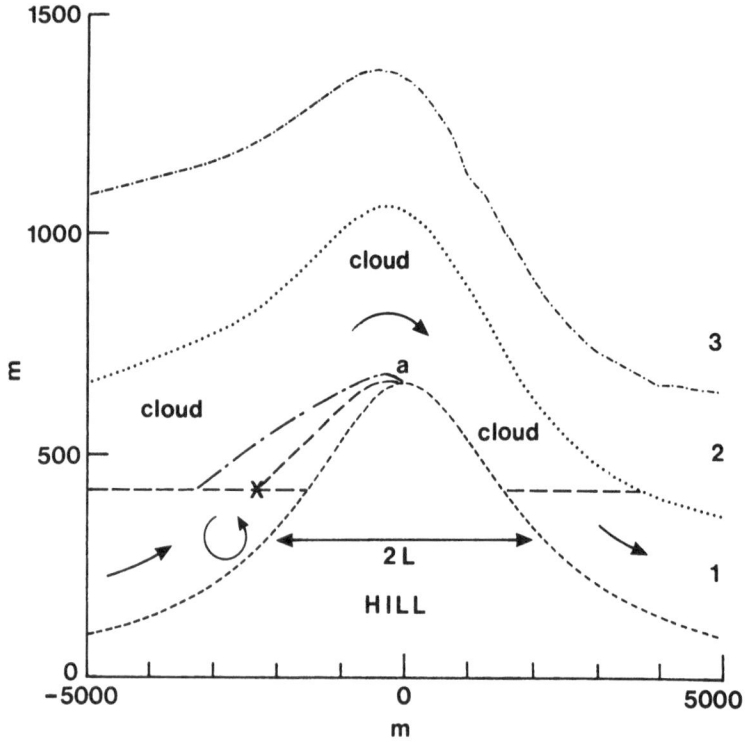

Figure 1. Schematic diagram showing turbulent deposition to a point 'a'
on the hill. The atmosphere has been divided into three
layers, the lowest of which contains the cap cloud. -.-
represents the outer boundary of the region whose droplets
may be deposited at 'a'. -- represents the average path to
'a'. Axes in metres.

liquid water content at A is calculated assuming no depletion (i.e. Z
must be > 10 m if Z_0 = 0.02 m) and hence we use the concentrations
predicted by the model 10 m above the ground.

3. CHEMISTRY MODEL

The chemical equations employed in these models are set out in Table 1.
The rate at which gases are absorbed or desorbed by droplets is
taken to be either infinite - assuming spontaneous equilibrium or
finite using the formulae developed by Schwartz (1983).
For gases in which equilibrium is almost instantaneously
established between gas and liquid phases, the partial pressure of each
gas, p_g at temperature T is given by:

$$P_g(t) = \frac{N(t)}{\frac{L(t).K}{1000.} + \frac{1.01 \times 10^5}{8.3.T}}$$

TABLE 1.

$H_2O \rightleftharpoons H^+ + OH^- \quad K_1 = 10^{-14} \exp{-6176.(1/T-1/298)}$

$SO_2(g) + H_2O \rightleftharpoons SO_2H_2O \quad K_2 = 1.23 \exp{3120.(1/T-1/298)}$

$CO_2(g) + H_2O \rightleftharpoons CO_2H_2O \quad K_3 = 3.11 \times 10^{-2} \exp{2423.(1/T-1/298)}$

$SO_2H_2O \rightleftharpoons H^+ + HSO_3^- \quad K_4 = 1.7 \times 10^{-2} \exp{2090.(1/T-1/298)}$

$CO_2H_2O \rightleftharpoons H^+ + HCO_3^- \quad K_5 = 4.3 \times 10^{-7} \exp{913(1/T-1/298)}$

$NH_3(g) + H_2O \rightleftharpoons NH_3.H_2O \quad K_6 = 1.7 \times 10^{-5} \exp{-4325.(1/T-1/298)}$

$NH_3.H_2O \rightleftharpoons NH_4^+ + OH^- \quad K_7 = 58.0 \exp{4035(1/T-1/298)}$

$H_2O_2(g) + H_2O \rightleftharpoons H_2O_2.H_2O \quad K_8 = 9.7 \times 10^4 \exp{6600.(1/T-1/298)}$

$O_3(g) + H_2O \rightleftharpoons O_3.H_2O \quad K_9 = 1.15 \times 10^{-2} \exp{2560.(1/T-1/298)}$

from Chameides (1984).

where $N(t)$ is the total number of moles in 1 m^3 of air both in the gas phase and in solution and K is the effective solubility of the gas, L is the liquid water content. It is assumed that ozone and carbon dioxide are always in equilibrium.

For highly soluble gases, ammonia and hydrogen peroxide and also for sulphur dioxide we assume a time dependency on gas takeup and use the equations derived by Schwartz (1983), e.g. for SO_2

$$\frac{dN_G(+)}{d+} = 4.2 \times 10^{-7} r \left(\frac{T}{0.22}\right)^{\frac{1}{2}} \left(\frac{1+3.3 \times 10^{-6}}{r}\right)^{-1} \left(\frac{[SO_2]_{aq} N}{\frac{K_2(1+K_4)}{[H^+]}8.3.T} - N_G N\right)$$

where N is the Avogadro number, r the radius of the droplet in metres and N_G the number of moles in the gas phase per cubic metre.

The equations in Table 1 are closed by the neutrality condition

$$[Na^+] + [H^+] + [NH_4^+] = [OH^-] + 2[SO_4^=] + [HSO_3^-] + [HCO_3^-] + [Cl^-]$$

and solved for $[H^+]$ after interval Δt.

Once the concentrations of the various species have been calculated the change in sulphate concentration is given by:

$$\frac{d[SO_4^=]}{dt} = K_H {}_{2O_2}.[HSO_3^-].[H_2O_2] + K_{O_3}.[HSO_3^-].[O_3]$$

180

where the rate constants are determined by the algorithms of Maahs (1983) and Martin and Damschen (1981).

The airflow model is valid for all hill sizes with aspect ratio H/L < 0.3 and L < 50 km (Carruthers, 1982). Hence the combined chemistry and microphysical model may be used to predict production/deposition rates for a wide variety of hills and mountains.

The Great Dun Fell hill is grass covered and experiment has suggested that a roughness length of 2 cm is appropriate. However, this can be changed in the model in order to describe other terrain eg. a forested hill. The rate of deposition will be higher over such a surface. If Z_0 is taken as 50 cm for such a hill then deposition rates in this model are increased typically by a factor of 2 or 3.

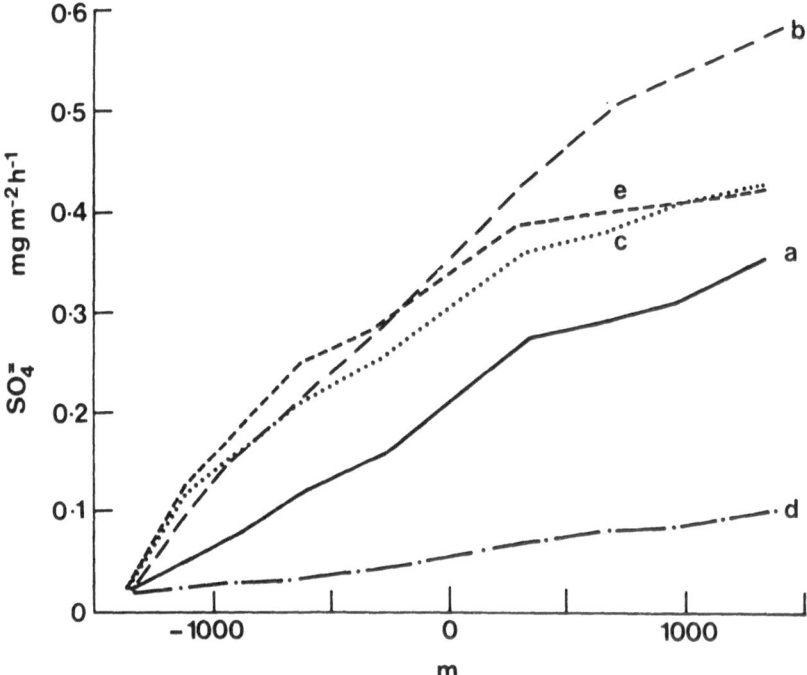

Figure 2a. Deposition rates of sulphate to the hill surface in mgm^{-2}h^{-1}. X axis in metres from centre of hill.
Initial gas concentrations:
curve 'a' 0.5 H$_2$O$_2$, 5.0 SO$_2$, 0 NH$_3$, 30 O$_3$
curve 'b'   "   ,   "   , 1.0 NH$_3$, "
curve 'c' 0.0 " ,   "   ,   "   , "
curve 'd'   "   ,   "   , 0.0 NH$_3$, "
curve 'e' 0.5 " , 20 SO$_2$,   "   , "

A full discussion of this model may be found in Hill et al 1986.
Figures 2a and 2b show the predicted rate of deposition of SO$_4^-$ and H$^+$ onto the surface of Great Dun Fell for different concentrations

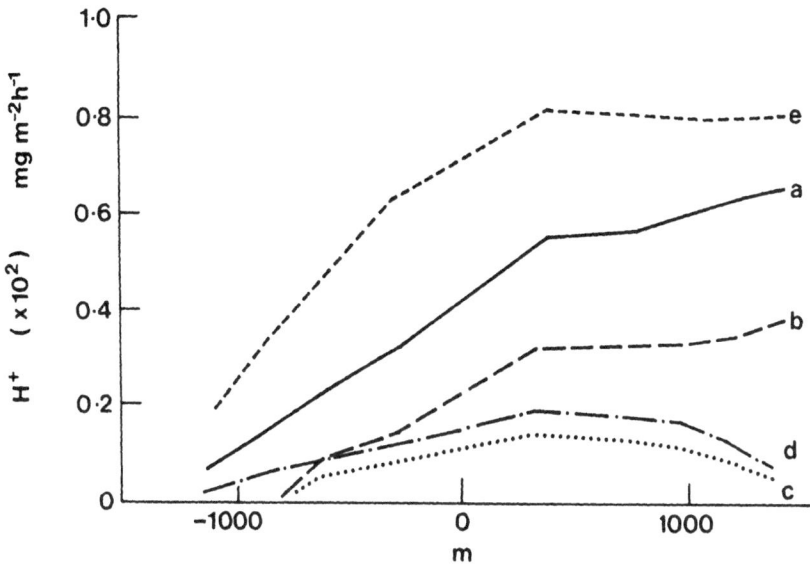

Figure 2b.   Deposition rate of hydrogen ions to the hill surface. See
figure 2a.
Deposition rate of sulphate to the hill surface for
three different atmospheric conditions. The input
chemistry is the same as for curve 'a' in figure 2a.

TABLE 2

Parameter	Value (initial if variable)
Stability of layer 1	0
Stability of layer 2	2×10^{-3} m^{-1}
Stability of layer 3	1×10^{-3} m^{-1}
Height of cloud base	400 m
Height of hill, H	665 m
Width at half height	2000 m
Roughness length, Z_0	0.02 m
Geostrophic wind, U_g	13 ms^{-1}
Temperature, T (at cloud base)	6°C
Height of inversion	400 m
Thickness of inversion	400 m
Ammonium sulphate	$0.4 \times 10^{-6} gm^{-3}$
Gaseous H_2O_2	0.5 ppbv
Gaseous SO_2	5 ppbv
Gaseous O_3	30 ppbv
Gaseous CO_2	320 ppmv
Gaseous NH_3	0 ppbv

182

of pollutant and oxidant. The atmospheric conditions supplied to the
model may be found in Table 2. The resulting flow is supercritical
with higher windspeeds on the leeside of the hill.

Curve a shows the result of using the input of Table 2. The
increasing deposition rate as the air flows over the hill is due to the
production of sulphate and the higher wind speeds on the leeside.
Curve b shows what happens if there is 1 ppbv of ammonia present. The
large increase in deposition rate is caused by two effects. Firstly
the higher pH (typically 4.5 - 5.5) means that the solubility of SO_2 is
increased and secondly, ozone becomes a much more effective oxidant
contributing about the same amount of sulphate as hydrogen peroxide.
Compare this with curve c in which ammonia is present, but there is no
hydrogen peroxide. Curve d shows the result of there being no ammonia
on the rate of production of sulphate by the ozone reaction. The pH
is now typically below 4.5 and the reaction is much slower.

Curve e has 20ppv of SO_2 present initially but is otherwise the
same as curve a. Note that although the production rate is initially
greater than in case a the reaction is limited by the hydrogen peroxide
concentration because ozone is relatively inefficient at low pH.

The curves in 2b show the corresponding H^+ ion deposition rates.
As expected, the runs with ammonia have lower deposition rates. The
highest rates occur when the SO_2 concentration is high.

In order to illustrate the effect of changes in atmospheric
stability we show the results of weakly stable upper air and a neutral
atmosphere with symmetrical flow (Figure 3). It can be seen that the
stability of the uper layers is important in determining the deposition
rate especially on the leeside. Different atmospheric conditions can
result in differences of a factor of 2.

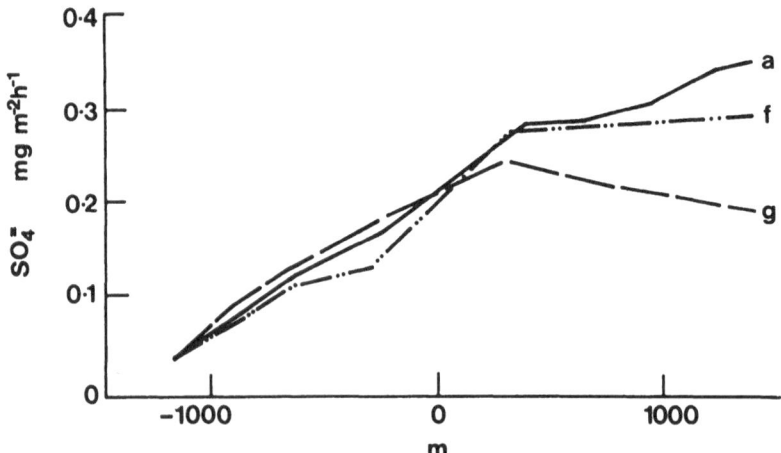

Figure 3. Curve 'a' is for a typical supercritical flow using the input
shown in Table 2. Curve 'f' represents one of a range of
curves for sub-critical flow patterns ($\mu_2 = 1.5 \times 10^{-3}$,
$\mu_3 = 1.5 \times 10^{-3}$ m^{-1} height of layer 2 is 1200m and thickness
300m). For curve 'g' all three layers are thermally
neutral. X axis in metres from centre of hill. Y axis is
deposition rate in mg m^{-2} n^{-1}.

4. WASHOUT MODEL

The other process by which cloudy water is deposited onto the hill
surface is precipitation. Because the average droplet lifetime in
deep cap cloud is typically ten to twenty minutes there is insufficient
time to generate precipitation sized droplets by coalescence. Instead
any rainout is likely to be due to the sweeping out of droplets by rain
falling form higher level cloud often associated with a front. This
process is frequently referred to as the 'seeder-feeder' process
(Bergeron 1965). The rainfall rate will therefore be enhanced over the
surface of the hill and the enhancement will be related to the path of
the raindrops falling through the cloud - being proportional to the
water content integrated along the path. The model is based on that
of Carruthers and Choularton 1983. We assume that, since raindrops
are typically 0.2 to 2 mm in radius, the collection efficiency for
droplets about 10 μm is 0.9 and use a Marshall-Palmer distribution for
the raindrops. The rainfall rate far upwind of the hill is taken from
observations. The liquid water distribution in the cap cloud before
washout is made dependent on streamline displacement for those
streamlines which are above cloud base.

A steady-state situation is envisaged in which cloud is being
continually swept out by the rain and replaced by fresh cloud due to
condensation in the air flowing over the hill. The path of the
droplet is determined by its fall speed and the flow speed of the air
over the hill. There are therefore two main factors which determine
the deposition pattern. Firstly, there is the path length of the rain
drop, determined by the hill length and the wind velocity and secondly
the rate at which water depleted by the washout process is replaced by
condensation. The condensation rate is determined by the updraught
speed and consequently hills of greater aspect ratio and hence higher
vertical winds close to the summit will have associated cap clouds
which are not so seriously depleted of water close to the summit. We
can summarise these arguments in terms of two length scales:

The horizontal drift of a raindrop radius r falling at speed V_r through
a cloud of depth Z_c

$$L_d \approx Z_c U_0 \Big/ V_r$$

The scale length of horizontal variation in liquid water content q

$$L_q \approx \frac{q}{\frac{\delta q}{\delta x}} \sim L$$

Where L is half length of the hill. for a short hill $L_q \lesssim L_d$.

To illustrate these effects Figures 4a and 4b show the rainfall
deposition patterns for the case of short hill (applicable to Great Dun
Fell) characteristic length 2000 m and for a long hill, length 20000m
All other parameters are kept constant.

184

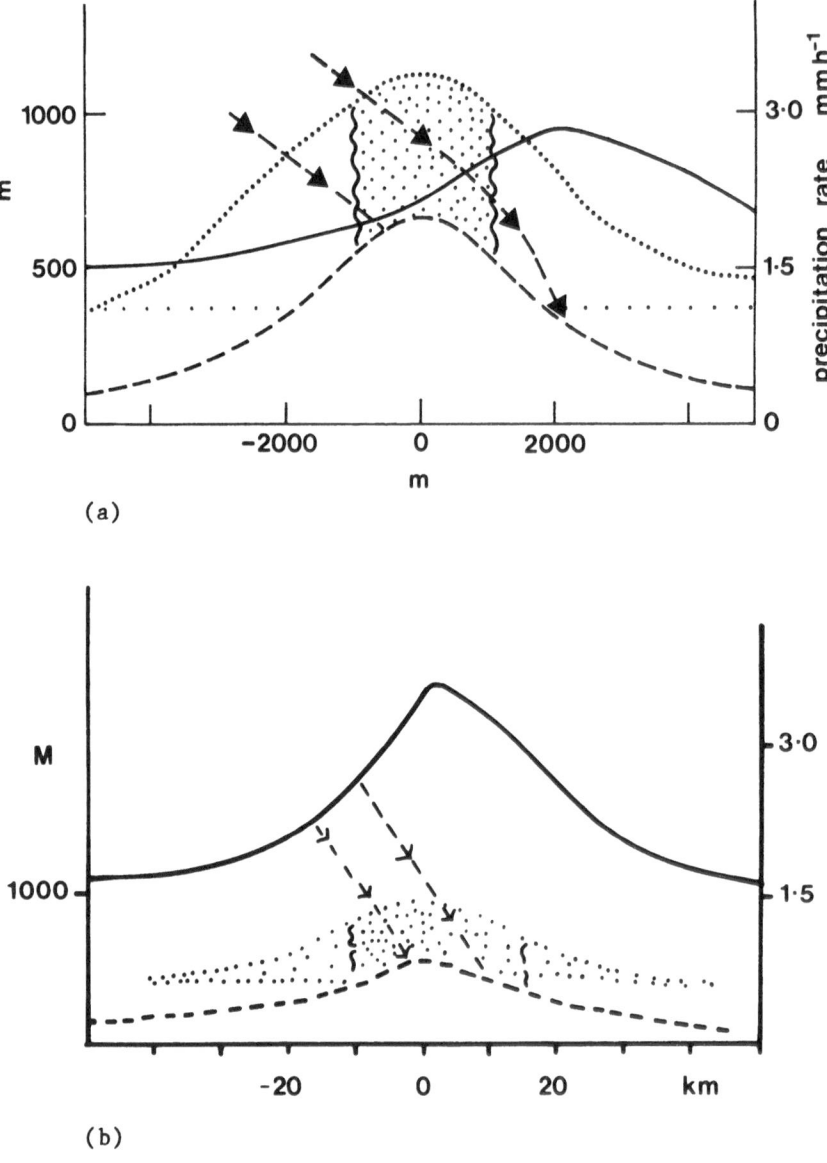

(a)

(b)

Figure 4. Schematic representations of washout processes for short and long hills respectively. Solid line shows typical precipitation rate at surface in mmh^{-1} as a function of position on the hill. Dotted area is region of maximum water content.

When the hill is steep, Figure 4a, the trajectory through the highest water conntent region is relatively short but the condensation rate is large due to the large vertical winds. Hence the greatest effect is the wind-drift of the air and the maximum deposition rate is displaced downstream.

Water is not so easily resupplied in the second case and the maximum enhancement occurs near the summit. The deposition magnitude will be greater than for a steep hill in the region of the summit because the maximum water content is relatively large and the effects of wind-drift are small.

If we now consider the deposition of the aerosol sulphate dissolved in the cloud water then the patterns for the short and long hills are typically represented in Figures 4a and 4b. For the short hill, the curve peaks downwind of the summit and is similar in shape to that in Figure 3a. However, for the long hill, Figure 4b, there is a faster decline in deposition rate downwind of the hill. This is because, although the water content is maintained to some extent by condensation, the aerosol sulphate is not replenished and hence declines continuously with position on the hill. This is not so crucial when the hill is short because the variation in the path length of the rain drops falling through the cloud is the dominant effect in this case. The model can similarly predict deposition rates for species such as NO_3^- and Cl^-.

5. COMBINING THE MODELS

It is to be expected that some chemistry will be taking place in the cap cloud through its depth as the air flows over the hill. We are interested in O_3 and H_2O_2 oxidising SO_2 to SO_4^-. It has been shown that H_2O_2 occurs in concentrations between 10 and 100 times smaller than O_3 in the free atmosphere and hence unless it can be replenished it will be exhausted in a few hundred seconds after condensation has taken place in a cloudy volume. Consequently over most of the cloud, the concentration of H_2O_2 may be relatively low. However, O_3 is in plentiful supply and oxidation can be expected to be taking place throughout the cloud. Note that close to the surface of the hill there will be a region in which H_2O_2 is still abundant because the cloud here has only recently formed and it is in this region that measurements of sulphate production are being made.

In order to investigate the effect of in-cloud sulphate production on the deposition pattern on the hill surface the production rate was parameterised. The cloud chemistry model was run along a streamline deep in the cloud from cloud base, over the hill and down the leeside in order to get representative rates of sulphate production for both oxidants on the time scales typical of cloud droplet lifetimes in the cloud (several 1000 seconds). Without in-cloud production of H_2O_2 oxidation is dominated by O_3 and a representative production rate of sulphate by oxidation is of the order of 1.0 ng m^{-3} s^{-1}. Figure 5 shows the deposition rates predicted for conditions typical of Great Dun Fell. We compare the rates predicted both with and without

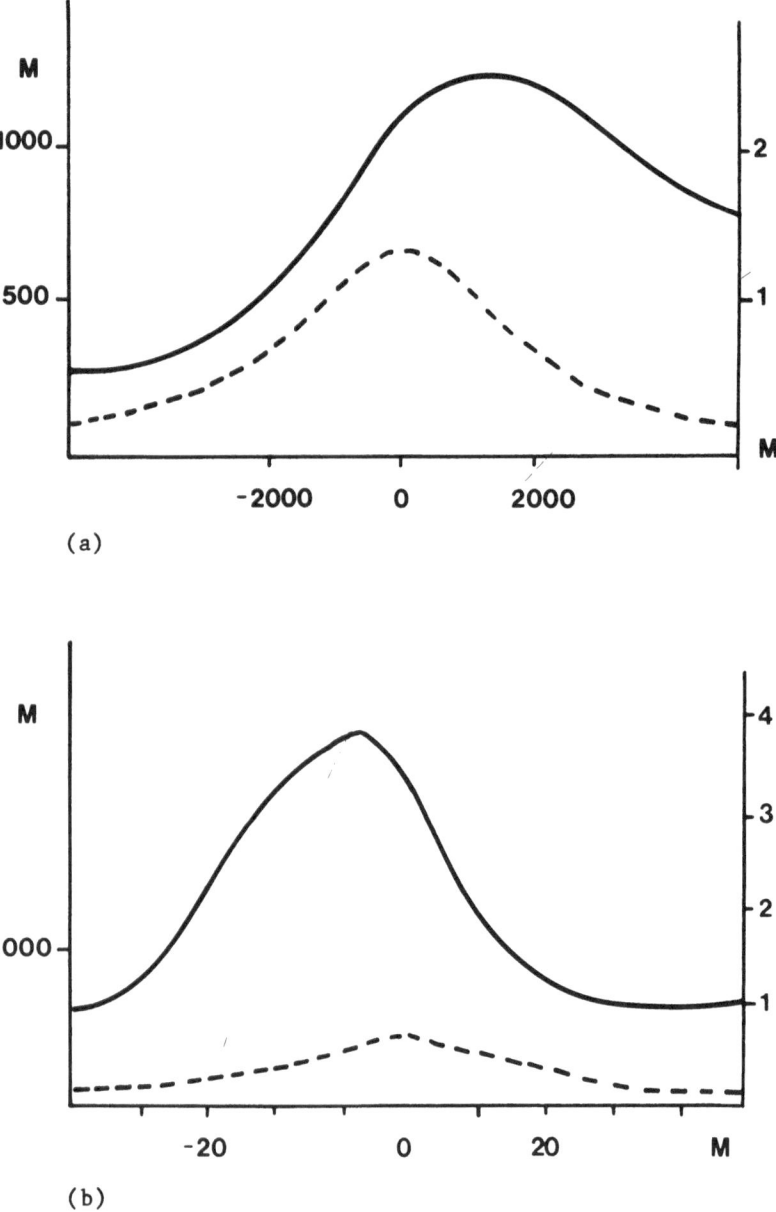

(a)

(b)

Figure 5. Schematic representation of rate of deposition of sulphate as a function of position on the hill for the case of a short hill and a long hill respectively. Typical deposition rates in $mgm^{-2}h^{-1}$.

sulphate production in the cloud.   It can be seen that in–cloud oxidation can significantly increase the sulphate deposition rate causing the rate to increase up the hill instead of levelling out and even decreasing towards the summit.   The precise oxidation rate is a function of many variables and it would not be suitable to include detailed chemistry when many simplifying assumptions have to be made about the physics.

6.   DISCUSSION

The models discussed in this paper predict that, during a precipitation event, the deposition due to rainout will be 5 to 10 times greater than turbulent or 'occult' deposition.   The pattern of deposition onto the surface of the hill, in both cases, is dependent on the atmospheric conditions and the rate of reaction within the cap cloud.   The latter is particularly important when rainout is absent.   However, as far as the washout of individual chemical species are concerned, the relative abundance in the rain falling into the cloud to that in the cap cloud itself may be more important.

The predictions of these models have been compared with field data with encouraging results.   The reader is referred to accompanying papers in this volume.

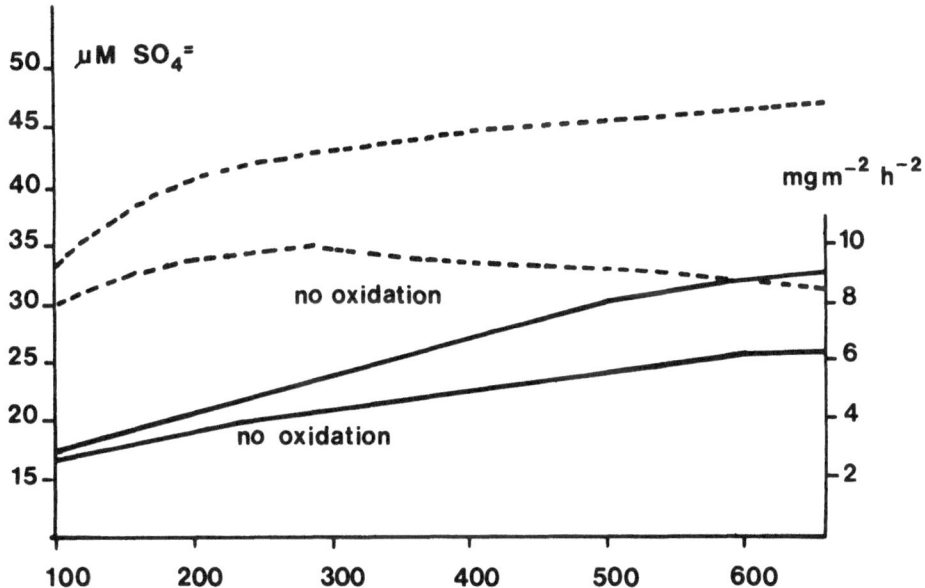

Figure 6. Comparison between sulphate precipitation concentration and deposition rate when no cloud chemistry taking place and when sulphate is being generated in the cloud as it flows over the hill.   Concentration in μM deposition in $mgm^{-2}h^{-1}$.

ACKNOWLEDGEMENTS

This work was part funded by the Department of the Environment. A J is in receipt of a Natural Environment Research Council studentship.

REFERENCES

Bergeron, T. 1965. On the low-level redistribution of atmospheric water caused by orography. Suppl. Proc. Int. Conf. Cloud Phys., Tokyo, May 1965, 96-1000.

Carruthers, D.J. 1982. PhD Thesis Univ. Manchester.

Carruthers, D.J. and Choularton, T.W. 1982. Air flow over hills of moderate slope. Q J Roy Met Society, 108, pp603-624.

Chameides, W.L. 1984. The photochemistry of a remote marine stratiform cloud. J. Geophys Res. 89, pp4739-4755.

Dollard, G.J., Unsworth, M.H. & Harvey, M.J. 1983. Pollutant transfer in upland regions by occult precipitation. Nature 302.

Hill, T.A., Choularton, T.W. & Penkett, S.A. 1986. A model of sulphate production in a cap cloud and subsequent turbulent deposition onto the hill surface. Atmos Environment (In press).

Jackson, P.S. & Hunt, J.C.R. 1975. Turbulent windflow over a low hill. Quart J.R. Met. Soc. 101, 929-955.

Maahs, H.G. 1983. Kinetics and mechanism of the oxidation of S(IV) by ozone in aqueous solution with particular reference to SO_2 conversion in nonurban tropospheric clouds. J Geophys Re, 88, pp 10721-10732.

Martin, L.R. & Damschen, D.E. 1981. Aqueous oxidation of sulphur dioxide by hydrogen peroxide at low pH. Atmos Environment 15, 1615-1621.

Middleton, P., Kiang, C.S. & Mohnen, V.A. 1980. Theoretical estimates of the relative importance of various urban sulphate aerosol production mechanisms. Atmos environment 14, pp463-472.

Penkett, S.A., Jones, B.M.R., Brice, K.A. & Eggleton, A.E.J. 1979. The importance of atmospheric ozone and hydrogen peroxide in oxidising sulphur dioxide in cloud and rainwater. Atmos Environment 13, pp123-127.

Schwartz, S.E. Gas- and aqueous-phase chemistry of OH_2 in liquid-culture clouds. Meeting of the American Chemical Society, Division of Environmental Chemistry, Washington, D.C. September 1983.

CLOUD CHEMISTRY RESEARCH AT GREAT DUN FELL

A S Chandler, T W Choularton, M J Gay, T A Hill, A Jones, A P
Morse and B J Tyler
UMIST
Manchester
UK

G J Dollard and B M R Jones
Environmental and Medical Services Division
AERE
Harwell
UK

S A Penkett
University of East Anglia
UK

ABSTRACT. A comprehensive series of experiments designed to
investigate the oxidation of SO_2 to sulphate by H_2O_2 and O_3 are being
performed in the cap cloud at Great Dun Fell. In this paper results of
the first set of experiments are presented. These took place during
November 1985.

The aim of these experiments was chiefly to monitor the H_2O_2
oxidation process by measuring its depletion with time within the cloud
in the presence of SO_2. Increases in sulphate content of the cloud
water were not observed during this experiment because H_2O_2 levels were
too low and oxidation by O_3 was inhibited by the low cloud water pH.
The concentrations of aqueous phase H_2O_2 measured were typically
100nmol and O_3 gas phase concentrations 20 ppbv.

It was found that, in the presence of SO_2, the concentration of
hydrogen peroxide declined much more rapidly with height above cloud
base than predicted by simple dilution by liquid water. Assuming this
to indicate reaction with SO_2, a comparison was made with the
predictions of the model of Hill, Choularton and Penkett (1986). It
was found that the rate of reaction was consistent with a value for the
second order rate constant $K_{H_2O_2}$ of $2 \pm 1 \times 10^5$ s^{-1} where $K_{H_2O_2}$ is
defined by:

$$\frac{d[S_{v1}]}{dt} = \frac{K_{H_2O_2} [H_2O_2][SO_2 . H_2O]}{0.1 + [H^+]} \quad \text{(Martin and Damschen 1981)}$$

M. H. Unsworth and D. Fowler (eds.), Acid Deposition at High Elevation Sites, 189–214.
© 1988 by Kluwer Academic Publishers.

This was determined with a cloud temperature of 8.5°C and a pH of 4.8. This is about 3x larger than the laboratory determined rate constant found in Martin and Damschen 1981 when corrected to the same temperature and pH.

On some occasions microphysical measurements in the cap cloud indicated that tropospheric air from above the cloud top was being entrained into the cloud. Increases in H_2O_2 concentrations with altitude within the cap cloud on these occasions showed that extra hydrogen peroxide was being simultaneously introduced to the cloud system. It is suggested that this entrainment process may play a very important role in SO_2 oxidation in clouds when the reaction is oxidant limited.

1. INTRODUCTION

A major collaborative experiment to investigate cloud chemical processes is being conducted at the University of Manchester Institute of Science and Technology (UMIST) field station on Great Dun Fell in Cumbria, England. Great Dun Fell (GDF) is 847 m above sea level and forms part of the long ridge of the Pennine hills, which run from the NW-SE down the centre of England. The prevailing south west winds blow almost perpendicular to the ridge, frequently forming cap clouds which envelope the site for parts of 250 days a^{-1}. The dynamics and microphysics of the cloud have been extensively studied, see for example Carruthers and Choularton (1982), Choularton et al. (1986). Additionally, these clouds make extensive contact with ground that is readily accesible, so that apparatus may be positioned at many different sites and the cloud examined at different stages in its history. These factors make the site ideal for the study of those aqueous phase chemical reactions which have a time constant of several hundred seconds or which produce measurable quantities of product on this timescale. The cloud is in effect a natural flow-through reactor which can be sampled at different stages of chemical state by collecting cloud water at different points on the hill in a line parallel to the prevailing wind. Experiments in wave clouds not in contact with the ground have been performed by Hegg and Hobbs (1982). Whilst these succeeded in detecting sulphate production, it was not possible to determine directly the mechanism of oxidation nor to monitor changes in chemical species within the cloud because of the great difficulty in making high resolution cloud chemical measurements from an aircraft.

During the past 10 years laboratory based evidence has accumulated to suggest that the oxidants H_2O_2 and O_3 are responsible for much of the atmospheric sulphate produced by SO_2 oxidation and that therefore aqueous phase oxidation contributes significantly to the acidity of precipitation (Penkett et al., 1979, Middleton et al., 1980, Chameides and Davies, 1982). This has been supported by recent studies of the chemical composition of cloud water from stratocumulus clouds, e.g. Kelly, Daum and Schwartz (1985) which have shown that there is a strong anti-correlation between the existence of SO_2 and H_2O_2 in aqueous

solution. In this paper we describe in some detail the results
obtained during 3 case studies performed during the autumn of 1985 and
designed to investigate the rate of consumption of aqueous phase H_2O_2
by SO_2 with the purpose of deriving a field reaction rate. At this
time of year gas phase H_2O_2 concentrations are normally very low, so
that the sulphate produced by this reaction would not be directly
measurable. However, H_2O_2 concentrations in solution can be measured
down to 0.05 μmol and hence changes in equivalent gas phase
concentration of about 0.5 pptv (1 pptv = 1 part in 10^{12} by volume) can
be observed. Hence, if changes in H_2O_2 are due to reduction by SO_2
then we have a very sensitive method of monitoring the reaction.

Interpretation of measurements in the field require detailed
knowledge of the likely microphysical and chemical history of the cloud
air intersecting the measurement points and it was for this reason that
the model of Hill et al. 1986 was developed. This model incorporated a
full description of the airflow over the hill using the treatment of
Carruthers and Choularton (1982). It also includes a detailed
description of the microphysical development as described by Carruthers
and Choularton (1986) and the sulphate production in the cloud droplets
by the oxidation of SO_2 by H_2O_2 and ozone. The method of
implementation is described later in the text.

2. THE EXPERIMENTAL SITE AND METHODS

The experiment was centred on 3 sites in the Great Dun Fell (GDF)
area. Figure 1 shows the deployment of the apparatus. Upwind of GDF
site 1 at Wharleycroft was our base laboratory in the Eden valley and
206 m above sea level. SO_2 and O_3 in the gas phase were measured in
the air stream entering the cloud along with wind direction, windspeed,
temperature and humidity. Site 2, the van site, was a mobile
laboratory situated on the front face of the hill about 675 m above sea
level. At this site gas phase SO_2 and O_3 were again measured. In
addition, the cloud liquid water content and droplet size distribution
were measured with a Knollenberg Forward Scattering Spectrometer Probe
(FSSP). Two passive cloud water collectors were deployed at this
site. These were of a similar design to those described by Kadlecek et
al. (1982) and strung with teflon fibre. These were used to collect
cloud water for the continuous monitoring of aqueous phase H_2O_2 using a
luminol technique (Ames 1983) and separate samples were stored for
later analysis by ion chromatography. Standard meteorological
measurements of windspeed, wind direction, temperature and humidity
were also made. Site 3, was at the summit of GDF. The same set of
measurements were made at the summit as were made at the van site. In
addition cloud water collectors were positioned at 8 other sites on the
hillside, these were used to provide additional batch samples for H_2O_2
measurements and ion chromatography. Site 3 was 1 km from site 2, and
5 km from site 1.

192

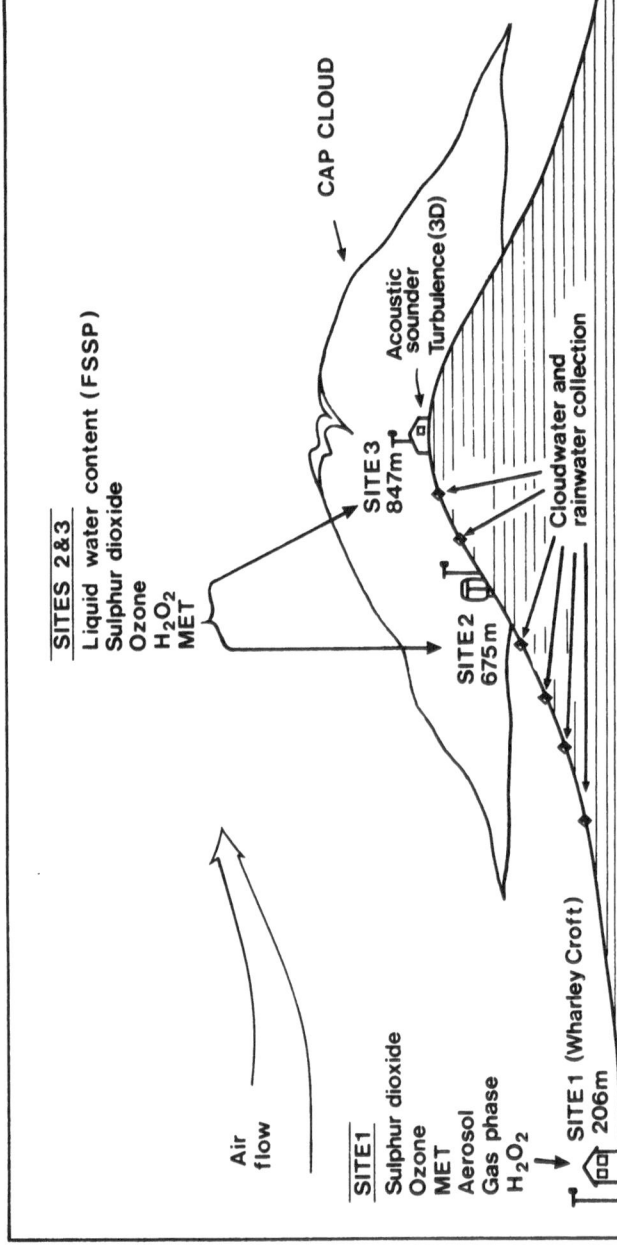

Figure 1.   Schematic diagram of the deployment of the instrumentation on Great
Dun Fell. Site 1 is the upstream site, site 2 the van and site 3 the
summit.

2.1 Measurements of Liquid Water Conent and Droplet Size Distribution

The microphysical measurements were made using Knollenberg Forward
Scattering Spectrometer Probes (FSS Probes) located at sites 2 and 3.
The techniques used and the precautions taken are described in detail
by Choularton et al, (1986).

During the field programme the 2 FSS Probes were run together side
by side at the summit of Great Dun Fell over an extended period. Based
on these comparisons and absolute calibrations performed in the
laboratory it is estimated that the liquid water contents quoted are
accurate to $\pm$ 10%.

2.2 The Chemical Measurements

2.2.1 Sulphur dioxide. SO_2 was measured using flame photometric
instruments (Melloy SA285E). Tests performed before purchase suggested
that the noise level of these instruments, when fitted with fast
response blocks, was below 0.2 ppbv (parts in 10^9 by volume).
Subsequent use in the field of these instruments showed the detection
limit to be about 1 ppbv. Long term drift of the baseline was a
significant problem and frequent checks of the zero level had to be
made either with a charcoal trap or with a tower containing glass beads
coated with KCO_3/glycerol solution. Calibration was performed with a
commercial calibrator which used a permeation tube to provide a steady
50 ppbv output flow. The 3 instruments were used at Wharleycroft, the
van and summit sites. During the experiment the Meloys were housed in
a thermostated environment which was maintained at 25°C which improved
the stability. The Meloy zero was checked every 10 min to counteract
any zero drift that occured. A Monitor Labs fluorescence instrument was
operated simultaneously and both sets of results agreed closely. The
sampling inlet for the SO_2 instrument was a 1.5 mm teflon tube with an
inverted 5 cm diameter funnel on the inlet end. For an inlet system of
this kind the flow at its mouth would be 0.1 cm sec^{-1} hence the intake
of liquid water was very small. In view of the above the SO_2
measurements quoted in the following sections are accurate to $\pm$ 0.5
ppbv.

2.2.2 The hydrogen peroxide measurements. H_2O_2 was normally measured
using the chemiluminescence observed when it reacts with luminol in the
presence of a microperoxidase (Sigma, type MP-11) catalyst.
Instruments were made to the design described by Ames (1983) which is
superior to the original luminol based technique described by Kok et
al. (1978). Ames described in detail the interferences encountered
with his instrument and points out that the more likely interference
(metal catalysed oxygen effect) lies well below the detection limit of
50nmol in normal circumstances. An instrument based upon the
fluorescence technique described by Lazrus et al. (1985) was also used
in these measurements. This latter instrument gave very similar values
to the luminol instruments for the peroxide concentrations when
compared. The linear range of the luminol instruments was 50nmol to
250μmol. When running in the field, the delay time between collection

and measurement was approximately 2-3 minutes when running continuously
and 30-60 minutes when analysing batch samples. The analysers were
calibrated each day using common, freshly prepared standards. All the
H_2O_2 analysers were operated using the same stock reagents.

The decay rates of the H_2O_2 in samples of cloud water has been
investigated for a number of case studies, in some cases starting with
fresh samples less than 2 minutes after collection. It was found to
vary and not to be related principally to the concentration of SO_2
present when the sample was collected. Around 50 measurements of the
decay rate of hydrogen peroxide have been made on stored cloud water
samples collected on different days and at different times of the
year. The decay rate was found to be between 8% and 12% h^{-1} and this
did not vary significantly with H_2O_2 concentration or from day to day.
Usually the rate of decay is not large enough to affect interpretation
of data from the batch samples.

Tests for organic peroxides in the samples were made using
catalyse to selectively destroy H_2O_2, followed by reanalysis (Lazrus et
al., 1985). These were found to contribute between 0 and 15%.

3. RESULTS

In the following paragraphs data are presented from 3 case studies
conducted during the autumn of 1985. Each illustrates different
experimental conditions and degree of complexity of the system. Case 1
concentrates on a period when a nearly adiabatic cap cloud formed and
H_2O_2 was consumed by reaction under conditions when there was ample SO_2
present. This represents the simplest system and the one whose results
are easiest to interpret. The results from this case study are used to
estimate a field reaction rate. Case 2 examines the effect of the
entrainment of tropospheric air from above cloud top on the H_2O_2
profile in the cap cloud when no significant SO_2 was present. In this
case changes in aqueous dilution and introduction of gaseous H_2O_2 along
with the entrained air. Case 3 also deals with a cap cloud which was
entraining free tropospheric air, but in this case large concentrations
of SO_2 were also sometimes present. The main features of the 3 case
studies are summarised in Table 1.

Table 1.

CASE	DATE	DURATION	WIND DIRECTION	WIND SPEED	CLOUD BASE	REMARKS
1	8.11.85	2.5 hr	SW	10 m s^{-1}	500 m ASL	2.5 ppbv SO_2
2	7.11.85	4.0 hr	WNW	8 m s^{-1}	> 600 m ASL	< 1 ppbv SO_2
3a	16.11.85	4.0 hr	SW	15 m s^{-1}	500 m ASL	5 ppbv SO_2
3b	16.11.85	1.0 hr	SW	8 m s^{-1}	$\cong$ 600 m ASL	< 1 ppbv SO_2

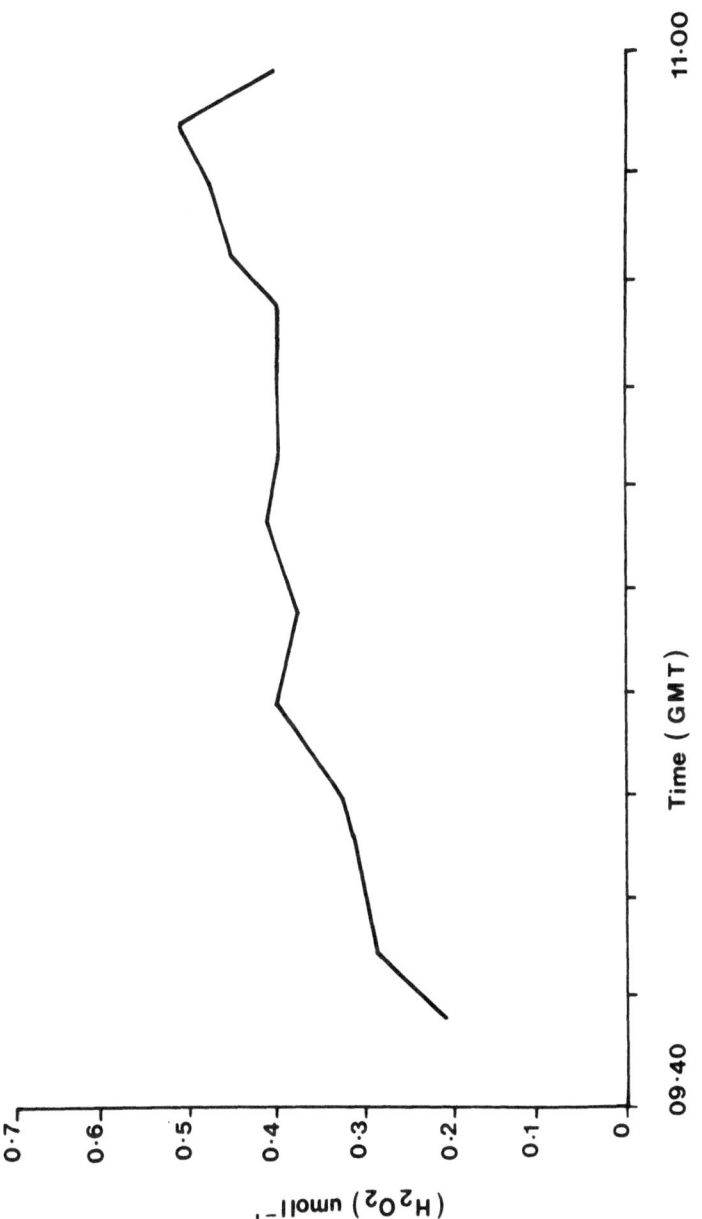

Figure 2. Aqueous phase H_2O_2 measurements at the van site (site 2) for 8 November 1985.

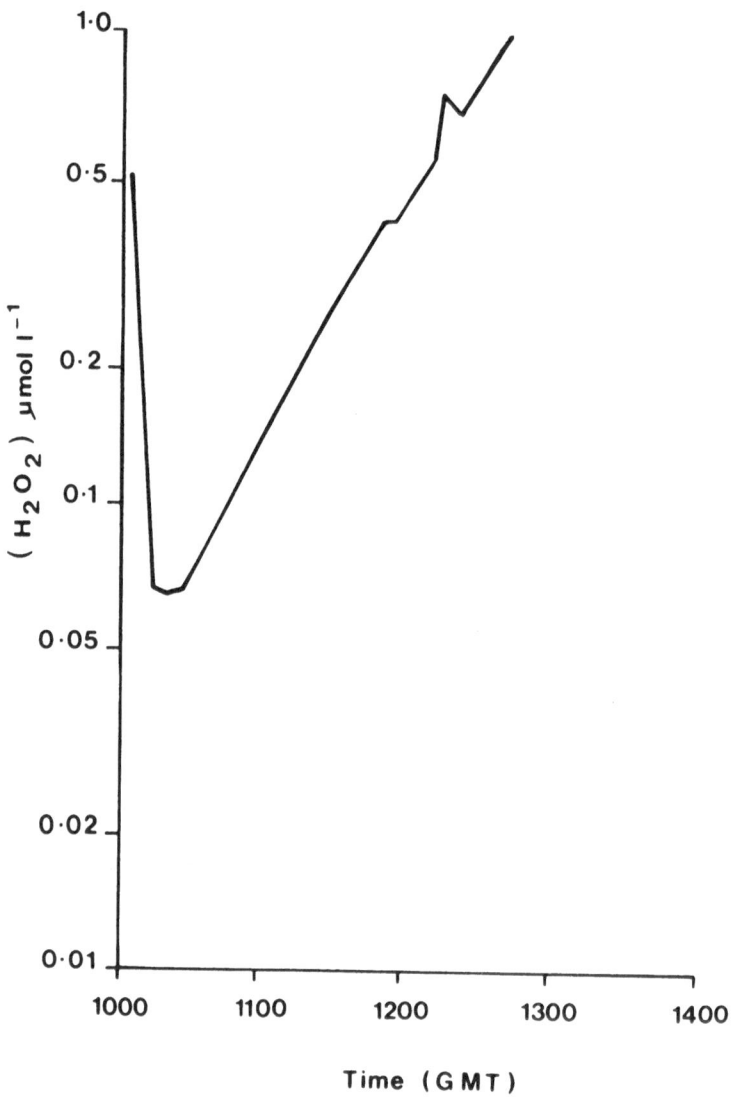

Figure 3. Aqueous phase H_2O_2 measurements at the summit site (site 3) for 8 November 1985.

3.1 Case 1, 8 November 1985

3.1.1 <u>Meteorological situation.</u> During the period when the data
described herein were gathered a SW flow covered the GDF area. The air
trajectory to GDF at 10.00 Z showed that the flow crossed an industrial
area around Merseyside about 200 km to the SW of GDF. The windspeed on
the summit of GDF was around 10 m s^{-1} from the SW throughout. The SO_2
trace at the summit indicated values of around 2 ppbv - 3 ppbv during
the period to 12 noon and around 1 ppbv or less later.

3.1.2 <u>Case study results.</u> During this experiment cloud water H_2O_2
measurements were made continuously at the van site and the summit
site. The results are shown in Figures 2 and 3. Data from the van
site are only available until 11.00 Z when a power failure occured
preventing any further data gathering at this site. Three sets of
batch samples of cloud water were collected: Run 1 approximately 10.00
Z to 11.00 Z, Run 2 11.00 Z to 12.00 Z and Run 3 12 noon to 13.00 Z.
These involved collecting samples from gauges 3 to 8 and each gauge
started collecting for the subsequent run as soon as its sample had
been collected. This process took about 30 minutes, so some overlap
existed between the collection periods for different gauges.
    Figure 4 shows the sulphate profiles up the hill, to the summit
(665 m above the Eden valley) obtained for the 3 runs. The decline in
concentration with height was due to dilution as the droplet radii
increase, most of the sulphate mass being incorporated into the
droplets at cloud base by nucleation scavenging.
    The concentrations of H_2O_2 present were too small for SO_2
oxidation to be directly observed by an increase in sulphate
concentration. The cloud water pH was about 4.8 at the summit and the
temperature was 8°C. Normalising with respect to Cl^- or NO_3^- ions as
shown in Figure 5, for run 3, confirms that sulphate production could
not be detected in the cloud water. Cl^- ion was chosen on this
occasion because it could be attributed to sea salt in the cloud
condensation nuclei. Normalisation on NO_3^- ion was chosen as a further
check because Cl^- ions occur predominantly in larger aerosol particles
unlike most of the sulphate mass. A slight but systematic upward trend
in NH_4^+ was also found in run 1 but not in run 2 where the scatter was
large. This may indicate that some sources of NH_3 gas exist on the
hillside but no direct measurements were available to confirm this.
Alternatively, the droplets may not have reached equilibrium with their
environment with respect to ammonia gas, see Hill (1986).

3.1.3 <u>Interpretation of the hydrogen peroxide measurements.</u> When data
collection started at the summit site at around 10.00 Z the adiabatic
increase in liquid water content between the van site and summit was
calculated to be 0.3 g m^{-3} compared to the observed change of about
0.28 g m^{-3} and so to within the limits of the accuracy of the FSSP ($\pm$
10%), this is close to adiabatic. During this period the H_2O_2
concentration at the summit was very much lower than at the van site,
even allowing for the effects of dilution, suggesting that H_2O_2 was
being removed by reaction with SO_2 which was present in considerable
excess (2-3 ppbv).

198

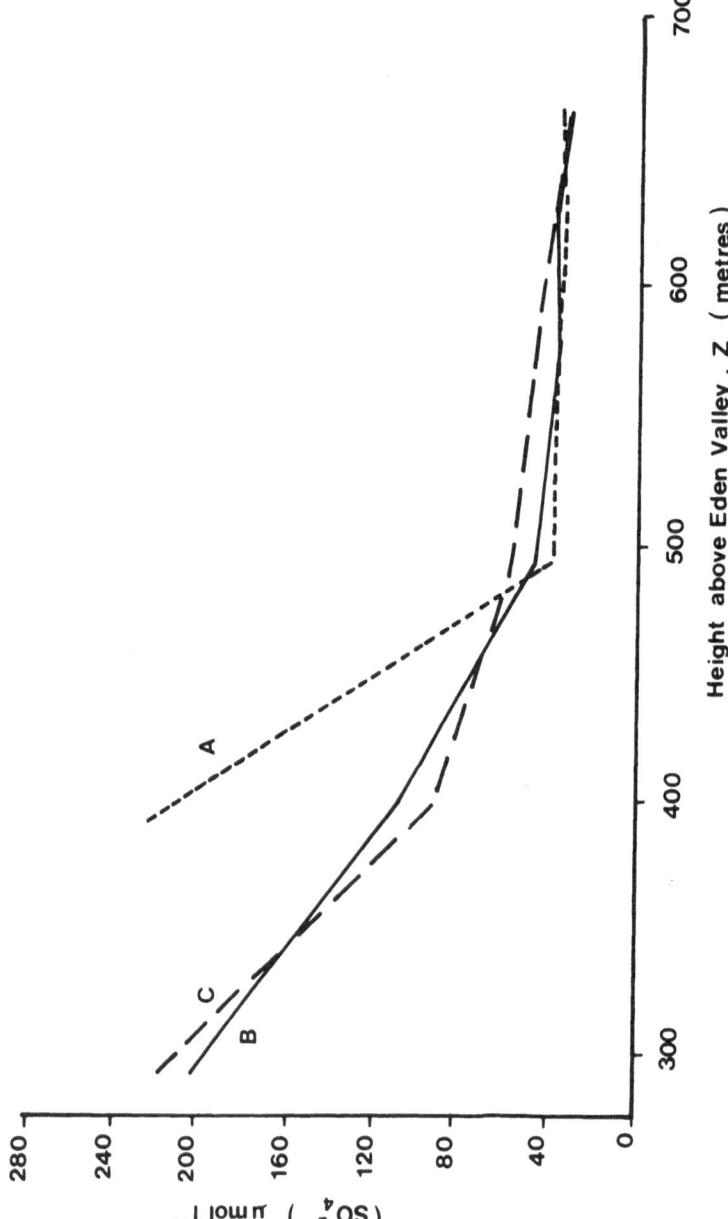

Figure 4. Concentrations of sulphate ion against height Z (metres) above the Eden Valley for the three runs on 8 November 1985. Curve A is for run 1, curve B for run 2 and curve C for run 3.

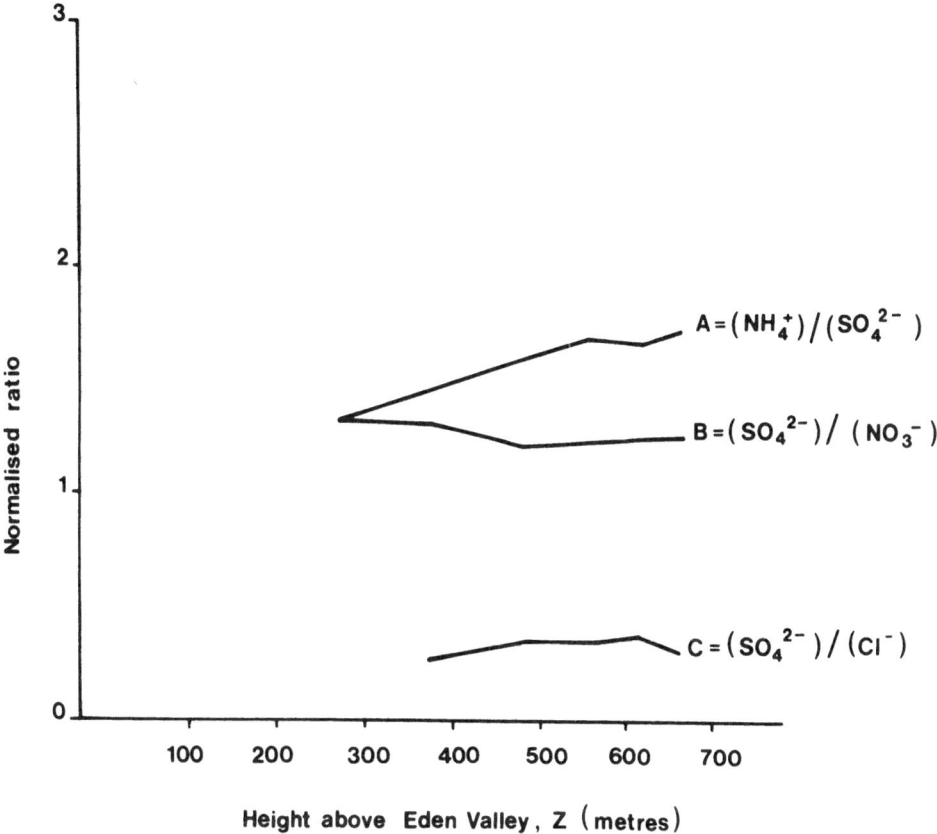

Figure 5.   The normalised profiles for sulphate and ammonium ions
against height Z above the Eden Valley on the 8 November
1985 for run 3.

From about 10.30 Z onwards the liquid water content at the van
site increased as cloud base lowered. At first the summit liquid water
content showed little change and so became increasingly subadiabatic.
This is usually caused by an increase in the rate of dry air
entrainment through the cloud top which mostly affects the cloud close
to the hill top and has been investigated in detail at this site during
the last few years, see for example, Choularton et al. (1986). After
11.00 Z the summit liquid water content decreased due to increasing
entrainment and after midday cloud base rose up the hill eventually
clearing the summit.

During this period the H_2O_2 concentration in the cloud water
collected at the summit increased faster than could be explained by the
decreasing liquid water content. This is believed, in the absence of
any suitable in-cloud production mechanism, to be due to the
introduction of extra H_2O_2 due to the mixing in of dry tropospheric air
from above cloud top. Recent airborne measurements in the US by

Calvert et al. (1985) have shown that whilst most primary pollutants, e.g. SO_2, are found in high concentrations in the boundary layer the concentration of H_2O_2 may be several times higher in the free troposphere. Further evidence to support this hypothesis is presented below.

3.1.4 Calculation of a field reaction second order rate constant for SO_2 and H_2O_2.

The model of Hill, Choularton and Penkett (1986) was used to investigate whether the observed decline in the H_2O_2 concentration was consistent with laboratory measurements of the reaction rate between SO_2 and H_2O_2. This model was especially designed for this purpose. It incorporates a full description of the airflow over the hill using the treatment of Carruthers and Choularton 1982 and allows for vertical mixing of the cloud air during transit. The chemical composition of the cloud passing a given point on the hill is calculated by integration of the microphysical and chemical equations along a computed representative trajectory from a point at cloud base. Hence, the expected change in H_2O_2 between two points or between below cloud base and one site can be calculated for a given initial concentration and a given oxidation rate constant. Full details of the chemistry in the model may be found in Hill et al. (1986). In this paper we compare a rate constant derived from field measurements with that derived by Martin and Damschen (1981) using laboratory data.

The data used in the following example were from the earlier part of the run when the summit liquid water content was close to the adiabatic level. The input to the model corresponded to the conditions observed at this time. The concentrations of gas phase H_2O_2 entering cloud base and, the rate constant for the reaction between the H_2O_2 and in aqueous solution, and bisulphite were adjusted so that the solution phase concentrations of H_2O_2 predicted at the van and summit sites agreed with those observed (see Figures 2 and 3), the liquid water contents being 0.3 g m^{-3} and 0.58 g m^{-3} respectively. The calculation was performed with 2.5 ± 0.5 ppbv of SO_2 entering cloud base. In the model the chemical calculations start when the liquid water content reaches 0.01 g m^{-3}. In order to maintain a pH close to that measured at the different sites (4.8 at the summit), sufficient NH_3 gas was assumed present in the pre-cloud air, 0.5 ppbv in this case. The cloud base temperature was set so as to reproduce the observed in-cloud temperatures (8°C at the summit). The time interval between droplets being collected by the fibre and entering the H_2O_2 instrument was estimated to be about 4 minutes, however, the microphysical and chemical parameters were approximately constant for time periods longer than this in the near adiabatic cloud consequently this is not expected to have a major effect on these calculations.

Taking account of the uncertainties in the measured quantities a reaction rate constant $K_{H_2O_2} = 2 \pm 1 \times 10^5$ s^{-1} where $K_{H_2O_2}$ is defined by:

$$\frac{d[S_{v1}]}{dt} = \frac{K_{H_2O_2} [H_2O_2][SO_2 \cdot H_2O]}{0.1 + [H^+]} \qquad \text{(Martin and Damschen 1981)}$$

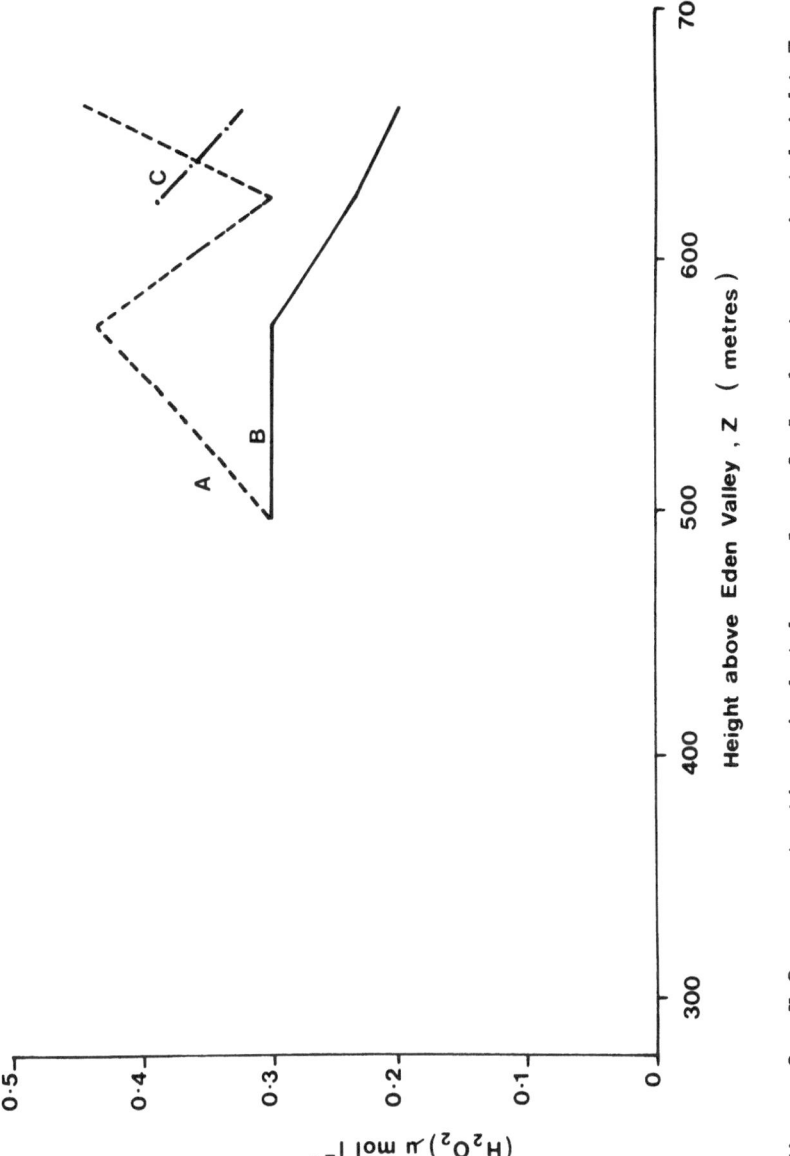

Figure 6. H_2O_2 concentrations in batch samples of cloud water against height Z above the Eden Valley for 7 November 1985 for the three runs. Curve A is for run 1, curve B for run 2 and curve C for run 3.

This was determined with an average cloud temperature of 8.5°C and an average cloud water pH of 4.8.

The result has been converted to this form which is only weakly pH dependent for ease of comparison with laboratory data in view of the large change in cloud water pH near to cloud base where the liquid water content is small and changing rapidly. This value is about 3x larger than the value quoted by Martin and Damschen (1981) when corrected to the same pH and temperature.

3.2 Case 2, 7 November

3.2.1 Meteorological conditions. A WNW airstream covered the GDF area during the period of this run with a ridge of high pressure approaching from the west. Air trajectory analysis confirmed that the flow was from the North Atlantic and consistent with this the air was clean with less that 1 ppbv of SO_2 throughout the period.

The summit site was in cloud from 14.00 Z to 18.00 Z and cloud base tended to rise during this period. The van was in cloud only from 14.30 Z to about 15.45 Z.

3.2.2 Case study results. Batch samples of cloud water were collected from gauges 5 to 8 (675 m to 847 m a.s.l.) for 3 periods: Run 1 14.20 Z to 15.30 Z; Run 2 15.40 Z to 16.35 Z; Run 3 16.35 Z to 17.23 Z. Figure 6 shows the concentrations of H_2O_2 found in the samples and the results may be compared with the continuous monitoring at the summit displayed in Figure 7. The batch sample results showed values about half those obtained during continuous monitoring at sites 2 and 3. This is attributed to decay of the stored samples, however the time interval between collection and analysis of each of the batch samples was identical and so relative values should be correct. These results are the only results presented in this paper affected by the decay of stored samples of H_2O_2.

The cloud samples were stored for later analysis for ionic species. As in case 1 no evidence was found for any sulphate production. The profiles are shown in Figure 8. The pH of the cloud water at the summit site was about 6.8.

3.2.3 Interpretation of the hydrogen peroxide profiles. During run 1 the H_2O_2 concentrations at the summit were higher than at the van site despite an increase in the liquid water content. The equivalent gas phase concentrations, calculated from the batch samples were 2.9 pptv (1 pptv = 1 part of H_2O_2 in 10^{12} parts of air by volume) at the summit and 1.3 pptv at the van site. In each of the other runs, however, a marked decline in the aqueous phase H_2O_2 occurred consistent with dilution.

It is suggested that the extra H_2O_2 observed at the summit site particularly during run 1, was due to the entrainment of free tropospheric air from above cloud top.

This interpretation is consistent with the microphysical data gathered during the experiment. During run 1 the liquid water content at the summit site was about 0.1 g m^{-3} higher than the van site. The

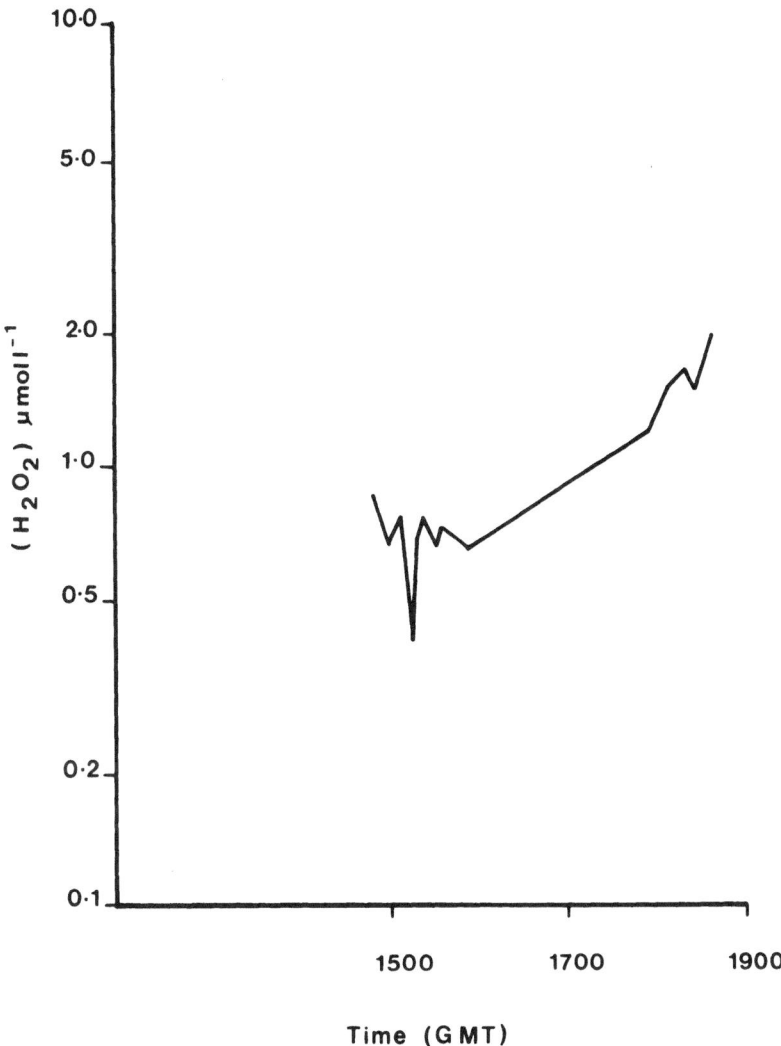

Figure 7. Continuous measurements of aqueous phase H_2O_2 concentration at the summit site (site 3) for 7 November 1985.

adiabatic change between the two was calculated to be 0.25 g m^{-3}. The mean droplet radius increased from 5.75 μm to 7.5 μm at the summit site explaining the marked reduction in sulphate concentration in the cloud water. The droplet number concentration dropped from 190 cm^{-3} to about 120 cm^{-3} at the summit site probably due to dry air entrainment completely evaporating a fraction of the droplets, as discussed by Choularton et al. (1986).

During runs 2 and 3 cloud base was higher up the hill and the effects of entrainment appeared to diminish. The reduction of H_2O_2 concentration with altitude due to dilution became dominant. Evidence

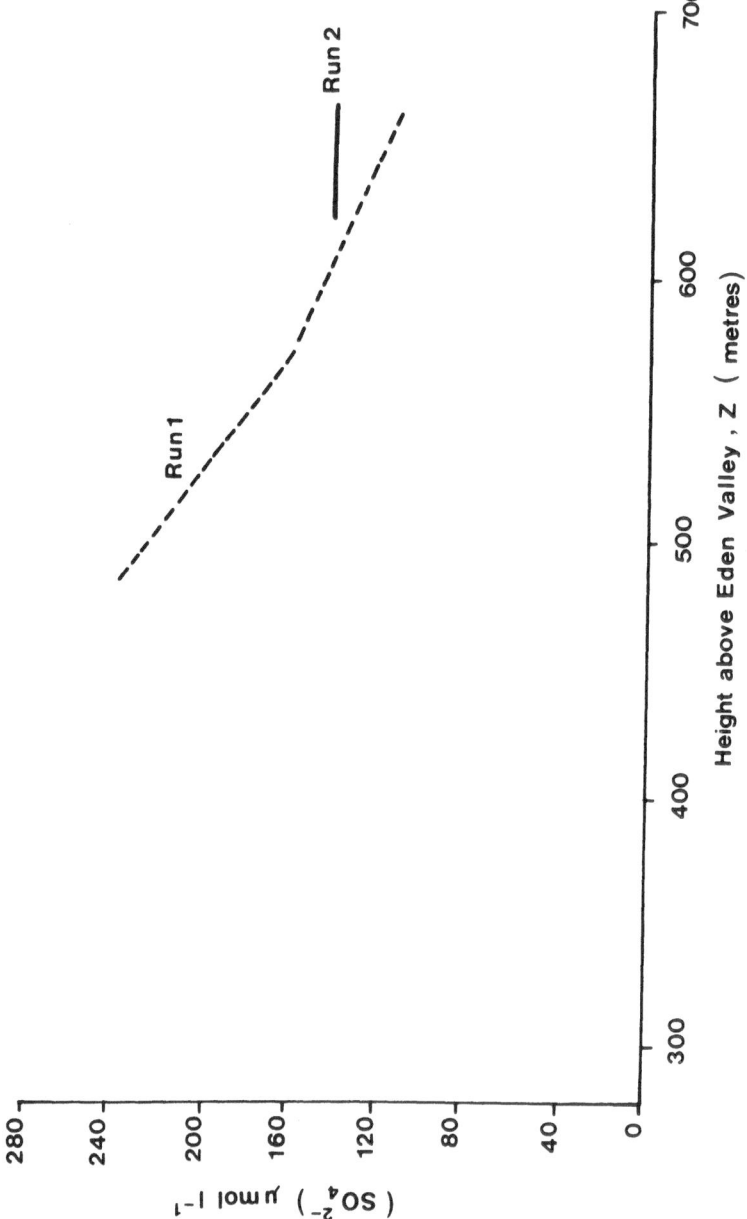

Figure 8. Sulphate ion concentration against height Z (metres) above the Eden Valley for the first two runs on 7 November 1985.

for this was found in the measured droplet number concentration at the summit which increased to values closer to those measured earlier at the van site, also the liquid water contents were closer to the adiabatic values estimated from the occasional approximate measurements of cloud base height.

In this case it was not possible to make comparisons with model predictions as no information could be deduced about the concentration of H_2O_2 in the free tropospheric air due to the decay of the stored samples. Further data of this kind is presented in case 3 and a more

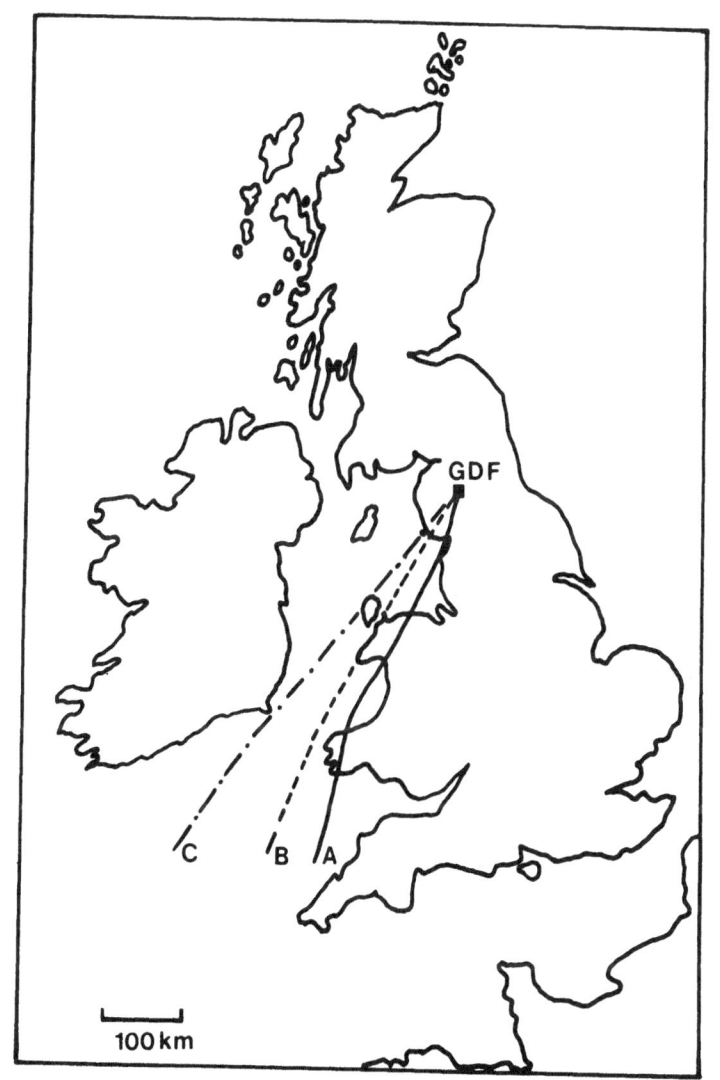

Figure 9. Air trajectories to Great Dun Fell for 16 November 1985 arriving at (A) 11 hrs (b) 13hrs and (c) 16 hrs.

206

Figure 10. The variation with time of SO_2 concentration in ppbv (A) at the summit, liquid water content, L g m^{-3}, at the summit (curve B), and at the van site (curve C) for 16 November 1985.

detailed interpretation is presented. The absence of SO_2 precluded H_2O_2 consumption by the $SO_2 - H_2O_2$ reaction.

3.3 Case 3, 16 November 1985

3.3.1 Meteorological situation. During the experiment a SW airflow covered the area with a cold front in the Irish Sea moving east. Figure 9 shows a series of air trajectories to the site during the course of the experiment. Two sets of batch samples were collected: Run 1 11.30 Z to 15.15 Z and Run 2 15.15 Z to 17.30 Z. Between these two runs the cold front passed over the site marked by a veer in the wind direction, introducing much cleaner air with an oceanic trajectory to GDF. This was also marked by a drop in windspeed from around 15 m s^{-1} to around 8 m s^{-1} at the summit.

3.3.2 Case study results. Figure 10 shows the variation in SO_2 measured at the summit and the liquid water contents measured at the summit and van sites. Examination of the liquid water content traces shows that dry air entrainment was affecting the clouds strongly near the summit (the adiabatic change between the two sites is about 0.3 g/m^3). This is supported by the ozone trace which showed a higher concentration at the summit than at the upstream site, i.e. about 20 ppbv at the summit and 10 ppbv at Wharleycroft. The concentration at the van site was close to that at the upstream site. Based on occasional measurements of cloud base the van site liquid water content was very similar to the adiabatic value. The pH of the cloud water at the hill summit varied from 4.7 at 14.20 Z falling to 4.2 at 15.30 Z and rising to 4.7 by 16.45 Z. The temperature was +0.5°C throughout at the hill summit.

3.3.3 Results of batch cloud water analyses. The batch cloud water samples collected during runs 1 and 2 were stored under refrigeration and analysed by ion chromatography. Figure 11 shows the sulphate profiles and these show a decrease with height following droplet growth as air ascended the hill. As with other case studies during the autumn period normalisation shows that there was no evidence of sulphate production. There was also no evidence of any relative increase in NH_4^+ with height.

During run 1 the sulphate concentration at the summit was considerably lower than at the van site despite the liquid water contents being almost equal. Table 2 summarises the time histories of droplet number concentration and mean radius at the two sites. During run 1 there was an increase in the mean radius between them which caused a dilution of the sulphate but there was also a reduction in the droplet concentration, due to evaporation, following dry air entrainment (data at the summit site were only available from 14.15 Z). It is this reduction in droplet number concentration which accounts for most of the reduction in liquid water content below the adiabatic value. The droplets that had been completely evaporated in this way were converted back to unactivated nuclei and these small aerosol particles would only be collected at low efficiency by the

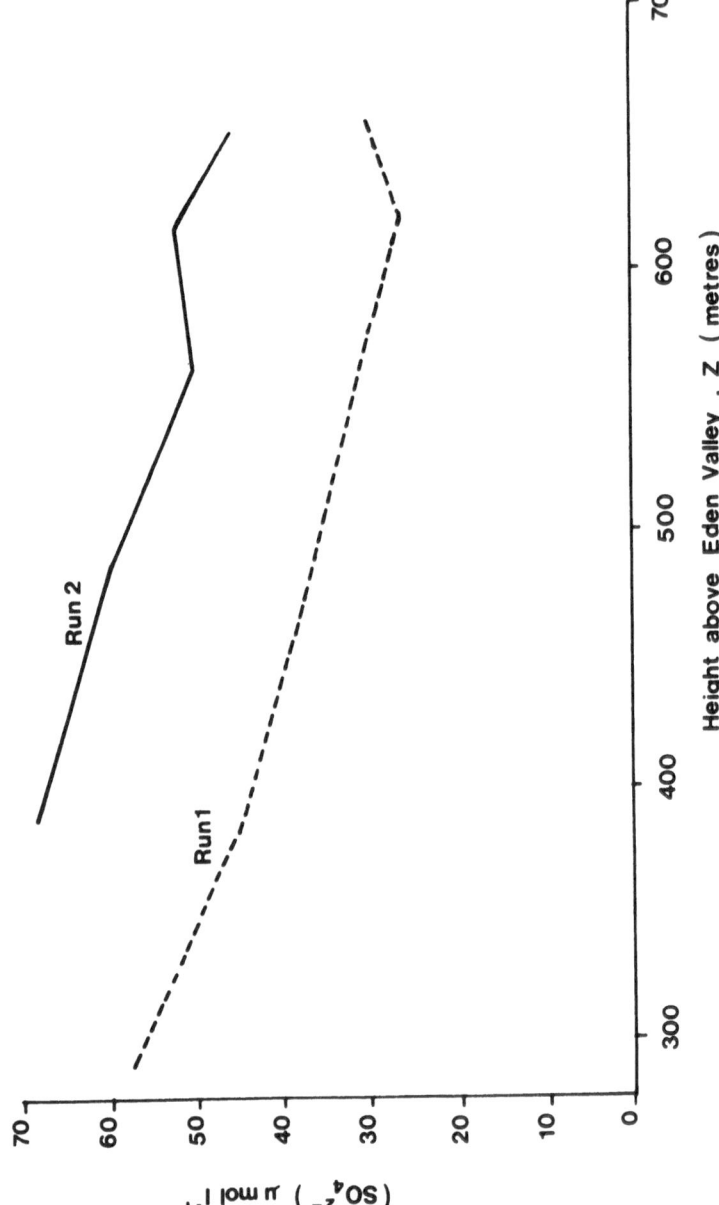

Figure 11.   The variation of sulphate concentration with height above the Eden Valley, Z, for run 1 and run 2 on 16 November 1985.

Table 2.

TIME	VAN SITE (SITE 2)			SUMMIT SITE (SITE 3)		
	L	N	R	L	N	R
14.30	0.32	310	6.2	0.27	250	6.5
15.00	0.40	270	6.5	0.36	170	7.0
15.30	0.18	160	5.0	0.27	120	6.6

Transit time between site 2 and site 3 was about 80 seconds.

cloud water collectors. A similar effect was found during other cloud
water collectors. A similar effect was found during other case
studies. This effect is discussed in recent studies of cloud
microphysics, e.g. Choularton et al. (1986) and Hill and Choularton
(1985). During run 2 the van site liquid water content was lower than
at the summit and so the same arguments do not apply.

During both runs the ratio of Na^+ to Cl^- ions was close to the sea
salt ratio of 0.85 by concentration and hence most of the Cl^- in the
samples was attributed to sea salt.

3.3.4 <u>Interpretation of the hydrogen peroxide profiles.</u> Figure 12
shows the H_2O_2 profiles for runs 1 and 2. In the case of Run 1 the
concentration of H_2O_2 decreased with altitude demonstrating that H_2O_2
was being consumed between the van site and the summit. About 6 ppbv
of SO_2 was measured in the gas phase at the same time.

During run 2, however, the SO_2 levels were much lower (close to
zero) particularly during the later part of the run and, in the absence
of any significant reaction with SO_2, the aqueous phase H_2O_2 level
increased markedly towards the summit of the hill. This was probably
due to dry air entrainment close to the summit introducing extra
gaseous H_2O_2. The interpretation of run 2 is complicated by the trend
caused by a rising cloud base during the experiment. In order to
quantify the suggestions the results were compared to our model
predictions.

3.4 Comparison of Field Data with Model Predictions

Using the model of Hill, Choularton and Penkett (1986), the model was
run with the following input parameters, run 1: cloud base 340 m above
the Eden valley floor, 5 ppbv SO_2, 10 ppbv O_3 and H_2O_2 entering cloud
base 2.0 pptv. The vertical structure of the atmosphere was estimated
using radiosonde data from Aughton, Longkesh and Shanwell stations.
The model was run with a variable entrainment rate into the cloud which
increased with height and was fixed to reproduce the observed liquid
water contents. No measurements were available of gas phase H_2O_2
entering the cloud from above and so the model was run with several
different assumed values. Figure 12a shows a comparison between the

210

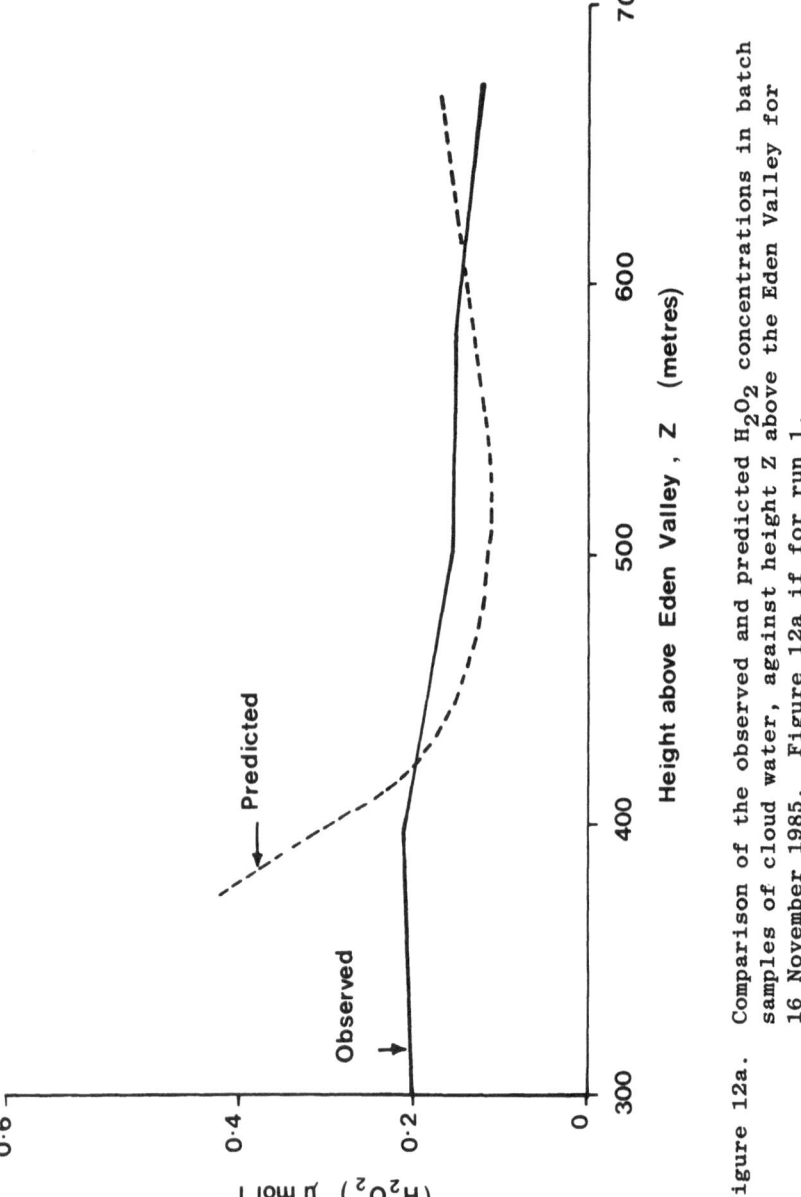

Figure 12a. Comparison of the observed and predicted H_2O_2 concentrations in batch samples of cloud water, against height Z above the Eden Valley for 16 November 1985. Figure 12a if for run 1.

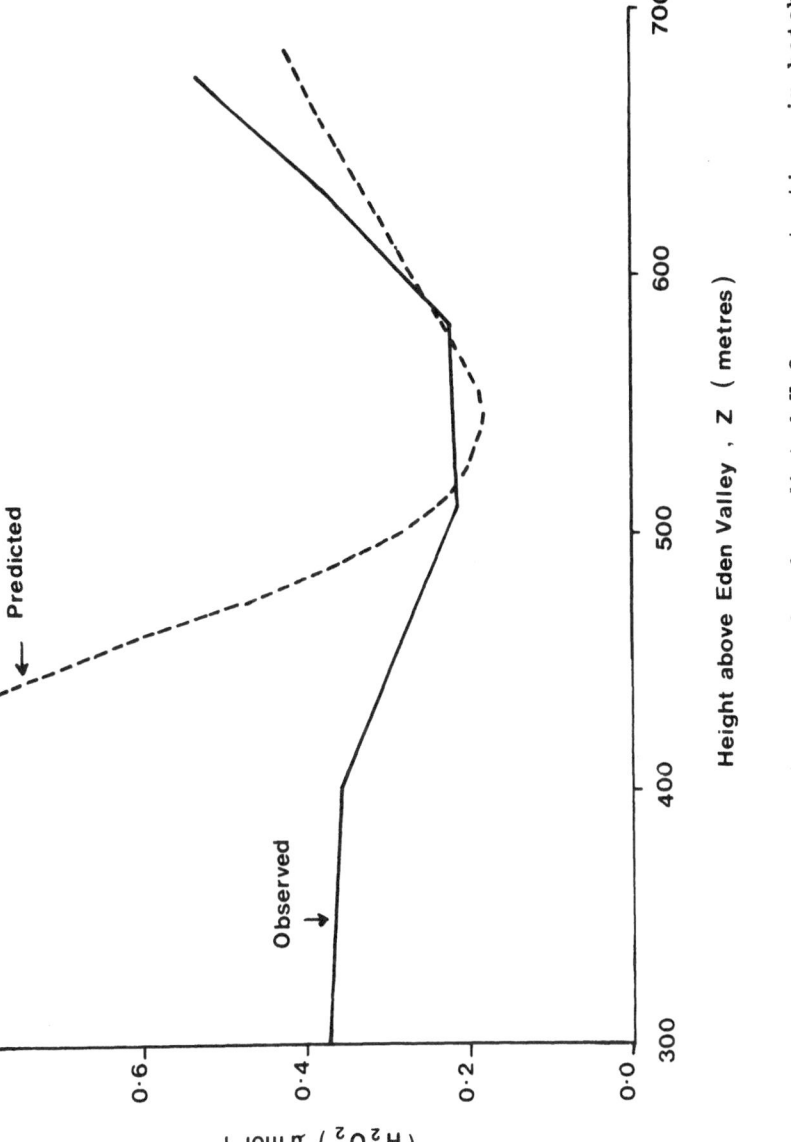

Figure 12b. Comparison of the observed and predicted H_2O_2 concentrations in batch samples of cloud water, against height Z above the Eden Valley for 16 November 1985. Figure 12b is for run 2.

data and the model predictions when H_2O_2 is present in the entrained air in a concentration of 6 pptv.

For run 2 cloud base was set at 440 m. The input gas concentrations were: 0 ppbv SO_2, 10 ppbv O_3 and 2 pptv H_2O_2 entering cloud base. The model was run as described above and the effect of varying cloud base was investigated. Comparison of the model and experimental data is presented in Figure 12b assuming there were 6 pptv of H_2O_2 in the entrained air.

In each case the level of agreement between model predictions and the observations supports the proposition that dry air entrainment introduces extra H_2O_2 and that, when present, SO_2 consumed the oxidant. Agreement between the model predictions and the data is poor close to cloud base. This is associated with the large fluctuations in liquid water content experienced here which are difficult to simulate in model calculations, see Choularton et al. (1986). Hence the batch samples which are long term averages do not show the high concentrations expected very close to cloud base. It should be stressed that only qualitative agreement can be achieved between the model and the observations when entrainment is occurring. However, the comparison clearly indicates the potential importance of entrainment introducing extra H_2O_2 to the cloud system.

4. CONCLUSIONS

The three case studies have shown that when significant concentrations of sulphur dioxide are present in the airstream the H_2O_2 dissolved in the cloud water is consumed at a rate consistent with that predicted by the model of Hill, Choularton and Penkett (1986) using the rate constant larger than that derived by Martin and Damschen (1981) from laboratory measurements. The field reaction rate constant, $K_{H_2O_2}$, is estimated as $2 + 1 \times 10^5$ s^{-1}.

When the cloud is affected by dry air entrainment from aloft, extra H_2O_2 is introduced. The results suggest that the concentration of H_2O_2 in the free tropospheric air entrained may be several times greater than in the boundary layer air in which the cloud forms. The reaction rate data suggest that the H_2O_2-SO_2 reaction is fast and is likely to be oxidant limited. Consequently in many conditions at low cloud water pH, the rate of sulphate production may be controlled by the rate at which H_2O_2 can be entrained into a cloud system as has previously been suggested (e.g. Kelly et al., 1985).

In these experiments carried out in November no significant sulphate production was observed. This was expected with the very low H_2O_2 concentrations observed (equivalent to 5 pptv), however, during the spring months H_2O_2 concentrations in the gas phase should be about 100 times higher, results from similar experiments conducted during these months are being analysed and will be reported.

ACKNOWLEDGEMENTS

This work is supported by the Department of the Environment and the Natural Environment Research Council.

REFERENCES

Ames, D.L. 1983. A method of determining hydrogen peroxide in cloud and rainwater. CERL note no. TPRD/L/2552, N83.

Calvert, J.G., Lazrus, A., Kok, G.L., Heikes, B.G., Walega, J.G., Lind, J. and Cantrell, C.A. 1985. Chemical mechanics of acid generation in the troposphere. Nature 317 , 27-35.

Carruthers, D.J. and Choularton, T.W. 1982. Airflow over hills of moderate slope. Quart. J. Meteorol. Soc. 108 , 603-624.

Carruthers, D.J. and Choularton, T.W. 1986. The microstructure of hill cap clouds. Ibid. 112 , 113-129.

Chameides, W.L. and Davies, D.D. 1982. The free radical chemistry of cloud droplets and its impact on the composition of rain. J. Geophys. Res. 87 , 4863-4877.

Choularton, T.W., Consterdine, I.E., Gardiner, B.A., Gay, M.J., Hill, M.K., Latham, J. and Stromberg, I.M. 1986. Field studies of the optical and microphysical characteristics of clouds enveloping Great Dun Fell. Quart. J. Meteorol. Soc. 112 , 131-148.

Hegg, D.A. and Hobbs, P.V. 1982. Measurements of sulphate production in natural clouds. Atmos. Env. 16 , 2663-2668.

Hill, T.A. and Choularton, T.W. 1985. An airborne study of the microphysical structure of cumulus clouds. Quart. J. Meteorol. Soc. 111 , 517-544.

Hill, T.A., Choularton, T.W. and Penkett, S.A. 1986. A model of sulphate production in a cap cloud and subsequent turbulent deposition onto the hill surface. Atmos. Env. 20 , 9, 1763-1771.

Hill, T.A. 1986. The role of entrainment in sulphate production in cumulus clouds. WMO Conf. Air Pollution Modelling and its Application, Leningrad, USSR.

Kadlecek, J.S., McLaren, N., Camarota, V., Mohnen, V. and Wilson, J. 1982. Cloud water chemistry at Whiteface Mountain. Proc. IVth Int. Conf. on Precipitation Scavenging, Santa Monica, CA. Vol. 1, 76-85.

Kelly, T.J., Daum, P.H. and Schwartz, S.E. 1985. Measurements of peroxides in cloud water and rain. J. Geophys. Res. 90 , 7861-7871.

Kok, G.L., Holler, T.P., Lopez, M.B., Nachtriels, H.A. and Yuan, M. 1978. Chemiluminescent method for hydrogen peroxide in the ambient atmosphere. Env. Sci. Tech. 12 (9), 1072-1076.

Lazrus, A.L., Kok, G.L., Lind, J.A., Gitlin, S.N., Heikes, B.G. and Shetter, R.E. 1985. Automated fluorometric method for hydrogen peroxide. Air. Analyt. Chem. 58 , 594-597.

Maahs, H.G. 1983. Kinetics and mechanism of the oxidation of S(IV) by ozone in aqueous solution with particular reference of sulphur dioxide conversion in non-urban tropospheric clouds. J. Geophys. Res. 88 , 10721-10732.

Martin, L.R. and Damschen, D.E. 1981. Aqueous phase oxidation of sulphur dioxide by hydrogen peroxide at low pH. Atmos. Env. **15** , 1615-1621.

Middleton, P., Kiang, C.S. and Mohnen, V.A. 1980. Theoretical estimates of the relative importance of various urban sulphate aerosol production mechanisms. Atmos. Env. **14** , 463-472.

Kelly, T.J., Daum, P.H. and Schwartz, S.E. 1985. Measurements of peroxides in cloud water and rain. J. Geophys. Res. **90** , 7861-7871.

Penkett, S.A., Jones, B.M.R., Brice, K.A. and Eggleton, A.E.J. 1979. The importance of atmospheric ozone and hydrogen peroxide in oxidising sulphur dioxide in cloud and rainwater. Atmos. Env. **13** , 123-127.

MEASUREMENTS OF AMBIENT SO_2 AND H_2O_2 AT GREAT DUN FELL AND EVIDENCE OF
THEIR REACTION IN CLOUD

G.J. Dollard and B.M.R. Jones
Environmental and Medical Sciences Division
AERE Harwell
Didcot
Oxfordshire
OX11 ORA

A.S. Chandler and M.J. Gay
Department of Physics
University of Manchester Institute of Science and Technology
P O Box 88
Manchester
M60 1QD

ABSTRACT. Details of field measurements of concentrations of SO_2 and
H_2O_2 in cloudwater at Great Dun Fell in Cumbria are given.
   Strong differences were found in H_2O_2 concentrations in cloudwater
collected during autumn and spring periods. Autumn values were
typically well below 1 μmol l^{-1} whereas springtime values were rarely
below 1 μmol l^{-1} with concentrations in excess of 100 μmol l^{-1}
recorded.
   Comparative measurements of aqueous concentrations of hydrogen
peroxide using a chemiluminescence analyser and a fluoresence based
analyser showed good agreement over a range of concentrations.
Residual signals, attributable to the presence of organic peroxides,
determined following catalase treatments, were found to constitute only
a small fraction of the total peroxide concentration.
   The second order rate constants for the reaction $H_2O_2 + HSO_3^-$ in
cloudwater, derived from the decay of H_2O_2 ranged from 4.2×10^2 to
1.84×10^4 l mol^{-1} s^{-1} for a pH range of $4.0 - 4.5$.

1. INTRODUCTION

For several years now the aqueous phase oxidation of sulphur dioxide by
hydrogen peroxide has been acknowledged as an important mechanism for
the production of acidity in clouds and rainwater (Penkett et al.,
1979, Middleton et al., 1980, Chameides and Davies, 1982). The
reaction between SO_2 and H_2O_2 in solution is particularly important,
unlike those with oxygen and ozone, the rate of reaction increases with

M. H. Unsworth and D. Fowler (eds.), Acid Deposition at High Elevation Sites, 215–229.

decreasing pH (Martin and Damschen, 1981). Below a pH of about 5.0 the hydrogen peroxide reaction will oxidise SO_2 faster than will ozone or reaction with oxygen in the absence of large concentrations of metal catalysts (Penkett et al., 1979).

Recent developments of sensitive instruments capable of detecting aqueous H_2O_2, e.g. Ames (1983), Lazrus et al. (1985) have facilitated field measurement of hydrogen peroxide in cloud and rainwater (Kadlecek et al., 1983, Kelly et al., 1985, Chandler et al., in prep.). Evidence has accumulated, in the form of an anti-correlation between SO_2 and H_2O_2, of the reaction of these species in cloud and rainwater.

Reaction rates have been determined for the SO_2 + H_2O_2 reaction in laboratory studies, e.g. Penkett et al. (1979), Martin and Damschen (1981) but it is only comparatively recently that field studies of the types described at this meeting have been instigated in order to generate data on field reaction rates.

It is important to obtain field based data on reaction rates as these contribute to a fuller understanding of the formation of acidity in clouds. This will allow a more realistic assessment of emission control strategies proposed for SO_2, as an attempt to reduce environmental damage in ecologically sensitive regions.

To obtain a comprehensive understanding of the formation of acidity from SO_2 oxidation, the reaction rates should be known for the range of air qualities likely to be encountered. As the detailed modelling study of Hough (1986) has shown, the source and quality of an air mass can markedly affect the chemistry of a cloud and influence the relative importance of oxidising species.

This paper will describe some of the measurements made at Great Dun Fell as part of a cooperative study with UMIST and ITE designed to yield information on the rate of reaction of SO_2 with H_2O_2 in cloudwater. The results presented relate principally to a recent field campaign during spring 1986.

2. MATERIALS AND METHODS

2.1 Experimental site

The experimental site and layout are described in detail elsewhere in this volume (Choularton, 1988). The measurements described here were made at the summit site some 847 metres above sea level. The equipment used was housed in a mobile laboratory which was connected to the main laboratory housing for power and data logging purposes.

Figure 1 represents schematically the arrangements of the principal items of equipment. In addition to the main measurements of SO_2, H_2O_2 and pH, ozone concentrations were monitored continuously, periodic hydrocarbon samples were taken and various meteorological parameters recorded as detailed in Figure 1.

The mobile laboratory was positioned near to the south westerly leading edge of the Fell allowing unperturbed measurements in the captive cap cloud.

2.2 Measurement of SO_2 concentrations

Gas phase concentrations of SO_2 were recorded continuously using a Meloy SA 285E flame photometric detector fitted with a fast response burner block which had a signal rise time to 95% of less than 30 seconds. As depicted in Figure 1 the inlet to the analyser was connected via teflon tubing to an inverted plastic funnel.

In order to maintain a satisfactory performance the instrument required thermostatic housing and regular zero and span checks. Zero checks were performed using a column of glass beads coated with a 0.8 M solution of KCO_3 in 10% glycerol. This was found to be a particularly effective blank system. Span checks were performed using a Meloy CS 10-3 permeation tube calibrator.

2.3 Hydrogen peroxide measurements

The majority of hydrogen peroxide determinations were made using a chemiluminescence instrument of the type described by Ames (1983), this monitors light from the chemiluminescent reaction between luminol (5-amino-2, 3-dihydro-1, 4-pthalazinedione) and hydrogen peroxide, a reaction catalysed by microperoxidase. Aqueous samples are mixed with EDTA to remove interfering transition elements. Other potential interferences have been evaluated by Ames (1983). Flow conditions and reagent strengths identical to those given by Ames (1983) were used.

Following publication of details of a fluorescence based analyser by Lazrus et al. (1985), a second device was built which utilised the reaction between parahydroxyphenylacetic acid and hydrogen peroxide in the presence of peroxidase, to form a fluorescent dimer. This was excited at 320 nM and the fluoresence measured at 400 nM on a commercial filter fluorimeter (Perkin Elmer Model LS 2B). Reagent strengths and flow conditions were the same as described by Lazrus et al. (1985). The fluoresence based instrument was used for some comparative tests with the chemiluminescence instrument.

To investigate the presence of organic peroxides, representative samples of cloudwater were treated with catalase as detailed by Lazrus et al. (1985), then reanalysed for any residual signal. Both hydrogen peroxide analysers were calibrated daily using fresh hydrogen peroxide standards. All reagents were made up in bulk in deionised water in order to minimize the effect of batch to batch variations in the activities of the enzyme reagents. Zero checks were made regularly during each sampling run.

Figure 1 illustrates the arrangements of the two analysers during their operation. Cloudwater was collected using a gauge of the design used on Whiteface Mountain by Kadlecek et al. (1983), positioned on the roof of the mobile laboratory. This gauge was strung with Teflon strands 50 cm long and 0.04 cm diameter. Cloudwater impacted on the vertical strands, was collected by a teflon funnel and taken directly into the laboratory. The water was directed routinely into the chemiluminescence instrument with any excess cloudwater flow collected for subsequent chemical analysis. When the fluorescence based analyser was operated the sample flow was split and directed through both

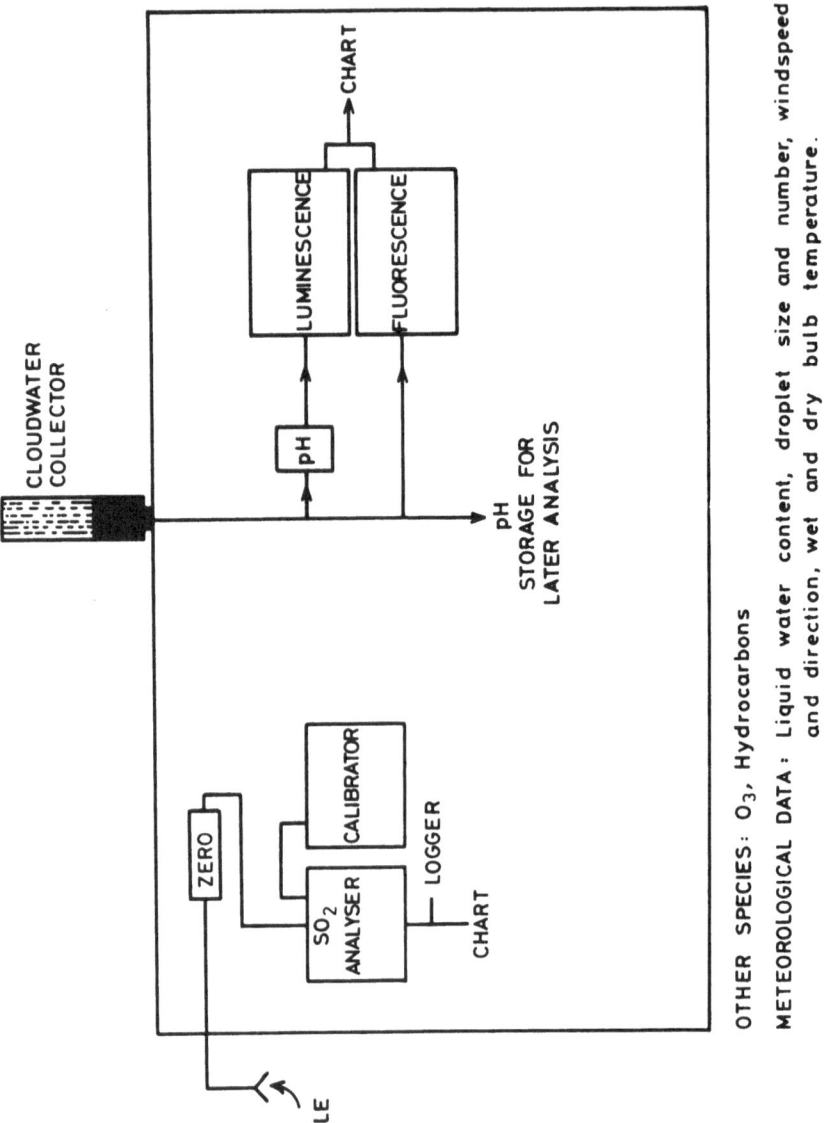

OTHER SPECIES: O₃, Hydrocarbons

METEOROLOGICAL DATA: Liquid water content, droplet size and number, windspeed and direction, wet and dry bulb temperature.

FIGURE 1. SCHEMATIC REPRESENTATION OF INSTRUMENT ARRANGEMENT

analysers simultaneously. The mixing requirements of the fluorescence technique resulted in a delay time of about 3 minutes between the two types of analyser.

The pH of the cloudwater samples were either monitored continuously at the sample inflow line to the chemiluminescence analyser using a micro combination pH probe or in the batch samples, collected for later chemical analysis, using a standard combination probe.

2.4 Derivation of second order rate constants

The reactions of interest in this matter are the following

$$SO_2\ g \xrightleftharpoons{H_2O} SO_2\ aq \tag{1}$$

$$SO_2\ aq \xrightleftharpoons{H_2O} HSO_3^- + H^+ \tag{2}$$

$$HSO_3^- + H_2O_2 \longrightarrow SO_4^{2-} + H^+ + H_2O \tag{3}$$

For reaction 1 the SO_2 aq term was derived from the Henry's Law relationship:

$$SO_2\ aq = K_H \cdot SO_2\ g \tag{4}$$

where K_H is the Henry's Law coefficient calculated using the expression, (NBS 1965):

$$K_H(5.0°C) = 1.23 \times [3120\ (1/T - 1/298)] = 2.61$$

The concentration of HSO_3^- in cloudwater was then determined for the equilibrium established in reaction 2:

$$K = \frac{[H^+]}{[SO_2\ aq]} \frac{[HSO_3^-]}{[H_2O]} \tag{5}$$

using the $[H^+]$ data from pH measurements and the constant K calculated using the relationship (NBS 1965):

$$K_{(5.0°C)} = 1.7 \times 10^{-2}\ exp\ [2090\ 1/T = 1/298)] = 0.028$$

By combination of equations 5 and 4 the bisulphite term may be calculated as

$$[HSO_3^-] = \frac{k\ (K_H\ [SO_{2g}])}{H^+} \tag{6}$$

As the sulphur species is always in stoichiometric excess the decay of H_2O_2 was assumed to follow pseudo first order kinetics and the 1st order rate constant was derived for H_2O_2 loss during transit up the Fell. It is assumed that levels of H_2O_2 entering the cloud do not vary substantially during the transit period.

For each of the case studies average transit times of droplets from cloud base to the summit or van sites were calculated using the model of Hill et al. (1986). This model computes the airflow pattern over the hill and calculates the evolution of the cloud droplet spectrum taking account of turbulent exchange in the boundary layer. As input, this model requires the vertical structure of the atmosphere which is divided into three layers of constant stability (Scorer parameter) μ_i, with the interfaces between the layers at $I(-\infty)$ and $J(-\infty)$ well upstream of the hill. This determines the pattern of the airflow over GDF. The surface geostrophic wind and height of cloud base are then supplied to calculate the transit times to the summit.

The first order rate constant for H_2O_2 decay during the transit time from cloud base to the summit was calculated and used to derive the second order rate constant. The generalised rate expression for reaction 3 is

$$\frac{d[SO_4^{2-}]}{dt} = K_2 \, [H_2O_2] \, [HSO_3^-] \tag{7}$$

where K_2 is the second order rate constant. It should be possible to derive K_2 in terms of SO_4^{2-} production, SO_2 loss and the decay of H_2O_2. In the present case we cannot use SO_2 data because of uncertainty associated with the calibration of the SO_2 analyser at the base monitoring site. The use of the SO_4^{2-} data is currently being evaluated, so here we derive K_2 on the basis of H_2O_2 loss in cloudwater.

On the basis that sulphate production is equivalent to H_2O_2 loss we can say

$$- \frac{d[H_2O_2]}{dt} = \frac{d[SO_4^{2-}]}{dt} \simeq K_1 \, [H_2O_2] \tag{8}$$

where K_1 is the pseudo first order rate constant, and from equation 7:

$$K_2 = \frac{K_1}{[HSO_3^-]} \quad 1 \text{ mol}^{-1} \text{ s}^{-1} \tag{9}$$

3. RESULTS AND DISCUSSION

3.1 Spring and autumn levels of H_2O_2 in cloudwater

Figure 2 summarises some data on seasonal levels of hydrogen peroxide measured in cloudwater at Great Dun Fell. The figures relate to peak values measured during hourly intervals.

The upper part of the figure illustrates data collected during three periods during autumn 1985. The highest peak concentrations measured were 7 to 8 μmol 1^{-1} however for the most part concentrations during this autumn period were well below 1 μmol 1^{-1}.

The lower part of the figure presents data from two sets of springtime measurements; it can be seen that concentrations were

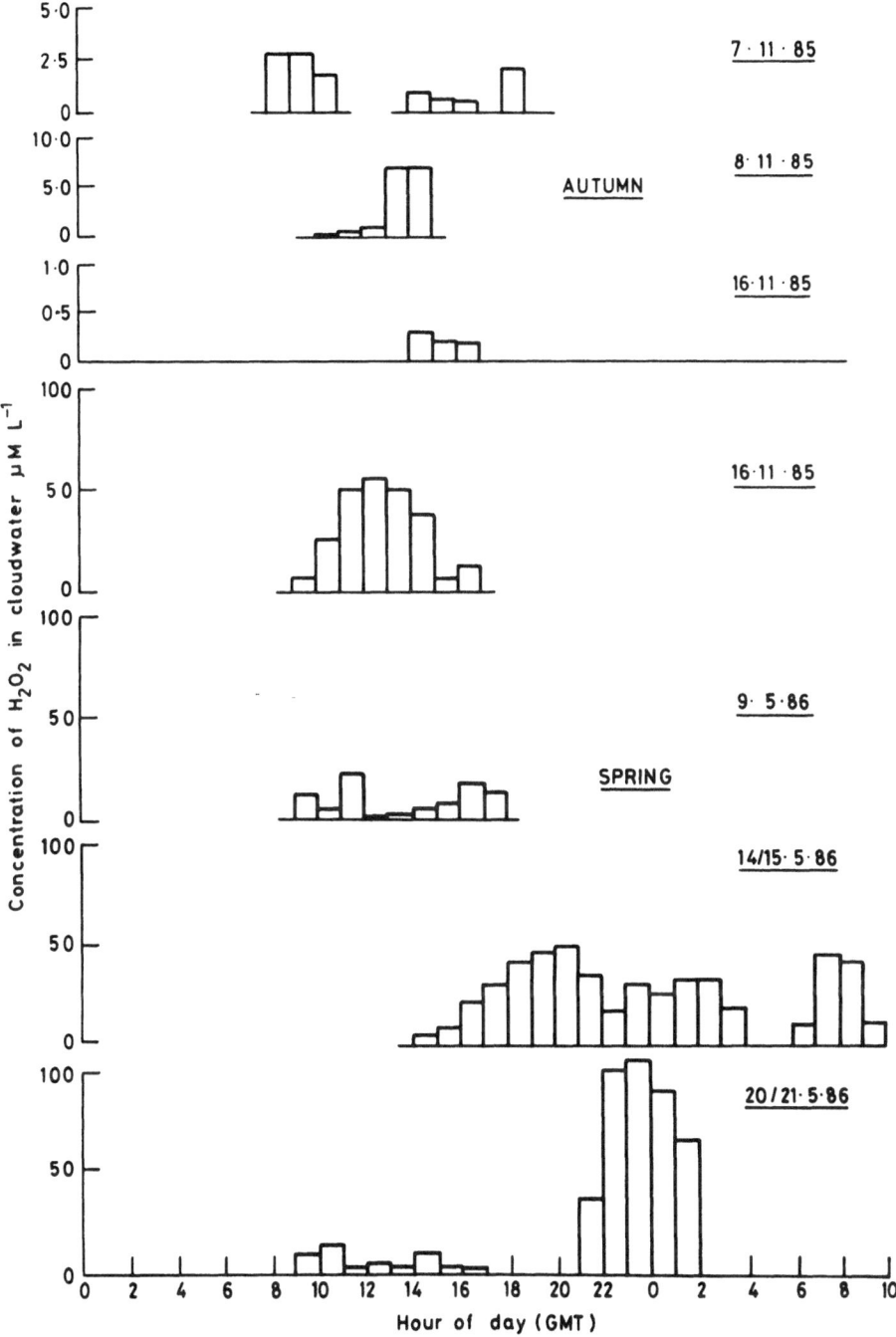

FIGURE 2. PEAK H₂O₂ CONCENTRATION (μM L⁻¹) IN CLOUDWATER AT GDF

considerably higher than during the autumn with peak levels in excess of 100 μmol 1^{-1}. Generally the H_2O_2 concentrations only rarely fell below 1 μmol 1^{-1}. Such seasonal differences would be consistent with the formation of hydrogen peroxide by photochemical processes involving free radical species.

In view of the photochemical origin of H_2O_2 precursors a diurnal variation in H_2O_2 concentrations might be expected. The data are not extensive enough to allow any identification of any such effect and it is interesting to note that very high concentrations of H_2O_2 were measured during the late evening period of the fourteenth and twentieth of June, nineteen eighty six.

The lower concentrations of H_2O_2 prevalent during the autumn in cloudwater samples may be of significance as lower autumn/winter concentrations of H_2O_2 could result in other species such as ozone assuming more importance in terms of SO_2 oxidation during such periods.

3.2 Comparison of hydrogen peroxide analysers

The two types of hydrogen peroxide analysers were operated simultaneously for periods during the autumn 1985 and spring 1986 campaigns at Great Dun Fell. Figure 3 shows a time trend plot for H_2O_2 in cloudwater. It is evident that the trends recorded by each analyser are very similar and allowing for the delay in the fluoresence analyser agreement is quite good. There is a period from about 1115 when the fluorescence instrument lost sensitivity, in this particular case this was due to a blockage in one of the flow lines which was not immediately apparent and was not fully cleared until about 1500.

Figure 4 summarises comparative measurements taken from the continuous traces made during spring 1985 plus some autumn 1985 data. The points marked as crosses relate to the autumn 1985 data. The dashed line represents the one to one fit and the solid line the fit to all the data points. Overall the fit is very good with a similar scatter over the whole range of concentration values. There is a slight indication of the luminol method giving higher values than the fluorescence based device. The enzyme catalysed luminol method of Ames (1983) compares more closely with the fluorescence method than does the copper catalysed system tested by other workers (Kadlecek et al., 1985, Lazrus et al., 1985).

From the comparisons carried out by Lazrus et al. (1985) there is a suggestion that positive interference due to organic peroxides could account for noted discrepancies between the chemiluminescence and fluorescence techniques. Previous tests carried out at Great Dun Fell have shown only a small contribution of organic peroxides (A. Chandler, unpublished data). Further checks were carried out as part of the instrument intercomparisons. The results are summarised in Table 1. It is evident that the residual signal attributed to organic peroxides constituted only a small fraction of the total and showed little correlation with the total peroxide originally present. An interesting feature of the data for 20.5.86 is the high value of 8.6% which was found in a sample taken immediately following a very heavy rain shower. There is an indication of higher organic peroxide

TABLE 1 Organic peroxides in cloudwater samples from spring 1986.

Date of sample	Original H_2O_2 concentration $\mu mol\ 1^{-1}$	% contribution from organics
9.5.86	12	1.4 *
	10	1.9 *
	15	0.6
	1.2	0.9
	3.6	0.6
	10	0.2
	8	0.3
14.5.86	4.5	0.9
	6	0.8
	20	0.2
	31	0.2
15.5.86	13	0.2 *
	22	< 0.1 *
	18	< 0.1 *
19.5.86	10	1.8
	30	0.2
	30	0.1
	8	0.7
	70	< 0.1
20.5.86	9	0.2
	1.4	8.6 **
	1.2	1.7 *
	110	0.2

*   precipitation during sample collection
** heavy precipitation during sample collection

224

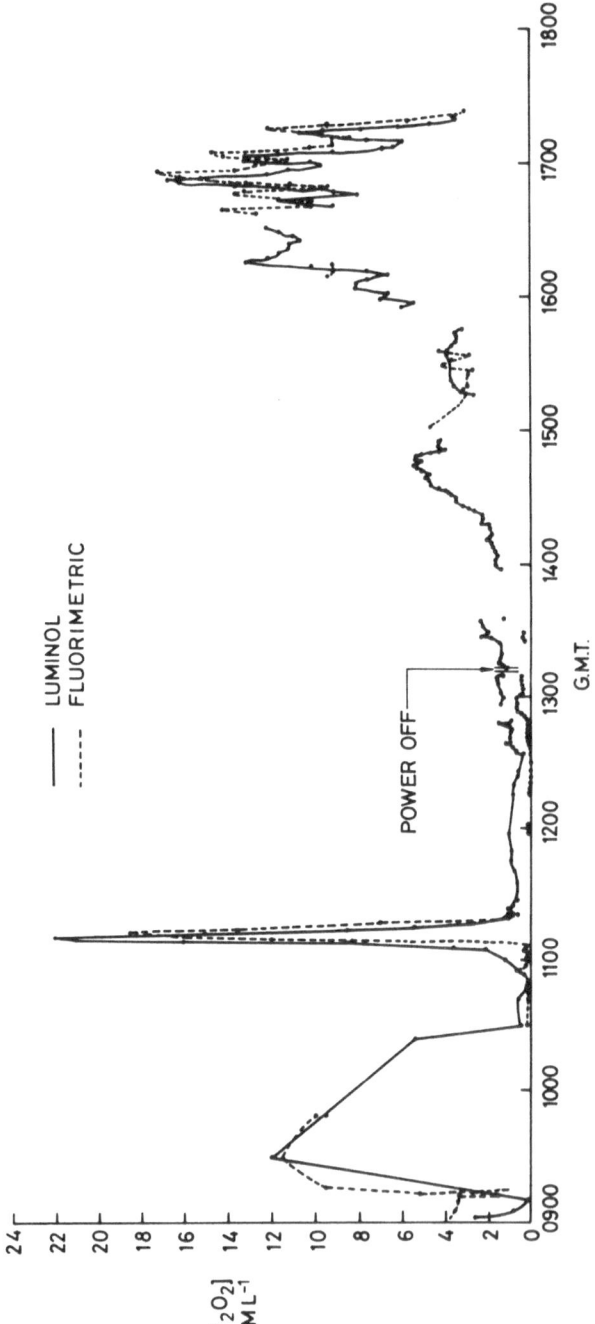

FIGURE 3. H₂O₂ CONCENTRATION AT G.D.F. SUMMIT 9/5/86

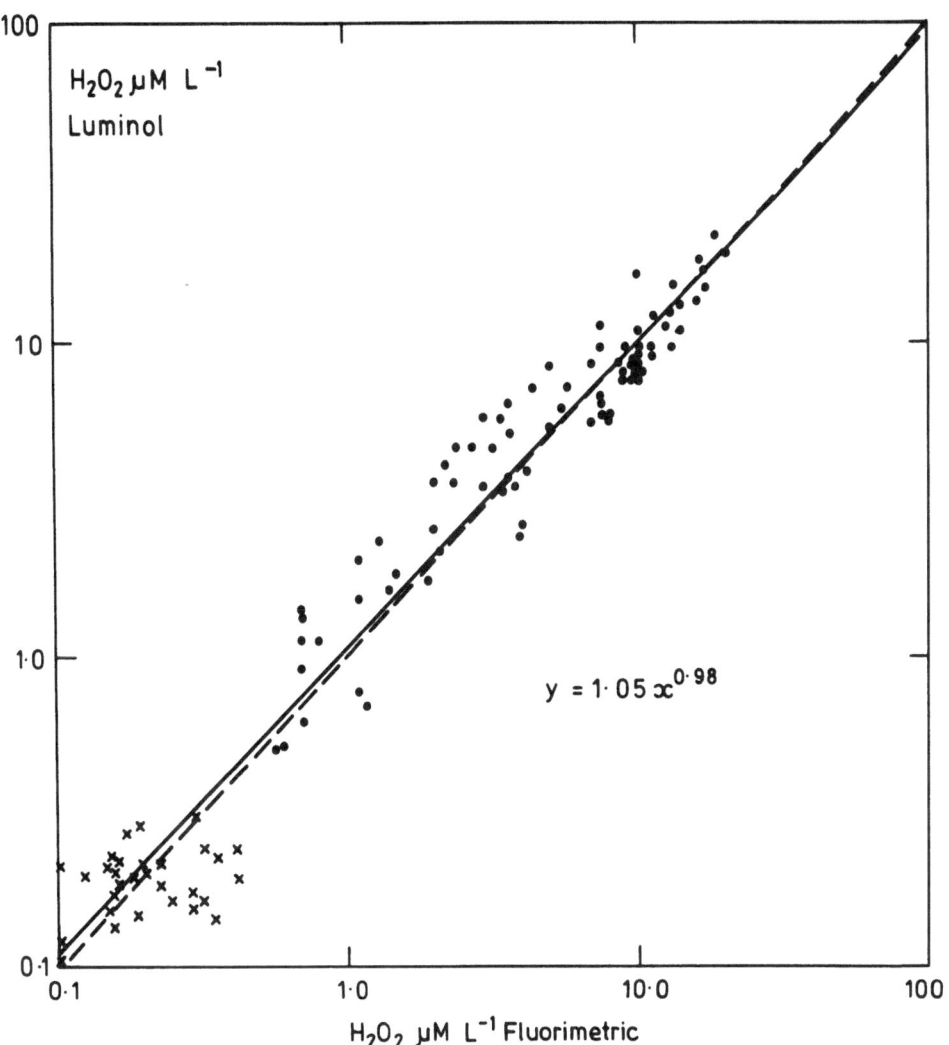

FIGURE 4. COMPARISON OF H₂O₂ ANALYSERS

226

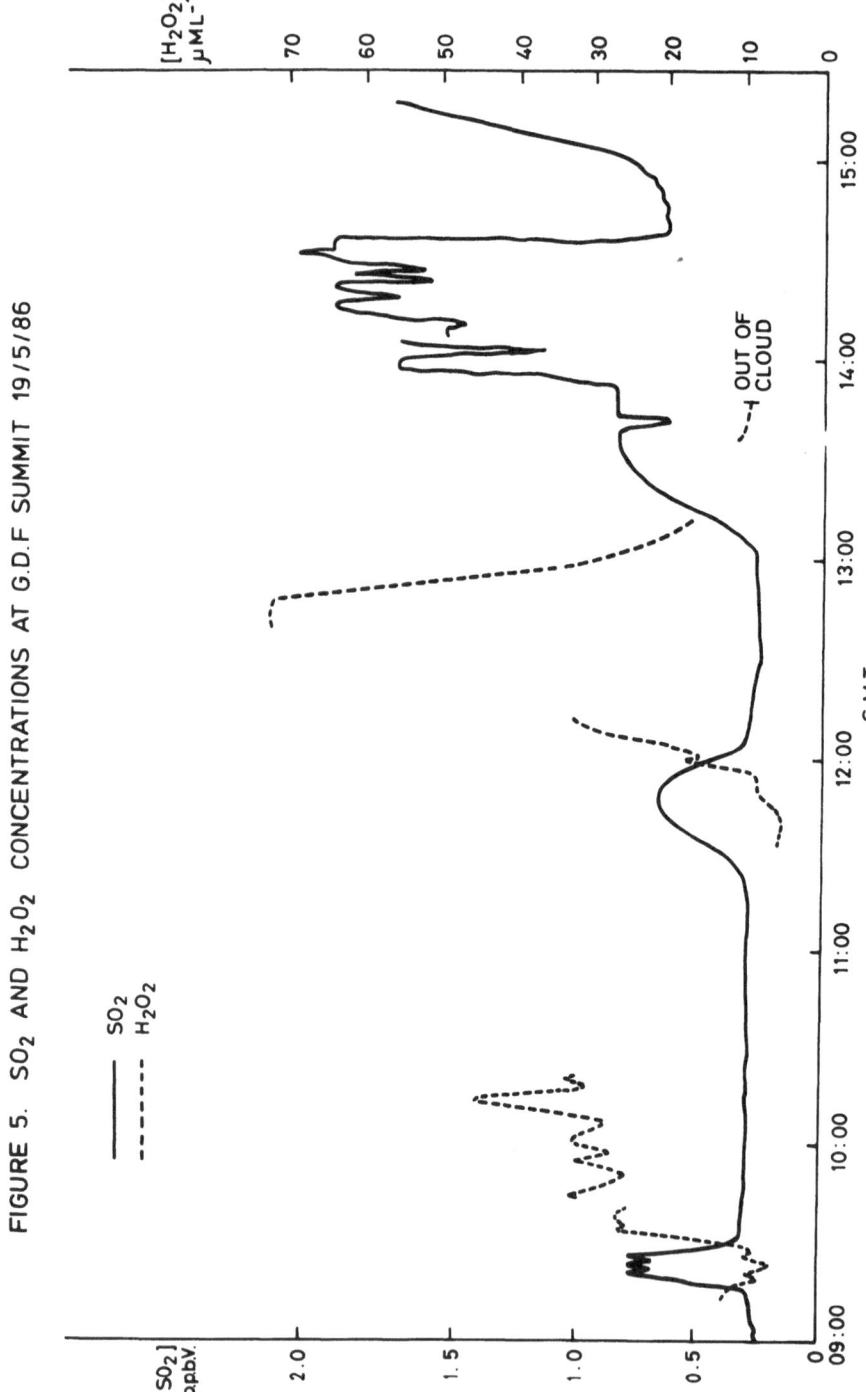

FIGURE 5. SO₂ AND H₂O₂ CONCENTRATIONS AT G.D.F SUMMIT 19/5/86

TABLE 2   Second order rate constants for SO_2 oxidation by H_2O_2 (5.0°C)

Date	Time GMT	Transit time seconds	$[H_2O_2]$ $\mu mol\ l^{-1}$	SO_2 loss ppbv	pH	K_2 $l\ mol^{-1}\ s^{-1}$
9.5.86	1200	160	15	9	4.0	2.6×10^3
19.5.86	0920	460	25	0.75	4.3	4.8×10^3
19.5.86	1200	300	35	0.75	4.4	7.1×10^3
19.5.86	1300	80	70	1	4.4	1.8×10^4
20.5.86	1700	280	5	8	4.5	420

228

concentrations in samples contaminated by rain on 9.5.86. However the
effect is not suggested in the data of 15.5.86. It is conceivable that
rain falling from above the cap cloud washed out greater concentrations
of organic peroxides or species capable of causing positive
interference, from a zone of photochemically active air above cloud
top.

3.3 Trends in SO_2 and H_2O_2 and derived rate constants

Figure 5 is an example of a time trend for SO_2 and H_2O_2 aq which were
used to derive second order rate constants.

On the 19.5.86 three periods were identified for further analysis,
these corresponded approximately to 0920, 1130 and 1300 hrs on Figure
5. Cloud was present continuously from the commencement of peroxide
analysis at 0910 hrs until clearing at about 1350 hrs. During this
period a great deal of structure is evident for both SO_2 and H_2O_2.

The second order rate constants derived from these curves, and
others not illustrated, are summarised below in Table 2. The values
obtained are similar to those derived from laboratory studies (Martin
and Damschen, 1981) ranging from 4.2 x 10^2 to 1.8 x 10^4 1 mol^{-1} s^{-1} for
a pH range of 4.0 to 4.5.

4. CONCLUSIONS

The ground-based study at Great Dun Fell has been successful in
generating data on seasonal trends in cloudwater hydrogen peroxide
concentrations and evidence for the reaction of SO_2 with H_2O_2 in cloud.

Examples of time trends for SO_2 and H_2O_2 presented here show a
great deal of structure and as a preliminary analysis have allowed the
derivation of rate constants for the reaction H_2O_2 + HSO_3^- under field
conditions.

Further analysis of the large body of data derived from the latest
field campain should significantly further improve our knowledge of the
oxidation of SO_2 in cloud systems.

ACKNOWLEDGEMENT

This work was supported by the Department of the Environment and the
Natural Environment Research Council.

REFERENCES

Ames, D.L. 1983. 'A method of determining hydrogen peroxide in cloud
    and rainwater'. Central Electricity Generating Board Report No.
    TPRD/L/2522/N83.
Chameides, W.L. and Davies, D.D. 1982. 'The free radical chemistry of
    cloud droplets and its impact upon the composition of rain'. J.
    Geophys. Res. 87 (C7), 4863-4877.

Chandler, A.S., Choularton, T.W., Dollard, G.J., Gay, M.J., Hill, T.A., Jones, A., Jones, B.M.R., Morse, A.P., Penkett, S.A. and Tyler, B.J. 1986. 'A field study of the cloud chemistry and cloud microphysics at Great Dun Fell' (in press, Atmospheric Environment).

Choularton, T.W. 1988. 'Cloud micro physical processes relevant to cloud chemistry'. This volume.

Hill, T.A., Choularton, T.W. and Penkett, S.A. 1986. 'A model of sulphate production in cap cloud and subsequent turbulent deposition onto the hill surface'. Atmos. Environ. 20 (9), 1763-1771.

Hough, A.M. 1986. 'A computer modelling study of the chemistry occurring during cloud formation over hills'. AERE report R-12180. The Harwell Laboratory, OX11 ORA England, pp. 43.

Kadlecek, J., McLaren, S., Camarota, N., Mohnen, V. and Wilson, J. 1983. 'Clcudwater chemistry at Whiteface Mountain'. Proc. IVth Int. Conf. on Precipitation Scavenging, Santa Monica A Vol 1.

Kelly, T.J., Daum, P.H. and Schwartz, S.E. 1985. 'Measurement of peroxides in cloudwater and rain'. J. Geophys. Res. 90 (D5), 7861-7871.

Lazrus, A.L., Kok, G.L., Gitlin, S.N., Lind, J.A. and McLaren, S.E. 1985. 'Automated fluorimetric method for hydrogen peroxide in atmospheric precipitation'. Anal. Chem. 57 , 917-922.

Martin, L.R. and Damschen, D.E. 1981. 'Aqueous oxidation of sulphur dioxide by hydrogen peroxide at low pH'. Atmos. Environ. 15 (9), 1615-1621.

Middleton, P., Kiang, C.S. and Mohnen, V.A. 1980. 'Theoretical estimates of the relative importance of various urban sulphate aerosol production mechanisms'. Atmos. Environ. 14 , 463-472.

National Bureau of Standards 1965. 'Selected values of chemical thermodynamic properties'. 1. NBS Tech. note 270-1, 124 pp.

Penkett, S.A., Jones, B.M.R., Brice, K.A. and Eggleton, A.E.J. 1979. 'The importance of atmospheric ozone and hydrogen peroxide in oxidising sulphur dioxide in cloud and rainwater. Atmos. Environ. 13 , 123-127.

WET DEPOSITION AND ALTITUDE, THE ROLE OF OROGRAPHIC CLOUD

D. Fowler, J.N. Cape, and I.D. Leith
Institute of Terrestrial Ecology
Bush Estate, Penicuik, Midlothian EH26 0QB, Scotland

T.W. Choularton, M.J. Gay, and A. Jones
Department of Pure & Applied Physics
University of Manchester Institute of Science and Technology
PO Box 88, Manchester M60 1QD

ABSTRACT. The effects of orography on the composition of rain and cloud water and patterns of wet deposition have been studied at Great Dun Fell in northern England. Measurements were made of rainfall and cloud water composition at 8 altitudes between 250 m and 847 m on the western slopes of the hill throughout the autumn of 1984 and spring 1985.
 With westerly airflow the hill summit is generally enveloped in orographic cloud during rain, and in these conditions the concentration of major ions in rain increases with altitude. On average (20 events) rainfall amount at 847 m (the summit) exceeds that at 250 m by a factor of 2 and concentrations of SO_4^{2-}, NO_3^-, Cl^-, NH_4^+ and H^+ are larger at the summit than at 250 m by factors of 2.2, 3.3, 2.9, 3.1 and 2.9 respectively. Wet deposition therefore increases rapidly with altitude at this site with amounts of major ions in precipitation deposited at the summit exceeding those at 250 m by about a factor of 5.
 The concentrations of all ions in cloudwater samples exceeded those in rain (at the summit by about a factor of 2), and the scavenging of the more polluted orographic cloud by falling rain is the cause of the increase in concentrations of major ions in rain with altitude. Concentrations of most ions in the orographhic cloud decreased with increasing altitude at rates close to those expected from adiabatic increases in cloud liquid water content between 250 m and 847 m.
 For easterly airflow or with blocked flow over the hill no changes in rainfall amount or composition were observed.

1. INTRODUCTION

During the last 15 years research into the meteorological and chemical processes involved in the atmospheric transport and chemical conversion

231

M. H. Unsworth and D. Fowler (eds.), Acid Deposition at High Elevation Sites, 231–257.
© *1988 by Kluwer Academic Publishers.*

of gaseous pollutants has identified the major processes controlling the long range transport of sulphur compounds in the atmosphere over Europe (Royal Society 1984).

Monitoring studies have steadily improved and the quality and quantity of data now provides the spatial patterns of wet deposition (BAPMON 1986). Within NW Europe and Scandinavia the areas receiving the largest amounts of wet deposited acidity are the high rainfall regions of SW Norway and western and northern Britain (Semb and Dovland 1986, Barrett et al 1983).

Networks of precipitation samplers throughout Europe and North America have been designed to obtain regionally representative measurements of the average composition of precipitation. To obtain the wet deposition fields from these data, the much more extensive data on precipitation amount from meteorological and hydrological precipitation samplers have been combined with the data from precipitation chemistry networks. Annual rainfall weighted average concentrations of SO_4^{2-}, H^+, NO_3^- and NH_4^+ have been generally shown to vary much less than precipitation amounts. However, there are areas of uncertainty, in particular the tops of hills where rainfall amounts and wet deposition are larger and where for practical reasons few measurements are made. Estimates of wet deposition on the high ground in NW Britain, Scandinavia and continental Europe, for which representative sites are not available, are made by assuming the composition of precipitation remains constant with altitude. As the areas experiencing the largest wet deposition are the upland districts which have been a focus of interest for freshwater acidification in Britain and Scandinavia, a measurement study of the variation in the composition of precipitation with altitude is necessary to define the inputs more precisely in these areas.

Measurements of the ionic composition of rainfall at different elevations in the Alps of France, in Austria and in western North America generally show smaller concentrations of the major ions SO_4^{2-}, NO_3^-, and NH_4^+ and H^+ at the higher elevations (3000 m) (Duncan et al 1986, Puxbaum 1988). On smaller mountains (500 to 1000 m) in NW Europe, orographic cloud frequently shrouds the summits and contains large concentrations of all major ions: SO_4^{2-}, NO_3^-, NH_4^+, H^+, Cl^- and Na^+. The low-level cloud is readily scavenged by rainfall from higher levels, enhancing rainfall by the seeder-feeder process (Bergeron 1965). Concentrations of the major ions in the feeder (lower) cloud are generally large compared with typical concentrations in rainfall, so that larger concentrations in precipitation reaching the ground may result. Wet deposition values predicted from a combination of regional rainfall chemistry measured at low level sites and rainfall amounts would under estimate inputs in such regions.

This study reports measurements of composition of precipitation and cloud water at a number of sites between 150 m above sea level and a hill summit 847 m above sea level in the NW of England. The study site includes a ridge lying NW - SE with the tops of individual hills at about 900 m, where annual rainfall amounts are in excess of 2000 mm, and a valley to the SW of this line of hills at 150 m above sea level with an annual rainfall of approximately 900 mm. The hill selected for

measurement (Great Dun Fell, GDF) at 847 m above sea level is in cloud
for some part of 250 days every year, and the road between the valley
site to the south-west and the summit provides access to a range of
sites for rain and cloud water collectors. In W and SW airflow the
seeder-feeder mechansim has been shown to be responsible for a
substantial enhancement of rainfall amounts at the higher levels on
these hills (Carruthers and Choularton 1983).

The objectives of this study were to obtain rainfall and cloud
water composition measurements at a range of sites along the SW slopes
of Great Dun Fell to show for individual rainfall events in south
westerly airflow:

(i) whether the concentrations of major ions in rainfall vary with
     altitude;
(ii) whether orographic cloud on Great Dun Fell significantly
     influences the composition of rainfall at the higher altitudes.

2. METHODS

Rainfall was sampled using 20 cm diameter Pyrex glass funnels mounted
1.5 m above ground at each of 8 locations between 250 m above sea level
in the Eden valley and 847 m at Great Dun Fell summit. The water
collected in polypropylene bottles beneath each funnel. The sites were
chosen to avoid, wherever possible, the influence of local
discontinuities in the topography although the 'local' exposure of
collectors varied considerably between the different collector sites.
A ground level Institute of Hydrology pattern rain collector (Institute
of Hydrology 1977) was also used at the summit. To provide an
indication of variability in precipitation amount and ionic composition
at one height, 8 identical collectors were arranged (at 50 m intervals
at 670 m above sea level) along a contour of the south west side of
Great Dun Fell. For similar reasons a series of 4 collectors were
also placed at 30 m intervals at a height of 807 m a.s.l where
windspeeds were generally large and collection efficiency problems were
anticipated. The arrangement of rain and cloud water samplers on the
hill-side is shown diagrammatically in Figure 1.

2.1 Cloud Water Collection

Cloud water samplers were located at 5 different altitudes on the hill
close to the location of precipitation samplers at the highest 5
levels. The 'harp' wire collectors were strung with 550 µm diameter
ETFE coated wire filaments, each separated at the top ring by a 3 mm
gap.

Cloud water collection is passive, droplets failing to follow the
streamline of airflow around the strings are intercepted by the strings
and run down into a polypropylene collector. Capture efficiency
increases with droplet size and wind velocity, but in practice, as
droplets that have been collected run down the strings and grow by
intercepting other drops, they may be blown off. It is therefore
difficult to predict from theoretical work the collection properties of

234

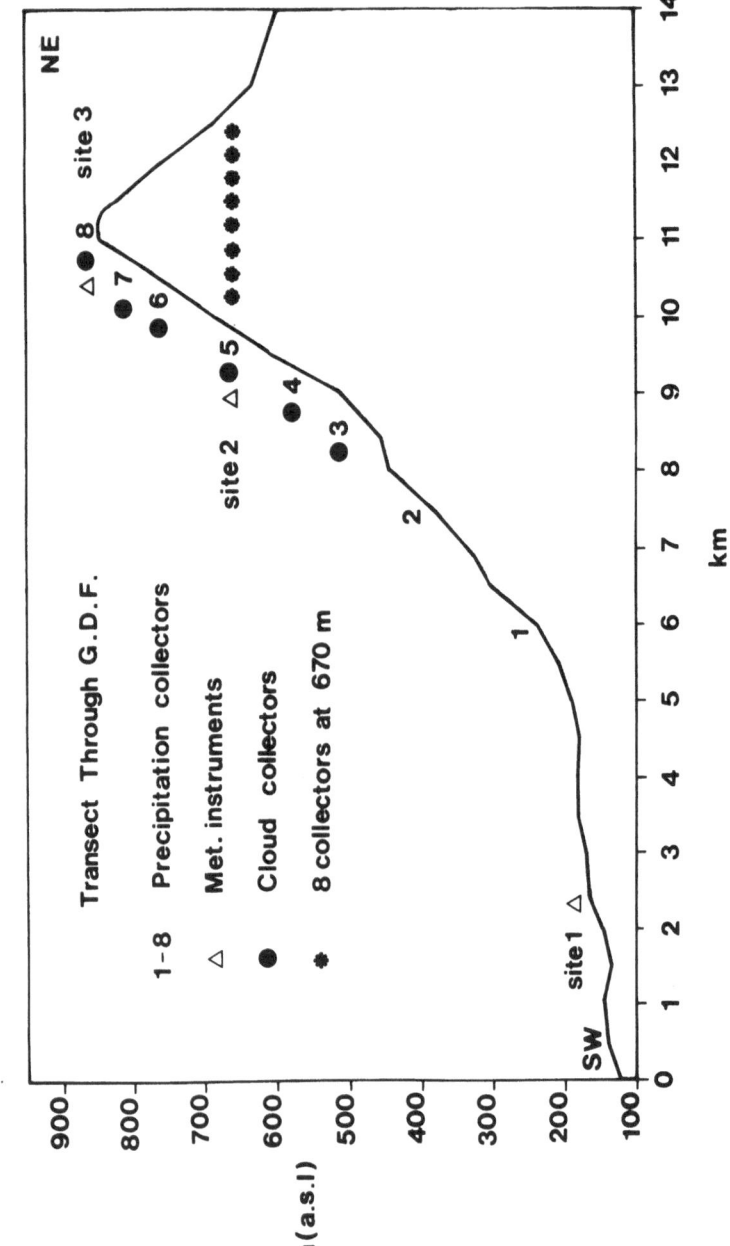

Figure 1. Schematic representation of Great Dun Fell showing locations of sampling points.

the gauges in field conditions. Field calibration of the passive cloud collectors was provided by Knollenberg forward scattering spectrometer probes (FSSP). The cloud water caught by the passive sampler was compared with the amount of cloud that would have passed through the vertical area cross-section of the cloud sampler. The FSSP provided cloud liquid water content and the mean drop size of the cloud water and a cup anemometer provided wind velocity. The periods of collection were between 20 and 200 minutes with average wind speeds from 10 ms^{-1} to 23 ms^{-1} and mean cloud drop sizes from 3.5 μm diameter. The collection efficiency ranged from 11-47% with a mean (±sd) of 28.9% ± 9.7%.

Precipitation and cloud water collectors were washed by spraying with de-ionised water prior to a collection period. Collection periods varied between 2 hours and 3 days, but on average were between 5 and 12 hours.

Precipitation and cloud water samples were stored at 4°C before analysis for SO_4^{2-}, NO_3^-, Cl^-, K^+, $Na^\pm$, Ca^{2+}, Mg^{2+}, and NH_4^+ by ion chromatography and for H^+ from pH measurements. Precision in these measurements was generally between 1 and 4% and the sums of cation and anion concentrations expressed in equivalents agreed within 5%. The measurement of precipitation amount was much less certain because of the exposed nature of many of the hillside collectors and the high wind speeds that are generally present at this site.

3. RESULTS

3.1 Precipitation Measurements, Autumn 1984

The first measurements of rainfall composition were made at the 8 hill sites, as described above, on 10 occasions between 24.10.84 and 5.12.84. The total amount of rainfall over this period was 62 mm and 99 mm at the lowest and highest collectors respectively. For individual events the variability between collectors on the hillside was large but a general increase in concentrations of the major ions towards the hill summit was observed for occasions with south-westerly airflow from the valley towards the summit.

Data for 4 and 5 December 1984 illustrate these general features. Rainfall amounts increased from 3 mm at an altitude of 433 m to 8 mm at the summit and concentrations of SO_4^{2-}, NO_3^-, NH_4^+ and H^+ all show an increase with altitude (Figure 2). Concentrations of SO_4^{2-} and NO_3^- increased between 433 m and 847 m by a factor of 1.5. For NH_4^+ the increase in concentration over the same height range is larger, about a factor of two. The acidity of rainfall also increases with altitude by about a factor of 3 between 433 m and 847 m, but changes in H^+ concentration are determined by the other ions in solution and may not be interpreted independently. The concentrations of Na^+, Cl^- and Mg^{2+} on this occasion did not change significantly between 433 m and 847 m despite a large change in rainfall amount. Increases with height in rainfall amount and ionic concentrations lead to a large increase in wet deposited SO_4^{2-}, NO_3^-, NH_4^+ and H^+ with altitude (Figure 3).

Figure 2.   Variation in concentrations of major ions in rain, as a function of altitude, 4-5 December 1984.

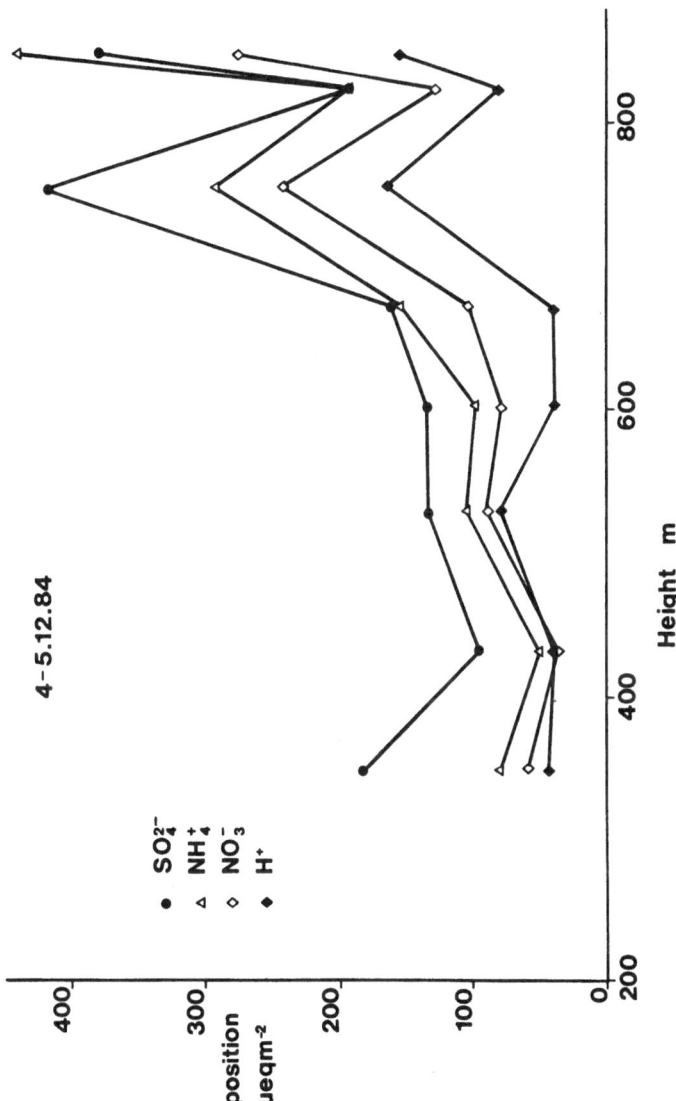

237

Figure 3. Variation in deposition of major ions in rain, as a function of altitude, 4–5 December 1984.

238

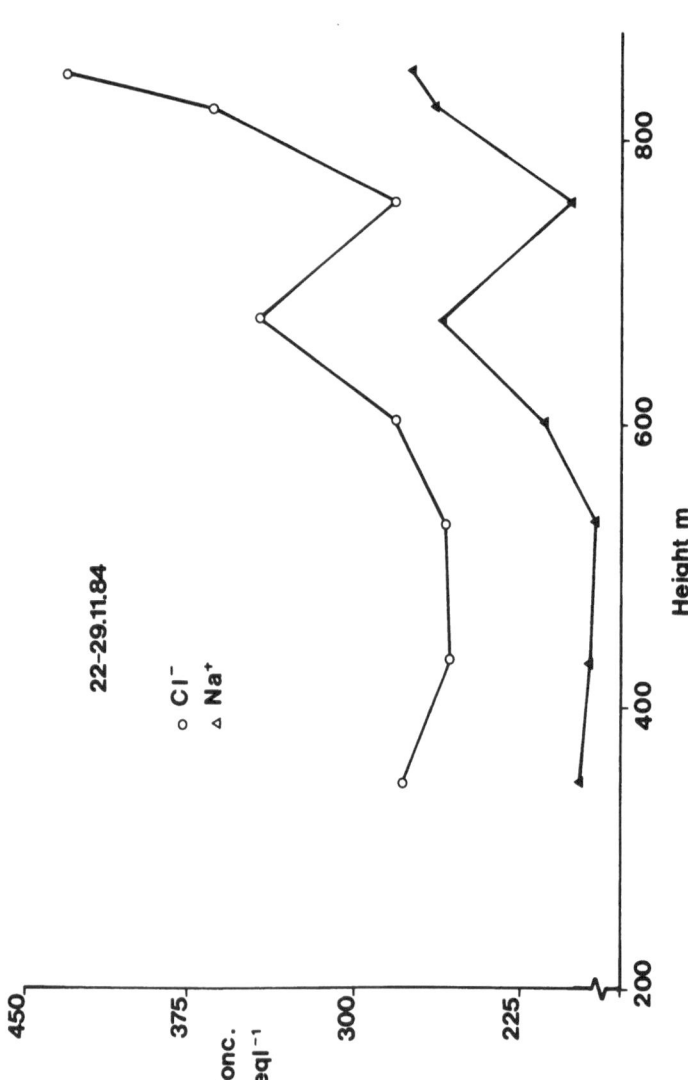

Figure 4. Variation in concentration of marine derived ions in rain, as a function of altitude, 22–29 November 1984.

All other episodes measured during Autumn 1984 showed similar increases in concentrations with altiude with airflow from the west. Strong south-westerly flow of maritime air generally show large concentrations of Cl^-, Na^+ and Mg^{2+} in rain and rather small SO_4^{2-}, NO_3^-, H^+ and NH_4^+ concentrations. Such an event is shown in Figure 4, in which the concentrations of all ions were larger at the hill summit. Precipitation amounts increased by about a factor of 2 between 347 m and 847 m and cl^- and Na^+ concentrations increased by a factor of about 1.5, providing a three-fold increase in deposition (Figure 5). The other occasions show similar results of increasing concentration and deposition with increasing altitude, although the scatter on individual profiles is large.

When the airflow was from the east or with blocked flow no increase in concentration or rainfall amount with altitude is observed.

Average properties of the measurements during the autumn of 1984 are provided in Figure 6 for which the rainfall weighted mean concentrations of each ion over the 6 weeks of sampling have been plotted for each collector on the hill. Concentrations of NO_3^-, SO_4^{2-}, H^+, Na^+, and Cl^- all increase with altitude, but only above 600 m are the changes significant. The concentrations increase by a factor of 1.5 between 530 m and 847 m. The increase in concentration which appears to be a consistent feature of the precipitation at the site, is always associated with orographic cloud on the hill.

3.2 Measurements of Precipitation and Cloud Water Composition, Spring 1985

Measurements of cloud water composition in the orographic cap-cloud and rainwater composition were made simultaneously during the Spring of 1985 to show whether it was the scavenging of low level cloud that was responsible for the increase in concentration of most ions in precipitation with altitude.

The measurements made between 16 April 1985 and 9 May 1985 included 7 occasions with significant amounts of precipitation at all levels with simultaneous samples of cloud water using the passive collectors described earlier. As in the Autumn of 1984, on all occasions with westerly or south-westerly airflow and the presence of a cap-cloud on Great Dun Fell, concentrations of the major ions were larger at the hill summit than at lower elevations.

3.3 Cloud Water Composition Measurements

Cloud water composition measurements were made initially at 3 sites (847 m, 808 m and 753 m) and, in general, concentrations for all ions were larger in cloud water than in rainwater. The difference varied between a factor of 1.5 and 8. A much larger data set (including samples from spring and autumn 1985 and later sampling) shows (Figure 7) that although the distributions of concentration (or SO_4^{2-} in this example) overlap, the concentrations of ions in cloud water samples are larger on average than in rain by a factor of 2.5. A paired t-test for concentrations of SO_4^{2-} in rainwater and cloud water showed that values

240

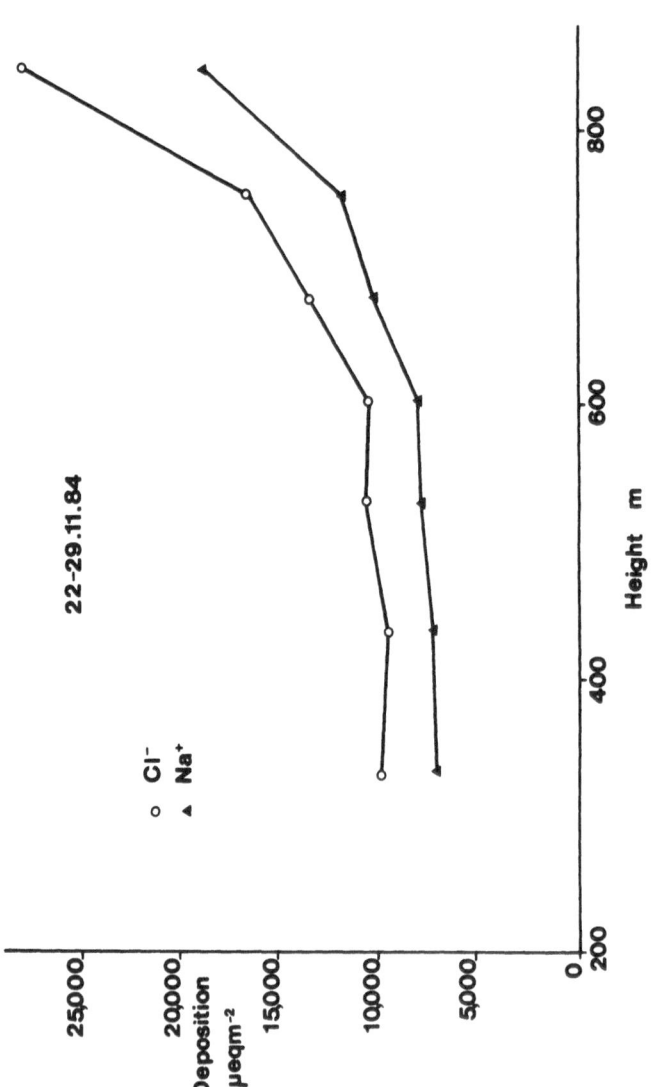

Figure 5. Variation in deposition of marine derived ions in rain, as a function of altitude, 22–29 November 1984.

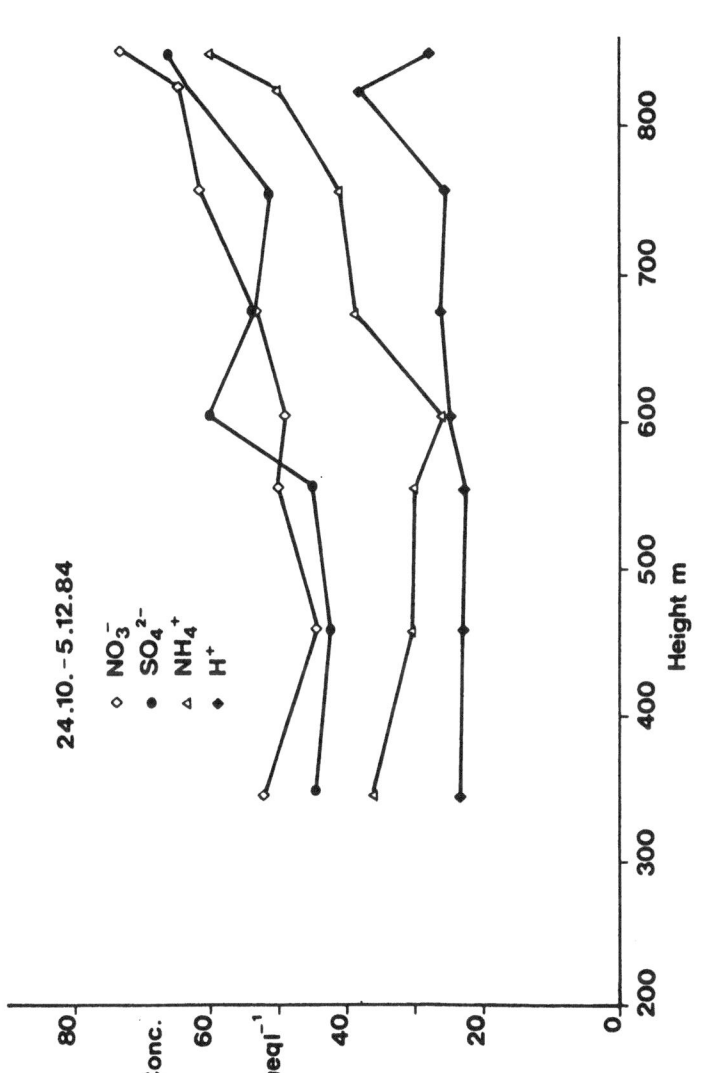

Figure 6.   Average concentrations (rainfall weighted) of major pollutant ions in rain, as a function of altitude, for the Autumn 1984 sampling period (24 October – 5 December).

242

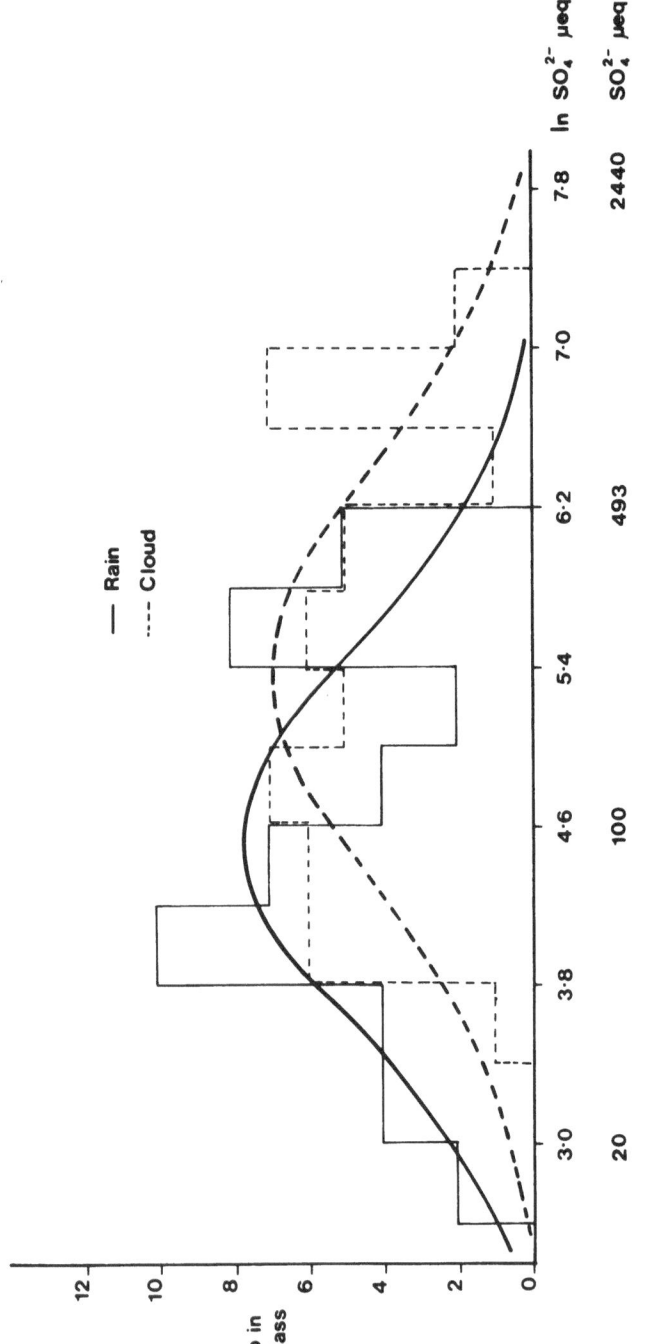

Figure 7.   Frequency distribution of sulphate concentrations in rain (solid lines) and cloud
(dashed lines) for simultaneous measurements, 1985.

differed significantly at the 0.1% level. The mean ratios of cloud/rain concentrations (in μeq 1^{-1}) for major ions are shown in Table 1 for 11 precipitation and cloud water events at the hill summit in Spring 1985.

Table 1. Ratio of concentrations in cloud/rain (μeq 1^{-1}).

H^+	NH_4^+	Cl^-	NO_3^-	SO_4^{2-}
3.9	2.4	2.6	2.8	2.0

Mean of 11 precipitation and cloud events at summit (847 m) during Spring 1985.

Concentrations of SO_4^{2-}, NO_3^-, Cl^-, H^+, NH_4^+, Na^+ and Mg^{2+} in cloud water are, however, strongly dependent on altitude. The largest concentrations occur at cloud base and decrease with height as cloud liquid water content increases with height and dilutes the concentrations present. For most occasions the lowest 100-200m of the cap-cloud at Great Dun Fell shows an adiabatic increase in liquid water content with height (Choularton et al., 1988), and in these conditions the change in concentration of major ions in cloud water with altitude may be predicted from a knowledge of the altitude of cloud base. The change in concentrations for SO_4^{2-}, NO_3^- and Cl^- for the 17 April 1985 when concentrations of SO_4^{2-} decreased from 238 to 141 μeq 1^{-1} between 755 m and 847 m, a decrease of 40% is shown in Figure 8. Cloud base measurements for the time of collection allow an estimate of the cloud liquid water content to be made for the 3 altitudes of the cloud water collectors, assuming the cloud to be adiabatic, and using these data the expected concentrations in solution may be estimated from the known concentration at the lowest sampling site. These are plotted in Figure 8, and the agreement between measured and estimated values is good. Other occasions with known (and fairly constant) cloud base and cloud water concentrations also show good agreement. There is therefore a pronounced and predictable vertical ionic concentration gradient in cloud water, with concentration decreasing with altitude.

The gauges below 500 m generally collect rain falling from higher levels which does not interact with the cap-cloud. Results from autumn 1984 show that at these levels the concentration changes little with altitude. At higher levels (500-847 m), rain falls through a cap-cloud which itself contains a large vertical gradient in concentration of the ions present.

3.4 Rainfall Composition

Rainfall amount and composition measurements from the Autumn 1984 at different altitudes show considerable scatter and there appeared to be

244

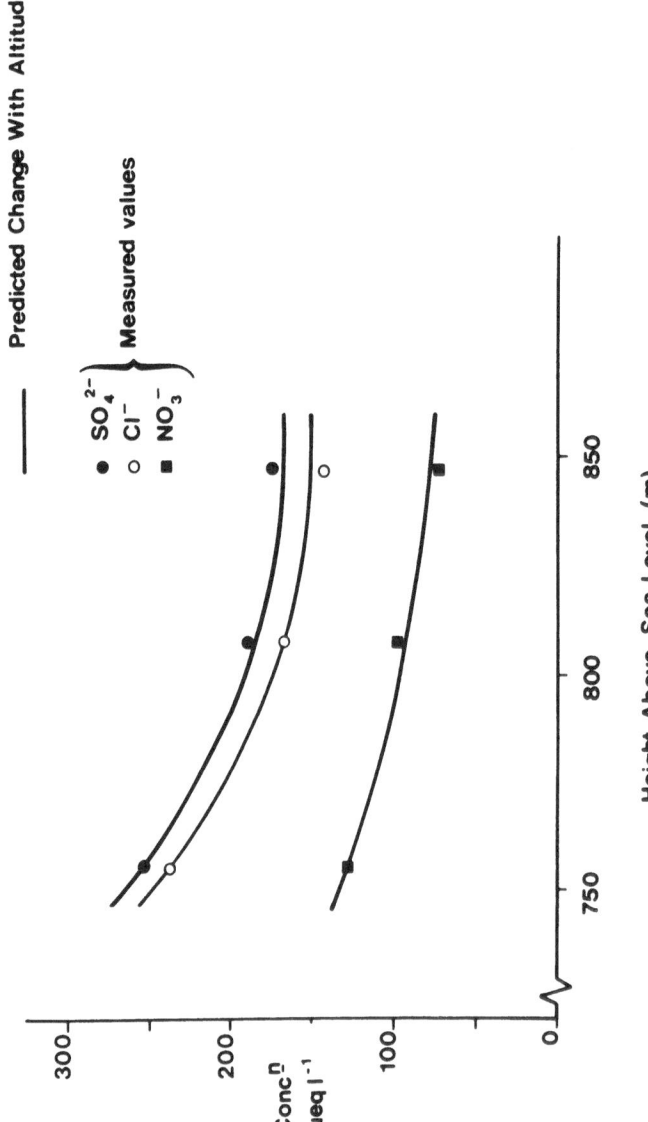

Figure 8.   Variation in concentration of major anions in cloud, as a function of altitude, compared with that predicted from a calculation of adiabatic water content.

particular problems at collectors Nos 7 and 8 close to the hill
summit. High windspeeds at these exposed locations cause aerodynamic
sorting of the raindrop size spectrum with many small drops being
carried over or around the funnel and only large drops entering. As
the chemical composition of rain is expected to vary with drop size in
these conditions (the large drops having smaller concentrations), this
will lead to a systematic underestimate of the concentrations of major
ions in rain at the higher levels. To provide information on the
variability of rainfall amount and composition at one altitude, 8
collectors were placed in 2 parallel lines with 50 m separating each
collector in the line and 20 m separating the lines, at the 670 m
contour on the hill. During Spring 1985, 8 rain events were measured,
varying between 0.7 mm and 11.0 mm, and for concentrations of SO_4^{2-},
Cl^- and NO_3^- ions the coefficient of variation was typically 10%.
Variability in rainfall amount was rather larger at about 20%. The
results of these measurements are summarised in Table 2. The
variability at higher, more exposed, collectors will no doubt be
greater, but practical problems such as finding the equipment on the
hill in cloud limited this aspect of the study. The collector at 823 m
frequently caught less rain that the summit collector during the Autumn
1984 measurements and a series of 4 collectors was placed along a
south-west to north-east line at approximately 30 m intervals to
provide a better estimate of both amount and composition of rain at
this level. Whenever possible the mean value for all collectors at a
given height was used to estimate ion concentrations and rainfall
amounts.

Two precipitation events during the Spring of 1985 which showed
contrasting chemical properties provide examples of the range of cloud
and rain compositions that is frequently encountered at Great Dun Fell.

The precipitation which fell during 20 and 21 April 1985 was
dominated by marine ions Na^+, Cl^- and Mg^{2+} (these contributing 75% of
the ions present, expressed in μeq l^{-1}). Concentrations of NO_3^- and
SO_4^{2-} were in the range 15-70 and 70-250 μeq l^{-1} respectively and all
ions in precipitation showed a marked increase in concentration with
altitude (Figure 9). For SO_4^{2-} and NO_3^- the increase was almost a
factor of 3 while for Cl^- it was closer to a factor of 2. The
differences in concentration of all ions between cloud water and
rainwater at the summit were (relative to other occasions) quite small,
less than a factor of 2. The 900 mb trajectories of the air prior to
precipitation showed an essentially northerly trajectory so that from
the originally maritime trajectory the air crossed only northern
Britain before reaching Great Dun Fell (Figure 10a). With such a
trajectory there is little opportunity for the composition of cloud and
rain to become dominated by the ions derived from the pollutants SO_2
and NO_x.

The 5th May 1985 provided precipitation associated with a 900 mb
trajectory taking the air over the major British source regions of SO_2
and NO_x before reaching Great Dun Fell (Figure 10b). On this occasion
the concentrations of all ions in rain increase with altitude, the
increases being typically a factor of 4 (Figure 11). The
concentrations of SO_4^{2-} and NO_3^- in cloud water at the summit are large

246

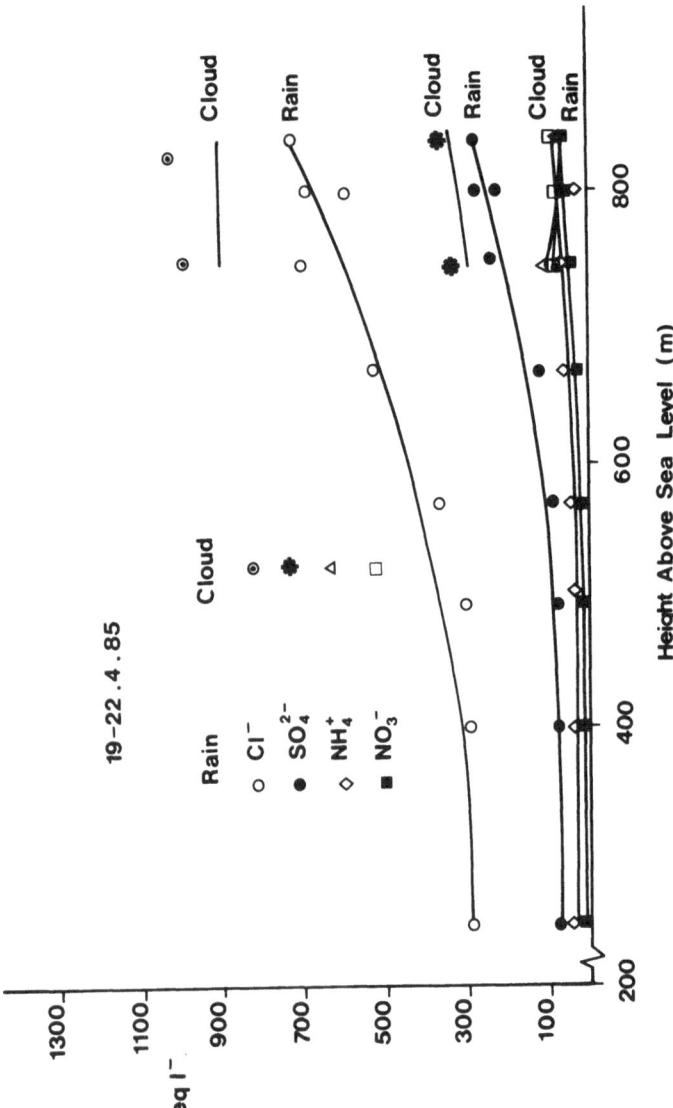

Figure 9. Variation in concentration of major pollutant ions in cloud and rain, as a function of altitude, 19–22 April 1985.

a)   **Trajectory for 900 mb at 3 hr intervals   20.4.85**

b)   **Trajectory for 900 mb at 3 hr intervals   5.5.85**

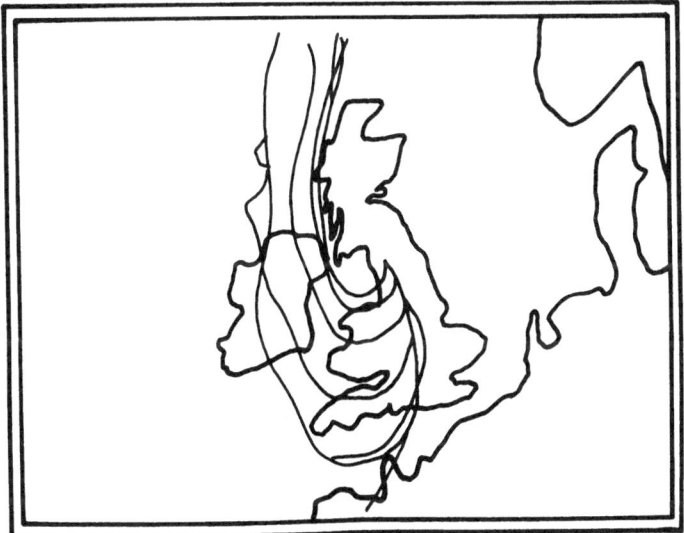

Figure 10.   (a) 900 mb trajectories of air masses arriving at Great Dun
              Fell 20–21 April 1985 (3-hour intervals)
              (b) 900 mb trajectories of air masses arriving at Great Dun
              Fell 5 May 1985 (3-hour intervals).
              Trajectories were calculated by D K J Keston, University of
              Edinburgh, Department of Meteorology.

248

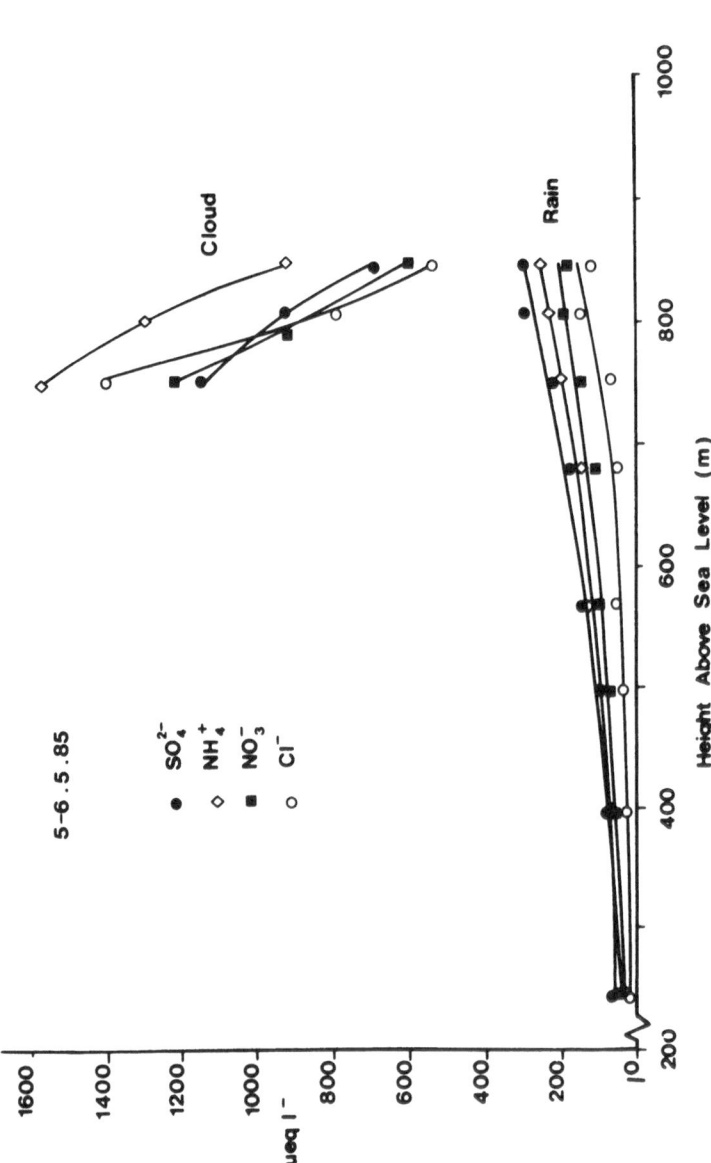

Figure 11.   Variation in concentration of major ions in cloud and rain, as a function of altitude, 5-6 May 1985.

Table 2. Measurements of rainfall and SO_4^{2-}, Cl^- and NO_3^- concentrations at 675 m above sea level on the SW slopes of Great Dun Fell using 8 collectors each separated by 40 m. the additional column provides the average rainfall amounts at 250 m and 810 m.

Date	Time	Mean Rain (mm) ±S.D.	SO_4^{2-} ±S.D.	Cl^- ±S.D.	NO_3^- ±S.D.	Rain (mm) (1) – (7)
16–17.4.85	1800–0900	1.63 ± 0.13	106 ± 18	89 ± 18	47 ± 3	1.02–3.63
19–22.4.85	1200–1500	5.88 ± 3.16	151 + 3.16	688 ± 141	54 ± 20	7.80–4.77
25–26.4.85	1800–1300	0.70 ± 0.15	245 ± 67	518 ± 332	96 ± 21	0.32–0.32
29.4.85	1000–1600	7.15 ± 1.93	92 ± 20	261 ± 144	23 ± 5	2.80–1.40
29–30.4.85	15600–1000	10.98 ± 1.93	34 ± 3	16 ± 13	10 ± 1	4.61–13.37
30–1.5.85	1800–1000	3.76 ± 0.98	43 ± 5	128 ± 37	9 ± 2	1.43–4.20
2–5.5.85	1800–1600	3.55 ± 1.17	350 ± 30	276 ± 62	183 ± 21	0.3–4.62 2
5–6.5.85	1600–0600	4.58 ± 1.43	196 ± 69	74 ± 49	139 ± 51	2.90–4.27

(600–900 μeq l^{-1}) and exceed concentrations in rain at the summit by about a factor of 3. The cloud/rain water concentration differences are much larger close to cloud base (factor of about 6).

The differences between these 2 occasions, and the link with trajectories suggests that cloud and rainwater chemistry at Great Dun Fell is largely controlled by the source of the air and the scale of sources of major air pollutants that lie along the trajectory. There is no evidence in any of the measurements to date that the mechanism of enhancement of SO_4^{2-} and NO_3^- deposition at this site is peculiar to local properties of the site. The valley upwind is a rural area with few sources of air pollutants and the only feature which appears to be responsible for the large increase in deposition is the presence of orographic cloud, which effectively transforms particulate pollutants from submicron aerosols into cloud droplets which are then scavenged relatively efficiently by falling rain.

3.5 Average Change in Rainfall Composition with Altitude

Rainfall and cloud water composition at different altitudes from the Spring of 1985 provides average concentrations at the hill summit that exceed the concentration in the valley by at least a factor of 2 for most ions (Table 3). The average data plotted for Cl^-, SO_4^{2-}, NO_3^-, H^+ and NH_4^+ in Figure 12 a–e. The larger rainfall at the higher levels (typically by a factor of 2) with the larger concentrations leads to wet deposition of most ions at the summit exceeding that at the valley site by a factor of at least 4. The widely applied technique for predicting deposition on hills from measurements at lower elevations is therefore likely to be seriously in error whenever hill cloud is common.

Table 3. Ratio of concentrations at summit/valley for major ions and rain amount

	H^+	NH_4	Cl^-	NO_3^-	SO_4^{2-}	RAIN
$\bar{x}$	2.9	3.1	2.9	2.3	2.2	2.0

20 precipitation events at G.D.F. with measurements at 8 levels (244–847 m) showing increase in concentration between valley and summit, 1984–85.

The rainfall events sampled in this study represent only a small proportion of the annual total. However, the sampled events showing the increase in concentration with altitude were all from the west or south-west, and these wind sectors provide the bulk of annual precipitation.

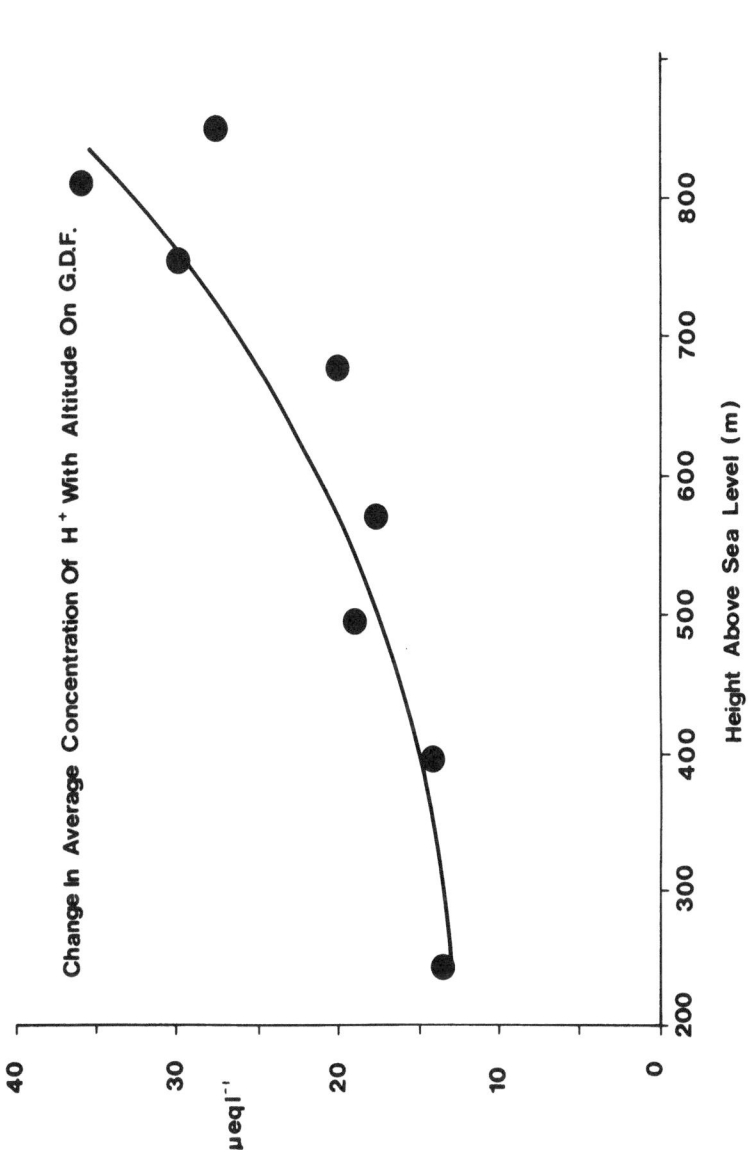

Figure 12a. Average concentration (rainfall weighted) of H^+ in rain, as a function of altitude, for the Spring 1985 sampling period (16 April – 9 May).

252

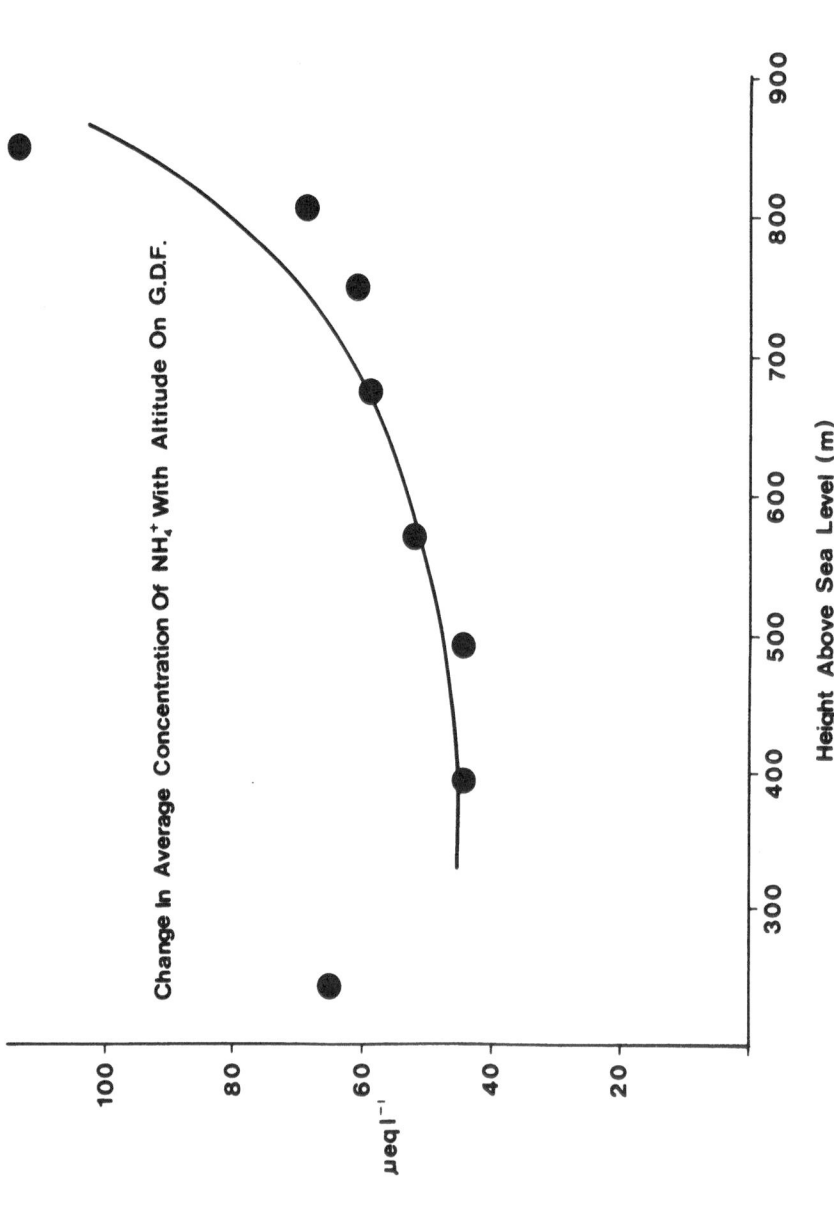

Figure 12b. Average concentration (rainfall weighted) of NH_4^+ in rain, as a function of altitude, for the Spring 1985 sampling period (16 April - 9 May).

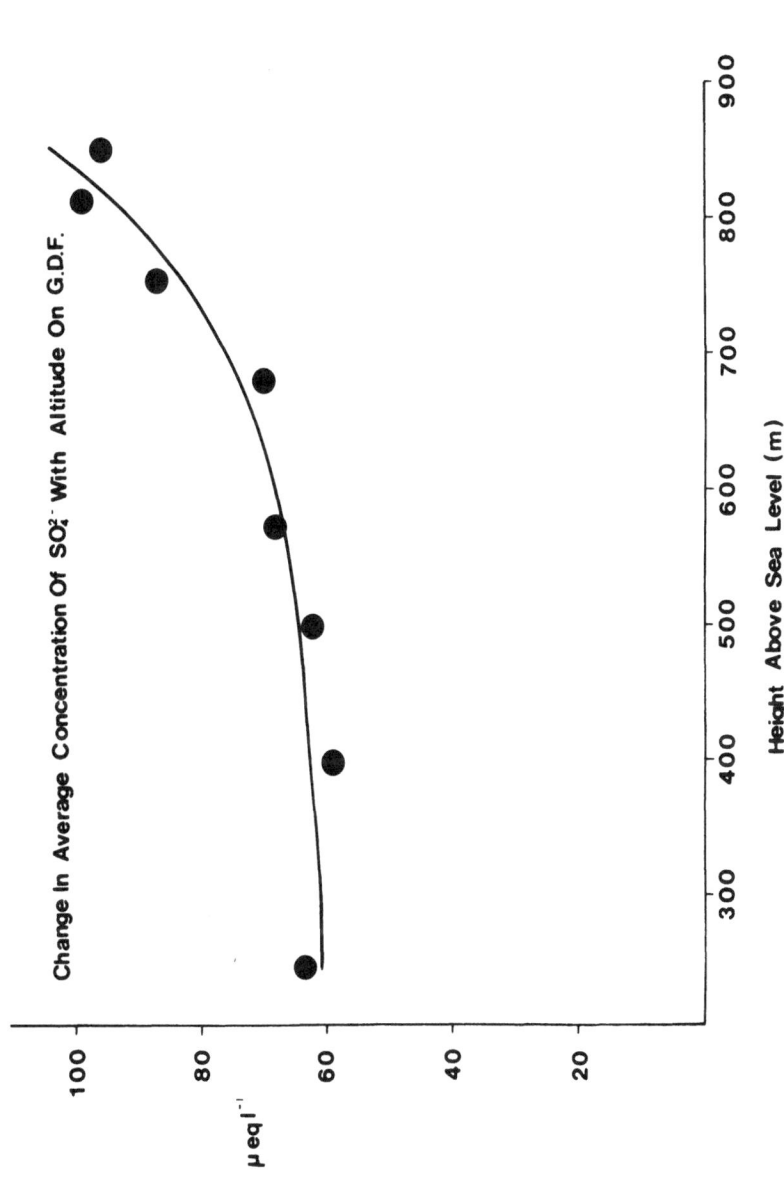

Figure 12c. Average concentration (rainfall weighted) of SO_4^{2-} in rain, as a function of altitude, for the Spring 1985 sampling period (16 April – 9 May).

254

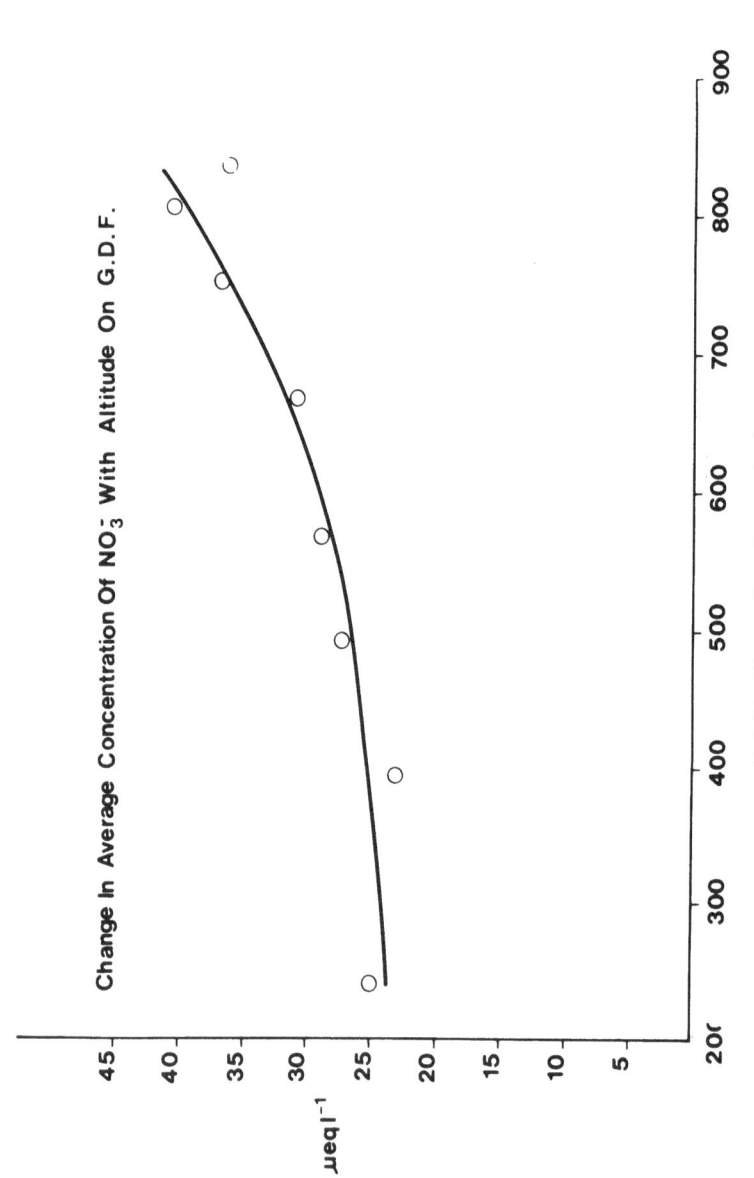

Figure 12d. Average concentration (rainfall weighted) of NO_3^- in rain, as a function of altitude, for the Spring 1985 sampling period (16 April – 9 May).

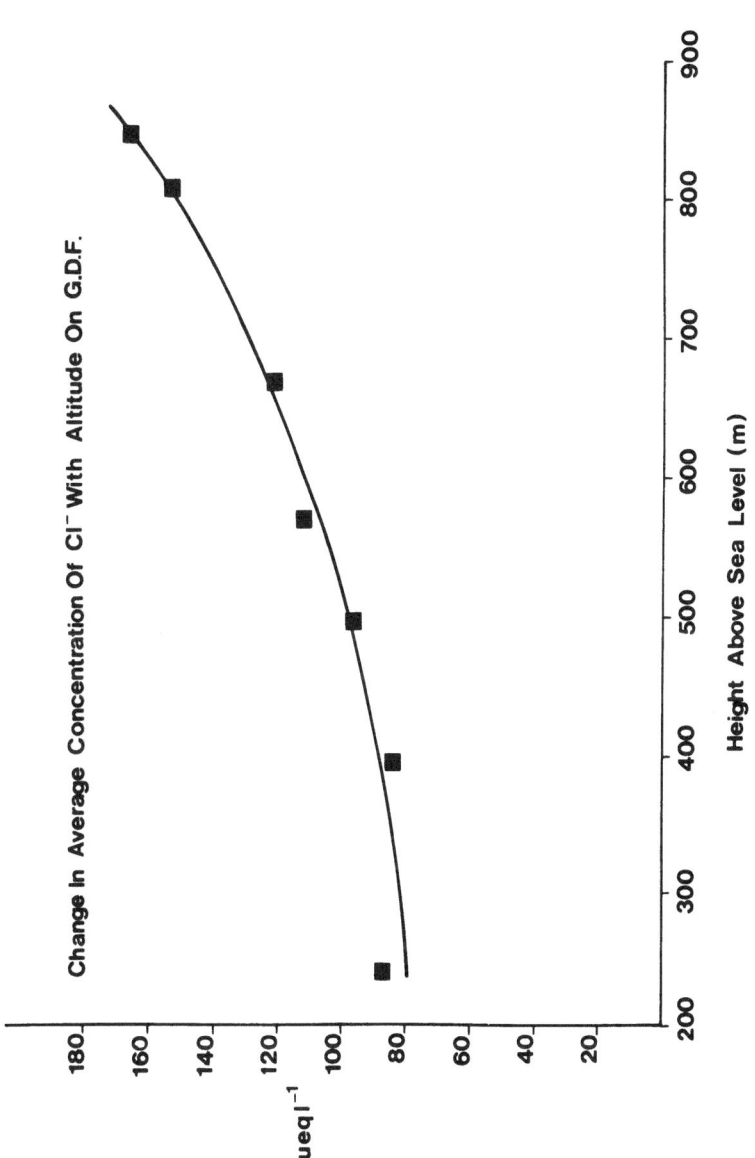

Figure 12e. Average concentration (rainfall weighted) of Cl^- in rain, as a function of altitude, for the Spring 1985 sampling period (16 April – 9 May).

The west facing hills in western Britain from Wales to the north west of Scotland, and the coastal mountains of Norway, are commonly shrouded by orographic cloud so that for these regions, which are already the areas of largest wet deposited SO_4^{2-}, NO_3^- and H^+ ions, the existing published values are underestimates. The work presented here may not be readily applied to other hills. The extent to which 'seeder-feeder' modification may be extended over upland Britain is uncertain. Such effects have so far been restricted to modelling studies (Hill et al 1987). These studies show that the effects observed here would be expected for most hills on which orographic clouds are a common feature, but that the rate of change of composition with altitude depends strongly on hill shape and atmospheric conditions. It is not possible, therefore, to extrapolate to other areas to gauge the magnitude of the underestimate in wet deposition. In order to extend the work to modify estimates of wet deposition in upland districts, further measurements are necessary of cloud and rain composition on a series of hills and over more extensive upland areas.

ACKNOWLEDGEMENTS

This work was supported by the UK Department of the Environment and the Natural Environment Research Council and forms a part of a larger collaborative study between ITE, UMIST, AERE (Harwell) and the University of East Anglia. The help from colleagues in all of these organizations is acknowledged. The trajectory analysis was done by Dr K J Weston of Edinburgh University, whose contribution to the work is gratefully acknowledged.

REFERENCES

BAPMON 1986. Global atmospheric background monitoring for environmental parameters. Vol II Precipitation Chemistry WMO/TD No 116

Barrett, C.F., Atkins, D.H.F., Cape, J.N., Fowler, D., Irwin J.E., Kallend, A.S., Martin, A., Pitman, M.J., Scriven, R.A. and Tuck, A.F. 1983. Acid deposition in the United Kingdom. 72pp. Published by the Warren Spring Laboratory, Stevenage, UK, SE1 2BX.

Barrett, C.F.D., Goldsmith, A.L., Hall, D.J. and Irwin, J.F. 1985. The variation of precipitation composition with altitude: a feasibility study. Stevenage: Warren Spring Laboratory LR 534 (AP)

Bergeron, T. 1965. On the low-level redistribution of atmospheric water caused by orography. Proc. Int. Conf. Cloud. Phys. Tokyo 1965 pp 96-100.

Carruthers, D.J. and Choularton, T.W. 1983. A model of the feeder-seeder mechanism of orographic rain including stratification and wind drift effects. Quart. J. Roy. Met. Soc. 109 , 575-588.

Choularton, T.W., Gay, M.J., Jones, A., Fowler, D., Cape, J.N. and Leith, I.D. 1988. The influence of altitude on wet deposition II. Comparison between field measurements at Great Dun Fell and the predictions of a seeder feeder model. Atmos. Environ. (in press).

Duncan, C.L., Welch, E.B. and Ausserer, W. 1986. The composition of precipitation at Snoqualmire Pass and Stevens Pass in the central Cascades of Washington State. Water Air and Soil Pollution 30, 217-229.

Hill, T.A., Jones, A. and Choularton, T.W. 1987. Modelling sulphate deposition onto hills by washout and turbulence. Q.J. Roy. Met. Soc. (in press).

Institute of Hydrology 1977. Selected measurement techniques in use at Plynlimon experimental catchments. Inst. Hydrology. Wallingford report no. 43.

Puxbaum, H. 1988. The chemistry of wet deposition in the Austrian Alps. In: Acid Deposition Processes at High Elevation Sites. (Editors Unsworth, M.H. and Fowler, D.). NATO-ARI, Edinburgh 1986 (in press).

The Royal Society, 1984. The Ecological Effects of Deposited Sulphur and Nitrogen Compounds. Phil. Trans. Roy. Soc. London (B) Vol. 305, pp 255-577.

Semb, A. and Dovland, H. 1986. Atmospheric deposition in Fenno-Scandia: Characteristics and trends. Water Air and Soil Pollution 30, 5-16.

TIME RESOLUTION IN PRECIPITATION AND CLOUD SAMPLING

A.R. Marsh, D.L. Ames, P.A. Clark, G.P. Gervat and W.J. McElroy
Central Electricity Research Laboratories
Leatherhead
Surrey
UK

ABSTRACT. Measurements of the composition of precipitation and clouds provide information which can be used to characterise deposition and aid the understanding of the wet chemical processes.

This paper reviews some of the measurements made by CERL in an attempt to characterise the processes determining the ionic composition of cloud and precipitation. The measurements described include the continuous chemical monitoring of rain composition, aircraft sampling of rain and precipitation and ground based cloud studies involving deliberate attempts to alter cloud chemistry.

A common theme throughout these experiments is the emphasis on temporal resolution of information. It is the correlation in time of ionic behaviour of precipitation and cloud water which is most useful both for interpreting the changes in chemical composition and for comparison with computer model predictions. The best opportunity at present for elucidating chemical processes in clouds appears to be that provided by those experiments in which cloud composition is altered directly by controlled emissions. These experiments are most easily conducted at well characterised hill top sites. However, they necessitate good temporal resolution of the ionic composition of clouds if useful information is to be obtained.

These studies demonstrate the importance of making detailed temporal measurements in the gas, aqueous and aerosol phases as a means of elucidating chemical transformation processes.

1. INTRODUCTION

This paper discusses field measurements of precipitation and cloud composition. A common feature of both types of measurement is the use of time as an aid to resolving physical and chemical processes.

Field measurements of the chemical composition of cloud and precipitation are usually made with one of two objectives in mind. The first objective might be described as characterising the deposition. Measurements are made to determine the average concentration and its variability. This, in turn can be used to calculate the flux of

M. H. Unsworth and D. Fowler (eds.), Acid Deposition at High Elevation Sites, 259–281.
© *1988 by Kluwer Academic Publishers.*

material and hence the deposition rates to catchment areas. The second objective in making field measurements of cloud and precipitation composition lies in an attempt to study the chemical processes that take place in the atmosphere.

The characterisation of deposition is important especially in upland areas where relatively little information is available, particularly when compared to the extensive precipitation data base available for urban and rural areas. However, to accomplish this objective it is usually sufficient, especially in the case of precipitation, to estimate average concentrations over reasonably long periods of time. Timescales used are typically annual, seasonal or monthly averages. That is not to say the measurements are only taken on that timescale. Modern networks usually operate on a 24 hour basis to allow comparison with air mass trajectories. In addition individual collectors may be reasonably sophisticated, with devices to distinguish between wet and dry deposition. But the result required is always an integral of the ionic composition over a relatively long period of time.

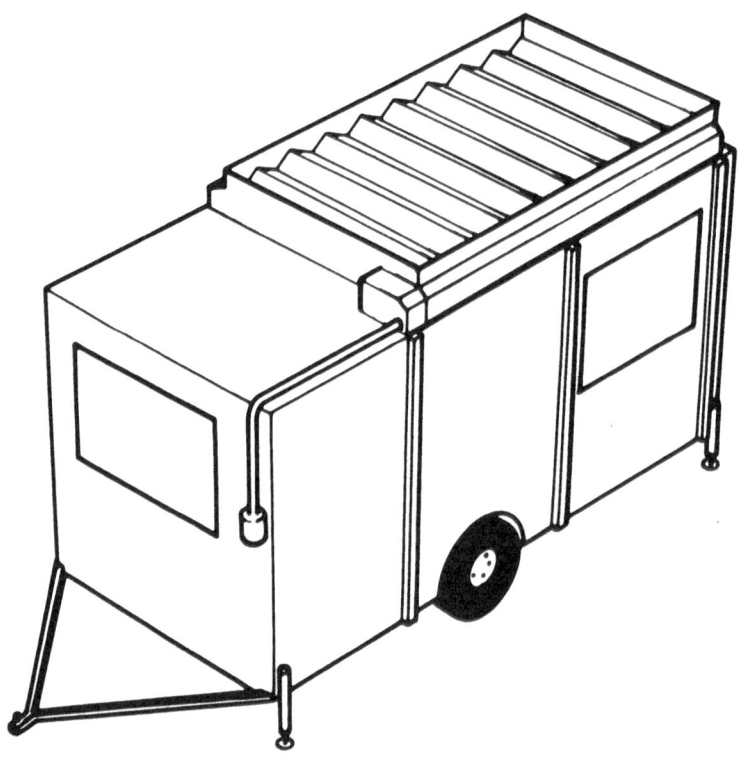

FIG. 1 CARAVAN BASED RAIN GAUGE

Although chemical analysis of bulk samples provides concentrations
of all the major ionic species present in precipitation, such
measurements do not provide any information on the processes which give
rise to the observed composition.

While there are a number of natural and international programmes
aimed at characterising wet deposition, few field studies have
attempted to investigate the chemical processes which determine
precipitation composition. Several years ago CERL initiated a research
programme to address this problem. The following sections review some
of the work carried out to date.

2. RAIN CARAVAN

About 6 years ago CERL initiated a project to build a rain collector
capable of the continuous analysis of the ionic composition of rain.
Figure 1 gives an illustration of the rain caravan. A detailed
description of the equipment is given by Ames et al., 1987. Basically,
the caravan provides a mobile laboratory which can be used in the field
as a base for gas and aerosol measurements in addition to precipitation
chemistry. The caravan serves as a wet-only collector with a protected
collection area on the roof. The area of 4 square metres provides
sufficient water for analysis even in light rainfalls. The collection
and analysis is computer controlled and can be activated by a variety
of sensors for rainfall. The computer system also controls the
calibration of analysis equipment and triggers an automatic flushing
gantry at regular intervals to supplement the wet-only protection
system and ensure clean collector surfaces. A schematic of the
equipment is shown in Figure 2. Rainfall is continuously analysed for
sulphate, nitrate, chloride, ammonium, sodium, calcium and hydrogen
ions by electrochemical and colorometric methods. Commercial gas
instrumentation is used for sulphur dioxide, oxides of nitrogen and
ozone while hydrogen peroxide is measured in both the gas and liquid
phases using a CERL constructed detector (Ames, 1983). Aerosol
analysis is by filter collection and ion chromatography and particle
size distributions are obtained from a Royco optical counter and a
charge mobility device (TSI EAA).

The advantages of such a continuous analysis device are
illustrated in Figures 3 and 4. Figure 3 shows very clearly the
variation of the concentration of ions with the rate of raining. This
event was recorded in Leatherhead during spring 1981. It is tempting
to deduce that the observed molecular composition resulted from a
mixture of sulphuric acid and ammonium nitrate with a little NaCl from
sea salt. However, this could well be a false deduction and the real
point is that all the ions are correlated in time and related to the
rate of raining. It is only by observing the relative temporal changes
in ionic composition that the various inputs giving rise to the ionic
composition can be inferred. In this case it can be deduced that there
is some acid sulphate input on the basis of the relative magnitude of
the ions. It is interesting to note that the major ion at the start of
the episode was ammonium. Several explanations are credible for this

262

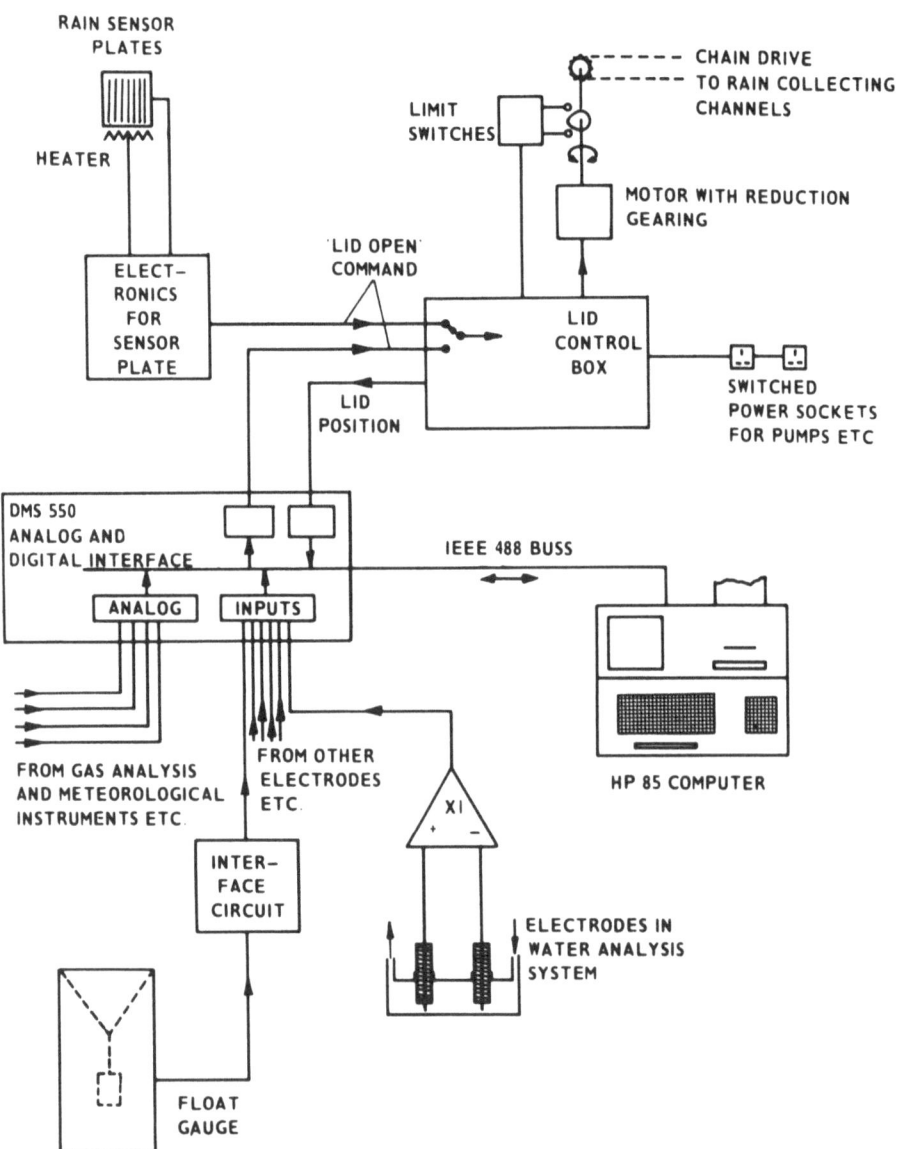

FIG. 2 THE RAIN DETECTION, LID CONTROL AND DATA ACQUISITION SYSTEM

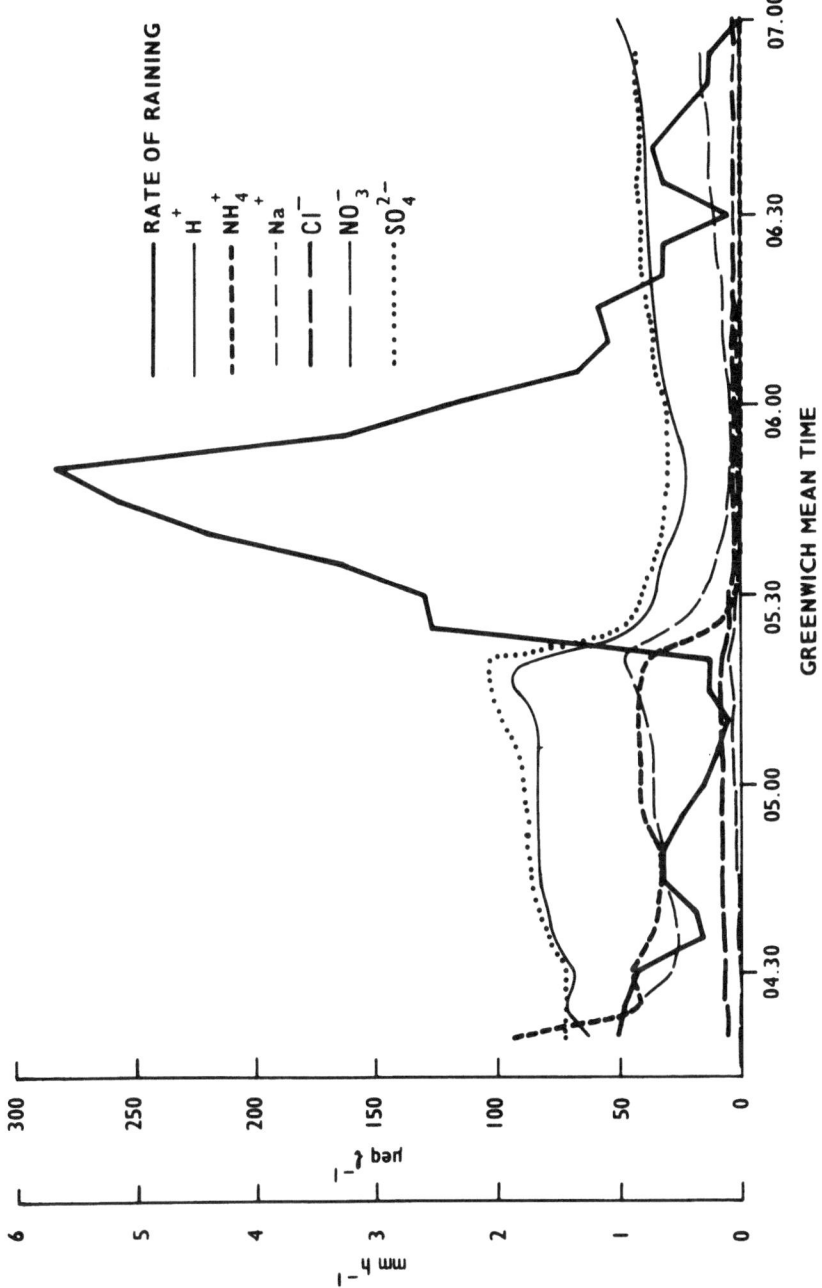

FIG. 3   RAIN CHEMISTRY DURING EPISODE 17B, 11th APRIL 1981

observation but washout of local ammonia can be considered a likely possibility.

Figure 4 shows another event recorded the following year in Leatherhead. The first hour of the record shows a fairly clear example of nitric acid contributing to the composition of precipitation. This can be seen from the correlation of the hydrogen and nitrate ions and their relative magnitude. It is generally considered that nitrate in precipitation arises from either gaseous nitric acid or nitrate aerosol, say ammonium nitrate, rather than oxidation of oxides of nitrogen species in solution. If the enhancement is due to the washout of nitric acid gas from the atmosphere below cloud then an estimate of the concentration of about 1 ppb is obtained, which is not an unreasonable figure given all the gross assumptions. Figure 4 again shows the reciprocal variation of composition with large changes in rainfall rate although the relationship is not always simple as can be seen from the second half of this event. It should be noted in passing that a discrete sampler would probably have missed this episode of nitric acid in the rain because the peak in the volume of rain occurred after the peak in nitric acid and the effect would have been greatly diluted in a bulk sample.

This nitric acid event is not important from a flux point of view since relatively little rain with this composition actually fell and further, the pH was not low enough to cause any direct effect. However, this event is significant from a chemical point of view in that it clearly demonstrates the presence of a particular species in the precipitation.

To fully understand quantitatively the origin of the nitric acid in the precipitation requires a great deal of information, not just the physical parameters concerning the rainfall and cloud, but chemical information concerning other species in the atmosphere such as ammonia. This underlines the importance of measuring many species in all three phases.

There is another way to use time resolved data. Species reflecting the photochemical activity of the atmosphere may have interesting diurnal cycles. If these species or their products are soluble and deposited by rain it is possible to build up a diurnal pattern. For precipitation this is much more complex than for gas phase species. Observations are of course only available for the duration of rain and to build up a complete 24 hour period may require a considerable number of days, each of which represents different conditions of wind direction, temperature, etc. Despite these complications a simple comparison of day/night is worth examining.

Two species can be readily examined. These are hydrogen peroxide and nitric acid. Figure 5 shows the diurnal pattern for hydrogen peroxide. This data was collected in central Wales during July 1985 within the catchment area of the Dinas hydro-electric power station. Nineteen eighty five was a particularly wet early summer and an impression of the diurnal cycle could be built up after only 2 weeks. Each point in Figure 5 represents a one minute average.

The highest values of hydrogen peroxide were observed during the daylight hours but a clear diurnal cycle is not apparent as relatively

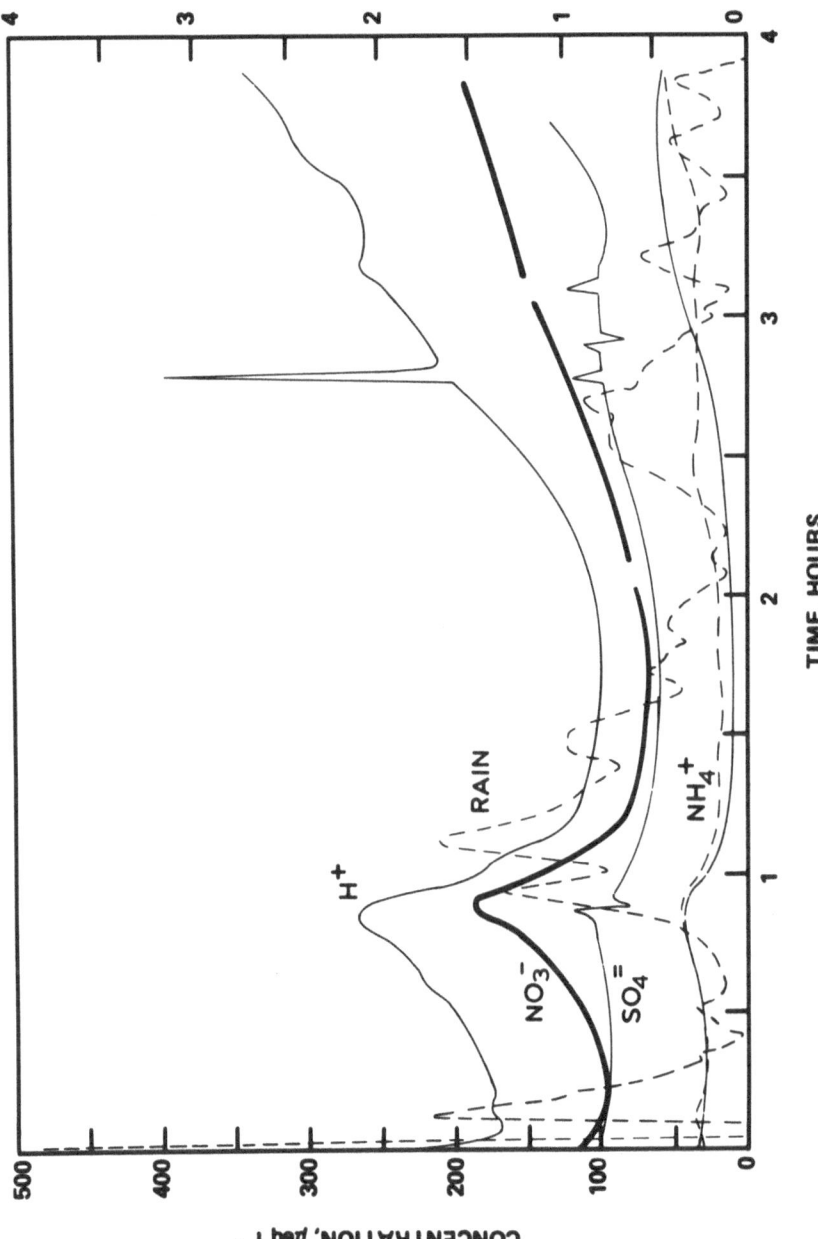

Fig.4 START 2032, 2 NOVEMBER 1982

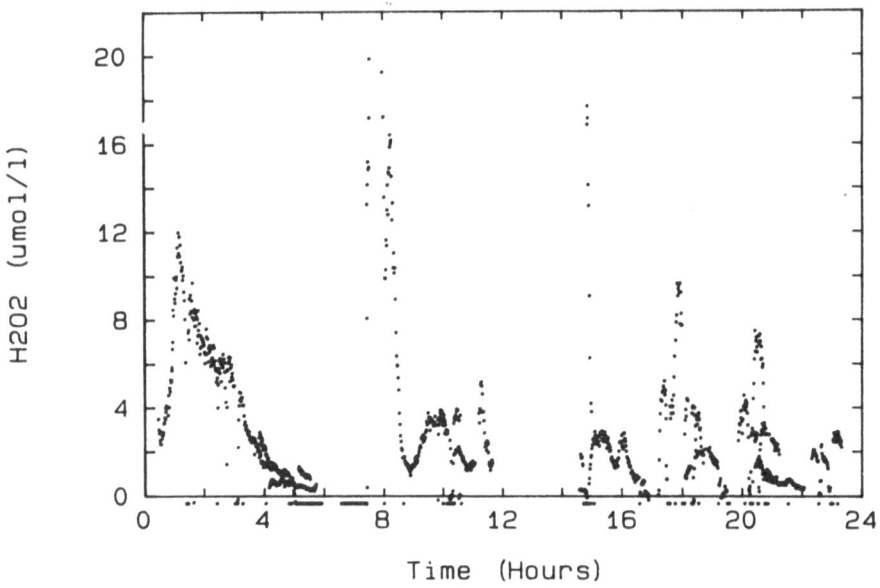

Fig.5 Diurnal Cycle H$_2$O$_2$.
DINAS, WALES, July 1985.

high values were also observed at night. Again assuming that hydrogen peroxide comes from washout below cloud, the events shown in Figure 5 crudely translate into average gas concentrations of the order of 1/3 ppb. It is, however, important to remember that the aqueous phase hydrogen peroxide observed at the ground represents the residual concentration after liquid phase reaction in the atmosphere has occurred. Consequently, the highest point observed in any event might be the best guide to concentrations in the atmosphere. Central Wales is relatively free from pollution and sulphur dioxide levels were < 2 ppb in the predominately westerly winds which gave rise to the data in Figure 5. Nonetheless, the variations in hydrogen peroxide seen in the precipitation reflect the composition of the atmosphere in a complex way and cannot easily be interpreted. However, the records clearly demonstrate comparable quantities of hydrogen peroxide throughout the 24 hour cycle.

Figure 6 shows the corresponding pattern for nitrate in precipitation. Hydrogen peroxide and nitric acid are similar in that they are photochemically produced soluble gases and therefore might have similar behaviour with respect to precipitation. However, two features complicate the comparison. The first is that nitric acid does not apparently undergo aqueous phase reactions and the nitrate ion concentration should not decay in the way that hydrogen peroxide in solution does in polluted atmospheres. The second feature is that, unlike hydrogen peroxide, nitric acid can be formed at night by reactions involving the nitrate radical and thus a profound diurnal cycle is not expected. This is borne out by the data shown in Figure 6.

The most striking feature of Figure 6 is the large initial values for the concentration of nitrate ion in precipitation. This occurs on virtually every occasion. The most probable explanation of this observation is washout of gaseous nitric acid below cloud.

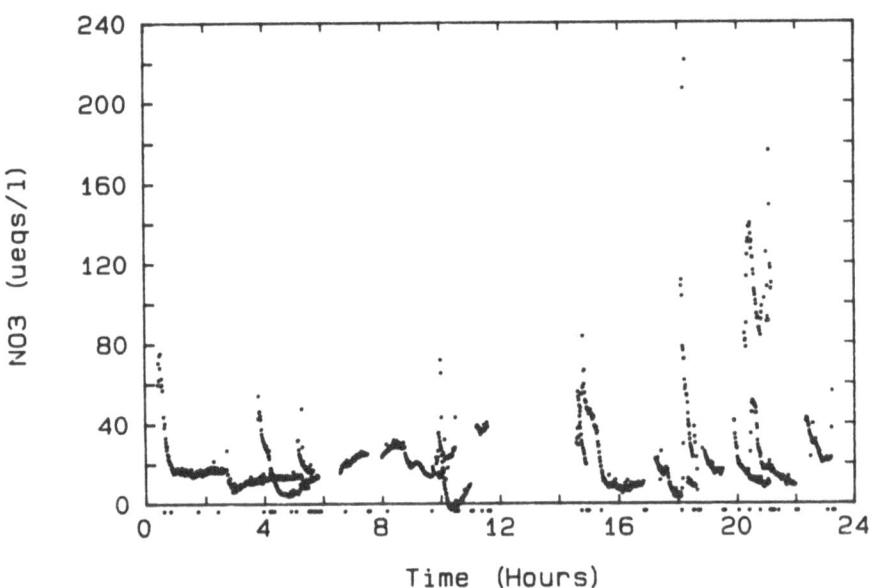

Fig.6 Diurnal Cycle NO_3^- ion.
DINAS, WALES, July 1985.

3. AIRBORNE SAMPLING

In order to confirm any deductions made from ground observations of precipitation it is necessary to make measurements aloft. Measurements in the "vertical" with respect to a ground observation site are very difficult. While an aircraft as a means of obtaining such information is far from ideal, CERL attempted a trial experiment in 1984. This attempt highlighted both anticipated and unexpected difficulties. Weather radar was not available and this is essential for a successful experiment. Some results obtained for the vertical profiles of sulphate and nitrate ions are shown in Figure 7. The data apparently show a decrease in concentration with height. However, there are several features not readily apparent in such a diagram. It is impossible to obtain a true vertical profile above a ground site since the trajectories of individually sized raindrops arriving at the ground will be different depending on their fall velocities. It is thus necessary to rely on the representativeness of samples taken in the vertical. Further, because the aircraft has to sample for a finite time, of the order of minutes, to obtain sufficient sample for chemical analysis, and the aircraft velocity is reasonably high, of the order of 70 m/s, the samples cover a large spacial area compared to the scale of heterogeneity within precipitation systems.

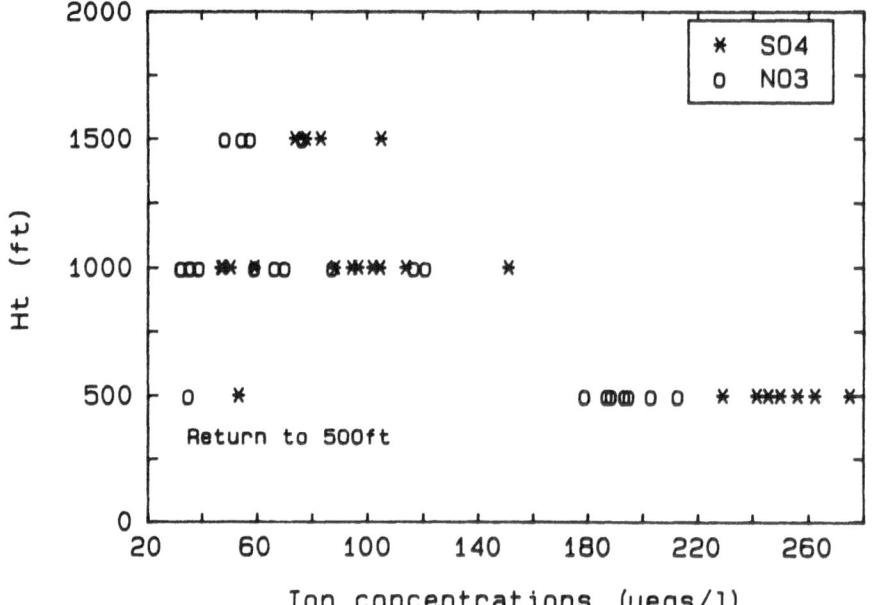

Fig.7 Vertical profile of rain
over rain caravan. Oct. 1984.

It is thus not possible to obtain a true "vertical" profile above any individual site with an aircraft since, of necessity, the aircraft flies at a sufficiently high velocity that a large spacial average is obtained at each height. By contrast, because of the relatively long sampling period required at each height, the total flying time is long and a "snapshot" of vertical concentration profiles cannot be obtained. The aircraft sampling illustrated in Figure 7 took of the order of one and a half hours, which is significant with respect to the duration of precipitation at ground level. This is illustrated in Figure 7 by the observations at 500 ft altitude. The samples collected initially at this height were in the range 180 to 280 µeq/l. However, after making measurements at height up to 1500 ft, returning to 500 ft yielded values in the range 40-60 µeq/l. Thus an aircraft is both too fast and too slow to study washout!

At least two interpretations of the data in Figure 7 are possible. The first is that washout of sulphur dioxide, (and oxidation), and nitric acid is occurring which accounts for the "observed" fall in concentration with height. Alternatively,

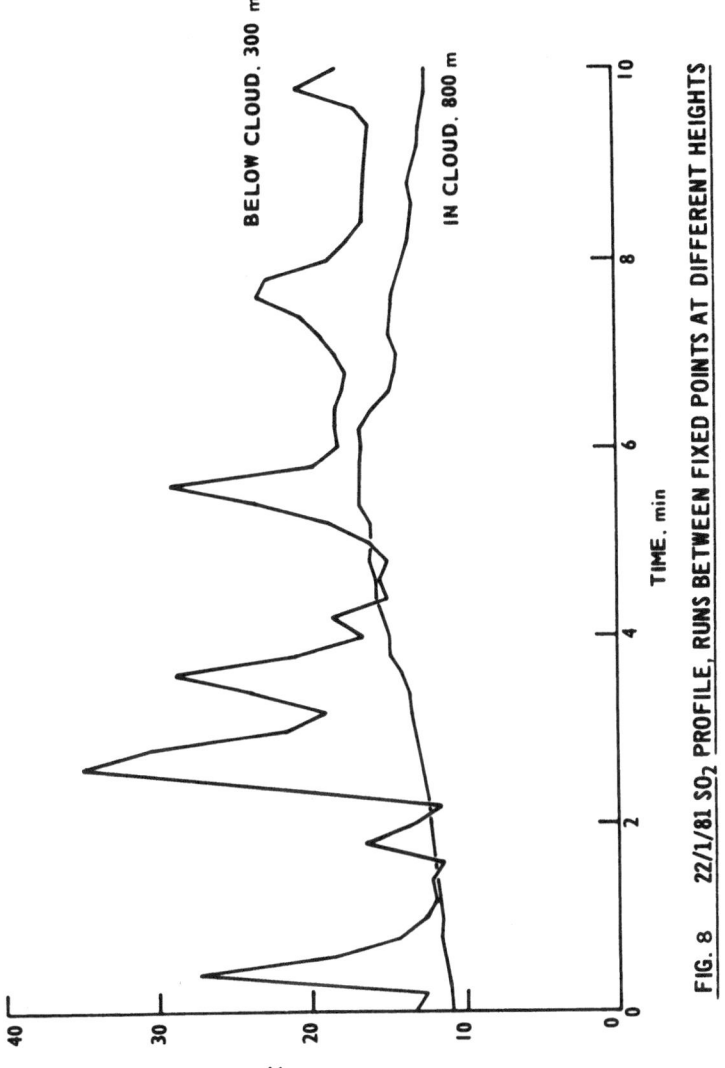

FIG. 8   22/1/81 SO$_2$ PROFILE, RUNS BETWEEN FIXED POINTS AT DIFFERENT HEIGHTS

evaporation of precipitation is occurring as the rain falls through unsaturated air. Quantitative results were not expected from this trial experiment, but it clearly demonstrates the complexities of designing an experiment to study washout in quantitative meteorological and chemical detail. This cannot be done on a routine basis and a large amount of sophisticated equipment both on the ground and in the air is required.

4. CLOUD SAMPLING

Aircraft can be used to examine clouds as well as precipitation. Some of the same problems encountered with precipitation are still relevant. Figure 8 shows the sulphur dioxide record for two transits at different heights, below cloud and in cloud. Some uptake of sulphur dioxide by the cloud water has apparently occurred. Again, the samples were taken along ground stationary flight legs rather than air stationary with respect to local wind. Even had air stationary sampling been possible, the problem of vertical velocities and trajectories of air parcels would remain to be resolved. A better approach would be in terms of budgets of material within a defined volume of air. This requires, in the case of sulphur dioxide, measurements of all three phases, gaseous sulphur dioxide, aerosol and cloud water sulphate.

The present generation of cloud water samplers for use in aircraft leaves much to be desired. They do not, in general, collect water quickly enough especially at low liquid water contents, even if concerns as to their integrity with respect to droplet sampling are ignored. Individual droplet size categories cannot be resolved by such samplers and requires a major advance in technology.

However, available cloud collectors can be used to good effect for budget studies especially if combined with tracer releases. Such experiments have been described by Clark et al., 1984. Figure 9 illustrates some of their results, not from the point of view of the budget study, but rather to demonstrate the response time problem of cloud sampling. It can be seen from Figure 9 that the response of the continuous cloud pH measurement compares unfavourably with that of the gas instrumentation. Generally insufficient water is collected to allow other continuous ionic analysis techniques used in the rain caravan to be applied to the aircraft cloud water samples. This means that only the gross features of the response of cloud chemistry to the air composition can be ascertained. Cloud samples require up to 5 minutes to collect, which at 80 m/s, (collection efficiency usually rises with velocity), represents one sample for 25 km of cloud. Thus only certain types of cloud such as stratus can be successfully sampled and all inhomogenieties within that distance will be averaged. The present generation of aircraft cloud collectors must be seen therefore as analoguous to the bucket rain gauge, providing only average concentration details and are not really a suitable tool for the investigation of chemical mechanisms except indirectly through budget type studies.

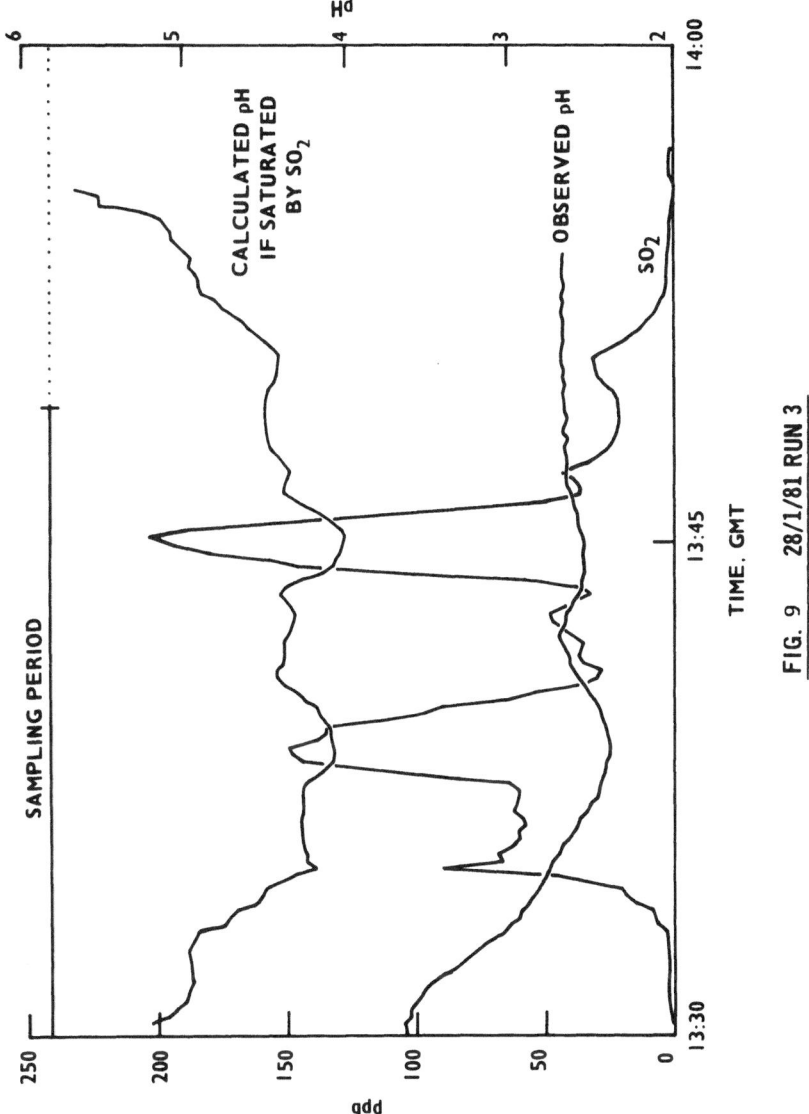

FIG. 9   28/1/81 RUN 3

5. CLOUD MODELLING

As intimated above, processes within clouds can be very fast, such as the oxidation of sulphur dioxide by hydrogen peroxide. This means that from a chemical process point of view the edges of clouds are of great interest. What happens chemically as the relative humidity in an air parcel rises as it approaches and is incorporated into the cloud state? Questions of this sort can be examined by computer modelling, as far as current knowledge will allow. Experiments to confirm or refute these expectations can then be planned in a quantitative manner.

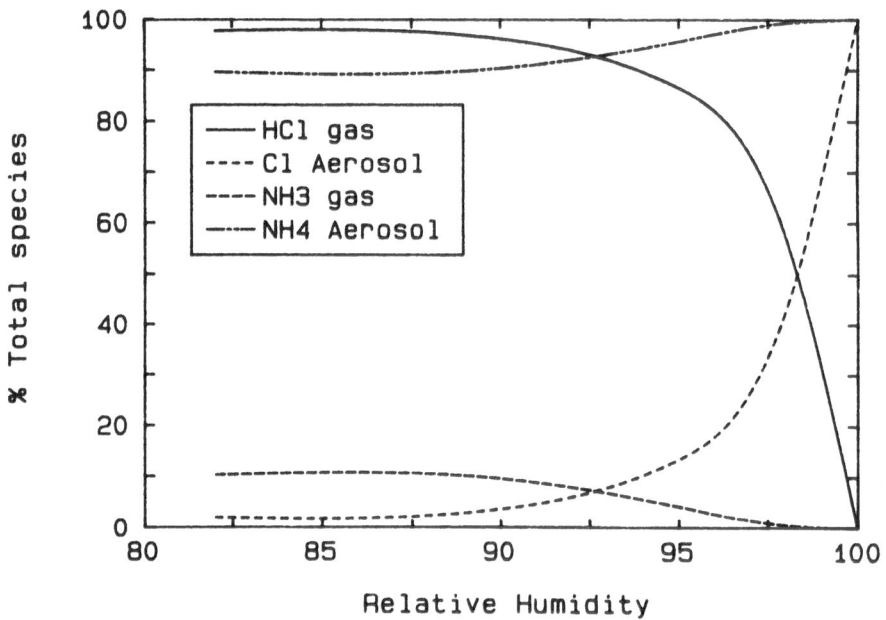

Fig.10 Mixed Ammonium Chloride Sulphate
Aerosol equilibrium

Figure 10 shows the calculated equilibrium behaviour of ammonium chloride sulphate internal mixture aerosol as the relative humidity rises (Marsh, 1982). At 80% relative humidity the bulk of the chloride should be in the gas phase as HCl with relatively little being incorporated into the predominately sulphate aerosol. Ammonia on the other hand should stay mainly in the aerosol with a low vapour pressure. As the humidity rises so the gaseous chloride and ammonia is incorporated into the aerosol droplet. Clearly profound changes in gas and aerosol composition could occur before cloud base is even reached. McElroy, 1987, has produced a computer model of cloud formation which takes into account as much of the aerosol, aqueous chemistry as is currently understood. The emphasis in the model is chemistry rather than meteorology and air parcels of a monodisperse aerosol are considered to move along a simple adiabatic path. Further details will be found in McElroy, 1987.

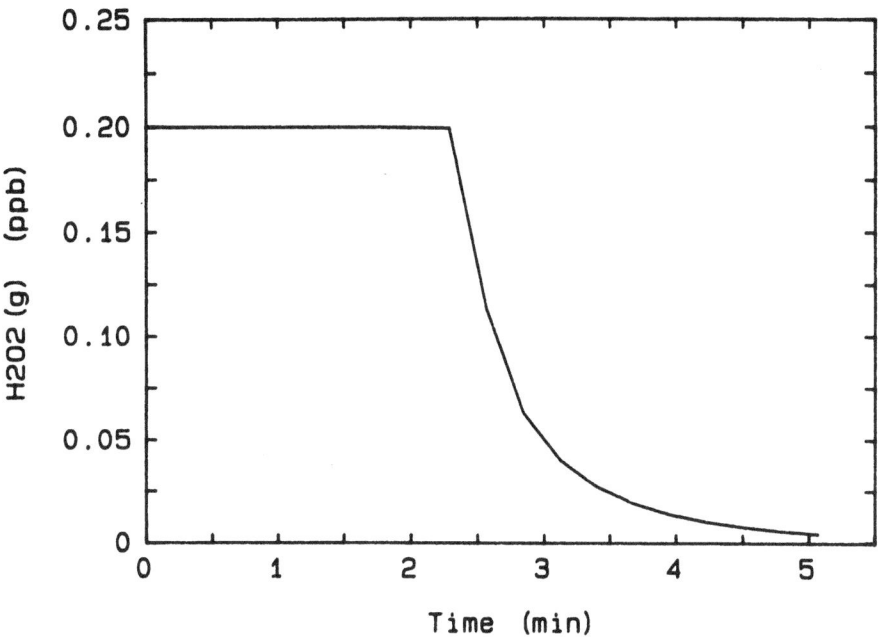

Fig.11a Model predictions for H_2O_2
in gas phase near cloud base

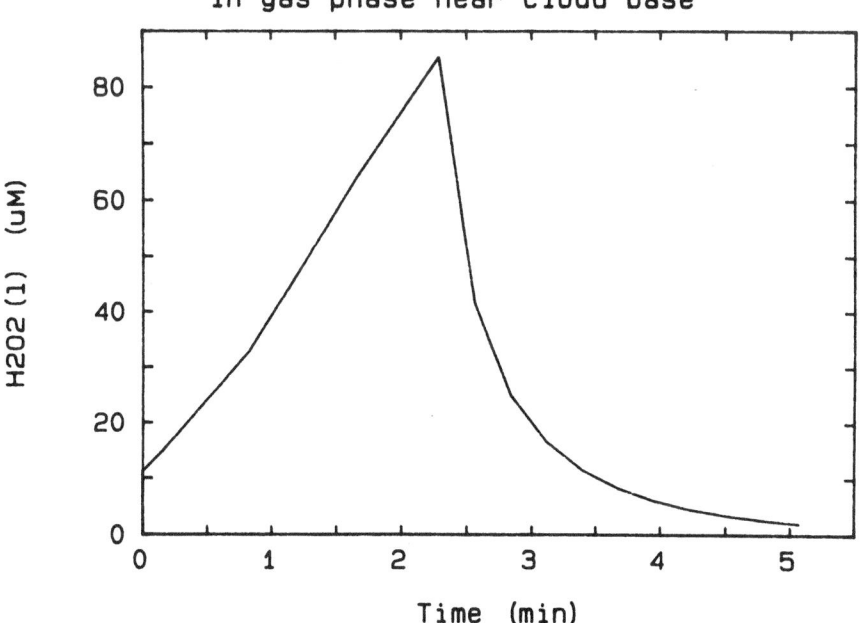

Fig.11b Model predictions for H_2O_2
in liquid phase near cloud base.

Such models are of course qualitative rather than quantitative, but they do provide a tool to investigate the likely relative importance of chemical processes. Figure 11 shows the behaviour of hydrogen peroxide with time as an air parcel approaches and is incorporated into cloud. The standard case boundary conditions used to generate Figure 11 are appropriate to ground based cloud experiments which will be discussed in the next section. The initial hydrogen peroxide concentration in the gas phase was taken to be 0.2 ppb. Figure 11 shows that the gas concentration is not significantly altered from this value until cloud base is reached after about 2.5 minutes. Initial relative humidity was taken to be 80% and Figure 11 shows that for the sea-salt aerosol used in this case, some of the hydrogen peroxide gas has been incorporated into the aerosol. As the relative humidity rises the ionic strength resistance to dissolution is reduced by droplet growth and the aerosol hydrogen peroxide concentration increases (McElroy, 1987). When cloud base is reached several effects influence the liquid phase concentration of hydrogen peroxide. The droplet grows rapidly leading to a fall in concentration. At the same time other soluble gases are being taken up in significant quantities by the greatly enlarged droplet, including sulphur-dioxide. This allows the reaction of hydrogen peroxide with dissolved sulphur dioxide to proceed and as a consequence the hydrogen peroxide concentration in solution falls even faster than straight dilution effects would predict.

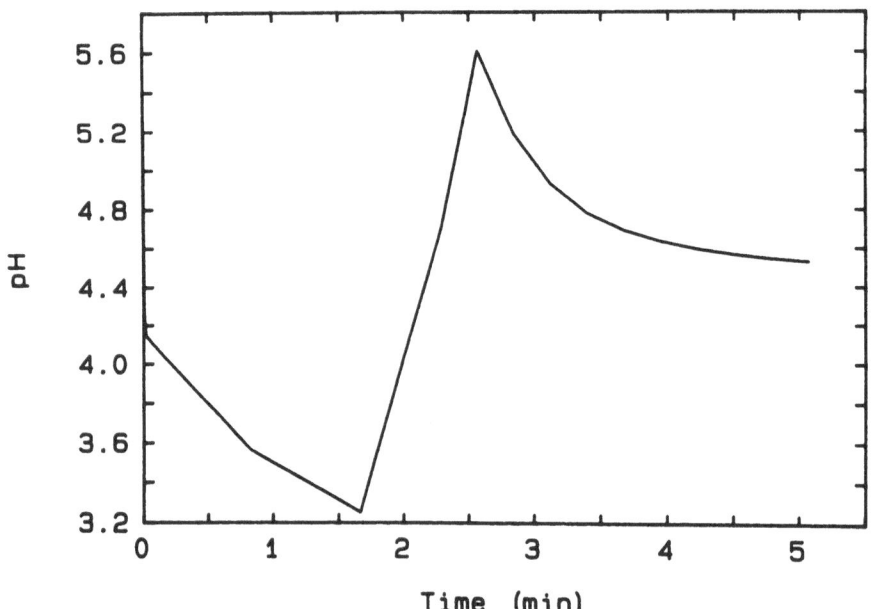

Fig.12 Model predictions for pH
near cloud base.

Figure 12 shows the corresponding pH prediction of the model. The dilution effects near cloud base after about 2.5 minutes dominate the record but the predicted decrease in pH due to sulphur dioxide oxidation in solution is clearly seen in the cloud itself. Indeed this oxidation reaction is also occurring in the aerosol before cloud droplet formation and is contributing to the lowering of pH as cloud base is approached. With the initial conditions used in the standard case pre-cloud reactions are not significant in total mass terms. However, if the concentration of hydrogen peroxide is an order of magnitude lower, 200 ppt, then all of the gas phase hydrogen peroxide is incorporated into the aerosol droplet before cloud formation, and can react there with sulphur dioxide. Such low concentrations have been observed in northern latitudes of the UK in winter (Gervat et al., 1986).

Such model calculations confirm the need to examine cloud chemistry in detail near to cloud. Given the difficulties of aircraft sampling of clouds, this is best achieved by ground based observations.

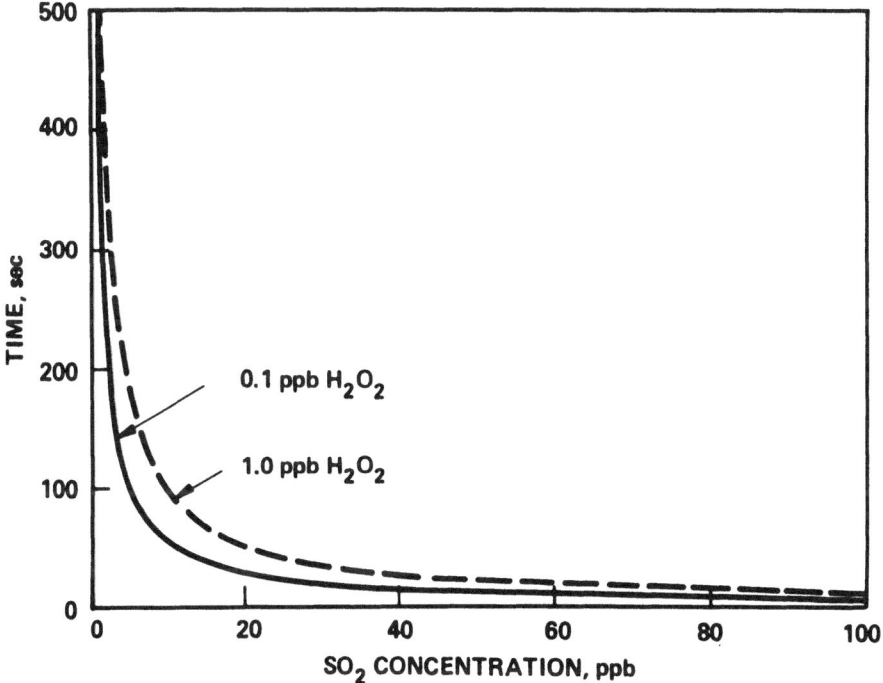

FIG.13 FIRST HALF LIFE OF HYDROGEN PEROXIDE IN SOLUTION

6. GROUND LEVEL CLOUD EXPERIMENTS

At first sight attempts to study clouds at elevated ground sites seems to have several disadvantages compared with aircraft studies. Certainly, cloud types that can be studied are restricted to intercepted cloud and orographic cloud. Further, there is probably no

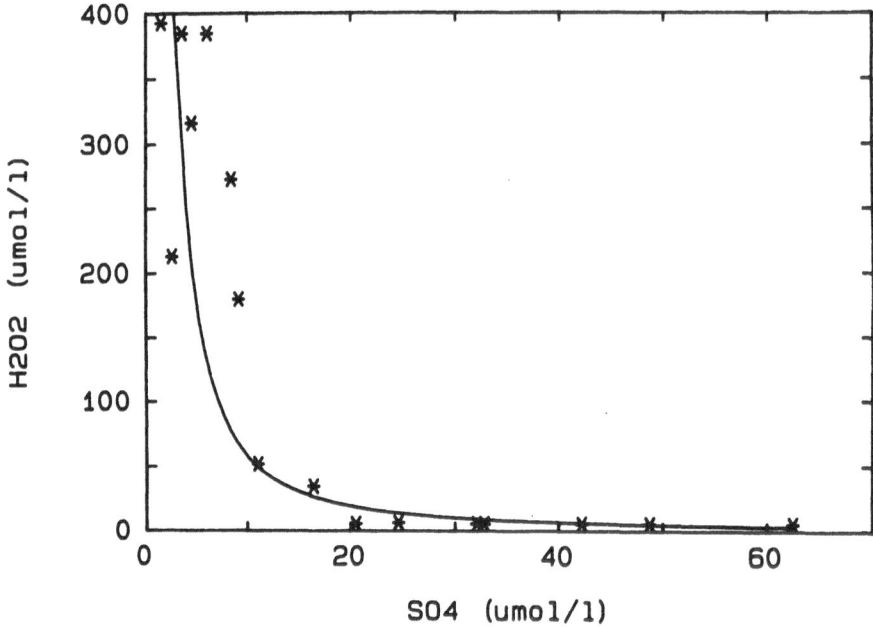

Fig.14 KEMA Observations of H_2O_2
and sulphate in cloud water.

access to cloud top, so that there will be difficulties in
characteristing air entrained from aloft. However, under restricted
conditions and for particular types of experiment the ground site has
some advantages.

The major advantage of measuring orographic cloud at ground level
on a hill-side is that to a first approximation, time equates with
distance up the hill. Since ground based cloud collectors are
relatively easily deployed on a hill-side, measurements at different
heights above cloud base represent cloud development history. Further,
for specific sites it can be the case that it is reasonably common that
the air flow over the hill follows a simple adiabatic lapse rate.
Under these conditions the major problems caused by entrainment are
avoided.

The order of the timescales of the sulphur dioxide oxidation
reaction with hydrogen peroxide are shown in Figure 13. Since air
transit times over simple hills is of the order of a few hundred
seconds, there is the possibility of studying this reaction at
relatively clean air sites where sulphur dioxide concentrations
are < 10 ppb. The general form of these expectations is confirmed by
aircraft results otained by KEMA in Holland. Figure 14 shows the
inverse relationship one would expect from "averaged" cloud water
composition obtained from aircraft samplers and a fast formation
reaction for sulphate. However, further interpretation of these
results is limited for all the reasons given hitherto.

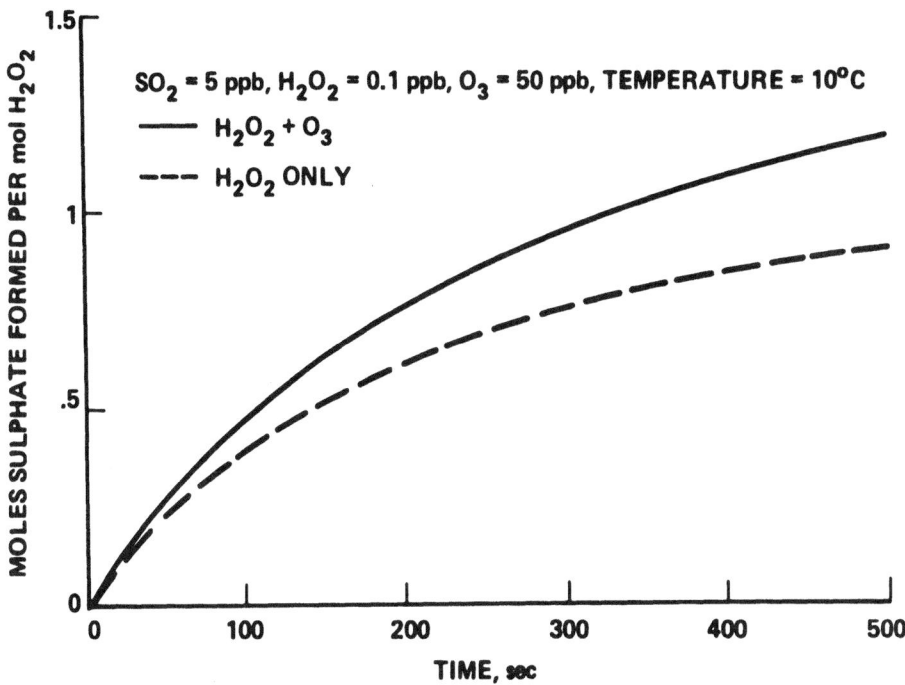

FIG.15 SULPHATE PRODUCTION

The reaction of sulphur dioxide with ozone can also be significant on the time scale of transit over simple hills depending on the pH of the cloud water. Figure 15 shows that the ozone reaction can contribute of the order of 10% of the sulphate formed within a few hundred seconds under suitable clean air conditions.

Great Dun Fell in Cumbria in the UK meets most of the requirements of a suitable hill top for ground level cloud chemistry experiments outlined above. In addition, attempts can be made to influence the cloud chemistry directly by altering one component of the atmosphere by a deliberate release of material upwind of a hill. For reference, an insoluble tracer can be used to characterise the dispersion of the release. CERL has developed continuous sulphur hexafluoride tracer detectors for use in aircraft budget studies, (cf Clark et al., 1984), and sulphur hexafluoride is ideal as a reference material since the emissions required are small. Simple dispersion calculations appropriate to a release point about 5 km upwind of the summit suggested that emission rates of the order of 2 kg/h sulphur hexafluoride would be required to achieve concentrations of the order of 100 ppt at the summit. This was confirmed to be of the right order by later experiments. Release of materials of chemical interest has to be larger to significantly exceed their background concentrations. Emission rates for sulphur dioxide to achieve a 10 ppb increase at the summit have to be of the order of 100 kg/h for an emission site about 5 km away.

This conceptual experiment can be further enhanced by another difference between aircraft and ground level sampling. Ground cloud collectors can be made to larger sizes than is possible on an aircraft. The increase in size that is possible more than outweighs the higher air velocity appropriate to the aircraft samplers and a greater volume of water can be collected on the ground than in the aircraft. This opens up the possibility of continuous ionic analysis. CERL have been able to exploit the expertise developed for this type of analysis in the rain caravan by applying it to ground based cloud collectors. Indeed, for quantitative understanding of such release experiments a continuous analysis technique is essential. If insufficient water is collected to supply analysis needs, it is very easy to combine two ground cloud collectors in parallel.

The idealised release experiment is shown schematically in Figure 16. In principle the greater the level of instrumentation that can be used in the experiment the more reliable and useful the resulting data. The advantage this experiment has over laboratory studies or chamber experiments is that there are no wall effects and the gaseous and aqueous radical concentrations are correctly "simulated"! Only one component of the atmosphere has been significantly altered. The first experiment planned was the study of sulphur dioxide oxidation. The objective was not to measure the reaction rate of dissolved sulphur dioxide with hydrogen peroxide, that is best determined in the laboratory under carefully controlled conditions. Rather, it is to see if the observed sulphur dioxide loss is fully understood and can be quantitatively accounted for by current understanding of cloud chemistry and that it is not inhibited or enhanced by any other process not allowed for in that understanding.

CERL together with UMIST have mounted a feasibility study and carried out three full scale experiments with sulphur dioxide at Great Dun Fell. Gervat et al., 1986 and Hill et al., 1986, papers presented at this NATO workshop, describe the results in detail. Suffice to say here that both expected and unexpected results were obtained.

7. FURTHER DEVELOPMENT

The existing rain caravan will be deployed at a variety of sites in the UK including forest and upland areas. It is an ideal tool for characterising air composition in all three phases at ground level. Some development work on the aerosol sampling techniques is in progress to improve their characterisation in terms of size categories and the gas analysis system is being further supplimented by hydrocarbon, ammonia and nitric acid measurements.

Development work is in hand on the aircraft cloud samplers to improve their efficiency but future experiments will be directed to budget and characterisation studies especially in "background" air. This includes both the "clean" boundary layer over the Atlantic and the free troposphere above the boundary layer. The cloud chemistry aspect of these studies forms part of a larger programme investigating photochemical oxidants and their precursors.

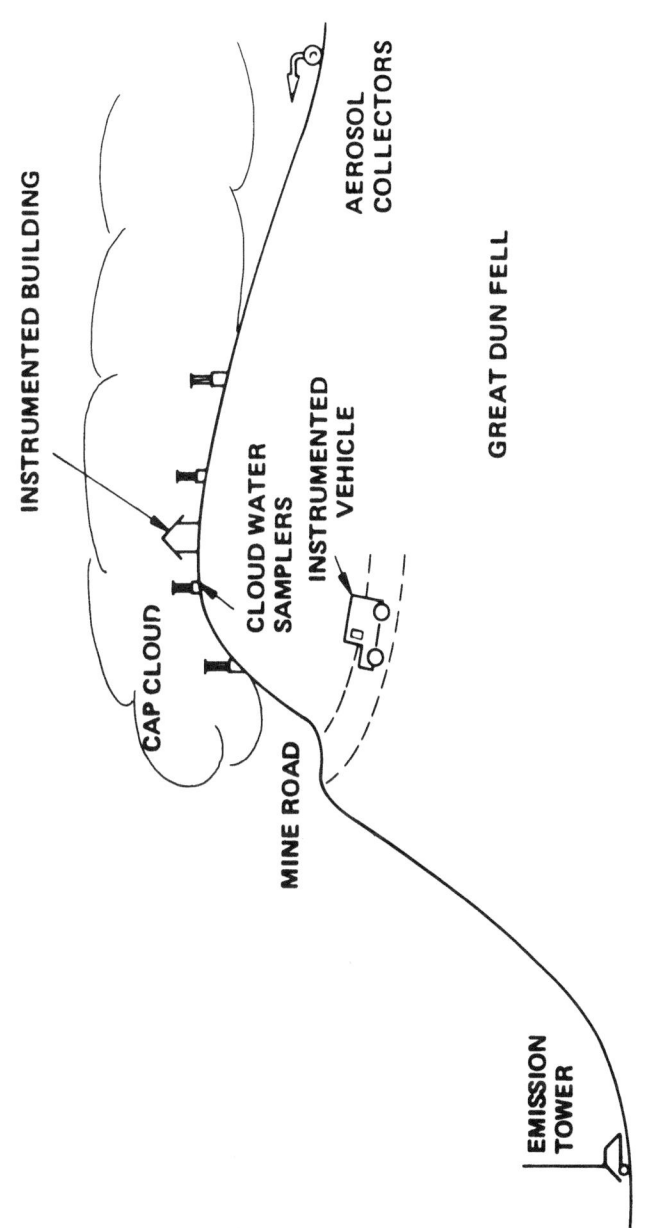

INSTRUMENTED BUILDING

CAP CLOUD

CLOUD WATER SAMPLERS

INSTRUMENTED VEHICLE

AEROSOL COLLECTORS

GREAT DUN FELL

MINE ROAD

EMISSION TOWER

FIG. 16 SCHEMATIC DIAGRAM OF FULL EMISSION EXPERIMENT

Plans for future ground level cloud experiments at hill top sites involve the release of other material both in conjunction with and independent of sulphur dioxide releases.

Table 1   Emissions to Influence Cloud Chemistry

Water	Air
Sulphur dioxide Ammonia Formaldehyde Chlorine Iron	Nitric oxide

Table 1 shows some possibilities. Two categories of release can be envisaged, those which mainly influence the aqueous chemistry and those which mainly influence the air chemistry. The former is of more immediate interest. Ammonia should have a profound effect on pH and hence on the rate of the ozone reaction with sulphur dioxide in solution. Formaldehyde would be interesting in its role as a possible inhibitor of sulphur dioxide. It might be possible to release aerosols and if so the role of transition metals in real clouds could be compared with expectations. Chlorine might be an interesting material to release in terms of its anticipated reaction with sulphur dioxide in solution. In a different vein it is worth considering the release of nitrogen dioxide at least in terms of confirming the expectation of little reaction in solution to form nitrate. However, such a release would greatly influence the gas chemistry via the photochemical stationary state. This could be better and more easily exploited by the release of nitric oxide which should influence ozone in the gas phase and hence corresponding affect the aqueous oxidation route of sulphur dioxide.

There are several exciting possibilities but there are several constraints, not least safety. On a practical level these are:-
(a)   the likely timescale reaction,
(b)   the background concentration, and
(c)   continuous analysis of species of interest in cloud water.
Obviously, a hill top experiment will only work if significant reaction occurs in the transit time available across the hill. Any emission has to significantly disturb the ambient concentration of the species of interest. This is no problem with the tracer sulphur hexafluoride but is really only just possible for sulphur dioxide at the Great Dun Fell site and would be totally impractical for ozone. While the continuous measurement of any species of interest in cloud water is not absolutely essential to the experiment, the experience to date (cf Gervat et al., 1986) shows that it is just this analysis which has clearly demonstrated the features of interest.

The interpretation of all of these experiments will require computer modelling of numerous aspects of atmospheric behaviour. A positive feedback system between experiments and computer modelling should help to elucidate chemical processes in the atmosphere.

REFERENCES

Ames, D.L., Roberts, L.E. and Webb, A.H. 1987. An automatic rain gauge for real time determination of rainwater chemistry. Atmos. Env. **21** , 1947-1956.

Ames, D.L. 1983. A method for determining hydrogen peroxide in cloud and rainwater. CEGB report TPRD/L/2552/N83.

Clark, P.A., Fletcher, I.S., Kallend, A.S., McElroy, W.J., Marsh, A.R.W. and Webb, A.H. 1984. Observations of cloud chemistry during long range transport. Atmos. Env. **18** , 1849.

Marsh, A.R.W. 1982. Vapour pressure of HCl above aerosols of mixed composition. CEGB report TPRD/L/2339/N82.

McElroy, W.J. 1987. The interactions of gases with aqueous aerosol particles. Part IV. The growth of cloud condensation nuclei coupled with aqueous chemical transformations in an adiabatically cooling parcel of moist air. CEGB report TPRD/L/3111/R87.

Gervat, G.P., Clark, P.A., Marsh, A.R.W., Choularton, T.W. and Gay, M.J. 1986. Controlled chemical kinetic experiments in cloud; a review of the CERL/UMIST Great Dun Fell project. NATO workshop: Acid deposition processes at high elevation sites, Sept. 1986, Edinburgh.

Hill, T.A., Jones, A. and Choularton, T.W. 1986. Modelling wet deposition of pollutants onto elevated topography. NATO workshop: Acid deposition processes at high elevation sites, Sept. 1986, Edinburgh.

CONTROLLED CHEMICAL KINETIC EXPERIMENTS IN CLOUD: A REVIEW OF THE
CERL/UMIST GREAT DUN FELL PROJECT

G.P. Gervat, P.A. Clark and A.R.W. Marsh
Central Electricity Research Laboratories (CERL)
Kelvin Avenue, Leatherhead, Surrey, UK

and

T.W. Choularton and M.J. Gay
Atmospheric Physics Research Group
University of Manchester Institute of Science and Technology
(UMIST), Manchester, UK

ABSTRACT. This paper describes a novel experimental technique for the
examination of chemical kinetic processes in clouds. This involves the
controlled release of a particular study gas and a tracer, upwind of a
cloud-capped hill. Atmospheric measurements before, in, and after the
cloud allow chemical changes within the cloud system to be monitored
and hence the determination of reaction rates in field conditions.

   Results from the application of this technique to the study of SO_2
oxidation reactions in clouds are presented. These are derived from a
series of experiments carried out at Great Dun Fell, Cumbria, over the
last two years. A pilot study in October 1984 was followed by three
full scale experiments involving the release of SO_2 and a tracer, SF_6.
Two were in spring (1985 and 1986) and one in winter (1985) when
photochemical activity was lower. Observed SO_2 oxidation rates
reflected the levels of photochemical activity.

   Plans for future experiments incorporating other releases are also
presented.

1. INTRODUCTION

Laboratory studies (e.g. Maahs, 1983) and aircraft measurements (e.g.
Bamber et al., 1984) have shown that oxidation reactions of SO_2 in
cloud droplets may be very rapid compared to those in the gas phase.
However, there are limitations in these techniques for understanding
the detailed processes taking place in clouds; it is difficult to fully
simulate clouds in the laboratory and aircraft measurement inevitably
have limited temporal resolution. In order to further understand the
oxidation processes and determine chemical reaction rates and effective
rate constants, a series of novel experiments has been carried out at a

M. H. Unsworth and D. Fowler (eds.), Acid Deposition at High Elevation Sites, 283–298.

284

site in northern England. This paper presents a brief overview of the project from the original concept to some of the most recent results. Plans for the future experiments of similar type are also outlined.

2. EXPERIMENTAL CONCEPT

On a cloud capped hill, air rising up the hillside cools adiabatically resulting in condensation on suitable nuclei. As cloud droplets pass up and over the hill gases dissolve in them, chemical reactions occur

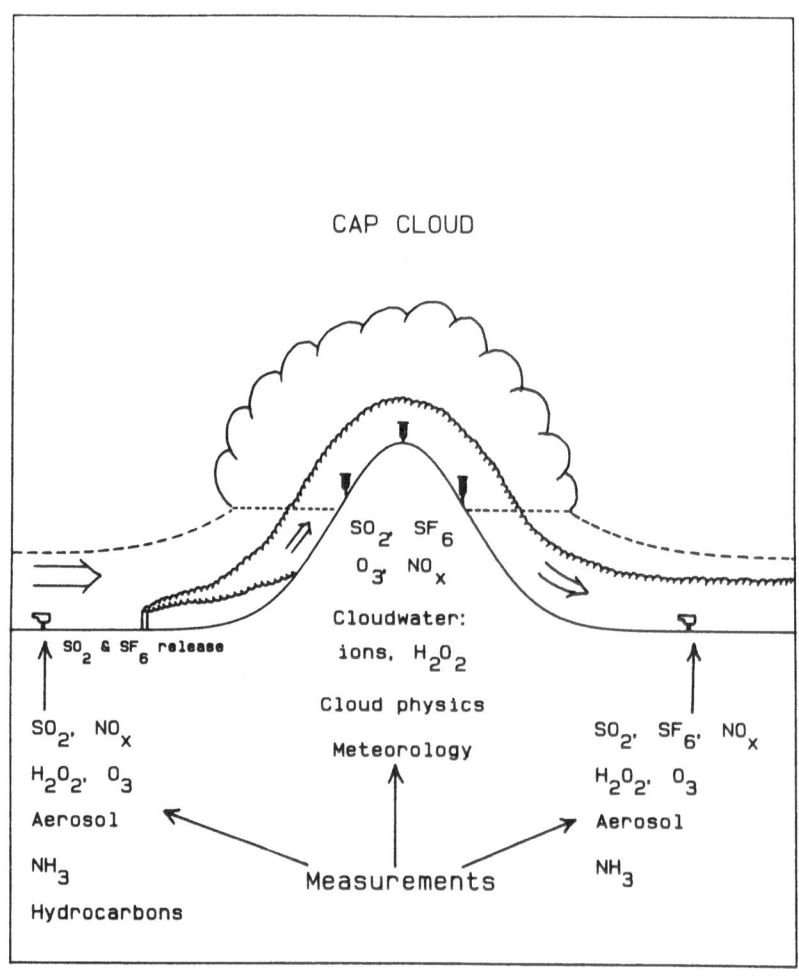

Fig. 1   The Ideal Experiment

and further particles may be incorporated. When the droplets evaporate on passing through the downwind edge of the cloud, material in the drops is returned to the gas and aerosol phases. Any chemistry occurring in the cloud modifies the downwind air composition compared with the upwind.

A series of experiments has been devised in which the upwind air is deliberately altered by releasing into it a species of interest. Changes in the cloudwater chemistry and air composition are then measured. Different release rates on one gas can, in principle, be used, or combinations of serveral gases released. The cloud-capped hill is treated as a large flow chamber with all parameters being approximately constant apart from those being deliberately altered. Figure 1 shows the ideal experiment. Air upwind, on, and downwind of the hill is fully characterised while cloud collectors stationed on the hill, along the line of release, enable progressive changes in cloud chemistry to be monitored.

The first gas releases have involved SO_2 along with an inert tracer, SF_6. Before carrying out these experiments, a feasibility study was undertaken to establish operational constraints.

3. FEASIBILITY STUDY

The feasibility study involved both a theoretical study and trial gas releases. It was necessary to know under what circumstances a release of SO_2 would produce a significant change in cloudwater sulphate concentrations and by how much ambient levels of SO_2 needed to be increased to produce that change.

The calculations were based on laboratory measurement of reaction rates applied to cloud of liquid water content 0.4 g m^{-3} and an initial pH of 5 at 10°C. In planning the experiment three years ago when considering the oxidation of SO_2 by H_2O_2, rate values from Penkett, Jones, Brice and Eggleton (1979) were used. However, the more recent rate data from McElroy (1986) confirm the original design criteria and have been used to derive Figure 2. This shows the calculated first half lives of hydrogen peroxide consumption as a function SO_2 concentration, for a gas phase H_2O_2 concentration of 1 ppbv entering cloud base. For values of H_2O_2 less than 1 ppb, the first half lives are almost identical to those for 1 ppbv H_2O_2, as long as the SO_2 concentration is greater than 1 ppb. While the half life is insensitive to the initial hydrogen peroxide concentration, at these levels it varies dramatically with SO_2 concentration, particularly if the concentration of SO_2 is less than about 15 ppb. Therefore, provided background concentrations of SO_2 are only a few ppbv, a small additional increase of ambient SO_2 concentration leads to a large change in H_2O_2 half life. Sulphate concentrations show an increase corresponding at least to the H_2O_2 consumed.

The reaction of SO_2 with O_3 was considered also using reaction rate data from Penkett et al. (1979). At pH values > 4.7 a significant contribution to sulphate production from this reaction was predicted, approximately 20% after 400 seconds. More recent determinations of

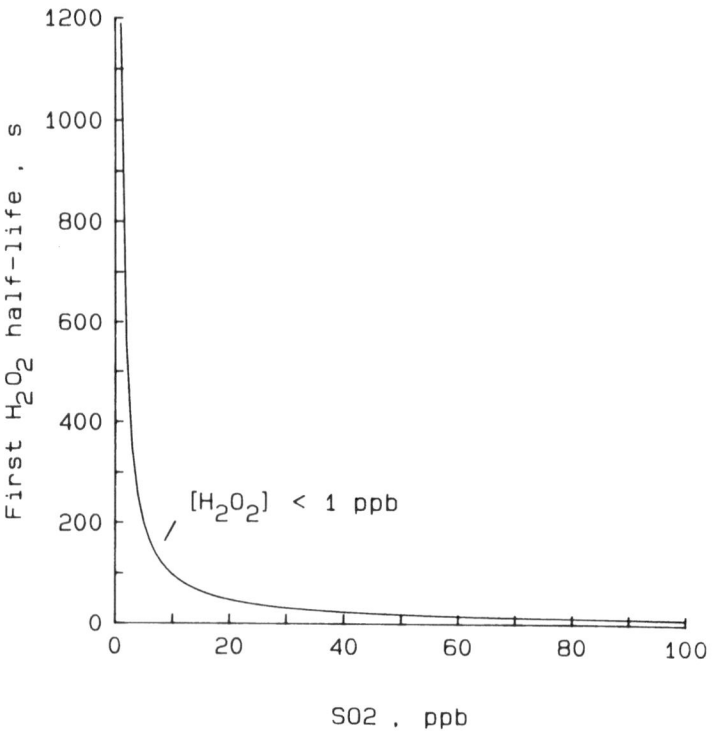

Fig. 2 Estimated first half-lives

of H_2O_2 with changing $[SO_2]$

reaction rate constants (Maahs, 1983) suggest that ozone, in particular, may be even more important at high pH than the earlier rate values suggested.

Great Dun Fell, in northern England, was chosen as the site for the experiments. At the summit, early measurements of background SO_2 in the appropriate wind sector for experiments (see later for details), showed levels ∿ 2 ppbv. Based on the calculations shown in Figure 2, it was decided to attempt to increase ambient SO_2 concentrations at cloud base by 10 ppbv. This was because the first half-life of H_2O_2 at the resultant SO_2 concentration of ∿ 12 ppbv would then be similar to, or shorter than, transit times in the cloud. In addition, 10 ppbv is a change easily characterised by current SO_2 analysers. It was estimated that to produce this increase from a source 5 km upwind a release rate of 100 kg h^{-1} would be required. The planned release points needed to be above the surface layer (about 10 m deep) to avoid significant depletion by dry deposition. To produce an increase of 100 ppt of SF_6 at cloud base (a suitable working level for the tracer detector), a release rate of 2.3 kg h^{-1} from the same site would be required under typical conditions. These estimates were based on simple Gaussian dispersion models.

In October 1984, a field trial was carried out at Great Dun Fell. The objective was to establish whether a gas release could be targetted at the summit. SF_6 was released from the top of a 10 m tower on several occasions from different sites in the nearby Eden Valley. The plume was successfully detected at and near the summit at approximately the predicted concentrations. These results indicated that it would be worthwhile attempting the full experiment.

4. EXPERIMENTAL DETAILS

Great Dun Fell (947 m asl) lies on the western edge of the northern Pennines on a limestone ridge running SE to NW, see Figure 3. Capping cloud covers the summit over 200 days each year. A surfaced road leads to the summit where electrical power is available. A track leads off the road along the southwest slope, to a mine.

Since the ridge runs SE to NW, release experiments are only possible for winds arriving from the south to west or north to east sectors. Measurements in winter 1984 (Gay and Gervat, unpublished work) showed that air arriving at the summit from directions with an easterly component could have SO_2 levels > 20 ppbv, probably due to emissions from Yorkshire or Tyneside. Winds arriving from the southwest were of maritime origin and SO_2 levels were correspondingly lower, < 5 ppbv. This, combined with the favourable location of the mine road, led to three gas release sites being chosen to cover winds arriving from SSW to WSW.

Since the field trial, three full scale experiments involving the release of SO_2 and SF_6 have been carried out at Great Dun Fell; two in 1985 and one in 1986. With each experiment the amount and sophistication of equipment used has increased. The deployment of equipment for the most recent campaign in Spring 1986, CUTE III (CERL UMIST Transformation Experiment III), is shown in Figure 3.

In CUTE III SO_2 and SF_6 were released from the site near the village of Knock. The SO_2 release system incorporated a hot water cylinder that evaporated liquified SO_2. The combined outputs of two small SF_6 cylinders were used for the SF_6 release system. The gases from both systems were piped to the top of a 20 m meteorological tower where they passed to the atmosphere. A wide band, (0.3–2.5 μm), solarimeter recorded insolation levels at this location.

At the background site, denuder tubes and double filter packs were used to determine NH_3, aerosol and trace acid gas concentrations. The denuder tubes were analysed for NH_3 on the same day at the summit building. The filter packs, incorporating one PTFE aerosol membrane pre-filter and a Whatman 41 absorbant filter impregnated with potassium carbonate, were analysed by ion chromatography after the campaign.

Further up the hill, on the mine road, the CERL Mobile Air Pollution Sampling Unit (MAPSU) repeatedly traversed the plume. The vehicle is equipped with an SF_6 analyser (Blackburn and Dear, 1981) to monitor the plume position, a Meloy SA285 flame photometric SO_2 analyser, a Monitor Laboratories NO_x analyser and a Bendix O_3 analyser.

288

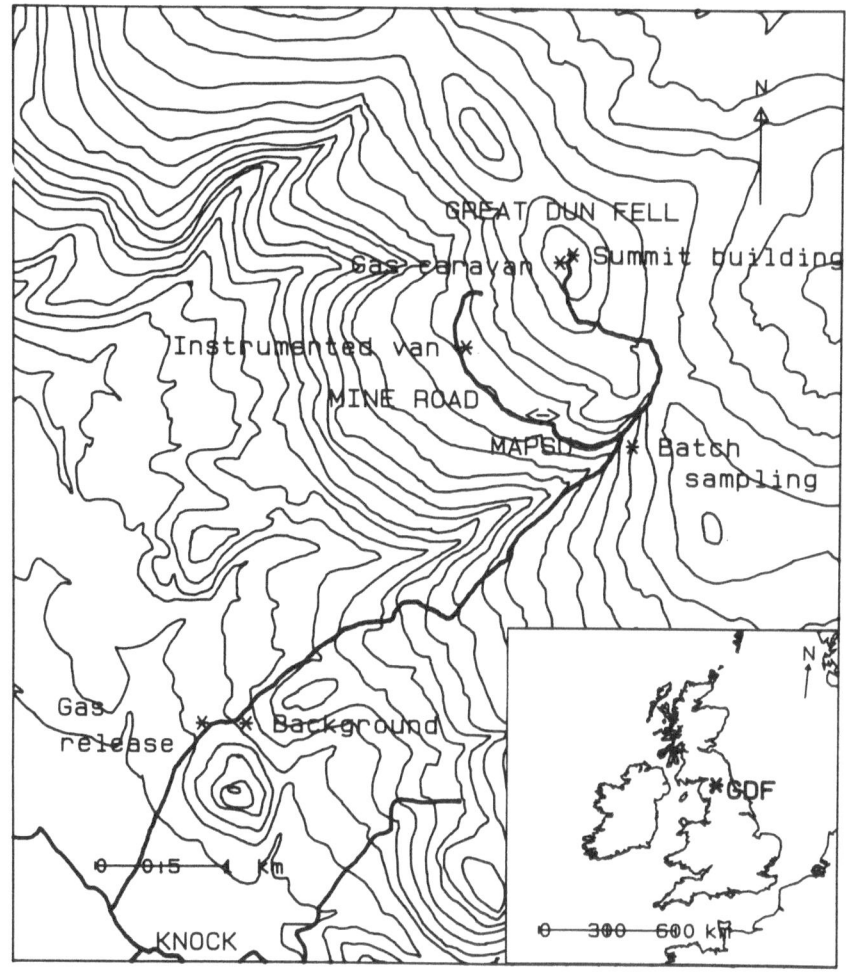

Fig. 3   Great Dun Fell Site
and Equipment Deployed
in Spring 1986, (CUTE III)

An instrumented van was sited to one side of the mine road near
the mine itself. It was equipped with two cloud collectors based on a
design of the Atmospheric Sciences Research Center, Albany. One was
used for collecting bulk samples. The output from the other fed a pH
electrode and a continuous H_2O_2 analyser (derived from the design by
Ames, 1983). An SF_6 analyser marked the passage of the plume and a
Meloy flame photometric detector provided ambient SO_2 concentrations.

Windspeed and direction, and wet and dry bulb temperatures were also measured. A Particle Measuring Systems Forward Scattering Spectrometer Probe (FSSP) provided cloud droplet spectra and hence cloud liquid water content.

At the southern end of the mine road another cloud collector provided bulk cloud samples. A battery powered bag sampler with a sampling time of about 5 mins, was operated concurrently and the bags were analysed for SF_6 concentration at the end of the experiment using one of the continuous SF_6 analysers.

At the summit, a Meloy flame photometric SO_2 analyser, two Severn Science Low-level SO_2 detectors and a further tracer detector were housed inside a caravan, and measured ambient concentrations. Two cloud collectors (the same design as those used on the mine road) were fixed to an instrument platform 4 m above the ground. Their combined outputs ran into the summit building where part of the flow was analysed continuously for H_2O_2, SO_4^{2-} using a colorimetric method and H^+ and NH_4^+ using ion-selective electrodes. The remaining flow was divided into 5-10 min bulk samples and retained for subsequent chemical analysis. (During part of the campaign when cloud liquid water content was high enough, sufficient sample was also available to allow continuous analysis of Na^+, Cl^- and NO_3^- composition using ion-selective electrodes). As at the mine road, a forward scattering spectrometer probe provided cloud droplet spectra and liquid water contents. Wind speed and direction, and wet and dry bulb temperatures were also measured.

At the base station, air samples were taken in stainless steel bottles for subsequent hydrocarbon analysis by gas chromatography (C_2-C_6). After the campaign, all cloud samples were returned to the participating institutions where they were analysed for major inorganic ions by ion chromatography and trace metals by inductively coupled plasma spectroscopy.

5. RESULTS AND DISCUSSION

5.1 Introduction

These experiments have produced large amounts of data, not all of which can be presented here. The results will therefore be described with a view to comparing seasonal effects.

Insolation levels are higher in spring than in winter: Figure 4 shows how the calculated energy available outside cloud for the photolysis of O_3 to produce O_D^1 radicals varied during the release days of the winter 1985 and spring 1985 and 1986 experiments. The differences of at least an order of magnitude in intensities between the winter and spring dates should have produced corresponding differences in photochemical activity. Therefore during a winter experiment it was expected that little extra sulphate would be produced in clouds from an SO_2 release. In the course of a spring release experiment, higher background ozone and H_2O_2 concentrations expected to lead to considerably greater sulphate production.

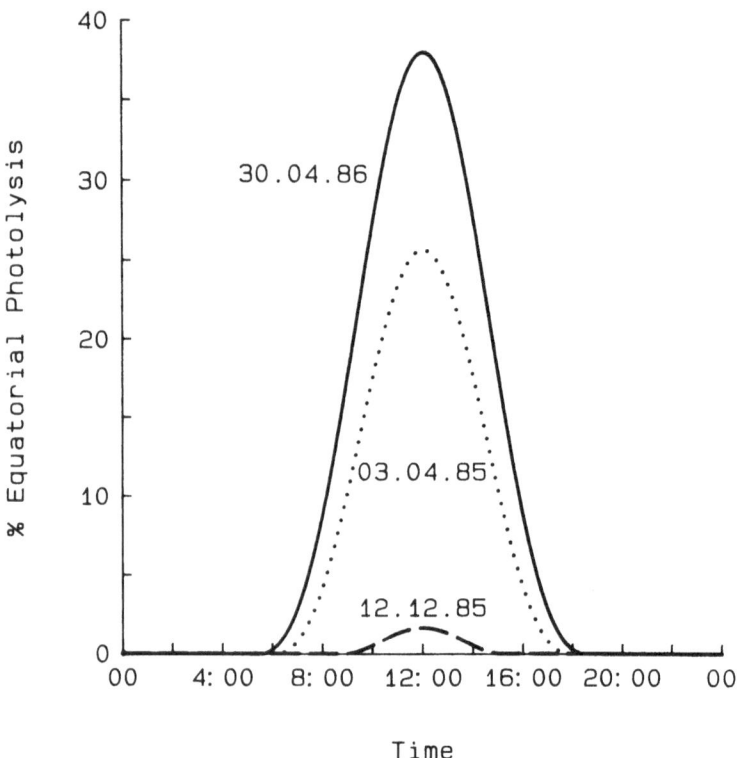

Fig. 4 Insolation Available for O$_3$

Photolysis at Great Dun Fell

5.2 Measurements in Winter

5.2.1 <u>Winter 1985, CUTE II</u>. The expectation of low sulphate production during a winter SO$_2$ release experiment was confirmed by the (CUTE II) experiment carried out on 12 December 1985. After allowing for changes in liquid water content, no measurable increase in sulphate concentration was observed at the time associated with the SO$_2$ release period (1242 to 1401 GMT). Some explanation of this can be gained by considering oxidant levels.

Ozone levels were between 20 and 26 ppbv during the course of the experiment. While, in principle, some oxidation of released SO$_2$ by ozone must have been occuring, sulphate production rates were so low that no consequent change in either aqueous sulphate concentrations or ambient ozone levels were observed.

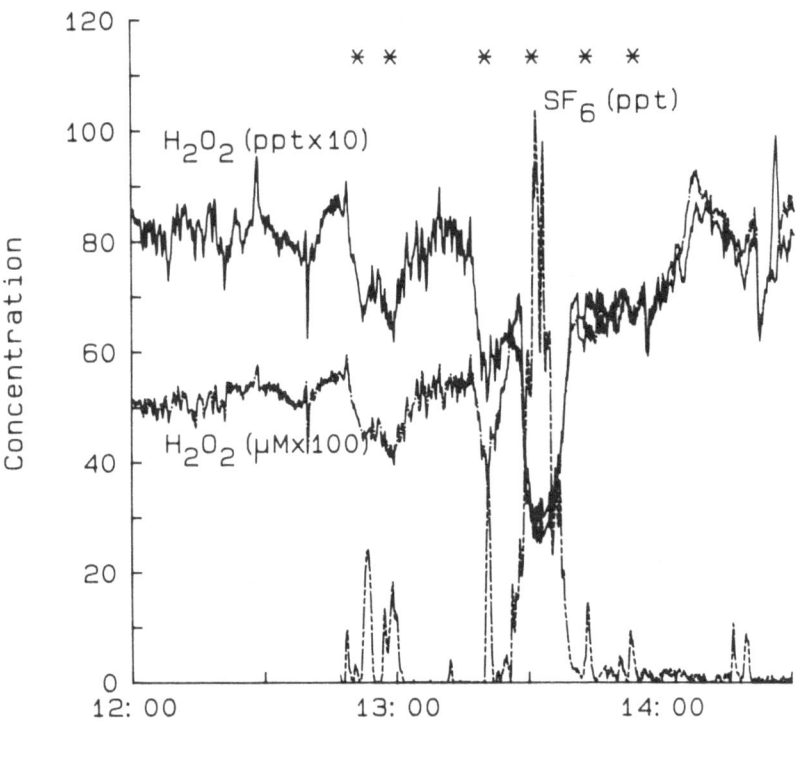

Fig. 5   SF$_6$ and SO$_2$ concentrations

at the summit, CUTE II, 12.12.85

Residual H$_2$O$_2$ concentrations in solution (Figure 5) were very low, 0.5 to 0.9 μM (equivalent to a total gas phase concentration ∿ 8 ppt), during the course of the experiment. Figure 5 does, however, show that there were very marked decreases in H$_2$O$_2$ concentration corresponding in time and relative magnitude to peaks in tracer (and hence SO$_2$). These indicate that despite the low levels of ambient H$_2$O$_2$, oxidation of SO$_2$ by H$_2$O$_2$ was occurring. The largest of the dips corresponds to a 0.4 μM loss of H$_2$O$_2$, equivalent to 4 pptv in the gas phase. A 0.4 μM change in sulphate concentration from one bulk sample to another is less than the changes due to natural variability in background aerosol composition. Consequently no change in SO$_4^{2-}$ concentrations due to H$_2$O$_2$ oxidation of released SO$_2$ was observed in the bulk samples.

The measurements, along with meteorological data, can be used to estimate a reaction rate constant. Consider the following simplified analysis of oxidation of SO$_2$ by H$_2$O$_2$. The decay of H$_2$O$_2$ can be expressed as,

$$\frac{d[H_2O_2]}{dt} = -k[H_2O_2][SO_2]$$

where k incorporates the aqueous reaction rate constant and solubility effects, and concentrations are <u>total</u> equivalent gas phase concentrations. For a volume of cloud moving from cloud base to the summit in a time τ and with SO_2 in large excess compared to H_2O_2 (and therefore with approximately constant concentration), this reduces to

$$\ln\left[\frac{[H_2O_2]_{init}}{[H_2O_2]_{final}}\right] = k\tau\,[SO_2]$$

A plot of the left hand side of this equation versus SO_2 concentration should yield a straight line of slope $k\tau$. Values of this function, derived from six of the peroxide dips marked on Figure 5, are plotted against SO_2 concentrations in Figure 6 along with a linear regression

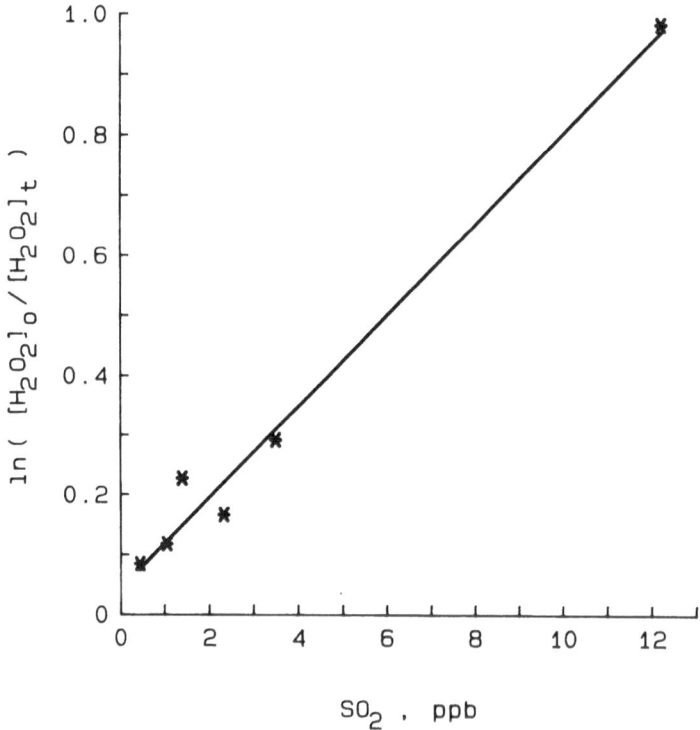

Fig. 6 H_2O_2 rate data, summit

CUTE II, 12.12.85

line. These points appear to lie on a reasonably straight line. Preliminary indications are that the derived aqueous rate constant is smaller (by up to a factor of 3) than those based on laboratory measurements, applied at the same pH and temperature (e.g. Martin and Damschen, 1981 and McElroy, 1986). A fuller explanation of the rate constant derivation is presented in Gervat, Clark, Marsh, Teasdale, Chandler, Choularton, Gay, Hill and Hill (in preparation). Limitations in the present method of analysis of the field data are currently being examined so that a more accurate value of the rate constant can be calculated.

5.3 Measurements in Spring

5.3.1 <u>Spring 1985, CUTE I.</u> It was anticipated that in spring, significant photochemical activity would lead to higher oxidant levels and hence greater in-cloud sulphate production during an SO_2 release experiment, than in winter. In the spring 1985 experiment (CUTE I, 30.04.85) H_2O_2 levels in solution were between 5 and 30 μM during the afternoon, equivalent to 0.1 to 0.4 ppbv in the gas phase. The extra

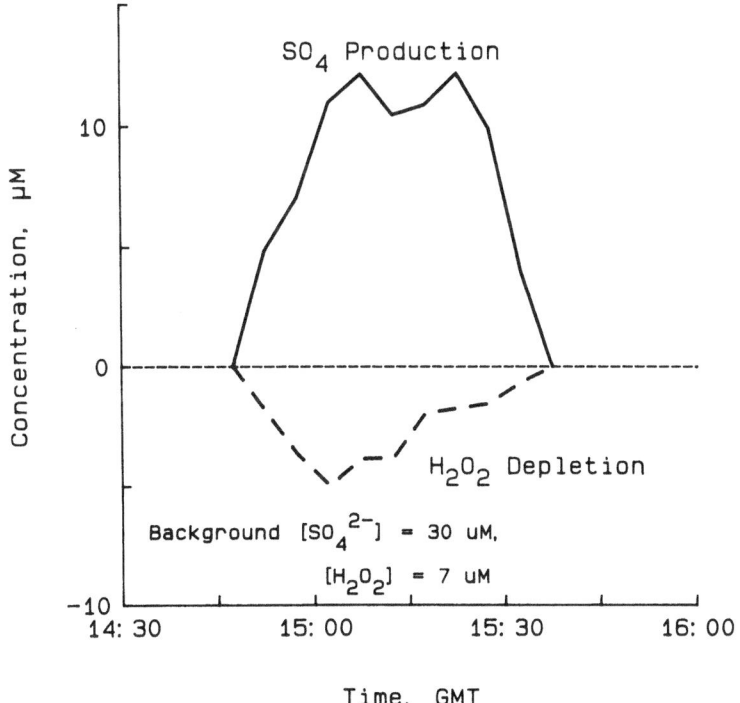

Fig. 7 Changes in cloudwater $[SO_4^{2-}]$ and $[H_2O_2]$
from background levels, Summit, 03.04.85

sulphate concentrations produced in cloudwater at the summit, above an estimated baseline of $\sim$ 30 μM, are shown in Figure 7. (The peak enhancement value of 12 μM corresponds to the period of maximum SO_2 release).

Interpretation of this sulphate peak is not simple because background air was not well characterised. Certainly, some oxidation of SO_2 by H_2O_2 was occurring as is shown by the average dip in H_2O_2 concentrations of 2.4 μM, from a 7 μM baseline during the same period. Some or all of the rest of the sulphate increase may have been due to changes in background aerosol. The most likely explanation is that the whole peak was due to in-cloud oxidation of SO_2 by H_2O_2 and O_3 in combination. This, combined with knowledge of cloudbase position and local windspeeds, leads to an estimated pseudo first order SO_2 oxidation rate of approximately 250% h^{-1} operating over a time period of 105 seconds.

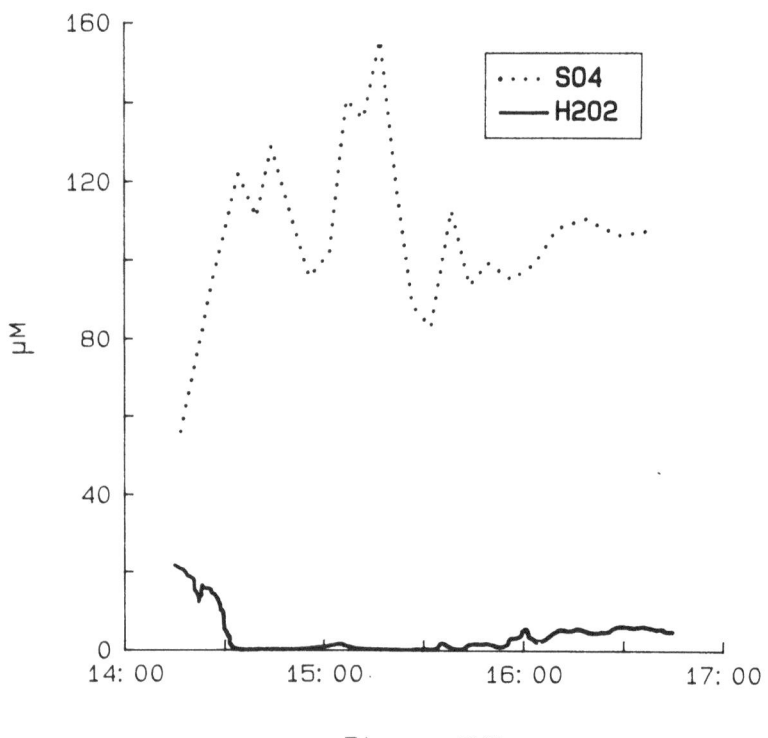

Fig. 8   [H2O2] and [SO4] in cloudwater

Mine Road Van, 30.04.86

5.3.1 <u>Spring 1986, CUTE III</u>. Results from the most recent experiment in Spring 1986 are, of necessity, only preliminary in nature. However, they are considered to be of such interest that they are now discussed in some detail.

On the 30 April 1986 SO_2 was released from 1413 to 1553 GMT. Figure 8 shows measurements of hydrogen peroxide and sulphate concentrations in cloud, at the instrumented van on the mine road. Strong sulphate peaks up to 60 µM above apparent background can be discerned during the time of SO_2 release and there are concurrent dips

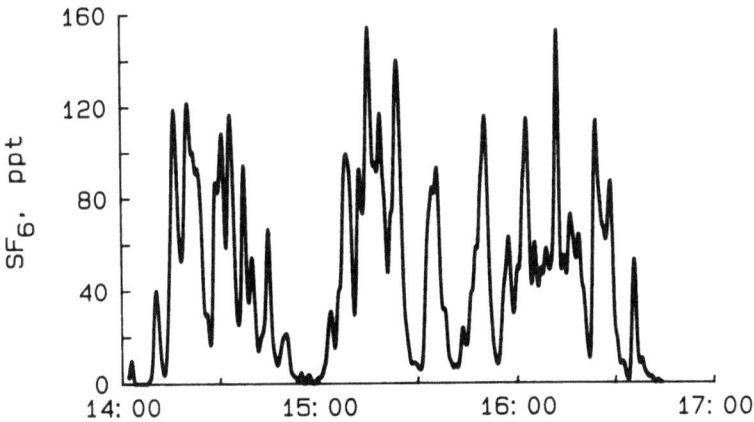

Fig. 9a SF_6 signal at the summit
CUTE III, 30.04.86

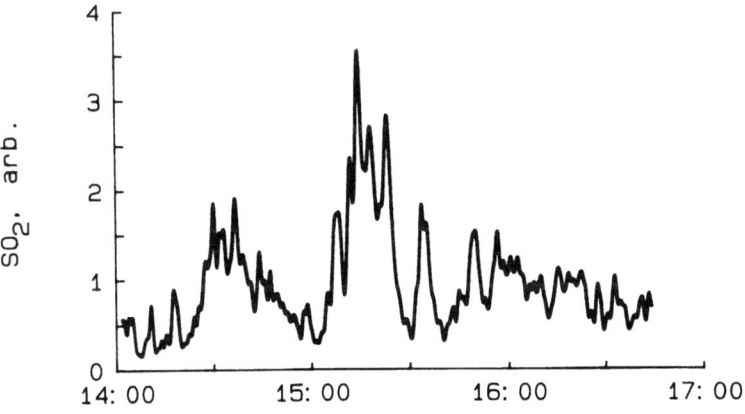

Fig. 9b SO_2 signal at the summit
CUTE III, 30.04.86

in H_2O_2 concentrations. These decreases in H_2O_2 concentration are so pronounced that from a background level of 18 μM at 1415 (0.14 ppb gas phase equivalent) the H_2O_2 in solution is virtually fully depleted until the SO_2 release rate declines after 1530. It is also apparent that there was only sufficient hydrogen peroxide in solution to account for a small proportion of the sulphate produced.

SF_6 and SO_2 measurements at the summit clearly show when the release plume was present (Figure 9). During the time of release of both gases, the shapes of both raw data traces are similar to each other both in overall pattern and in detail. Calibration of the signals is not yet complete but initial estimates suggest that more depletion of SO_2 occurred under these spring conditions than was the case in winter 1985 (CUTE II).

Changes in concentration of some of the species in cloudwater at the summit are shown in Figure 10. The sulphate concentrations are derived from both the in-situ continuous colorimetric analysis and subsequent ion chromatographic analysis of bulk samples. The shapes and magnitudes of both traces correspond closely and, as at the mine road, there are three major and possibly several minor peaks.

H_2O_2 values at the summit were initially lower than at the mine road, $\sim$ 10 μM. This was partly due to dilution by increased cloud liquid water content. Available H_2O_2 was again almost fully depleted during the time of the sulphate peaks. There does not appear to have been enough peroxide to have produced the peaks in sulphate concentration of up to 60 μM unless extra H_2O_2 had been entrained into the cloud or generated within the droplets. However, the droplet pH was high enough (around 5) to suggest that O_3 could contribute significantly.

The ammonium concentrations are of particular interest. They conflict with simple ideas which explain cloud water ammonium just in terms of dissolution of upwind NH_3 together with NH_4^+ which were well correlated with peaks in SO_4^{2-} due to the release of SO_2. Ammonia is a highly soluble gas and any present in the gas phase upwind of the cloud would have almost completely dissolved in the cloud droplets at the pH observed. The production of extra sulphate in the droplets as a result of the SO_2 release should not have affected the ammonium ion concentration. Further examination of this phenomenon is in progress.

This paper has presented some of the more interesting results of each of the three experimental campaigns. In each campaign SO_2 has been successfully emitted and changes in cloud chemistry observed. The experiments clearly validate the principle of SO_2 release as a method of determining SO_2 oxidation rates in clouds. Full interpretation clearly needs detailed computer models of air flow and cloud chemistry. This work is in progress and will be reported later.

6. FUTURE PLANS

While the experiments carried out so far have involved the release of SO_2 with SF_6 as a tracer, numerous other species could also be

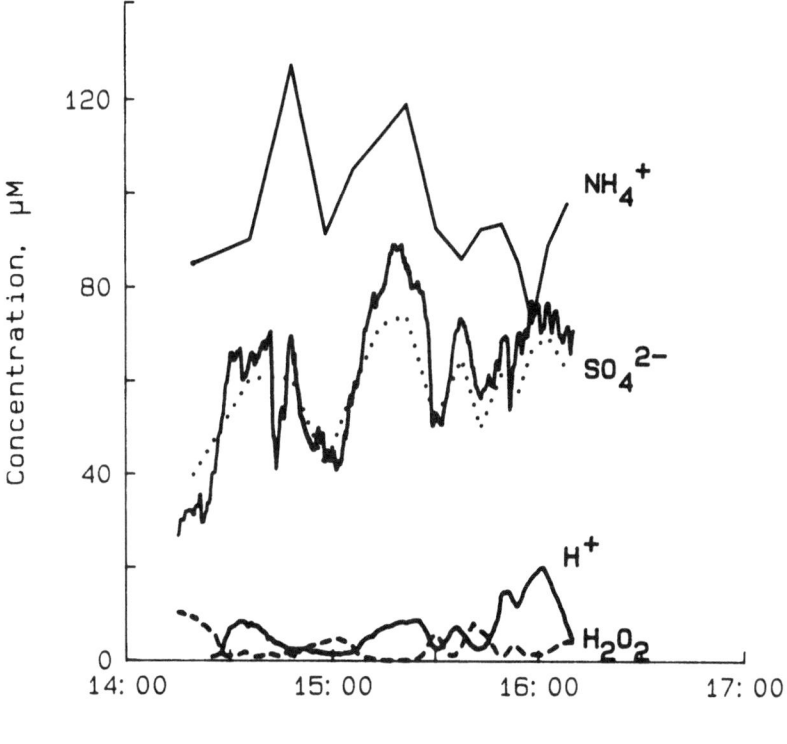

Fig. 10   Cloudwater composition

at the Summit, 30.04.86

released, separately or in combination. Several species are presently being actively considered to form the basis of future experiments; these include NH_3, NO_x, Cl_2 and formaldehyde.

7.   CONCLUSIONS

A novel experiment technique has been developed to establish chemical reaction rates and effective rate constants in ground-level clouds. This technique was first shown to be feasible by laboratory calculations and a field trial before three successful field experiments were carried out at Great Dun Fell. These large scale experiments have involved the application of new experimental methods, such as the real time continuous analysis of ions and H_2O_2 in cloudwater.

H_2O_2 depletion by reaction with SO_2 was observed and measured in winter even though background levels of H_2O_2 were low, less than 1 μM

298

in solution. In spring, when photochemical activity is higher, significant peaks in cloud sulphate concentrations were produced in both the 1985 and 1986 campaigns. In the spring 1985 experiment, the observed sulphate peak corresponded to a quasi-linear average SO_2 oxidation rate of 250% h^{-1} acting over 105 s. Peaks produced in both the spring 1985 and the spring 1986 experiments cannot be accounted for by the decline of H_2O_2 concentrations in solution alone. NH_4^+ concentrations appear to have been increased as a result of the SO_2 release, for reasons which are not readily explicable.

These three experiments have shown this new technique to be a successful way of examining the chemical kinetics of in-cloud oxidation processes. Further analysis of existing data and future experiments should provide further valuable information.

ACKNOWLEDGEMENTS

This work was carried out by staff of the Central Electricity Research Laboratories and members of the Cloud Physics Group, UMIST. Thanks are due to many colleagues at both establishments; at one time or another, over 25 people have been involved. This paper is publishsed by permission of the Central Electricity Generating Board and UMIST.

REFERENCES

Ames, D.L. 1983. A method for determining hydrogen peroxide in cloud and rainwater, CERL Internal Publication, TPRD/L/2552/N83.

Bamber, D.J., Clark, P.A., Glover, G.M., Healey, P.G.W., Kallend, A.S., Marsh, A.R.W., Tuck, A.F. and Vaughan, G. 1984. Air sampling flights round the British Isles at low altitudes: SO_2 oxidation and removal rates. Atmospheric Environment 18 No. 9, 1777-1790.

Blackburn, A.J. and Dear, D.J.A. 1981. A continuous monitor for sulphur hexafluoride and perfluoromethyl-cyclohexane in air, CERL Laboratory Publication, RD/L/216/R81.

Gervat, G.P., Clark, P.A., Marsh, A.R.W., Teasdale, I., Chandler, A.S., Choularton, T.W., Gay, M.H., Hill, M.K. and Hill, T.M. in preparation. Controlled chemical kinetic experiments in cloud; the oxidation of SO_2, for submission as a Letter to Nature.

McElroy, W.J. 1986. Sources of hydrogen peroxide in cloudwater. Atmospheric Environment 20 , 427-438.

Maahs, H.G. 1983. Kinetics and Mechanism of the Oxidation of S(IV) by ozone in aqueous solution with particular reference to sulphur dioxide conversion in non-urban tropospheric clouds. J. Geophys. Res. 88 , 10721-10732.

Martin, L.R. and Damschen, D.E. 1981. Aqueous oxidation of sulphur dioxide by hydrogen peroxide at low pH. Atmospheric Environment 15 (9), 1615-1621.

Penkett, S.A., Jones, B.M.R., Brice, K.A. and Eggleton, A.E.J. 1979. The importance of atmospheric ozone and hydrogen peroxide in oxidizing sulphur dioxide in cloud and rainwater. Atmospheric Environment 13 , 123-137.

PHYSICS OF CLOUDWATER DEPOSITION AND EVAPORATION AT
CASTLELAW, S.E. SCOTLAND.

R. Milne, A. Crossley & M.H. Unsworth
Institute of Terrestrial Ecology
Bush Estate
Penicuik
Midlothian EH26 OQB
Scotland

ABSTRACT. Measurements of cloud and rain water deposition at a hilltop
site in SE Scotland are presented. Evaporation rates of pure water
from plant surfaces during a cloud event are given and the influence of
such evaporation on acidity enhancement at surfaces is estimated. The
possibility of acidity enehancement on cloud water gauges is noted. A
method of scaling the rate of cloud water capture measured by a gauge
to that occuring on a natural surface is proposed.

1.   INTRODUCTION AND METHODS

The requirement to monitor the duration and chemistry of rain and cloud
in upland areas, together with the associated weather conditions,
presents a number of problems. A major limitation on the routine
collection of samples of cloudwater is the difficulty of excluding
rainfall from a cloudwater collector. Rainwater usually has much lower
chemical concentrations than cloudwater due to the physical processes
causing the growth of raindrops. Correction of measured chemical
concentrations in mixed samples is not possible using water from an
adjacent raingauge as the capture characteristics of the two collectors
differ in such a way that chemical concentrations also differ. It is
therefore necessary to monitor continuously the occurrence of both
cloud and rain and hence eliminate mixed samples in the cloudwater
collector. Mains power is usually unavailable at upland sites
therefore passive or battery-operated equipment must be used.
    At Castlelaw (10 km S. of Edinburgh, 480m a.s.l. and 300m above
the surrounding countryside) the requirement was to record the weather
conditions during typical cloud and mist events with a view to

i)   describing the general air mass and its possible source
ii) estimating deposition and evaporation fluxes of water and
iii) recording the occurrence of rain and cloud separately.

M. H. Unsworth and D. Fowler (eds.), Acid Deposition at High Elevation Sites, 299–307.

An automatic weather station (A.W.S.) was chosen as the solution. The variables measured were:-

1. Net radiation
2. Total incoming solar radiation
3. Windspeed
4. Wind direction
5. Air temperature
6. Air humidity
7. Rainfall rate
8. Cloudgauge capture rate.

The raingauge was installed with its top at ground level and surrounded by a 1.3m x 1.3m grid to Rodda's (1968) design to minimise wind induced undercatch. The cloudgauge was to the same design as that used to collect cloudwater for chemical analysis with the addition that the outflow was monitored by a tipping bucker raingauuge. The cloudwater collector was based on the conical design of Dollard et al (1983) but with five improvements:

i)   the size was changed to give a 330mm high by 200mm diam. cone which increased the collection rate,
ii) the wire frame was coated with low density polythene to eliminate the Cl^- leaching from the original PVC covering,
iii) ETFE coated copper wires were used instead of nylon to improve strength and reduce dry deposition of nitric acid vapour to the strings,
iv) the addition of an upper polypropylene spacer designed to allow use of a single length of wire stringing which could be readily removed for cleaning, and,
v)   the addition of a 1.2m diameter polypropylene-faced lid, chosen to exclude raindrops larger than 0.5 mm when the windspeed was less than 5 m/s.

The A.W.S. measured air temperature and humidity in a naturally ventilated psychrometer using dry and wet bulb platinum resistance thermometers. All the sensors used produced electrical outputs that were recorded at five minute intervals on magnetic tape, the whole system being run from rechargeable batteries. The data on the magnetic tapes were transferred to a desk-top computer for analysis. Unsworth (1984) has shown that deposition and evaporation of water from cloud or mist to vegetation depends critically on net radiation, air humidity and wind speed. In cloud relative humidity is greater than 90% but wet and dry bulb psychrometers are least accurate (± 5% RH) in this range. Thus when the magnetic tape data logger was replaced by new equipment, an additional measurement of air humidity by an alternative method was added. This sensor used a polymer film element, the electrical capacitance of which varies linearly with the relative humidity of the surrounding air and has a quoted accuracy of ± 2% RH.

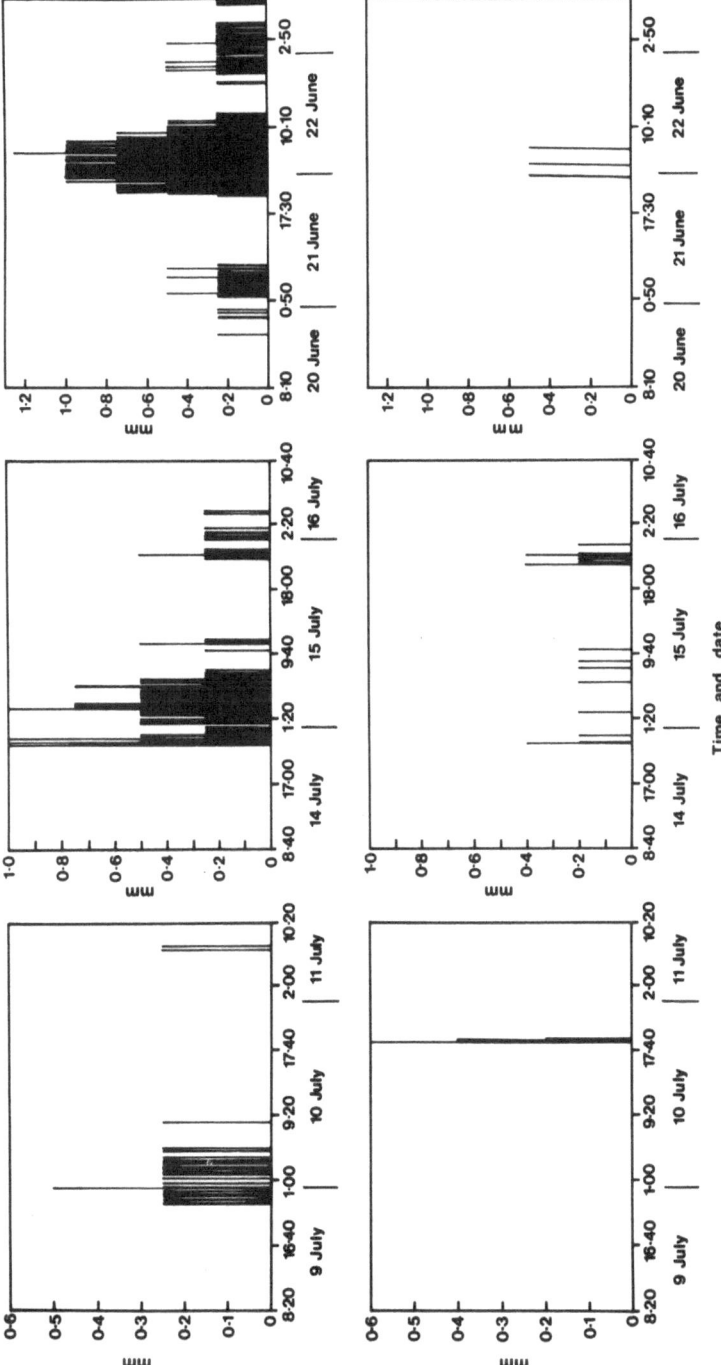

Figure 1. Occurrence of cloud and rain on three separate occasions at Castlelaw, SE Scotland as indicated by strung cloud water gauge and ground level rain gauge. 1a – event when collected cloud water not mixed with rain, 1b – event when collected cloud water mixed with rain, 1c – event when water collected on rain gauge was likely to be cloud water only, i.e. cloud only in cloudwater gauge.

2. OCCURRENCE OF CLOUD AND RAIN

Water samples for chemical analysis were taken over 2-3 day periods from cloud and rain collectors. Water samples from the cloud collector were classified as pure cloud or mixed cloud and rain by using the data from the logger. Each block of data covering the period of water sample collection was analysed to give:

1. Number of hours of cloud only,
2. Number of hours of rain only,
3. Number of hours with rain and cloud and windspeed greater than 5 m/s, (i.e. when the cloudguage lid did not provide full protection against rain contamination).
4. Number of hours with rain and cloud but windspeed less than 5 m/s, (i.e. cloudgauge protected against rain).
5. Number of dry hours, and,
6. Total time of sample collection.

Samples of pure cloudwater could then be chosen e.g. in Fig. 1a sample C66 was pure cloudwater because cloud and rain did not occur together but in Fig. 1b sample C68 cloudwater was contaminated, as rain fell while cloud was present. Fig. 1c is an example where, in dense cloud, water was detected by the raingauge but was probably due to condensation or deposition on the raingauge funnel rather than rain. A summary of 200 days data showed that cloud deposition alone accounted for 347 hours, rain for 87 hours, mixed cloud and rain for 112 hours leaving 4254 hours of dry deposition. During this period 16.2 litres (540 mm) were collected in the raingauge and between 34.5 and 46.2 litres in the cloudgauge (690-924 mm to capture cross-section), depending on how water collected in this gauge was classified between rain and cloud during mixed events when the windspeed was greater than 5 m/s. These collected cloudwater volumes are strongly dependant on the gauge design and therefore are of no direct use in defining deposition fluxes to natural surfaces.

3. CANOPY WATER EXCHANGE DURING CLOUD

Unsworth (1984) has discussed the possibility of acidity enhancement occurring on plant surfaces in cloud due to re-evaporation of pure water. The enhancement ratio is given by:

$$c_l/c_c = D/(D-E) \tag{1}$$

where c_l is the concentration of an acid substance on the leaf, c_c is the concentration in the cloud, D is the deposition rate of cloud water and E is the evaporation rate of pure water from the leaf.

Shuttleworth (1977) and others have shown that water is deposited on surfaces at a rate which can be estimated from

$$D = W/r_a = C_d u W \tag{2}$$

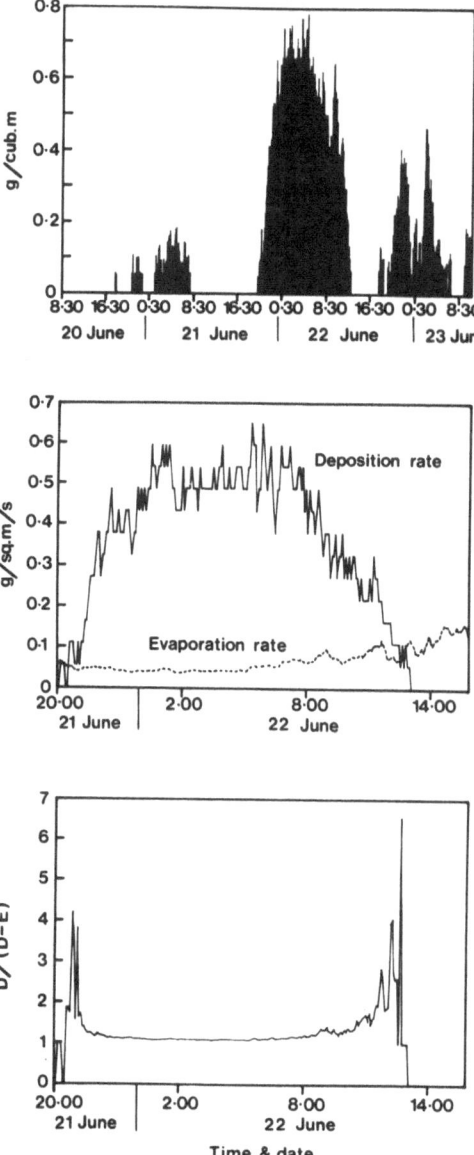

Figure 2.   a.   Hourly estimates of cloud water content at Castlelaw, SE
Scotland on 20–23 June 1986 estimated from water collected
on strung cloud water gauge.   b. Hourly estimates of
deposition and evaporation rate at a typical forest
subjected to the cloud event in Fig. 2a.   c. Enhancement of
acidity in surface water at forest of Fig. 2b.

where W is the liquid water content of the cloud (LWC), r_a is the aerodynamic resistance to transfer of momentum to the canopy, u is wind velocity and C_d is a drag coefficient.

Evaporation from a continuous or near-continuous water film on a surface can be estimated from the Penman-Monteith equation:

$$E = \frac{1}{\lambda} \quad \frac{A + \rho c_p(e_s - e_a)/r_a}{\Delta + \gamma} \tag{3}$$

where $(e_s - e_a)$ is the water vapour pressure deficit, A is the available energy (approximately equal to net radiation), λ is the latent heat of vaporisation of water, Δ is the rate of change of saturation vapour pressure with temperature, ρc_p is the volumetric specific heat of dry air and γ is the psychrometric constant. The assumption is made in the equation that momentum and water vapour are each exchanged at the surface through a resistance of value, r_a.

From these equations it can be seen that, to estimate typical enhancement ratios in cloud, it is necessary to measure liquid water content, net radiation and vapour pressure deficit during the presence of cloud. The cloud and rain gauges indicated the occurrence of cloud only events within Sample C59 on the 20 to 23 June 1986 (Fig. 1c). The net radiation and vapour pressure deficit for this period were recorded on the A.W.S. and the liquid water content of the cloud over each five minute logger sample period was calculated using the bucket tip count of the cloud gauge, the wind velocity and the capture efficiency of the gauge (Fig. 2a). The cloud detector was calibrated at Great Dun Fell by placing it alongside a Knollenburg Forward Scattering Spectrometer Probe (FSSP) (measuring liquid water content and droplet size spectrum) and comparing the cloud gauge catch with the volume of cloudwater that passed through its area cross section. Comparisons were made in cloud on 12 occasions, with windspeeds from 10 to 33 m/s and mean drop size from 3.5 um to 6.9 um. The gauge efficiency ranged from 11% to 46% with 29% being typical (Fowler 1986, pers. comm.).

Sampled period C59 is typical of the cloud events occuring at Castlelaw which generally happen during the night with some extension at either end from evening and into morning. The vapour pressure deficit (v.p.d.) recorded during cloud was small (i.e. high humidity), but during the day energy from the sun cleared the cloud and the v.p.d. increased. On 22 June however, there was insufficient energy for this to occur and vapour pressure remained near the saturation value throughout the day. The cloud water content estimates show the variation as cloudbase decreases through the night and increases again to midday. A rough check on these water content values was obtained by the visual observation that at the end of the sample period (11.00 hours 23 June) cloudbase was about 100 meters below the measurement site which, assuming that LWC in the lowest layers of the cloud increases adiabatically with height, indicates a value of 0.2 g m^{-3} similar to that estimated in Fig. 2a.

If it is assumed that the cloud event occuring on the night of 21/22 June (Fig. 2a) is fairly typical for upland sites, we may investigate the acid enhancement effect by considering what would happen if this event occurred over a typical coniferous forest. Jarvis et al (1976) have published drag coefficients, C_d, for various forests and using a typical value of 0.09 with the data recorded during the above event (in Eqns. 2 & 3) allows the forest deposition and evaporation rates to be calculated (Fig. 2b). Deposition rate increases to about 0.65 g^{-2} s^{-1} as cloud water content reaches its maximum while evaporation rate remains low during the night, increasing slowly from dawn with the increasing net radiation. The small residual evaporation during the night when net radiation is zero, is driven by turbulent transport to the air which is slightly unsaturated. The small vapour pressure deficit recorded is within the error of the wet bulb depression sensor in the psychrometer, illustrating the need for more accurate humidity measurements. Such improved values will be available from the thin-film capacitance humidity sensor when difficulties with calibration have been overcome.

Using these deposition and evaporation rates, the acidity enhancement ratio was calculated from Eqn. 1 for each five minute period during the event (Fig. 2c). These results confirm that significant enhancement of the acidity on forest surfaces can occur, although the greatest effect seems to be limited to the beginning and end of cloud events. Note that the apparent 10% enhancement throughout the night is within the possible error due to the psychrometer inaccuracy. During the morning when net radiation is available the evaporation increases and several hours of 100% or greater ($D \geq 2$) enhancement of acidity on surfaces occur. It is interesting to observe that after deposition has ceased, the mechanism for acidity enhancement described here will alter when increases in acidity will occur solely by the continued evaporation of pure water from the liquid remaining on the foliage. This increase will continue until a final acidity is reached which will depend on several factors e.g. initial acidity, evaporation rate, ion exchange with plant and the solution chemistry within the liquid film.

In order to estimate the net effect on the tree of the enhanced acidity in the surface liquid film during cloud events it will be necessary to interpret data from an extended period and from several sites, especially forests.

4.   ACIDITY ENHANCEMENT ON CLOUD GAUGES

To calculate the concentration on foliage it is necessary to multiply enhancement ratio by cloud ionic concentration. Cloud ion concentrations are usually estimated using the water samples from the strung collectors but it should be noted that concentration enhancement probably also takes place in the water on the strings. As evaporation occurs concurrently with deposition, the removal of water at the base is not important as there will always be some water on the strings, just as in a canopy where there is always considered to be a near-complete water film, although stemflow and drip is occuring. To estimate the magnitude of acidity enhancement on collector/gauges an effective drag coefficient for the gauge was calculated by reworking

306

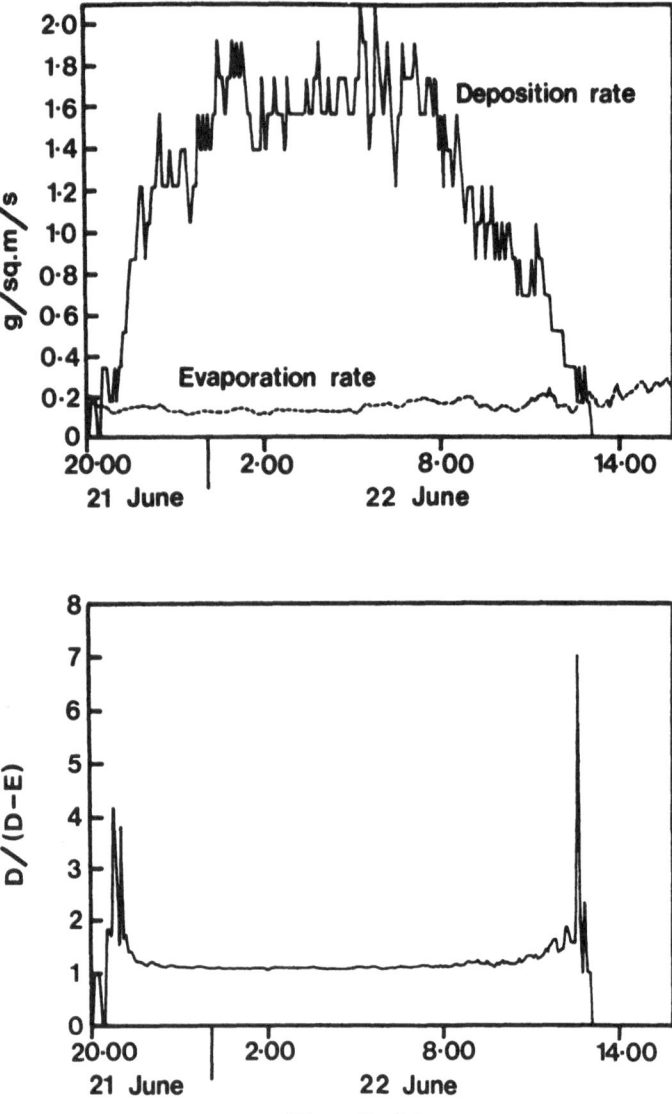

Figure 3. a. Hourly estimates of deposition and evaporation rates for strung cloud water gauge during event on 20–23 June 1986 at Castlelaw, SE Scotland (Fig. 2a). b. Enhancement of acidity of water collected by strung cloud water gauge during cloud event shown in Fig. 2a.

the capture efficiency data. If the cloud is assumed to transfer water droplets to the gauge by turbulent transport and the effective capture area is the triangular cross-section then its drag coefficient is equal to the capture efficiency (0.29 for the cloud detector). Applying Eqns. 1, 2 & 3 as before but with C_d = 0.29 allows estimates of gauge deposition rate, evaporation rate and enhancement ratio (Figs. 3a & 3b). For the gauge the deposition and evaporation rates are both greater than those for a typical forest hence giving lower but not insignificant enhancement ratios (1.1-1.5) especially in the morning when net radiation is available. Further investigation of this method of estimating enhancement of ion concentrations on strung collector/gauges is necessary, e.g. by estimating the drag coefficient of the gauge using alternative methods.

5. CLOUDWATER FLUXES TO NATURAL SURFACES

From Equation 2 it can be seen that deposition flux is directly proportional to C_d and hence it is possible to scale the collection on the cloudgauge (C_d = 0.29) to natural surfaces. Thus the grass at Castlelaw ($C_d \sim 0.01$) would have received over the 200 day observation period 24-32mm from cloud while a forest ($C_d \sim 0.09$) experiencing the same weather would have received 216-289mm. With rainfall of 540mm (collected at Castlelaw) this indicates that, excluding input from snow about, 33% of the volume of water input to forests may be from clouds.

REFERENCES

Dollard, G.J., Unsworth, M.H. & Harvey, M.J. 1983. Pollutant transfer in upland regions by occult precipitation. Nature **302** , 241-242.

Jarvis, P.G., James, G.B. & Landsberg, J.J. 1976. Coniferous Forest. In: Vegetation and the Atmosphere, Vol. 2, J.L. Monteith (Ed.), p171-236. Academic Press.

Rodda, J.C. 1968. The rainfall measurement problem. IASH Pub. No. 78. (General Assembly, Berne, 1967), 215-231.

Shuttleworth, W.J. 1977. The exchange of wind-driven fog and mist between vegetation and the atmosphere. Boundary Layer Meterol. **12** , 463-489.

Unsworth, M.H. 1984. Evaporation from forests in cloud enhances the effects of acid deposition. Nature **312** , 262-264.

A COMPARISON OF METHODS FOR ESTIMATING CLOUD WATER DEPOSITION
TO A NEW HAMPSHIRE (U.S.A.) SUBALPINE FOREST

Gary M. Lovett
Institute of Ecosystem Studies
The New York Botanical Garden
Mary Flagler Cary Arboretum
Box AB
Millbrook, New York 12545
USA

ABSTRACT. Deposition of cloud droplets is thought to be a major input
mechanism for water, plant nutrients, and airborne pollutants in the
high-elevation forests of the eastern U.S. This paper discusses three
methods for determining amounts of cloud water deposition. The data
used to illustrate the methods were all taken in the subalpine balsam
fir forests of Mt. Moosilauke, New Hampshire, USA.
    The first method uses a resistance model of the droplet deposition
process based on the capture characteristics of individual canopy
components. The cloud water deposition estimate of 27 to 125 cm for
the June–October growing season depends strongly on estimates of mean
meteorological and cloud characteristics. The second method uses data
on rainfall, stemflow and throughfall to calculate a water balance for
the canopy. The deposition estimates for the June–October period
range from 24 to 41 cm and depend on the estimate of interception loss
from the canopy. The third method uses measurement of SO_4^{2-}
concentrations in spring seeps to calculate the SO_4^{2-} mass balance for
the drainage areas. The estimates of cloud water deposition for the
June–October period range from 0 to 22 cm depending on assumptions
about evapotranspiration and the concentration of SO_4^{2-} in cloud
water. To calculate chemical deposition, these water deposition
estimates must be multiplied by measured chemical concentrations in
cloud water.
    Because the data were gathered from disparate studies, comparisons
between the methods are difficult. However, it is suggested that the
canopy water balance method may be the most appropriate means of
measuring cloud water deposition on scales of space and time relevant
to ecological studies.

1.   INTRODUCTION

Deposition of cloud droplets onto vegetation surfaces has received

M. H. Unsworth and D. Fowler (eds.), Acid Deposition at High Elevation Sites, 309–320.

considerable attention as a mechanism for the transfer of water, plant
nutrients, and pollutants to high elevation forests in the eastern
U.S.A. This attention has been heightened recently by the realization
that red spruce (Picea rubens) is undergoing a severe decline in the
same forests. The cause of this decline is unknown, but air pollution
stress is suspected, along with stresses from drought and low
temperatures (Woodman and Cowling, 1986).

Despite the concern, estimates of the amount of cloud water
deposition are still crude. Methods have included artificial surface
collectors (e.g. Schlesinger and Reiners, 1974; Vogelmann et al.,
1968; Scherbatskoy and Bliss, 1984) and inferential (sensu Hicks,
1986) methods (Lovett et al., 1982). Several research groups have
developed programs to measure chemical concentrations in cloud water,
but few estimates of deposition fluxes have been reported.

This paper will discuss three ways to estimate cloud water
deposition to forests. For all of the methods, the objective is to
estimate water deposition; these must then be multiplied by
appropriate chemical concentrations in cloud water to estimate chemical
deposition. The illustration and comparison of these methods will
draw on data published over the last 10 years by researchers working on
Mt. Moosilauke, New Hampshire, U.S.A. (44°01'N, 71°50'W) in association
with Dr. William Reiners. These studies are concerned primarily with
the upper subalpine forest zone, an area of nearly continuous forest
cover existing in the elevational range of roughly 1200 to 1450 m.The
dominant tree species in this zone is balsam fir (Abies balsamea, 85%
of the basal area) with scattered individuals of red spruce and paper
birch (Betula papyrifera var. cordifolia). The climate of this zone
is cold, wet, and windy, and the soils are thin and rocky. The
vegetation and environment of this zone were described in more detail
by Reiners and Lang (1979).

Although all of the studies cited below were performed on Mt.
Moosilauke, they were not performed at the same time nor in the same
place on the mountain. Therefore, comparisons between the methods are
at best approximate. The primary purpose of this paper is to
demonstrate the methods and discuss their applicability to ecological
studies. The estimates of cloud water deposition will be restricted
to the period June through October, which is the season in which all of
the studies were performed.

2. DEPOSITION MODEL

Lovett et al., (1982) estimated cloud water deposition for a stand of
balsam fir at 1220 m elevation using a micrometeorological modeling
approach. The model was described in detail by Lovett (1984).
Briefly, cloud deposition fluxes are calculated in the model from a
cloud water gradient and a network of resistances. The resistances
parameterize cloud droplet transport into the forest and through the
boundary layers of the individual canopy components (shoots, branches
and trunks). Wind tunnel studies using fluorescein-tagged droplets
provided the data used to calculate the boundary layer resistances

Thorne et al., 1982). Wind speed, cloud liquid water content, and
the droplet size distribution at the canopy top were input data for the
model, along with the vertical distribution of canopy surface area.
The model was tested against cloud drip collections made during cloud
events of known meteorological conditions in an instrumented balsam fir
stand and was found to agree within 20% (Lovett 1984).

Extrapolation of the model predictions to longer time scales
requires average meteorological and cloud characteristics which are not
accurately known. Estimates of mean wind speed, cloud liquid water
content (LWC), and cloud immersion time are particularly crucial
because the predicted cloud water deposition is nearly directly
proportional to changes in these parameters. Using measurements made
on Mt. Moosilauke and taken from the literature, Lovett et al., (1982)
estimated mean values for the variables as: wind speed, 4 m s^{-1};
liquid water content 0.4 gm^{-3}; and cloud immersion 40% of the time.
Unfortunately, no better estimates are available today. Using these
estimates and a droplet diameter mode of 10 µm in the current version
of the model gives a cloud deposition of 64 cm for the June-October
period. This figures represents "gross" cloud water deposition;
i.e., before interception losses in the canopy. Decreasing the wind
speed, LWC, and cloud immersion time by 25% (to 3 m s^{-1}; 0.3 g m^{-3},
and 30% immersion time) decreases the estimated cloud deposition to 27
cm, and increasing them all by 25% (to 5 m s^{-1}, 0.5 g m^{-3}, and 50%)
increases the estimated deposition to 125 cm. Thus relatively small
changes in the three most crucial input parameters, when taken
together, can produce a nearly 5-fold range in cloud deposition
estimates.

The uncertainty in the model estimates appears therefore to be
strongly dependent on the uncertainty in the input parameters.
Because these parameters are likely to be extremely variable in both
space and time, good estimates will require long-term monitoring at
many locations in the mountain landscape, a difficult task because the
measurements require meteorological towers and fragile
instrumentation. This illustrates the major problem with this method
for estimating deposition: although high-quality point measurements
may be made, extrapolation to broader spatial and temporal scales
requires enormous effort and expense.

3. WATER BALANCE IN THE CANOPY

Water that enters a forest canopy as rain (R) or cloud water deposition
(C) is either evaporated as interception loss (I) or drips to the
forest floor as stemflow (S) or throughfall (T). Thus, the canopy
water balance can be expressed as:

$$C = S + T + I - R \tag{1}$$

Measurements of rainfall, stemflow, and throughfall were reported
by Olson et al. (1981) for three balsam fir stands at 1250 m elevation
on Mt Moosilauke. The sampled plots were each 20 x 20 m, and were

chosen to maximize differences in canopy structure. Collections were made after each discrete rain event during the period 14 July–23 October 1975.

The results of the study show that S + T exceeded R by 18–29% in the three stands, and the authors attributed the excess to cloud water deposition. Because interception loss occurs, this is a low estimate of cloud water deposition. Interception losses in mature coniferous forests are typically 15–25% of incident rainfall (Olson et al., 1981, Rutter 1975). One might speculate that in these high-elevation balsam fir forests, interception losses might be relatively high, because the forests are wet for long periods of time with cloud water, and the wind speeds are high. Nevertheless, this information provides ranges of estimates for (S + T)/R and I/R, which can be used with equation 1 to estimate the ratio of cloud water deposition to rainfall (C/R):

$$C/R = (S + T)/R + I/R - 1 \qquad (2)$$

In Eq. 2 we normalize the hydrologic variables to rainfall because we expect the normalized values to be more general (representative of average conditions) than the un-normalized values would be. If we assume 1) that the above values of (S + T)/R and I/R are typical for this elevation and 2) that the average annual precipitation of 180 cm (Dingman, 1981) is spread evenly through the year (giving 75 cm for June through October), we can estimate a range of values of cloud water deposition from these data. Table 1 shows values of C calculated from equation 2. The calculated cloud water deposition ranges from 25 to 41 cm, depending on the values chosen for I/R and (S + T)/R.

TABLE 1. Cloud water deposition (cm) for an average June–October period calculated from equation 2, assuming R = 75 cm.

		(S+T)/R	
		1.18	1.29
	.15	25	33
I/R	.20	29	37
	.25	32	41

Equations 1 or 2 provide a simple and direct estimate of cloud water deposition using easily obtainable data (rainfall, throughfall, and stemflow). One obvious drawback is that interception loss must be estimated. This can be done by monitoring wind speed, temperature, relative humidity, and net radiation above the canopy and calculating evaporation rates when the canopy is wet. This requires considerable effort, although not as much as measuring the cloud characteristics necessary for the modeling approach discussed above. Even without the correction for interception loss, measurement of R, S and T provides a firm lower limit for cloud water deposition. For R = 75 cm, (S+T)/R = 1.18, and I/R = 0 (ignoring interception loss), this lower limit is 14 cm.

Spatial variability in cloud water deposition within a forest stand is reflected in the spatial variability of water flux in throughfall and stemflow. This necessitates multiple samples to characterize the mean deposition for a stand. Replication of the samples within and among stands is inexpensive and, in contrast to micrometeorological approaches, is not limited by complexities of terrain or vegetation cover.

4. STREAMWATER CHEMISTRY

4.1 Approach

It is, in theory, possible to extend the water balance concept to broader areas by measuring hydrological fluxes on whole watersheds. However, this requires streamwater gauging stations that are expensive to install and maintain. If some assumptions about the chemistry and hydrology of the watershed can be tolerated, it is possible to use streamwater chemistry to estimate cloud water deposition by employing a "tracer" element which is present in rain, cloud, and streamwater. The assumptions involve the application of the principle of conservation of mass to water and the tracer:

1) Hydrologic Mass Balance. Rain and cloud are the only water inputs to the watersheds of the streams sampled, and evapotranspiration and streamflow are the only outputs. Deep seepage is negligible.

2) Tracer Mass Balance. Sources and sinks of the tracer within the watershed are negligible compared to measured inputs and outputs.

If these assumptions hold, then the flow-weighted concentration of the tracer in streamwater (X_s) is equal to the total deposition of the tracer to the watershed (D_T) divided by the streamflow (S) for the period:

$$X_s = \frac{D_T}{S} \qquad (3)$$

If there is no deep seepage, the streamflow can be calculated as:

$$S = R + C - E \qquad (4)$$

where E is evapotranspiration from the watershed.

The total deposition of the tracer to the watershed is the sum of wet deposition (D_w), cloud water deposition (D_c), and dry deposition (D_d):

$$D_T = D_w + D_c + D_d = RX_r + CX_c + D_d \qquad (5)$$

where X_r and X_c are volume-weighted mean concentrations of the tracer in rain and cloud water, respectively. We can combine equations 3, 4 and 5 and solve for C as follows:

$$C = \frac{X_s (R-E) - RX_r - D_d}{X_c - X_s} \qquad (6)$$

Equation 6 shows that cloud water deposition can be estimated from streamwater concentrations if the amount of rain, evapotranspiration, dry deposition, and the concentration of the tracer in rain and cloud water are known. As mentioned above, X_s, X_r and X_c should all be weighted so as to produce an accurate flux when multiplied by S, R and C respectively. Weighting X_r by rain amount is not difficult, but this exercise assumes that S and C are not measured. Therefore, this method will be inaccurate to the extent that simple mean values of X_s and X_c differ from their appropriately weighted counterparts.

4.2 Application to Mt. Moosilauke

4.2.1 Choice of Tracer.
Concentrations of a number of macronutrient ions in rain, cloud, and streamwater have been measured on Mt. Moosilauke. Choosing one to use as a tracer involves several considerations. First, there must be minimal sources and sinks in the watershed. This requirement excludes NO_3^- and NH_4^+ which are strongly taken up by the biota in many watersheds (Vitousek 1977) and also the cations Na, Ca, K, and Mg, which are released from minerals in the soil by weathering. Of the ions commonly measured in nutrient cycling studies, Cl^- and SO_4^{2-} remain possibilities. Chloride has little interaction with the biota, and is not released from common rock minerals. While S is an essential plant nutrient, the amounts delivered to these systems in acidic deposition overwhelm the requirements of the plants and the sorption capacity of the soil, so that only a small change in SO_4^{2-} concentration is seen as water moves from throughfall to streams (Cronan 1980). No S-bearing rocks have been found on Mt. Moosilauke.

A second requirement for a good tracer is that it be much more concentrated in cloud water than in rain. This is illustrated by Figure 1, in which the ordinate represents the percent change in tracer concentration in streamwater that could be expected from 35 cm of cloud water input relative to no cloud input, and the abscissa represents the ratio of concentration of the tracer in cloud and rain (X_c/X_r). This calculation is shown in Figure 1 for two values of E, 17 and 42 cm. Reiners et al., (1984) derived an equation for actual evapotranspiration (AET) as a function of elevation on Mt. Moosilauke which indicates 42 cm yr^{-1} AET at 1220 m. This represents 23% of the precipitation at that elevation. Because most, but not all, of the annual AET is likely to occur during the June-October growing season, the AET for that period should lie between 42 cm and 17 cm, which is 23% of the average precipitation for the growing season.

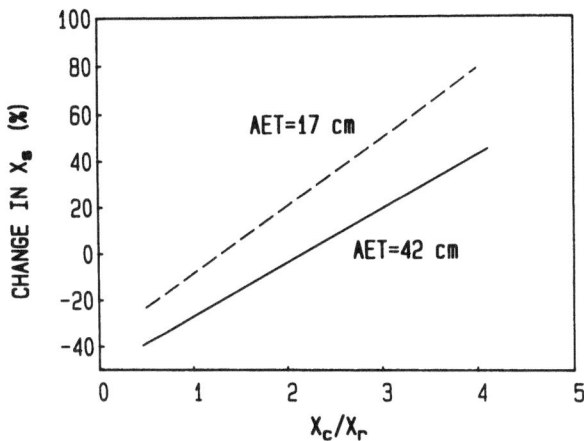

Figure 1. Change in tracer concentration in streamwater (X_s) expected from 35 cm cloud water deposition relative to no cloud deposition, given various ratios of tracer concentration in cloud water (X_c) and rain (X_r).

Figure 1 illustrates that the change in streamwater concentration that can be expected as a result of cloud water deposition is small unless concentration of the tracer in cloud is much higher than the concentration in rain. In order to see a 30% increase in streamwater concentration over what would be expected for rain alone, the ratio X_c/X_s should approach 3. A change of 30% in the mean X_s is probably the minimum detectable, given the variability in X_s in space and time. Because of evapotranspiration and the relatively small amounts of water deposition from clouds relative to rain, the cloud deposition "signal" is muted in streamwater.

The data of Weathers et al. (this volume) indicate that X_c/X_r ratios for Cl^- are typically 1-2, while those for SO_4^{2-} are 3-4. This suggests that SO_4^{2-} is likely to be the best tracer to use for calculation of cloud water deposition.

4.2.2 Calculation of Deposition. Cloud water deposition can be calculated from equation 6 if R, E, X_s, X_r, X_c, and the dry deposition flux (D_d) are known. Cronan (1980) reported concentrations of SO_4^{2-} averaging 132 μeq L^{-1} for several spring seeps in the balsam fir zone on Mt. Moosilauke during the growing season of 1975-77. Cronan used a composite value for rainwater concentration (X_r), but Olson et al. (1981) measured SO_4^{2-} in bulk precipitation and found the volume-weighted mean of 40 μeq L^{-1} during the 1975 growing season. Concentration of SO_4^{2-} in cloud water on Mt. Moosilauke was reported by Lovett et al. (1982) to be 342 μeq L^{-1} on Mt. Moosilauke. This mean value was based on only 10 samples and had a standard error of 234 μeq L^{-1}, but the more extensive data of Weathers et al. (this volume) for nearby Mt. Washington, New Hampshire indicate a mean of 268 μeq L^{-1}. Because some of these samples may have been contaminated by rain, this may be a low estimate of the mean.

No direct measurements of the dry deposition of SO_4^{2-} have been made for Mt. Moosilauke. However, Eaton et al. (1978) used a watershed mass balance to calculate a dry deposition of 6.1 kg S ha^{-1} yr^{-1} for the Hubbard Brook Experimental Forest about 10 km away (and 500-800 m lower in elevation). This is supported by estimates made from atmospheric sampling in the eastern U.S., which indicate an expected dry deposition of slightly less than 7 kg S ha^{-1} yr^{-1} for this area (NRC 1983, pp. 92-93). Both of these estimates include dry deposition of SO_2 and SO_4^{2-}. If we assume a dry deposition of 6.1 kg S ha^{-1} yr^{-1}, spread evenly through the year, with all SO_2 converted to SO_4^{2-} in the ecosystem, then the June-October dry deposition of SO_4^{2-} is 7.6 kg ha^{-1}.

Table 2 shows the cloud water deposition calculated from equation 6 using these data and assuming 75 cm precipitation during the period June through October. Two estimates of X_c (268 and 342 ueq L^{-1}) and E (17 and 42 cm) are used in an attempt to bracket the correct value. The values for cloud deposition range from < 0 to 22 cm, substantially lower than the values presented for the modeling and canopy water balance methods.

The negative values shown in Table 2 are unreasonable, but they indicate that if 42 cm AET is assumed, then wet and dry deposition can more than account for the SO_4^{2-} measured in streamwater. The estimates in Table 2 are very uncertain because they rely on scant data measured in different places and at different times. These calculations might best be considered an illustration of the method (i.e. eq. 6) rather than an estimate of deposition. Clearly, a concerted effort to measure all of the variables in equation 6 at the same place and time would be necessary for a fair test of this method.

TABLE 2. Values of cloud water deposition (cm) for the June-October growing season calculated using equation 6 using two values of SO_4^{2-} concentration in cloud water (X_c) and two values of evapotranspiration (E). Other constants used in equation 6: R = 75 cm, X_s = 132 ueq L^{-1}, X_r = 40 ueq L^{-1}, and D_d = 7.6 kg ha^{-1}.

	X_c (ueq L^{-1})	
	268	342
E(cm) 17	22	14
E(cm) 42	< 0	< 0

5. DISCUSSION

All of the methods discussed above require concentrations of solutes in cloud water in order to estimate the deposition of those solutes to high-elevation forests. However, measurement of cloud water chemistry is insufficient to infer deposition rates, and a method such as those illustrated above must be employed. Estimating deposition at a single point in space and time with reasonable certainty may be possible, but very high uncertainties are associated with broader-scale estimates.

In comparing cloud water deposition to rainfall, it should be noted that measurement of precipitation in windy environments is itself a problem. Care must be taken to shield the precipitation collector from high winds which will tend to decrease the collection efficiency. Both the canopy water balance and stream chemistry methods discussed here rely on accurate measurements of rainfall amount.

The ranges of values discussed above for the three techniques are: model, 15-125 cm; canopy water balance, 25-41 cm; and streamwater sulfate, 0-22 cm. These ranges are not true uncertainties, but rather are an attempt to bracket the estimates from each method by manipulating the parameters within reasonable bounds. Because the data are drawn from several different studies, the comparability of the ranges is questionable and should be interpreted with caution. It cannot be determined which technique is most accurate, because no "standard" is available for comparison. In fact, cloud water deposition could vary over this entire range (0 to 125 cm) in different areas of the mountain and in different years. Modeling analyses indicate that spatial gradients in wind speed, cloud liquid water content, cloud immersion time, and canopy structure will result in large variability of cloud water deposition over a mountain landscape (Lovett 1984, Lovett and Reiners 1986). The canopy water balance data of Olson et al. (1981) indicate that, at least for that site in that year, the lower limit for cloud water deposition (the minimum observed enhancement of SF + TF relative to rain) is 14 cm. This minimum value is the firmest value deducible from these studies.

These estimates can be compared to other values of cloud water deposition reported for the mountains of northern New England. Schlesinger and Reiners (1974), using collectors exposed near the summit (1372 m) of Mt. Moosilauke, found that water deposition into a collector surmounted by artificial foliage was 4.5 times the deposition in a nearby open bucket. This implies that cloud water deposition is 350% of rainfall at this site, but no attempt was made to extrapolate to real forest canopies or to less-exposed sites. Using our nominal value of 75 cm rainfall (June-October), this suggests 263 cm of cloud water deposition. Vogelmann et al., (1968), using rain collectors surmounted by wire screens in a forest clearing, found cloud deposition to be 67% of rainfall at 1100 m on Camel's Hump, Vermont. Again, no attempt was made to extrapolate this result to a real canopy. Using our 75 cm rainfall amount, this suggests 50 cm of cloud water deposition. Scherbatskoy and Bliss (1983) used artificial collectors and an empirically derived collector-to-canopy extrapolation factor to estimate cloud water deposition of 50-85% of rainfall in the elevational range 1110 to 1220 m on Madonna Mountain, Vermont. This suggests 38-64 cm of cloud water deposition at our site.

Although the estimates discussed in this paper are not directly comparable, it is possible to compare their potential usefulness in studies of atmospheric deposition at high elevation sites. The deposition modeling approach requires extensive monitoring of data that are difficult to collect (especially cloud liquid water content, droplet size distribution, and cloud immersion time). The model is extremely sensitive to the accuracy of input data and it requires faith in assumptions about the applicability of the micrometeorological measurements and models in mountainous terrain with heterogeneous vegetation. This is probably not a practical method for near-term (i.e. 5 yrs) estimation of cloud water deposition on the time and space scales of hectares and seasons.

Measuring streamwater chemistry is somewhat easier and naturally integrates deposition over broad spatial scales. However, the assumptions about deep seepage of water and lack of sources and sinks of the tracer in the watershed are difficult to verify. If dry deposition of the tracer is important, as it appears to be for SO_4^{2-}, measurement of that flux is at least as difficult as measurement of cloud water deposition, so nothing is gained in terms of ease of measurement. This method could be improved by finding a tracer for which dry deposition is likely to be negligible, but adding this condition to the two mentioned above (non-reactiveness in the ecosystem and high ratio of cloud/rain concentrations) will make the choice of tracer very difficult. Accurate estimates of evapotranspiration may also be problematic. Using ratios of tracers, such as SO_4^{2-}/Cl^-, may help solve those problems.

The canopy water balance method integrates over a smaller spatial scale than does the streamwater method. However, the water balance measurements are direct, are not technologically demanding, do not rely on tenuous assumptions, and can integrate over long periods. The main drawback is that an estimate of interception loss is necessary. If done with micrometeorological methods, this could be difficult. However, other simpler methods may suffice. For instance, if a tracer can be found that 1) has approximately the same concentration in rain and cloud, and 2) does not interact significantly with the canopy, the concentrations of this tracer in below-canopy precipitation could permit calculation of interception loss. Once again, using a ratio of tracers may help. Note again that a lower limit can be calculated even without the correction for interception loss.

In summary, the canopy water balance method appears to have the most promise in terms of ease of application, scale of measurements, and accuracy. A comparative experiment in which all methods were tried at a single site would permit better comparison. Nonetheless, it is clear that methods exist that can provide deposition estimates at ecologically relevant scales of space and time; these measurements should be given a priority equal to cloud chemistry measurements in studies of atmospheric deposition to mountainous regions.

ACKNOWLEDGEMENTS

I am very grateful to Christopher Cronan, David Hollinger, Gerald Lang, Richard Olson, and especially William Reiners, whose careful studies on Mt. Moosilauke permitted the analysis reported in this paper. I thank K.C. Weathers and G.G. Parker for helpful reviews of the manuscript. Research support for the author was provided by the Electric Power Research Institute ("Integrated Forest Study on the Effects of Atmospheric Deposition", contract RP2621 through Oak Ridge National Laboratory) and the Mary Flagler Cary Charitable Trust. This is a contribution to the program of the Institute of Ecosystem Studies of the New York Botanical Garden.

REFERENCES

Cronan, C.S. 1980. Solution chemistry of a New Hampshire subalpine ecosystem: a biogeochemical analysis. Oikos **34**, 272–281.

Dingman, S.L. 1981. Elevation: a major influence on the hydrology of New Hampshire and Vermont, USA. Hydrol. Sci. Bull. **26**, 399–413.

Eaton, J.S., Likens, G.E. and Borman, F.H. 1978. The input of gaseous and particulate S to a forest ecosystem. Tellus **30**, 546–551.

Hicks, B.B. 1986. Measuring dry deposition: A re-assessment of the state of the art. Water Air Soil Poll. **30**, 75–90.

Lovett, G.M., Reiners, W.A. and Olson, R.K. 1982. Cloud droplet deposition in subalpine balsam fir forests: hydrologic and chemical inputs. Science **218**, 1303–1304.

Lovett, G.M. 1984. Rates and mechanisms of cloud water deposition to a subalpine balsam fir forest. Atmospheric Environment **18**, 361–367.

Lovett, G.M. and Reiners, W.A. 1986. Canopy structure and cloud water deposition in a subalpine coniferous forest. Tellus **36**, 319–327.

National Research Council. 1983. Acid deposition: Atmospheric Processes in Eastern North America. National Academy Press, Washington, D.C.

Olson, R.K., Reiners, W.A., Cronan, C.S. and Lang, G.E. 1981. The chemistry and flux of throughfall and stemflow in subalpine balsam fir forests. Holarctic Ecology, **4**, 291–300.

Reiners, W.A. and Lang G.E. 1979. Vegetational patterns and processes in the balsam fir zone, White Mountains, New Hampshire. Ecology **60**, 403–417.

Reiners, W.A., Hollinger, D.Y. and Lang, G.E. 1984. Temperature and evapotranspiration gradients of the White Mountains, New Hampshire, USA. Arctic and Alpine Res. **16**, 31–36.

Rutter, A.J. 1975. The Hydrologic Cycle in Vegetation. pp. 115–154 In J.L. Monteith (ed.). Vegetation and the Atmosphere. Vol. I: Principles. Academic Press, London, 278 pp.

Scherbatskoy, T. and Bliss, M. 1984. Occurrence of acidic rain and cloud water in high elevation ecosystems in the Green Mountains of Vermont. pp. 449–463. In: P.J. Samson (ed.). The Meteorology of Acidic Deposition. Air Pollution Control Association, Hartford, Connecticut.

Schlesinger, W.H. and Reiners, W.A. 1974. Deposition of water and cations on artificial foliar collectors in fir krummholz of New England mountains. Ecology 55, 378–386.

Thorne, P.G., Lovett, G.M. and Reiners, W.A. 1982. Experimental determination of droplet impaction on canopy components of balsam fir. Journal of Applied Meteorology 21, 1413–1416,

Vitousek, P.M. 1977. The regulation of element concentrations in mountain streams of the Northeastern United States. Ecolo. Monogr. 47, 65–87.

Vogelmann, H.W., Siccama, T., Leedy, D. and Ovitt, D.C. 1968. Precipitation from fog moisture in the Green Mountains of Vermont. Ecology 49, 1205–1207.

Woodman, J.N. and Cowling, E.B. 1986. Airborne chemicals and forest health. Environ. Sci. Tech. 21, 120–126.

A COMPARISON OF ATMOSPHERIC EXPOSURE CONDITIONS AT HIGH- AND LOW-ELEVATION FORESTS IN THE SOUTHERN APPALACHIAN MOUNTAIN RANGE

S.E. Lindberg[1], D. Silsbee[2], D.A. Schaefer[1], J.G. Owens[1] and W. Petty[3]
Environmental Sciences Division, Oak Ridge National Laboratory, Oak Ridge, Tennessee 37831, USA[1], Uplands Research Laboratory, Great Smoky Mountains National Park, Tennessee, USA[2], Grinnell College, Grinnel, Iowa, USA[3].

ABSTRACT. Two research sites were established at 300 m and 1800 m elevations in the southern Appalachian Mountains for the study of effects of atmospheric deposition on forest element cycles. Meteorological and chemical data are collected continuously and on an event basis to compare the rates of wet and dry deposition to indigenous conifer forests. Climatic data confirm the expected differences in atmospheric exposure conditions between sites: precipitation, wind speed, and cloud/fog immersion time increase with elevation by factors of 2 to 50. Chemical data collected during the winter indicate comparable concentrations of most constituents in air and rain, while cloudwater contains higher concentrations of acidity and acid anions than does fog water. All of these factors combine to create much higher deposition loading to the mountain site. Differences in dry deposition are reflected in significantly higher net throughfall fluxes in the high elevation spruce stand.

1. INTRODUCTION

The role of the atmosphere in forest element cycling has received increased interest since the discovery of widespread damage to forests in Europe and North America. The importance of atmospheric deposition in forest decline is unknown; however, there is circumstantial evidence that deposition may be involved. One common characteristic of many reported declines is their tendency to occur initially, more frequently, and more severely at high elevation sites (Schutt and Cowling, 1985). Forests at high elevations must adapt to generally more severe climatologic conditions than their lower elevation counterparts. In addition, these same conditions are expected to result in higher atmospheric deposition rates than occur in forests at lower elevations (Lovett, 1984a).

Mountainous terrain is conducive to orographic precipitation, high wind speeds, and cloud immersion, and is commonly populated by

M. H. Unsworth and D. Fowler (eds.), Acid Deposition at High Elevation Sites, 321–344.
© *1988 by Kluwer Academic Publishers.*

322

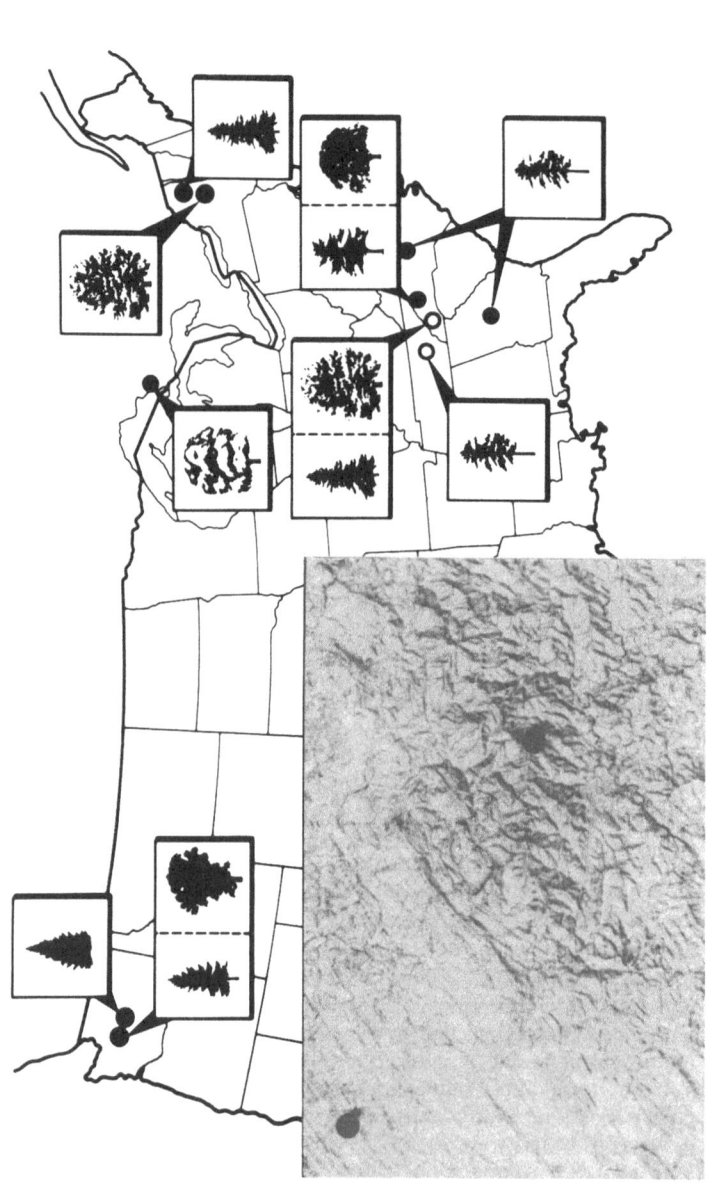

Figure 1. Location of Integrated Forest Study sites in North America showing the local terrain of the sites discussed in this paper (inset). The intensive deposition sites include the following (clockwise from left): Douglas fir, Thompson Forest, Washington; mixed deciduous, Huntingdon Forest, New York; spruce/fir, Whiteface Mountain, New York; loblolly pine, Duke Forest, North Carolina; red spruce, Smoky Mountains, North Carolina; white pine, Coweeta, North Carolina; loblolly pine, Grant Forest, Georgia; and loblolly pine, Oak Ridge, Tennessee. There is an additional site not shown: Norway spruce, Nordmoen, Norway. The sites described in this paper are indicated by the open circles.

coniferous vegetation having relatively high surface area throughout the year. It the concentrations of elements in precipitation and air are comparable at high and low elevations, these physical factors should result in higher atmospheric loading to high elevation forests. Reports of positive gradients of Pb in forest floor samples from low to high elevations in the northeastern United States suggest that the elevation effect on deposition is measurable (Johnson et al., 1982).

During 1986, we established atmospheric deposition research sites in one European and several North American forests as part of the Integrated Forest Study on the Effects of Atmospheric Deposition (IFS) (Johnson and Lindberg, 1986). These sites are operated by local scientists and include basic nutrient cycling and soil chemistry studies in addition to detailed research on deposition rates and processes. The North American sites are located primarily in the United States and include 6 major forest types falling into three general elevation ranges: 300 m, 500 - 1100 m, and 1500 - 1800 m (Fig. 1). We report here preliminary results from sites at high- and low-elevations in the southern Appalachian Mountains (open circles in Fig. 1).

2. SITE DESCRIPTIONS AND METHODS

Measurements of atmospheric chemistry and meteorology were taken in and above forests at Oak Ridge, Tennessee, and at the Great Smoky Mountains National Park, North Carolina (Fig. 1 inset) during January through July, 1986. The site in the Smokies is located at Nolan Divide on Clingmans Dome Mt. at an elevation of approximately 1800m on a slope of 15° with a southwest aspect. Overstory vegetation is dominated by old-growth (200-300 years) red spruce (Picea rubens). The primary understory consists of patches of Fraser fir (Abies fraseri) and red spruce regeneration. The overstory canopy is patchy, with large areas of dense mature spruce interspersed with smaller areas of regrowth in fir stands recently killed by the Balsam wooly aphid. The height of the dominant spruce canopy ranges from 23-32m. Recent measurements of the total above-ground biomass at the research plots range from 2.3 to 2×10^5 kg ha^{-1}. Preliminary estimates of the surface area index of the canopy are in the range of 8 to 14 m^2 per m^2.

The Oak Ridge site is located in a forest plantation at an elevation of approximately 300m in rolling terrain. Overstory vegetation is loblolly pine (Pinus taeda) of approximately 30 years age. The canopy is relatively uniform in height and density, with a mean height of 31m. Total above-ground biomass in the research plots at the site is on the order of 1.2×10^5 kg ha^{-1}. Preliminary estimates of the surface area index of the canopy are in the range of 5 to 6 m^2 per m^2.

All of the deposition sites in the IFS project are similarly instrumented for the collection of routine meteorological data and event measurements of air quality, deposited particles, wet deposition (above and below the canopy), and cloud/fog water. Each research site consists of a central tower surrounded by two ~0.1 ha forest plots,

324

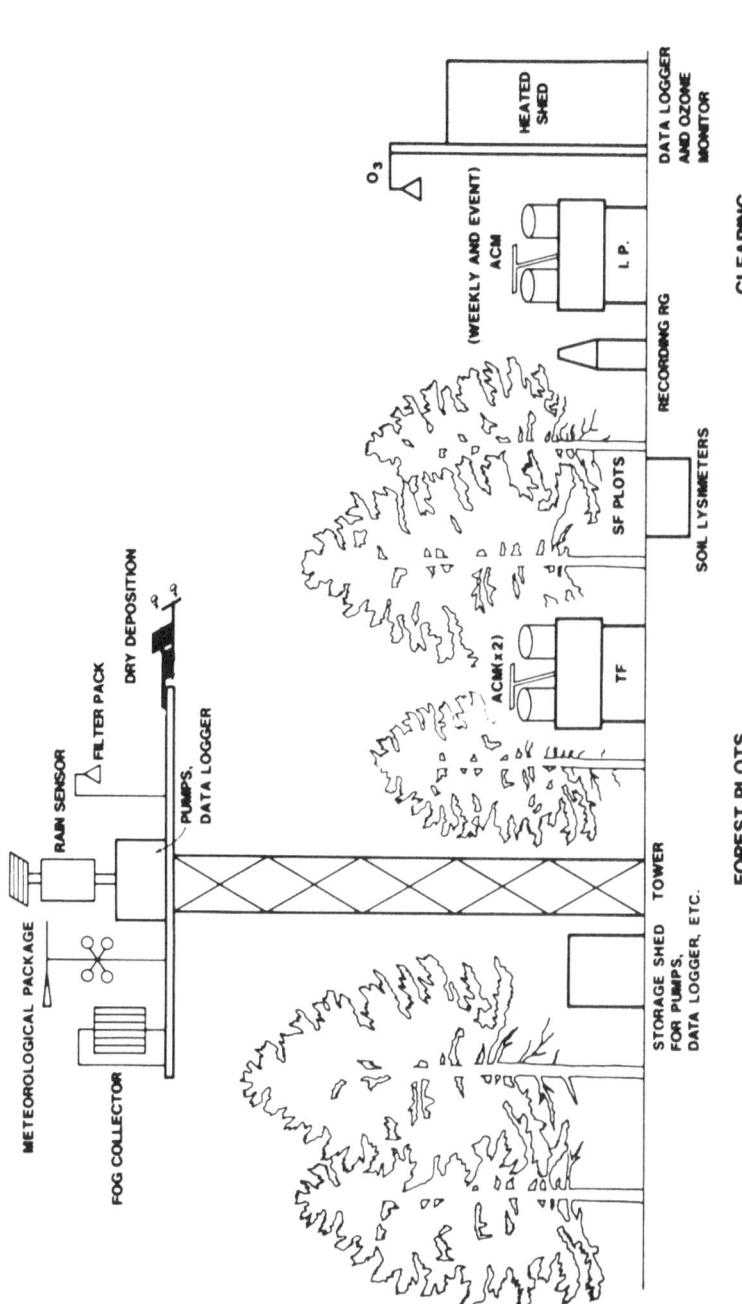

Figure 2. Schematic representation of collection equipment at each intensive deposition sampling site. (Abbreviations: TF = throughfall, I.P. = incident precipitation, ACM = Aerochem Metrics Inc. wet-only sampler, RG = rain gauge, SF = stemflow.

which contain from two to six throughfall/stemflow collectors plus the nutrient cycling apparatus, and an adjacent forest clearing (Fig. 2). Details on the equipment and overall sampling approach have been published (Johnson and Lindberg, 1986; Hosker and Womack, 1985; Hicks et al., 1985; Lindberg et al., 1986; Lindberg and Lovett, 1985; Lovett and Lindberg, 1984; Lovett, 1981).

Briefly, meteorological data on wind speed and direction, solar radiation, precipitation, temperature, relative humidity, and surface wetness are collected with standard sensors mounted 5 to 10m above the canopy (wetness sensors are 1m above the canopy) on towers. Data are collected continuously using microprocessor-controlled data loggers which calculate and store hourly mean values.

Suspended particles, SO_2 and HNO_3 vapor concentrations are measured above the canopy using filter packs fitted with teflon, nylon, and carbonate-treated cellulose filters. Ozone is monitored in nearby clearings using a UV absorption analyzer and separate data logger system. Dry deposited coarse particles are collected above the canopy using inert surfaces in automatic collectors. Cloudwater at high elevation sites is collected using a combination of active (McFarland and Ortiz, 1984) and passive collectors, and low elevation radiation fogs are sampled with the active collectors. Samples are generally collected above the canopy. Incident precipitation is collected in nearby clearings and below the forest canopy using standard wet-only collectors, and stemflow is collected using plastic collars. All analyses are by standard atomic absorption, automated colorimetric, and ion chromatographic methods.

The sampling approach is to collect all samples on an event basis, pairing dry/wet periods whenever possible. In practice, a typical sampling period consists of a 30 to 60 h dry period followed by a precipitation event, each sampled separately for all of the above parameters. The meteorological and ozone data are collected continuously. Cloud water is collected by event, with subsamples taken during events as necessary. Attempts are made to sample all precipitation and as much of the intervening dry periods as possible, while cloud water and fogs are samples as frequently as conditions permit. Wet deposition above and below the canopy is determined from continuous hydrologic measurements and event chemistry. Cloud water and fog input is determined from meteorological and chemistry data using the model of Lovett (1983) and on-site estimates of canopy structure. Dry deposition is determined by a number of independent methods including inferential air concentration methods, measured net throughfall fluxes, and mechanistic models (Baldocchi, 1986; Lovett and Lindberg, 1984; Lindberg et al., 1986).

3. RESULTS AND DISCUSSION

3.1 Climatology and Meteorology

Complete meteorological data are available from both sites for May through July, 1986, and for certain parameters starting in January,

1986. The elevational influence on climatology is clearly reflected in the average meteorological conditions at each site (Table 1). In

Table 1. Average meteorological conditions at high- and low-elevation forest plots in the southern Appalachian Mountains for the period May through July, 1986.

MEANS

Site	Elevation (m)	Temp. (°C)	RH[a] (%)	Wind Speed (ms⁻¹)	PAR[b] (daylight hours) (wm⁻²)	Primary Wind Direction (degrees)	Primary Wind Frequency (%)
Oak Ridge	300	23	72	1.4	350	225-270	33
Smokies	1800	14	82	3.7	410	225-270	28

TOTALS

Site	Precip. (cm)	Precip. Duration (hours)	Precip. Duration (% of period)	Surface Wetness Duration (hours)	Surface Wetness Duration (% of period)	Cloud/Fog Immersion Jan-Apr. (hours)	Cloud/Fog Immersion May-Jly. (hours)	Cloud/Fog Immersion Jan-Apr. (% of period)	Cloud/Fog Immersion May-Jly. (% of period)
Oak Ridge	17	100	4	780	35	57	120	2	5
Smokies	38	150	7	600	27	640	200	25	9

[a] RH = relative humidity,
[b] PAR = Photosynthetically active radiation.

general, all measured parameters increase with elevation: humidity, wind speed, rainfall, rain duration, cloud/fog immersion time, and solar radiation, with gradients ranging from ~ 15% for solar radiation and humidity to a factor of 10 for cloud/fog immersion (Jan.-Aprl. period). The difference in cloud/fog immersion diminished from the winter to the summer period as cloud base moved higher up the mountain. Only temperature as expected, and surface wetness duration decreased with elevation. The difference in wetness largely reflects a higher frequency of dew formation at the low elevation site where nocturnal winds are minimal. Additional data on precipitation from an intermediate site clearly illustrate the influence of orographic effects on rainfall for both winter and summer periods (Fig. 3).

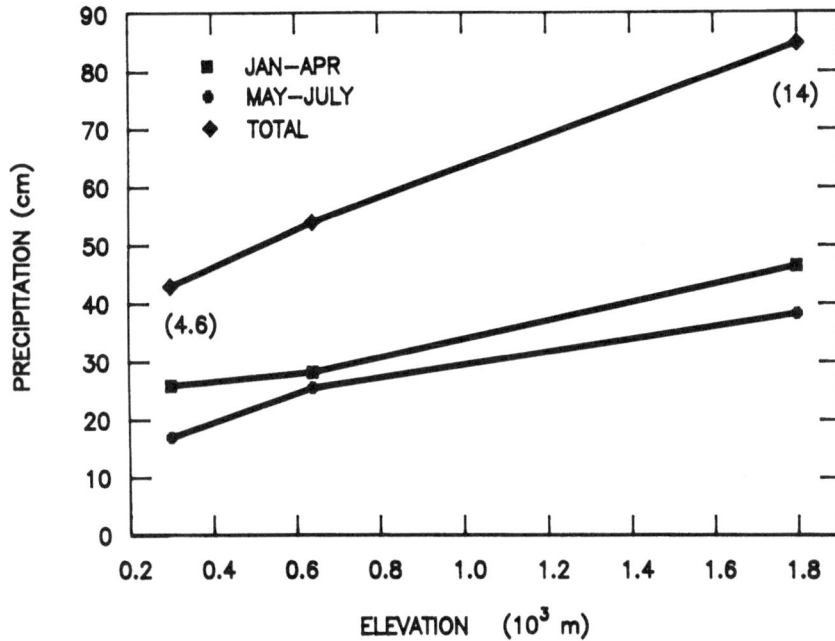

Figure 3. The influence of elevation on seasonal precipitation at three sites in the southern Appalachian Mountains. The locations are Oak Ridge (300m), Uplands (640m), and Smokies (1800m). The Uplands data are from the U.S. National Atmospheric Deposition Program site operated by the National Park Service at the base of the Smoky Mountains, ~13 km north of the Smokies site. The numbers shown below the upper curve represent the amount of precipitation which occurred as snow or sleet at two of the sites.

The influence of elevation on wind direction and speed is shown in more detail in Fig. 4. The prevailing winds in this region are from the southwest as reflected in the plot and Table 1 for both sites. The major difference between the sites is in the frequency of northerly winds (sectors 360 and 45), 15% at the Oak Ridge site vs 36% at the Smokies site. This difference may reflect a nocturnal downslope drainage flow as the Smokies site has a south to southwest aspect. However, segregating the data into day/night periods indicated that the northerly flow was more frequent during the day than at night. Hence, this difference may be due to a combination of gross terrain effects and to the generally counter-clockwise rotation of wind direction with altitude.

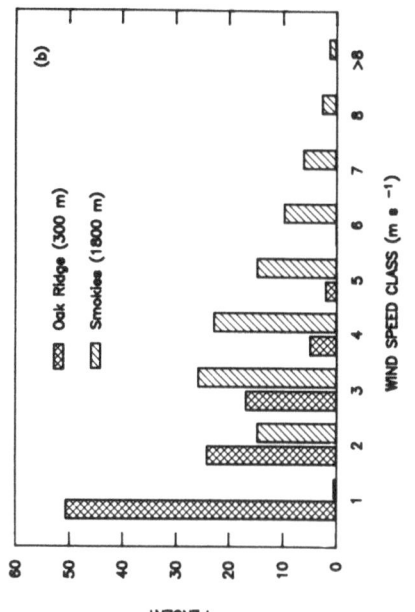

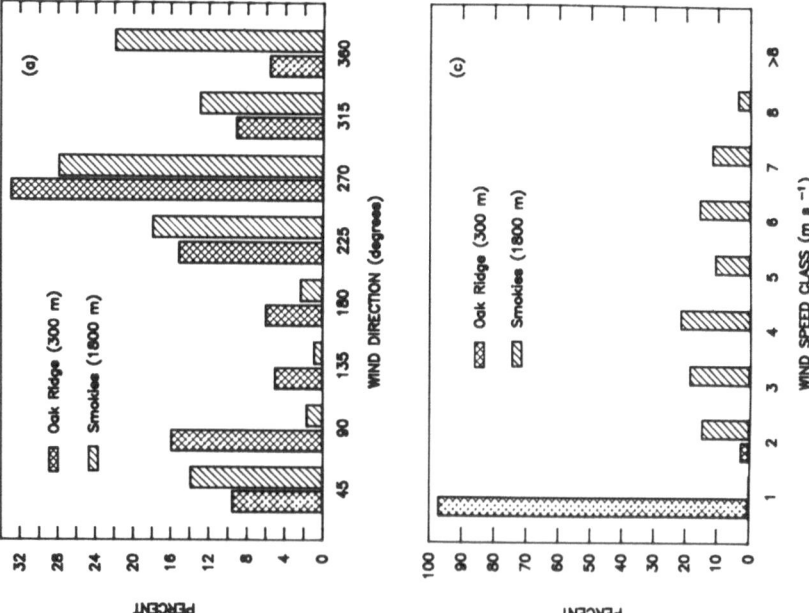

Figure 4. Frequency histograms of wind direction and speed at the high and low elevation sites for the period May through July, 1986. The data are plotted at the frequency (%) of the values less than or equal to the stated value on the X axis. Plot (a) is for wind direction, while the lower plots illustrate the wind speed frequencies for all periods (b) and for in-cloud/fog periods (c).

The wind speed frequency distributions indicate very different wind regimes at each site. For 50% of the time, wind speeds at the low elevation are 1 m s^{-1} or less, while at the mountain site nearly 60% of the values exceed 3 m s^{-1} (Fig. 4b). The most striking difference is the significantly higher wind speed during cloud/fog immersion periods (Fig. 4c). Low elevation radiation fog periods are characterized by generally calm winds, while in the mountains over 40% of the in-cloud periods occur with wind speeds greater than 4 m s^{-1}. These differences have a profound influence on the relative importance of occult deposition at each site as discussed below.

3.2 Occult Deposition and Cloud/Fog Chemistry

Many of the characteristics of the high elevation Smokies site should result in significantly higher deposition rates of fog/cloud associated ions to this site than to the Oak Ridge site. To illustrate the relative influence of these parameters, the cloudwater deposition model of Lovett (1984) was used to estimate the occult deposition of H$^+$ to model forests at each site under a range of conditions characteristic of each site (Fig. 5). In general, as the parameters are increased from values representative of low elevations to those representative of mountainous sites, the predicted ion flux increases, with the most influential parameters being H$^+$ ion concentration, liquid water content, immersion time, and wind speed, as expected.

Also shown in Fig. 5 are estimates of the probable ranges in the flux of H$^+$ by cloud/fog impaction, based on conditions measured at each site during immersion periods and on the chemistry of samples collected to date. These values (0.001 meq m^{-2} month^{-1} at Oak Ridge and 11 to 17 meq m^{-2} month^{-1} at Smokies) are significantly to somewhat lower than values estimated to a high elevation fir forest in New Hampshire (Lovett et al., 1982). Such analyses clearly indicate that wind driven clouds at the high elevation site are capable of depositing significantly more free acidity than are radiation fogs at the low elevation site (by nearly 4 orders of magnitude). This results not only from substantial differences in the physical factors which influence occult deposition as indicated in Fig. 5, but also from significant differences in the chemistry of cloud/fog water measured at the sites (Fig. 6).

Cloud water collected at the mountain site is an acidic solution, dominated by HNO$_3$ and H$_2$SO$_4$ [H$^+$/(SO$_4{}^{2-}$ + NO$_3{}^-$) equivalent ratio of 0.6]. Fog collected at the low elevation site is a neutral solution, dominated by Ca^{2+} and NH$_4{}^+$ salts of SO$_4{}^{2-}$ and HCO$_3{}^-$ [H$^+$/SO$_4{}^{2-}$ + NO$_3{}^-$) equivalent ratios of 0.005]. Unlike cloudwater, ground-level radiation fogs form under stagnant nighttime conditions that lead to effective neutralization of scavenged acidity by basic constituents with significant ground-level sources (e.g. NH$_3$ and CaCO$_3$). Preliminary estimates of occult deposition of other ions to each site are as follows (shown as Smokies/Oak Ridge inmeq m^{-2} month^{-1}): SO$_4{}^{2-}$ = 15/0.13; NO$_3{}^-$ = 3.5/0.04; NH$_4{}^+$ 5.8/0.13; Ca^{++} = 1.7/0.12). Because of the patchy nature of the canopy and the steep slope at the Smokies site, the actual ratios could be much higher. Lovett (in press) has

330

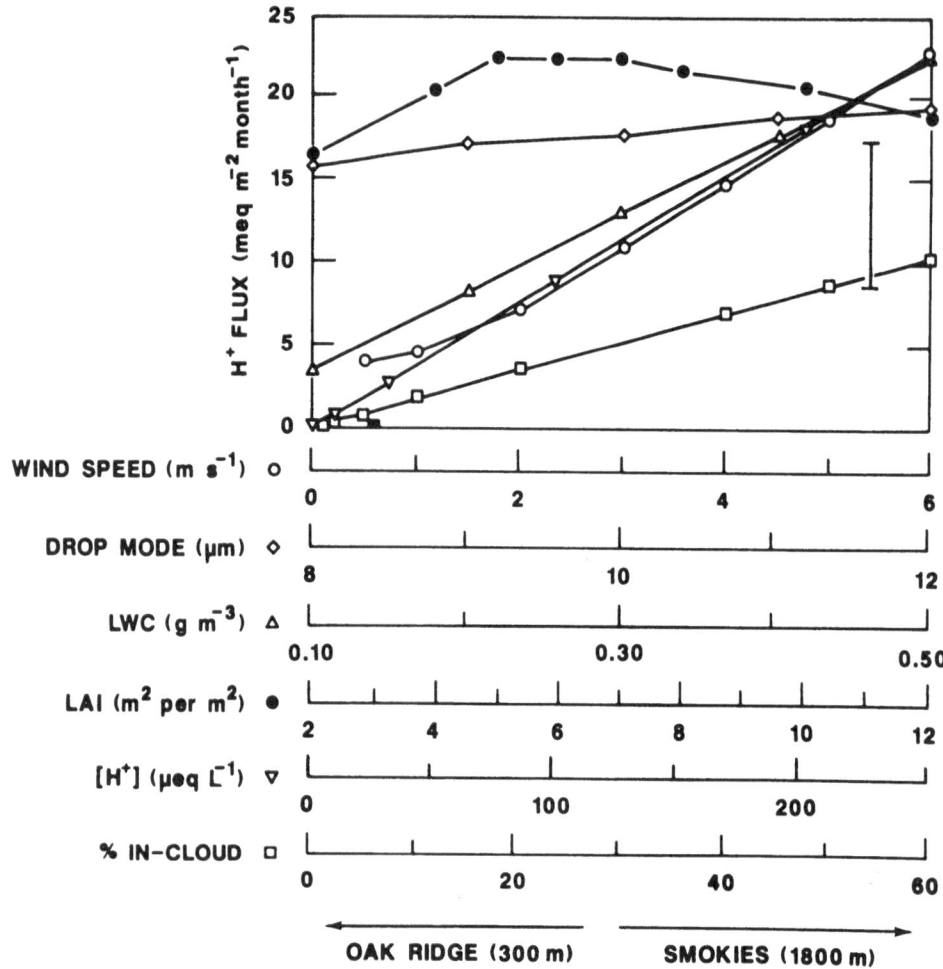

Figure 5. The influence of elevation on predicted cloud/fog water
deposition of H+. The curves were generated using the
cloudwater impaction model of Lovett (1984) by varying each
parameter individually while holding all others constant.
When constant, the values were as follows: LWC (liquid water
content) = 0.4 g m^{-3}; wind speed = 4.8 m s^{-1}; temperature =
10°C; relative humidity = 100%; net radiation = 0 Ly min^{-1};
cloud/fog immersion frequency = 25%; [H+] = 200 µeq l^{-1} (pH =
3.7). The curve for LAI (leaf area index) was taken directly
from Lovett (in press) for a model subalpine fir canopy.
The vertical bar on the right represents the estimated range
in H+ deposition to the high elevation site, while the
estimate for the low elevation site is shown by the filled
square at the lower left.

considered the maximum possible effect of canopy gaps on cloud water
deposition. He estimated the magnitude of this effect to be on the
order of a 7-fold increase incloud water deposition due to edge
effects.

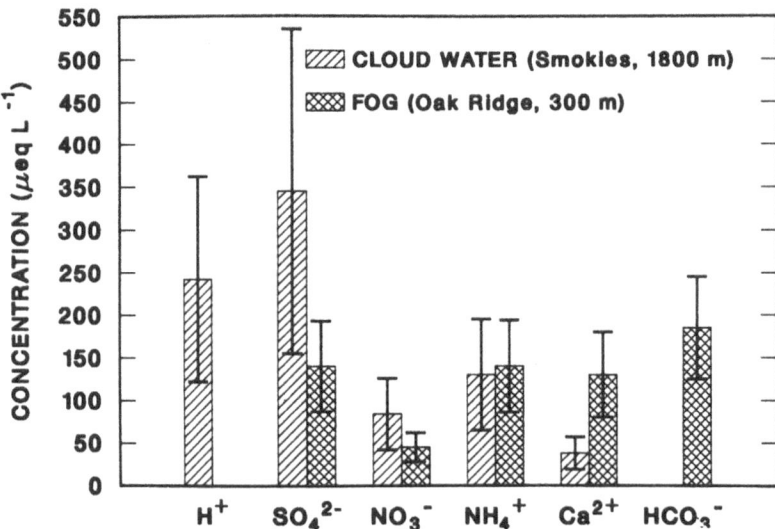

Figure 6. Mean ion concentrations (and standard errors) in cloud water
collected at the high elevation Smokies site and in fog
collected at the low elevation Oak Ridge site from 8/85 to
8/86.

3.3 Atmospheric Chemistry and Dry Deposition

Many of the same characteristics which result in enhanced occult
deposition at the high elevation site can similarly influence dry
deposition. Wind speed, atmospheric turbulence, and canopy
structure/surface area, which generally increase from the low to the
high elevation site, are of particular importance for the deposition of
many constituents (Hosker and Lindberg, 1972). As the magnitude of
these parameters increase, the predicted dry deposition rates to plant
canopies increase for coarse particles (Davidson et al., 1982), fine
aerosols (Wiman and Agren, 1985), and gases (Baldocchi, 1986), assuming
constant air concentrations. These theoretical results are generally
supported by the field and wind tunnel studies of these authors and
others (eg. Sehmel, 1980). Many of the other factors which influence
dry deposition such as humidity, solar radiation, leaf conductance, and
capture efficiency of canopy elements, also tend to increase with
elevation (Lovett, 1985).

Significant differences in air chemistry between elevations also

Table 2. Atmospheric chemistry data collected during matched dry
periods at low- (Oak Ridge, 300m) and high- (Smokies, 1800m)
elevation sites in the southern Appalachians.

		ATMOSPHERIC CONCENTRATIONS (μg m^{-3})						
SITE	PERIOD	SO_2	HNO_3	SO_4^{2-}	NO_3^-	H^+	NH_4^+	Ca^{2+}
Oak Ridge	3/25–3/31	4.3	3.1	5.9	0.42	0.003	1.3	0.85
Smokies	3/25–3/26	4.3	3.8	5.1	0.16	0.026	1.6	0.27
Oak Ridge	4/7–4/14	8.6	3.9	5.4	0.24	0.001	1.2	1.1
Smokies	4/10–4/14	9.3	3.9	4.4	0.94	0.001	1.5	0.64
Oak Ridge	4/26–4/28	9.8	4.4	8.6	0.65	0.006	2.3	1.3
Smokies	4/25–4/28	6.7	6.2	4.9	0.53	0.003	1.6	0.88
		MEANS (±SE) FOR ALL DATA COLLECTED						
Oak Ridge	Jan.–April	16 (3)	3.2 (0.4)	5.1 (0.7)	0.7 (0.06)	0.007 (.002)	1.4 (0.2)	0.68 (0.13)
Smokies	Jan.–April	5.8 (1.4)	4.2 (0.7)	4.1 (0.7)	0.45 (0.18)	0.008 (.006)	1.3 (0.2)	0.5 (0.16)

		MONTHLY MEAN (±SE) OZONE CONCENTRATIONS (ppb)[a]		
		May	June	July
Oak Ridge	(all hours)	33 (32)	30 (19)	37 (26)
Smokies	(all hours)	49 (9.4)	53 (16)	67 (17)
Oak Ridge	(daytime)	46	41	48
Smokies	(daytime)	48	50	61
Oak Ridge	(nighttime)	24	20	23
Smokies	(nighttime)	50	56	64

[a] Ozone data for Smokies collected by the National Park Service at
Cove Mt. (elevation 1400m, 19km North of Smokies site).

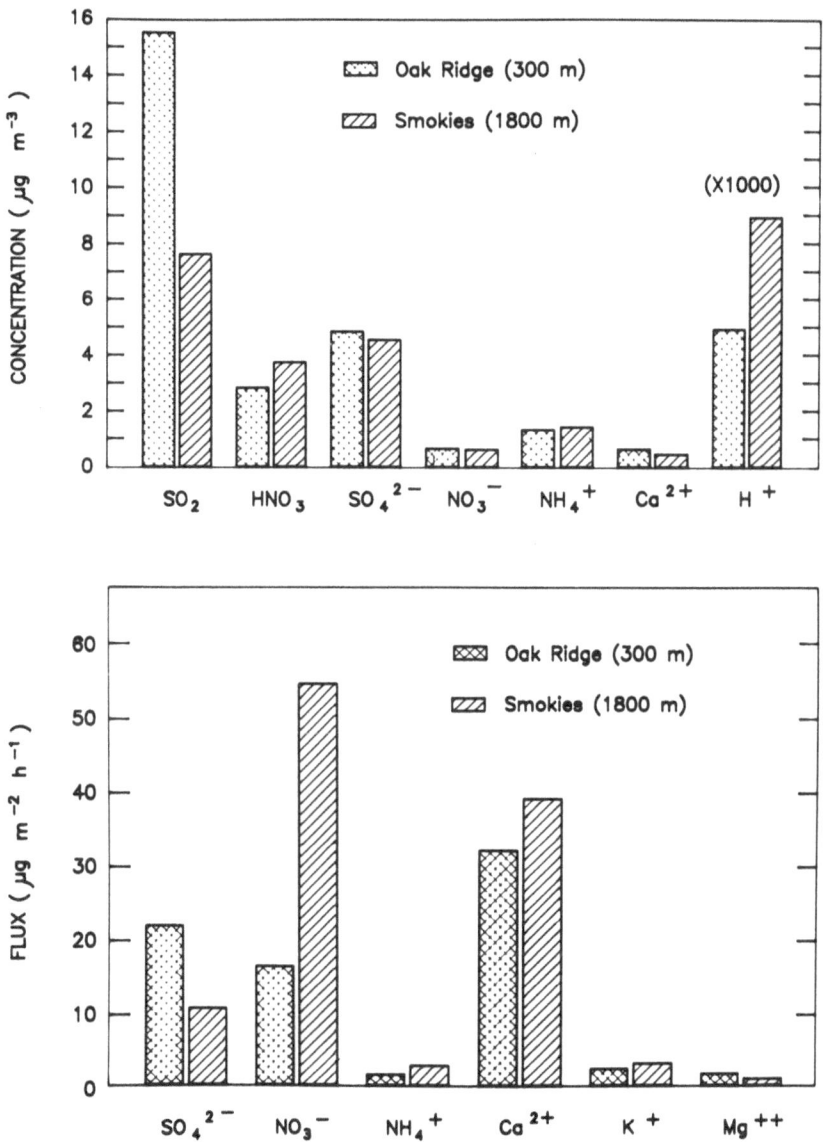

Figure 7. Mean concentrations of several atmospheric particles and gases (lower graph) and mean dry deposition rates of coarse particles to inert surfaces (upper graph).

effect dry deposition rates. Limited data from the Oak Ridge and Smokies sites for the winter/spring period, suggests that the atmospheric concentrations of most constituents do not differ drastically (Table 2). Based on data from the same three dry periods and on overall means, it appears that air concentrations of several particulate species are comparable. However, both gaseous and particulate sulfur and particulate Ca^{2+} concentrations are consistently lower at the high elevation site, while those of HNO_3 vapor are consistently higher. The trend in S compounds is expected because their primary emission sources are in the valley. This is also true for airborne Ca^{2+} which originates primarily from resuspended soil and road dust. All of the concentrations at the Oak Ridge site are comparable to or somewhat lower than the mean winter/spring concentrations recently measured at Walker Branch Watershed (WBW), a deciduous forest ~10km east of this site (Lindberg et al., 1986).

The trend in particle concentrations is reflected in the mean coarse particle deposition rates to inert surfaces at each site (Fig. 7), with the exception of NO_3^- which shows a substantially higher flux at the mountain site for all sample periods. The deposition rates measured at these sites are generally lower than rates previously measured at WBW, particularly for SO_4^{2-} (mean = 72 ± 12 µg m^{-2} h^{-1} during the winter). The WBW site is located within 4 km of an emission source of sulfur and nitrogen oxide. The mean NO_3^- flux at WBW (46 ± 5 µg m^{-2} h^{-1} during the winter) is also much higher than that measured at the Oak Ridge site, but ~20% lower than that at the Smokies site (Lindberg and Lovett, 1985).

The large difference in the mean NO_3^- flux to identical inert surfaces between the high and low elevation sites could be due to wind speed. Similarly, differences in coarse particle NO_3^- concentrations, which are not efficiently measured by filters, could explain the trend. These differences could also be explained by artifact NO_3^- from HNO_3 vapor deposited to the plates. We have measured negligible fluxes of SO_2 to both wet and dry plates in exposure chambers (Lindberg et al., 1984), and plan similar experiments with HNO_3 for the future. Even if these NO_3^- fluxes are controlled primarily by HNO_3, the data suggest significantly higher total NO_3^- deposition rates at the mountain site relative to the lowland pine site.

The concentrations of HNO_3 vapor and O_3 are both elevated at the mountain site (Table 2) relative to the Oak Ridge site, possibly reflecting the regional area-source nature of their precursors, combined with higher biogenic emissions and solar radiation at the mountain site. We are not aware of other reports of increasing HNO_3 concentrations with elevation, but this phenomenon is known for O_3 (Singh et al., 1978). While the filter pack method for HNO_3 suffers some artifacts in urban, polluted conditions, good agreement has been shown between filter pack and denuder systems at Whiteface Mt. (Kelly et al., 1985).

Another interesting difference between the sites is the absence of nocturnal depletion of O_3 at the mountain site (cf. day/night means in Table 2), suggesting lower levels of oxidant scavengeres in the air and minimal nighttime dry deposition. A similar trend has been reported at a mountain site in Europe (Broder and Gygax, 1985). Ozone and HNO_3 vapor concentrations measured at Whiteface Mountain, New York (1500m) during July, 1982 were generally lower than measured in the Smokies, except during "episodes" when concentrations of 60-80 ppb for O_3 and 2.6-7.9 μg m^{-3} for HNO_3 were measured (Kelly et al., 1985).

Although the limited data preclude detailed analysis, these measurements suggest greater dry deposition of NO_3^- to the high elevation site during this period. Higher wind speeds and canopy surface area combined with similar concentrations of most of the other species, except the S compounds, are expected to yield generally higher dry deposition fluxes than measured at the low elevation site. These differences are possibly being manifested in measurably higher throughfall fluxes of many ions at the mountain site, as discussed below.

Table 3. Precipitation chemistry for events sampled from the same frontal storm systems at low- (Oak Ridge, 300m) and high- (Smokies, 1800m) elevation sites in the southern Appalachians from January through April, 1986.

SITE	DATE	PRECIPITATION (cm)	SO_4^{2-}	NO_3^-	Cl^-	H^+	NH_4^+	K^+	Ca^{2+}	Mg^{2+}	Na^+
			CONCENTRATIONS (μeq l^{-1})								
Oak Ridge	1/25-26	1.5 (snow)	46	41	7.6	65	16	0.77	3	0.8	3
Smokies	1/25-27	2.9 (snow)	64	35	3.3	27	20	6.6	4	0.8	7.4
Oak Ridge	2/4-5	1.7	37	9.5	9	34	12	0.13	3	0.8	8.7
Smokies	2/5-6	2.2	24	7.9	4.5	28	6.7	2	4	0.8	4.4
Oak Ridge	2/10-11	1.4	53	30	4.8	48	14	0.13	10	0.8	2.2
Smokies	2/10-22	1.8	39	9.2	2	47	16	0.12	1	0.8	1.3
Oak Ridge	4/28	1.1	47	20	11	48	23	1.5	39	3.3	4.4
Smokies	4/28	0.9	34	11	3.9	35	19	1.5	10	1.6	2.6
		MEANS ($\pm SE$) FOR ALL DATA COLLECTED									
Oak Ridge	(Jan.- April)		54 (7)	31 (8)	9.3 (1.0)	50 (4)	33 (12)	1.2 (0.3)	20 (6)	2.7 (0.6)	7.6 (2.6)
Smokies	(Jan.- April)		68 (20)	27 (8)	10 (5)	53 (13)	28 (9)	3.1 (1)	16 (8)	3.7 (2)	5.7 (1.4)

3.4 Precipitation Chemistry and Wet Deposition

Wet deposition is the product of total precipitation and ion concentrations in rain. Hence, if the chemistry of precipitation collected at each site is comparable, orographic effects at the mountain site will result in proportionally higher input. Based on data from the winter/spring sampling period, it appears that rain chemistry is similar at both sites, with few consistent trends seen in the concentrations measured in the four events sampled from the same storm systems at each site (Table 3). Higher rainfall at the Smokies site might result in lower concentrations due to dilution. However, comparison of rainfall weighted mean concentrations at each site do not indicate obvious differences, with the exception of Ca^{2+} (values shown as the ratio of weighted means, Oak Ridge/Smokies): $SO_4^{2-} = 0.92$, $NO_3^- = 0.90$, $Cl^- = 0.93$, $H^+ = 1.2$, $NH_4^+ = 1.0$, $Ca^{2+} = 2.3$. The concentrations measured at these sites are generally similar to those reported for Walker Branch during the winter/spring season (Lindberg et al., 1984), except for the N compounds (see below).

The estimated total wet deposition to the two sites for January through April indicates the strong influence of orographic precipitation on wet inputs in the mountains (Fig. 8). The monthly mean wet deposition rates at the Smokies site are 50 to 100% higher than at the Oak Ridge site, except for Ca^{2+} which is 20% lower due to

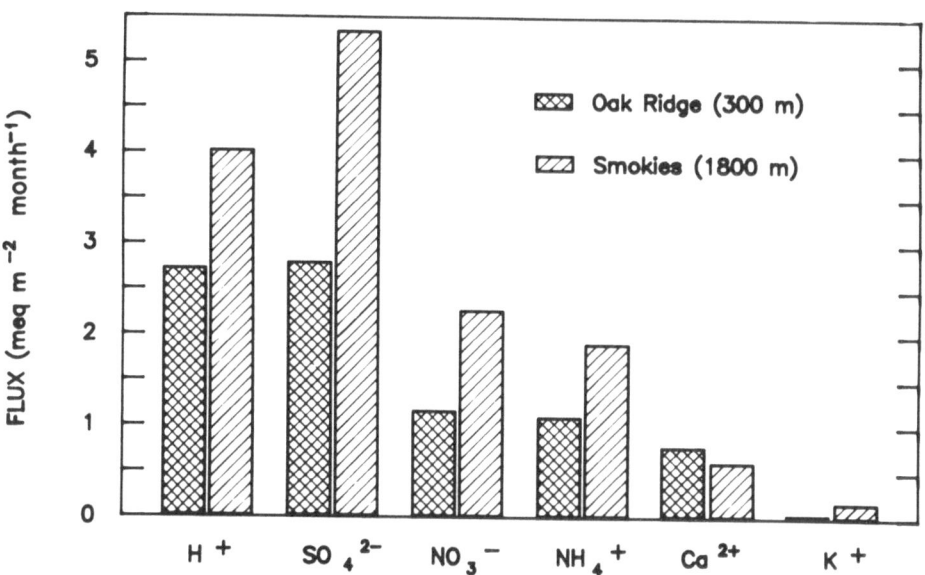

Figure 8. The estimated monthly mean wet deposition of several ions to high and low elevation sites in the southern Appalachian Mountains during January through April, 1986. Deposition was calculated from weighted mean ion concentrations and the measured total precipitation over the period.

its higher concentrations in rain at Oak Ridge. The primary source of Ca^{2+} in both rain and dry deposition is large-particle soil and road dust, and its removal by rain is probably controlled by below cloud processes. Thus its primary source at ground level is reflected in its higher concentration in rain at the low elevation site. The primary emission sources of S are also in the valley, but its concentration in rain is strongly controlled by in-cloud processes such that its distribution in rain tends to be more regionally uniform.

It is difficult to compare these wet deposition rates with those measured previously in this area because of the severe drought during 1986; rainfall at both sites is 50% below normal. Thus, the wet deposition rates measured at Walker Branch during the winters of 1981-1983 are generally higher than those in Fig. 8: $H^+ = 5.5$ meq m^{-2} month^{-1}, $SO_4^{2-} = 5.9$, $NO_3^- = 1.4$, $NH_4^+ = 0.72$, $Ca^{2+} = 0.78$, $K^+ = 0.08$ (Lindberg et al., 1986). Interestingly, the fluxes of both N species are generally higher in the 1986 period than earlier, despite the drought. This is a result of measurably higher (by 50 to 100%) concentrations in rain during this period. The limited data preclude determination of any real temporal trends in wet deposition of N at this time.

3.5 Throughfall as an Indicator of Total Deposition

Preliminary analysis of the data from the first four months of this 4-year study suggests that occult, wet, and dry deposition of major nutrient and pollutant ions are substantially higher to forests in the mountains than to forests at the low elevation site. This agrees with expectations based on physical principles and with measurements at other high elevation sites (eg. papers in Pruppacher et al., 1983). The limited duration of the data reported here prevents anything more than approximations of the total deposition to these sites. This is particularly true for dry deposition because of the lack of models applicable in complex terrain, although some are in development at this time (Hicks and Myers, this volume). However, it may be possible to infer the magnitude of dry and total deposition to each site by analysis of throughfall fluxes, as reported elsewhere (Lakhani and Miller, 1980; Lovett and Lindberg, 1984). Despite some uncertainty in estimating dry deposition from net throughfall, the values do provide an accurate measure of the ion transfer to the forest floor for comparison between sites. For forested areas during periods free from significant litterfall, such as the January to April data from these sites, throughfall represents the major pathway of the ion flux to the ground.

While the chemistry of precipitation deposited to these forest canopies was comparable during this period, that of throughfall was not. Throughfall chemistry from three matching rain events summarised in Table 3 and the mean throughfall concentrations for all data are substantially higher for samples collected at the mountain site than for those collected at the low-elevation site (Table 4). This is particularly true for the ions whose main sources in throughfall are thought to be dry deposition washoff, based on work in other forest

Table 4. Throughfall chemistry for events sampled from the same rain-
producing weather systems at low- (Oak Ridge, 300m) and high-
(Smokies, 1800m) elevation sites in the southern Appalachians
from January through April, 1986.

SITE	DATE	THROUGH-FALL (cm)	SO_4^{2-}	NO_3^-	Cl^-	H^+	NH_4^+	K^+	Ca^{2+}	Mg^{2+}	Na^+
			THROUGHFALL CONCENTRATIONS ($\mu eq\ l^{-1}$)								
Oak Ridge	2/4-5	1.5	62	22	19	52	12	8.0	21	7.1	14
Smokies	2/5-6	1.4	230	230	100	240	84	26.0	120	37	77
Oak Ridge	2/10-11	1.2	140	56	18	80	16	14.0	42	14	13
Smokies	2/10-11	1.1	120	72	14	120	14	9.1	33	9	15
Oak Ridge	4/28	1.0	86	48	34	53	40	51	71	32	6.8
Smokies	4/28	0.43	160	260	40	220	81	49	130	31	13
			MEANS (±SE) FOR ALL DATA COLLECTED[a]								
Oak Ridge	(Jan.- Apr.)		110 (6)	76 (8)	27 (3)	84 (4)	36 (8)	20 (5)	61 (6)	19 (2)	13 (2)
Smokies	(Jan.- Apr.)		200 (30)	240 (50)	57 (13)	200 (30)	77 (14)	26 (6)	130 (20)	36 (7)	39 (11

[a]Includes liquid precipitation periods only.

systems: SO_4^{2-}, NO_3^-, H^+, and Ca^{2+} (Parker et al., 1980; Lovett and
Lindberg, 1984; Lindberg et al., 1986). The mean concentrations of
these ions in Smokies throughfall exceed those at the Oak Ridge site by
80% for SO_4^{2-} to over 200% (H^+ and Ca^{2+}) to 320% for NO_3^-, while the
concentration of K^+, which is controlled by foliar leaching in
deciduous forests (Lovett and Lindberg, 1984), is only 30% higher in
Smokies throughfall. The influence of differences in tree physiology
and morphology on these values cannot be determined at this time. For
example, the higher canopy surface area of the spruce stand can enhance
both dry deposition washoff and foliar leaching.
    Differences in throughfall concentrations reflect trends in both
the incoming rain chemistry and in canopy washoff/leaching effects.
However, net throughfall flux illustrates the canopy effect alone, and
is expressed as the difference between the throughfall flux of an ion

during an event minus the flux of the ion in incident precipitation.
For the three storms simultaneously sampled at each site, the total net
fluxes of all ions except K^+ are measurably higher at the Smokies site
than at the Oak Ridge site, despite a 25% lower total throughfall
volume (Fig. 9). The ratios of total net fluxes (as Smokies/Oak
Ridge) range from 2.6-3.1 for SO_4^{2-}, Cl^-, and Ca^{2+}, to 7.0-8.5 for NO_3^-
and H^+, while that for K^+ is 0.8. If the results we have reported for
other forest types in the Walker Branch Watershed (Lovett et al., 1985)
are applicable to these sites, then these limited data confirm the much
higher dry deposition rates at the high elevation site expected from
physical principles.

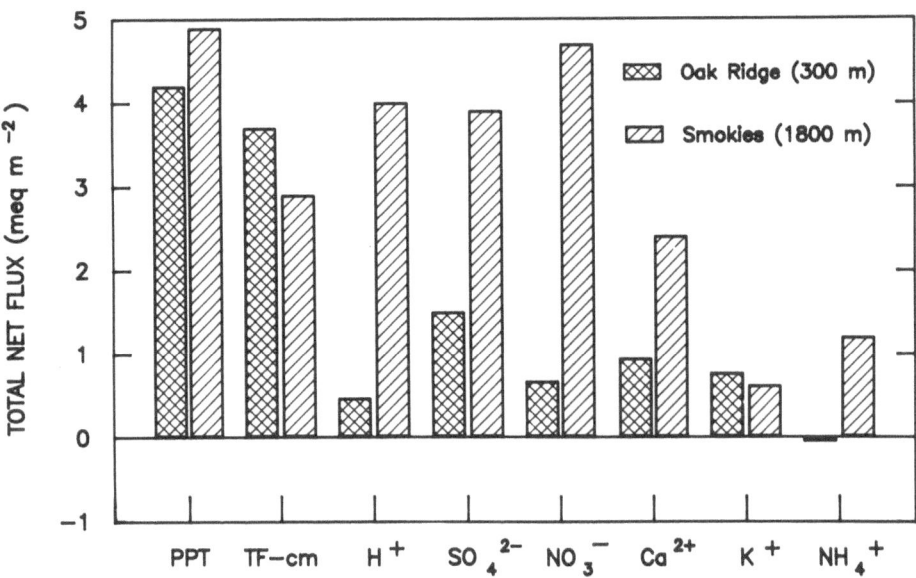

Figure 9. Total net throughfall fluxes at high and low elevation sites
in the southern Appalachians for three events sampled from
the same storm systems from February to April, 1986. The
"PPT" column represents the total rainfall measured, and
"TF-cm" is the total throughfall amount, both in cm as
indicated by the values on the y axis.

The model we published to separate net throughfall into its
component parts relies on positive correlations between the length of
the antecedent dry period and dry deposition washoff, and between
rainfall amount and foliar leaching; these coefficients were
significant for the Walker Branch data (Lovett and Lindberg, 1984).
Although we cannot yet apply this model to the spruce canopy because of
limited date, we have some recent evidence supporting the influence of

340

dry deposition on net throughfall at the Smokies site.   In July, 1986,
we performed an intensive study of Pb deposition during which all rain
events and preceding dry periods were sampled.   For the five matched
dry/wet periods, there was a positive correlation between antecedent
dry duration and the net throughfall flux of Pb in the following rain
event.   Including the effect of the concentrations of Pb in air during
each dry period improved the relationship considerably, suggesting an
obvious influence of dry deposited Pb on net throughfall (Fig. 10).
We have previously shown than Pb in throughfall below a deciduous
canopy is derived primarily from dry deposition (Lindberg and Harriss,
1981).

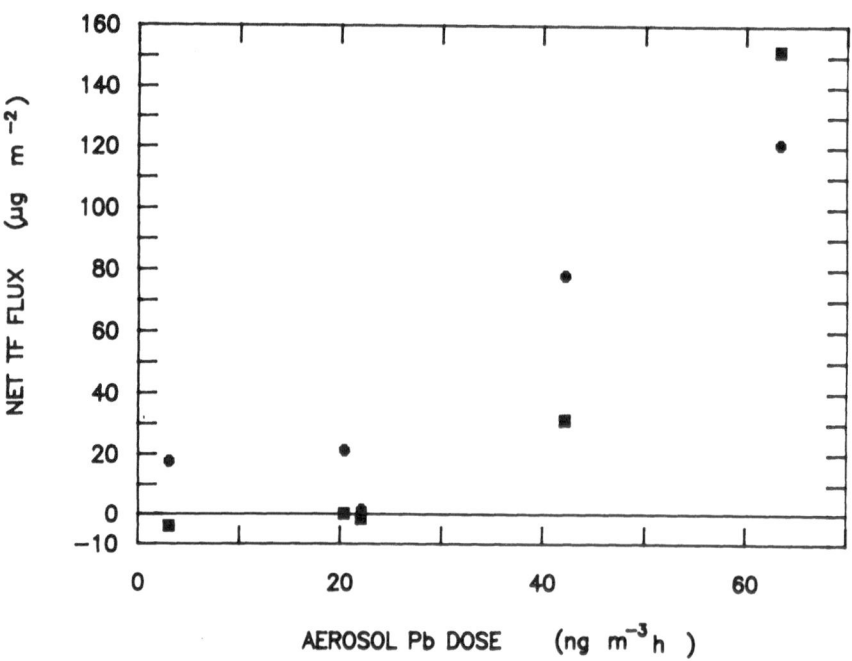

Figure 10 The relationship between the net flux of Pb in throughfall
         and the aerosol dose of Pb during July, 1986 at a high
         elevation spruce stand in the southern Appalachians.   (Dose
         = the Pb air concentration during the dry period prior to
         each storm event multiplied by the dry period duration).
         The different symbols represent the fluxes measured at two
         adjacent forest plots.

    The data from the April 28 storm sampled at the Smokies site
provide the opportunity to further test this hypothesis, as it was
preceded by a 150-h dry period which was sampled for air
concentrations, and which was free from cloudwater impaction.   Because
HNO_3 probably dominates the dry deposition of NO_3^- to this forest, as

elsewhere (Lindberg et al., 1986), we can estimate the deposition velocity of HNO_3 during this dry period from the net flux of NO_3^-. We assume no foliar leaching of NO_3^-, which seems reasonable for nitrogen limited systems, but questionable for N saturated forests (Lindberg et al., in press). Ignoring particle NO_3^-, which comprised only 7% of the airborne NO_3^-,(Table 2), the net flux of 1 meq NO_3^- m^{-2} for this event (Tables 3 and 4) yields a mean deposition velocity for HNO_3 of 1.8 cm s^{-1}. This is a conceivable value considering the physical conditions at the site, and clearly indicates that dry deposition of HNO_3 could support this net flux of NO_3^-. In fact, because dry deposited HNO_3 may be absorbed by the canopy to some extent (Lovett et al., 1985), this estimate is probably low.

If the storms sampled at each site are representative of the population of events over the winter period, weighted mean concentrations can be used to estimate the total flux to the forest floor at each site for selected ions. This is a reasonable approach for the Oak Ridge site as > 90% of the events were collected, but tenuous for the Smokies site where only 15% of the total throughfall volume has been analyzed for chemistry. Hence, until more complete data are available, the following values for total ion fluxes to the soil at the high elevation site must be regarded as crude estimates. In addition, stemflow has not been included due to uncertainty in water fluxes. Stemflow could potentially increase these fluxes by 2 to 5%.

The estimates are useful primarily for establishing the possible magnitude of the differences in total fluxes to each site, bearing in mind the possible roles of in-canopy absorption (primarily of NO_3^- and H^+, meaning that throughfall underestimates the actual flux to the canopy) and of canopy leaching (primarily of SO_4^{2-} and Ca^{2+}, meaning that throughfall overestimates flux to the canopy). The estimated monthly fluxes to the forest floor for the January to April period at the mountain site are 2 to 5 times higher than those at the low elevation site (values shown as Smokies/Oak Ridge in meq m^{-2} month^{-1}): SO_4^{2-} = 10/5, NO_3^- = 9/2, H^+ = 9/4, and Ca^{2+} = 4/2. The estimates for the Smokies site are conservative in that cloud water drip was not sampled during this period. The calculations of cloud impaction in Fig. 5 suggest that for H^+ total fluxes could be ~2 times higher.

4. CONCLUSIONS

These data support the hypothesis that atmospheric inputs to high elevation forests are enhanced over those to forests in lowlands due to natural physical phenomena such as high winds, cloud immersion, and orographic precipitation. Based on our estimates, the increases can be on the order of a factor of 2-5 or more for forests along an elevation gradient of 1500m in 100 km in the southern Appalachian Mountains. Differences in the measured wet deposition above and below the forest canopies suggest that dry deposition and cloud water impaction are primarily responsible for the enhancement. As the air quality in these environments is increasingly impacted by industrial and automotive emissions, native vegetation will be subject to stresses in addition to those imposed by the often harsh climates of these zones. Well-designed long-term monitoring programs in these areas are crucial to our understanding of man's influence on atmosphere/biosphere interactions.

ACKNOWLEDGEMENTS

We wish to thank M. Ross-Todd for the ozone data from the Oak Ridge site, and J. Elwood and C. Garten for comments on the manuscript. Research supported by the Electric Power Research Institute and the Office of Health and Environmental Research, U.S. Department of Energy, under contract DE-AC05-84OR21400 with Martin Marietta Energy Systems, Inc. Publication Number 2803, Environmental Sciences Division.

REFERENCES

Baldocchi, D. 1986. A multi-layer, resistance analog model for estimating SO_2 deposition to a deciduous oak forest canopy. Atmospheric Turbulence and Diffusion Laboratory Contribution 86-14, Oak Ridge, TN.

Broder, B. and H.A. Gygax. 1985. The influence of locally induced wind systems on the effectiveness of nocturnal dry deposition of ozone. Atmospheric. Environ. 19, 1627-1637.

Davidson, C.I., J.M. Miller, and M.A. Pleskow. 1982. The influence of surface structure on predicted particle dry deposition to natural grass canopies. Water Air Soil Pollut. 18, 25-43.

Hicks, B.B., D.D. Baldocchi, R.P. Hosker, B.A. Hutchison, D.R. Matt, R.T. McMillin, and L.C. Satterfield. 1985. On the use of monitored air concentrations to infer dry deposition. NOAA Technical Mem. ERL ARL-141.

Hicks, B.B. and Myers, T.P. Measuring and modeling dry deposition in mountainous area (this volume).

Hoffman, W.A., S.E. Lindberg, and R.R. Turner. 1980. Precipitation acidity: The role of the forest canopy in acid exchange. J. Environ. Qual. 9, 95-100.

Hosker, R.P. Jr. and Lindberg, S.E. 1982. Review: Atmospheric deposition and plant assimilation of gases and particles. Atmos. Environ. 16 , 889-910.

Hosker, R.P. Jr. and J.D. Womack. 1985. Simple meteorological and chemical filterpack system for estimating dry deposition of gaseous pollutants. Proceedings Fifth Symposium on recent advances in measurement of air pollutants, Raleigh, NC, May 14-16, 1985.

Johnson, A.H., T.G. Siccama, and A.J. Friedland. 1982. Spatial and temporal patterns of Pb accumulation in the forest floor in the northeastern U.S. J. Envir. Qual. 11 , 577-580.

Johnson, D.W. and S.E. Lindberg. 1986. Project summary: Integrated forest study of effects of atmospheric deposition. Oak Ridge National Laboratory Report.

Kelly, T.J., Tanner, R.L., Newman, L., Glavin, T. and Kadlecek, J.
    1984. 'Trace gas and aerosol measurements at a remote site in the
    northeast US'. Atmos. Environ. 18 , 2565-2576.
Lakhani, K.H., and H.G. Miller. 1980. Assessing the contribution of
    crown leaching to the element content of rainwater beneath trees.
    pp. 161-172. In Effects of Acid Precipitation on Terrestrial
    Ecosystems. T.C. Hutchinson and M. Havas, eds., Plenum Publ.
    Corp., New York. 666 pp.
Lindberg, S.E. and G.M. Lovett. 1985. Field measurements of particle
    dry deposition rates to foliage and inert surfaces in a forest
    canopy. Envir. Sci. Technol. 19 , 238-244.
Lindberg, S.E. and R.C. Harriss. 1981. The role of atmospheric
    deposition in an eastern U.S. deciduous forest. Water, Air, Soil
    Pollut. 15, 13-31.
Lindberg, S.E., G.M. Lovett, and K.J. Meiwis. Deposition and canopy
    interactions of airborne nitrate. In T.C. Hutchinson (ed.)
    Proceedings Advanced NATO Workshop on Effects of Acidic Deposition
    on Ecosystems, Springer-Verlag, NY (in press).
Lindberg, S.E., Lovett, G.M. and Coe, J.M. 1984. 'Acid deposition/
    forest canopy interactions'. Final report RP-1907-1. Electric
    Power Research Institute, Palo Alto, California.
Lindberg, S.E., G.M. Lovett, D.R. Richter, and D.W. Johnson. 1986.
    Atmospheric deposition and canopy interaction of major ions in a
    forest. Science 231 , 141-145.
Lovett, G.M. and Lindberg, S.E. 1984. Dry deposition and canopy
    exchange in a mixed oak forest determined from analysis of
    throughfall. J. Appl. Ecol. 21 , 1013-1028.
Lovett, G.L. Canopy structure and cloud water deposition in subalpine
    coniferous forests. Tellus (in press).
Lovett, G.M. 1984a. Pollutant deposition in mountainous terrain. In
    P.S. White, The southern Appalachian spruce-fir ecosystem. U.S.
    Dept. of Interior, National Park Service Research Management
    Report (in press).
Lovett, G.M., Reiners, W.A., and Olson, R.K. 1982. Cloud droplet
    deposition in subalpine balsam fir forests: Hydrological and
    chemical inputs. Science 218 , 1303-1304.
Lovett, G.M. 1981. Forest structure and atmospheric interactions:
    predictive models for subalpine balsam fir forests. Ph.D thesis,
    Dartmouth College, Hanover, New Hampshire.
Lovett, G.M. 1984b. Rates and mechanisms of cloud water deposition to a
    subalpine balsam fir forest. Atmos. Environ. 18 , 361-367.
McFarland, A.R., and C.A. Ortiz. 1984. Characterization of the mesh
    impaction fog sampler. Report 84-RD-95 of the Texas Engineering
    Experiment Station, College Station, TX.
Parker, G., S.E. Lindberg, and J.M. Kelly. 1980. Atmosphere canopy
    interactions in southeastern U.S. forested watersheds. In D.S.
    Shriner, C.R. Richmond, and S.E. Lindberg (eds.), Atmospheric
    Sulfur Deposition, pp. 477-493. Ann Arbor Science Publishers, Ann
    Arbor, MI, 568 pp.

344

Pruppacher, H.R., R.G. Semonin, and W.G.N. Slinn (eds.). 1983.
    <u>Precipitation Scavenging, Dry Deposition, and Resuspension.</u>
    Elsevier Science Publ., New York.
Schutt, P. and E.B. Cowling. 1985. Waldsterben, a general decline of
    forests in central Europe: symptoms, development, and possible
    causes. <u>Plant Disease</u> **69** , 548-558.
Sehmel, G.A. 1980. Particle and gas dry deposition: A review. <u>Atmos.
    Environ.</u> **14** , 983-1012.
Singh, H.B., F.L. Ludwig, and W.B. Johnson. 1978. Tropospheric ozone:
    Concentrations and variabilities in clean remote atmospheres.
    <u>Atmos. Environ.</u> **12** , 2185-2196.
Wiman, B.L.B., and G.I. Agren. 1985. Aerosol depletion and deposition
    in forests - A model analysis. <u>Atmos. Environ.</u> **19** , 335-347.

CHEMICAL CONCENTRATIONS IN CLOUD WATER FROM FOUR SITES IN THE EASTERN UNITED STATES

Kathleen C. Weathers[1], Gene E. Likens[1], F. Herbert Bormann[2], John S. Eaton[1], Kenneth D. Kimball[3], James N. Galloway[4], Thomas G. Siccama[2] and Daniel Smiley[5].

[1]Institute of Ecosystem Studies, Millbrook, New York 12545, USA
[2]Yale University, New Haven, Connecticut 06511, USA
[3]Appalachian Mountain Club, Gorham, New Hampshire 03581, USA
[4]University of Virginia, Charlottesville, Virginia 22903, USA
[5]Mohonk Preserve, New Paltz, New York 12561, USA

ABSTRACT. Event samples of cloud and fog water were collected in 1984 and 1985 as part of the Cloud Water Project, a large-scale network designed to chemically analyse cloud and rain water from ten sites in North America. The data presented here are from four sites in the eastern United States that ranged in elevation from 5 m to 1534 m, and in geographic location from Virginia to Maine.

Cloud water from Loft Mt. (Shenandoah National Park), VA (990 m); Mohonk Mt., NY (467 m); Lakes-of-the-Clouds (Mt. Washington), NH (1,534 m); and Bar Harbor, ME (5 m) was found to be acidic (pH range 2.4-5.5) and dominated by H^+, NH_4^+, $SO_4^=$, and NO_3^-. Sulphate:nitrate ratios in cloud water for all four sites were similar to sulphate: nitrate ratios in rain water, but varied considerably by event. Very acidic cloud water samples ($\leq$ pH 3.4) were common for all sites: Virginia (31% of all events), Maine (31% of all events), Mohonk (28% of all events) and Mt. Washington (22% of all events).

Although cloud water chemistry at these four sites was similar, deposition rates are likely to be quite different from site to site. For example, wind speeds vary typically from < 1 m/s (Maine) to greater than 40 m/s (Mt. Washington) and vegetation types range from lowgrowing coniferous (Mt. Washington), to broad-leaved tree species (Virginia and Mohonk).

1. INTRODUCTION

In a preliminary study, cloud water collections were made near the summit of Mt. Washington, New Hampshire (Lakes-of-the-Clouds [LC]) from 1981-1983. Average values for cloud water from LC were higher in dissolved substances and lower in pH than rain water collected from the Hubbard Brook Experimental Forest, approximately 80 km SW of the LC

345

M. H. Unsworth and D. Fowler (eds.), Acid Deposition at High Elevation Sites, 345–357.
© 1988 by Kluwer Academic Publishers.

site (Bormann, 1983). These data, coupled with a few other studies suggesting the same phenomenon (Mrose, 1966; Okita, 1968; Falconer and Falconer, 1980) led us to initiate the Cloud Water Project (CWP) to evaluate the potential stress on terrestrial ecosystems subjected to cloud and fog.

The project was designed to gather information on the chemical concentrations of selected elements in cloud, fog and rain water from several sites in North America. The CWP established ten coastal and montane sites that ranged in elevation from 5 m to 1534 m. General criteria for site selection included a high frequency of cloud or fog events, distance from local sources of pollution (e.g. power plants, automobile traffic, and agricultural areas) and availability of field technicians to initiate collections and maintain CWP equipment (Weathers et al., 1986; Weathers et al., in review). The data presented here are from four CWP sites in the eastern United States.

2. METHODS

A CWP active collector was used to collect samples of cloud and fog water. A battery-driven fan at the rear of the collector pulls atmospheric droplets (50% efficiency for 5 μm diameter droplets) through a ventral opening and across a cartridge of 0.78 mm diameter Teflon strands at a velocity of 7.6 m sec^{-1}. The droplets impact on the strands, form larger droplets and drip into a polyethylene bottle below. The collector has been described elsewhere in detail by Daube et al. (1987).

Cloud collections were initiated when a prominent, stationary object one kilometer distant (e.g. tower, tree etc.) was obscured from view consistently for fifteen minutes. These conditions constituted an event. The sampling time for each cloud water event was set at five hours from initiation of collection, primarily because of the power limitations of the batteries (few of the CWP sites had AC power in close proximity). Sampling times varied, however, since (a) not all events lasted for five hours and (b) at the Maine site, fog events often occurred from late evening to early morning hours, hence collections were made for the duration of the night.

If clouds lifted during the event and visibility improved for fifteen minutes or more, sampling was discontinued.

Rain water was collected (wet only) using a standard Hubbard Brook-type rain collector (Likens et al., 1967). Both cloud and rain collectors were covered between events to exclude dry deposition.

Field operators used standardized protocol for sample and equipment handling, and for measuring pH. Shipping bottles were acid washed (50% HCl), rinsed copiously with deionized water at the Institute of Ecosystem Studies (IES), and sent to the sites. Collectors, collection bottles, and tubing were cleaned initially at IES, and were rinsed copiously with deionized water after each collection at field laboratories. Samples of collection-equipment rinsewater and site deionized water were sent to IES for chemical analysis during the collection season.

Field operators completed a Sample Description Form for each event noting date; sample type; time collection began and ended; beginning and ending temperature, windspeed and wind direction (at sites where meteorological equipment was available); and general information, including whether the vegetation was wet as a result of the cloud-fog event.

Samples were measured for pH (Thomas general purpose glass and Fisher calomel reference electrodes) in the field, usually within one hour of collection. Periodically, inter-laboratory precision was checked by sending reference samples to each of the field technicians for pH measurement. Intra-laboratory accuracy was determined by having the field operators measure pH on two separate aliquots of the same sample for each event they collected. If the two pH measurements differed by more than 0.05 pH units, the aliquots were discarded and two new aliquots of the sample were measured.

Samples were stored in the dark and kept cool until they were shipped to IES in Millbrook, New York for chemical analyses. Where volume was sufficient (> 50 ml), samples were analysed for Ca^{++}, Mg^{++}, K^+, Na^+, NH_4^+, $SO_4=$, NO_3^-, and Cl^- and measured for pH in the IES laboratory.

Sulphate and nitrate were determined by ion exchange chromatography (Dionex 21 I and Dionex 14) (Small et al., 1975), chloride and ammonium using standard Technicon autoanalyser techiques (NH_4^+ = Indophenol method, Cl^- = Mecuric thiocyanate method), potassium and sodium by flame atomic absorption (Perkin Elmer 2380) (Slaven, 1968), and calcium and magnesium by inductively coupled plasma (Perkin-Elmer 6000) emission spectroscopy. pH was measured with a Fisher Accumet 610A pH meter.

3. COLLECTION SITES

The four CWP sites considered here are: Loft Mt., Virginia (VA); Mohonk Mountain, New York (MK); Lakes-of-the-Clouds, New Hampshire (LC); and Bar Harbor, Maine (ME). The sites are described elsewhere (Weathers et al., in review, Dauble et al., 1987), but in general, vary in elevation from 5 m (ME) to 1534 m (LC), and in vegetation type from low-growing, spruce/fir krummholz (alpine vegetation) (LC) to northern hardwood (VA) (Table 1). Three of the sites were located in montane areas, including VA (Blue Ridge Mountains, central VA), MK (Shawangunk Mountains, eastern NY), and LC (White Mountains, northern NH). Sampling season varied somewhat between sites and was dependent primarily upon weather: rime ice collection using the CWP collector was not possible, therefore, since rime ice is often the form of cloud precipitation from September until May at LC, sampling was conducted from June-September at LC. At sites where rime ice was not a problem, the sampling season was extended beyond June-September: at the ME site collections were made from June-November, from June-December at VA, and year-round at MK.

Table 1   Cloud Water Project.   Site and Sampling Information

| Site | Sampling Period* | | Elevation | Vegetation Type |
	1984	1985	(m)	
Loft Mt., VA	16 June–13 Aug	1 July–30 Nov	500	Northern Hardwood
Mohonk Mt., NY	18 June–3 Dec	1 Jan–12 Nov	467	Mixed Hardwood, Conifer
Lakes-of-the-Clouds, NH	18 June–12 Sept	16 June–6 Sept	1534	Spruce/Fir Krummholz Alpine Tundra
Bar Harbor, ME	8 June–29 Nov	5 June–13 Oct	5	Conifer

*Dates when first and last samples were collected.

4.   RESULTS AND DISCUSSION

Cloud water samples from these four CWP sites were acidic, with
generally higher concentrations of ions relative to rain water from the
same site.   Sulphate, nitrate, ammonium and hydrogen ions constituted
from 75 to 94% of the overall mean ionic composition for all four
sites.   The percent contribution of various ions was similar for all
sites (Fig. 1).   However, sodium and chloride from the Maine site,

AVERAGE PERCENT CONTRIBUTION of IONS for CLOUD WATER (µeq/l)

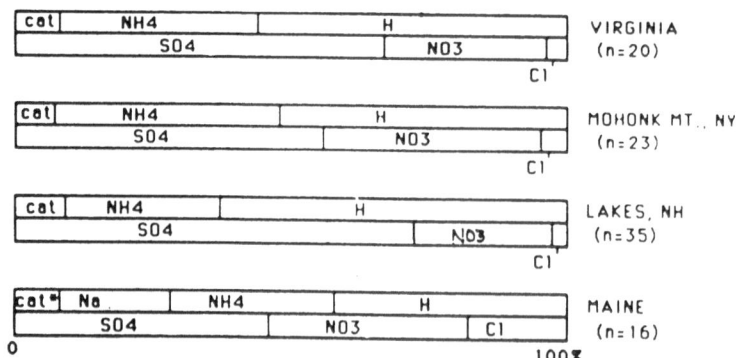

Figure 1   Average percent contribution of ions for cloud water
(unweighted) (µeq/l).   Cat = Ca^{++}, Mg^{++}, K^+, Na^+.   Cat*
=Ca^{++}, Mg^{++}, K^+.

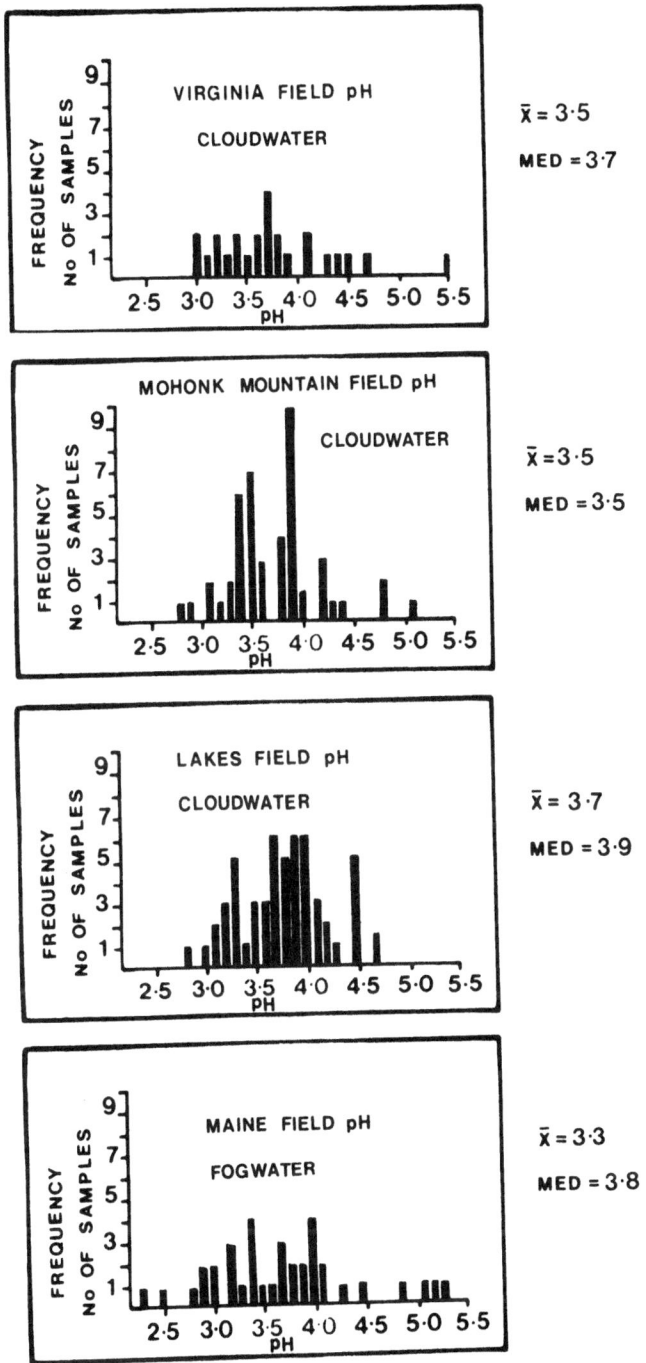

Figure 2 Field pH measurements of cloud water.

350

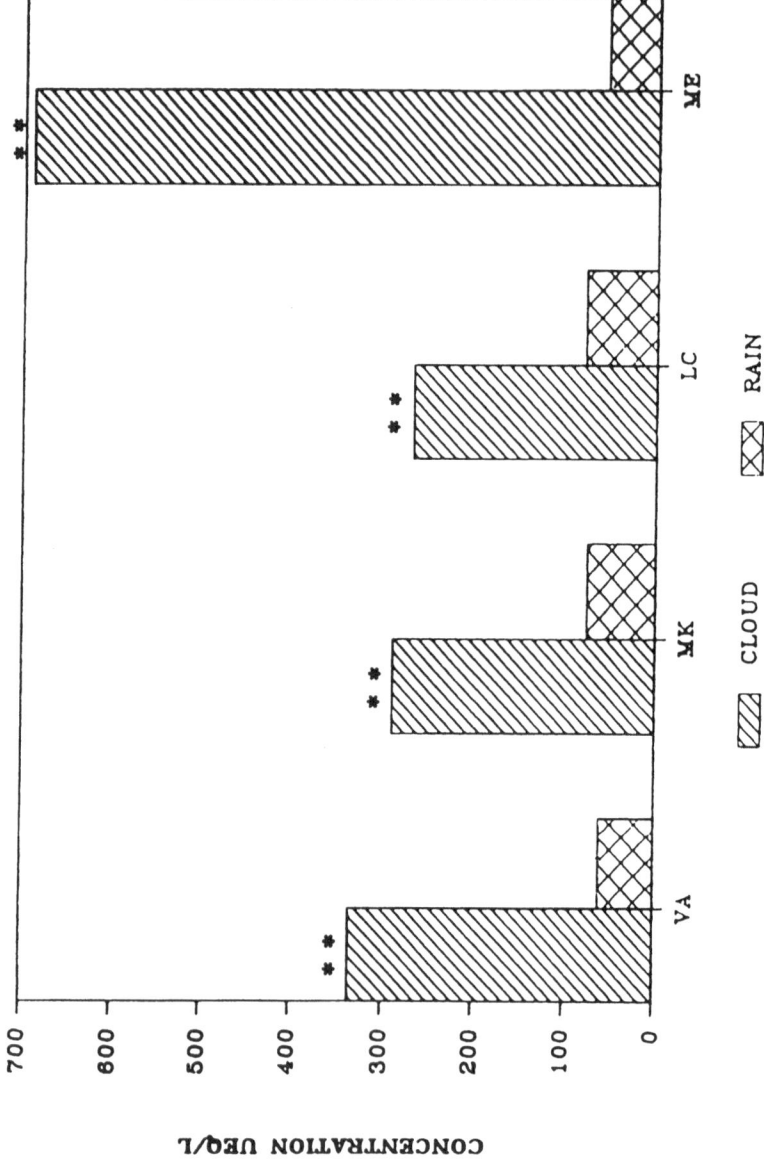

Figure 3. Mean concentration for cloud and rain water (sulphate). ** = p $\leq$ 0.01.

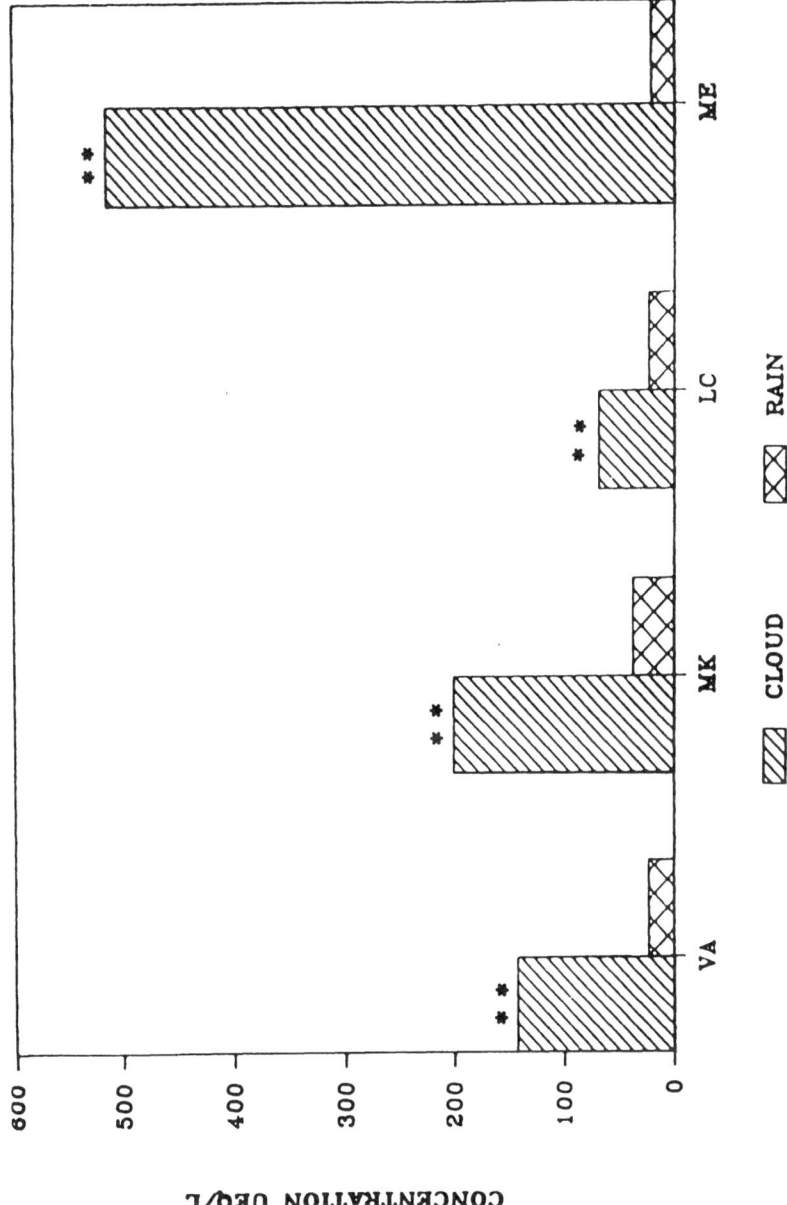

Figure 4. Mean concentration for cloud and rain water (nitrate). ** = p ≤ 0.01.

352

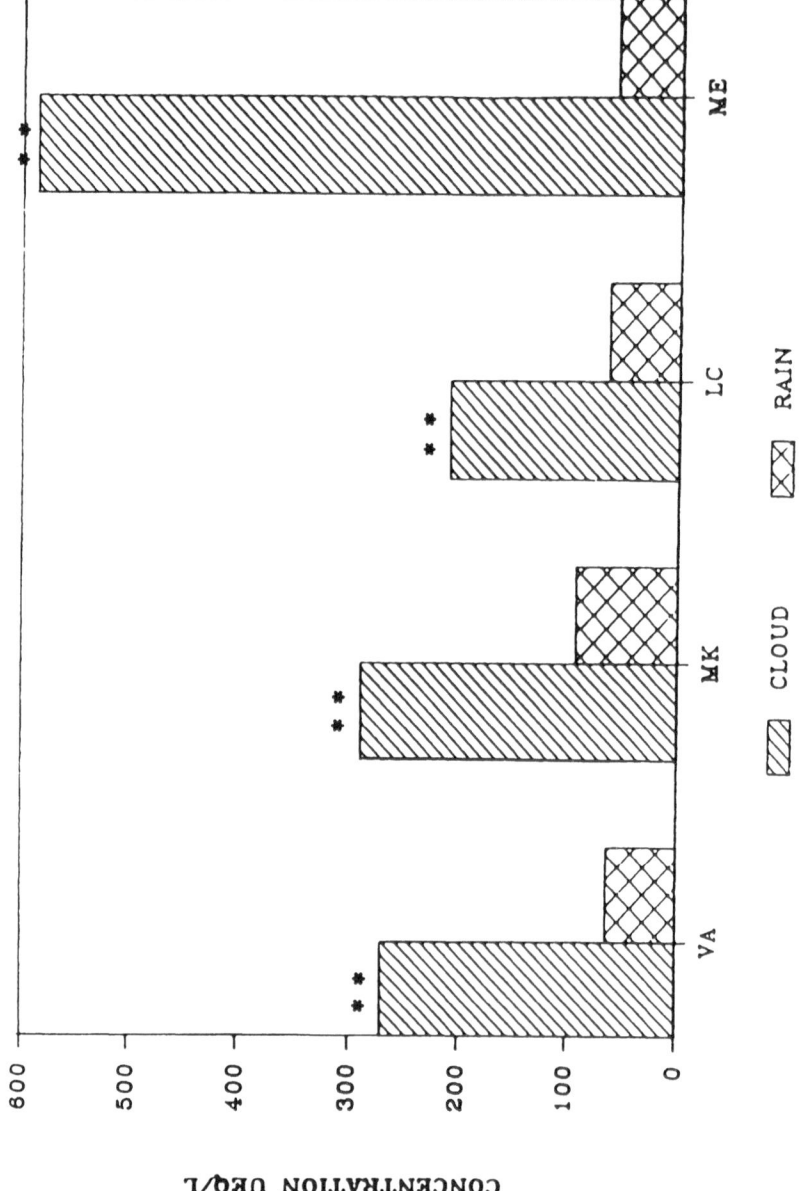

Figure 5.   Mean concentration for cloud and rain water (hydrogen).   ** = p ≤ 0.01.

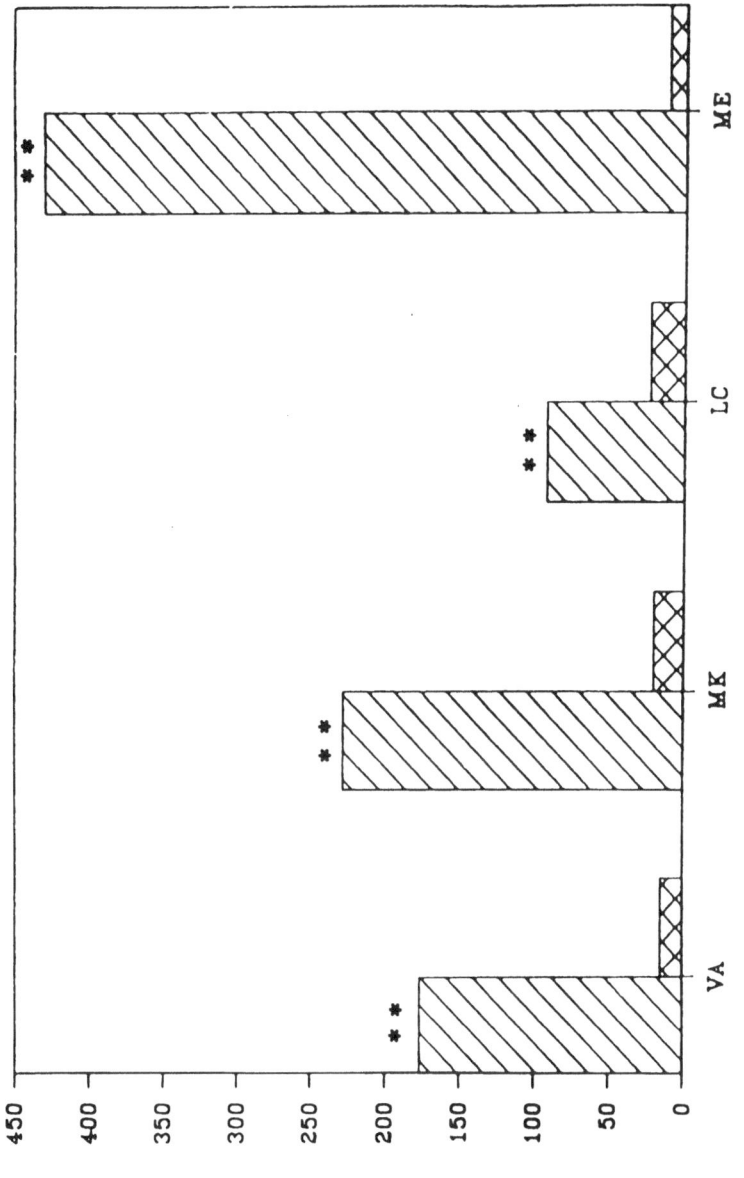

Figure 6. Mean concentration for cloud and rain water (ammonium). ** = p ≤ 0.01.

354

which is closest to the ocean and therefore most influenced by seasalt,
were more dominant ions than in the other sites. For VA, MK and LC,
base cations (Ca^{++}, Mg^{++}, K^+ and Na^+) were of little importance to the
total ionic strength of the samples.

Frequency of occurrence of field pH values measured at all sites
is illustrated in Figure 2. These data include all samples, regardless
of whether the volume was sufficient for complete chemical analysis.
Mean and median values show cloud water from these four sites to be
very acidic and well below the pH of rain (4.1) (Galloway et al., 1984)
for the eastern United States. Although the range in pH is wide, with
2.4 (ME) the lowest pH measured, and 5.5 (VA) the highest pH measured,
there are few samples at the extremes. Inter-site pH values were
similar: Maine (mean = 3.3, median = 3.8), MK (mean = 3.5, median =
3.5), VA (mean = 3.5, median = 3.7) and LC (mean = 3.7, median = 3.9),
as was the frequency of events collected that had pHs $\leq$ 3.4: VA (32%
of all events, n = 25), ME (31% of all events, n = 36), MK (28% of all
events, n = 46), and LC (22% of all events, n = 54).

To examine cloud vs rain, we performed t-tests on arithmetic mean
concentrations of ions for cloud and rain from each of the sites. We
are able to reject the null hypothesis that cloud was equal rain (p $\leq$
0.0001 for H^+, NH_4^+, NO_3^- and $SO_4=$ at all four sites (Figs. 3-6). At
the VA and ME sites we were also able to reject the null hypothesis for
Ca^{++}, Mg^{++}, K^+, Na^+ and Cl^-.

AVERAGE SO4=/NO3⁻ RATIOS
FOR CLOUD AND RAIN WATER

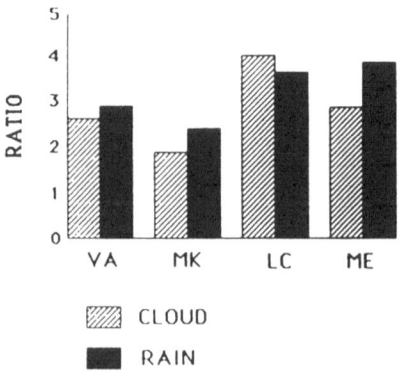

Figure 7 Average sulphate:nitrate ratios for cloud and rain water.

Although mean cloud and rain concentrations were significantly
different, sulphate:nitrate ratios for cloud were similar to $SO_4=:NO_3^-$
ratios for rain from the same sites (Fig. 7). On average, sulphate is
more concentrated than nitrate at all sites for both cloud water and
rain water. However, there is considerable intra-site variability for
cloud water (Fig. 8). The MK site had the lowest $SO_4=:NO_3^-$ ratio, due
presumably to its proximity to urban pollution sources (e.g.

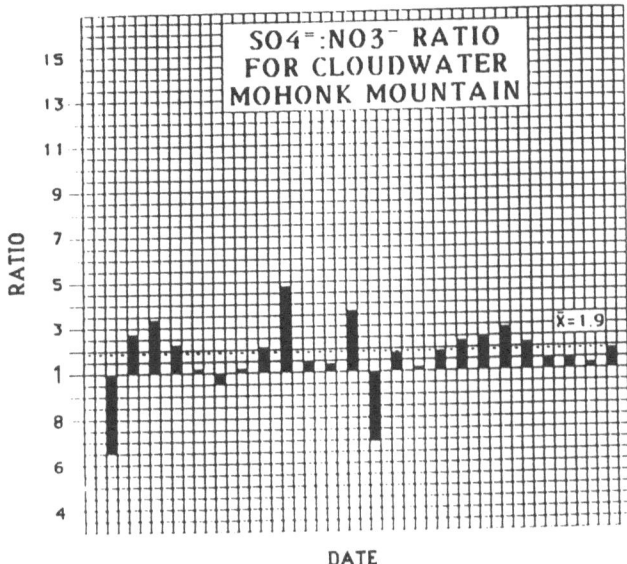

Figure 8   Sulphate:nitrate ratio for cloudwater from Mohonk Mountain,
New York.

automobiles in New York City).   The $SO_4=:NO_3^-$ ratios for cloud water in
the eastern United States differ from observations made in California
by Waldman et al. (1985), where sulphate:nitrate ratios for fog are
generally less than 1.
     These data show that, on average, ionic concentrations in cloud
water are acidic and similar in magnitudes at four sites in the eastern
United States.   A critical question remains, however: what is the
deposition rate of cloud/fog water to the surrounding vegetation and
other surfaces?   Only a few researchers have attempted to answer that
question (Waldmann et al., 1984; Lovett et al., 1982; Hori et al.,
1968), and no in-depth field study to validate cloud water deposition
models has been made.   The information needed to assess annual rates of
deposition to a forest would include: liquid water content of the fog,
total immersion time in cloud/fog, vegetation type—size,
architecture-windspeed throughout the stand, and, for deciduous
species, total leaf time (number of days of the year the deciduous
canopy has leaves).   Although we cannot address the question of
deposition rates from these data, we can make some speculations.
Waldman (1985) has identified windspeed and receptor surface geometry
as two of the main criteria that determine deposition of fog in windy
environments.   If we consider average windspeed alone (LC > MK
& VA > ME) and assume comparable forest canopy capture efficiencies
between all four of our sites, we would expect LC to have the greatest
deposition rate by virtue of the fact that it has the highest overall
windspeed.   If we add an additional parameter--vegetation type (LC =

coniferous, MK & VA = deciduous and coniferous, and ME = coniferous)--
and assume that coniferous vegetation is more efficient at scavenging
cloud droplets than deciduous vegetation, we might still estimate
highest deposition rates at LC, but, it becomes increasingly difficult
to speculate about the relative deposition rates at the other three
sites. Maine has the lowest average windspeeds, but is also
characterised by coniferous vegetation while VA and MK have higher
windspeeds and deciduous vegetation. The problem of quantifying and
comparing deposition rates becomes very complex when comparing various
sites with different characteristics. The challenge is clear relative
to quantification of cloud and fog water deposition on natural
landscapes.

From the data we have presented for four CWP sites, it is clear
that while chemical concentrations of cloud water are similar between
the sites, deposition rates may vary considerably. To make estimates
of the input of mineral acids and nutrients to specific ecosystem via
cloud water deposition, future research should be directed toward
making the measurements necessary to enable quantification of
deposition of cloud water to forest canopies.

ACKNOWLEDGEMENT

This research was funded in part by grants from the Andrew W. Mellon
Foundation, and through a cooperative agreement with the United Statdes
Environmental Protection Agency (813-934010 as part of the Mountain
Cloud chemistry programme, Dr Volker A. Mohnen, Principal Invesigator.
We thank B. Daube, W. Malpass, J. Di Mauro, M. Keifer, R. Mannix, J.
Coury, C. Wilson, D. Cass, J. Andersen, S. Eliassen and P. Huth for
field assistance, and the Appalachian Mountain Club, Shenandoah
National Park, and the College-of-the-Atlantic for use of their
facilities. This is a contribution to the programme of the Institute
of Ecosystem Studies, The New York Botanical Garden and the Hubbard
Brook Ecosystem Study.

REFERENCES

Bormann, F.H. 1983. Factors confounding evaluation of air pollution
    stress on forests: Pollution input and ecosystem complexity.
    Proceedings of the symposium "Acid deposition a challenge for
    Europe". (Edited by H. Ott and H. Stangler, preliminary
    edition). Commission of the European Communities.
    XII/ENV/45/83. Brussels, Belgium.
Daube, B.C. Jr., Kimball, K.D., Lamar, P.A. and Weathers, K.C. 1987.
    Two new cloud water collector designs that reduce rain
    contamination. Atmos. Environ. 21: 893-900.
Falconer, R.E. and Falconer, P.D. 1980. Determination of cloud water
    acidity at a mountain observatory in the Adirondack Mountains of
    New York State. J. Geophys. Res. 85 , 7465-7470.

Galloway, J.N., Likens, G.E. and Hawley, M. 1984. Acid precipitation: Natural versus anthropogenic components. Science **226** , 829-831.

Hori, T. (ed.) 1953. Studies on fogs in relation to fog preventing forests. Tanne Trading Co., Japan.

Likens, G.E., Bormann, F.H., Johnson, N.M. and Pierce, R.S. 1967. The calcium, magnesium, potassium and sodium budgets for a small forested ecosystem. Ecology **48** , 772-785.

Lovett, G.M., Reiners, W.A. and Olson, R.K. 1982. Cloud droplet deposition in subalpine balsam fir forests: Hydrological and chemical inputs. Science **218** , 1303-1304.

Mrose, H. 1966. Measurements of pH, and chemical analyses of rain-, snow-, and fogwater. Tellus XVIII, 266-270.

Okita, T. 1968. Concentration of sulphate and other inorganic materials in fog and cloud water and aerosol. J. Meteorol. Soc. of Japan **46** , 120-127.

Slavin, W. 1968. Atomic absorption spectroscopy. John Wiley Interscience, New York.

Small, H., Stevens, T.S. and Bauman, 1975. Novel ion exchange chromatographic method using conductimetric detection. Anal. Chem. **47** , 1801-1809.

Waldman, J.M., Munger, J.W., Jacob, D.J., Flagan, R.C., Morgan, J.J. and Hoffman, M.R. 1982. Chemical composition of acid fog. Science **218** , 677-679.

Waldman, J.M. 1985. Depositional aspects of pollution in fog. Ph.D. thesis, California Institute of Technology Report # AC-11-85.

Weathers, K.C., Likens, G.E., Bormann, F.H., Eaton, J.S., Bowden, W.B., Andersen, J., Cass, D., Galloway, J.N., Keene, W.C., Kimball, K.D., Smiley, D. and Huth, P. 1986. A regional acidic cloud/fog event in the eastern United States. Nature **319** , 657-658.

Weathers, K.C., Likens, G.E., Bormann, F.H., Bicknell, S., Bormann, B.T., Daube, B.C. Jr., Eaton, J.S., Galloway, J.N., Kadlecek, J.A., Keene, W.C., Kimball, K.D., Lugo, A., McDowell, W.H., Siccama, T.G., Smiley, D. and Tarrant, R. Cloud water chemistry from ten sites in North America (in review).

MEASUREMENTS OF THE PROPERTIES OF HIGH ELEVATION FOG IN QUEBEC, CANADA

Robert S. Schemenauer
Atmospheric Environment Service
Downsview, Ontario M3H 5T4
Canada

Peter H. Schuepp and Selim Kermasha
MacDonald College, McGill University
Ste. Anne de Bellevue, Quebec H9H 1CO
Canada

Pilar Cereceda
Pontificia Universidad Catolica de Chile
Casilla 114-D, Santiago, Chile

ABSTRACT. The Chemistry of High Elevation Fog (CHEF) programme is operational twelve months a year on two mountains (Mt. Tremblant and Roundtop Mtn.) in Quebec, Canada and for four months on a third mountain (Mt. Épaule), all having maximum elevations of 970 m. Measurements include: fog water chemistry, precipitation chemistry, ozone concentrations and meteorological observations. Field observations indicate that the summit of Roundtop Mtn. is in cloud about 44% of the time. The mountain (above 457 m) has an estimated forest exposure to acidic fog of 10^5 km^2 h y^{-1}. In autumn 1985 the daily median Roundtop summit fog pH was 3.8 (mean 3.7), and the median precipitation pH 4.55 (mean 4.3). At Mt. Tremblant the values were 3.9 (mean 3.8) and 4.3 (mean 4.25) respectively. In spring 1986 the daily median fog pH on Mt. Tremblant was 4.0 (mean 4.0) and the precipitation pH 4.5 (mean 4.4). The $NO_3^-/SO_4^=$ equivalents ratios in the spring fog water and precipitation were in the 0.4 to 0.5 range with concentrations of NO_3^- and $SO_4^=$ enhanced by a factor of 5 to 10 in the fog water.

1. INTRODUCTION

Concern over the possibility of a major forest decline problem has been slow to develop in Canada. This is not because of any lack of awareness of the nature and extent of the problem in Europe, but rather because field observations supporting a serious health problem with Canadian forests were absent in the early 1980s. Morrison (1984) reviewed the literature on acid deposition effects in forest ecosystems

M. H. Unsworth and D. Fowler (eds.), Acid Deposition at High Elevation Sites, 359–374.

and concluded that laboratory and isolated field studies had demonstrated some effects but large scale damage to forests from acid rain was unproven. The situation has changed significantly, however, in the last several years as a result of repeated reports of forest dieback problems at higher elevations in the eastern United States and as a result of a survey of hardwood forest damage (primarily sugar maple, <u>Acer</u> <u>saccharum</u> Marsh) in the province of Quebec in 1985. The survey showed substantial areas of damage in an area (~3 x 10^4 km^2) south of the St. Lawrence river. In some portions surveyed, > 50% of the trees showed a noticeable loss of foliage and dead branches. The impact of this problem on the maple syrup industry has resulted in heightened public and scientific awareness of potential damage to our forests from all causes including atmospheric deposition.

This paper describes results from the Chemistry of High Elevation Fog (CHEF) programme. CHEF was designed to examine the nature and extent of wet and dry atmospheric deposition to high elevation eastern Canadian forests.

Schemenauer (1986) has extensively reviewed the background of CHEF and its close links with the United States Mountain Cloud Chemistry Project (MCCP). The principle objective of MCCP (and CHEF) is to provide air chemistry and meteorological support to studies of the effects of atmospheric deposition on higher elevation forests in eastern North America. A secondary objective is to monitor selected gaseous and particulate air pollutants and the concentrations of pollution related ions in cloud water over an extended period of time in order to estimate time trends in such data.

Cowling (1985) in his comments on the review paper of McLaughlin (1985) states that with the exception of some direct effects of ozone on forests and of the combination of ozone and acid mist "all the presently available evidence indicating a possible role of airborne chemicals in the current 'declines' of forests in Europe and North America is circumstantial". CHEF has adopted the approach of looking first at fog water chemistry (from 1985), then at ozone concentrations (from May 1986) and finally at other atmospheric chemicals as resources permit in keeping with Cowling's conclusions.

2. SAMPLING PROTOCOLS AND INSTRUMENTATION

CHEF operated two sites in 1985: Mont Tremblant (46°13'N, 74°33'W) with a maximum elevation of 970 m and the main observing location at 860 m; and Roundtop Mountain (45°05'N, 72°33'W) with a maximum elevation of 970 m and the main observing location at 845 m. For the summer of 1986 a third site was added at the Montmorency Forest (47°19'N, 71°10'W) again with a maximum elevation of 970 m which was also the main sampling location. Data discussed here will only be from the first two sites.

Each site has a building equipped for year-around operation. A roof-top platform contains standard instrumentation to measure wind speed, wind direction, temperature, relative humidity, pressure, solar radiation and precipitation amount (in summer). A pair of passive fog

water collectors and a pair of passive collectors for precipitation
chemistry are also roof-mounted as is the inlet for the ozone
analyser. The fog water collector is of the Whiteface Mtn. (ASRC)
design. It has ~ 370 vertical Teflon fibres ~45 cm long with a
diameter of 0.53 mm spaced ~3 mm apart in two concentric circles
resulting in a cross sectional surface area of ~880 cm^2. The
precipitation sample comes from a 50 cm diameter plastic container in
which a sterile polyethylene bag is placed prior to each precipitation
event, rain or snow.

An operator visits the main sampling building and each
sub-location on a daily basis. The meteorological and ozone data are
recorded continuously on data loggers with a 15 min averaging period.
The wet chemistry samples are collected on an event or sub-event basis
depending on sample volume and operator availability. During the
normal work day sample periods can be as short as one hour. Overnight
samples can extend for 16 to 24 hours. CHEF is designed to sample
every fog (cloud on the mountain) and precipitation event during the
course of the year and at several different elevations at each site if
practical. Valley precipitation, at ~250 m, is continuously monitored
for comparison purposes.

The data of Schemenauer (1986) indicate that the summit (970 m) of
Roundtop Mtn. is in cloud ~44% of the time, the ridge location (845 m)
~38% and a lower ridge (530 m) ~23%. This results in a substantial
number of sampling opportunities and as will be seen below a
substantial forest exposure to acidic fog.

3. ANALYSIS

The pH measurements on the samples were made in the laboratory with a
Corning model 145 pH meter. Measurements were made two weeks (or more)
after collection following storage and shipping at ~4°C. The
laboratory values agreed well ($\pm$ 0.1 pH units) with the values measured
in the field by the site operator shortly after collection.

The samples were analysed for concentrations of anions (Cl^-, NO_3^-,
$SO_4^=$) and cations (Na^+, NH_4^+, K^+, Ca^{++}, Mg^{++}) by High Performance
Liquid Chromatography utilizing an IC-PAK Waters column and a Waters
pump model 590. The samples were filtered using a Millipore HV 0.45 μm
filter and injected with an automatic injector (Waters model WISP
710). The conductivity detector was a Waters model 430 and the
chromatograms produced by a Shimadzu C-R3A chromatopac integrator.
Detection limits were estimated to be 0.02 ppm for Cl^-, Na^+ and NH_4^+,
0.03 ppm for NO_3^-, $SO_4^=$, K^+ and Mg^{++} and 0.04 ppm for Ca^{++}. The ion
chromatograph became operational in late spring of 1986. Only a
portion of the analyses from 1986 are available at this time.

Both the fog water samples and the precipitation samples were
collected by exposing a pair of identical collectors for the same time
interval. This resulted in a set of simultaneous sample pairs which
could be screened for possible contamination. In practice, 7% of the
samples might be rejected due to notes made by the site operator of
possible contamination, usually wind blown material in the

precipitation samples. About 3% of the samples are rejected due to a significant difference in the concentration of a major ion between sample pairs. As an example of the agreement in analysis between the accepted sample pairs, the 17 fog sample pairs discussed in the next section had a median difference in $SO_4=$ concentration of 4 µeq 1^{-1} (0.20 ppm) compared to a sample median and median daily median $SO_4=$ concentration of 199 µeq 1^{-1}. The median difference in NO_3^- concentration between sample pairs was 1.9 µeq 1^{-1} compared to a sample median NO_3^- concentration of 65 µeq 1^{-1} and a median daily median NO_3^- concentration of 119 µeq 1^{-1}. The median difference in concentrations between the accepted pairs of simultaneous samples is about 2% of the reported median concentrations for the species.

4. FOG AND PRECIPITATION CHEMISTRY

In this section the chemistry of the fog water and precipitation samples from the spring of 1986 at Mt. Tremblant will be compared to measurements from late summer and fall of 1985 at Mr. Tremblant and Roundtop Mtn. In addition, the compatibility of the measurements with those of a detailed aircraft and surface cloud chemistry project in the winter of 1984 will be explored. Only the H^+, NO_3^-, $SO_4=$ and NH_4^+ concentrations will be discussed at this time.

4.1 Spring 1986 Surface Data

The Mt. Tremblant data from March, April and May of 1986 are given in Table I. The samples are divided into 3 categories: fog; precipitation (almost all rain); and mixed fog and precipitation samples from the Teflon string collector. The data are presented in 3 ways to illustrate the effects of the type of presentation on the data. Mean values are presented in Table Ia), medians of all the samples in Ib) and medians of the daily median values in Ic). Isaac and Daum (1986) argue that the latter presentation is the most reasonable for this type of data since it removes any weighting effect due to some days having more samples than others. The one negative aspect of this approach is that the number of sample days is less than the number of samples which can be a problem in small data sets. The major conclusions from the Table remain unchanged as one moves from Ia) to Ic) but there are some noticeable shifts in emphasis. For example, the difference between fog and precipitation H^+ concentration is similar but the fog plus precipitation category moves closer to the precipitation only value. Also the difference in $NO_3^-/SO_4=$ ratios between fog and precipitation is increased. In the discussion below we will make use of the values in Table Ic).

The median daily median fog water pH value is 3.98 compared to a summit precipitation pH of 4.49. This represents 3.3 times as much H^+ in the fog water, 104 versus 32 µeq 1^{-1}. The valley precipitation has essentially the same pH (4.46) as the precipitation at the summit (4.49). The median daily median mixed fog and precipitation pH value, 4.35, is closer to that of precipitation alone and argues that when

Table I a)

Location	Elevation (m)	Number of Samples	H⁺ (µeq ℓ⁻¹)	pH from H⁺	$\frac{NO_3^-}{SO_4^=}$	Sample Composition
Pic White	860	17	108 ± 62	3.97	0.52	fog
" "	860	32	93 ± 96	4.03	0.51	fog + precip
" "	860	25	39 ± 38	4.41	0.57	precip
Valley	275	27	48 ± 39	4.32	0.57	precip

Summary of the data collected at the two sampling locations at Mt. Tremblant during March, April and May 1986. Each "sample" is the mean of a simultaneous sample pair. Data are means of all the samples. The equivalents ratio for NO_3^- to $SO_4^=$ is given.

Table I b)

Location	Elevation (m)	Number of Samples	median H⁺ (µeq ℓ⁻¹)	pH	$\frac{NO_3^-}{SO_4^=}$	Sample Composition
Pic White	860	17	113	3.95	0.57	fog
" "	860	32	54	4.27	0.32	fog + precip
" "	860	25	30	4.52	0.41	precip
Valley	275	27	35	4.46	0.43	precip

As for I a) but medians of all the samples.

Table I c)

Location	Elevation (m)	Number of Samples	median day H⁺ (µeq ℓ⁻¹)	pH	$\frac{NO_3}{SO_4}$	Sample Composition
Pic White	860	17 (7)	104	3.98	0.53	fog
" "	860	32 (25)	45	4.35	0.33	fog + precip
" "	860	25 (24)	32	4.49	0.40	precip
Valley	275	27 (27)	35	4.46	0.43	precip

As for I a) but medians of the daily median values. The sample number in parentheses results from grouping the data by day.

rain and fog occur simultaneously the major portion of the water in the passive Teflon string collector is from the rain. This agrees with the conclusion of Schemenauer (1986). A "hat" type cover is not used to shield the fog water collector due to the possibilities of contamination due to blow-off and run-off. A sample is considered mixed when there is evidence during the collector exposure period of rain from the operator's observations, the recording rain gauge or the precipitation chemistry sampler.

The median daily median $NO_3^-/SO_4^=$ ratio is higher (0.53) in fog than in rain (0.40) at the same altitude. This difference is larger than if one examines the means, 0.52 versus 0.57 and in the opposite direction. The valley precipitation values are similar (slightly higher) to those on Pic White. The median daily median fog plus precipitation value of 0.33 is lower than that of fog or precipitaton alone. It is difficult to draw conclusions regarding these differences due to the natural variability of the fog and precipitation events. Castillo (1984) also found this for summer cloud water $NO_3^-/SO_4^=$ values at Whiteface Mtn., N.Y. The variability is well illustrated by Table II in which all of the fog samples are listed for this period. Fog samples from March and April are few in number because most were mixed with precipitation. Since airmass origins can change seasonally and annually it points to the value of having a long record in order to state median values with some degree of confidence. The alternative is to take the case study approach and examine one episode in depth in order to try to understand the processes active at the time. An example might be 23 May 1986 where 5 consecutive fog samples were obtained and the pH dropped from 4.80 to 3.95 while the $NO_3^-/SO_4^=$ ratio seemed to rise and then fall.

A summary of the July to December 1985 data from the Mt. Tremblant and Roundtop Mtn. sites is given in Table IIIa) (mean values) and Table IIIb) (medians of daily medians). Most of the fog only values are from September and October. The median pH (3.88) of the autumn fog samples at Pic White on Mt. Tremblant is slightly lower than that for the spring of 1986 discussed above (3.98). The median values for the small number of samples at the summit (3.80) and the high elevation ridge (3.96) of Roundtop Mtn. in the autumn also fall in the same range. The Mt. Tremblant precipitation values for the autumn at the summit (4.30) and the valley (4.17) are also slightly lower than the corresponding values for the spring (4.49 and 4.46). The Roundtop Mtn. values for autumn 1985 at the summit (4.55) and valley (4.21) locations are again comparable to the spring data.

This look at the data from the two sites in the autumn of 1985 and the Mt. Tremblant site in two seasons (autumn 1985 and spring 1986) shows that the fog water acidity is consistently about 3 times higher than precipitation acidity on the upper part of the mountains. In addition, the autumn data in particular suggests lower pH values for precipitation in the valley as opposed to near the summit. This may be largely due to evaporation but scavenging may play a role. The $NO_3^-/SO_4^=$ ratios for the spring data fall in the 0.4 to 0.5 range. The general agreement in fog water (and precipitation) acidity levels in the autumn and spring is perhaps understandable in terms of an almost

Table II

Date 1986	Sample #	pH	H^+ (μeq ℓ^{-1})	$\dfrac{NO_3^-}{SO_4^=}$	Comments
March 4/5	5/6	3.69	205	0.53	24 h
April 7/8	7/8	3.98	104	0.75	24 h
May 6	21/22	3.76	174	0.33	1100-1200 EST
6	23/24	3.70	200	0.36	1200-1400
6	25/43	3.83	147	0.49	1400-1500
6	26/27	3.93	118	0.61	1500-1530
7	39/40	4.24	58	0.77	1300-1400 EST
7	41/42	3.90	125	0.66	1400-1500
8	46/51	4.10	79	0.23	1130-1300 EST
8	52/53	3.76	172	0.21	1300-1400
8	54/55	3.79	164	0.19	1400-1530
23	120/121	4.80	16	0.57	0930-1030 EST
23	122/123	4.59	26	0.70	1030-1130
23	124/125	4.36	44	0.78	1130-1230
23	126/127	4.17	67	0.76	1230-1400
23	128/129	3.95	113	0.69	1400-1515
24	138/139	4.54	29	0.18	
		3.97	108.3± 62.2	0.52± 0.22	N=17

Listing of the fog samples collected at Mt. Tremblant during March, April and May 1986. Each value in columns 3, 4 and 5 is the mean of a simultaneous sample pair. The mean pH is from the mean H^+ concentration. The $NO_3^-/SO_4^=$ ratio is from concentrations in μeq ℓ^{-1}.

366

Table III a)

Location	Elevation (m)	Number of Samples	$\overline{H^+}$ (μeq l^{-1})	$\overline{pH}$ from $\overline{H^+}$	Sample Composition
Mt.Tremblant Pic White	860	20	153.2± 95.9	3.82	fog
" Mid-level	590	2	73.4± 3.6	4.13	fog
" Pic White	860	35	87.3± 83.6	4.06	fog + precip
" Mid-level	590	16	80.1± 62.5	4.10	fog + precip
" Pic White	860	49	55.9± 37.9	4.25	precip
" Valley	275	25	74.1± 45.2	4.13	precip
Roundtop Summit	970	4	196.2±182.1	3.71	fog
" Ridge	850	2	110.4± 43.8	3.96	fog
" Summit	970	8	84.3± 33.8	4.07	fog + precip
" Ridge	850	3	46.5± 26.7	4.33	fog + precip
" Summit	970	24	48.1± 47.4	4.32	precip
" Ridge	850	9	92.3± 22.7	4.04	precip
" Valley	245	49	77.3± 64.8	4.11	precip

Mean values of H$^+$ concentration for the data collected at the different sampling locations at the Mt. Tremblant and Roundtop Mtn. Sites from July to December 1985. Each precipitation sample is the mean of a simultaneous sample pair; most fog samples were single samples. From Schemenauer (1986).

Table III b)

Location		Elevation (m)	Number of Samples	median H$^+$ (μeq l^{-1})	pH from med. H$^+$	Sample Composition
Mt.Tremblant	Pic White	860	20 (15)	132	3.88	fog
"	Mid-level	590	2 (2)	73	4.14	fog
"	Pic White	860	35 (32)	73	4.14	fog + precip
"	Mid-level	590	16 (16)	69	4.16	fog + precip
"	Pic White	860	49 (48)	50	4.30	precip
"	Valley	275	25 (25)	68	4.17	precip
Roundtop	Summit	970	4 (4)	160	3.80	fog
"	Ridge	850	2 (2)	110	3.96	fog
"	Summit	970	8 (8)	82	4.09	fog + precip
"	Ridge	850	3 (3)	48	4.32	fog + precip
"	Summit	970	24 (24)	28	4.55	precip
"	Ridge	850	9 (9)	98	4.01	precip
"	Valley	245	49 (48)	61	4.21	precip

As for Table IIIa) except the H$^+$ concentrations are medians of the daily median values. The sample number in parentheses results from grouping the data by day.

constant molar ratio of SO_x to NO_x emissions throughout the year in the northeastern United States (Summers and Barrie, 1986), and an approximate equivalence in autumn and spring in the incoming solar radiation which in part controls the production rates of NO_3^- and $SO_4^=$.

4.2 Comparison with Aircraft Data

Isaac and Daum (1986) report on the results of an extensive aircraft and surface observation study conducted near North Bay, Ontario during January and February 1984. This location is about 400 km due west of Mt. Tremblant. Both locations are in areas which are 100 km or more from major pollution sources. Table IV compares some of the winter data from North Bay to the autumn and spring data from Mt. Tremblant.

Table IV

Species	Season	Cloud/Fog Water	Precipitation High Elevation	Precipitation Low Elevation
pH	autumn	3.88	4.30	4.17
	winter	3.59	4.55	4.22
	spring	3.98	4.49	4.46
$SO_4^=$ (μeq l^{-1})	autumn	-	-	-
	winter	255.0	22.5	44.0
	spring	199	32	39
NO_3^- (μeq l^{-1})	autumn	-	-	-
	winter	164.0	22.4	91.0
	spring	119	16	22
$NO_3^-/SO_4^=$	autumn	-	-	-
	winter	0.70	0.62	1.40
	spring	0.53	0.40	0.43

Median daily median values for three seasons of the year. The autumn and spring CHEF project values are for Mt. Tremblant, Quebec. The winter values are from Isaac and Daum (1986) for the North Bay, Ontario aircraft and ground data.

The aircraft cloud water samples were collected at a median altitude (1.2 km) comparable to that on the mountain (0.9 km) but the aircraft precipitation samples were collected somewhat higher (1.9 km).

The basic agreement between the data sets is very good. Cloudwater pH values are lower than in precipitation, and precipitation aloft (CHEF Pic White) has higher pH values than that at low elevation (ground surface North Bay; valley CHEF). Sulphate and nitrate concentrations are 5 to 10 times higher in cloudwater than in precipitation, and concentrations at low elevations are higher than those aloft. The equivalents ratio of $NO_3^-/SO_4^=$ is lower in spring than winter for cloud/fog water and the spring precipitation values are lower both aloft and at low elevations. At North Bay most of the precipitation fell as snow while at Mt. Tremblant almost all of the precipitation was rain. If snow is an efficient scavenger of HNO_3 (which is present in high concentrations at the ground in winter, Anlauf et al. 1986) then this would explain the high NO_3^- concentration in low level precipitation in winter as well as the high $NO_3^-/SO_4^=$ value.

Banic et al. (1986) looked at the problem of whether cloud water chemical composition reflects the aerosol and gas composition of the air below cloud base or whether in-cloud aqeuous phase reactions significantly modify cloud water composition. This was done in part by examining ratios of certain primary soluble ions. Table V compares some of the North Bay aircraft and ground data from summer 1982 and winter 1984 to the spring 1986 Mt. Tremblant CHEF data. The spring $NO_3^-/SO_4^=$ and $H^+/SO_4^=$ ratios are similar to but generally not quite as high as for the summer and winter data. Leaitch and Strapp (1987) have argued that in the winter NO_3^- was dominant in the determination of cloud water acidity and that the NO_3^- and acidity levels in cloud water were higher than could be accounted for by simple dissolution of ground aerosol and acidic gases, perhaps the result of in-cloud HNO_3 production mechanisms. The quick look at the aircraft and CHEF data sets presented here suggests that when fully analysed the CHEF data will be able to shed considerable light on these processes on an annual basis.

5. FOREST EXPOSURE

The survey of sugar maple damage symptoms published as a map by the Quebec Ministry of Energy and Resources in 1985 includes the area of the province surrounding the CHEF Roundtop Mtn. site. We have taken a preliminary look at the influence topography might play in determining where the damage areas are located. The hypothesis is that if the dieback of sugar maples and other deciduous species in mountainous areas is related to the input of some atmospheric pollutant such as acidic fog/cloud water or ozone then the dose received by the trees should depend on such parameters as altitude, wind direction, wind speed, slope orientation and perhaps solar exposure. Indeed, these parameters will determine to a large degree the weather related stresses on the trees even in the absence of significant pollutant concentrations. It may be very difficult to partition the two contributions without a careful look for episodic (pollutant and meteorological) damage within the canopy.

A study area of approximately 165 km^2 was chosen around the Roundtop Mtn. site. It extends from 45°00'N to 45°10'N and 72°30'W to 72°40'W. The percentages of land area in various elevation ranges are shown in Table VI. Only areas above 457 m (1500 ft) are shown. This

Table V

Species	Season	Cloud/Fog Water	Precipitation High Elevation	Precipitation Low Elevation	Surface Air
$NO_3^-/SO_4^=$	summer	0.9	0.5	0.7	0.3
	winter rain	0.7	0.8	0.9	0.3
	winter snow	0.8	0.8	2	0.5
	spring	0.5	0.4	0.4	-
$H^+/SO_4^=$	summer	0.8	1	1	0.5
	winter rain	0.8	1.7	1	0.6
	winter snow	1	0.8	2	0.8
	spring	0.5	0.8	0.7	-
$NH_4^+/SO_4^=$	summer	0.5	0.3	0.4	0.9
	winter rain	0.3	0.6	0.2	0.8
	winter snow	0.5	0.4	0.4	0.6
	spring	0.6	0.5	0.4	-

Ion ratios (equivalents) in cloud/fog water and precipitation for the summer and winter (from Banic et al., 1986) at North Bay, Ontario and for the spring (this paper, median daily median values) at Mt. Tremblant, Quebec.

Table VI

Elevation Range (m)	Contoured Area (km^2)	Representative Slope (°)	Slope Area (km^2)	Time in Cloud (%)	Forest Exposure (km^2h y^{-1})
457-610	23.47	18	24.68	25	54,050
610-686	7.60	20	8.09	30	21,260
686-762	4.77	22	5.14	34	15,310
762-838	2.60	25	2.87	37	9,300
838-970	0.78	28	0.88	42	3,240
			41.7		103,160

Estimated yearly forest exposure to acidic fog water at Roundtop Mountain. The time in cloud is calculated from the data of Schemenauer (1986).

corresponds to terrain that is in cloud 25% or more of the time.

In Table VI, the areas in each altitude range corrected for slope are calculated and combined with the estimated time in cloud to produce an annual forest exposure to acidic fog water. The lowest altitude range 457-610 m has an exposure of 54,050 km^2 h y^{-1} and the highest range 838-970 m an exposure of 3,240 km^2 h y^{-1}. Even though the time in cloud increases with altitude, the decrease in available surface area dominates the calculation. The total forest exposure (above 457 m) in the 165 km^2 study area is 103,160 km^2 h y^{-1}. When one considers the potential multiplying effect introduced by the ~250,000 km^2 area in southern Quebec that the CHEF data might be applied to, the total forest exposure to acidic fog becomes very large. Certainly not all of this area has 25% of the land above 457 m but a large portion does. Also the percentage of time elevated areas are in cloud may vary but based on the three CHEF sites the variation in percentage will not be large. Using the same assumptions as above for the area of Quebec south of Chicoutimi but including the Gaspé gives a forest exposure of ~1.6×10^8 km^2 h y^{-1}. Even assuming lower values for areas of elevated land and cloud frequencies would leave one with a value of the order of 10^8 km^2 h y^{-1} for southern Quebec. Using a relatively conservative estimate of fog water deposition rate to a forested area, 0.2 mm h^{-1}, after Lovett (1984), produces an annual water deposition of 2×10^{10} m^3 y^{-1}. This is obviously a substantial amount of water. It amounts to ~10% of the annual precipitation in this area. Future calculations making use of the regional and seasonal variability of fog water chemistry from the CHEF data and a detailed look at the provincial topography will attempt to translate these values into annual deposition figures for the major ionic species.

In examining the sugar maple dieback areas at our Roundtop Mtn. study site, except for a small area of moderate (26–50% of the trees affected) damage close to the town of Sutton, the major damage appears to be generally on ridges (at higher elevations) with less severe damage areas occurring in valleys. The relationship of these damage areas to the prevailing wind direction, the steering effects of the main valley, the flow over and around ridge lines, slope, slope solar exposure, etc. are currently being examined.

Detailed calculations of the amount of fog water deposited to a forested mountain slope are very difficult to make. Calculations such as the above based on a deposition value from Lovett's (1984) model are useful approximations. But eventually better ways of transferring measurements of fog liquid water content and fog water chemistry into forest deposition values will have to be found. The model, even if initialised with actual field measurements, will always have great difficulty in dealing with deposition to complex mountainous terrain. Ultimately some form of basin water balance approach may have to be utilized, however, this would be a major effort requiring considerable resources. It may be that an approach that examines the role of a tree as a collector of wet and dry deposition under actual conditions will eventually allow extrapolations from passive fog and precipitation collectors to be made to entire forested areas. Schuepp et al. (1986) examine some aspects of this problem and indicate that the technique shows promise of success.

6. DISCUSSION

The Chemistry of High Elevation Fog (CHEF) programme is still in the initial stages of data collection, analysis and interpretation. The field aspects of the programme ran semi-continuously from July to October 1985 and have operated continuously since that time. This paper has demonstrated that acidity levels in fog and precipitation samples at mountain sites in Quebec, Canada are similar in two seasons, autumn 1985 and spring 1986. In addition, ion analysis of the spring 1986 data has been shown to be consistent with that from a comprehensive aircraft and ground-based field programme in the winter of 1984.

Calculations of forest exposure to acidic fog suggest that the near vicinity of Roundtop Mtn. receives roughly 10^5 km^2 h y^{-1}. Southern Quebec as a whole may receive of the order of 10^8 km^2 h y^{-1}. Translating these values, which indicate how much forest area is in fog, to water and ionic deposition values is a major challenge and is one of the areas in which efforts are continuing. Another related problem, on which work has begun at one site, is an examination of the detailed topography and meteorology to see if a better understanding of these topics will lead to some insight as to why small scale damage patterns appear where they do on the mountain ranges. If the forest decline observed in sugar maples in Quebec is due to atmospheric deposition or to meteorological stresses then some correlation would be expected.

At this time the expectations are that CHEF will operate until the end of 1990 and provide a data base in excess of 5 years in length. This should allow for an assessment of seasonal, annual and geographical changes in fog water and precipitation chemistry, meteorology and gaseous and particulate air pollutants. Continuous ozone measurements began in May 1986 and other parameters will be added as resources permit. Close cooperation with forest scientists will ensure that the data obtained is relevant to current and future forest decline research.

ACKNOWLEDGEMENTS

We would like to express our thanks to the site operators at Mt. Tremblant, Kevin Parker and Francine Juillet, whose efforts in acquiring the samples are much appreciated. Our thanks also to Mohammed Wasey and Steve Bacic for their technical guidance and to Carol Sguigna for typing the manuscript.

REFERENCES

Anlauf, K.G., Bottenheim, J.W., Brice, K.A. and Wiebe, H.A. 1986. 'A comparison of summer and winter measurements of atmospheric nitrogen and sulphur compounds'. J. Water Air Soil Poll. 30 , 153-160.
Banic, C.M., Leaitch, W.R., Joe, P.I., Schemenauer, R.S., Strapp, J.W., Tremblay, A. and Isaac, G.A. 1986. 'Airborne studies of chemical and physical processes in clouds in Ontario, Canada'. 2nd Intl. Spec. Conf., Meteorology of Acidic Deposition, March, Albany, New York, 302-316.
Castillo, R. 1984. 'The seasonal comparison of ion concentrations: cloud water and precipitation event samples at Whiteface Mountain, New York'. J. Rech. Atmos. 18 , 167-172.
Cowling, E.G. 1985. 'Critical review discussion paper on "Effects of air pollution on forests". J. Air Poll. Cont. Assoc. 35 , 916-919.
Isaac, G.A. and Daum, P.H. 1987. 'A winter study of air, cloud and precipitation chemistry in Ontario, Canada'. Atmos. Environ. 21, 1587-1600.
Leaitch, W.R. Bottenheim, J.W. and Strapp, J.W. 1987. 'Possible contribution of N_2O_5 scavenging to HNO_3 observed in winter stratiform cloudwater'. Submitted to J. Geophys. Res.
Lovett, G.M. 1984. 'Rates and mechanisms of cloud water deposition to a subalpine balsam fir forest'. Atmos. Environ. 18 , 361-371.
McLaughlin, S.B. 1985. 'Effects of air pollution on forests - a critical review'. J. Air Poll. Cont. Assoc. 35 , 512-534.
Morrison, I.K. 1984. 'ACID RAIN a review of literature on acid deposition effects in forest ecosystems'. Forestry Abst. 45 , 483-506.

Schemenauer, R.S. 1986. 'Acidic deposition to forests: The 1985 Chemistry of High Elevation Fog (CHEF) project'. Atmos. Ocean 24 , 303-328.

Schuepp, P.H., McGerrigle, D.N., Leighton, H.G., Paquette, G., Schemenauer, R.S. and Kermasha, S. 1986. 'Observations on wet and dry deposition to foliage at a high elevation site'. Accepted, NATO Advanced Research Workshop on Acid Deposition to High Elevation Sites. Edinburgh, Sept. 8-12, 1986. D. Reidel Publ.

Summers, P.W. and Barrie, L.A. 1986. 'The spatial and temporal variation of the sulphate to nitrate ratio in precipitation in eastern North America'. J. Water Air Soil Poll. 30 , 275-283.

EXPERIMENT ALSDAN (1985): SOME PRELIMINARY RESULTS OF HEIGHT RESOLVED
MEASUREMENTS OF TRACE GASES, AEROSOL COMPOSITION, CLOUD- AND
PRECIPITATION WATER

L. Kins
Institute of Meteorology
University of Munich
Barbarastr. 16
D-8000 Munchen 40
Federal Republic of Germany

K.P. Müller, and D.H. Ehhalt
Nuclear Research Center (KFA) Jülich
Institute of Chemistry 3
P.O. Box 1913, D-5170 Jülich
Federal Republic of Germany

and

F.X. Meixner
Fraunhofer-Institute of Atmospheric Environmental Research
Kreuzeckbahnstr. 19, D-1800 Garmisch-Partenkirchen
Federal Republic of Germany

ABSTRACT. In- and below-cloud scavenging experiments were performed
at Deuselbach (49.8°N, 7.2°E), a rural site in West Germany, in
April/May and October/November 1985. Cloud water samples were
collected on two platforms of a TV-tower on top of a hill (861 m and
821 m a.s.l.), while rain was sequentially sampled at 754 m, 480 m,
and 430 m a.s.l. Cloud- and rainwater samples were analyzed for
conductivity, major anions, cations, and H_2O_2. We also monitored the
mixing ratios of SO_2, NO, NO_2, O_3 in air and standard meteorological
parameters at 821 m and 480 m a.s.l. At the same locations, HNO_3,
HCl, NH_3, and aerosol were measured by discontinuous sampling. The
experiments were generally limited to meteorologically simple
precipitation events, like frontal rains. A preliminary analysis of
the data shows that the trace substance concentrations in cloudwater
at the cloud base always exceed those of rainwater by a factor of 1.5
- 4. Also within the cloud a concentration gradient is observed with
higher concentrations at the cloud base. Aerosol particles (r > 0.5
µm) seem to be effectively scavenged by rain droplets, while
scavenging of trace gases was barely detectable except in the case of
HNO_3.

M. H. Unsworth and D. Fowler (eds.), Acid Deposition at High Elevation Sites, 375–394.
© 1988 by Kluwer Academic Publishers.

1. INTRODUCTION

Scavenging by hydrometeors is an important removal path in the atmospheric cycle of many trace substances. Several physical processes contribute to the wet removal of gases and aerosols: (a) condensation of water vapour on pre-existing particles ("nucleation scavenging"), (b) uptake of particles and gases by cloud droplets ("in-cloud scavenging"), and (c) scavenging of aerosols and gases by raindrops falling from the cloud base to the earth's surface ("below-cloud scavenging") (Hales, 1972). These mechanisms and subsequent chemical reactions within the hydrometeors determine the chemical composition of cloud- and rain water (e.g. Hegg et al., 1984).

An experimental approach to study the scavenging processes within and below clouds should include (a) the simultaneous collection of cloud- and rain water, (b) the determination of the spatial distribution of gases and aerosols before, during and after a particular rain event, and (c) measurements and evaluation of the meteorological parameters associated with the rain event.

Since the scale of the wet removal processes ranges from the molecular to the micro- (0.2 - 2 km) and even to the meso-scale (20 - 200 km) (Hobbs, 1981), a given field program can investigate only certain aspects of scavenging and only within simple cloud systems and precipitation fields. For example clouds and precipitation associated with warm fronts are believed to be essentially uniform (Hobbs, 1981) at least in the horizontal scale. Therefore, case studies of frontal rain from stratiform clouds should reduce the task of measurements of trace gases, and aerosol compounds in air and cloud- and rain water to a one-dimensional (vertical) problem.

To realize such experiment, several German and Dutch groups joint forces to launch a field program which was dubbed ALSDAN, the German acronym for "Auswaschen löslicher Spurenstoffe durch atmosphärischen Niederschlag" i.e. "Scavenging of Soluble Trace Substances by Atmospheric Precipitation" (Kins et al., 1986 a). Eleven precipitation events were investigated in April/May and October/November 1985. In this paper we present the design of the field experiment, we describe the methods and techniques utilized to measure trace substances in the different phases during a precipitation event, and we present as an example the results from a single event on 1/2 November, 1985.

2. EXPERIMENTAL

2.1 Design of the field experiment

A suitable site to realize our concept was found at Deuselbach, F.R.G. (49°N, 7.3°E). A schematic diagram of the site and the experimental set up in Fig. 1. The entire location is situated in the Hunsrück mountains, a hill (500 - 750 m a.s.l.) region in Central Europe.

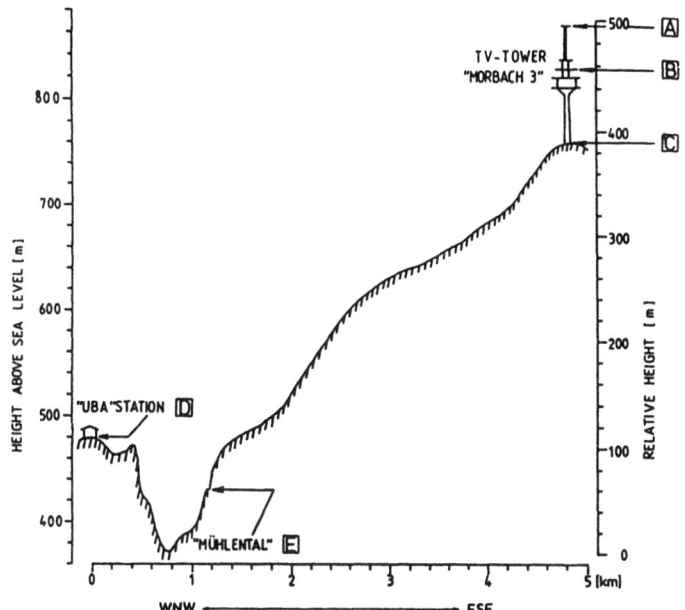

Ⓐ TV tower top (861m, a.s.l) :

windspeed, wind direction, air temperature, relative humidity, cloud water sampling[a]

Ⓑ TV tower platform (821m, a.s.l.) :

SO_2, NO, NO_2, O_3 (continous recording); HNO_3, HCl, NH_3, SO_2 (discontinous sampling) liquid water content, bulk aerosol samples[a]

Ⓒ TV tower base (754m, a.s.l.) :

sequential rain sampling[a]

Ⓓ "UBA"-station (480m, a.s.l.) :

SO_2, NO, NO_2, O_3, CO_2 (continous recording); HNO_3, HCl, NH_3, SO_2 (discontinous sampling) aerosols: bulk samples[a], H_2SO_4-, $SO_4^=$-recording, impactor samples[a], size distribution (optical) sequential rain sampling[a], windspeed, -direction, air temperature, relative humidity

Ⓔ "Mühlental"-site (430m, a.s.l.) :

HNO_3, HCl, NH_3, SO_2 (discontinous sampling); bulk aerosol samples[a], sequential rain sampling[a]

[a] lab analysis : H^+, NH_4^+, Ca, Mg, Cl^-, NO_3^-, $SO_4^=$, $HCOO^-$, CH_3COO^-, $(COO^-)_2$, $[SO_3^=, H_2O_2]$

Figure 1. Design and set up of the experiments of the ALSDAN field program at the Deuselbach site.

There, in spring and autumn, precipitation is normally associated with cyclonic frontal systems of the zonal westerlies. Experience has shown, that in these seasons and in the case of precipitating stratiform cloud systems the TV tower (at least its top) is covered by clouds. Therefore, it was possible to take simultaneously cloud water samples at the top of the TV tower and height resolved rain samples at the TV tower base, at the station of the Umwelt-Bundesamt (UBA), and at the "Mühlental"-site. The vertical distribution of trace gases and aerosol compounds was determined by measurements at a TV tower platform, the "UBA"-station, and the "Mühlental"-site. Some of the wet chemical analysis of cloud-, rain-, and aerosol samples could be performed in the laboratory of the "UBA"-station.

From the observed temporal changes of the concentrations in air, and cloud- and rain water at different altitudes it should be possible to infer in- and below cloud scavenging of trace gases and aerosols. Moreover, since all measurements were made simultaneously and height-resolved, the vertical dependence of individual scavenging phenomena could be investigated and the spatial variability of a rainband moving over the sampling site could be studied.

2.2 Instrumentation, sampling and chemical analysis

A schematic description of our instrumentation at different heights is shown in Fig. 1. The investigated trace substances, the methods and technqiues utilized to measure them are listed in Tab. 1. Some more detailed information about sampling and detection methods are given below.

2.2.1 Sampling of cloud water. A new active collector was developed to sample airborne fog-cloud droplets by impaction on an array of TFE-strings (Kins et al., 1986b). The design of the sampler is shown in Fig. 2. The strings (Ø 0.5 mm) are offset such that any straight trajectory is blocked and the collector cross section is once fully covered with strings. Due to the short length of the strings the droplets are exposed to the ambient air stream only for about 1 min. on their way down along the strings to the 50 ml polythene sampling bottle. Air is drawn through the sampler by a fan providing a minimum flow speed of 1.4 m/s. At that speed a sample of 5 ml volume can be collected in 12 min., assuming a liquid water content of 0.5 g/m^3. A windvane mounted onto the sampler turns the sampler's inlet into the wind direction. During dry periods the sampler's inlet is closed by a lid. At the beginning of a cloud event it can be opened either manually or by a sensor which is activated by cloud droplets only. Thus contamination by impaction of aerosol particles during dry periods could be minimized.

The sampler's collection efficiency was determined by wind tunnel experiments (Kins et al., 1986b). In a saturated air stream, suitable jets generated water droplets of 1 to 40 μm diameter, the size range of actual cloud droplets (Pruppacher and Klett, 1978). Droplet size spectra before and behind the sampler were measured by a light scattering probe (Dibelius and Voß, 1983) which left air flow and droplet distribution in the tunnel undisturbed. At a mean liquid water content of 0.22 g/m^3 and a flow speed of 1.4 m/s provided by the fan, the cut-off diameter, diameter above which 50% of the droplets are collected, is 14 μm.

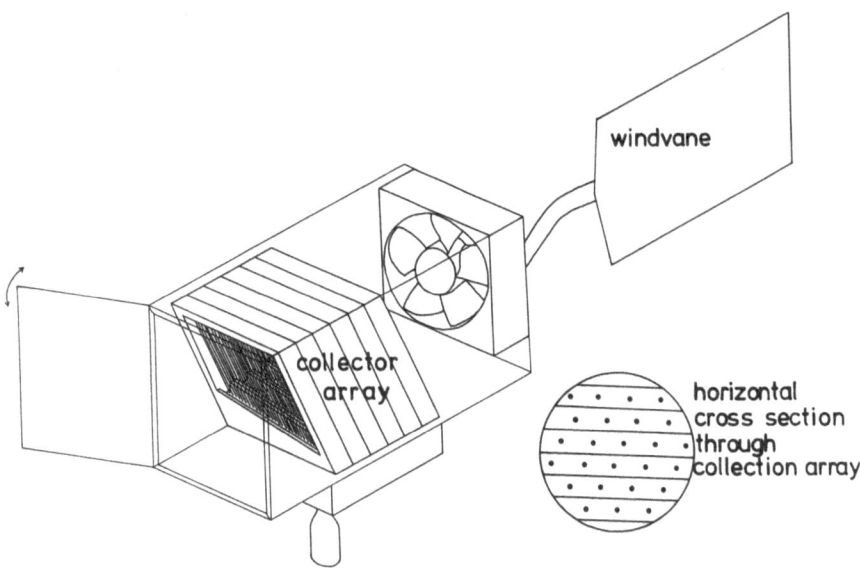

Figure 2.   Schematic diagram of the ground-based active collector for cloud- and fog water.

During the measurements on the TV-tower the sampling characteristics was controlled by the ambient wind speed which varied between 8 and 13 m/s. Under these circumstances the cut-off diameter is about 4-5 µm.

Droplets up to 500 µm diameter could be collected efficiently at these high wind speeds by the cloud water sampler. A rain sensor similar to that used at the sequential samplers was mounted on the TV-tower to registrate the occurence of rain droplets.

During the first experimental period of ALSDAN (April/May 1985) only one cloud water sampler was operated at the top of the TV tower (861 m a.s.l.). Since the corresponding measurements indicated a vertical gradient of cloud water concentrations (Kins et al., 1986a), a second sampler was installed in October/November 1985 on a further platform of the tower (821 m a.s.l.).

2.2.2 Liquid water content. The liquid water content of clouds was estimated by the amount of water sampled by the cloud water collectors, using the actual flow through the sampler and the measured collection efficiency.

Cloud water was also occasionally sampled at the platform 40 m below the tower's top by a high volume collector which consisted of a Schleicher+Schüll[r] paper filter (0 90 mm) mounted in a light-weight filter holder. The filter was first primed by running for 5 min. Then the filter pack was weighed and repeatedly run again. After sampling (19 - 20 min.) the filter pack was weighed again on a electronic analytical balance (Sartorius) which was installed inside the TV tower. From the mass difference and the sampled volume the liquid water content could be inferred.

2.2.3 **Sequential rain sampling.** The sequential rain samplers were improved versions of earlier instruments (Asman et al., 1983; Kins, 1982). A schematic diagram of the sampler is shown in Fig. 3. With the onset of precipitation the sampler is automatically opened triggered by a sensor which is activated by rain drops only. Precipitation water is then sampled via a polythylene funnel ($\emptyset$ 30 cm) into small polypropylene bottles. The precipitation was apportioned into samples of equal volume (19 ml = 0.1 mm of rain) by means of a commercial fraction collector (LKB, Redirac). The times of opening/closing of the sampler's lid and of each sample change are recorded by a micro-computer. Therefore, the duration of a rain event and the rain intensity is documented.

2.2.4 **Sampling of aerosol particles.** As indicated in Fig. 1, bulk samples of atmospheric aerosol particles were taken at the platform B of the TV tower (821 m), at the "UBA"-station (480 m), and at the "Mühlental"-site (430 m). PTFE-filters were used, applying a high volume filter pack technique (Meixner et al., 1985).

The size distribution of the aerosol composition was deduced using a 8-stage Anderson type impactor (size range: 0.43 µm to 11 µm, sampling period: $\geq$ 8 hours). Another physical characterization of the aerosols was obtained by quasi-continuous (integration time: 1 hour) monitoring with a laser aerosol spectrometer (modified version of a Particle Measurement Systems[r] instrument, model 250X), which provided the number density distribution of aerosol particles in 12 classes (0.2 - 12 µm) of aerodynamic diameter.

2.2.5 **Trace gases.** Trace gases were regularly measured at the "UBA"-station (480 m) and the TV tower platform (821 m), while measurements of selected compounds were also performed at the "Mühlental"-site (s. below). NO and NO_2 was continuously monitored by chemiluminescence instruments. Ozone mixing ratio was continuously measured by UV-absorption instruments. A continuous coulometric and a fluorescence instrument as well as discontinuous filter sampling wereapplied to measure SO_2 mixing ratio. Prior and after each measurement campaign all instruments were intercompared and calibrated with standard gax mixtures at the Deuselbach experimental site.

Inlet systems of 3 - 7 m tubing (stainless steel, PTFE) were used at the TV tower and the "UBA"-station. To avoid wall reactions both inlet systems were well ventilated (several m^3h^{-1}) and precautions were taken to exclude co-sampling of coarse aerosols, rain and cloud droplets.

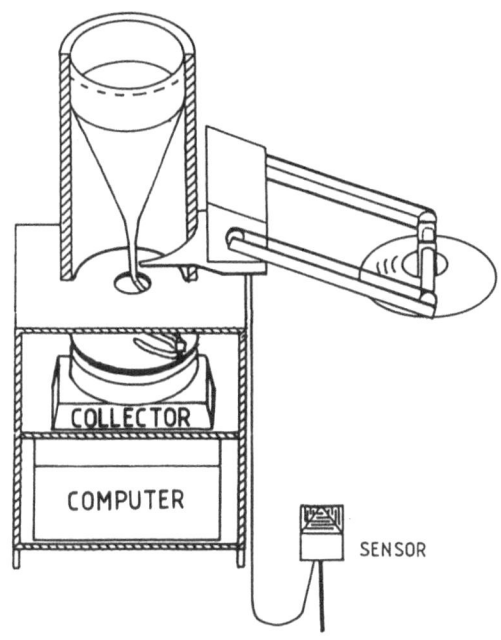

Figure 3.   Schematic diagram of the sequential rain sampler.

Discontinuous sampling of HNO3 was performed either by filter pack or by denuder techniques without inlet systems. The denuder systems (also installed at the "Mühlental"-site) furthermore provided discontinuous sampling of NH3 and HCl. After sampling, the filters (HNO3 filter pack) are sent frozen to KFA Jülich (F.R.G.), the denuder tubes to ECN Petten (NL) for wet chemical analysis (s. below).

2.2.6 Wet chemical analysis. Already during a rain event the cloud- and rain water samples were transported to the laboratory of the "UBA"-station. There, in a clean air bench, each sample was partitioned in several aliquots. Still in the "UBA"-laboratory pH and electrical conductivity were measured in a thermostated micro flow-cell (500 µl), while the content of S(IV) and H2O2 of another aliquot (3 ml) was determined by chemiluminescence techniques (Jaeschke, 1985). The rest of each sample was stored at -18°C awaiting the analysis at KFA Jülich and ECN Petten some weeks later. Analysis of ions (H+, NH4, Cl-, NO3-, SO4=) were performed by ion chromatography (Müller, 1984; Slanina et al., 1981).
    Extracts of the filters and the impactor plates were analyzed for the same ions as the cloud- and rain water samples.

382

2.3 Meteorology measurements and supporting data

Even in the case of simple precipitation systems a fair amount of
meteorological information is necessary to characterize the conditions
of the lower troposphere in terms of frontal activity and homogeneity
of air masses. Therefore, at both platforms (s. Fig. 1), the TV tower
top and the "UBA"-station, windspeed and -direction, air temperature,
and relative humidity were continuously monitored.

A further advantage of the Deuselbach experimental site is the
dense net of routine meteorological and aeronomic measurements within
a circle of 30 km. Radiosonde (00, 06, 12 UT), weather radar (on
request), ceilometer (i.e. height of cloud layers, continuously) and
the usual set of meteorological data reported by several weather
stations (Trier, Offenbach, Idar-Oberstein) and air bases (Hahn,
Bitburg) were available. Individual short term forcasting was
provided on request by the nearest office of the German Weather
Service (30 km) (Trier).

3.   PRELIMINARY RESULTS

A total of 11 events was investigated during the whole span of the
ALSDAN field program. To illustrate the kind of results obtained, the
data from a single event of 1/2 November, 1985 are briefly discussed.

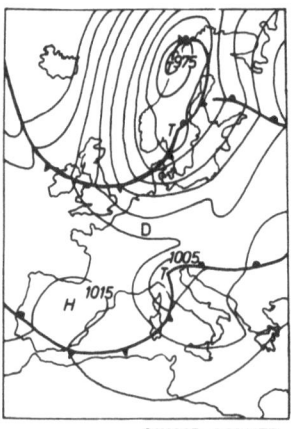

2.11.1985 1:00 MEZ

Figure 4.   Surface pressure chart of 2 November 1985, 00 GMT. The
symbol "D" denotes the location of Deuselbach site.

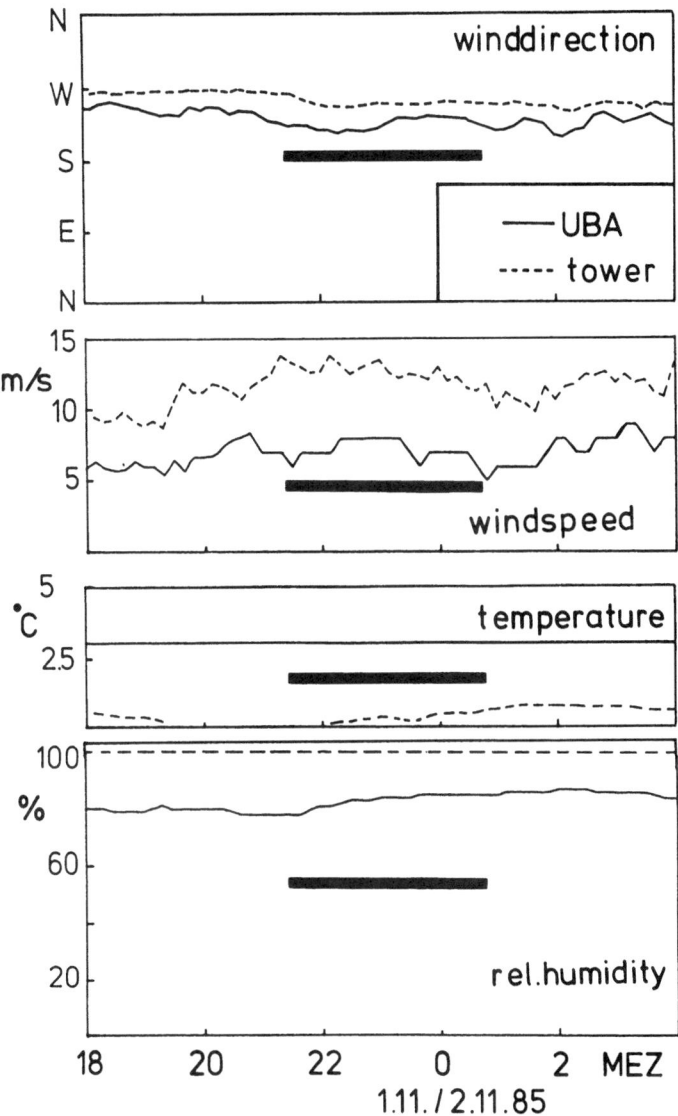

Figure 5. Results of the meteorological measurements during the
precipitation even 1/2 November 1985. Measurements
performed at the TV tower top (861 m a.s.l.) are indicated
by dashed, those of the "UBA"-station (480 m a.s.l.) by
straight lines. The time of the precipitation event is
given by black bars, rain intensities are shown in Fig.
11.

384

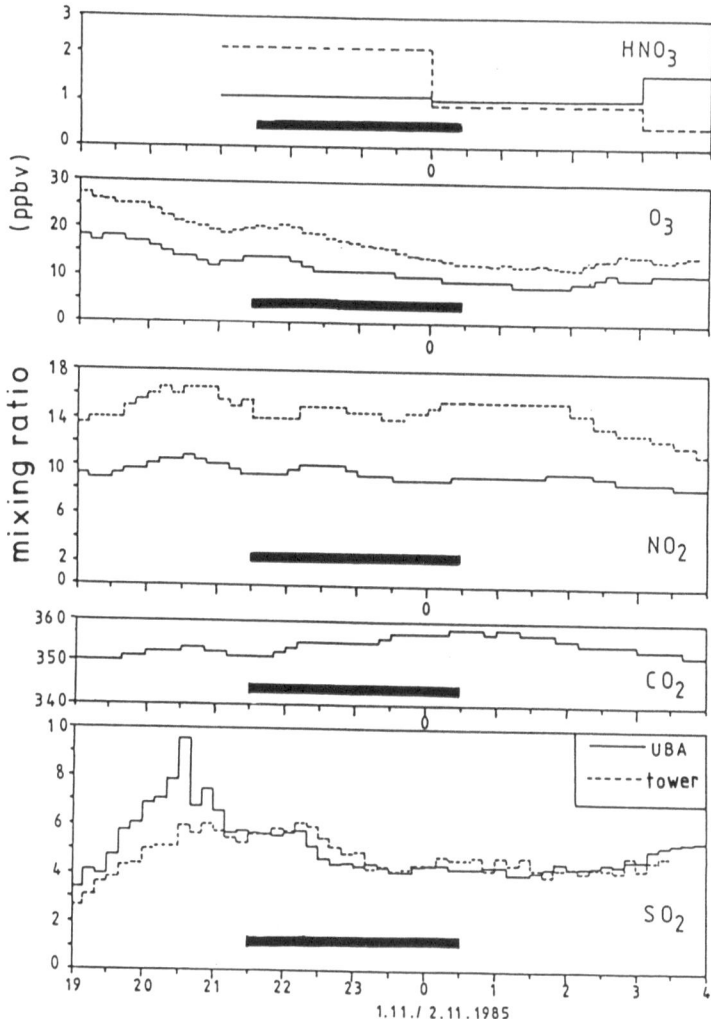

Figure 6.   Results of trace gas measurements during the precipitation
event 1/2 November 1985. TV tower platform (821 m
a.s.l.): dashed lines, "UBA"-station (480 m a.s.l.):
straight lines. The precipitation event is indicated by
black bars.

3.1 Weather situation and meteorological measurements

On 2 November, 01 CET, the upper air chart (500 mb) showed a pronounced zonal flow from WSW over Mid-Europe. The ground-level weather situation is characterized by the surface pressure chart given in Fig. 4. While on 1 November between 01 CET and 13 CET a weak convergence line crossed Deuselbach area, the precipitation during the night 1/2 November was not accompanied by the passage of a front (see Fig. 4). Light to moderate rainfall at 21:30-00:30 CET originated from stratocumulus clouds. We observed the cloud base first at 800 m a.s.l. (21:00-23:30 CET) and later at 550 m (23:00-02:00 CET).

The meteorological measurements at Deuselbach are shown in Fig. 5. Nearly constant values of wind speed and -direction, temperature and relative humidity were observed on both levels (821 m and 480 m), indicating that the meteorological situation remained rather uniform before, during, and after the precipitation event.

3.2 Trace gas and aerosol data

The temporal variation of HNO_3, O_3, NO_2, and SO_2 is shown in Fig. 6, that of the NH_4^+, Cl^-, NO_3^-, and $SO_4^=$ content of bulk aerosol samples in Fig. 7, and that of the denuder results in Fig. 8. During this event, the NO mixing ratio remained always below 0.1 ppb the detection limit of the instruments.

At the "UBA"-station, the SO_2 mixing ratio showed a maximum prior to the rain event and at that time exceeded that measured on the TV tower platform. From the begin of rainfall till the end of the observational period the SO_2 mixing ratios at both levels did not show much differences. A certain decrease of the mixing ratio was observed at 22:15 CET at the "UBA"-station, followed 20 min. later at the TV tower. This decrease is not considered a result of scavenging for the following reasons:

- rain had started 1/2 h respectively 1 h before the SO_2 mixing ratio began to decrease
- the sampling site at the TV tower was inside the cloud layer, and there SO_2 would have been scavenged by cloud droplets
- at the "UBA"-station below-cloud scavenging by falling rain droplets was the removal mechanism for SO_2.

It seems not likely that these two different scavenging mechanisms would result in the observed parallel trend of the mixing ratio at both levels. That and the decrease of the O_3- and the increase of the CO_2-mixing ratio at nearly the same time rather point to a more local change in air mass.

NO_2 showed nearly constant mixing ratios at both levels and a remarkably similar trend. This indicates, that the uptake of NO_2 by cloud- and rainwater was of minor importance, as would be expected from the dissolution kinetics of nitrogen oxides (Schwartz and White, 1983).

The HNO_3 mixing ratio was significantly reduced after the rain at the TV tower platform (more than 50%). In contrast a nearly constant mixing ratio was observed at the "UBA"-station, either for the filterpack or the denuder measurements.

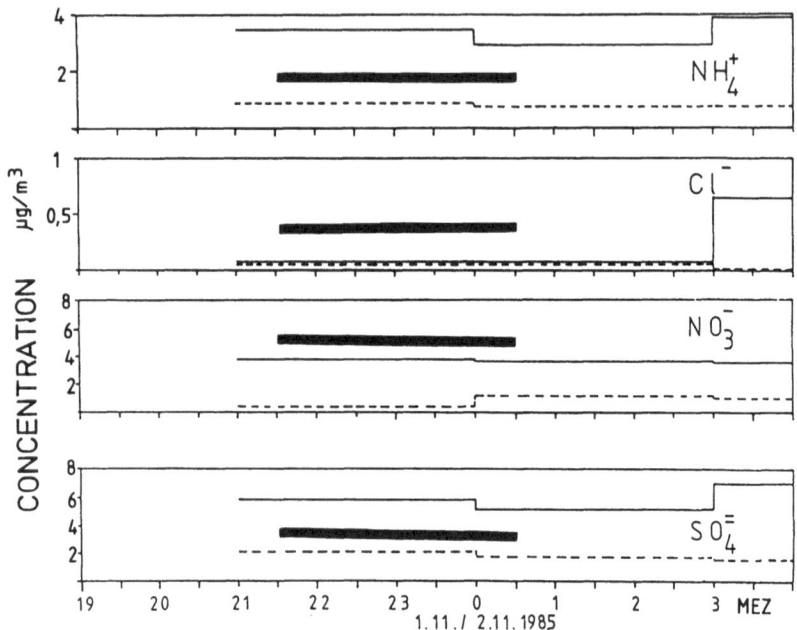

Figure 7.   Results of (bulk) aerosol measurements during the
             precipitation event (1/2 November 1985). TV tower
             platform (821 m a.s.l.): dashed lines, "UBA"-station (480
             m a.s.l.): straight lines. The precipitation event is
             indicated by black bars.

        A reduction of NO_3^-, $SO_4^=$, Cl^- and NH_4^- ions in the aerosol after
the precipitation could be hardly detected.
        As the temporal resolution of these measurements was either 3 h
for the filterpack or 6 h for the denuder method, a short-term
decrease of the aerosol particles during the rain would not be
detected by these measurements.
        Such a decrease of aerosol particles during the rain event could
be demonstrated by the measurements of the laser aerosol
spectrometer. The results of these measurements (Fig. 9) before,
during and immediately after the rain event indicate an effective
reduction of particles with diameters > 5 μm during the rain and a
fast increase of these particles after the rain.

3.3 Cloud- and rain water data

The results of the cloud- and rain water measurements are given in
Fig. 10 and Fig. 11. The most striking feature, common to all species

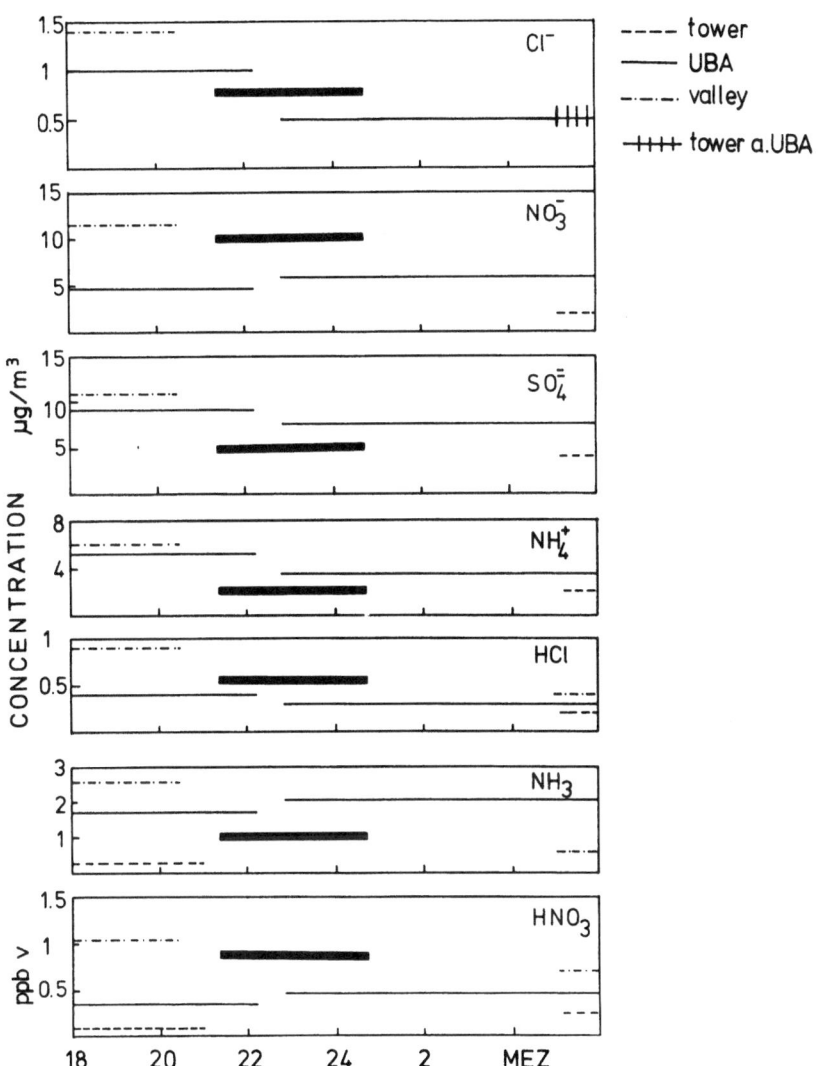

Figure 8.   Results of denuder measurements during the precipitation
event 1/2 November 1985. 4 upper panels show the aerosol
results, the lower panels show the gas phase
measurements. TV tower platform (821 m a.s.a.): dashed
lines, "UBA"-station (480 m a.s.l.): straight lines.
"Mühlental"-site (430 m a.s.l.): dashed-dotted lines.

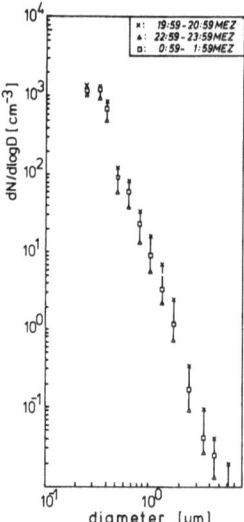

Figure 9.   Results of the laser aerosol spectrometer measurements
            (480 m a.s.l.) during the precipitation event 1/2 November
            1985.

is the large difference between cloud- and rain water. Cloud water
concentrations exceeded those found in rain water by a factor of 3 to
4. Moreover there is a significant difference even in the cloud water
concentrations obtained at the TV tower top (861 m) and the platform
which is only 40 m lower. At the platform, concentrations were
enhanced by 6 - 22%, which points to a concentration gradient inside
the cloud.

The rain water concentration of all investigated ions revealed
nearly the same temporal variation during the event. With the
exception of the data obtained at the TV tower's base (754 m a.s.l.) a
concentration minimum was observed just in the middle of the rainfall
event. NO_3^- -concentrations and conductivity at the TV tower's base
were significantly lower than those found at the other levels
throughout the event. To be precise at the beginning and the end of
the event different concentrations were found at the three sampling
sites with lowest values at the TV tower, while in the middle of the
event the concentrations were more or less equal. This may point to a
certain contribution of below-cloud scavenging to the rain water
concentrations at the "UBA"-station and the "Mühlental"-site,
respectively.

The most important results common to all rain water measurements,
demonstrated here by means of a single event can be summarized as
follows:

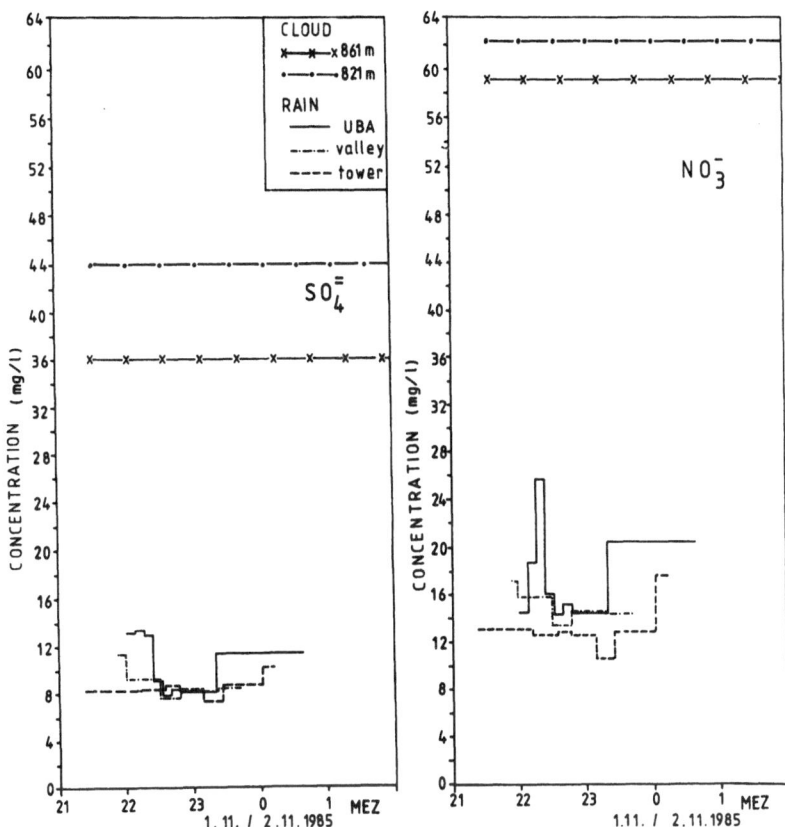

Figure 10.   Results of cloud- and rain water measurements during
precipitation event 1/2 November 1985.

-   Significant concentration differences of rain water collected in
    different heights occurred, with few exceptions, at the very
    beginning or the end of a precipitation event.
-   Concentrations of cloud water, collected near the cloud base
    exceeded rain water concentrations by a factor of 1.5 - 4.
-   There is a significant concentration decrease with height inside
    the clouds.
-   Trace gas and aerosol below and in cloud showed little decrease
    during cloud or precipitation events.
-   Measurements of the particle size distribution showed a decrease
    of particles with diameters 0.5 μm during rain events.

390

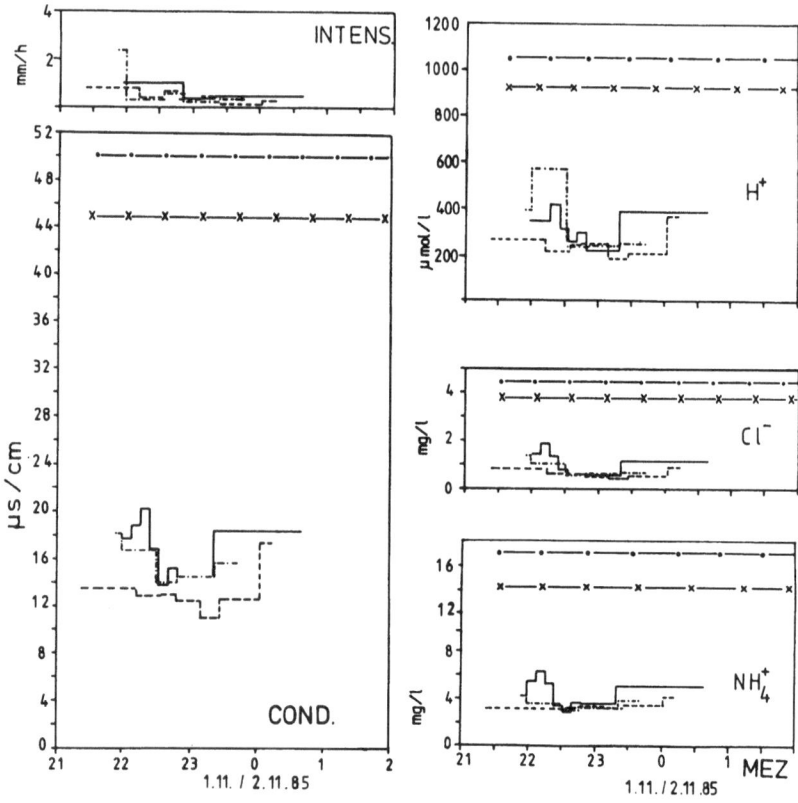

Figure 11.   Results of cloud- and rain water measurements during
precipitation event 1/2 November 1985.

4.   DISCUSSION AND CONCLUSION

The most striking feature of our measurements is the large difference
between cloud- and rain water concentrations. As the chemical
composition of the rain water is the result of enrichment of trace
substances and of the growth mechanism of the droplets within the
whole cloud, this observation can only be explained by assuming a
concentration gradient inside the clouds with decreasing
concentrations with increasing height.

Before a possible explanation for such a gradient is given, some
hints on the cloud characteristics and the rain formation during
ourmeasurements seem to be necessary.

As in the example given above, all of our examined precipitation
events are from status or stratocumulus clouds. Because of the low

temperatures (-2° to 13°C) the upper parts of these clouds must have consisted of supercooled water droplets and ice particles. In clouds like that rain droplets form either by aggregation of ice crystals or by accretion of supercooled water to ice crystals (s. Pruppacher and Klett, 1978).

For this kind of clouds sulfur chemistry has been modelled by Hegg et al., (1984). Indeed, their results are able to explain some of our observations.

In their kinetic model for sulfur chemistry in warm-frontal rainbands Hegg et al., (1984) presented a comprehensive study of in-cloud and below-cloud scavenging of SO_2 and $SO_4=$ particles within horizontally homogenous and wide spread rain bands. They apply a detailed treatment of cloud microphysical processes featuring the socalled "seeder-feeder" mechanism. In this process ice particles from a "seeder" cloud grow as they fall through a lower-level "feeder" cloud: the "seeder" zone aloft consists of weak convective cells that produce ice particles that fall through a zone of stable uplift ("feeder" zone) where they grow either by vapor deposition, riming (i.e., collection of supercooled cloud droplets), or a combination of these two mechanisms.

The model is applied to two different types of rainbands. The first is characterized by weak vertical velocities ($0.1-0.2$ ms^{-1}) in the feeder cloud, the second by rather strong vertical velocities (≈ 0.7 ms^{-1}).

Beside characteristic differences in the hydrometer fields both rainband types show maximum sulfate concentrations near the cloud base. For rainband "type 1" the region of maximum sulfate content has a greater vertical extension (≈ 1 km) than for rainband "type 2" (≈ 100 m).

In the "type 1" rainband most of the precipitation mass is formed by vapour deposition on snow particles because of the low liquid water content of these clouds. Such growth transfers no liquid-phase sulfate to the snow particles, only dry scavenging can enhance sulfate content of the precipitation.

For the rainband "type 2" cloud systems, a substantial fraction of the precipitation mass reaching the ground is attributable to the riming of snow particles during their fall through the feeder zone. Since only in the lowest part of the feeder zone, and in a very limited layer of thickness enhanced sulfate concentrations of cloud water droplets occur, the main growth of the snow particles will be due to the collection of supercooled cloud droplets of lower sulfate content. So, for both rainband types the sulfate concentration of the rain water will be lower than that of the cloud droplets at the cloud base.

As the assumptions of the microphysical structure of the (model-) clouds are in general comparable to most of the clouds observed during our field campaign, the remarkable difference between the concentrations of rainwater and cloud water collected near the cloud base has a reasonable explanation. Further detailed evaluation of available synoptic data (visible and IR sattelite imageries, weather radar imageries, radiosonde data, synoptic fields of cloud top temperatures and vertical velocities) will contribute to answer this question.

ACKNOWLEDGEMENTS

The authors would like to thank A. Waayers, A. van Wensveen, A. van Westen, C., Schoonebeek, F. Bakker, P. Fonteijn, G. Kos, F.P. Jongejan, M. Langen, N. Beltz, H. Obenland, and H. Franken for chemical analysis and assistance during the field experiments. The continued engagement of the staff and personnel of the TV tower "Morbach 3", "Wetteramt Trier", "Geophysikalischer MeBzug 300, Idar-Oberstein", and "Umweltbundesamt, MeBstelle Deuselbach" is gratefully acknowledged. Most of the synoptic data were provided by the German Weather Service.

REFERENCES

Asman, W.A.H., Jonker, P., Slanina, J., Baard, J. 1983. "Interpretation of sequential rain sampling results". In: H.R. Pruppacher, R.G. Semonin, and W.G.N. Slinn (eds.), Precipitation Scavenging, Dry Deposition, and Resuspension, pp. 265-274, Elsevier New York Amserdam Oxford.

Dibelius, G., Voss, H. 1983. VDI-Berichte No. 487 131, VDI-Verlag, Düsseldorf, F.R.G.

Hales, J.M. 1972. "Fundamentals of the theory of gas scavenging by rain", Atmos. Environ. 6, 635-659.

Hegg, D.A. Hobbs, P.V., Radke, L.F. 1984. "Measurements of the scavenging of sulfate and nitrate in clouds", Atmos. Environ. 18, 1939-1946.

Hegg, D.A., Rutledge, S.A., Hobbs, P.V. 1984. "A Numerical model for sulfur chemistry in warm-frontal rainbands", J. of Geophys. Res. 89, 1939-1946.

Hobbs, P.V. 1981. "Scales involved in the formation and organization of clouds and precipitation". In: P.V. Hobbs and A. Deepak (eds.), Clouds, their formation optical properties and effects, pp. 1-14, Academic Press New York London Toronto.

Jaeschke, W. 1985. "Multiphase atmospheric chemistry". In: W. Jaeschke (ed.), Chemistry of Multiphase Atmospheric Systems, pp. 3-40, Springer Verlag Heidelberg New York.

Kins, L. 1982. Differentialanalysen der chemischen Zusammensetzung von Einzelniederschlägen, Master, thesis, 99 p., J.W.v. Goethe University Frankfurt am Main, F.R.G.

Kins, L., Meixner, F.X., Ehhalt, D.H., Müller, K.P., Slanina, J., Moels, J.J., Beltz, N., Jaeschke, W., Rumpel, K.J. 1986a. "Scavenging of trace substances by clouds and precipitation". In: Preprints, American Chemical Society, Div. Pet. Chem. 31, No. 2, pp. 457-463.

Kins, L., Junkermann, W., Meixner, F.X., Müller, K.P., Ehhalt, D.H., 1986b. "Development and calibration of a ground-based active collector for cloud- and fogwater". In: Preprints, American Chemical Society, Div. Pet. Chem. **31**, No. 2, pp 527-532.

Meixner, F.X., Jaeschke, W. 1981. The detection of low atmospheric SO$_2$ concentrations with a chemiluminescence technique, Int. J. Environ. Anal. Chem. **10**, 51-67.

Meixner, F.X., Müller, K.P., Aheimer, G., Höfken, K.D. 1985. "Measurements of gaseous nitric acid and particulate nitrate". In: F.A.A.M. Leeuw, N.D. van Egmond (eds.), Proceedings of the COST action 611 meeting "Pollutant cycles and transport: modelling and field experiments", pp. 103-119, RIVM Bilthoven, The Netherlands

Meixner, F.X., Franken, H.H. 1986. "Emission and deposition of NH$_3$ in Europe", Contribution of the KFA Jülich, Final report to the EC, Organisatie voor Toegepast Natuurwetenschappelijk Onderzoek (TNO), Delft, The Netherlands

Müller, K.P. 1984. "Ionen-Chromatographie in Niederschlagswasser", Fres. Z. Anal. Chem. **317**, 345-346

Pruppacher, H.R., Klett, J.D. 1978. Microphysics of Clouds and Precipitation, D. Reidel Publishing Company, Dordrecht Boston London

Rumpel, K.J. 1979. "Kontinuierliche Messung mit dem COSO 2. In: Berichte des Umweltbundesamtes, pp. 55-64, Umweltbundesamt Berlin, F.R.G.

Schwartz, S.E., White, W.H. 1983. "Kinetics of reactive dissolution of nitrogen oxides into aqueous solution". In: S.E. Schwartz (ed.), Trace Atmospheric Constituents, pp. 1-116, Advances in Environmental Science and Technology, Vol. 12, J. Wiley & Sons, New York

Schwartz, S.E. 1984. "Gas-aqueous reactions of sulfur and nitrogen oxides in liquid water clouds". In: J.G. Calvert (ed.), SO$_2$, NO and NO$_2$ Oxidation mechanisms: Atmospheric Considerations, pp. 173-208, Butterworth Boston

Slanina, J., Can Lamoen-Doornencal, L., Lingerak, W.A., Meilof, W., Klockow, D., Niessner, R. 1980. "Application of thermodenuder analyzer to the determination of H$_2$SO$_4$, HNO$_3$ and NH$_3$ in air", Report ECN-80-113, Stichting Energieonderzoek Centrum Nederland (ECN), Petten, The Netherlands

Slanina, J., Bakker, F.P., Jongejan, P.A.C., van Lamoen, L., Moels, J.J. 1981. "Fast determination of anions by computerized ion chromatography coupled with selected detectors", Anal. Chem. Acta **130**, 1-8

Slanina, J. 1982. "Measurements of strong acids and the corresponding ammonium salts in the Netherlands". In: VDI Berichte **429**, p. 177-182, VDI Verlag Düsseldorf, F.R.G.

Slanina, J., Schoonebeek, C.A.M., Klockow, D., Miessner, R. 1985. "Determination of sulfuric acid and ammonium sulfates by means of a computer controlled thermodenuder system", Anal. Chem. **75**, 1955-1960

Submitted:

Doyle, K.M., T.J. Fahey and R.D. Paratley. Subalpine heathland of the Mahoosuc Range, Maine. Bull. Torrey Bot. Club (submitted).

Fahey, T.J. and J.B. Yavitt. Soil solution chemistry in lodgepole pine (Pinus contorta ssp. latifolia) ecosystems, southeastern Wyoming. Biogeochemistry (submitted).

Fahey, T.J., J.B. Yavitt and G. Joyce. Precipitation and throughfall chemistry in lodgepole pine ecosystems, southeastern Wyoming. Can. J. Forest Res. (submitted).

Hughes, J.W. and T.J. Fahey. Seed dispersal and seedling establishment following large-scale disturbance of northern hardwood ecosystems. Journal of Ecology (submitted).

Hughes, J.W. and T.J. Fahey. Population persistence and demographics of a woodland herb, Aster acuminatus. Amer. J. Bot. (submitted).

Nyberg, R.C. and T.J. Fahey. Soil hydrology in lodgepole pine ecosystems, southeastern Wyoming. Soil Sci. Soc. Amer. J. (submitted).

CHEMISTRY OF CLOUD WATER AND PRECIPITATION AT AN ALPINE MOUNTAIN STATION

K. Munzert
Fraunhofer-Institute for Atmospheric Environmental Research
Kreuzeckbahnstr. 19
D-8100 Garmisch-Partenkirchen
F.R. Germany

ABSTRACT. Cloud water and precipitation samples were monitored using passive cloud water collectors and precipitation gauges in the Bavarian alps for 14 months. The measurements show that the samples from the passive cloud water collector contained larger concentrations of all major ions than precipitation samples, typically by a factor between two and three. These measurements lead to under estimates of the actual cloudwater/precipitation ratios for individual ions, as the passive collectors at these windy sites also collect precipitation. The passive cloud water collectors also collect during 'dry periods', and the resulting 'deposit' may represent almost half of the collection during cloud and precipitation events.

1. MEASURING SITE AND COLLECTION DEVICE

The Fraunhofer-Institute for Atmospheric Environmental Research operates a network of two valley and four mountain top stations in the Bavarian alps. At the station on Wank peak at an altitude of 1780 m asl, which serves as a background air pollution monitoring station of the World Meteorological Organization, meteorological data and all relevant trace constituents, gases and aerosols are measured. Precipitation samples have been collected and analysed for the main chemical constituents since 1966. In spring 1985 the collection of cloud water with a passive fog collector began.

The instrument, which were used in 1985-86 for the fog collector intercomparison at Frankfurt/Main, FRG, uses a roof to prevent the collection of precipitation. The collecting grid is made of stainless steel wires with a diameter of .25 mm and the mesh width is 1.8 mm. The total area of the cylindrical sampling net is 1500 cm^2, so that the vertical collecting cross section is 475 cm^2.

M. H. Unsworth and D. Fowler (eds.), Acid Deposition at High Elevation Sites, 395–401.

2. METEOROLOGICAL CHARACTERISTICS OF SAMPLING PERIODS

The characteristics of the collection periods for cloud water, and precipitation, are listed in Table I. On average, the 104 exposure intervals from March 1985 to May 1986 last for 3 days with variations from 1.5 to 6 days depending on synoptic situation. The mountain station is within clouds usually for 1.5 days per week. The volume of the collected cloud water is denoted in ml and also in equivalent mm precipitation. This value is reduced to the horizontal cross-section of the 300 cm^2 funnel. The high volumes of the fog collector samples compared to those of precipitation shows that under high wind speeds the roof is too small to prevent precipitation collection. Another problem for a separated collection of cloud water is the considerable portion of snowfall with more than 50% of all precipitation events. The occurrence of fog, that means for a mountain peak station within clouds, is similar to the times of precipitation. For that reason no pure cloud water samples not affected by precipitation were collected.

3. CONDUCTIVITY AND PH-VALUE

The differences between cloud water (cldw) and precipitation (prec) acidity (Table II) show that the fog collector does not simply measure precipitation. The averaged proton concentration of all cloud water samples is a factor of 2 higher than that of precipitation. The ratio of the electrical conductivities reaches 2.6. Despite the lower pH values in cloud water, the relative fraction of H^+ compared to other ionic species is lower. The dynamic range from minimum to maximum proton concentrations is 4 times broader than the variability of proton concentration in precipitation whereas the relative deviations of the conductivities are equal.

All data for the 15 months sampling period are combined, because no statistically significant annual variations were observed. The greatest seasonal differences between winter and summer months are smaller than the change from spring 1985 to spring 1986. The time trend over 15 months shows increasing concentrations, especially for NH_4^+, for both cloud water and precipitation, leading to higher pH values. The conductivities remain roughly constant during the period as a result of the higher equivalent conductance of H^+ relative to NH_4^+. These 15 months are however too short for the proof of a statistically significant trend of such strongly scattering magnitudes.

4. IONIC COMPOSITION OF CLOUD AND PRECIPITATION

The mean concentrations of ammonium and all other analysed ions are shown in Table III. The cloud water values in the upper lines are compared with those of precipitation samples below. For both sample types the concentrations of the major ions $SO_4^=$, NO_3^- and NH_4^+ amounts to 87% of the analysed ions'. The large proportion of ammonium is believed to originate from local agricultural acitivity. The variations of the individual constituents indicated by s/x is much smaller for precipitation than for cloud water. The median values for cloud water compared to those of precipitation are shifted for a factor 3.3 to higher concentrations. The skewness of the distributions with many small values and few peaks is indicated by the difference between median and mean x and the different spreads from median to 10% representing 90% value. Skewness is most significant for NO_3^- and NH_4^+ and better developed for cloud water than for precipitation.

5. IONIC COMPOSITION

The ionic concentrations expressed as equivalents are shown in Table IIIb. The fraction of acid denoted by H^+ concentration amounts to 9.4% of the sum of analysed ions for precipitation. As already indicated in the data of pH and conductivity, this share is significantly lower in the case of cloud water. The sulphate contribution dominates in the cloud water samples and the ratio $SO_4^=/NO_3^-$ decreases from 2.3 to 1.5 for precipitation.

The total ionic sum of all, even of the non analysed ion species, can be deduced from electrical conductivity and pH value. If equilibrium of CO_2 between atmosphere and liquid phase is assumed, the concentrations of HCO_3^- and $CO_3^=$ are governed by pH. The atmospheric CO_2 mixing ratio of 350 ppm buffers the liquid samples to a pH of 5.63. The measured electrical conductivity is reduced by the conductivities of H^+, OH^-, HCO_3^- and $CO_3^=$ determined from their concentrations and ionic equivalent conductances. From this reduced conductivity the ionic sum of mineral compounds is estimated. For the analysed non H^+ ions and an average ionic composition a mean ionic equivalent conductance of 70 $\mu S.cm^{-1}/mequiv.l^{-1}$ is derived.

The contribution of about 85% of the analysed ion species to this calculated total sum of 705 $\mu equiv/l$ for cloud water representing 253 $\mu equiv/l$ for precipitation is given at the bottom of Table III. The balance cations/anions is achieved if the not analysed magnesium is included. The concentrations of Ca and Mg are raised due to the erosion processes at the limestone of the mountains consisting of Ca-carbonate and Mg-carbonate (dolomite). The calculated ionic sum seems to be too high for about 5% to 10% compared with the analysed ionic sum.

6. TOTAL IONIC SUM AND ACID FRACTION

The frequency distribution of all estimated ionic sums and the enrichment factor deduced from the ratio of the ionic sums in cloud water and precipitation can be taken from Table IV. The relative variabilities s/x are equal for cloud water and precipitation in contrast to the deviations of the individual ion species. This means, that even though the ionic composition of individual samples varies strongly, the total ionic sum provides a more even distribution. The individual ionic species in cloud water vary inversely and reduce the scatter of the ionic sums. The enrichment factor which averages at 2.9 varies only slightly. Its relative deviation is much less (cv 52%) than that of the individual ionic sum.

The acid fraction is determined as the portion of free acid in the reduced ionic sum without H^+, OH^-, HCO_3^- and $CO_3^=$. The free acid is defined as H^+ concentration without the CO_2 buffer protons originating from the dissociation of HCO_3^- and $CO_3^=$. For samples with a pH value greater than 5.63, free base is required. This means, that the sum of (cations – H^+) is greater than the sum of (anions – OH^- – HCO_3^- – $CO_3^=$). The sign of acid fraction was chosen so that '–' denoted free acid. Corresponding to the results in Tables II and IIIb, the share of free acid in precipitation's ionic sum is a factor 1.5 higher than acid fraction in cloud water. Free base is observed in 15% of the precipitation events (17% of all cloud water samples). The maximum values of acid, i.e. base fraction reaches +5%.

In the last line of this table the total deposition to the fog collector is computed. Most of the deposition occurs in single events. The arithmetic mean corresponds to the 72% value of the cumulative frequency distribution and the spread from median to 90 percentile is 3.5 times greater than the spread median to 10 percentile.

7. DEPOSITION RATES UNDER FAIR WEATHER

The combined deposition of cloud and rain water precipitation not deflected by the roof must be compared with the dry deposition rates given in Table V. Here in the upper lines the characteristics of fog sampler collecting periods with no measureable liquid gain are represented. The 'dry only' periods last much longer than the cloud water collecting exposures. In the case of a sample period with no fog or if a small volume is already evaporated, the collecting grid and the funnel was washed with 250 ml distilled water. The pH value calculated from averaged H^+ of these samples is 5.6 and is the buffer pH for atmospheric CO_2. The mean acid fraction of the dissolved neutral salts is +1% with only a slight variation from –4% to +5%. The daily deposition of aerosol is in contrast to moist deposition of fog evenly distributed around 43 µequiv/day. This deposition originates from an aerosol concentration of 6.4 µg/m^3, containing .099 µequiv/m^3 of sulphate, nitrate and ammonium, and a contribution from the gases SO_2, NO_2, HNO_3 and NH_3.

The data shows that about one half of the deposited material comes from dry deposition. The question arises, whether a passive fog sampler is a cloud water collector or just a combined aerosol gas and precipitation sampling device. This problem will be investigated using an active collector installed in spring 1986 and running only during fog.

8. CONCLUSIONS

A passive fog water collector at an Alpine mountain station provides valuable information on the relative chemical content of precipitation and combined samples of cloud water and rain. The passive fog water collector suffers from the major problem of collecting precipitation, as well as cloud drops, but the combined cloud and rain samples may be representative at least qualitatively of the cloud and rain collected by conifer trees. This study emphasizes the need for active cloud water collectors at mountain sites. The problems of collection of snow crystals or rime remain unsolved.

REFERENCES

Texte 13/86 Tagungsband: Atmospharische Prozesse
    IMA Querschnittsseminar, Teil I, II Umweltbundesamt Berlin, 1986.
Proceedings: Physico-Chemical Behaviour of Atmospheric Pollutants,
    Third European Symposium, Varese, 1984.
Winkler, P. 1984. An instrument to determine the wet deposition of
    acid. WMO-TECOMAC Symposium Den Haag.

TABLE I.   Statistical characteristics (cumulative frequency distribution) of cloud water and precipitation samples.

10%	25%	med	75%	90%	x	s	n	
45.	50.	73.	100.	127.	84.	41.	104	duration of expos, h
60.	134.	296.	606.	1218.	464.	457.	104	cloud water, ml
3.	5.	15.	25.	37.	18.	16.	104	fog duration, h
1.4	4.6	11.5	18.4	36.0	16.1	20.7	104	precipitation, mm
3.	6.	13.	22.	38.	17.	16.	104	precipitation, h

(40% rain, 14% mixed, 46% snow)

TABLE II.   Comparison between cloud water and precipitation pH and conductivity: x pH - arithmetic mean of pH values
    x pH (H^+) pH calculated from averaged H^+ concentrations

10%	25%	med	75%	90%	x	s	n	
3.93	4.43	4.88	5.35	5.97	4.93	0.74	104	cldw, pH
					4.46			cldw, pH (H^+)
4.29	4.61	4.99	5.37	5.74	5.01	0.54	100	prec, pH
					4.72			prec, pH (H^+)
17.	25.	39.	64.	129.	59.	54.	104	cldw conduct, µS/cm
7.	10.	14.	26.	51.	23.	23.	99	prec

TABLE IIIa Comparison of main ionic constituents in cloud water and precipitation.

10%	25%	med	75%	90%	x	s	n		
0.74	3.41	7.47	11.89	19.61	9.93	13.96	73	cldw	SO_4^-, mg/l
0.17	1.07	2.26	4.18	6.63	2.89	2.44	92	prec	
1.08	1.79	2.76	5.05	11.27	5.58	9.14	79	cldw	NO_3^-, mg/l
0.69	1.23	1.76	2.58	4.19	2.41	2.22	92	prec	
0.19	0.69	1.73	3.60	6.28	3.02	4.29	72	cldw	NH_4^+, mg/l
0.00	0.17	0.45	1.44	2.13	0.90	1.20	88	prec	
0.17	0.49	0.68	1.07	1.41	0.77	0.47	28	cldw	Cl^-, mg/l
0.00	0.14	0.29	0.63	1.18	0.45	0.47	85	prec	
0.09	0.14	0.35	0.71	0.85	0.61	1.07	29	cldw	Na^+, mg/l
0.01	0.05	0.09	0.15	0.24	0.12	0.12	88	prec	
0.00	0.05	0.13	0.27	0.40	0.25	0.49	29	cldw	K^+, mg/l
0.00	0.00	0.02	0.07	0.15	0.05	0.09	89	prec	
0.24	0.29	0.58	1.17	3.44	1.09	1.19	73	cldw	Ca^{++}, mg/l
0.03	0.09	0.14	0.39	0.94	0.32	0.38	89	prec	

TABLE IIIb. Chemical composition of cloudwater and precipitation.

The first two %-lines relate to the sum of analysed ions, the lat two to the total ionic sum calculated from conductivity and pH.

	H^+	SO_4	NO_3^-	NH_4^+	Cl^-,Na^+,K^+,Ca^{++}	
cloudwater	0.03	9.93	5.58	3.02	2.72	mg/l
	35.	207.	90.	167.	109.	µequiv/l
	5.7	34.0	14.8	27.5	18.0	%
precipit.	0.02	2.89	2.41	0.90	0.94	mg/l
	19.	60.	39.	50.	35.	µequiv/l
	9.4	29.6	19.1	24.6	17.3	%

contribution to total ionic sum

	cations	anions	cat+anions	total ion sum
cloudwater	41.0%	45.2%	86.2%	705 µequiv/l
precipit.	36.1%	44.2%	80.3%	253 µequiv/l

TABLE IV   Total ionic sum in cloud water and precipitation and
            enrichment factor as ratio ionic sum cloud water/ionic sum
            precipitation. The acid fraction gives the portion of free
            acid of the total ionic sum: - denoted free acid, + denotes
            free base(pH > 5.63). The last line gives the total
            deposition of the fog sampler.

10%	25%	med	75%	90%	x	s	n	
198.	285.	457.	803.	1361.	705.	709.	103	cldw ion sum,µequiv/l
83.	116.	165.	274.	476.	253.	264.	99	prec
1.46	1.84	2.54	3.67	4.94	2.91	1.50	99	cldw/prec enrichm fact
-15.	-10.	-5.	-1.	+1.	-6.	6.	103	cldw acid fract, %
-20.	-14.	-7.	-3.	+1.	-9.	9.	99	prec
10.	18.	52.	98.	196.	90.	124.	103	deposit, µequiv/d

TABLE V   Survey on fog sampler exposures during dry periods. The
           dry deposited matter is washed off using distilled water.

10%	25%	med	75%	90%	x	s	n	
69.	72.	93.	119.	150.	102.	39.	22	duraction of expos, h
4.89	5.20	5.96	6.46	6.80	5.98	0.68	21	pH
18.	25.	42.	71.	92.	50.	28.	21	conduct, µS/cm
399.	894.	1537.	2619.	3528.	1840.	1165.	21	ion sum, µequiv
-4.	-1.	+1.	+4.	+5.	+1.	4.	21	acid fract, %
7.	26.	43.	59.	73.	44.	24.	21	deposit, µequiv/d

MEASUREMENTS OF THE CHEMICAL COMPOSITION IN CLOUD AND FOGWATER

G. Schmitt
Institute of Meteorology and Geophysics
University of Frankfurt
Feldbergstr. 47
6000 Frankfurt/Main

ABSTRACT. Measurements of the chemical composition of fogwater are presented. The samples were collected by two different methods. The sampling site is located at the mountain "Kleiner Feldberg/Taunus" (800 m asl) 25 km northwest of Frankfurt/Main. The analysis of fogwater shows extremely high trace element concentrations. Compared to rain water the concentrations in fog are 3 - 15 times higher. Element specific enrichment factors are given. In the single events the pH-values show a wide variation between pH 2.4 - 7.0. Characteristic distributions of short time fluctuation of elements during single fog events are given. During the episodes the variability of element concentration in fog is mainly determined by the actual amount of the LWC. A decrease of the LWC by a factor of 2 leads to a two fold increase of the element concentrations. Especially at the beginning and the end of a fog event the concentrations increase, when the liquid water content is reduced.

1. INTRODUCTION

Up to now the knowledge about acid fog and the chemical composition of fogdroplets is based on only a few measurements, which are often not comparable due to different sampling methods. Nevertheless fog is one important pathway of removal of trace elements from the atmosphere. The interception of fogdroplets can moisten canopy surfaces, thus increasing their effectiveness in capturing trace gases and particles from the air. In forest areas with high fog frequency the total amount of the intercepted fogwater can equal the yearly precipitation rates (Grunow 1955). Therefore the mechanism of interception of fog droplets may be of great importance with respect to the atmospheric input into forest ecosystem.

In Middle Europe the temporal distribution of the fog events shows a strong maximum during the months October and November. In this time most of the fog events were observed. Usually the number of days with fog increases with height above 400 m. At 600 m above sealevel nearly 120 days/year are observed, at 800 m more than 200 days with fog/year are registrated. At the station "Kleiner Feldberg/Taunus" the average

403

M. H. Unsworth and D. Fowler (eds.), Acid Deposition at High Elevation Sites, 403–417.
© 1988 by Kluwer Academic Publishers.

mean value (30 years) is 235 days/year. The regional distribution of
the fog frequency shows a similar distribution as the forest declines
do. Most of the damages are observed in areas with a high fog
frequency.

For these reasons fogwater chemistry is becoming an increasingly
important field of atmospheric research. Besides the measurements of
the chemical composition of fog main interests on our investigations
are focussed on the temporal variations of concentrations of trace
species in fogwater during individual events. Therefore an active
sampling system was developed, which allows to collect several samples
during individual fog episodes.

2. SAMPLING METHOD

Most fogwater sampling systems are based on the principle of inertial
impaction. By forcing the airstream to move around a barrier, fog
droplets greater than a given radius will not be able to follow the
streamlines. These droplets are impacted at the barrier. Collection
efficiency and cut-off depend on the relative velocity between the
airstream and the barrier as well as the dimensions of the barrier
itself. With increasing relative velocity, the cut-off permits smaller
droplets to be collected (Fuchs 1964).

In general the principal ways to collect fogwater can be
distinguished into two main categories. The passive sampling methods
use the natural relative velocity between the fog droplets and a
stationary sampler (Black and Landsberg 1983, Mohnen 1980). In the
opposite the active collection systems take the advantage of an
increased relative velocity by rotating a collector through the fog or
by sucking the fog droplets through the collector by means of a fan
(Hoffmann et al. 1983, Katz 1980).

2.1 Passive Method

Figure 1 shows a sketch of the developed passive sampling system. The
collector consists of a teflon support structure and teflon strings of
0.3 mm in diameter, mounted every 3 mm around the periphery. The fog
droplets are impacted at the strings and grow up to larger drops.
Finally they run down the strings and are collected in a 100 ml
polyethelene sampling bottle. At normal wind speeds, all droplets
greater than 5 μm in diameter are impacted (Georgii and Schmitt 1985).

In general a sampling time of 2 hours is necessary to collect a
volume of 15-25 ml, sufficient for the chemical analysis. The whole
sampling system is constructed in a way that only during fog events the
collector is exposed. For the remaining time the collector is kept in
a metal cylinder as protection against contamination by rain and dry
deposition. A fog sensor was constructed, comparing the actual
temperature and the dewpoint. As soon as the actual temperature
reaches the dewpoint, adequate to an increase of the relative humidity
up to 100%, a motor (M) is moving the collector (C) out of its shelter
(G). In case of rain, detected by an additional sensor, the collector
is moved back again into its shelter.

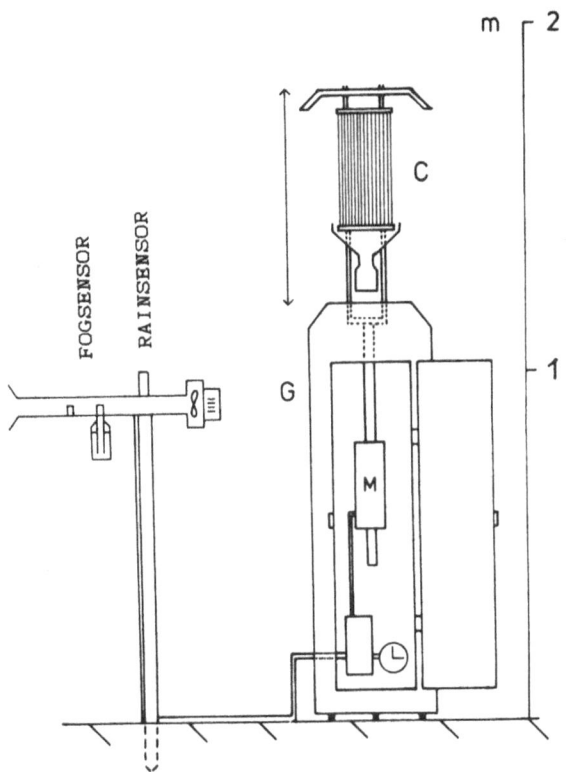

Figure 1   Passive fogwater sampling system
           A fog sensor, comparing the actual temperature and the
           dewpoint, gives a signal to a motor (M), which moves the
           collector (C) out

2.2   Active Method

Compared to the passive method, the active system simulates higher
relative velocities between fog droplets and collector.   The sampling
system uses a powerful electric motor for rotating an aluminium arm of
2 meter length horizontally through the fog.   At both ends of the arm
collectors are installed (Fig. 2a).   A sketch of the collector in
detail is shown in Figure 2b.   Seven teflon bars, each 10 mm in
diameter are mounted in three rows.   Each bar has a sampling slot of 3
mm width (Fig. 2c).   The fog droplets are impacted in the sampling
slots and are driven into the bottles.   The main advantage is that
previously impacted drops in the slots cannot be lost due to the high
speed of the airstream.   Also the evaporation of the sampled drops is
reduced.   The velocity of rotation can be adjusted electronically from
0 to 300 rpm, so that the velocity at the collectors reaches up to 30
m/sec.   At this speed all droplets greater than 5 μm are impacted in

406

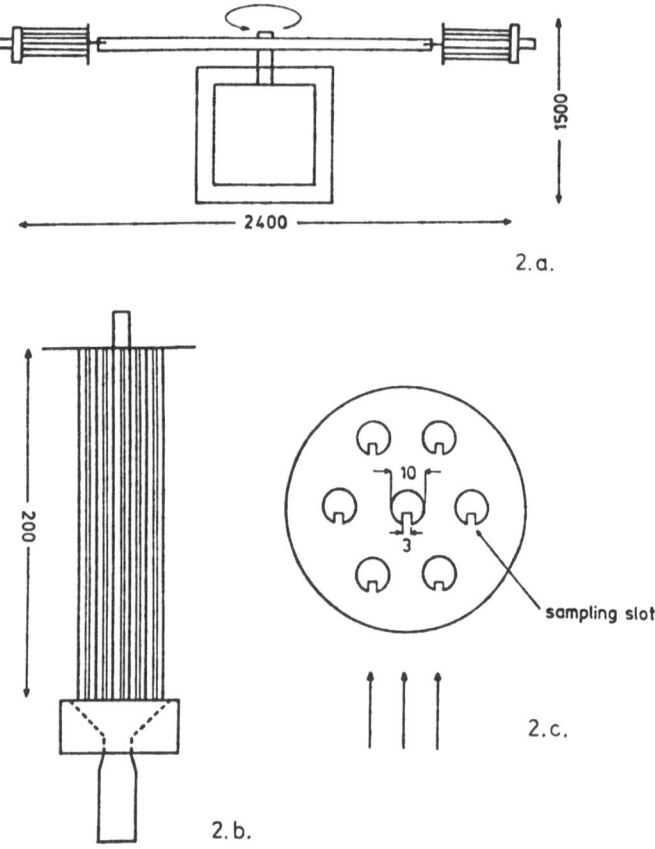

Figure 2. Active fogwater sampling system.
          a) general view, b) collector in detail, c) profile of the
          sampling slots.

the sampling slots.   Using different relative velocities, the cut-off
can be varied.   Thereby it is possible to sample different droplet
size fractions to obtain information about the chemical composition
within these fractions.
     Sampling time necessary to receive sufficient volume of 20 ml
fogwater range from 10 to 20 minutes, depending on the liquid water
content of the fog.   Using an active sampling system allows
investigations on the temporal fluctuations of concentrations during
individual fog events with a sufficient time scale (15-30 min.).
Detailed information about the two sampling methods are given in
Georgii et al. (1986).

3. RESULTS

Both sampling systems are used for collecting cloud and fogwater at the
station on top of the mountain "Kleiner Feldberg/Taunus" at an altitude
of 800 m above sea level.   The station is situated 25 km northwest of
Frankfurt/Main.   All fog water samples are analysed for pH and
electrical conductivity.   The compounds $SO_4^=$, NO_3^-, Cl^-, NH_4^+, Na
and K are measured by ion chromatography (IC).   The analysis of the
elements Ca, Mg and the heavy metals Pb, Fe and Cd are done by means of
atomic absorption spectroscopy (AAS).

3.1 pH in Fogwater

The pH values in fog are plotted as frequency distribution.   The
calculation includes 250 single events.   The cell limits used are 0.5
pH units.   The pH values, measured in rainwater at the same station,
are also indicated (Fig. 3).
    Compared to rainwater the pH in fog is decreased.   In fog 4% of
the pH values are below 3.0.   The lowest pH found in the two years was
2.38, but also high values between pH 5 and pH 7 were observed.   Most
of all fogwater samples (70%) show pH values in the range from pH 3.5
to 4.5.   In fog the mean value is pH 3.78, whereas in rain water the
average value is pH 4.30.   The presentation also demonstrates the

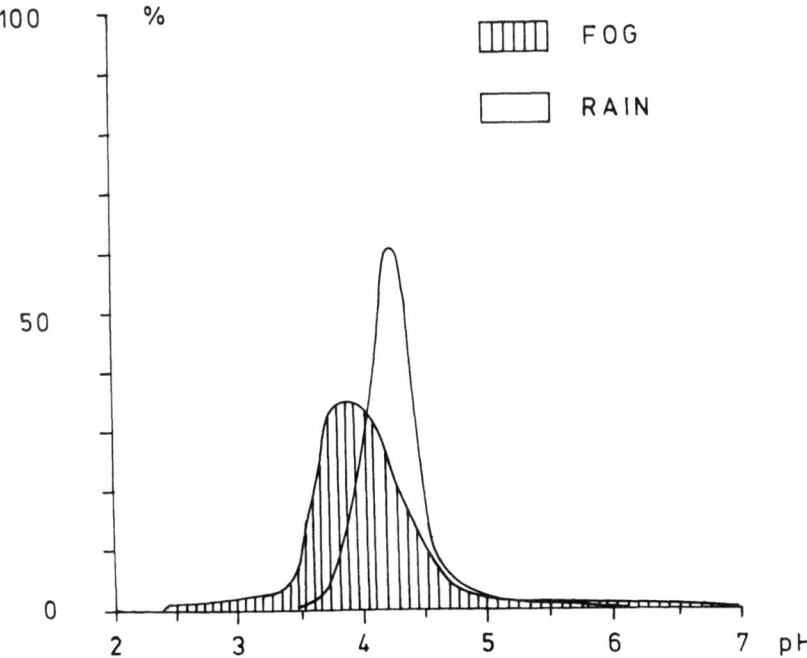

Figure 3.   Frequency distribution of the pH-values in fog and rainwater
at the station "Kleiner Feldberg" (autumn 1983 - spring
1986).

408

extremely wide variation of the pH range, corresponding to a range in
H^+ by more than four orders of magnitude. This is indicated by the
lower levels of the frequency. In rainwater the pH only ranges from
pH 3.0 to pH 6.50.

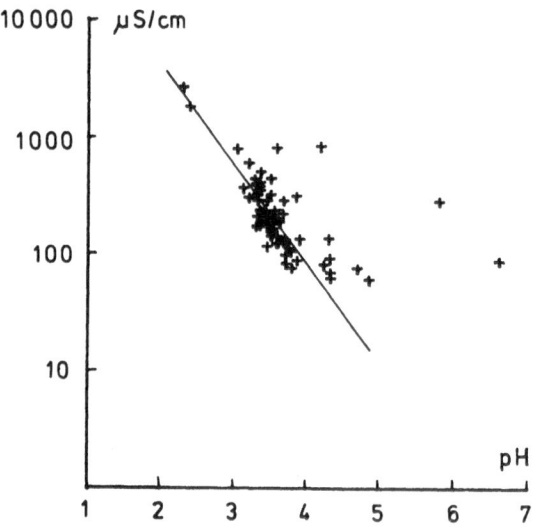

Figure 4. Electrical conductivity as function of pH in fogwater at
the station "Kleiner Feldberg" (autumn 1983 – spring 1986).

3.2 Electrical Conductivity

Besides the pH-values the electrical conductivities are measured. Due
to the small sampling volumes a micro-electrode type WTW LDM/S 5807/02
was used. Figure 4 demonstrates the results of the electrical
conductivity in dependence of the pH-values.

Extremely high electrical conductivities in the range of more than
2000 μS/cm sometimes were observed. These high values only occurred in
samples with very low pH-values. Most values however range from 100 to
300 μS/cm. Although in a few cases some high conductivities exist
along with relatively high pH, the correlation between the electrical
conductivity and the concentrations of the H^+-ions, respectively pH,
shows a very good agreement. The correlation coefficient is R = 0.96.

3.3 Element Concentrations in Fogwater

The concentrations of the anions and the ammoniumions in fog are
compared to the concentrations in rainwater. The deltoides give the
range of concentration. The peaks represent the maximum and minimum
concentations, whereas the average values are indicated by the
crossbeams (Fig. 5).

The highest concentration is found for the compound $SO_4^=$,which
reaches up to 87 mg S/l. The average value is 13.8 mg/l.
Concentrations less than 2.3 mg S/l in fog are not found. In rainwater

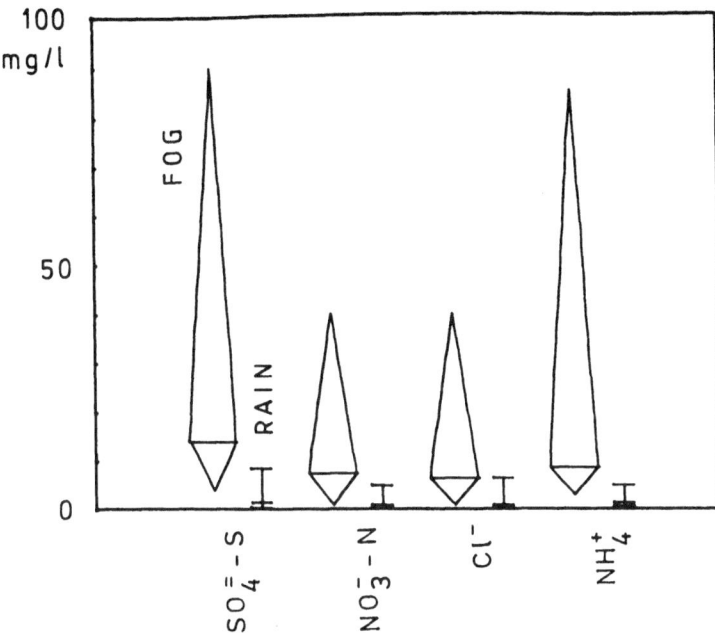

Figure 5. Mean concentrations of the anions and NH_4^+ for the period
autumn 1983 – autumn 1985 in fogwater and rain at the
station "Kleiner Feldberg"

the mean value is 1.5 mg/l. Compared to rain the concentration is
9 times higher. Similar relations are found for nitrate and chloride.
For NO_3^- the mean value is 7.5 mg N/l with highest concentrations up to
40 mg/l. For the compound Cl^- the measured concentrations range from
0.7 up to 40 mg/l with a mean value of 6.5 mg/l. These compounds also
show an enrichment factor of 8-9 with respect to rainwater. The result
of the high anion concentrations along with the relatively high
pH-values can be explained by the high contributions of ammonium. The
mean concentrations of NH_4^+ is 14 mg/l. The maximum reaches up to 85
mg/l. The lowest concentration measured in fog was 2.8 mg NH_2^+/l.
Table 1 indicates the mean concentrations of the heavy metals Pb,
Fe and Cd and the compounds Na, K, Ca, Mg. Additionally the minimum
and maximum concentrations in fog and the mean concentrations in
rainwater are presented.
For all compounds the concentration in fog is much higher than in
rain. In fog the lowest concentations are in the same range as the
mean values in rainwater. For Pb and Ca the minimum concentrations in
fog are even higher than the average values in rain.
Extremely high concentrations were measured for Pb and Fe. The
concentrations can reach more than 1.5 mg/l. Highest variability is
registrated for sodium. Between the mean concentration and the maximum
a factor 20 is calculated. For the heavy metals lead shows the
greatest variation in the range of 10. Again these results demonstrate

the high variations of the element concentrations in the individual fog
events as described for the pH-values.

Table 1.   Mean concentrations, minimum and maximum in fogwater and
            mean concentrations in rain at the station "Kleiner
            Feldberg".

| | fogwater | | | rainwater |
| | mean | max | min | mean |
	(............mg/l ....)			(mg/L)
Pb	0.203	2.000	0.016	0.013
Fe	0.439	1.600	0.053	0.068
Cd	0.0047	0.020	0.0004	0.0013
Na	2.7	41.8	0.2	0.29
K	1.1	5.1	0.2	0.19
Ca	2.4	8.8	0.8	0.2
Mg	0.3	3.2	0.1	0.09

Table 2.   Mean specific enrichment factors of elements in fogwater.

H^+-Ions	11	K	6
NH_4^+	9	Ca	12
$SO_4^=$-S	9	Mg	3
NO_3^--N	9	Pb	15
Cl^-	8	Fe	7
Na	9	Cd	4

From these data of the averaged concentrations in fog and
rainwater enrichment factors were calculated. Table 2 demonstrates the
increasing of the element concentrations in fog compared to rain. For
all compounds the enrichment factors show positive values, that means
all elements in fog show higher concentrations than in rainwater.

It seems, that different enrichment factors for the different
compounds exist. The highest factors were calculated for the elements
Pb, Ca and the H^+-ions. In fog the mean concentrations of these
elements are enhanced by a factor of 11-15.

While the anions and ammonium show a similar increase by a factor
of 8-9, lower enrichment factors were calculated for Mg, K and Cd.
Here the increase is in the range of 3-7.

Similar results are also found by investigations of other authors
(Winkler, 1986). At the station Schauinsland/Breisgau (1284 m) of the
German Environmental Protection Agency (UBA) analysis of cloud and
fogwater show comparable specific enrichment factors (UBA, 1984).

Up to now the elementspecific enrichment factors cannot be explained completely. One reason may be different anthropogenic sources or different physico-chemical processes, such as certain reactions in the gaseous phase.

3.4 Ionic Balance

The ionic budget will be defined as the sum of the concentration of all dissolved anions and cations. For all fogwater samples ionic balances were calculated. For the estimation the anions include the measured compounds sulphate, nitrate and chloride, whereas the sum of the cations is indicated by the H^+-ions, NH_4^+, Na, Mg, Ca and K:

$$H^+ + NH_4^+ + Mg^{++} + Ca^{++} + Na^+ + K^+ = SO_4^= + NO_3^- + Cl^- \quad (1)$$

These compounds represent the major ions in fogwater. The results are shown in Figure 6. For the calculation all concentrations are used in equivalent units. In most of all cases the total sum of the cations and anions is less than 3 meq/l. For these samples the relation between the sums of the anions and cations is nearly 1.

The obtained results show a good correlation. The correlation coefficient is R = 0.93.

Only at higher absolute concentrations (about 3 meq/l) the sum of the anions is higher than the corresponding sum of the positive ions.

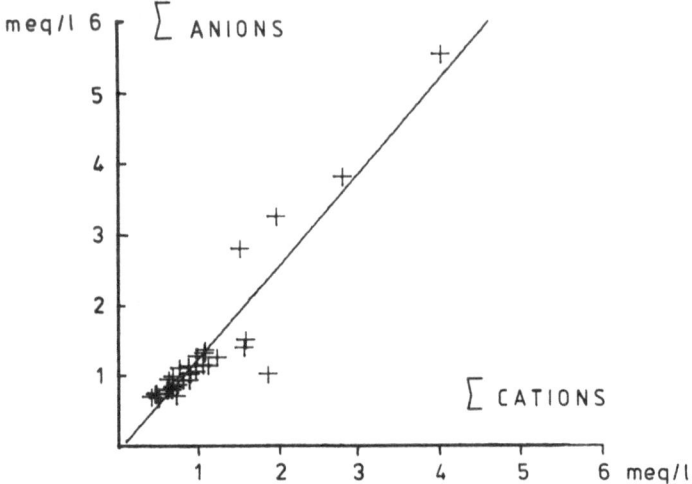

Figure 6. Correlation between the sums of cations and anios in fogwater.

Figure 9 gives information about the percentual distributions of the analysed ions in fogwater. The diagram shows the mean values for the whole sampling period from autumn 1983 - spring 1986.

412

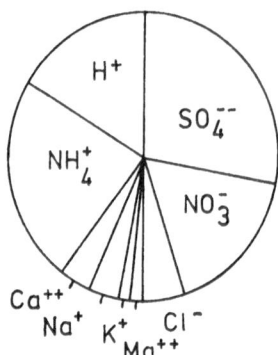

Figure 7. Percentual distribution of the measured ions in fogwater.

As one can see, for the anions sulphate is the dominant compound.
The contribution reaches up to 27%. Nitrate and chloride play a minor
role. Their fraction of the total ionic composition only contribute to
17% and 6%. Due to the similar enrichment factors of the negative ions
the ionic composition of fog shows an identical distribution as
rainwater. For the cations the highest contribution is measured for
NH_4^+. Compared to rainwater the value is slightly increased. In rain
we find a contribution of 22%, whereas in fog NH_4^+ has a fraction of
24%. The contribution of the H^+-ions is 16%. For Na, K, Ca and Mg the
mean contributions are in the range of 1-4% of the total sum of the
measured ions in fog.
    It has to be mentioned, that the results reported in Figure 7 are
mean values averaged over the whole sampling period. In the individual
samples the percentage distribution of the element can change in a wide
range.

3.5 Temporal Distribution of Concentration During Single Fog Events

As an impressive example not for very high concentrations, but for the
typical temporal fluctuation of the concentrations during a single fog
event the distribution of the anions, ammonium, lead, electrical
conductivity, pH and liquid water content (LWC) are shown for a case of
a fog period collected at the seventh of October 1985 (Fig. 8). The
samples are taken at half hour intervals. Sampling started when the
visible range was less than 300 m.
    All element concentrations show a distinct equivalent
distribution. At the beginning all measured compounds have high
concentrations. In the following samples the concentrations decrease.
For the following samples the concentrations increase again and reach
in some cases ($SO_4^=$ -S, Cl^-, Pb) the highest values of the whole fog
event. Before the fog is disappearing, within the last samples again
increased concentrations were observed. For the compound ammonium at
the end of the fog episode the highest concentration was measured
(14.45 mg/l), whereas in the first sample we found 12.3 mg/l.

For the analysed heavy metals, as example the distribution of lead is given. The measured concentrations range from 120 µg/ml to more than 300 µg/ml.

While the same temporal distribution appears for the electrical conductivity (150 - 380 µS/cm), the pH-values show an inverse behaviour. Samples with high element concentrations have low pH-values, decreasing pH-values are associated to increasing element concentrations. Nevertheless there is no explanation for the first pH-values being relatively high in this case.

It has to be pointed out, that for all investigated fog events we find the highest element concentrations in the beginning phase and at the end as a characteristic distribution for fog episodes (Schmitt, 1986).

In addition Figure 8 demonstrates the results of the samples taken by the passive sampling system at the same time indicated by the dashed

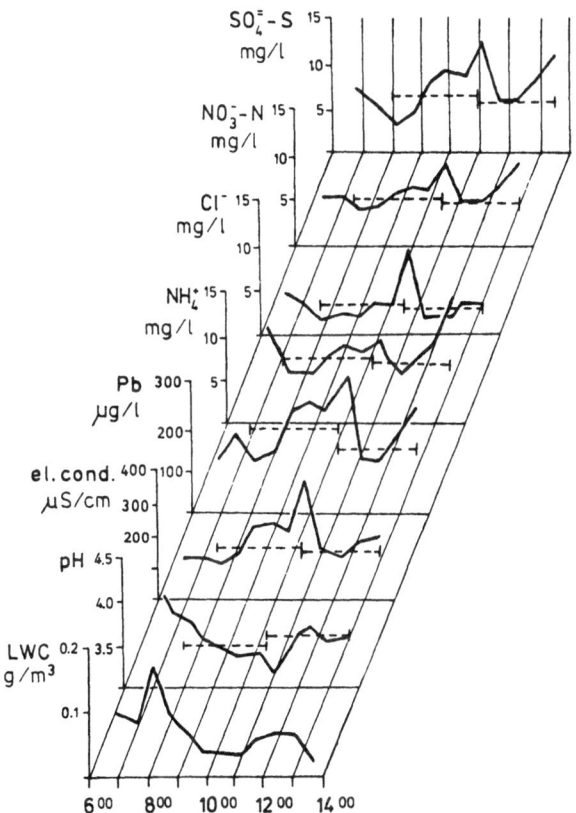

Figure 8.   Temporal distributionn of the anions, ammonium, pH, Pb, electrical conductivity and LWC during a single fog event (7.10.1985) at the station "Kleiner Fledberg" (solid lines: active system, dashed lines: passive system).

414

lines. During the whole event only two samples could be collected.
The sampling time was 3 hours for each sample. The measured element
concentrations are in the same range as the samples collected by the
active sampling system, but due to the longer sampling time the results
produce only an integrated mean value of the temporal variation of the
concentrations. Therefore investigations on short time fluctuations of
the concentrations within individual fog events passive collection
systems are not satisfactory.

The observed variations of the concentrations and the pH-values
strongly depend on the actual amounts of the LWC. In general the LWC
in fog ranges from 0.1-0.5 g/m^3. The values of the LWC were
approximated by dividing the volume of the sampled fog water by the
volume of air passed by the rotating collectors. The presentation
shows, that low element concentrations and therefore high pH-values are
only observed, when the LWC increases. In the opposite at low LWC the
solution is concentrated leading to increasing concentrations.

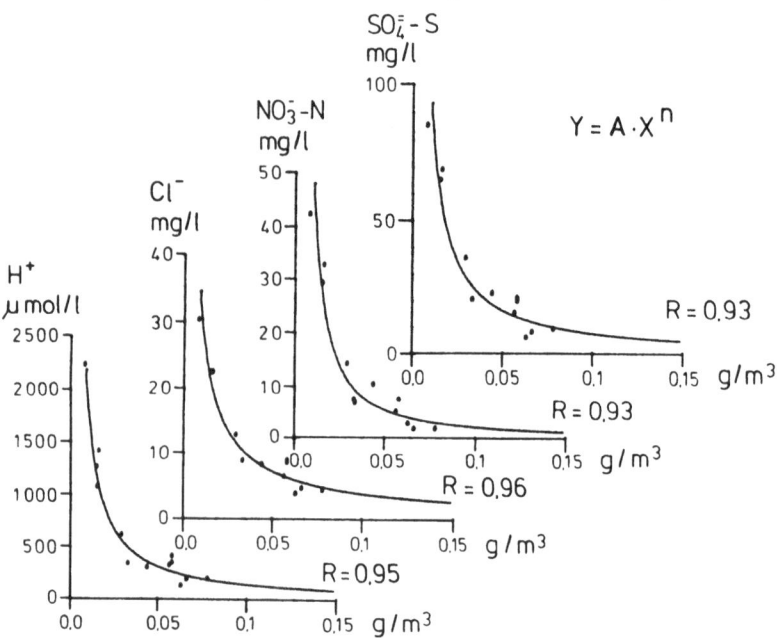

Figure 9.   Concentration of the anions and H$^+$ as function of LWC
during a single fog event (15.10.1984) at the station
"Kleiner Feldberg".

Generally high concentrations in fog compared to rain can be
explained by different reasons. Near the ground, where fog mostly
arises, the concentrations of most of the important pollutants are
enhanced compared to higher levels of the atmosphere, where clouds are
formed. In fog the scavenging coefficient may be higher than in
clouds, because in fog a larger fraction of small droplets can be found
(Juisto 1983). Thereby the mean ratio of surface to volume of droplets

is increased compared to clouds (Pruppacher and Klett 1978). Due to these differences, the incorporation and absorption of trace gases and aerosol particles is considered to be more effective in fog than in clouds. Finally in fog the concentrations will be increased due to the lower values of the LWC.

For further estimations we assume the concentration of the trace substances in the atmosphere and the scavenging coefficient to be nearly constant during the time for the fog event. As a result the element concentration in the fog water depends only on the actual values of the LWC. On the basis of the measured concentrations and the estimated LWC correlation coefficients were calculated. In Figure 9 the concentrations of the anions and H^+-concentration are plotted as a function of the LWC for a fog period in autumn 1984.

The absolute amounts of concentrations are much higher than in the case previously described before. The obviously strong dependence is given by correlation coefficients with high values of $R > +0.9$. The estimations also indicate the extremely high concentrations, when the LWC is reduced. By decreasing the LWC for 50% the concentrations will be increased by a factor of 2. For the investigations on forest decline this effect has to be taken into account. Due to short time reduced LWC, such we find characteristical for the beginning and at the end of a fog period, extremely high concentrations will occur. On the surface of the vegetation, when highly concentrated fog droplets were intercepted, the same effects may result in damage on the vegetation.

By the data received from these measurements also a rough estimation on the abundant aerosol concentration can be produced. Compared to aerosol filter measurements the sample volume is defined by the total volume of air, passed by the collectors during rotation. The product of the concentration K and the collected fogwater (Vw) is independent of the LWC. Dividing this product by the sample volume (Va) leads to concentrations of the elements (M) in the atmosphere:

$$M = \frac{K \cdot Vw}{Va} \tag{2}$$

M = element concentration in the atmosphere (mg/m^3)
K = concentration in fog water (mg/l)
Vw = volume of the sampled fogwater (l)
Va = volume of the filtered air (m^3)

Figure 10 shows the calculations for the anions, lead and cadmium. The estimated concentrations of the anions range from 0.2 - 1.0 $\mu g/m^3$. For the compounds lead and cadmium the concentrations vary in a small range from 10 - 20 ng/m^3 respectively 0.5 - 1.0 ng/m^3.

Compared to aerosol filter measurements at this station these results are in the same order of magnitudes (Georgii et al. 1982). The absolute amounts are always smaller, because the calculations by equation 2 only take into account element masses associated with the sampled fog droplets. Due to the independence on the LWC the concentration of M should be nearly constant during the fog event.

Nevertheless the temporal distribution of M shows a slight decrease
during the fog event. This effect may be the result of removal of the
elements from the atmosphere.

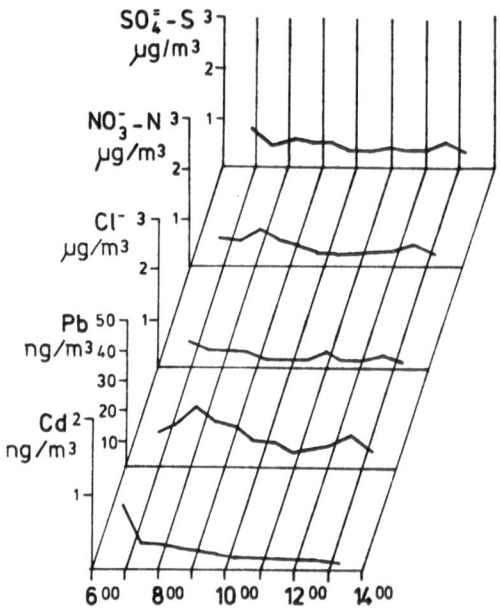

Figure 10   Temporal distribution of the element concentration M
             calculated by equation 2 for a single fog event (7.10.1985)
             at the station "Kleiner Feldberg".

4.   CONCLUSIONS

The measurements reported herein emphasize that the investigations on
fogwater chemistry are necessary for a complete understanding of the
chemical processes in the atmosphere. Compared to rainwater in general
the element concentrations in fog are increased. The enrichment is
depending on the different compounds. For the measured elements
enrichment factors in the range of 3-15 were observed. During
individual fog episodes the variability of element concentration in fog
is mainly determined by the actual amounts of the LWC. Extremely high
concentrations are registered, when the LWC is reduced. Especially at
the beginning and the end of a fog event the concentrations increase.
A decrease of the LWC by half leads to an increase of the element
concentrations by a factor of 2. The same effects occur on vegetation
surfaces, when intercepted fog droplets are evaporating. These results
demonstrate the important role of fogwater chemistry with respect to
investigations on forest decline.

A detailed research programme in this field is planned in the future.

ACKNOWLEDGEMENTS

The investigations are sponsored by the Umweltbundesamt, Berlin under contract number 104 02 715.

REFERENCES

Black, H. and Landsberg, H. 1983. A method for continuous records of the pH of low-level clouds. Final Rep. Contr. DE-AS05 ER 6000044 College Park, Maryland.

Fuchs, N.A. 1964. The mechanics of aerosols. Pergamon Press, Oxford.

Georgii, H.W. and Schmitt, G. 1985. 'Ergebnisse der Nebelwasseranalyse', Staub Reinh. der Luft, Nr. 6 , pp. 260-264.

Georgii, H.W., Perseke, C. and Rohbock, E. 1982. Feststellung der Deposition von sauren und langzeitwirksamen Luftverunreinigungen aus Belastungsgebieten; Abschlussbericht Forschungsprojekt 104 02 600 im Auftrag des Umweltbundesamtes, Eigenverlag des Universtäts- instituts für Meteorologie und Geophysik, Frankfurt am Main, 205 p.

Georgii, H.W., Grosch, S. and Schmitt, G. 1986. Feststellung der Schadstoffbelastung von Waldgebieten in der Bundesrebuplik Deutschland durch trockene und nasse Deposition; Abschlussbericht Teil A Forschungsprojekt 104 02 715 im Auftrag des Umweltbundesamtes, Eigenverlag des Universtäts instituts für Meteorologie und Geophysik, Frankfurt am Main, 235 p.

Grunow, J. 1955. 'Der Nebelniederschlag im Bergwald'. Forstw. Gbl. 74 .

Hoffmann, M. et al. 1983. 'Design and calibration of a rotating arm collector for ambient fog sampling'. Precipitation Scavenging, Dry Deposition and Resuspension, ed. H.R. Pruppacher et al., Elsevier Publ., New York, Vol. 1.

Jiusto, J.E. 1983. Clouds, their formation, optical properties and effects, Academic Press, New York.

Katz, U. 1980. 'A droplet impactor to collect liquid water from laboratory clouds for chemical analysis'. Comm. a la VIII Conf. Intern. sur la Physique des Nuages, Vol. 2.

Mohnen, V. 1980. 'Cloud water collection from aircraft'. Atmospheric Technology 12 .

Pruppacher, H.R. and Klett, J.D. 1980. Microphysics of clouds and precipitation, Dordrecht, Reidel Publ.

Schmitt, G. 1986. 'Temporal distribution of trace element concentrations in fogwater during individual events. In: Atmospheric pollutants in forest areas, ed. H.W. Georgii, Reidel Publishing Company, Dordrecht, pp. 129-141.

UBA 1984. Monatsberichte des Umweltbundesamtes 7 .

Winkler, P. 'Observation on fog composition in Hamburg. In: Atmospheric pollutants in forest areas, ed. H.W. Georgii, Reidel Publishing Company, Dordrecht, pp. 143-151.

CHEMICAL COMPOSITION OF WET DEPOSITION IN THE EASTERN ALPINE REGION

H Puxbaum, W Vitovec and A Kovar
Institute for Analytical Chemistry
Technical University Vienna
A-1060 Wien
Getreidemarkt 9
Austria

ABSTRACT. In the northern and central part of the eastern Alps a "wet only" precipitation network is operating collecting samples on a daily basis. In this presentation the results of 1984/85 for 9 stations situated at an elevation from 400 - 1700 m are summarised. For the stations at the northern part of the Alps a concentration gradient from west to east is found for the components NH_4^+ and SO_4^{2-}. The wet deposition of ionic components is significantly higher in the summer season as compared to the winter season. The wet deposition of the main ionic components is occurring in an episodic mode. The input of acidity is highly episodic at all stations.

1. INTRODUCTION

A major part of the eastern Alps is situated in Austria consisting of three main formations, the Northern and Southern "Lime-Alps" and the "Central Alps". As a consequence of the altitude gradient humid air masses which are frequently transported towards the Alps during westerly and north-westerly weather situations cause intensive rain- and snowfall. The main fraction of the humidity is deposited in the northern Alps and the northern slopes of the central Alps, where yearly wet deposition patterns appear to be height dependent and ranging up to 2500 mm of water equivalents per year (Böhm, 1986). By contrast the southern parts of the Central Alps receive less than half of the wet deposition at comparable altitudes. In this presentation the results for the collection period 1984/85 from a "wet deposition" network operating in the Austrian Alps are summarised and some conclusions concerning the episodic nature of the deposited main constituents are drawn.

M. H. Unsworth and D. Fowler (eds.), Acid Deposition at High Elevation Sites, 419–430.
© 1988 by Kluwer Academic Publishers.

2. EXPERIMENTAL

2.1 Precipitation Sampling Network

In the northern and central part of the Austrian Alps a "wet only" precipitation network is operating, collecting snow and rain samples on a daily basis since 1982/83. For the sampling period 1984/85 nine stations situated at an elevation from 400 - 1700 m delivered daily samples. The location of the stations is presented in Figure 1.

A short description of the sampling site characteristics is given in Table 1.

The stations are generally situated outside of the centre of villages some distance from the main motorways in the cases of the valleys with dense traffic. The stations Achenkirch, St. Koloman, Wergenweng and Innervillgraten are situated more remote from larger local sources and can be regarded as "regional" background stations. The daily control of the samples is carried out by local observers usually in the morning hours. The collection procedure follows the recommendations of the Austrian Ministry for Health and Environment (1984). The samples are stored in polyethylene bottles at 0 - 4°C and transported in biweekly intervals to the central laboratories in Vienna. There the samples were stored at 0 - 4°C again and analysed generally within 1 - 4 weeks after receipt.

Samples for all stations were analysed at the Institute for Analytical Chemistry at the Technical University of Vienna (IAC-TU Wien) with the exception of the stations Achenkirch and St. Koloman, which are part of the "EMEP" network. The respective samples were analysed by the group "Air hygiene" at the Umweltbundesamt in Vienna (UBA).

All stations are equipped with "WADOS" (wet and dry only samplers) (Fa. Kroneis, Vienna). General features of the samplers are: diameter of wet deposition collection funnel - 250 mm, suitable for wintertime operation, wintertime proven lid mechanics, sensitive wet deposition (conductometric) sensor with two heating steps. Based on intercomparisons with standard "Hellmann" gauges the sampling efficiency for rain water is 90% on the average for a three months observation period.

2.2 Analysis

In all samples conductivity (25°C) and pH were determined by standard methods (Ministry for Health and Environment, 1984). Anions were determined by "suppressed mode" ion chromatography using a Dionex D10 ion chromatograph. Monovalent cations were determined at the "IAC" by "single column" ion chromatograph on a WESCAN "ICM" analyser under the conditions described by Tsitouridou and Puxbaum (1987). For the determination of the divalent cations (Ca^{2+}, Mg^{2+}) atomic absorption spectrometry was used.

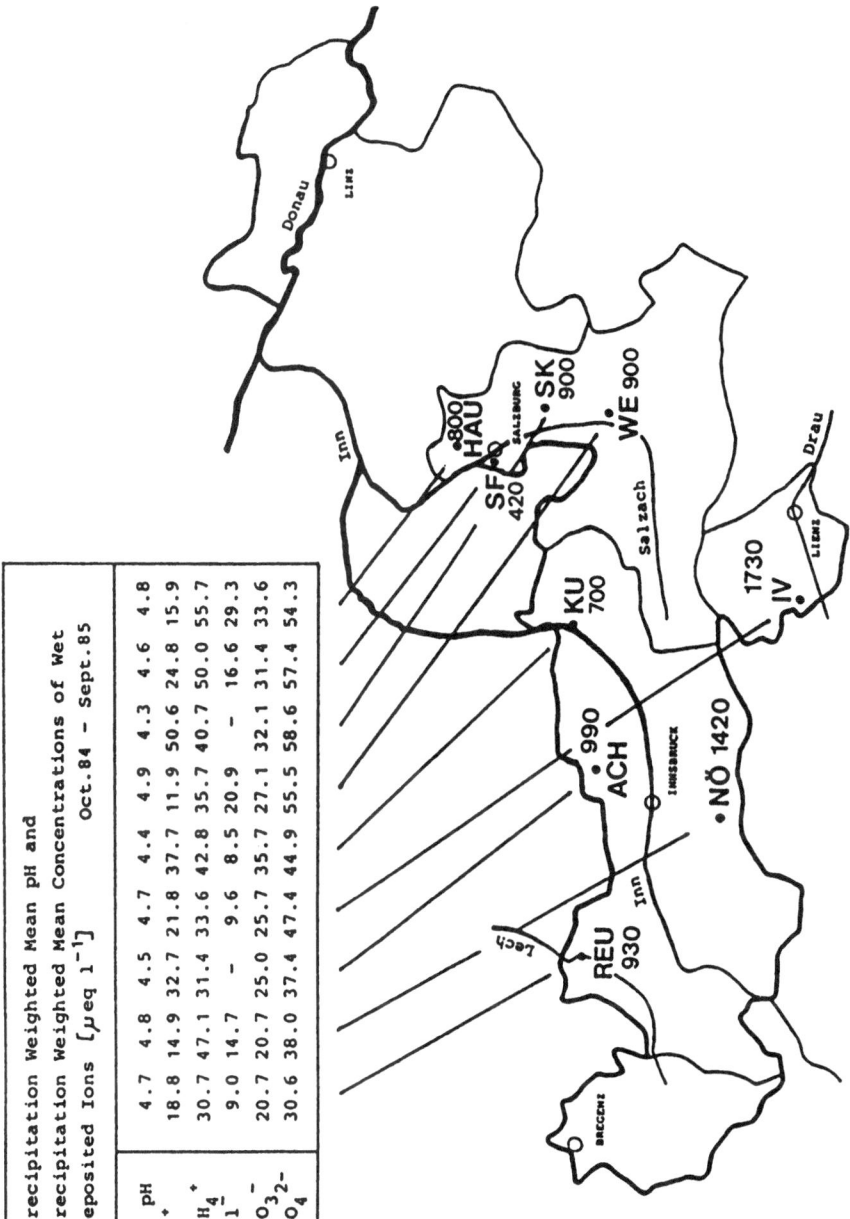

Figure 1. Weighted mean pH and concentrations of deposited ions.

Table 1.   Sampling site characteristics -

Type A: Northern Alps, stations in valleys and plains open
         to the north or northwest.
Type B: Central Alps, stations in "inner alpine" regions.

Stations	Elevation m ASL	Region	Characteristics
Type A:			
A-west:			
Reutte	930	Lechtal	Valley with some local industry.
Achenkirch	990	Achental	Narrow valley, no major local influence.
Kufstein	700	Inntal	Industrialised valley, heavy traffic and densely populated.
A-east:			
Salzburg Fl.	420	Salzachtal, Flachgau	Suburban terrain north of Salzburg.
St. Koloman	900	Tennengau	Mountain slope directly exposed to upslope rain.
Haunsbergh	800	Salzachtal, Flachgau	Hilly terrain north of Salzburg.
Type B:			
Werfenweng	900	Pongau	Narrow side valley south of the chains of the northern alps.
Nösslach	1420	Brenner	Sparsly populated valley, northern slopes of central alps, influenced by high traffic density.
Inner-villgraten	1730	Osttirol	Narrow side valley south of the main chains of the central alps, remote from major local sources.

3. RESULTS AND DISCUSSION

3.1 Concentration and Deposition

The regional distribution of the concentrations annual weighted mean
values of the main ionic components is illustrated in Figure 1. The
results indicate that for the stations of type A (northern alpine
station) a concentration gradient from west to east is observed for the
components NH_4^+, Cl^-, NO_3^-, SO_4^{2-}. For acidity in rainfall no spatial
trends but relatively high fluctuations have been observed. We have
grouped the stations into three classes (westerly and easterly stations
of type A and stations of type B - Table 1, 2). Compared to the
stations type A west, the stations type A east exhibit higher
concentrations for the components SO_4^{2-}, NO_3^-, Cl^- and NH_4. However,
the differences are significant only for the components NH_4^+ and SO_4^{2-}
(Table 2). The wet precipitation at stations type B show intermediate
concentrations for the component NH_4^+, Cl and SO_4 but less for nitrate
as compared to stations type A west and east. The differences are not
significant in all cases.

Table 2.   Annual precipitation weighted mean concentrations (μeq/1),
October 84 – September 85, at three classes of stations.

A_{west}: Reutte, Achenkirch, Kufstein
B   : Nösslach, Innvervillgraten, Wergenweng
A_{east}: Salzburg Flughafen, St. Koloman, Haunsberg

Significance of differences: + level 5%, n.s. not
significant.

	A_{west}	B	A_{east}	A_{west}: A_{east}	B: A_{west}	B: A_{east}
H^+	29.7	16.2	30.4	n.s.	n.s.	n.s.
NH_4^+	35.8	38.8	48.8	+	n.s.	n.s.
NO_3^-	27.1	24.5	32.4	n.s.	n.s.	n.s.
SO_4^{2-}	37.6	47.0	56.8	+	n.s.	n.s.

An explanation for the concentration differences might be the fact
that the Salzach valley (stations: Salzburg-Flughafen, Haunsberg, St.
Koloman) is a broad, relatively dense populated open valley extending
towards the city of Munich. The "Upper Inn-valley" (station: Kufstein)
is also a relatively densely populated and industrialised area, whereas
the valleys leading to Reutte and Achenkirch are sparsely populated and
narrow. Thus regional emissions might be responsible for the
differences of deposition of wet deposition main constituents at the
stations of type A.

The amount of rainfall at the stations of type A is in a comparable range (Figure 2). Also with the exception of the station Salzburg-Flughafen all stations of type A are situated at comparable elevations. Therefore the orographic effect on the concentrations of rain constituents seems to be of less importance for the Type A stations compared. The stations of type B receive less wet precipitation on a yearly basis.

The regional pattern of the wet deposited main components; NO_3^-, SO_4^{2-}, Cl^-, H^+ and NH_4^+ accumulated during one year (1984/85) is given in Figure 2.

According to the concentration trend the deposition of the ionic main constituents NH_4^+, Cl^-, NO_3^- and SO_4^{2-} is more pronounced at the easterly station compared to the westerly stations of type A. The inneralpine valley stations in Tyrol (Nösslach and Innervillgraten) receive lower deposition for the components NH_4^+, Cl^-, NO_3^- and SO_4^{2-} as compared to the stations at the northerly slopes.

Seasonal differences in the ion concentrations for the summer and winter half year periods are shown in Table 3. No clear trends can be deduced from these results. A marked difference between summer and winter period concentrations is found for all major ions at the site Innervillgraten where the winter period ionic concentrations were considerably lower as compared to the summer period values. This might indicate that in the winter period the transport of pollutants to regions south of the Central Alps is more restricted as compared to the summer period.

The amount of wet precipitation in the summer period is a factor of 2,2 - 3,5 higher than in the winter period (Table 4). As a result of this effect for all major ions and all stations the wet deposition is significantly higher in the summer period. The trends of the monthly averaged concentrations show size dependent patterns. As observed elsewhere in Europe (Cape et al., 1984 or Granat, 1978) a marked spring maximum for the ions $\overline{H^+}$, $\overline{NH_4^+}$, NO_3^- and SO_4^{2-}, but not for Cl^-, is found at the type A west sites (Figure 3a). At the stations of type A east (Figure 3b) a spring maximum is observed in March for the ions H^+ and NO_3^-, whereas a broader maximum from March through May appears for the ions SO_4^{2-} and NH_4^+. For the stations type B the seasonal trends of the ionic concentrations are not so clear (Figure 3c).

3.2 Episodicities

At all sites the main fraction of the wet precipitation of the ionic main constituents was deposited during heavy rain events in episodic mode. Smith and Hunt (1978) provide an explanation for the episodic nature of wet deposition observed in different regions of Europe and gave an arbitrary definition for the term "episodicity". We will use the term "episodicity" in the same way as the ratio expressed as a percentage of the number of "episode days" to the annual number of wet days. "Episode days" are days with the highest wet deposition of a certain component which, when summed make up 30% of the total annual wet deposition for the respective component. In Table 5 the

425

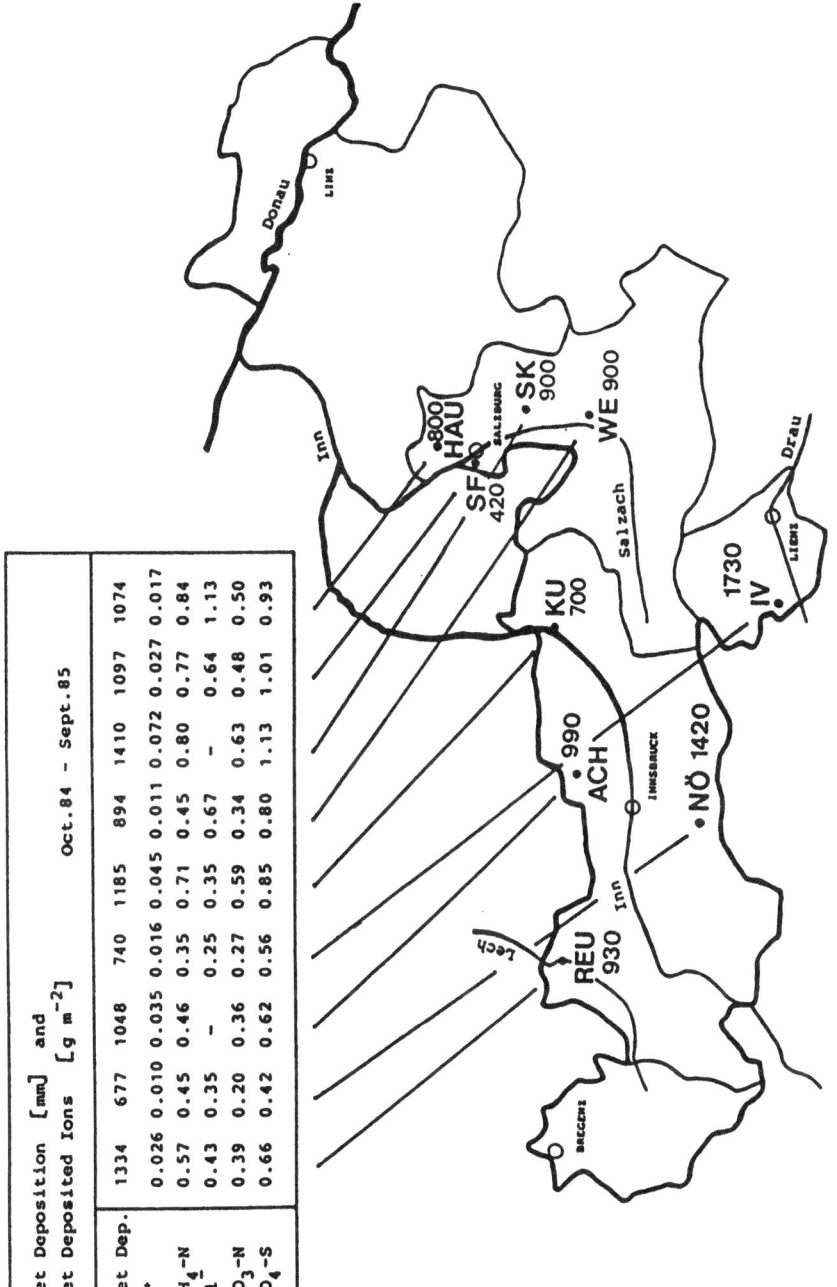

Wet Deposition [mm] and Wet Deposited Ions [g m⁻²]									Oct.84 – Sept.85	
Wet Dep.	1334	677	1048	740	1185	894	1410	1097	1074
H⁺	0.026	0.010	0.035	0.016	0.045	0.011	0.072	0.027	0.017
NH₄–N	0.57	0.45	0.46	0.35	0.71	0.45	0.80	0.77	0.84
Cl⁻	0.43	0.35	–	0.25	0.35	0.67	–	0.64	1.13
NO₃–N	0.39	0.20	0.36	0.27	0.59	0.34	0.63	0.48	0.50
SO₄–S	0.66	0.42	0.62	0.56	0.85	0.80	1.13	1.01	0.93

Figure 2. Wet deposition.

Table 3: Winter (October 84 - March 85) and Summer (April 85 - September 85) period weighted mean values of ionic concentrations (μeq/l). S/W Summer- vs. Wintertime deposition ratio

	Reutte		Nößlach		Achenkirch		Innervillgraten		Kufstein		Werfenweng		St.Koloman		Salzburg Airport		Haunsberg	
	W	S	W	S	W	S	W	S	W	S	W	S	W	S	W	S	W	S
pH	4.8	4.7	5.2	4.8	4.6	4.4	4.9	4.6	4.7	4.3	4.9	4.9	4.3	4.3	4.6	4.6	4.6	4.8
H+	16.9	19.8	6.9	14.9	24.8	37.7	12.9	24.8	19.8	45.6	11.9	11.9	46.6	52.6	26.8	23.8	22.8	13.9
S/W	1.2		2.2		1.5		1.9		2.3		0.9		1.1		0.9		0.6	
NH$_4$-N	14.3	37.1	40.0	49.3	21.4	35.0	13.6	42.8	30.7	47.1	23.6	37.8	27.1	45.7	85.0	40.0	84.2	46.4
S/W	2.6		1.2		1.6		3.1		1.5		1.6		1.7		0.5		0.6	
Cl-	7.3	9.9	6.8	17.2	-	-	4.5	11.6	5.6	8.5	10.2	24.3	-	-	25.9	13.8	30.7	29.0
S/W	1.4		2.5		-		2.6		1.5		2.4		-		0.5		0.9	
NO$_3$-N	19.3	21.4	26.4	17.1	25.0	25.0	8.6	32.8	33.6	36.4	24.3	27.8	37.1	30.0	40.7	29.3	42.8	30.7
S/W	1.1		0.6		1.0		3.8		1.1		1.1		0.8		0.7		0.7	
SO$_4$-S	21.8	34.3	33.1	40.0	42.4	34.9	13.1	62.4	33.7	48.0	63.6	65.5	50.5	61.8	73.6	52.4	58.6	52.4
S/W	1.6		1.2		0.8		4.8		1.4		1.0		1.2		0.7		0.9	

Table 4: Winter (October 84 - March 85) and Summer period (April 85 - September 85) wet deposition of rain (mm) and ions (g/m²).
S/W Summer- vs. Wintertime deposition ratio

	Reutte		Nößlach		Achenkirch		Innervillgraten		Kufstein		Werfenweng		St.Koloman		Salzburg Airport		Haunsberg	
	W	S	W	S	W	S	W	S	W	S	W	S	W	S	W	S	W	S
Wet Dep.	412	922	150	527	281	767	227	514	358	827	203	690	379	1031	256	841	247	827
S/W		2.2		3.5		2.7		2.3		2.3		3.4		2.7		3.3		3.3
H+	0.007	0.018	0.001	0.008	0.007	0.029	0.003	0.013	0.007	0.038	0.003	0.008	0.018	0.054	0.007	0.020	0.006	0.012
S/W		2.6		8.0		4.1		4.3		5.4		2.7		3.0		2.9		2.0
NH4-N	0.084	0.48	0.084	0.36	0.084	0.38	0.042	0.31	0.15	0.55	0.070	0.36	0.14	0.66	0.31	0.48	0.29	0.53
S/W		5.7		4.3		4.5		7.4		3.7		5.1		4.7		1.5		1.8
Cl-	0.11	0.32	0.035	0.32	-	-	0.035	0.21	0.071	0.25	0.071	0.60	-	-	0.25	0.43	0.28	0.85
S/W		2.9		9.1		-		6.0		3.5		8.5		-		1.7		3.0
NO3-N	0.11	0.28	0.056	0.13	0.098	0.27	0.028	0.24	0.17	0.42	0.070	0.27	0.20	0.43	0.14	0.35	0.15	0.35
S/W		2.5		2.3		2.8		8.6		2.5		3.9		2.2		2.5		2.3
SO4-S	0.14	0.51	0.080	0.34	0.19	0.43	0.048	0.51	0.19	0.64	0.080	0.72	0.30	1.03	0.30	0.71	0.24	0.69
S/W		3.6		4.3		2.3		10.6		3.4		9.0		3.4		2.4		2.9

428

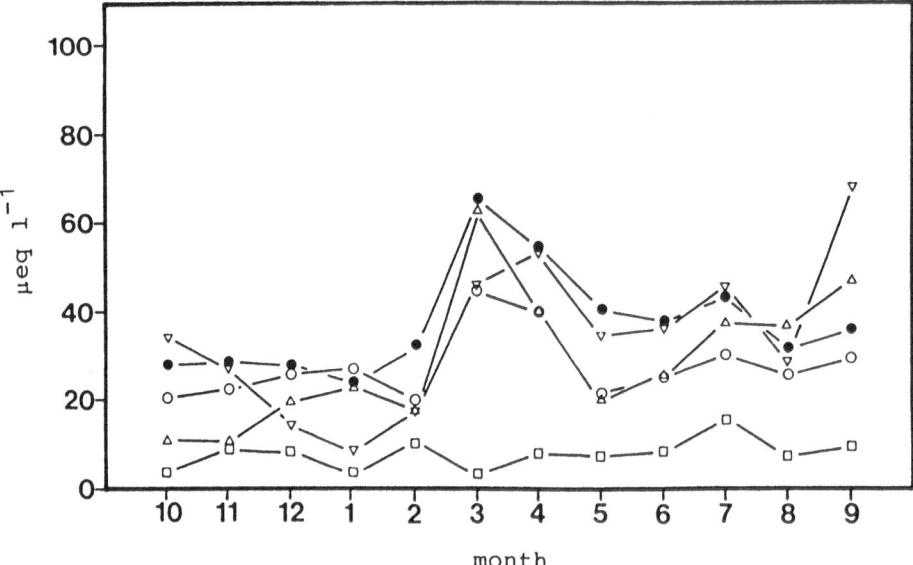

Figure 3a Seasonal variation of ionic main constituents
concentrations – Station type A west – period 1984/85
$\triangle$...H$^+$, $\triangledown$...NH$_4^+$, $\square$...Cl$^-$, O...No$_3^-$, ●...SO$_4^{2-}$.

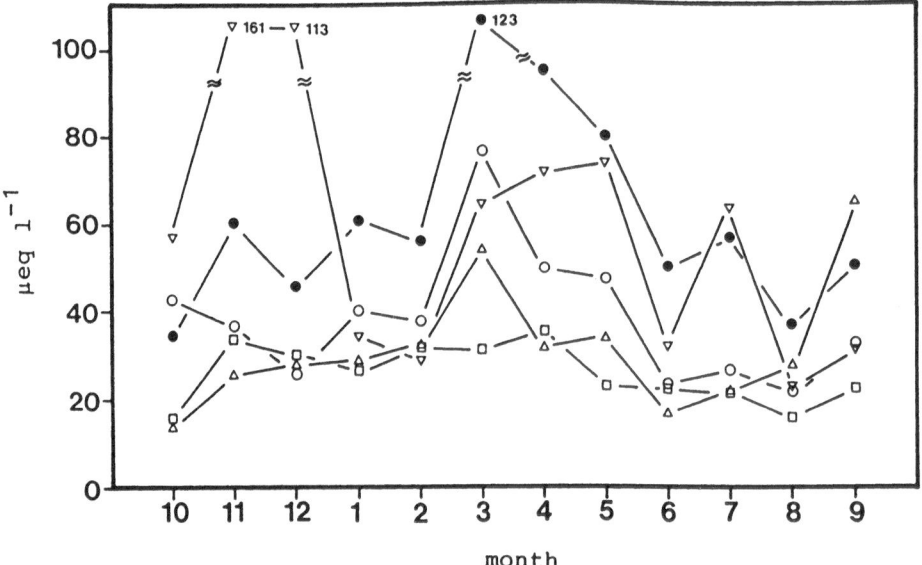

Figure 3b Seasonal variation of ionic main consituents concentrations
– Station type A east – period 1984/85 $\triangle$...H$^+$, $\triangledown$...NH$_4^+$,
$\square$...Cl$^-$, O...NO$_3^-$, ●...SO$_4^{2-}$.

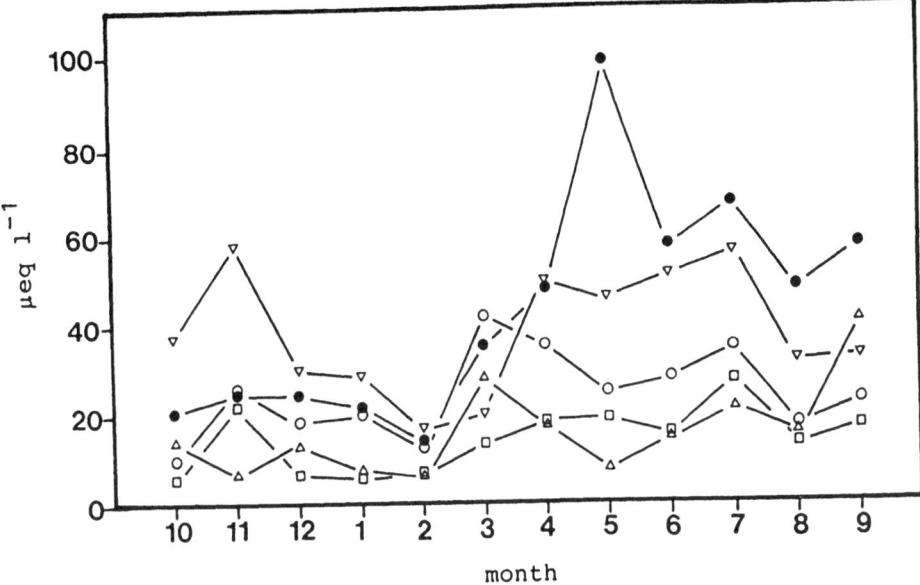

Figure 3c   Seasonal variation of ionic main constituents
concentrations – Station type B $\triangle$...H^+, $\triangledown$...NH_4^+, $\square$...Cl^-,
O...NO_3^-, $\bullet$...SO_4^{2-}.

Table 5: Episodicities

	Reutte	Nößlach	Achenkirch	Innervillgraten	Kufstein	Werfenweng	St.Koloman	Salzburg Airport	Haunsberg
SO_4-S	7.5	7.5	7.5	4.7	6.7	6.2	5.5	6.7	9.7
NH_4-N	5.6	6.7	5.1	5.2	5.9	3.8	3.8	5.4	7.7
NO_3-N	8.4	11.0	3.5	4.0	6.5	7.3	5.0	7.2	9.8
H^+	2.7	3.5	4.0	4.5	4.1	3.6	3.4	3.6	3.1

"episodicities" of the main ions are listed for the different stations of the sampling network.

In contrast to the presentation of Smith and Hunt (1978) where the eastern alpine region was not classified as a highly episodic region it is evident from Table 5 that the wet deposition of acidity is highly episodic at all stations under consideration. For wet deposition SO_4^{2-} and the NO_3^--episodicities are varying at the different stations from a high episodic to the episodic state. In one case (nitrate at the station Nösslach) the deposition appears to be unepisodic. This station is situated in a valley with intensive highway traffic which might be the cause for a different behaviour of the nitrate deposition. The wet deposition network is operated by the environmental departments of the local governments of Tyrol and Salzburg.

REFERENCES

Böhm, R. 1986. Der Sonnblick, Osterr. Bundesverlag Wien.
Cape, J.N., Fowler, D., Kinnaird, J.W., Paterson, I.S., Leith, I.D. and Nicholson, I.A. 1984. Chemical composition of rainfall and wet deposition over northern Britain. Atmos. Environ. **18** , 1921-1932.
Granat, L. 1978. Sulphate in precipitation as observed by the European Atmospheric Chemistry Network. Atmos. Environ. **12** , 413-424.
Ministry for Health and Environment 1984. Richtlinie 11 "Immissions-messung des nassen Niederschlages und des sedimentierten Staubes", Wien.
Smith, F.B. and Hunt, R.D. 1978. Meteorological aspects of the transport of pollution over long distances. Atmos. Environ. **12** , 461-477.
Tsitouridou, R. and Puxbaum, H. 1987. Application of a portable ion chromatogrphy for field site measurements of the ionic composition of fog water and atmospheric aerosols. Int. J. Environ. Anal. Chem. **31** , 11-22.

CONCENTRATION GRADIENTS IN ATMOSPHERIC PRECIPITATION IN AREAS OF HIGH
ANNUAL PRECIPITATION

L. Granat
Department of Meteorology
University of Stockholm
Arrhenius Laboratory
S-106 91 Stockholm
Sweden

ABSTRACT. The chemical composition of atmospheric precipitation was
studied in and around some areas with high annual precipitation. The
concentration in areas of high precipitation was estimated by
extrapolation from areas with less precipitation and compared with
measured values. Two studies in southern Sweden indicated a decrease
in concentration of only about 5% or less compared to extrapolated
values when the precipitation amount increased by about 60%. Two
studies of accumulated snow in the mountaineous north-western part near
to the Atlantic indicated a decrease in deposition with increasing
annual precipitation due to a rapid decrease of concentration in a
westward direction. The concentration in the mountains might be
approximately estimated by an extrapolation from regular measuring
sites - it will at least not be underestimated. The results from the
northern study are considered to be valid only in the mountaineous area
near the Atlantic where the increased amount of precipitation is due to
the influence of the Atlantic with its relatively clean air.

1. INTRODUCTION

The starting point for discussion in this paper is that elevated areas
often receive more precipitation than nearby lower areas. In some
areas, for instance on the Swedish west-coast only a quite modest
elevation of about 150 m at some 50 km from the coast induces
substantially increased precipitation while the other area of high
precipitation is found in north-western Sweden along the border to
Norway and near the Atlantic with altitudes ranging from 1000 to 2000 m
and an annual precipitation of 1 to 2 m. Precipitation chemistry is
not usually measured in these areas of high precipitation and an
estimate of wet deposition must then be based either on an
extrapolation from one side or an interpolation between regular
measuring sites on both sides of such an area. It is therefore of
interest to examine available chemistry data with a better spatial

431

M. H. Unsworth and D. Fowler (eds.), Acid Deposition at High Elevation Sites, 431–441.
© *1988 by Kluwer Academic Publishers.*

resolution and try to find some guidelines for extrapolation/
interpolation from regular precipitation chemistry measuring sites to
areas with high precipitation. Below we will examine two studies each
of 12 months duration in southern Sweden and two snow surveys in
northern and north-western Sweden.

The results from the southern and north-western part of Sweden
show as we shall see below different relations between gradients of
precipitation and concentration and it was therefore interesting to
include the studies in southern Sweden although they do not by
themselves qualify for the theme of this conference with respect to
high elevation.

2. COLLECTION AND ANALYSIS OF PRECIPITATION

The bulk collector for use in summer is made of an about 200 mm
diameter funnel connected to a 5-1 bottle. A smaller funnel with a
long sprout is inserted in the bottom of the wider funnel and gives a
good evaporation shield. A small net is also inserted. All material
is of polyethylene. The bottle and part of the funnel are wrapped in
aluminium foil (to keep collected water dark and prevent growth of
algae). In winter a collector is used which can accommodate a full
month of precipitation as snow and prevent evaporation of melted
water. The shape has varied somewhat during the years but consisted
essentially of a long polyethylene bag fixed to an about 200 mm
diameter plastic ring (to define the collection area) mounted about 2 m
above the ground. A restriction at the bottom of the bag acts as an
evaporation shield. In the presently used winter collector the bottom
of the bag is connected to a funnel and bottle of the same kind as used
in summer. The different constructions are considered to give the same
sampling results but the later version is more easy to handle. The
wet-only collector is cylindrical in shape with a lid that swings away
to the side during precipitation. It is used both in winter and
summer. Collected snow is slowly melting by thermostated heat. All
collectors are designed and made at the Department of Meteorology,
University of Stockholm.

The collectors are preferably located in small openings in
forested regions well away from dust producing areas such as roads and
farming fields. The location of collectors in forested areas is
thought to minimize the influence of wind on sampling efficiency,
particularly during winter and may also keep the dry deposition to the
collector low. In the temporary studies in southern Sweden (A and B in
Figures 4 and 5) one and usually three collectors, respectively were
used at each site. In the presently running regular network the bulk
collectors are used in pairs with a distance usually of a few hundred
metres as an efficient countermeasure of occasional very local
influence. Eighteen wet only collectors are used in the present
network co-located with bulk collectors at five sites. No wet-only
collectors were used in the previous studies.

Samples are analysed for nine major components. The analysis of
sulphate and nitrate changed at the end of 1984. Prior to that

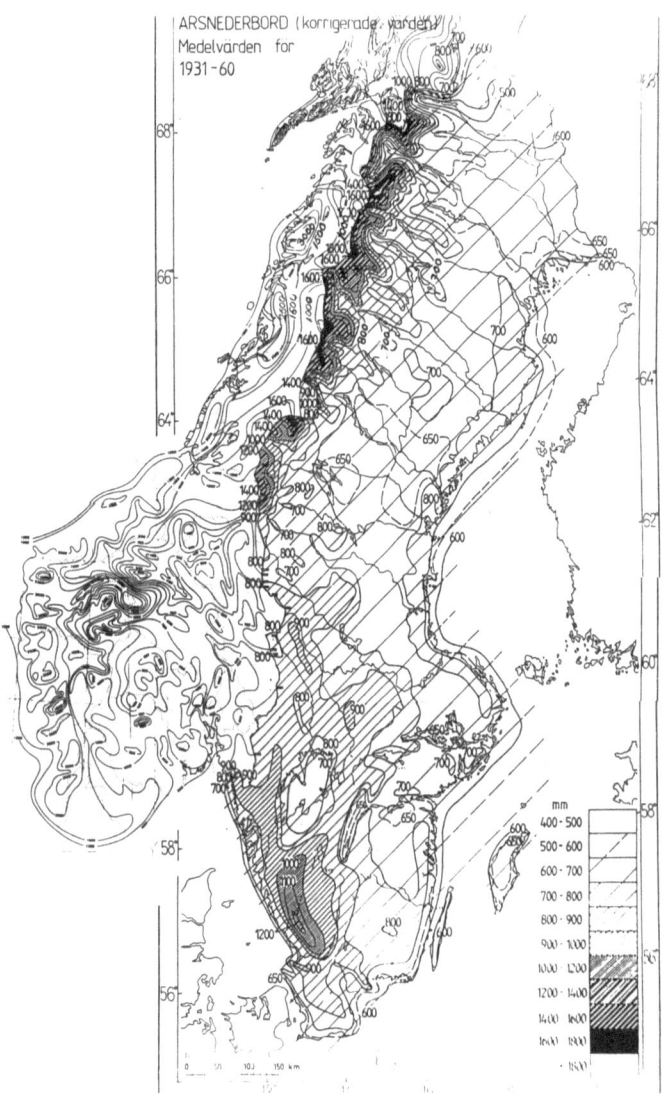

Figure 1.   Yearly amount of precipitation. Data obtained from the
             Norwegian and the Swedish Meteorological Networks and are
             so called "reference values" based on measurements during
             the period 1931-1960. Det Norske Meteorologiske Institut
             (B.Aune) and Sveriges Meteorologiska och Hydrologiska
             Institut, (Bertil Eriksson).

sulphate was analysed colorimetrically by precipitation of barium
sulphate with thorin as indicator and nitrite colorimetrically as an
azo dye after reduction to nitrate. After 1984 both components were
analysed by ion cromatography. During an overlapping period the
colorimetric and IC methods were compared and the agreement was found
to be very good. Amount of precipitation is obtained both from our
collectors (by weighing samples) and from the meteorological networks
in Sweden and Norway from which we here use the 'reference values'
which were collected during the period 1931 to 1960. The Swedish
values have been corrected for a number of sampling errors including
wind errors. In the following evaluation no distinction has been made
between precipitation data obtained from our collectors and the
reference values. Comparison from recent years data has shown an
agreement better than 5% for a yearly average of all stations.

3. RESULTS

Data from obviously contaminated samples and other outliers have been
removed. Excess sulphate (non-seasalt sulphate) was calculated as

$$(Ex\ SO_4) = 0.120\ (Na)$$

where values within parenthesis are concentrations in μ equivalents/l.
Weighted mean values are calculated for each site that was in operation
during the two studies in southern Sweden and for the three last years
of network operation. In Figure 1 is given the yearly precipitation
(reference values) based on both Swedish and Norwegian summaries. The
regular precipitation chemistry network provides information from a
number of points covering most of Sweden up to an altitude of about
600 m. No regular sites are thus operated above the timberline (which
ranges from 600 - 800 m). Concentration of excess sulphate and nitrate
- being the most important contributors to acid precipitation - are
shown in Figures 2 and 3 which includes both the volume weighted mean
values at each site and concentration fields evaluated subjectively
based on an objective analysis. During this evaluation conditions in
neighbouring countries have been taken into account but not variations
in altitude or amount of precipitation. Results with regard to excess
sulphate from the studies with better spatial resolution are given in
Figures 4 and 5.
    Nitrate concentration from snow analysis is shown as a function of
distance between the Atlantic and the Baltic. Amount of precipitation
is estimated from data in Figure 1.

4. DISCUSSION AND CONCLUSIONS

In Case A (Figure 4) amount of precipitation increases about 60% from
the inland towards the maximum near the coast. The corresponding
decrease in excess sulphate concentration is less than a few percent.
The concentration in the area of maximum precipitation can therefore be

quite accurately estimated from a linear extrapolation from inland
sites - if no measurements were available in the former area. In Case
B (Figure 5) precipitation also increases about 60% above inland values
while concentration decreases less than a few percent.

Similar results were also found for other major components such as
hydrogen ion, nitrate and ammonium, although the picture is not quite
as clear as for excess sulphate due to a more pronounced east-west
gradient especially for the latter two elements.

The two studies in the south and south-west of Sweden thus
indicate that the concentration of the major acidifying and
neutralizing components in an area with high yearly amount of
precipitation can be estimated by extrapolation from nearby areas which
receive less precipitation. From this also follows that a large-scale
concentration field can be evaluated directly from the measurements
without consideration of differences in yearly amount of precipitation
between measuring sites.

Conditions in the mountainous north-western part of Sweden are
less investigated both due to the experimental difficulties and small
incentive as the area is considered to be comparatively unpolluted.
The only information on precipitation chemistry in this region is
obtained from snow surveys by Wright and Dovland (1978), and our own
measurements (i.e. Ross and Granat, 1986). The main focus of our snow
surveys have been to map concentration gradients from the coast of the
Baltic to the inland and in the very north of Sweden and Finland but
there are also a few transects across the mountains in a west-east
direction at the Norwegian border near the Atlantic which is of special
interest in the present context (Figure 6). Snow samples taken in
duplicate below the timberline usually agree very well - almost within
the precision of the chemical analysis (see for instance Ross and
Granat, 1986). Concentration in snow samples taken at regular sites
agree within 10-20% of results obtained with collectors for the same
period. The scatter shown in Figure 6 can to a large degree be related
to the separation between sampling points - in some cases up to 100 km
in north-south direction. However, above the timber line the results
start to scatter even between nearby samples obviously depending on the
very uneven distribution of the snow. The results are therefore less
certain than those obtained below the timberline. Amount of
precipitation could not at all be estimated from the snow samples since
the snow depth is so variable. Instead, the meteorological reference
value has been used as an indication of the relative spatial change in
amount of precipitation during the period when the snow had
accumulated. We have chosen to present nitrate rather than sulphate
data because calculation of excess sulphate may be uncertain near the
coast. In Figure 6 data from three different snow surveys have been
plotted although only two of them cover high elevation sites. Data
from the regular network have also been included as a general
reference. It should here be observed that samples taken near the
coast of the Baltic represent shorter accumulation periods and may
therefore not be directly comparable to samples further to the west and
also that the data from the network are for the whole year. Near the
border to Norway precipitation amount increases from some 700-800

436

mm/year to more than 2 m with large variations due to topography. In the same area the concentration continues to decrease such that even deposition decreases despite the rapid increase in amount of precipitation although the uncertainty in the data does not permit definite quantitative estimates. The reason for the rapid decreases in concentration near the coast is evidently due to the fact that this area receives precipitation from the Atlantic from air parcels which

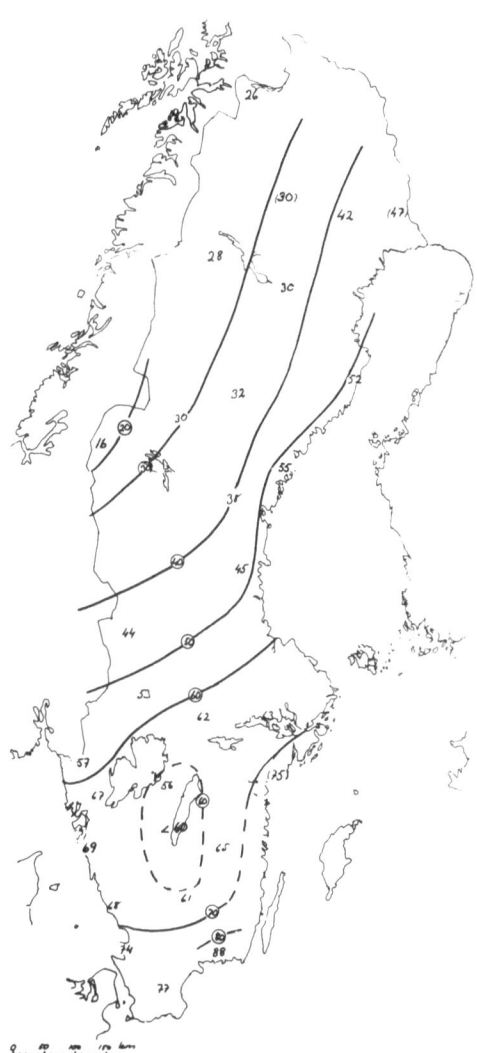

Figure 2. Weighted mean concentration of excess sulphate (in µeq/1) based on measurements in the MISU/PMK network 1983-1985.

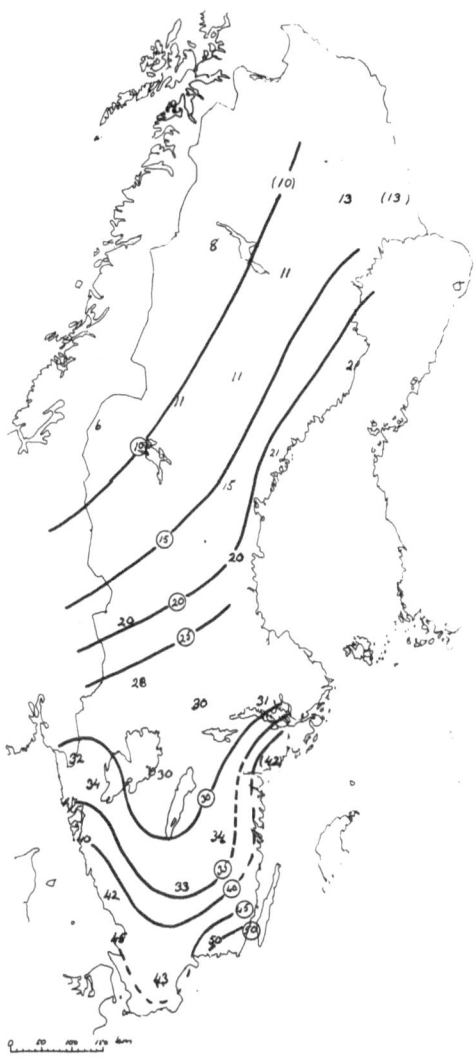

Figure 3. Weighted mean concentration of nitrate (in µeq/1) based on measurements in the MISU/PMK network 1983-1985.

438

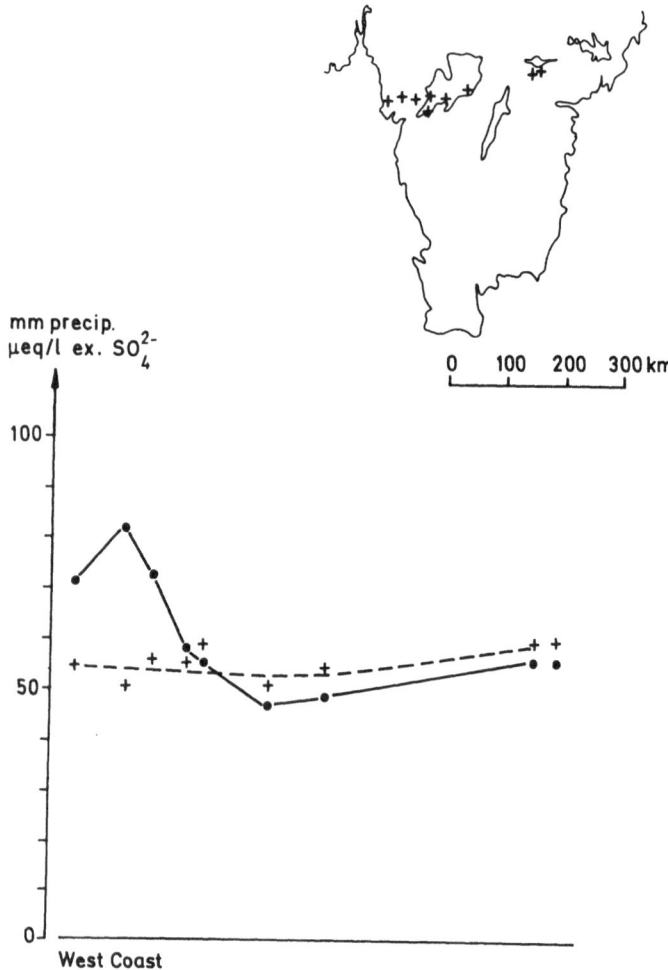

Figure 4. Concentration of excess sulphate ("+"; µeq/l) and monthly amount of precipitation ("."; mm) along a transect across southern Sweden. In the text referred to as Case A.

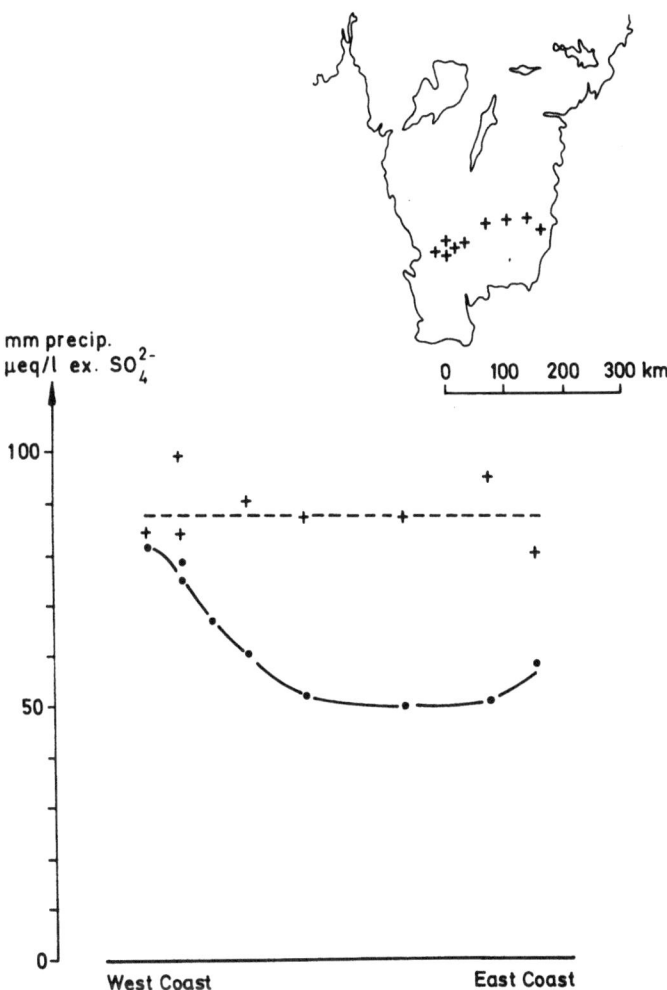

Figure 5. Concentration of excess sulphate ("+"; µeq/1) and monthly
amount of precipitation ("."; mm) along a transect in
south-west Sweden. In the text referred to as Case B.

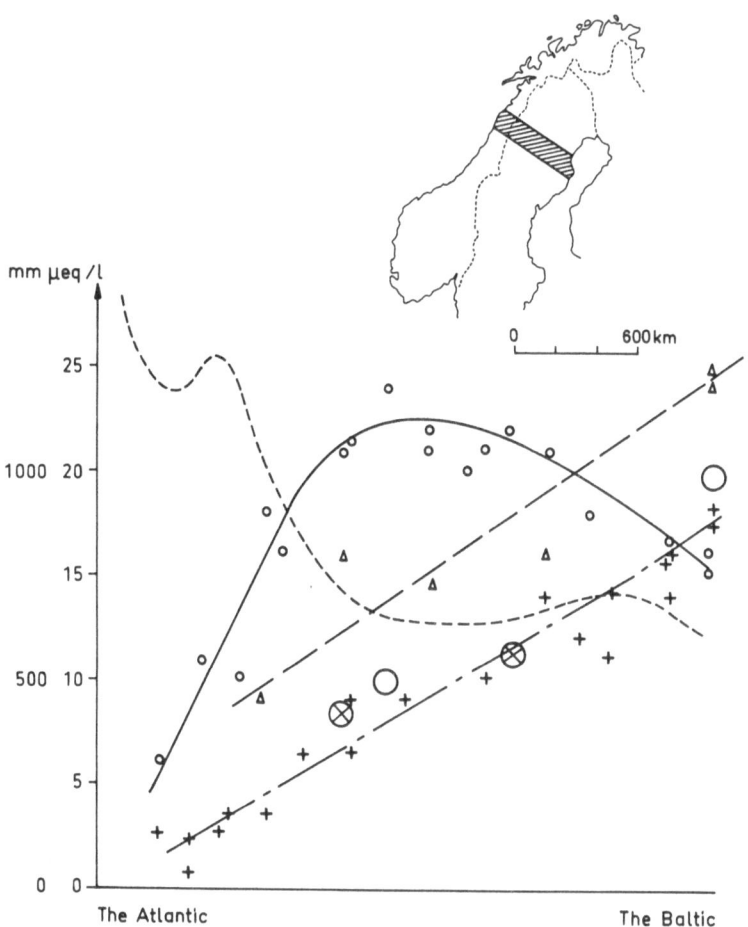

Figure 6. Gradients in northern Sweden of NO_3^- concentration in
accumulated snow (+: March 1981, o: March 1982, △: March
1984) and from the precipitation chemistry network as mean
weighted concentration for the period 1983-1985 (⊗: regular
sites, ○: estimated by interpolation) together with yearly
amount of precipitation (---: reference values in mm/a).

are relatively free of pollutants. It has been estimated (personal
communication, Christer Persson and Sven Kindell, Swedish
Meteorological and Hydrological Institute) that the mountain ridge
between Sweden and Norway, receives more than 90% of the winter
precipitation from westerly winds while this value has decreased to

about 40% near the coast of the Baltic. In southern Sweden, on the other hand, both easterly and westerly winds bring in both precipitable water and pollutants.

The results presented here indicate that concentrations can be estimated in areas of high precipitation by interpolation from nearby sites which receive less precipitation with an error of about 5% or less. For the deposition estimates accurate data on precipitation amount with good spatial resolution are required. The procedure might be less accurate in the mountainous area in NW Sweden near the coast and might give an overestimate especially when it is based on extrapolation from measurements in Sweden alone.

The information available from Sweden does thus - so far - not indicate that high elevation sites (above the altitude of the highest regular sampling sites) receive wet deposition of acidic and other pollutants in larger amounts than has been estimated from the regular network. It remains, however, to be seen if dry deposition or deposition of cloud or fog may alter this picture.

ACKNOWLEDGEMENT

Financial support for the precipitation chemistry investigations was obtained from the National Swedish Environment Protection Board, the National Swedish Enviornmental Programme (PMK) and the University of Stockholm.

REFERENCES

Ross, H.B. and Granat, L. 1986. Deposition of atmospheric trace metals in northern Sweden as measured in the snow pack. Tellus 1986, 38B, 27-43.
Wright, R.F. and Dovland, H. 1978. Regional surveys of the chemistry of the snow pack in Norway late winter 1973, 1974, 1975 and 1976. Atmos. Environ. 12 , 1755-1768.

FOG CHEMISTRY AND DEPOSITION IN THE PO VALLEY, ITALY

S. Fuzzi
Istituto FISBAT - C.N.R.
Via de'Castagnoli 1,
40126 Bologna
Italy

ABSTRACT. In the Po Valley, the combined effect of high fog frequency (up to 30% of the time during the fall-winter months) and the high ionic concentration of the solution droplets (ionic strength up to 20 meq 1^{-1}) result in wet deposition rates for chemical substances which have an important impact on the environment. An evaluation of this phenomenon is presented within the context of an international programme aimed at the study of the multiphase physico chemical processes taking place in the radiation fog.

1. INTRODUCTION

Since 1982, an experimental fog chemistry programme has been established at the FISBAT field station of S. Pietro Capofiume in the Po Valley, Italy. Very low pH values, down to 2.2, were in fact measured in samples of fog liquid water collected during intense radiation fog episodes (Fuzzi et al., 1983). The particular interest in fog as regards this area of Italy is due to the unusually high local occurrence, which can reach up to 30% of the time in the fall-winter season, from October through March (ITAV, 1978). The reported early findings, and the growing scientific interest in the acid deposition issue, encouraged us to undertake a wider research project, aimed at the evaluation of the possible effects of the fog chemical composition on the natural environment and at investigating the mechanisms responsible for the high acid content of the Po Valley fogs. It was clear that such an enterprise was not an easy one, because of the wide range of competencies involved: analytical chemistry, thermodynamics, fog microphysics, meteorology, aerosol sampling technology and the need for instrumental capabilities not all present at our institute. A collaboration has therefore been set up with several European and American institutes interested in a field study on radiation fog chemistry in the Po Valley.

The general idea behind the whole project is related to the particular meteorological and microphysical nature of radiation fog, which is formed in high pressure and clear sky conditions by the steady

443

M. H. Unsworth and D. Fowler (eds.), Acid Deposition at High Elevation Sites, 443–452.
© 1988 by Kluwer Academic Publishers.

cooling of air close to the ground. These conditions of high atmospheric stability imply a minimal air mass exchange, so that it is reasonable, in this case, to apply a mass balance criteria between gas, liquid and particulate phases. In this way, it is possible to use radiation fog (a cloud in contact with the ground) as a natural laboratory for the study of the physico chemical transformations in the atmospheric dispersed liquid phase (droplet phase). The relatively high pollution conditions in the Po Valley, the largest agricultural as well as industrial area in Italy, are, paradoxically, an advantage for a study on fog chemistry.

The groups involved in the Po Valley fog project are from the following institutes: Atlanta University (AU), Atmospheric Sciences Research Center (ASRC), Boris Kidric Institute (KIBK), Centro Italiano Studi Esperienze (CISE), ENEL Centro Ricerche Termiche Nucleari (CRTN), Georgia Institute of Technology (GIT), Lawrence Berkeley Laboratory (LBL), National Center for Atmospheric Research (NCAR), Presidio Multizonale USL 28 (PMP), Technical University of Vienna (TUV), University of Frankfurt (UF), University of Padova (UP), University of Vienna (UV).

2. THE FISBAT FIELD SITE

The FISBAT field site (SPC in Fig. 1) is located near the village of S. Pietro Capofiume, in the central part of the eastern Po Valley, at 10 m a.s.l. with the typical climatic conditions of a continental region. The surrounding area is part of a large agricultural region. The high fog frequency is mainly due to the orography of the valley, closed on three sides by high mountain ranges. Under high pressure

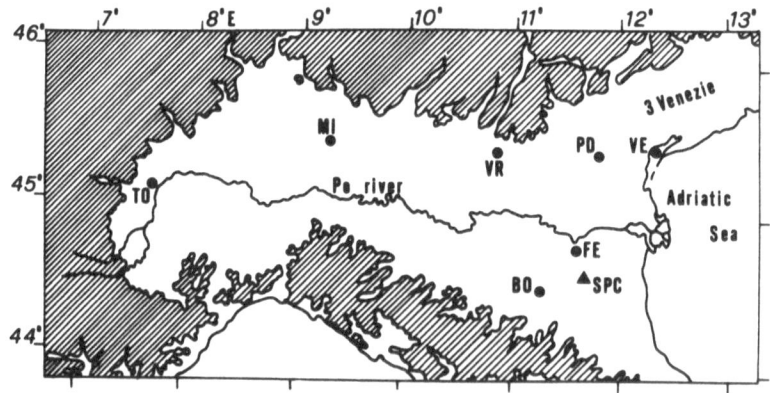

Figure 1.   Map of the Po Valley. The shaded area refers to elevations greater than 500 m (Alps to the north and west, Appennines to the south). The triangle indicates the location of the FISBAT field station of S. Pietro Capofiume (SPC).

conditions, the establishment of strong temperature inversions is favoured, so that water vapour saturation is often reached in the low levels of the atmosphere (Hesse, 1954). The station is run by the FISBAT Institute in collaboration with the Department of Agriculture of Regione Emilia Romagna. A meteorological station is routinely operated by in situ pesonnel, analysis and forecasting charts and Meteosat II satellite pictures are available at the site. The experimental field is an area of 4 hectares with several power distribution points and boxes for instrumentation sheltering. A 50 metre tower equipped with an elevator is also located in the field. Several different research projects are normally carried out through the year.

3. DESCRIPTION AND FIRST RESULTS OF THE EXPERIMENTAL FOG CHEMISTRY
   PROJECT

Two major campaigns have been carried out up to now: November 84 and November 85. A third campaign is already scheduled for November 86. The part of the experimental field equipped for the fog chemistry study is schematically shown in Figure 2. The group responsible for each

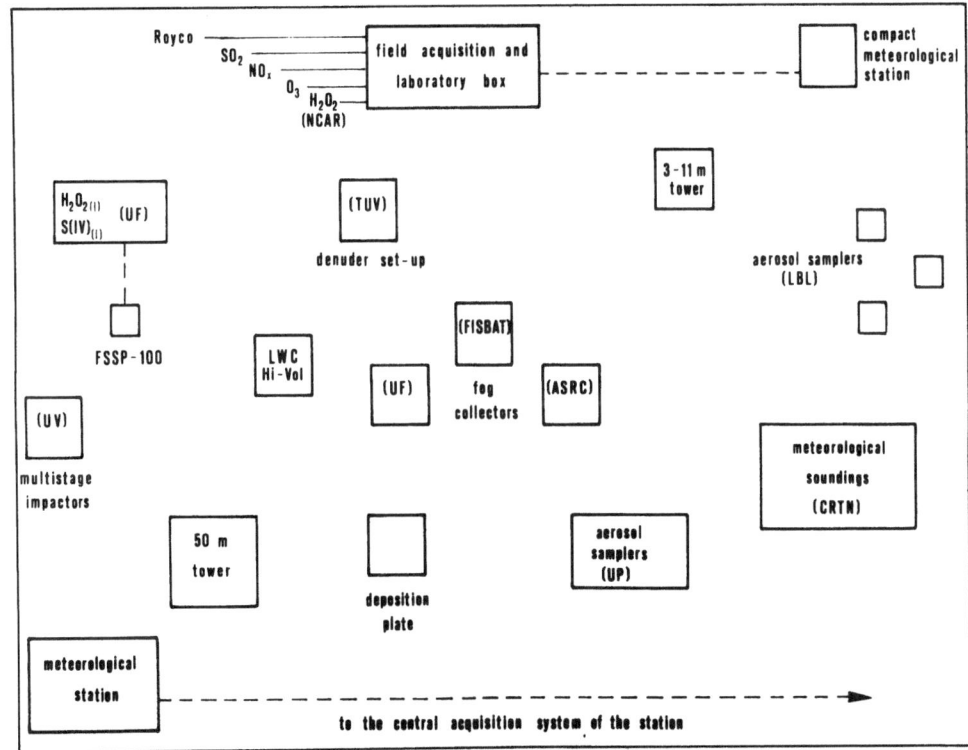

Figure 2.   Schematic diagram of the instrumented field network for the Po Valley fog project. For the identification of the groups see text.

Table 1   Mean concentration and range (in brackets) of the different components in fog water samples collected during two field campaigns: November 84 (87 samples) and November 85 (63 samples).

Components		Range of concentration and mean concentration	
		November 84	November 85
H^+	(μeq 1^{-1})	0.4 - 1148	0.2 - 3388
		(263)	(320)
NH_4^+	"	279 - 4056	519 - 5896
		(1629)	(1582)
K^+	"	6 - 103	14 - 302
		(18)	(47)
Na^+	"	1 - 190	12 - 201
		(34)	(40)
Ca^{2+}	"	11 - 454	12 - 235
		(45)	(59)
Mg^{2+}	"	2 - 98	6 - 82
		(14)	(20)
F^-[a]	"	9 - 135	23 - 467
		(57)	(104)
Cl^-	"	48 - 331	53 - 1404
		(147)	(201)
NO_2^-	"	0 - 104	0 - 26
		(7)	(5)
NO_3^-	"	23 - 3340	172 - 2270
		(995)	(555)
SO_4^{2-}	"	150 - 2066	326 - 3063
		(732)	(908)
HCHO	(μM 1^{-1})	23 - 105	1 - 53
		(64)	(13)
S(IV)	"	n.m.	77 - 764
		-	(310)
H_2O_2	"	< 0.1	0.1 - 2
		-	(1.2)
Fe	"	1 - 18	1 - 27
		(6)	(8)
Mn	"	0 - 4	0.4 - 4
		(1.1)	(1.6)

n.m. = not measured
[a]some weak organic acids co-elute with F^-

measurement is also indicated. A detailed description of all experimental operations has been reported elsewhere (Fuzzi, 1986a).

As an initial indication, a summary of the fog water chemical composition for the two periods is reported in Table 1. As can be seen, approximately 90% of the total ionic strength of the fog water solutions is accounted for by four major ions: H^+, NH_4^+, NO_3^-, $SO_4^=$. NO_3^- and $SO_4^=$ are the principal acidifying anions, whereas NH_4^+ is the main neutralizer of the atmospheric acidity. The equivalent ratio $NH_4^+ + H^+/NO_3^- + SO_4^=$, always very close to one (Figure 3), indicates that fog water acidity principally derives from the fraction of HNO_3 and H_2SO_4 not neutralized by ammonia. Particular care was therefore devoted to the investigation of the $SO_2/SO_4^=$ and $NO_x/HNO_3/NO_3^-$ systems.

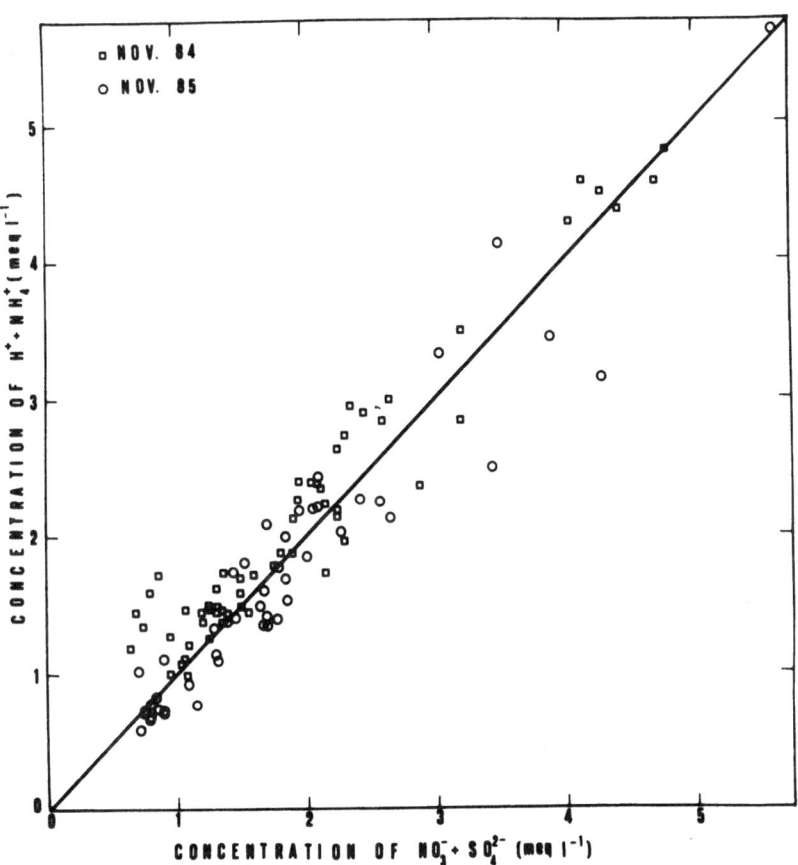

Figure 3.   Plot of $NH_4^+ + H^+$ equivalent concentration versus $NO_3^- + SO_4^=$. The solid line indicates 1:1 ratio.

Some initial evidence was obtained from the November 84 data (Fuzzi et al., 1986). The main points can be summarised as:
a) a major contribution by the atmospheric particulate (nucleation scavenging) has been found for the sulphate accumulation in fog water, as theoretically predicted by Stelson et al. (1985);

b) excess of nitrate is frequently present in the liquid phase, with
respect to clear air aerosol.

The former point suggests a lack of oxidants in the winter fogs:
ozone concentration is always close to the detection limit of the
instrumentation (ca. 5 ppbv); low gas phase H_2O_2 concentrations were
always detected in pre-fog periods (maximum of 150 pptv, with an
average of less than 50 pptv) (Lind and Lazrus, 1986). The only
efficient mechanism of SO_2 oxidation seems to be the transition metal
catalysed reaction. At the same time, the second observation indicates
the importance of the $NO_x/HNO_3/NO_3^-$ in the Po Valley fogs. The excess
nitrate detected in the liquid phase could either be due to HNO_3
incorporation from the gas phase at the early stage of fog formation,
or to heterogeneous transformation processes taking place in the
dispersed system (Heikes and Thompson, 1983). The already processed
data do not allow us to discriminate between these two processes and we
are looking for new evidence from the November 85 results and from our
next campaign.

A particular point which has been stressed in several papers
(Fuzzi et al., 1983; 1984; Fuzzi, 1986b) is the importance of the
connection between the chemical processes occurring in fog and
microphysical variations. One of the major difficulties is
represented by the different time scale of microphysical variations
(order of one minute or less) and the resolution of chemical samplings,
both for particulate aerosol (few hours) and fog droplet (10 to 60
minutes). Furthermore, the lack of a reliable method for accurate
measurements of the fog liquid water content, which is the
macroscopical parameter reflecting the droplet size distribution,
represents a major uncertainty in the evaluation of the chemical
results.

For the next field experiment (November 86), the instrumentation
set-up and the sampling and analytical procedures are being completely
redesigned in the light of the already acquired evidence.

4. FOG DEPOSITION

The high fog occurrence in the Po Valley and the high ionic loading of
the fog water solutions could, in principle, produce a relevant
deposition of chemical substances. Preliminary results from
measurements of cloud and fog hydrological input to the environment
made in the USA (Lovett et al., 1982; Lovett, 1984) and the UK (Dollard
et al., 1983), testify to the importance of this particular kind of
deposition, which has been defined as "occult precipitation", insofar
as it is not recorded by standard rain gauges and, therefore, not taken
into account in wet deposition budgets.

Fog deposition takes place either in form of gravitational
settling or turbulent impaction, depending on the droplet size and wind
speed. The principal characteristic of radiation fog is, as stated
before, the extremely low wind speed involved, usually less than 1 m
s^{-1}. Given the characteristic size of the fog droplets in our
conditions (mean volume radius in the range of 7-15 μm (Fuzzi et al.,

Table 2   Comparison between calculated deposition due to fog and measured bulk precipitation deposition both on an annual basis (1984) and for the period January – March and October – December (fog season in the Po Valley). Total water deposition is also reported.

Ion	Bulk precipitation yearly deposition (P_y) (Kg Km^{-2})	Bulk precipitation deposition Jan–Mar/Oct–Dec (P_s) (Kg Km^{-2})	Fog deposition (Kg Km^{-2})	F/P_s (%)	F/P_y (%)
H$^+$	—	—	0.7	—	—
NH$_4^+$	2795	641	84.9	13.2	3.0
NO$_3^-$	5559	1375	166.2	12.1	3.0
SO$_4^=$	5387	2403	126.3	5.3	2.3
Water deposition (mm)	662	305	4.2	1.4	0.6

1980)), it is expected that most of the water deposition to the ground is due to sedimentation, with impaction playing only a minor role (Lovett, 1984).

Some measurements of fog deposition were thus performed during the field experiments by using a polyethylene flat funnel device (Fuzzi et al., 1985). The funnel was exposed at the onset of dense fog and taken in immediately after dissipation. An average fog water deposition rate of 7.98 g m^{-2} h^{-1} was measured for five individual fog episodes. The calculated mean deposition velocity (deposition rate/LWC) is consistent with terminal fall velocity of 12 µm droplets, typical of a mature fog. The average deposition rate for the major ions was then deduced from the amount and chemical composition of the bulk deposited water. It is not possible to compare the chemical composition of airborne fog droplets with that of deposited water, since the latter is collected on a several hours time basis. Calculated deposition rates for major ions resulted (in µeq m^{-2} h^{-1}): H$^+$ = 1.38, NH$_4^+$ = 9.41, NO$_3^-$ = 5.36, SO$_4^=$ = 5.26.

The effectiveness of fog in removing chemical substances from the air is indicated by the high percentage of atmospheric particulate deposited compared to the atmospheric loading before the onset of fog: values ranging from 24 to 78% by mass were measured.

From our measurements and from statistical data available from the National Weather Service (ITAV, 1978), an average of 500 hours of dense fog in the period October–March was calculated; fog occurrence during the rest of the year is negligible. It was therefore possible to evaluate the annual deposition due to fog by multiplying the average deposition rates reported above, times the total fog hours. Comparison between the deposition due to fog and that due to precipitation collected at the same station (Gruppo di Studio sulle Caratteristiche Chimiche delle Precipitazioni Atmosferiche del Nord Italia, 1985) are reported in Table 2. H$^+$ deposition due to precipitation is not reported in the table because neutralisation of acidity is likely to occur in bulk samples, mainly due to deposition of mineral dust in the open collector during the "dry" periods. It can be seen that, although quite small on a yearly basis, the contribution of fog chemical deposition during the fall–winter season is by no means negligible.

It has also to be considered that the reported values are representative for a flat surface, whereas a much higher amount of fog water can be captured by vegetation (Wattle et al., 1984). Moreover, since the environmental effects of acid deposition are more closely related to H$^+$ concentration that to the total H$^+$ deposited (Evans et al., 1984), it is also more appropriate to consider these fog events as individual intense pollution episodes, instead of averaging the deposition values in terms of long term budgets. Wisniewski (1982) also noted that although only a small amount of moisture is usually associated with fogs, this amount is sufficient to entirely cover the surfaces exposed to the atmosphere, without any appreciable run-off. Thus, the contact time with the fog droplet solution is also much longer than in the case of precipitation. In these conditions, previously dry deposited substances are also dissolved, therefore increasing the potential effects of fog droplet deposition.

5. CONCLUSIONS

Radiation fog, a common phenomenon in the Po Valley, has been proven to be a convenient natural laboratory for the study of multiphase physico chemical processes responsible for for water chemical composition. The international project described in this paper has already provided some initial results and new experiments are scheduled for the near future. The role of radiation fog as an efficient agent for deposition of chemical substances has also been stressed.

ACKNOWLEDGEMENTS

All groups and individuals who participated in the cooperative fog field programme are gratefully acknowledged for their efforts. Particular thanks are due to P. Mandrioli, manager of the FISBAT field station and to my co-workers M.C. Facchini, G. Nardini and G. Orsi. The research programme is sponsored by "Progetto Finalizzato Energetica 2, Sottoprogetto Ambiente e Salute" and "Progetto Strategico Clima ed Ambiente dell'Area Mediterranea", both of the Italian National Research Council.

REFERENCES

Dollard, G.J., Unsworth, M.H. and Harvey, M.J. 1983. Pollutant transfer in upland regions by occult precipitations. Nature 302, 241-243.

Evans, L.S., Raynor, G.S. and Jones, D.M.A. 1984. Frequency distributions for durations and volumes of rainfalls in the eastern United States in relation to acidic precipitation. Water Air Soil Pollut. 23, 187-195.

Fuzzi, S., Gazzi, M., Pesci, C. and Vicentini, V. 1980. A linear impactor for fog droplet sampling. Atmos. Environ. 14, 797-801.

Fuzzi, S., Orsi, G. and Mariotti, M. 1983. Radiation fog liquid water acidity at a field station in the Po Valley. J. Aerosol Sci. 14, 135-138.

Fuzzi, S., Castillo, R.A., Jiusto, J.E. and Lala, G.G. 1984. Chemical composition of radiation fog water at Albany, New York and its relationship to fog microphysics. J. Geophys. Res. 89, 7159-7164.

Fuzzi, S., Orsi, G. and Mariotti, M. 1985. Wet deposition due to fog in the Po Valley, Italy. J. Atmos. Chem. 3, 289-296.

Fuzzi, S. (ed.) 1986a. Heterogeneous Atmospheric Chemistry Project. FISBAT Report, Bologna, Italy.

Fuzzi, S. 1986b. Fog chemistry and microphysics. In: Chemistry of Multiphase Atmospheric Systems (W. Jaeschke ed.), pp. 213-226, Reidel, Dordrecht.

Fuzzi, S., Orsi, G., Nardini, G., McLaren, S., McLaren, E. and Mariotti, M. 1986. Heterogeneous processes in the Po Valley (Italy) radiation fog. Submitted to J. Geophys. Res..

452

Gruppo di Studio sulle Caratteristiche Chimiche delle Precipitazioni Atmosferiche del Nord Italia 1985. Deposizioni atmosferiche sul Nord Italia. Acqua e Aria **8**, 721–735.

Heikes, B.G. and Thompson, A.M. 1983. Effects of heterogeneous processes on NO_3, HONO and HNO_3 chemistry in the troposphere. J. Geophys. Res. **88**, 10883–10895.

Hesse, W. 1954. Nebel und Sichtverhältnisse in Ober- und Mittelitalien. Pure Appl. Geophys. **27**, 179–189.

ITAV 1978. Visibilità orizzontale inferiore a 1000 m sulle pianure dell'Italia settentrionale. Aeronautica Militare, Servizio Meteorologico, Report CDU 551.582.3, Rome, Italy.

Lind, J. and Lazrus, A. 1986. Measurements of hydrogen perioxide vapour at S. Pietro Capofiume. In: Heterogeneous Atmospheric Chemistry Project (S. Fuzzi, ed.), FISBAT Report, pp. 27–31, Bologna, Italy.

Lovett, G.M., Reiners, W.A. and Olson, R.K. 1982. Cloud droplet deposition in subalpine balsam fir forest: hydrological and chemical inputs. Science **218**, 1303–1304.

Lovett, G.M. 1984. Rates and mechanisms of cloud water deposition to a balsam fir forest. Atmos. Environ. **18**, 361–371.

Stelson, A.W., Kiang, C.S. and Fuzzi, S. 1985. Relative times for sulphate accumulation in fog. Proceedings, Heterogeneous Processes in Source-Dominated Atmospheres, October 8–11, New York.

Wattle, B.J., Mack, E.J., Piliè, R.J. and Hanley, J.T. 1984. The role of vegetation in the low-level water budget in fog. Arvin Calspan Advanced Technology Center, Report DAAG29–82–0022, Buffalo, New York.

Wisniewski, J. 1982. The potential acidity associated with dews, frosts and fogs. Water Air Soil Pollut. **17**, 361–377.

PARTICLES IN OROGRAPHIC CLOUD AND THE IMPLICATIONS OF THEIR TRANSFER TO PLANT SURFACES

A. Crossley
Institute of Terrestrial Ecology
Bush Estate
Penicuik
Midlothian EH26 OQB
UK

ABSTRACT. Continuous bulk collections of cloud and rainwater have been made at Castlelaw Hill (480 m), S.E. Scotland since April 1986. The cloud gauges were passive 'harp wire' devices strung with either nylon or Ethyltetrafluoroethylene (ETFE) and set out open or protected from rainfall to a limited extent by polypropylene lids. Automatic weather station data was also taken at the site, together with detector activity on the frequency, intensity and duration of cloud and rain events. Particulate contents in cloud and rainwater were determined gravimetrically after filtration through 0.2 μm pore size PTFE membranes. The cloudwater frequently contained higher concentrations of particulate matter (up to 50 mg 1^{-1}) than the rainwater although this may have been enhanced by dry deposition onto the collecting filaments, especially the ETFE. The particles included biological debris (pollen, spores, plant fragments), mineral fragments from wind blown soil and dust, but were dominated by pulverised fuel ash (less than 10 μm diameter) and carbon residues.
 Scanning electron microscope examinations of Sitka spruce foliage at high altitude (600 m and subject to cloud deposition) and low altitude (300 m no cloud) sites revealed a faster 'weathering' of the needle epicuticular wax at the upper site together with greater accumulation of particles.

1. INTRODUCTION

The recent observations regarding forest decline in Europe and North America have indicated that there is an increase in damage level with increase in altitude. (Johnson and Siccama, 1983, 1984; Schuut and Cowling, 1985). This has stimulated research into possible causes which include direct interception of cloudwater by the high elevation vegetation. Studies on cloud and rainwater chemistry in Germany (Georgii and Schmitt, 1985; Grunow, 1955), Britain (Dollard et al., 1983; Gervat, 1985; Fowler et al., 1986), North America (Falconer and Falconer, 1980; Waldman et al., 1982; Daum et al., 1984), and Japan

453

M. H. Unsworth and D. Fowler (eds.), Acid Deposition at High Elevation Sites, 453–464.
© 1988 by Kluwer Academic Publishers.

(Okita, 1968) reveal that the major ion concentrations in cloudwater can be many times that found in rainwater.

These concentrations may be high enough to cause direct foliar injury or may be enhanced by evaporation processes (Unsworth, 1984; Milne, Crossley and Unsworth, this publication) to produce highly concentrated and potentially damaging ionic solutions on the leaf surface.

During the course of such a study our observations have indicated that as well as the high ion concentration in cloudwater there were also substantial levels of particulate matter.

Although this material contained biological debris (pollen, spores and plant fragments) and mineral fragments, it was dominated by pulverised fuel ash (PFA) residues which are typically produced by large coal- and oil-fired power stations.

These particles were generally less than 5 μm in diameter and as such could have been transported over hundreds of kilometers in the atmosphere (Mamane, 1986). Dry particles of this size would only be inefficiently deposited onto conifer foliage (Chamberlain, 1975; Chamberlain and Little, 1981).

Studies on airborne particles of power plant (Mamane, 1986; Eltgroth and Hobbs, 1979) or urban and natural origin (Orsini et al., 1986) have not addressed the implications of their transfer to plant surfaces. The limited publications on the topic deal with dry deposition of fly ash (e.g. Mishra, 1986), or with the chemical changes in leaf tissue after removal of surface deposits (Murray, 1984).

The preliminary study reported here was initiated to examine the possible role of cloudwater in the transfer of particles to plant surfaces and to evaluate the effects that this might have on the surface integrity.

2. METHODS

Continuous collections of cloud and rainwater bulk samples have been made at Castlelaw Hill (480 m) in the Pentlands, S. Scotland. The cloudwater collector was a development from that used by Dollard et al. (1983) resulting in a passive harp wire device strung with either Teflon fluorocarbon (ETFE) coated copper wire (commercially available as wire wrap, Trade Mark Tefzel) diameter 0.55 mm or monofilament nylon, diameter 0.65 mm. The accumulation of impacted cloud droplets running off the filaments was collected by a glass funnel and directed into a polypropylene bottle. To reduce rainfall contamination of the cloudwater sample a polypropylene faced lid 1200 mm in diameter was supported over the cloud gauge. This excluded raindrops larger than 0.5 mm in diameter at windspeeds up to approximately 5 m sec^{-1}.

Rainwater collections were made using a 200 mm diameter glass funnel set 2 m above the ground and the liquid was again collected in a polypropylene bottle.

Identical collectors were coupled to tipping bucket rain gauges with their outputs recorded on a datalogger to monitor the frequency, duration and intensity of cloud and rain deposition.

Bulk collections were taken every 2–3 days and the type of collection was assessed by comparison with the data logger data on meteorological parameters and detector activities. This enables the samples to be categorised in to cloud only, rain only, cloud and rain (with windspeed above 5 m sec^{-1}), cloud and rain (with windspeed less than 5 m sec^{-1}) and dry deposition. Dry deposition samples were taken by spraying the gauges with de-ionised water and collecting a fixed volume of washing run off.

The pH of the samples was determined immediately after collection before storage at 4°C in a dark cold room prior to chemical and particle analysis.

The particulate loadings were determined gravimetrically by filtering 100 mls of sample (when available) through 0.2 μm pore size, pre-weighed, pre-dried, PTFE filter membranes (Gelman Sciences Ltd). After drying over phosphorus pentoxide for 1 week the filters were reweighed. PTFE was chosen as a result of trials which indicated that wetting agents and high affinity for water vapour precluded the use of the more standard cellulose nitrate and nylon filters respectively.

Selected examples of cloudwater were examined under the optical microscope to determine the particle types that were present and assess whether image analysis techniques could be used to categorise and quantify the particles.

To assess the accumulation and possible effects of particles on vegetation, a simple field study was carried out at Dunslair Heights near Peebles, S. Scotland.

At this site there was mixed coniferous forest (dominated by Sitka spruce and larch) planted from a height of 200 m up to the summit at 600 m. At two elevations (300 and 600 m), Sitka spruce trees were selected for foliage sampling. In this initial study only one tree at each site was selected; the tree at the lower site was a naturally regenerated, 12-year-old individual with older, taller trees on its leeward side. The tree from the upper site was 24 years old but, as a result of its poorer growth, was of the same height as the tree at the lower site.

At both elevations, shoots bearing current, first and second year needle classes were harvested monthly between August and November 1985. Individual needles were carefully removed from the shoots and secured, abaxial surface uppermost, to Scanning Electron Microscope (SEM) stubs using double sided adhesive tape. After drying for 1 week in a dessicator containing P_2O_5, the stubs were coated with 25 nm of gold using a sputter coater with a water-cooled target and specimen stage.

The stubs were examined in a Cambridge Stereoscan Mk IIA. Regions of the needles were selected at random and photographed at magnifications of x1000 and x2000, operating in twin channel mode to produce positive and negative recordings. A twin projector system was used to evaluate the surface structure of the needles and the accumulation of particulate matter. The initial comparison was carried out by simultaneous projection of SEM + ve images of upper site and lower site needle surfaces, the two images were compared and ranked visually for better wax structure. The relative abundance of particles was scored for regular spherical particles (flyash) or irregular particles (biological and mineral fragments).

Each comparison was pre-coded so that the examination was carried out 'blind'.

3. RESULTS

The data for the cloud and rainwater detectors combined with the meteorological records over a 200 day assessment at Castlelaw Hill between 13.3.86 and 16.11.86 are summarised in Table 1.

The greatest proportion (88.5%) of the 200 day monitoring period was taken up by dry deposition. Of the wet deposition categories, cloud only, was surprisingly dominant at 63%.

The data logger results recorded that these events typically occurred overnight with some extension at either end into the evening and morning and this would account for the unexpectedly high cloud activity that had not been apparent from daytime observations. The mixed cloud and rain events at high windspeed accounted for 20% of the wet deposition but these data must be treated with caution as the high cloudwater volume due to rain capture at high windspeeds will have

TABLE 1. The duration and volume via five categories of cloud and rainwater depostion at Castlelaw Hill, 1986.

	Hours	% Total Hours	% Wet Dep Hours	Volume (litres)	Deposition ml m^{-2} yr^{-1}
Cloud only	346.2	7.28	63	34.2	3186*
Rain only	86.5	1.82	16	6.1	347
Cloud & Rain Wind < 5 m sec^{-1}	111.5	2.34	20	19.3	1525+
Cloud & Rain Wind < 5 m sec^{-1}	4	0.09	1	0.8	55+
Dry Deposition	4211.5	88.47		0.0	
TOTAL	4760				5113

* Using a cloud detector cross sectional area of 0.051 m^2

+ Combined cloud and rain calculations

distorted the true picture. There were only a few occasions (< 1% of wet deposition) when the site experienced mixed cloud and rain events with windspeeds less than 5 m sec^{-1} resulting in fully protected cloud samples due to the 1.2 m lid. This low figure was not surprising considering the mean windspeed for the Castlelaw site over the period was in excess of 10 m sec^{-1}.

Rain only collections accounted for only 15% of the wet deposition yet this figure when converted to mm yr^{-1} (837) agrees well with previously estimated rainfall for the area (Barrett et al., 1983).

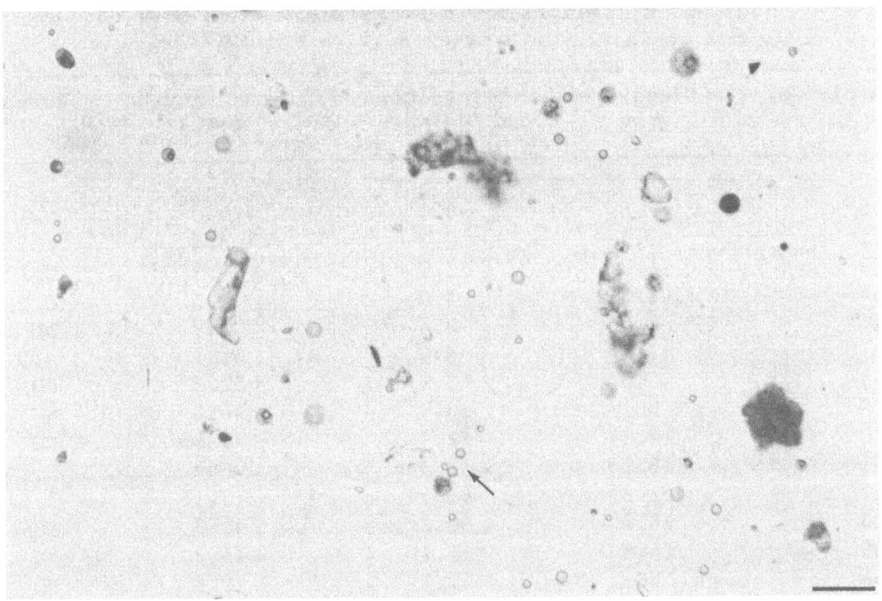

Figure 1.   Light micrograph of particles observed in cloudwater. The
            material is a mixture of biological debris, mineral
            fragments and pulverised fuel ash (arrowed), bar = 20 μm.

The particulate content of a selection of Castlelaw cloud and rain samples is given in Table 2. Samples of less than 100 mls could not be assessed gravimetrically without the risk of large errors. As a consequence many of the rainfall samples have no data. The mean values for the wet deposition samples clearly show higher concentrations of particles in the cloudwater than in rain and constantly higher values for the ETFE than for the nylon strung gauges. (Note: open gauges did not have the 1.2 m lid for rain protection.) The particles in the samples (Figure 1) were a mixture of biological debris (pollen, spores, diatoms, plant fragments etc.), soil-derived mineral fragments (often very angular) and pollutant residues of carbon aggregates and pulverised fuel ash (PFA). The PFA was a very recognisable particle type with a perfect spherical form. Image analysis techniques could be easily applied to separate out this feature from the rest of the material using an X/Y dimension ratio of approximately 1. Biological material was identified by eye leaving a mixture of mineral fragments and unidentified material.

An observation from Table 2 was the high level of particulate material collected in the washings from the gauges after a dry period especially on the ETFE filaments. Most of these observed particles were less than 10 μm in diameter and as such would not normally be

TABLE 2. Particle concentrations mgl^{-1} in rain and cloudwater collected at Castlelaw between 30.5.86 and 30.7.86.

Sample No.	Cloud Open ETFE	Cloud Open Nylon	Cloud Lidded ETFE	Rain
C50	7.8	6.0	4.8	0.4
C51	3.3	2.2	3.4	ND
C53	18.5	22.0	23.3	0.1
C54	1.5	1.0	1.1	0.5
C55	0.5	0.8	0.9	ND
C57	11.25	8.6	16.2	0.3
C58	33.6	10.9	32.2	ND
C59	ND	18.9	20.0	ND
C60	12.3	12.3	1.6	ND
C61	116.9*	35.3*	ND*	0.1*
C62	261.8*	68.5*	213.1*	3.8*
C63	26.7	30.2	44.8	ND
C65	15.8	11.7	15.3	ND
C66	ND	11.1	12.8	ND
C67	13.6	3.8	15.3	ND
C68	4.3	5.4	15.4	1.1
C70	ND	7.7	8.0	ND
C71	1.0	ND	5.1	ND
C72	0.7	6.0	ND	1.7
C73	3.4	4.7	5.5	0.3
C74	ND	ND	ND	ND
Excluding dry deposition				
Mean	10.3	9.6	12.6	0.6
Max	33.6	30.2	44.8	1.7
Min	0.5	0.8	0.9	0.1

*   Dry deposition measurement

ND No data

expected to be deposited very efficiently onto the filaments. When examined in the SEM the surface roughness of the two filament types were comparable. A hypothesis that the filaments develop an electrostatic charge as a result of fast airflow through the gauges might explain the enhanced particle capture especially for a very poor conductor such as ETFE.

The SEM observations for the needle surfaces revealed distinct differences between the upper and lower sites for both epicuticular wax quality and particle accumulation.

The epicuticular wax of the _Picea sitchensis_ needle samples studied had two forms. The inter stomatal regions were covered by flat amorphous or plate-like wax, whereas the wax immediately adjacent to and filling the stomatal apertures had a fibrillar form. This gave the needles a pronounced longitudinal glaucous banding when viewed with the naked eye. These bands became less distinct with increasing needle age.

The wax structure at the upper and lower sites were indistinguishable immediately after needle elongation (Figure 2a and 2b) with a good porous wax matrix in the stomatal plug. After twelve months exposure there was a marked deterioration of the upper site wax with large areas of fibrillar wax fused or abraded into smooth amorphous coverings with frequent inclusions of particles (Figure 2c). Many of these particles were readily recognised as PFA spheres. The lower site by comparison was less damaged although fibril fusion or abrasion was evident (Figure 2d).

After at least two years exposure the needles of the upper site had little or no recognisable fibrillar wax and extensive accumulation of pollutant, mineral and biological particles (Figure 2e). At the lower site wax fusion was also extensive although there were areas that still retained the fibrillar wax form. Particles were more abundant than observed after one year's exposure, especially the irregular type.

The results of this visual comparison for wax structure and particle abundance are summarised for two sampling dates in Figure 3. The interpretation of this histogram should be that a bar equally dispersed about the 0 axis indicates that the two sites were indistinguishable and if there was a marked shift of the bar to one site, this indicated that there was better wax structure or a relatively higher abundance of particles at that site. Figure 3 clearly indicates that the greatest separation of the two sites on the basis of particle abundance and wax degradation occurred after 12 months exposure of the needles.

4. DISCUSSION

The use of ground based cloudwater collecting devices has indicated that at Castlelaw Hill, with a relatively low elevation of 480 m, cloudwater impaction is the dominant wet deposition pathway. Where cloudwater interception is likely to be very efficient, as in afforested upland regions, the potential input of pollutants to the ecosystem will be greatly increased.

600m   **300m**

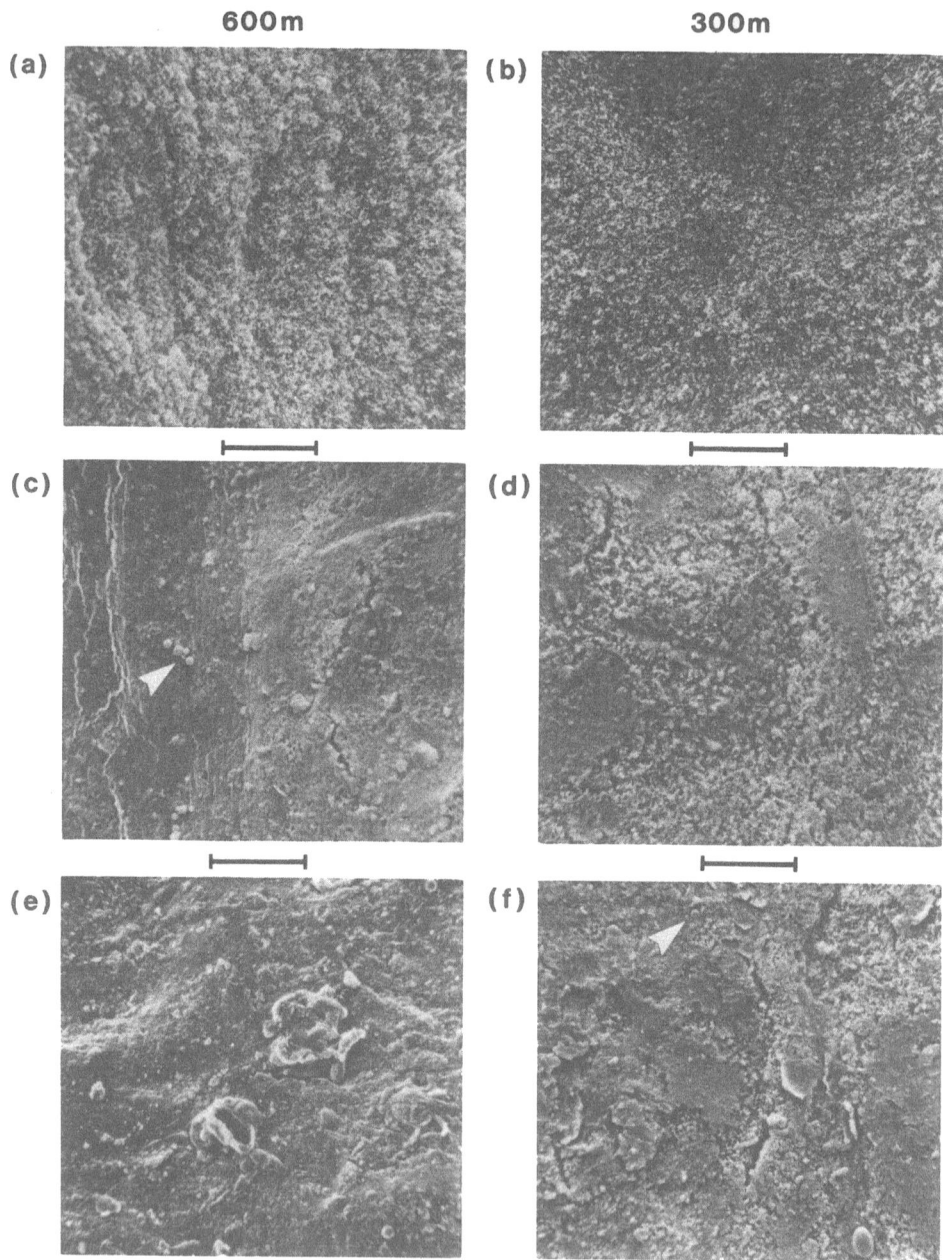

Figure 2.   Scanning electron microscope micrographs illustrating the
             typical surface characteristics for current (a + b) 1st
             year (c + d) and third year (e + f) Sitka spruce needle
             classes harvested from a high elevation (600 m) and low
             elevation (300 m) site at Dunslair Heights, Peeblesshire,
             Scotland.   Bar = 20 μm, PFA spheres arrowed.

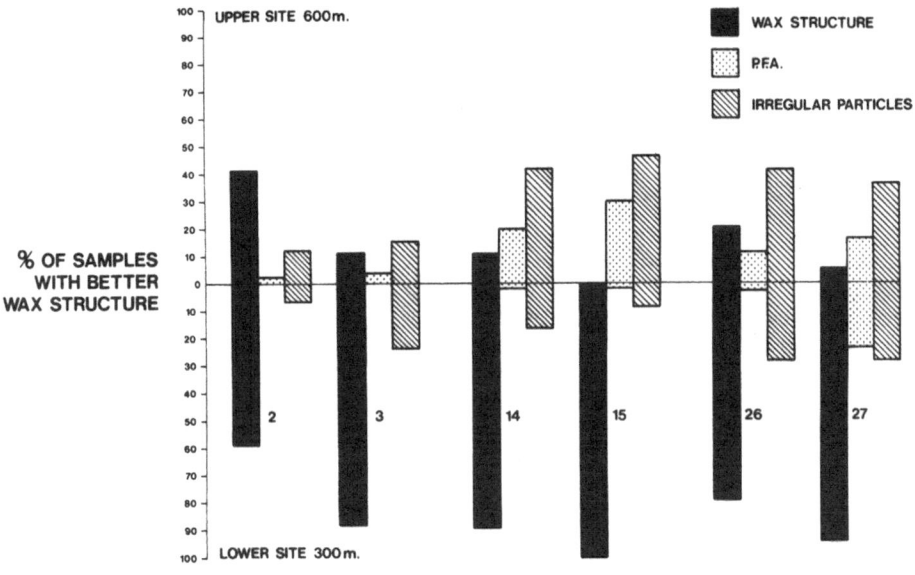

Figure 3.   Histogram of relative particle abundance and comparative
            wax structure between high elevation (600 m) and low
            elevation (300 m) sites at Dunslair Height, Peeblesshire.

    In Britain the cloudwater is frequently polluted (Dollard et al.,
1983; Gervat, 1985) with high concentrations of all the major ions and
also high concentrations of particulates. Direct foliar damage by
acidity could be expected below pH 3 (Jacobson, 1980). Cloudwater
often has low pH levels (see Gervat, 1985), and as such could be
regarded as injurious, especially so if the ion concentrations are
further enhanced by evaporative effects (Unsworth, 1984; Milne and
Crossley, this publication).

    If the cloud liquid water content was assumed to be 0.1 g m^{-3},
then the observed particle concentrations in cloudwater would have
arisen by 2 µg of particles from each 1 m^3 of air being incorporated
into cloud drops.

    Since the annual mean concentrations of particles in urban air of
Britain are typically 17 µg m^{-3}, which appear low by comparison with
those presented for Brazil by Orsini et al. (1986), it is not
surprising that we and others measuring cloudwater in Britain
frequently observe large concentrations of particles (Gervat, 1985;
Coutts, pers. comm.; Harvey, pers. comm.).

    The implications of particle transfer to foliage could be examined
from two aspects. Firstly, the particles themselves may contain
reactive species that are not readily released into solution but could

have damaging effects on contact with plant surfaces. Secondly, the presence of particles on the wax coating of a plant leaf will alter the physical properties of the surface. The epicuticular and cuticular wax barrier at the leaf surface is normally very hydrophobic, especially the fine fibrils of wax typically found around and in the stomatal apertures. This results in high contact angles for water drops on the surface (Cape, 1983) reducing the liquid/wax interface and increasing the 'run-off'. Changes in the chemistry of the wax or its physical integrity from fibrillar to fused forms reduces the contact angle. Where particles may have been deposited in droplets via cloud or rain deposition, subsequent drying of the water before run off would leave the particles on the surface, held by electrostatic forces. The particles may in themselves be chemically inert and have no direct effect on the wax, but their presence changes the physical integrity of the epicuticular wax. This renders the surface more wettable and less likely to allow water run off, and thus compounding the subsequent accumulation of more particulate material. Eventually the leaf surface is so contaminated that it is completely wettable allowing long residence times for aqueous solutions of pollutant ions. This could result in greater potential for ion exchange/leaching across the cuticle, resulting in foliar damage. Figures 2 and 3 clearly illustrate the effect of increased altitude resulting in increased particle accumulation, but do not enable the cause of the epicuticular wax damage to be determined. The increased 'weathering' of the epicuticular wax at the upper site may simply reflect the greater exposure, although Crossley and Fowler (1986) indicated that this was not the sole factor for Scots pine wax. It is not possible at this stage to ascertain whether the accumulation of particles caused the fusion of the epicuticular wax fibrils (either directly or indirectly) or whether greater 'weathering' of the upper site wax resulted in increased particle accumulation.

There is however no doubt that the needles of conifers growing at higher altitudes in Britain are subjected to much greater pollutant inputs as a result of direct interception of cloud droplets, and this, combined with exposure factors could lead to visible foliar damage.

ACKNOWLEDGEMENTS

This work was funded by the Commission of the European Communities and the Natural Environment Research Council. I would like to thank Dr R Milne for his assistance in the development and interpretation of the A.W.S. data and Miss D Wilson for Scanning Electron Microscopy.

REFERENCES

Cape, J.N. 1983. Contact angles of water droplets on needles of Scots pine (Pinus sylvestris) growing in polluted atmosphere. New Phytol. **93** , 293-299.

Chamberlain, A.C. 1975. The movement of particles in plant communities, Vegetation and the Atmosphere, I. Principles (Ed. by J.L. Monteith, pp. 155-1203. Academic Press, London.

Chamberlain, A.C. and Little P. 1981. Transport and capture of particles by vegetation. Plants and their Atmospheric Environment (Ed. by J. Grace, E.D. Ford and P.G. Jarvis), pp. 147-173. Blackwell, Oxford.

Crossley, A., Fowler, D. 1986. The weathering of Scots pine epicuticular wax in polluted and clean air. New Phytol. 103 , 207-218.

Daum, P.H., Schwartz, S.E. and Newman, L. 1984. Acidic and related constituents in liquid water stratiform clouds. Journal of Geophysical Research 89, D1 ., 1447-1458.

Dollard, G.J., Unsworth, M.H. and Harvey, M.J. 1983. Pollutant transfer in upland regions by occult precipitation. Nature 302 , 241-242.

Falconer, R.E., Falconer, P.D. 1980. Determination of cloud water acidity at a mountain observatory in the Adirondack Mountains of New York State. J. Geophys. Res. 85 , 7465-7470.

Fowler, D., Cape, J.N., Leith, I.D., Paterson, I.S. 1986. Acid deposition in Cumbria. In: Pollution in Cumbria (ed.) P. Ineson. ITE Symposium, ISSN 0263-8614, No. 16.

Georgii, H.W., Schmitt, G. 1985. Methoden und ergebnisse der nebelangalyse. Staub Reinhaltung der luft 45(6) , 260-264.

Gervat, G.P. 1985. Clouds at ground level: samples from the southern Pennines. Central Electricity Research Laboratories Report No. TPRD/L/2700/N84.

Grunow, J. 1955. Die Niederschlag in Bergwald. Forstwiss Zentbl. 74 , 21-36.

Hobbs, P.C., Hegg, D.A., Eltgroth, M.W. and Radke, L.F. 1979. Evolution of particles in the plume of coal-fired power plants. I. Deductions from field measurements. Atmospheric Environment 13 , 935-951.

Jacobson, J.S. 1980. The influence of rainfall composition on the yield and quality of agricultural crops. In: Ecological Impact of Acid Precipitation, (eds. D. Drablos and A. Tollan. SNSF, As-NLH, Norway.

Johnson, A.H., Siccama, T.G. 1983. Acid deposition and forest decline. Environ. Sci. and Tech. 17 , 294-305.

Johnson, A.H., Siccama, T.G. 1984. Decline of red spruce in the northern Appalachians: assessing the possible role of acid deposition. TAPPI, 68-72.

Mamane, Y., Miller, J.L., Dzubay, T.G. 1986. Characterisation of individual fly ash particles emitted from coal- and oil-fired power plants. Atmospheric Environment 20(11) , 2125-2135.

Mishra, L.C., Shukla, K.N. 1986. Effects of fly ash deposition on growth, metabolism and dry matter production of maize and soybean. Environmental Pollution (Series A) 42 , 1-13.

Murray, F. 1984. The accumulation by plants of emissions from a coal-fired power plant. Atmospheric Environment 18(8) , 1705-1709.

464

Okita, T. 1968. Concentration of sulphate and other inorganic materials in fog and cloudwater and in aerosol. J. Met. Soc. Japan **46** , 120–127.

Orsini, C.Q., Tabacniks, M.H., Artaxo, P., Andrade, M.F., Kerr, A.S. 1986. Characteristics of fine and coarse particles of natural and urban aerosols of Brazil. Atmospheric Environment **20(11)** , 2259–2269.

Schutt, P., Cowling, E.B. 1985. Waldsterben – a general decline of forests in Central Europe: symptoms, development, and possible causes of a beginning breakdown of forest ecosystems. Plant Disease.

Unsworth, M.H. 1984. Evaporation from forests in cloud enhances the effects of acid deposition. Nature **312** , 262–264.

Waldman, J.D., Munger, J.W., Daniel, J.J., Flagan, R.C., Morgan, J.J. and Hoffman, M.R. 1982. Chemical composition of acid fog. Science **218** , 677–679.

HIGH EFFICIENCY ANNULAR DENUDERS FOR THE DETERMINATION OF SPECIES RESPONSIBLE FOR ATMOSPHERIC ACIDITY

A Liberti, A. Febo and M Possanzini
Istituto sull'Inquinamenta Atmosferico del C.N.R.
Area della Ricerca di Roma
Via Salaria Km 29,300 - C.P. 10
00016 Monterotondo Stazione (Roma)
Italy

ABSTRACT. Basic principles relating to the applications of high efficiency annular denuders for the sampling of gases and aerosols from the atmospheric environment are presented, and the analytical procedures for the simultaneous sampling and determination of nitric and nitrous acid, sulphur dioxide, ammonia and aerosolic species (ammonium, sulphate and nitrate ions) are reported. The described procedure permits us to obtain a full picture of most species responsible for acid deposition in the environment.

1. INTRODUCTION

Minor components of the atmosphere such as ammonia, nitrous acid, nitric acid, nitrogen oxide and sulphur dioxide play a very important role in air chemistry, and their determination is essential to clarify several environmental problems. Their measurement involves however several analytical problems and the sampling of gases and aerosols such as $SO_4^=$, NO_3^-, and Cl^- might be affected by a variety of artifacts when conventional procedures, based on separating the particulate material by collection on a filter with subsequent downstream detection of gaseous compounds in an absorbing medium, are used. Some of these artifacts, which arise from a number of different interactions which might occur in gas and in solid phases are: the retention of gaseous compounds on the filter owing to adsorption or reaction with aerosol particles already collected; the reaction of collected particulate nitrate and chloride with co-collected strong acids to release volatile acids; and volatilisation from the filter of ammonium salts which have an appreciable vapour pressure to form gaseous species.

To overcome problems associated with gas-particle interactions which can be responsible for positive and negative interferences, the use of diffusion tubes (denuders) has been suggested not only to discriminate between gaseous and aerosol species, but also to fractionate species with different chemical properties. Denuders are

465

M. H. Unsworth and D. Fowler (eds.), Acid Deposition at High Elevation Sites, 465–478.
© 1988 by Kluwer Academic Publishers.

cylindrical glass tubes, whose walls have been coated with a layer of substance capable of reacting with a selected gas. When a laminar airstream passes through a denuder, the walls act as an irreversible sink for that gas while particles larger than 0.01 μm in diameter proceed unaffected. The principle upon which a denuder operates is the difference of the diffusion coefficient of a gas and a submicron particle. As the value of the former is 3-6 orders of magnitude larger than the latter's the gas is absorbed on the wall, whereas particles, which proceed unaffected through the diffusion tube, can be collected at the exit on a back-up filter.

It is well known that cylindrical denuders have some limitations, the main one being the operational flowrate, which has to be very low in order to have a larger sorption efficiency; this requires a long sampling time to have a sufficient amount of material for chemical analysis. The use of several cylindrical tubes in parallel configuration has been proposed but this assembly has not been found very convenient.

The adoption of denuders with annular geometry, where air is sampled through the annular space determined by two coaxial glass cylinders, supplies to air chemistry a very efficient sampling device for the determination of air components in traces. Such geometry permits operation at high flowrate with high efficiency, and by coupling denuders, which may act specifically on various atmospheric species, it is possible to obtain quite a detailed picture of most components in the gas and aerosol phases.

The theory of annular denuders and their application to the determination of important species is presented.

2. THEORY

In a laminar airstream passing through a cylindrical denuder the fractional penetration of a species, expressed as the exit (C) to inlet (C_0) concentration ratio, may be described approximately by the following equation:

$$\frac{C}{C_0} \cong 0.819 \exp. \ (-14.63\Delta) \tag{1}$$

where

$$\Delta = \frac{\pi D 1}{4F} = \frac{D 1}{\gamma R_e d} \tag{2}$$

D is the diffusion coefficient of the species, $\underline{1}$ the length of the tube, γ the kinematic viscosity of air, d the internal diameter of the tube, R_e the Reynolds number and F the air flowrate. Reynolds number, which is an index of flow laminarity, must be below 2000; equation 1 is valid when this condition is fulfilled.

In the case of an annular denuder, the parameter Δ is given by the expression (the subscript 'a' refers to the annular denuder):

$$\Delta_a = \frac{\pi D l}{4F} \frac{d_1 + d_2}{d_2 - d_1} \tag{3}$$

where d_1 and d_2 are the inside and outside diameters of the annulus, respectively. By comparing equation 2 with equation 3, it can be seen that at given F/l ratios, values of Δ_a much larger than those corresponding to Δ can be obtained. In fact, Δ_a may also be optimised by reducing the equivalent diameter of the annulus $d_2 - d_1$ and increasing its inside diameter. Moreover, large flow rates do not alter the laminar regime in the annular geometry since R_e is inversely proportional to $d_1 + d_2$.

Thus, the condition $R_e \leqslant 2000$ can in any case be ensured by increasing the denuder diameter. The fractional penetration of a species passing through an annular denuder can be expressed by an empirical equation of the same form as equation 1, i.e.

$$\frac{C}{C_o} \simeq 0.82 \exp. (-22.53 \Delta_a) \tag{4}$$

By comparing the two expressions 1 and 4, the result is that, in the case of a cylindrical denuder, the only parameter governing the mass transfer to the wall is the tube length/air flow ratio whereas in an annular denuder C/C_o depends also on the annulus diameter and the width of the annular section. By arranging expressions 1 and 4 the following relationship is obtained

$$(\frac{F}{l})_a \simeq 1.54 \frac{d_1 + d_2}{d_2 - d_1} (\frac{F}{l}) \tag{5}$$

Equation 5 clearly demonstrates that the ratio F/l for an annular denuder is larger than that obtainable with a cylindrical diffusion tube. For instance, to obtain an efficiency better than 95% with a 0.3 cm i.d. SO_2 cylindrical denuder of 50 cm length, the airflow should not exceed 1.7 L min^{-1} (R_e = 791), while the same efficiency is obtained when an airflow of 20 L min^{-1} passes through an annular denuder 20 cm long, d_1 = 3.0 cm, $d_2 - d_1$ = 0.3 cm (R_e = 442). These performances indicate the possibility of using HEADs (high efficiency annular denuders) for short term sampling of atmospheric species at sub-ppm level.

3. EXPERIMENTAL

A schematic diagram of an annular denuder is shown in Figure 1. The walls of the denuder are frosted in order to make the coating easier; the coating operation is carried out by either sucking into the annular space an ethanol or water-ethanol solution of the selected species, or by pouring a few millilitres of the solution and by rotating the denuder to ensure a uniform wetting of the surface. The excess solution is drained out and the walls are dried by blowing through a clean air stream. The denuders are then sealed at both ends with threaded teflon caps until they are used.

468

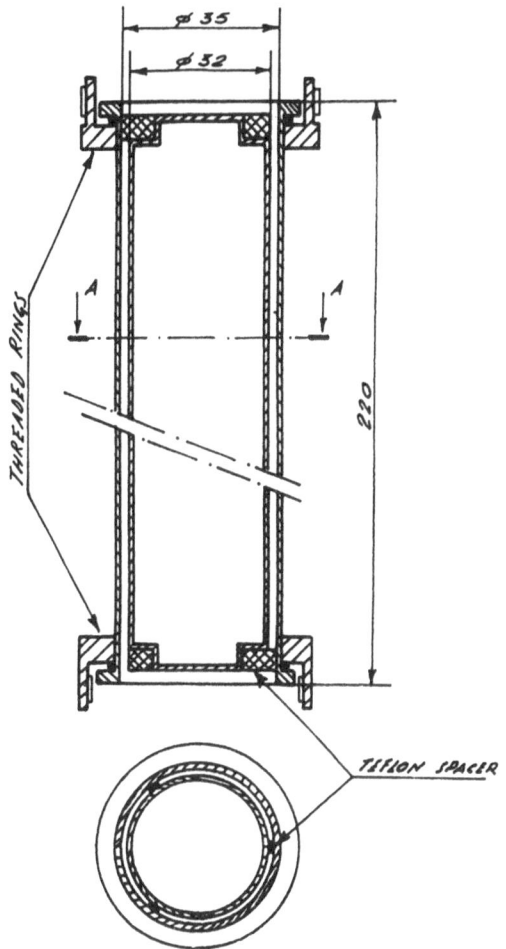

Figure 1.   Schematic diagram of an annular denuder. Units are mm.

    HEADs permit sampling without bubbling air into an absorbing
solution, eliminate gas-particle interactions, perform the sampling
with a high efficiency by using a high flow rate up to 20 1/m, and
correspondingly decrease the limit of detection of a particular
species. Table I lists some applications of annular denuders for the
sampling of specific species. The absorbing layer indicated has been,
in several cases, developed by investigating in a cylindrical denuder
reported in the references.
    One of the more interesting features of HEAD is the possibility of
setting three or more denuders in series to sample from air most
species responsible for atmospheric acidity.

TABLE 1. Denuders used for selective sampling

Compounds	Absorbing layer	References
SO_2	tetrachloromercurate	Possanzini et al., 1983
	lead dioxide	Durham et al., 1978
	sodium carbonate	Tanner et al., 1980
NH_3	oxalic acid	Ferm, 1979
	tungstic acid	McClenny et al., 1982
	citric acid	Bos, 1980
	phosphorous acid	Stevens et al., 1978
HNO_3	sodium carbonate	Ferm, 1986
	tungstic acid	McClenny et al., 1982
	aluminum sulphate	Lindquist, 1985
	sodium fluoride	Slanina et al., 1981
	magnesium oxide	Forrest et al., 1982
HNO_2	sodium carbonate	Ferm and Sjodin, 1985
HF	sodium bicarbonate	Weinstein and Mandl, 1971
HCl	sodium nitrate	Bailey et al., 1976
	sodium fluoride	Niessner and Klockow, 1981
	sodium carbonate	De Santis et al., 1985
HCHO	bisulphite-triethanolamine	Cecchini et al., 1985
NO_2	potassium iodide	Possanzini et al., 1984
	manganese dioxide	Adams et al., 1986
	alkaline guaiacol	Buttini et al., 1987
Aniline	oxalic acid	De Santis and Perino, 1987
Tetraalkyl lead	iodine monochloride	Febo et al., 1986

The determination of species of relevant interest in air chemistry are reported in the following sections.

3.1 Ammonia and Ammonium Ion

Ammonia is a minor constituent of the atmosphere, as it is usually found at concentrations ranging from 1 to 25 ppb. It however plays an important role in determining atmospheric acidity, as its neutralising action is the most probable pathway for acidity removal before any deposition may occur. Acid substances introduced or formed in the atmosphere can be neutralised by ammonia leading to aerosols containing ammonium nitrate, sulphate and bisulphate. It has been suggested that ammonia in vapour phase is in equilibrium with ammonium nitrate only, and it has been observed by Stelson (1979) that the $[NH_3]$ $[HNO_3]$ product agrees well with the calculated equilibrium constant. A similar result has been obtained by Larson and Taylor (1983) during daytime sampling. During nighttime sampling, NH_3 concentrations were much higher and HNO_3 much lower than those predicted by phase equilibria. This was probably the effect of the relative humidity, which is an additional critical parameter in determining the equilibrium concentration.

Measurements of ammonia and ammonium ion can be carried out with high reliability by HEAD by using the sampling train shown in Figure 2A. It consists of two denuders coated with 1% citric acid in ethanol connected together through a filter which collects ammonium salts. The second denuder is used to measure ammonia released from ammonium salts collected on the filter and its contribution to the ammonium concentration is taken into account.

3.2 Nitric and Nitrous Acid

The determination of both acids is of a great environmental importance in terms of acid deposition and evaluation of photochemical smog. The determination of gas-phase nitric acid in the atmosphere is important to the understanding of the contribution of NO_x emissions to rain or fog chemistry and to visibility. Gaseous nitrous acid is considered to play an important role, mainly due to its ability to produce OH radicals through direct photolysis. It has been found that a sodium carbonate layer is a perfect sorbent for both acids, and the measurement of nitrate and nitrite carried out by ion chromatography on the eluted solutions permits determination of these compounds. Allowance has to be made however for species which may yield nitrite as well as nitrate in order to have consistent results. They are NO_2 and PAN, the former being in a concentration almost one order of magnitude higher than HNO_2, and the latter in the range of ppbv. The evaluation of these interferences has been studied by Liberti et al. (in press); the determination of HNO_2 can be carried out by considering that the NO_2 and PAN removal efficiency of a sodium carbonate denuder is in the range of a fewpercent. Hence the mass distribution of these interferents along successive denuder sections may be regarded to be the same. Then by operating two denuders in series, in the first there is the quantitative removal of HNO_2; this value has to be corrected for the nitrite amount found in the second denuder (Figure 2B).

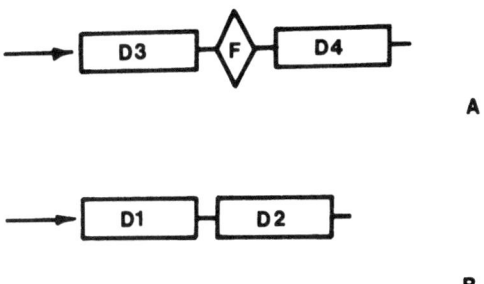

Figure 2. Denuder assembly for sampling ammonia and ammonium ion (A),
           nitric and nitrous acid (B) in the atmosphere.
           D_3, D_4 citric acid coated denuders
           F   Teflon filter
           D_1, D_2 sodium carbonate coated denuders.

3.3 Annular Denuder-filter Pack Assembly for the Measurement of
     Gaseous and Aerosol Species

The possibility of coupling in series by means of flanges and threaded
locking rings, annular denuders with different characteristics permits
the construction of a sampling train for simultaneous collection of
gases HNO_3, HNO_2, SO_2, NH_3 and aerosol material $SO_4^=$, NO_3^-, NH_4^+ by
operating with a gas flow at 15 L min^{-1}. As it is shown in Figure 3,
it consists of:

a) Two annular denuders (22 cm long, $d_2 - d_1 = 0.3$ cm, $d_1 = 3.0$ cm)
    coated with Na_2CO_3 for the collection of HNO_3, HNO_2 and SO_2. The
    second denuder is used to correct for interferents in the
    measurements of the above species.
b) A hollow tube of Teflon, 5.5 cm long and 3.6 cm in diameter,
    inserted between the Na_2CO_3 denuders. This expansion coupler
    produces similar fluid dynamic conditions at both denuder inlets,
    where particle losses by impaction may be considered the same.
c) Two shorter annular denuders (13 cm long, $d_2 - d_1 = 0.3$ cm, $d_1 =$
    3.0 cm). The first denuder collects gaseous ammonia while the
    second one, placed downstream of the filter pack, is intended to
    recover NH_3 released by the Teflon filter.
d) A dual filter holder, constructed of Teflon, which accommodates a
    47 mm diameter 10 μm Teflon filter, which retains HNO_3 resulting
    from the dissociation of ammonium nitrate collected on the Teflon
    filter. By adding the amounts of nitrate and ammonium found on

the nylon filter and the back-up denuder, respectively, to those collected on the Teflon filter, total nitrate and ammonium content in particulate matter can be determined.

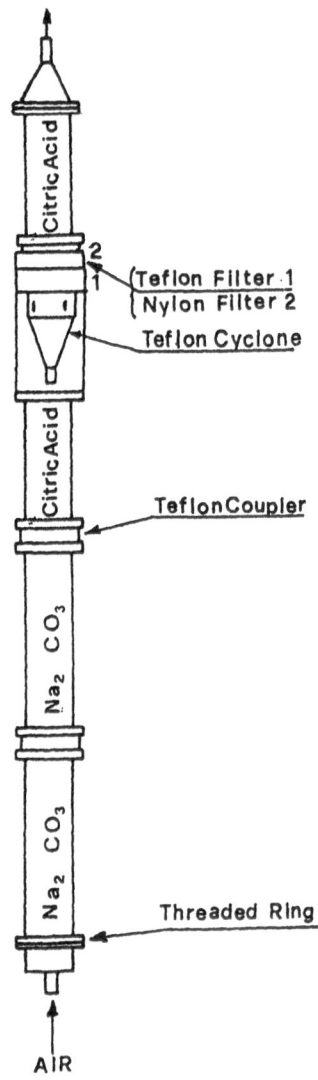

Figure 3.   Annular denuder system for collecting gaseous acids, ammonia and related aerosol species in the atmosphere.

The sampling train is made to operate in a vertical position to eliminate gravitational settling and is set into a case, where heating may be supplied. This precaution is required in the presence of high humidity which alters the absorbing layers of the denuder.

Table II reports daily concentrations of gaseous pollutants and related ionic species in aerosol measured in a suburban area 30 km from Rome.

3.4 Estimation of the Tropospheric Concentration of the Hydroxyl Radical Through HEADs Measurement

The dominant role played by the hydroxyl radical in tropospheric chemistry makes it essential to have better definition of its concentration. This determination can be carried out by direct measurement, by utilizing a physical characteristic or chemical property of the radical in the troposphere, but also indirectly by the measurement of tropospheric constituents. In the former case, highly sophisticated optical methods are employed, such as laser induced fluorescence, long-path absorption of laser radiation, and others, whilst indirect determination can be carried out by measurement of atmospheric trace constituents taking part in reactions influenced by HO radicals.

Photochemical models of varying sophistication have been used to derive tropospheric concentrations of the HO radical. A rather simple approach arises from the determination of atmospheric nitrogen oxides and acids. Nitrous acid concentration is affected by the hydroxyl radical, as its photolysis is a source of this radical

$$HNO_2 + hv \longrightarrow HO+NO \tag{6}$$

Formation of HNO_2 in the atmosphere can occur according to the following reactions

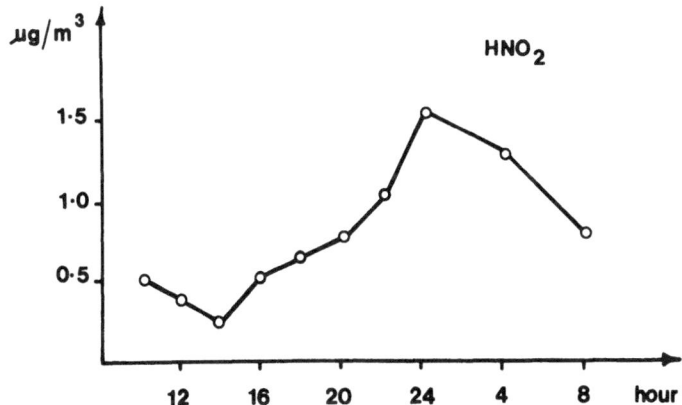

Figure 4.   Diurnal variation of HNO_2 atmospheric levels measured on a two-hour basis. Rome, February 26-27, 1985.

TABLE II. Amount of ionic species measured by the annular denuder method (ADM) during 12–24 h sampling at 15 L min⁻¹ and relative atmospheric levels in the gas and particle phases.

DATE 1985	Denuder							Filter				Den 4	Gas phase				Particle phase		
	1			2			3	1			2								
	NO_3^-	$SO_4^{=}$	NO_2^-	NO_3^-	$SO_4^{=}$	NO_2^-	NH_4^+	NO_3^-	$SO_4^{=}$	NH_4^+	NO_3^-	NH_4	HNO_3	SO_2	HNO_2	NH_3	NO_3^-	$SO_4^{=}$	NH_4
2-6	3.0	149	52.0	1.0	3.1	8.0	69.7	54.6	73.6	42.5	18.9	15.0	0.09	4.2	1.9	3.0	3.2	3.2	2.5
2-8	2.0	165	32.1	0.7	2.4	3.0	88.3	30.0	51.6	24.2	7.2	10.6	0.10	9.0	2.4	7.4	3.1	4.3	2.9
2-12	2.7	134	37.6	1.0	2.8	0.6	80.2	39.5	39.5	26.8	7.4	11.6	0.14	7.2	3.1	6.7	3.9	3.3	3.2
2-14	3.8	171	23.2	1.9	3.9	7.0	90.1	51.8	73.8	45.1	18.2	23.4	0.10	6.2	0.9	5.0	3.9	4.1	3.8
2-15	3.2	139	37.0	2.6	3.8	6.4	118.8	49.0	92.8	50.3	24.5	21.7	0.03	5.6	2.0	7.5	4.6	5.8	4.5
2-25	7.8	186	17.3	2.0	2.4	4.1	104.1	64.0	86.4	54.0	10.5	14.4	0.48	9.9	1.0	8.7	6.2	7.2	5.7
2-26	5.4	197	55.5	1.9	4.0	7.7	149.9	70.2	84.6	62.2	29.0	27.8	0.19	7.2	2.7	8.3	5.5	4.7	5.0
2-27	14.2	170	44.0	2.6	2.2	6.5	107.4	31.1	96.8	33.4	3.9	4.1	0.52	5.1	1.7	4.9	1.6	4.4	1.7
3-29	6.5	102	21.5	1.4	3.1	6.5	25.5	40.4	99.0	62.7	5.8	5.5	0.23	2.9	0.7	1.2	2.1	4.5	3.1
4-2	5.7	126	51.4	2.6	1.0	4.8	58.5	63.1	62.1	60.6	10.5	8.6	0.13	3.4	2.2	2.5	3.2	2.7	3.0

ION CONTENT (μg) — ATMOSPHERIC CONCENTRATIONS (μg/m³)

TABLE III.   Concentrations of OH radicals calculated from Ez. (9) on the basis of 4-h HNO_2 and NO measurements.

Day	Time	HNO_2 nmole m^{-3}	NO nmole m^{-3}	$J_{HNO_2}10^3$ sec^{-1}	OH concentration 10^6 molecules cm^{-3}
2-4	10,30-14,30	5.1	80	1.0	12
2-5	9,30-13,30	20.2	358	0.6	6
2-7	9,30-13,30	15.3	360	0.6	4
3-17	14,30-18,30	5.3	40	0.6	14
4-2	13,30-17,30	4.9	44	1.1	26
4-15	9,30-13,30	16.4	290	1.1	12
4-23	9,30-13,30	4.9	200	1.1	5
4-23	13,30-17,30	2.1	45	1.1	11

$$NO + NO_2 + H_2O \longrightarrow 2HNO_3 \qquad (7)$$

$$2NO_2 + H_2O \longrightarrow HNO_2 + HNO_3 \qquad (8)$$

As a consequence of these reactions, the concentration of HNO_2 is not constant but changes noticeably with time in a trend which is shown in Figure 4. The concentration is higher at night, and decreases in the day. As reaction (7) has been investigated by Chan et al., experimental results relative to the measurement of nitrous acid in the atmosphere have to be discussed by comparing them with the calculated equilibrium values. It has been found that HNO_2 in air is always lower than the equilibrium value, by day and also at night when its concentration is higher. As the major differences are observed during daytime, they have to be attributed to the photolytic process of HNO_2. Since OH and NO are in photochemical equilibrium with HNO_2 during the day, the concentration of hydroxyl radical can be estimated if HNO_2 and NO concentrations are measured, and the photolysis constant of HNO_2 is calculated in the measured radiation conditions

$$(HNO_2) = \frac{k \ (OH) \ (NO)}{J_{HNO_2}} \qquad (9)$$

k is the constant for the backward reaction (Equation 6), $(5,14 \times 10^{-12}$ $cm^3 \ molec^{-1} \ sec^{-1})$. J_{HNO_2} is the photolysis constant obtained from the data of Stockwell and Calvert (1978) by averaging the volumetric UV radiation intensity in the time interval selected for the measurements of HNO_2 and NO. HNO_2 was measured by sampling air on a 4 hour period as above described, and nitrogen oxide concentrations were obtained by integrating the response of a commercial chemiluminescence analyser. Some results are reported in Table III. Hydroxyl radical concentrations reported in Table III are somewhat higher than reported by various authors (e.g. Hewitt and Harrison, 1985); improvements in the analytical methods for the measurement of HNO_2 and NO will allow us to obtain more accurate figures.

REFERENCES

Adams, K.M., Japar, S.M. and Pierson, W.R. 1986. Development of a MnO_2-coated cylindrical denuder for removing NO_2 from atmospheric samples. Atmospheric Environment **20** , 1211-1215.

Bailey, R.R., Field, P.E. and Wightman, J.P. 1976. Determination of low concentrations of hydrogen chloride in moist air. Analytical Chemistry **48** , 1818-1819.

Bos, R. 1980. Automatic measurement of atmospheric ammonia. Journal of Air Pollution Control Association **30** , 1222-1224.

Buttini, P., Di Palo, V. and Possanzini, M. 1987. Coupling of denuder and ion-chromatographic techniques for NO_2 trace level determination in air. Science of the Total Environment **61** , 59-72.

Cecchini, F., Febo, A. and Possanzini, M. 1985. High-efficiency annular denuder for formaldehyde monitoring. Analytical Letters 18 (A6), 681–693.

Chan, W.H., Hordstrom, R.J., Calvert, J.G. and Shaw, J.H. 1976. Kinetic study of HONO-formation and decay reactions in gaseous mixtures of HONO, NO, NO$_2$, H$_2$O and N$_2$. Environmental Science & Technology 10 , 674–682.

De Santis, F., Liberti, A. and Rotatori, M. 1985. Determinazione dell' acido cloridrico nell'atmosfera mediante campionamento con tubi di diffusione ad alta effeicienza. Acqua e Aria 6 , 529–534.

De Santis, F. and Perrino, C. 1986. Personal sampling of aniline in working sites by using high-efficiency annular denuders. Annali di Chimica 76 , 355–364.

Durham, J.L., Wilson, W.E. and Bailey, E.B. 1978. Application of an SO$_2$-denuder for continuous measurement of sulphur in submicrometric aerosols. Atmospheric Environment 12 , 883–886.

Febo, A., Di Palo, V. and Possanzini, M. 1986. The determination of tetraalkyl lead in air by a denuder diffusion technique. Science of the Total Environment 48 , 187–194.

Ferm, M. 1979. Method for determination of atmospheric ammonia. Atmospheric Environment 13 , 1385–1393.

Ferm, M. and Sjodin, A. 1985. A sodium carbonate-coated denuder for determination of nitrous acid in the atmosphere. Atmospheric Environment 19 , 979–983.

Ferm, M.. 1986. A Na$_2$CO$_3$-coated denuder and filter for determination of gaseous HNO$_3$ and particulate NO$_3$ in the atmosphere. Atmospheric Environment 20 , 1193–1201.

Forrest, J., Spandau, D.J., Tanner, R.L. and Newman, L. 1982. Determination of atmospheric nitrate and nitric acid employing a diffusion denuder with a filter pack. Atmospheric Environment 16 , 1473–1485.

Hewitt, C.N. and Harrison, R.M. 1985. Tropospheric concentrations of the hydroxyl radical-A review. Atmospheric Environment 19 , 545–554.

Larson, T.V. and Taylor, G.S. 1983. On the evaporation of ammonium nitrate aerosol. Atmospheric Environment 17 , 2489–2495.

Liberti, A., Allegrini, I., De Santis, F., Di Palo, V., Febo, A., Perrino, C. and Possanzini, M. 1987. Annular denuder method for sampling reactive gases and aerosols in the atmosphere. Science of the Total Environment 67 , 1–16.

Lindquist, F. 1985. Determination of nitric acid in ambient air by gas chromatography-photoionization detection after collection in a denuder. Journal of the Air Pollution Control Association 35 , 19–23.

McClenny, W.A., Galley, P.C., Braman, R.S. and Shelley, T.J. 1982. Tungstic acid technique for monitoring nitric acid and ammonia in ambient air. Analytical Chemistry 54 , 365–369.

Niessner, R. and Klockow, D. 1981. A new approach to the determination of atmospheric strong acids. Proc. 9th Annual Conference of the Association for Aerosol Research, Duisburg, F.R.G., Sept. 23–25.

Possanzini, M., Febo, A. and Liberti, A. 1983. New design of a high-performance denuder for the sampling of atmospheric pollutants. Atmospheric Environment 17 , 2605-2610.

Possanzini, M., Febo, A. and Cecchini, F. 1984. Development of a KI annular denuder for NO_2 collection. Analytical Letters 17 (A10), 887-896.

Slanina, J., Lamonen-Doornenbal, L., Lingerak, W.A., Meilof, W., Klockow, D. and Niessner, R. 1981. Application of a thermo-denuder analyser to the determination of H_2SO_4, HNO_3 and NH_3 in air. Intern. J. Environ. Anal. Chem. 9 , 59-70.

Stelson, A.W., Friedlander, S.K. and Seinfeld, J.H. 1979. A note on the equilibrium relationship between ammonia and nitric acid and particulate ammonium nitrate. Atmospheric Environment 13 , 369-371.

Stevens, R.K., Dzubay, T.G., Russwurm, G. and Rickel, D. 1978. Sampling and analysis of atmospheric sulphates and related species. Atmospheric Environment 12 , 55-68.

Stockwell, W.R. and Calvert, J.G. 1978. The near ultraviolet absorption spectrum of gaseous HONO and N_2O_3. Journal of Photochemistry 8 , 193-203.

Tanner, R.L., D'Ottavio, T., Garber, R. and Newman, L. 1980. Determination of ambient aerosol sulphur using a continuous flame photometric detection system. I. Sampling system for aerosol sulphate and sulphuric acid. Atmospheric Environment 14 , 121-127.

Weinstein, L.H. and Mandl, R.H. 1971. The separation and collection of gaseous and particulate fluorides. VDI Berichte 164 , 53-63.

A FOG CHAMBER AND WIND TUNNEL FACILITY FOR CALIBRATION OF CLOUD WATER
COLLECTORS

R K A M Mallant
Netherlands Energy Research Foundation ECN
P O Box 1
1755 ZG Petten
The Netherlands

ABSTRACT. The interest in cloud water chemistry and the role of
intercepted cloud water in forest die-back has led to the construction
and use of various droplet collectors. Most of these collectors have
not been calibrated adequately. In this paper a facility for testing
of cloud and fog water collectors is described. Water droplets of
known composition and size are used in the calibration process. To
overcome errors due to evaporation of the droplets after generation,
the air is pre-humidified and thermostated. By using different types
of generators, various polydisperse droplet spectra can be obtained.

1. INTRODUCTION

The interest in the role of clouds in atmospheric chemistry and the
role of fog and intercepted clouds in forest die-back has led to the
development of many different cloud and fog water collectors. Some of
the researchers involved are not interested in the collection
efficiency for different drop sizes (Mack and Pilie, 1975; Scott, 1978;
Black and Landsberg, 1983; Schripff et al., 1984), others rely on
theory (Brewer et al., 1983; Fuzzi et al., 1984; Georgii and Schmitt,
1985). In few cases the collectors were laboratory calibrated with
solid aerosol (Jacob et al., 1984) or oil droplets. Jacob et al.
(1982) tested a rotating arm collector in a chamber in which water
droplets were sprayed. Using the difference in droplet spectrum
before and during operation of the collector, a lower cut-off diameter
was deduced, but no real collection efficiency was established. Mohnen
(1980) used water droplets to calibrate his cloud water collector. The
droplets were not sprayed in pre-humidified air. The short residence
time and high liquid water content reduced errors due to evaporation.

It is preferable to use a stable cloud or fog to determine the
efficiency of a collector, instead of using a solid aerosol or oil
droplets. Also, theory often can not be used to calculate the
efficiency. Jacob et al. (1984) calculated a D_{50} (diameter at 50%
collection efficiency) of 8 micrometers for their rotating arm

M. H. Unsworth and D. Fowler (eds.), Acid Deposition at High Elevation Sites, 479–490.
© 1988 by Kluwer Academic Publishers.

collector using a theory for impaction of particles on a cylinder. Calibration with sodium fluorescein yielded a D_{50} of 20 micrometers. The discrepancy between theory and practice is attributed to the fact that the collection surface in fact is not a cylinder. For several reasons theory will also not be applicable to determine the collection efficiency for fog droplets of thin wires (Jacob et al., 1984). One reason is that the dimensions and roughness, and thereby the impaction characteristics, of a thin wire are significantly altered as soon as collected droplets start to pile up and form large water drops. This process cannot be simulated by calibration with a solid aerosol.

In two occasions fog water collectors were compared in the field (Hering and Blumenthal, 1985; Jaeschke et al., 1986). Such a comparison is very useful, but care should be taken in using the results to establish if evaporation or condensation in a specific collector occurs. At first, the composition of the fog water before collection is not known. Therefore, the reference must be an other collecting device. Secondly, if the functioning of a collector is judged by the difference in composition of samples obtained with that and other collectors, one should bear in mind that each collector has its own lower and upper cut-off diameters which, combined with a drop size dependent chemical composition of the fog water, may cause the observed differences.

Testing with a stable fog in the laboratory will not only give information about the collection efficiency for a certain size spectrum, but it will also reveal errors that may occur in the collection of fog and handling of the sample. These errors may be due to evaporation of water after collection, or contamination of the sample by materials of which the collector is constructed. To reveal errors due to evaporation or condensation (in case of adiabatic cooling), it is necessary to simulate natural conditions, i.e. when a relative low amount of liquid water is available, compared to the amount of water vapour.

Although this paper is focussed on the use of fog for testing of collectors, the same fog generation technique as described here can also be applied in other experiments. The current awareness that intercepted clouds may play a considerable role in forest die-back has initiated investigations into the effects of acidic fog on trees (Wood and Bormann, 1974, 1975; Wedding and Ligotke, 1979; Hindawi et al., 1980; Scherbatskoy and Klein, 1983; Granett and Musselman, 1984; Krause et al., 1985; Skeffington and Roberts, 1985). Most of these experiments are lacking in precautions to avoid evaporation of the droplets. So, in this type of experiment too, there is a need for a good technique to create a stable fog of known composition. In the experiments performed by Mallant et al. (1986) and Masuch et al. (1986) special attention was given to this aspect.

Another application for the description fog generation system is the study of cloud chemistry in the laboratory.

2. PROBLEM

Conducting experiments in which fog is involved is a precarious matter. The water vapour content of saturated air at 20°C is 17.3 g/m³ (CRC Handbook of Chemistry and Physics, 1983). The Liquid Water Content (LWC) of natural fog is typically in the order of 0.1 g/m³ (Jiusto, 1981), and consequently less than 1% of the water vapour concentration. This, combined with the fact that the slope of the saturation curve at 20°C is about 1 gram/m³/°C, makes that minor changes in temperature (and therefore in the degree of saturation) cause drastic effects on the LWC. Consequently, heat fluxes must be kept at a minimum and the air must be humidified.

In addition to the problem of creating an environment in which relative small amounts of liquid water can be maintained, there are additional problems, related to the type of experiments in which the fog will be used. Examples of these experiments are cloud chemistry studies, plant effect studies, or tests of fog water collectors.

In case the fog generation system is meant to be used in cloud chemistry experiments or plant effect studies the following parameters must be controllable and measurable:
1. The size spectrum.
2. The Liquid Water Content (LWC).
3. The chemical composition of the fog water.

These requirements can be met by using both pneumatic and/or ultrasonic nebulizers for droplet generation. Although LaMer-type generators offer the possibility to create monodisperse droplets, they have the disadvantage that the chemical composition of the droplets is not so easily controlled. Spinning disc generators produce droplets that are too large (D 20 micrometer), and the output is not sufficient in case the air-throughput in the system is large. Vibrating orifice generators also suffer from too small production rates.

In case the set-up is meant to test cloud or fog water collectors the following requirements should also be met:
4. The collector should not have a too drastic effect on the fog, i.e. the volume throughput in the chamber should be large with respect to the volume of air aspirated by the collector. Rotating collectors should not be able to impart their rotation to the air in the fog chamber, which would cause impaction of droplets to the walls. This can be realized by a rapid exchange of air in the chamber.
5. Many of the designed fog water collectors rely on wind speeds in their collection process. Cloud water collectors need even higher wind-speeds, up to 70 m/sec. Therefore, the set-up must comprise a windtunnel.

The use of monodisperse droplets would facilitate the calibration of fog water collectors. However, as stated above, all available techniques that are normally used to produce monodisperse droplets, cannot be used in case large amounts of fog are needed. The use of polydisperse droplets creates its own typical problems. The collection efficiency, expressed as the fraction of LWC collected, gives no information about the relation between collection efficiency and drop size. For some samplers, i.e. passive string-type collectors, the efficiency can be determined by comparing the droplet spectra that are measured in front and behind the strings. If this procedure cannot be followed for other collectors, another possibility remains. By using several spectra with varying mean diameters, the collection efficiency as function of diameter can be calculated as follows:

- Divide the diameter range of interest in intervals D_i (i = 1 - N_1).
- Let the collector have efficiencies E_i for droplets in the different intervals D_i.
- Determine the mass of liquid water $M_{i,j}$ within each interval D_i of droplet spectra S_j (j = 1 - N_2), as measured by an optical probe.
- The (measurement) fraction F_j of the LWC of spectrum S_j as sampled by the collector then equals:

$$F_j = \text{sum } (E_i * M_{i,j}) \text{ for } i = 1 \rightarrow N_i$$

This equation holds N_1 unknown efficiencies E_i. These E_i can be calculated by making $N_2 = N_1$, i.e. by generating N_1 different drop size spectra that fit within the diameter range of interest. In practice this calibration procedure also has its drawbacks. For instance, N_2 may be small, e.g. 4 or 5. As a result the relation between droplet diameter and collection efficiency will not be known very accurately.

3. MATERIALS AND METHODS

In earlier experiments in which the effects of H_2O_2-containing acidic fog on young trees were studied, it was found that continuous flow systems in which the entering air is humidified in a packed bed are very suitable to conduct fog experiments (Mallant et al., 1986; Masuch et al., 1986). In these experiments fog was generated by means of a pneumatic nebulizer (Plomp, 1985). Packed-tower type humidifiers are easily constructed and operated. The air is not only humidified, but also attains the temperature of the water in the top of the tower. By controlling the water temperature such that it equals that of the laboratory air, changes of the temperature and humidity of the air in the rest of the set-up can be avoided. The system was scaled up to build the test facility.

The set-up is called CHIEF (Chamber for Investigations with Equilibrated Fog). Air flow rates are up to 4500 m^3/hr, and LWC ranges from 0.01-1 g/m^3. The set-up is as follows (Figure 1): outdoor air is blown in by a controllable 15 kW blower. This air enters into the lower part of a humidifier, which consists of a tower sized 1.15 x 15 x 3 meter. Part of this tower (2 m^3) is filled with about 90,000 ceramic Raschig rings of 2.5 cm. Water is poured over this column at a rate of 2.5 litre/second. The water is heated by a controllable 50 kW electric heater. The air passes through the top section of the humidifier where devices for fog generation are placed. Then, the fog enters a cylindrical vessel of 2.5 m diameter and 4 m height (20 m^3). In this chamber the droplets are given time to mix (and if necessary to

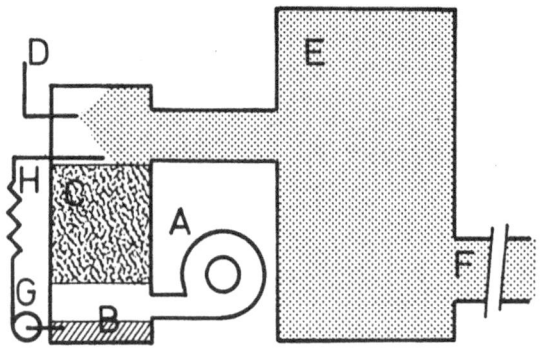

A = Blower

B = Reservoir

C = Humidifier

D = Fog generator

E = Fog Chamber

F = Tunnel (5 m)

G = Water pump

H = Heater

Figure 1.   Schematic drawing of CHIEF.

equilibriate) with the air. Also, fog water collectors can be placed
in this vessel. The fog leaves the vessel through a 5 metre long 50 x
50 cm channel. The wind-speed can be varied from 1-5 m/sec. The fog
is discharged in the outdoor air. With the use of conical inserts,
the wind-speed can be increased to meet requirements to test passive
fog water collectors or cloud water collectors. As a consequence of
the fact that CHIEF is also intended to be suited to study fast
chemical reactions in clouds, it was decided not to recirculate the
air, but to continuously use fresh, unsaturated air. This defines a
lower limit on the outdoor air humidity and temperature at which CHIEF
can operate. These worst conditions are: temperature of the incoming
(outdoor) air 10°C, at 50% R.H., temperature of the air leaving the
humidifier 20°C at 100% R.H., and a throughput of 4500 m³ air per hour.
     A theory of mass transfer in packed towers is used to calculate
the dimensions of the tower (see appendix). Since the driving force
for the evaporation of water in the humidifier is the difference in
water vapour pressure of the air and the surface layer of the water, it
follows theoretically that 100% saturation cannot be reached. In
practice, first results have shown that evaporation losses of fog water
due to this deviation are limited to a few percent.
     Special attention should be given to the amount of water flowing
through the tower. The water temperature in the tower decreases due to
evaporative cooling. Consequently, the water vapour pressure is a
function of height in the tower. This can be taken into account in the
calculations of the dimensions of the humidifier. It is much easier
however, to take the amount of water large enough to limit the decrease
in temperature to a few degrees under extreme conditions.
     Even when all possible precautions to avoid evaporation of water
droplets have been made, it is useful to check whether the composition
of the fog water after spraying remains unchanged. Therefore, it is
necessary to obtain a sample of the fog water. In case of cloud

chemistry experiments, sampling of the water is obligatory. For this purpose, small cyclones made of glass or polyethylene were constructed. These cyclines have a modified Stairmand design, and the sampling efficiency is calculated according to the Barth theory (Barth, 1956; Hochstrasser, 1976). In laboratory investigations into the effects of fog on trees and in cloud chemistry experiments, these cyclones have proven to operate satisfactorily. However, in these experiments, the LWC was high ($>$ 1 g/m^3). Tests in CHIEF must reveal the suitability of cyclones for the collection of less dense clouds.

CHIEF is equipped with a Particle Measuring Systems Inc. FSSP-100 optical probe to determine drop size distributions. Temperature and dew point temperature are monitored by means of an Endress & Hauser HMT 240 probe. After calibration this probe has proven to be able to measure the difference in temperature and dew point temperature at an accuracy of 0.1 °C.

4. RESULTS

The difference in temperature and dew point temperatures of the air in CHIEF lies within the detection limit of 0.1 °C (droplets generators inactive). According to the design specifications of CHIEF, the dew point depression should be less than 0.02 °C, which is far less than the detection limit of the instrument. If CHIEF does not comply with these specifications, then the dew point depression in the worst possible case could equal the detection limit of 0.1 °C. In this hypothetical case, it would still be possible to evaporate about 0.1 g water per m^3, leading to a considerable evaporation of the sprayed fog water. This can only be established by comparing the chemical composition of both spray solution and fog water collected by a reference instrument that is not likely to give evaporation or condensation, e.g. a string-type collector. The use of such a reference method in CHIEF is much more acceptable than in natural fogs. Although not much is known about the size dependency of the chemical composition of fog droplets, one may expect that this dependency will be almost or perhaps completely absent in CHIEF, while such a dependency applies very likely to natural fogs.

The size distribution of droplets generated by an Heyer USE 77 ultrasonic nebulizer in given in Figure 2. The LWC, inferred from the mass distribution measured by the FSSP-100, is 0.03 g/m^3. The mass median diameter (MMD) is 7.4 micrometer, and the geometric standard deviation (GSD) is 1.35. The mass distribution of a Spraying Systems 1/4J-1a pneumatic atomizer is given in Figure 3. The LWC is 0.15 g/m^3. The concentrations of fog droplets is 532 cm^{-3}. The mass median diameter is 11.8 micrometer, and the geometric standard deviation is 1.59. Both the LWC and the MMD of this nebulizer can be influenced by varying the pressure and the liquid flow. The fog produced is very stable; fluctuations in LWC and MMD are less than 5% over a period of one hour.

The results of two initial experiments with a cyclone in CHIEF are given in Table 1. The cyclone is a 20 cm diameter version of the small

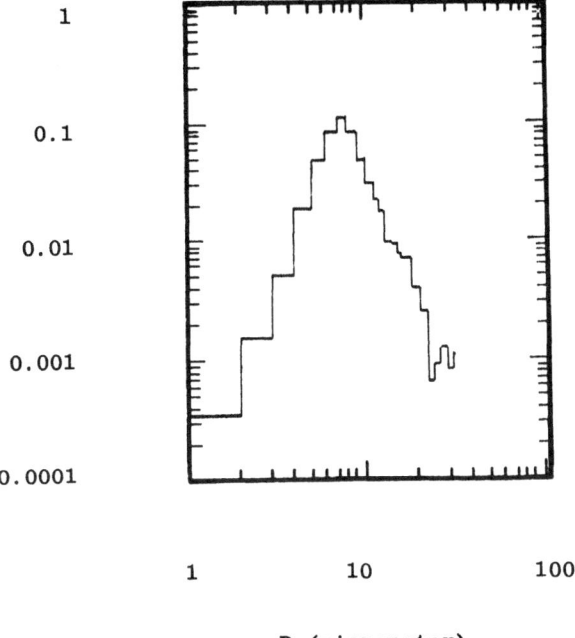

D (micrometer)

Figure 2. Mass distribution of fog droplets generated by an
ultrasonic nebulizer (Heyer USE 77).
Mass Median Diameter = 7.4 micrometer
Geom. Standard Dev.   = 1.35
Liquid Water Cont.   = .03 g/m³

cyclones mentioned earlier in this paper. It sampled 2 m³/min, and is
designed to have a 50% (lower) cut-off diameter of 5 micrometer. The
purpose of the tests was to see if evaporation or condensation occurs
in the cyclone. Therefore, the spray solution contained 5.9 ppm K⁺.
The composition of fog water collected with a string type collector was
chosen as the reference for the real fog water composition. From the
sample obtained by this reference method, it was concluded that 20%
respectively 30% of the sprayed solution had evaporated after spraying
in CHIEF. In the first experiment the LWC varied from 0.05 to 0.4
g/m³, with a mean value of 0.2 g/m³. The fluctuations in LWC were
intentional, they served to investigate the working range of the
Sonicore nebulizer. The MMD in this experiment varied from 11-15

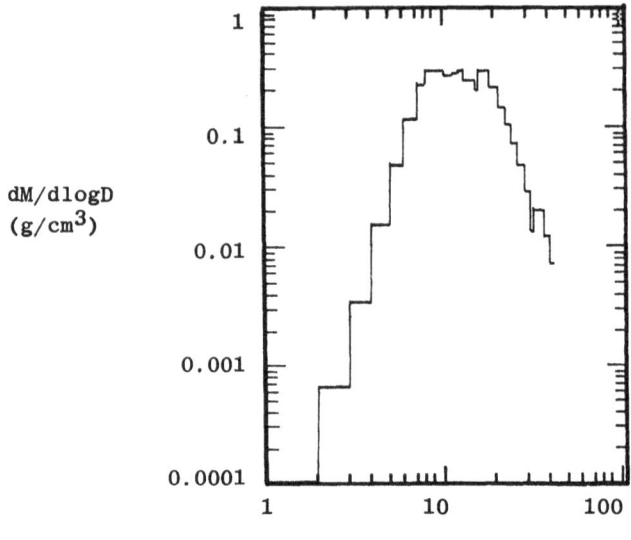

D (micrometer)

Figure 3.   Mass distribution of fog droplets generated by a
            pressurized nebulizer (Spraying Systems 1/4J-1A).
            Mass Median Diameter = 11.8 micrometer
            Geom. Standard Dev.   =   1.59
            Liquid Water Cont.   =   .15 g/m^3

micrometer. In the second experiment, the LWC was kept more constant
at 0.2 g/m^3, and the MMD was 12.7 micrometer. The composition of the
samples obtained by the cyclone was compared with that of the reference
method. The samples obtained with the cyclone showed an increase of
the K$^+$ concentration of 12% for the first, and of 3% for the second
experiment.

5. DISCUSSION

The facility described in this paper produces a stable fog which can be
used to test fog and cloud water collectors. Both types of atomizers
that are available give useful mass distributions. The Heyer USE 77
produces a relative narrow spectrum, which cannot be influenced by
operating conditions of the nebulizer. Relative to the Heyer, the
Sonicore 035H atomizer produces broader spectra, with larger mass
median diameters, that can be varied by changing pressure and liquid
flow. The MMD of the spectra presented here are small compared to
those measured in a radiation fog by Fuzzi et al. (1984) (MMD ranging
from 20-30 micrometers). The reason is that the author first attempted
to find atomizers that produce small droplets. It will be no problem

Table 1. Comparison of K^+ concentrations in fog water collected by a cyclone and in the original spray solution, respectively in water collected by a reference (string-type) collector. See text for operational conditions.

Date 27/8/1986

Sample No.	$\dfrac{[K^+](\text{cycl.})}{[K^+](\text{spray.})}$	$\dfrac{[K^+](\text{cycl.})}{[K^+](\text{ref.})}$
1	1.43	1.14
2	1.64	1.31
3	1.41	1.13
4	1.64	1.31
5	1.26	1.01
6	1.33	1.06
7	1.22	0.98
8	1.25	1.00
9	1.43	1.14
Mean and standard deviation:		1.12 ± 0.12

Date 28/8/1986

Sample No.	$\dfrac{[K^+](\text{cycl.})}{[K^+](\text{spray.})}$	$\dfrac{[K^+](\text{cycl.})}{[K^+](\text{ref.})}$
1	1.37	0.96
2	1.59	1.12
3	1.52	1.07
4	1.39	0.98
5	1.44	1.01
Mean and standard deviation:		1.03 ± 0.07

to find atomizers that generate larger drops.

Some first experiments with a cyclone showed that errors due to evaporation probably are small. The LWC of 0.2 g/m^3 used in this test is not unrealistically high, especially not when the high water vapour content (WVC) of 19.4 g/m^3 (22 °C) is taken in consideration. LWC values of natural fogs often are lower than the value of 0.2 g/m^3, but in most cases the temperature and therefore the WVC is lower too. No attempt was made to determine the overall efficiency of the cyclone. This will be done in future.

From samples of the fog in CHIEF obtained by a string-type collector, it can be concluded that some evaporation still occurs. In future a more accurate temperature controller will be installed. Also the string-type collector will be equipped with an on-line method, e.g. a conductivity cell, to check if the collected water differs from the spray solution. This facilitates adjustments of the temperature settings.

ACKNOWLEDGEMENTS

The author would like to acknowledge P.B. Fonteijn for his fine work in constructing CHIEF, G.P.A. Kos for his contribution in the experiments in CHIEF, and the members of the former Aerosol Research Group for their contributions in the discussion leading to the design of CHIEF.

REFERENCES

Barth, W. 1956. Design and Layout of the Cyclone seperator on the Basis of New Investigations. Brennstoff-Warme-Kraft **8** : 1.

Black, H.D. and Landsberg, H.E. 1983. A Method for Continuous Records of the pH of Low-Level Clouds, Rep. no. SR-83-17, Univ. of Maryland, Dept of Meteor., College Park, Maryland.

Brewer, R.L., Gordon, R.J. anad Shepard, L.S. 1983. Chemistry of mist and fog from the Los Angeles urban area. Atm. Environ. **17** (11): 2267.

CRC Handbook of Chemistry and Physics 64th Edition, 1983.

Fuzzi, S., Castillo, R.A., Jiusto, J.E. and Lala, G.G. 1984. Chemical composition of radiation fog water at Albany, New York, and its relationship to fog microphysics. J. Geoph. Res. **89** : 7159.

Georgii, H.W. and Schmitt, G. 1985. Methoden und Ergebnisse der Nebelanalyse. Staub-Reinhalt. Luft **6** : 260.

Grannett, A.L. and Musselman, R.C. 1984. Simulated acidic fog injures lettuce. Atmos. Environ. **18** (4): 887.

Hering, S.V. and Blumenthal, D.L. 1985. Fog Sampler Intercomparison Study, Final report ST1 11 90063-308-FR Sonoma Techn. Inc., Santa Rosa, California 95401.

Hindawi, I.J., Rea, J.A. and Griffis, W.L. 1980. Response of Bush bean exposed to acid mist. Amer. J. Bot. **67** (2): 168.

Hochstrasser, J.M. 1976. The investigation and development of cyclone dust collector theories for application to miniature cyclone presamplers. Ph.D. Thesis, Univ. of Cincinnati.

Jaeschke, W. et al. 1986. Unpublished study at the 'Kleinen Feldberg', FRG.

Jacob, D.J., Wang, R.T. and Flagan, R.C. 1984. Fogwater Collector Design and Characterisation. Environ. Sci. Technol. **18** : 827.

Jacob, D.J., Flagan, R.C., Waldman, J.M. and Hoffmann, M.R. 1982. Design and calibration of a rotating arm collector for ambient fog sampling. Precipitation Scavenging, Dry Deposition, and Resuspension Proc. of the 4th Int. Conf., Santa Monica, CA, Nov. 1982 **1** : 123.

Jiusto, J.E. 1981. Clouds: Their Formation, Optical Properties, and Effects. p. 187, Acad. Press Inc.

Krause, G.H.M., Jung, K.D. and Prinz, B. 1985. Experimentelle Untersuchungen zur Aufklärung der neuartigen Waldschäden in der Bundesrepublik Deutschland). VDI Berichte **560** : 627.

Mack, E. and Pilie, R. 1975. Fog Water Collector. U.S. Patent 3889532.

Mallant, R.K.A.M., Slanina, J., Masuch, G. and Kettrup, A. 1986. Experiments on H_2O_2-Containing Fog Exposures on Young Trees. Aerosols. Editors S.D. Lee, T. Schneider, L.D. Grant and P.J. Verkerk. Lewis Publ., Chelsea, Michigan. pp. 901–910.

Masuch, G., Kettrup, A., Mallant, R.K.A.M. and Slanina, J. 1986. Effects of H_2O_2-Containing Acidic Fog on Young Trees. Int. Journ. of Envirnmental Analytical Chemistry 27 : 183.

Mohnen, V.A. 1980. Cloud Water Collection from Aircraft. Atmosph. Technol. 12 : 20.

Plomp, A. 1985. Performance of commercial atomisers for therapeutic and diagnostic use as compared to a newly developed atomiser. Proc. of ECLASS 85: 3rd Int. Conf. on Liquid Atomisation and Spray systems-85. Imperial College London, July 8–10. p. IIIB/4/1.

Scherbatskoy, T. and Klein, R.M. 1983. Response of Spruce and Birch Foliage to Leaching by Acidic Mists. J. Environ. Qual 12 (2): 189.

Schrimpff, E., Klemm, O., Eiden, R., Frevert. T. and Herrmann, R. 1984. Anwendung eines Grunow-Nebelniederschlägen. Staub-Reinhalt. Luft 44 : 72.

Scott, W.D. 1978. The pH of Cloud Water and the Production of Sulphate. Atmospheric Environment 12 : 917.

Skeffington, R.A. and Roberts, T.M. 1985. The effects of ozone and acid mist on Scots pine samplings. Oecologia 65 : 201.

Wedding, J.B. and Ligotke, M. 1979. Effects of Sulphuric Acid Mist on Plant Canopies. Environ. Sci. Technol. 13 (7): 875.

Wood, T. and Bormann, F.H. 1974. The effects on an artificial acid mist upon the growth of Betula alleghaniensis Britt. Environ. Pollut. 7 : 259.

Wood, T. and Bormann, F.H. 1975. Increases in Foliar Leaching Caused by Acidification of an Artificial Mist. Ambio 4 (4): 169.

APPENDIX

The calculation of mass transfer in a packed tower was made following Strauss (1975). First, the diameter of the tower was calculated using an equation based on operation at 50% of the flooding rate. This equation contains a number of partly empirical parameters in which properties of the packing material, gas and liquid are combined. Using a liquid flow rate of 1 $kg/sec/m^2$, a diameter of 1.28 m was calculated. This implies a total liquid flow rate of 1.3 litre/sec. The worst conditions for operation described before (outside air at 10°C and 50% R.H., final temperature 20°C and 100% R.H.) imply an evaporation rate of water of 15.6 g/sec at an air flow rate of 1.25 m^3/sec. The power consumption for evaporation alone already equals 35 kW. Heating the air for 10°C requires an additional 15 kW. If all this energy is to be withdrawn from the water flowing in the column at a rate of 1.3 litre/sec, then this water cools about 9°C. This was considered to be too much, so water flow rate was doubled. This does not have a large influence on the percentage of flooding which now comes at an acceptable 55%. The height of one transfer unit according to Sherwood and Pigford (1952) or Perry and Chilton (1973) is calculated to be 23 cm. The number of transfer units required to reach a defined saturation can be calculated according to Rietema (1976) using:

$$N = \ln \left[(C_*-C_0)/(C_*-C_f) \right] \quad \text{in which:}$$

C_* = concentration of water vapour at the water surface ($17.3 \ g/m^3$).
C_0 = initial water vapour concentration (inlet air; $4.7 \ g/m^3$).
C_f = desired final water vapour concentration after humidifier.

(Here it is obvious that this theory is in fact not suited to calculated humidifier performance as the concentration of vapour at the water surface, C_*, decreases when the water reaches the lower part of the column, due to cooling).

It will be clear that reaching 100% saturation is impossible according to this equation. We settled for 99.9%, which implies that another $0.0173 \ g/m^3$ of water can be evaporated. (The difference in temperature and dew point temperature then is about $0.016°C$). At typical LWCs in the order of $0.1 \ g/m^3$ this would lead to an evaporation of 17.3% of the fog water. In practice, this will be less if the worst conditions for operation are not met. Also it can be overcome by adjusting the temperature of the water slightly above that of the surrounding air.

The number of transfer units calculated to reach 99.9% is 6.6, resulting in a column height of 1.52 m.

REFERENCES

Perry, R.H. and Chilton, C.H. 1973. Chemical Engineer's Handbook 5th edition. McGraw-Hill Kogakusha Ltd, Tokyo.
Rietema, K. 1975. 'Fysische Transport - en Overdrachtsverschijnselen' Prisma-Technica 60, p. 251.
Sherwood, T.K. and Pigford, R.L. 1952. Absorbtion and Extraction. McGraw-Hill, New York.
Strauss, W. 1975. 'Industrial Gas Cleaning' 2nd edition, Pergamon Press.

CHEMICAL COMPOSITION OF BULK ATMOSPHERIC DEPOSITION TO SNOW AT COL DE LA BRENVA (MT BLANC AREA)

F Ronseaux and R J Delmas
Laboratoire de Glaciologie et Géophysique de l'Environment
B.P. 96
38402 St Martin d'Hères Cedex
France

ABSTRACT. Two 10-m firn cores were drilled in June 1984 at Col de la Brenva, a high elevation site (4350 m a.s.l.) located just below the summit of Mt Blanc (4807 m) in the French Alps. At this elevation, summer melting does not occur and the chemical composition of snow impurities reflects the background aerosol composition on a regional scale. One of the firn cores was carefully subsampled in ninety 10 cm long sections in clean air conditions and the meltwater samples analysed by ion chromatography for common ions (Na^+, K^+, NH_4^+, Mg^{++}, Ca^{++}, $SO_4^=$, NO_3^- and Cl^-). Acidity/alkalinity was accurately titrated. Snow layer dating indicates that the studied time period covers about 2.5 years. Along the cores, several dusty bands corresponding to short Saharan dust events were discovered. Sulphate and calcium dominate the ionic composition. A tentative interpretation concerning the initial composition of the alpine aerosol at high elevation sites is made. In June 1986, a 70 m-deep firn core was drilled on a nearby high plateau. The chemical analylsis of this new core will provide an environmental record for at least the last four decades, a time period over which the impact of atmospheric pollution has increased most rapidly.

1. INTRODUCTION

The chemical analysis of polar ice samples has already provided valuable information on the global atmospheric environment. Polar ice sheets are generally remote from most aerosol sources and the chemical composition recorded in the snow deposited in these areas reflects background conditions. On the other hand, alpine snow fields located at high elevations in central Europe may provide information on the impurity content of the atmosphere over a densely populated and industrialised continent. The exchange of pollutants between the various European countries is a matter of growing international controversy. Mt Blanc, the highest summit of the Alps, located on the border between France, Italy and Switzerland could be an excellent

M. H. Unsworth and D. Fowler (eds.), Acid Deposition at High Elevation Sites, 491–510.

Figure 1. Map of the French Alps showing the areas of highest
elevations. The Mt Blanc summit (4807 m) is located at the
border between France, Italy and Switzerland (upper part of
the map).

observatory for investigating the effects of the long range transport
of natural and anthropogenic substances on the chemical composition of
the aerosol in the free atmosphere of the European continent (Figure
1). Surprisingly, published scientific data for very high elevation
alpine precipitation is still incomplete. Ten years ago our laboratory
started glaciochemical measurements on snow and firn samples collected
on the very summit of Mt Blanc (Batifol and Boutron, 1984) and on a
flat snowfield located at Col du Dome at a lower elevation (4350 m)
(Delmas and Aristarain, 1978; Briat, 1978). Deep ice cores from
another most promising central Alpine site, Colle Gnifetti (4400 m), in
Switzerland, have demonstrated of going back over several centuries in
time (Haeberli et al., 1983).

Over recent years, environmental problems have become increasingly
acute in western Europe. Rapid and accurate glaciochemical analytical
methods have recently been developed in our laboratory, especially for
polar programmes. We therefore decided in 1984 to undertake a
comprehensive investigation on the chemical composition of Alpine
precipitation. The first step was to locate a suitable site for deep
drilling in the Mt Blanc area. A virgin site was chosen and a field
campaign organised to drill a firn core down to several metres for
preliminary glaciochemical investigations.

This paper reports on the data obtained from the samples collected on this occasion, preceded by an exhaustive description of the methods and techniques used for these glaciochemical studies.

2. SAMPLING SITE AND DRILLING

The samples were taken in June 1984 in the Mt Blanc range (Figure 1) at a location called Col de la Brenva (elevation 4350 m a.s.l.). The col is oriented east-west on a ridge joining Mont Maudit (4465 m) and Mont Blanc (4807 m) (Figure 2). It is not on the route most commonly used by climbers attempting the Mt Blanc summit.

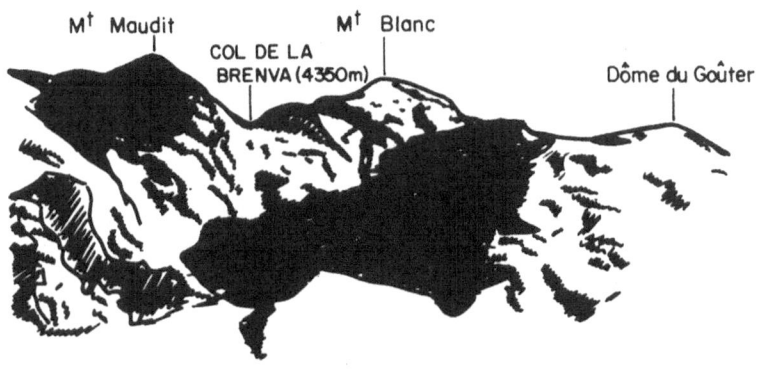

Figure 2. Drawing showing the upper part of the Mt Blanc area and the study site (Col de la Brenva). The summit of Mt Blanc (4807 m, in the background) is higher than that of Mt Maudit (4465 m).

In the area, the wind regime is generally from the west so that Col de la Brenva is frequently crossed by air masses in the west-east direction. Southerly or easterly regimes are however also observed. The surrounding mountains (north, south and south-west directions) are constituted by crystalline rock (the Alpine arch). However the pre-Alp summits, at lower elevations, are calcareous.

At these elevations, snow accumulation rates are generally around 3 metres per year, the snow accumulating mostly in early spring. At high elevations, precipitation tends to occur at the beginning of depressions whereas the opposite is observed at lower elevations influenced more by the ends of meteorological disturbances.

Dust falls occur principally in spring and summer (Haeberli, 1977). In addition to events of a local or regional origin, dust falls from the Sahara desert (located nearly 1500 km to the south) are sometimes observed.

The mean annual temperature at Col de la Brenva is -13°C. Despite this relatively low snowpack temperature, the possibility of some

494

summer melting had to be considered. Percolation is an important
phenomenon which can, if present, limit the quality of chemical depth
profiles. In the firn core studied, we observed thin ice layers
(< 5 cm), some corresponding to the winter-season (in this case, caused
by wind). No percolation columns were observed in the snow pack.
Moreover the temperature depth profile at Col de la Brenva is not
typical of temperate firn such as is observed for instance at 3950 m
for the Glacier des Violettes (Figure 3) where summer percolation is
frequent. It can be calculated that the melting of only 0.5 m of snow
(density 0.4) is able to raise the mean temperature of 15 m of firn up
to nearly 0°C (as is the case for the Glacier des Violettes). It may
therefore be assumed that the environmental record at the Col de la
Brenva has remained undisturbed since its deposition. The results
presented below will show that this assumption is reasonable and that
this site is well suited for glaciochemical studies.

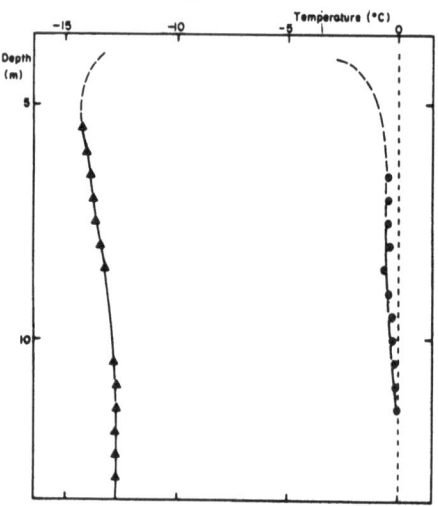

Figure 3.   Temperature vs depth profiles obtained at Col de la Brenva
             (triangles) and at another site (Glacier des Violettes,
             circles) where summer percolation occurs.

3. DATING OF THE SNOW LAYERS

The firn core was dated using three different methods: visual
stratigraphy, palynology and reference dust layers.

- In cold alpine glaciers, summer and winter snow layers can generally
   be easily identified: winter snow accumulation is fine grained
   whereas summer snow is large grained with frequent thin melt
   layers. The alternating summer and winter layers enable a
   relatively accurate (to within half a year) dating of our firn core
   as a function of depth. In order to reach a more refined
   chronology, two additional methods were used.

- The composition of the pollen grains deposited on snow varies from one season to another, according to the vegetation acitivity in the surrounding Alpine region (J L Borel, pers. comm.). This enables an accurate dating of the snow layers corresponding to these seasons. This method, which to our knowledge has never before been used for dating snow layers, appears to be very promising.

- Reference horizons of various origins (radioactive levels, volcanic acid layers, dust layers, etc.) is a method commonly used for the dating of ice cores. Total beta radioactivity is particularly useful for the years following large atmospheric nuclear tests (Jouzel et al., 1977). Over recent years, the atmospheric radioactivity has nearly returned to its natural level so that no clear reference horizons are available for dating purposes.

    Haeberli et al. (1983) used mineral dust atmospheric events to date the deep ice core from Colle Gnifetti. We have also detected several well marked dust events recorded in our 10 m firn core, probably corresponding to the long range transport of Saharan dust. Unfortunately their exact date of arrival over the Alps is unknown in most cases.

    The combined utilisation of these three techniques led to conclude that our firn core covers the time period spanning from winter 1981-82 to early summer 1984 (date of the drilling), i.e. nearly 30 months.

4. SUB-CORING AND LABORATORY ANALYSES

The frozen firn core was stored in a freezer until sample cleaning and analysis operations. In spite of the precautions taken in the field, the core surface cannot be considered as contamination-free. It had to be re-cored in a clean air cabinet installed in a cold room (-15°C). This is done with the aid of a mini-drilling set-up (Figure 4) according to a protocol described in detail elsewhere (Legrand et al., 1984; Legrand and Delmas, 1987). The sub-cores obtained are then melted in sealed plastic bags in order to minimize the absorption of ammonia or of other gaseous contaminants by the meltwater. Ninety samples were prepared from the 10 cm long firn core, giving a mean length of 10 cm of firn per sample. The sampling frequency was therefore approximately 36 samples per year.

    This decontamination procedure was carefully tested by measuring the impurity content of an artificial ice core prepared from deionized water, recored with the aid of the minidrill and melted in one of the plastic bags used for storing and meltwater samples (blank value). The subcoring operation was started only after having reached negligible blank test values. Between two series of chemical determinations, the samples were kept in a frozen state.

    Before chemical determinations, the samples were filtered in order to remove insoluble dust particles. However an aliquot of each sample was taken just before filtration for further alkaline metal determinations. The filtration operation has been suspected to

496

significantly increase the ammonium content of the samples, so ammonium (and consequently also sodium and potassium) were measured on unfiltered samples.

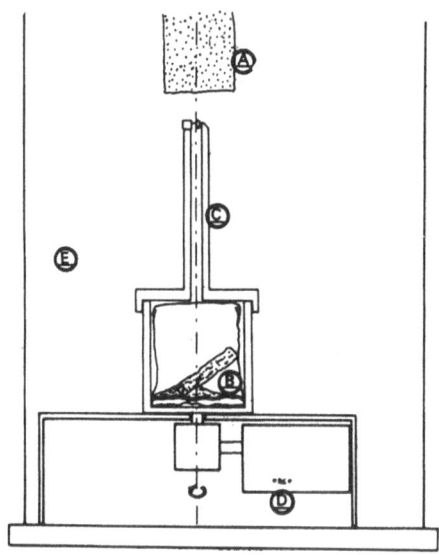

Figure 4. Firn recording set-up (mini-drill): A: firn core, B: sub-sample, C: PTFE auger equipped with molybdenum bits, D: electrical motor, E: dust-free enclosure.

All major ions, except H^+ were determined by ion chromatography using a Biotronik apparatus for alkaline cations ($NH_4^+ - Na^+ - K^+$) and a Dionex 2010 i equipment for alkaline earth cations (Ca^{++} and Mg^{++}) and anions (Cl^-, NO_3^- and $SO_4^=$). Preconcentration columns were used in order to improve measurement sensitivity.

Each ion chromatographic determination necessitates the addition (using a syringe) of 3 to 5 ml of liquid. The chromatograms are recorded on a 2-way recorder (Figure 5).

Calibration curves were obtained using diluted standard solutions (either Merck (R) or Riedel de Haen (R)). Ultrapure water was used throughout. The calibration curve (based on 5 reference concentrations) was checked before and after each measurement series. The mean errors were as follows:

Na^+ : 1.7%, K^+ : 3.5%, NH_4^+ : 3%, Mg^{++} : 2%, Ca^{++} : 3%, Cl^- : 7%, NO_3^- : 3.6% and $SO_4^=$: 3.4%.

A few calcium measurements were also performed by atomic absorption spectrophotometry (AA) to compare results obtained using two different techniques. Filtered samples were used. The results are reported in

Table 1. The agreement between the two sets of measurements is satisfactory:

(1) $Ca_{AA} = 0.94 \cdot Ca_{IC} - 2.0$ with $r = 0.983$.

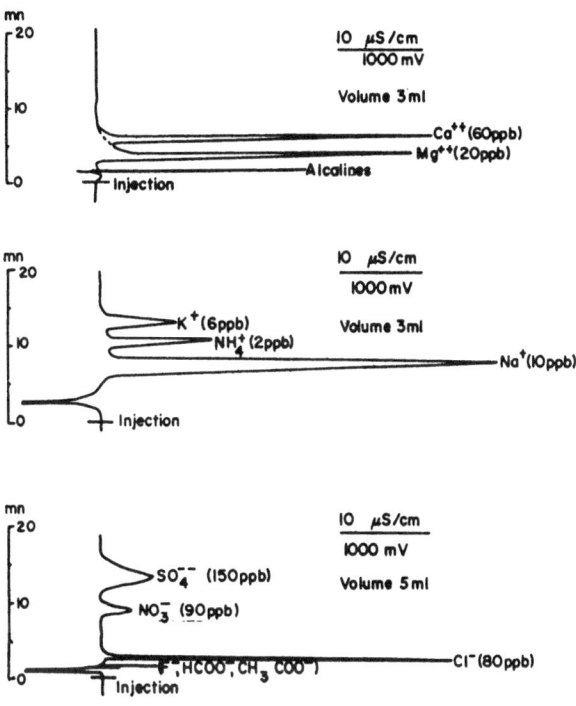

Figure 5.   Typical ion-chromatograms obtained for alkaline-earth metals, alkaline metals (middle) and anions (bottom) in meltwater-samples form the Col de la Brenva firn core.

Table 1.   Comparison of calcium measurements to atomic absorption (AA) and ion chromatography (IC).

IC	0.0	0.0	4.6	33.1	108	112.8	143.2
AA	0.0	0.4	2.5	27.5	76.5	112.7	141.6

PH was not measured. Acidity was determined using a titration method based on the addition of known quantities of a diluted acid solution (H_2SO_4 5.10^{-3} N). The advantage of this method is to entirely eliminate the influence of dissolved CO_2 by lowering the pH of the sample (at pH < 4.6 the dissociation of carbonic acid is negligible). When the sample is initially alkaline, the first additions of the

498

titrant serve to neutralise the solution, which then becomes
progressively acidic. Twelve successive additions are carried out for
each titration. The final linear regression giving the acidity of the
sample is similar to the Gran's plot method. The detailed protocole is
described by Legrand et al. (1982). A Mettler titration system (DK 31
A/D converter and DV 11 buret drive with analog output) is used. The
pH electrode is from Microelectrodes Ltd. The titration vessel is
maintained at a constant temperature (+ 0.05 °C) using a thermostat
controlled water bath.

The method has already been successfully employed for several
thousand polar meltwater samples. For samples collected at
mid-latitudes and in particular in the Alps, the interpretation of this
acid titration is somewhat more complicated since one part of the
titrant may be consumed by the destruction of carbonated particles (if
any). In fact only a few samples were found to contain (bi)
carbonates. Our titration method therefore gives a measure of strong
acidity (acids with pK lower than 3.6, see Legrand et al., 1982), with
a mean accuracy of 2%.

5. RESULTS AND DISCUSSION

5.1 General

The results for the soluble impurities concentrations are given in
Table 2 and the ion budget Σ_i as a function of depth in Figure 6.
The overall ion-balance is well respected (Figure 6). The general
imbalance ΔC is <1% when calculated with respect to the mean ionic
budget (the sum of the anions and cations). For 70% of the individual

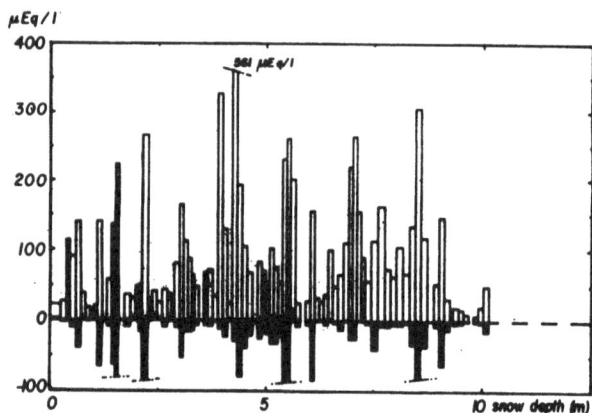

Figure 6.   Ion budget (Σ cations + Σ anions) in the 90 samples of the
10 m fir core from Col de la Brenva (open bars). The ionic
imbalance (Σ cations - Σ anions) is given below (solid
bars).

sample, it is less than 15%. If ΔC is calculated as:

$$C = \text{cations} - \text{anions} = [Na^+] + [K^+] + [NH_4^+] + [Ca^{++}] + [Mg^{++}] + [H^+] - [Cl^-] - [SO_4^=] - [NO_3^-] - [HCO_3^-]$$

(concentrations expressed in µeq l^{-1})

Then the mean value of the imbalance (ΔC) is found equal to 0.8 µeq l^{-1}, i.e. 0.8% of the ion budget. The imbalance observed for 30% of the samples is generally negative, showing that, in this case, a cation has probably been missed. It is worth emphasizing, however, that anions and alkaline earth metals were measured after meltwater filtration and, alkaline metals on unfiltered samples, and that the acidity was determined after the destruction of bicarbonates by the acid used in our titration method.

Table 2.   Mean concentrations of major ions in the soluble fraction of snow meltwater collected at Col de la Brenva for 62 samples in ionic balance. Eleven samples were found to be alkaline.

	Concentrations		Standard deviation	Variation range
	ng g^{-1}	µeq l^{-1}	µeq l^{-1}	µeq l^{-1}
Na$^+$	16.1	0.7	1.5	9.5 - 0.1
K$^+$	19.6	0.5	1.1	5.2 - 0.04
NH$_4^+$	36.2	2.0	1.1	5.2 - 0.2
Mg^{++}	63.2	5.2	6.6	33.1 - 0.1
Ca^{++}	595	29.7	41.2	214.8 - 0.2
H$^+$	8.4	8.4	6.4	21.3 - 0.2
Total cations		**45.5**		
Cl$^-$	63.8	1.8	1.4	5.7 - 0.2
NO$_3^-$	298	4.8	5.9	40.7 - 0.3
SO$_4^=$	914	38.1	43.2	249.0 - 1.0
Total anions		**44.7**		
HCO$_3^-$(alkaline samples)	5.3		6.76	21.9 - 0.8

Finally, for 70% of the samples, the low values of ΔC is a good indication that no major ion has been missed in the ion budget calculation. Calcium and sulphate ions are largely dominant over the studied profile but this is apparently due to the sporadic deposition of concentrated (up to 26.5 mg l^{-1}) dust layers superimposed on a general background where the concentrations are one order of magnitude lower.

500

Taking Na as the reference element for sea salt, it can be deduced
from Table 2 that this aerosol source contributes very little to the
impurity deposition in the study area: the mean value for this element
is 0.7 μeq l^{-1} part of which is also of crustal origin.

The measurements indicate that the snow is slightly acidic (mean
value 7.4 μeq l^{-1} corresponding to a calculated pH-value of about 5.1)
but a few snow layers show a slight alkaline reaction. Among the 90
samples analysed in this study about 18% were found to be alkaline and
had high calcium contents, 9% were found to be acidic and had nearly no
calcium, and 73% contained both calcium and acidity. The apparent
acid-base equilibrium existing in the snow of this area will be
discussed in detail below.

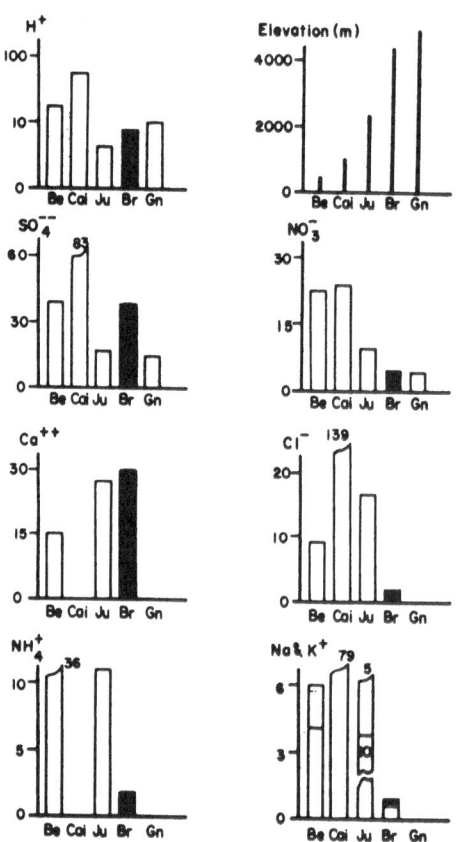

Figure 7   Chemical concentration in precipitation collected at five
European locations of various elevations. Be: Bern,
Switzerland, 500 m, (Fuhrer, 1984); Cai: Cairngorm
Mountains, Scotland, 1100 m (Davies et al.); Ju:
Jungfraujoch, 3500 m (Fuhrer, 1984); Br: Col de la Brenva,
4350 m (this study); Gn: Colle Gnifetti, 4400 m (Schotterer
et al., 1984).

5.2 Comparison With Other Sites

The impurity content of the snow deposited at Col de la Brenva has been compared with (1) some available glaciochemical data for other very high elevation glaciers (Table 3) and (2) precipitation chemical compositions for European sites located at various elevations (Figure 7).
The transport of sea-salt particles to upper tropospheric levels decreases rapidly as a function of elevation so that the amount not only of calcium, but also of sodium (except in polar areas) depends mainly on the deposition of crustal weathering material. The Na and Ca values listed in Table 3 indicated that this continental influence is significantly stronger in the Alps than at all the other locations. Ammonium, an ion formed from the local or regional emissions of NH_3 is apparently one order of magnitude less at remote areas such as South Pole, Crête (Greenland) or Mt Logan (Yukon) than in the Alps. The continental input (as indexed by Ca) is particularly important at high elevations (Figure 7) suggesting some airborne transport in the mid troposphere and not only a contribution from the surrounding ground at lower elevations. On the other hand, sulphate and overall nitrate concentrations are higher at low elevation sites, which indicates probable sources at ground level. Finally, comparison of the acidity values is generally difficult since the methods used for determining this parameter vary and the significance of the measured value may be different depending on the technique. At upper sites, the acidities are however significantly less than at low altitudes, probably in relation to the sulphate and nitrate concentrations. Note that at Col du Dôme, Delmas and Aristarain (1978) found mean annual values in the range 10-14 μeq 1^{-1} for mid seventies alkalinity (the titration method used in the latter study was very similar to that of the present study).

5.3 Seasonal Variations

Seasonal patterns in the deposition of impurities may provide useful information as to their origin. For instance, it is generally found at low altitudes that atmospheric pollutants peak in winter whereas dust derived compounds are present in greater amounts in summer precipitation. We shall focus our attention on only three parameters for which the seasonal variations are particularly well marked: the Ca/Mg equivalent concentration ratio, the concentration of NH_4^+ and the acidity (Figure 8). Summer layers are defined by meltlayers.

5.3.1 Ca/Mg ratio. The Ca/Mg ratio exhibits large variations, in the range $\overline{0.6 - 7.9}$ (Figure 8). Minima (around 2) are observed in winter snow, maxima (5) for summer layers. As these ratios are calculated on the soluble part of the crustal material, they cannot be compared with standard rock compositions. This seasonal variation may be linked with the larger amount of $CaSO_4$ in summer than in winter snow.

Table 3. Sodium, ammonium and calcium content in snow at very high elevation sites.

	South Pole	Crête (Greenland)	Mt Logan (Yukon)	Quelccaya (Peru)	Colle Gnifetti (Swiss Alps)	Col du Dôme	Mt Blanc summit French Alps	Col de la Brenva
Elevation	2800	3170	5300	5670	4450	4280	4807	4350
Na	0.6	0.3-1	0.062	~1	~2	4.2	2.0	0.7
Ca	0.04	0.4	-	-	-	9.8	5.0	30.4
NH_4	0.2	0.3	0.26	-	~6	-	-	2
(μEquiv. l^{-1})								
Ref.	(1)(2)	(3)(4)	(5)	(6)	(7)	(8)	(8)	(9)

(1) Legrand and Delmas, 1984
(2) Boutron, 1979a
(3) Hammer, 1978
(4) Boutron, 1979b

(5) Delmas et al., 1985
(6) Lyons et al., 1985
(7) Schotterer et al., 1984
(8) Batifol and Boutron, 1984
(9) this study

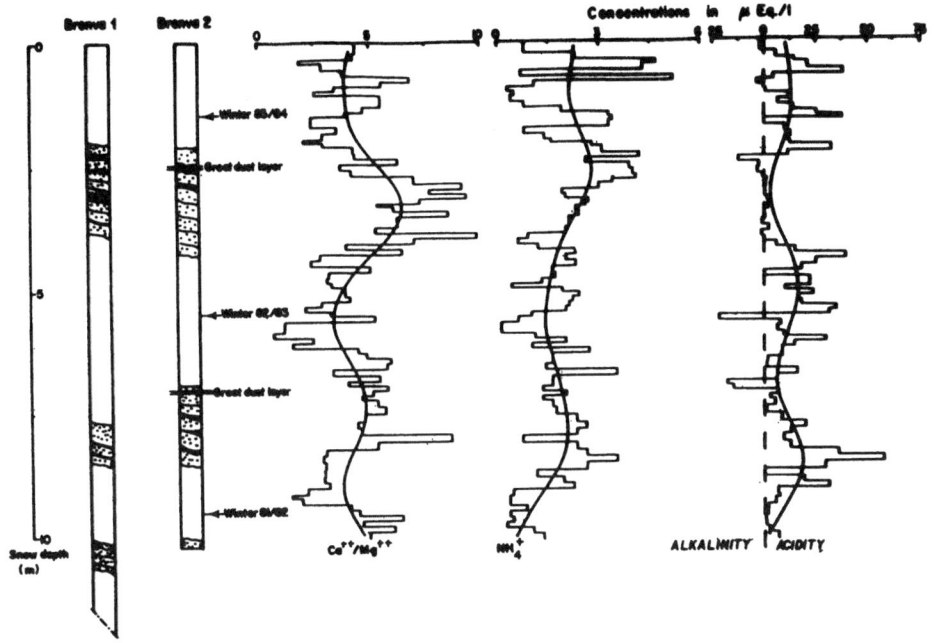

Figure 8   Variations of acidity/alkalinity values, ammonium
            concentrations and Ca/Mg equivalent ratios as a function of
            depth. The dating of the firn cores is also given.

5.3.2 Ammonium. The concentrations of ammonium are apparently higher
in summer that in winter snow layers, varying by about one order of
magnitude. However the studied time period is too short to draw a firm
conclusion. Our results are not in contradiction with the conclusion
of Asman and Janssen (1986) who indicate that NH3 production is
generally minimum in winter, in relation to agricultural practices and
weather conditions.

5.3.3 Acidity. Our acidity profile exhibits a clear seasonal pattern
with values near zero in summer and up to 40 - 50 μeq l⁻¹ in winter.
The measured acidity being the result of an atmospheric balance between
acid and alkaline compounds, this observation can be interpreted in
different manners (see below). The "negative" acidity values are found
in summer layers, particularly in frozen meltlayers containing dust
bands.

5.4 Interpretation of the Results

5.4.1 Insoluble particulate matter. The sea salt contribution being
particularly low at the study site, the aerosol deposited has mostly a
continental origin. This deposit of soil-derived material is generally
poorly soluble.

We shall consider first the case of aluminosilicates (clays). The equilibrium to take into account has been proposed by Stumm and Morgan (1970), (equation 3):

$$CAT-Al-Silicate + H_2CO_3 + H_2O \rightleftharpoons HCO_3^- + H_2SiO_4 + CAT + Aluminosilicate$$

In this reaction the cation (CAT = Na^+ - K^+ - Ca^{++} - Mg^{++}) may be freed, depending on the pH of the solution. At pH levels of the natural precipitation (in the range 5 - 6) these metals are only partly released. However, as already pointed out, the alkaline metals Na^+, K^+ and NH_4^+ were measured by ion chromatography on unfiltered samples. The acid eluent (pH $\sim$ 2.3) used in this analytical method dissolves a great part of natural particulate matter and totally releases the alkaline metals bound to aluminosilicates (Legrand and Delmas, 1987). Both Na^+ and K^+ are found to be highly correlated (r = 0.85, see Table 4), showing that they have the same (crustal) origin.

Alkaline earth metals were measured on filtered samples. For these metals the acidity conditions of equation 1 are natural and the release of Ca^{++} and Mg^{++} from aluminosilicates is likely to be low. In this case, we can assume that the major part of calcium and magnesium in solution is due to soluble compounds such as nitrates, sulphates or carbonates. These two metallic ions account for 33% and 6% of the ionic budget, respectively. The excellent correlation between them (r = 0.95, Table 4) suggests that they have common origins which may be not only crustal (calcium and magnesium carbonates and calcium nitrate) but also anthropogenic (sulphates). Carbonates are important since they can neutralise part of the natural acidity:

(4) $(Ca, Mg) CO_3 + H^+ \longrightarrow CO_2 + (Ca, Mg)^{++} + H_2O$

This reaction could occur during aerosol transport in the cloud droplets or just after sample melting. From the large excess of sulphate in comparison to nitrate ($SO_4^=/NO_3^- = 8$), we deduce that the acid involved in equation 4 is essentially H_2SO_4.

5.4.2 <u>The acid-base equilibrium of snow</u>. As shown in the preceding paragraph, carbonated dust plays a major role in the acid-base equilibrium occurring in meltwater samples. On the other hand, the very weak contribution of ammonium to the ionic budget ($\sim$ 2%) indicates that ammonia is a negligible natural acidity neutralising agent at these elevations. As ammonium chloride and nitrate are unstable in the atmosphere at extremely low NH_3 vapour pressures, it is reasonable to consider that ammonium bisulphate (NH_4HSO_4) is the most probable NH_4-containing compound of the background atmosphere in the area.

We have tried to reach a deeper understanding of the intricate acid-base equilibrium of snow. The following interpretation, based on a proposition first made by Legrand (1985) for Antarctic snow, must be considered as provisional for Alpine precipitation since based on an insufficient number of samples. These three distinct categories of samples have been considered:

A – poorly buffered acid samples (low dust contents),
B – moderately impurity-loaded samples,
C – highly impurity-loaded, alkaline or nearly alkaline samples,
    coresponding to dust bands.

The ionic-balance diagrams for these three different cases are reported
in Figures 9, 11, and 13.

Table 4   Correlation table between the measured ion concentrations.
           Only the values of r which may be considered as significant
           are given (i.e. higher than 0.3, according to Pearson's
           criteria).

	Na^+	NH_4^+	K^+	Cl^-	NO_3^-	SO_4^{--}	H^+	Mg^{++}	Ca^{++}
Na^+	–	–	.85	–	–	–	–	–	–
NH_4^+		–	–	–	–	–	–	–	–
K^+			–	–	–	–	–	–	–
Cl^-				–	.62	.61	–	.67	.60
NO_3^-					–	.91	–	.87	.89
SO_4^{--}						–	–	.91	.96
H^+							–	–	–
Mg^{++}								–	.95
Ca^{++}									–

Cases A:

This case, representative of only a limited number (5) of samples,
allows us to describe atmospheric chemistry features in background
winter conditions when pollution is high and dust fallout unimportant
(< 0.1 mg g^{-1}). The sum of the concentrations of soluble alkaline and
alkaline earth metals is lower than 2 μeq l^{-1} and the ion balance is
very similar to those found in polar regions (Legrand and Delmas, 1984)
(Figure 10). The acidity (12 μeq l^{-1} corresponding to a pH of 4.8) is
relatively weak, 76% due to H_2SO_4, with HNO_3 and HCl accounting for
only 16 and 8% respectively. No significant amount of organic acids
(in particular formic acid) was detected by ion chromatography.
    The $SO_4^=/NO_3^-$ ratio is equal to 5.6 (concentrations expressed in
μeq l^{-1}) a value significantly higher than in rainwater from northern

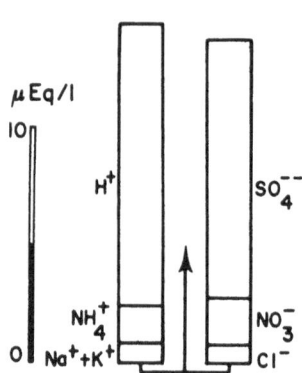

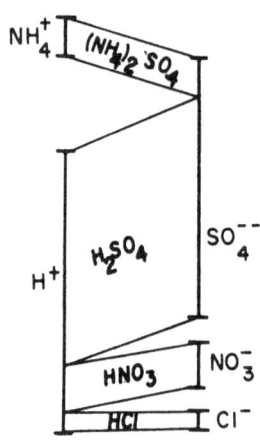

Figure 9 Ion balance for
samples with low ion
contents (Cases A, see
text).

Figure 10 Tentative aerosol
composition
reconstruction from
measurements (for cases
A, see text).

mid-latitudes, where values in the range 2-3 are more common (Likens et al., 1979). Very recently, Laird et al. (1986) have found values around 3 for the same ratio in snow from the western mountains in the USA. This could be an indication of the high degree of pollution of the European mid-troposphere by sulphur pollutants.

Cases C:

Eleven samples were found to contain high amounts of dust. Generally their ionic budget is very high (130 μeq l^{-1}) and the insoluble impurity content in the range 10-30 μeq l^{-1}. Ca^{++} and $SO_4^=$ are the most important ions, with mean concentrations of 52 μeq l^{-1} (Figure 11). Alkaline metal concentrations remain low ($\sim$ 1 μeq l^{-1}) with K^+ concentrations higher than Na^+. These samples are found in late-winter or spring snow layers.

The alkalinity measured for these samples corresponds to a mean pH of 6.2, suggesting that they contain bicarbonates (from calcium and magnesium carbonates). The residual alkalinity we have titrated is the result of a neutralisation reaction (equation 4 in the preceding section). We have assumed a $CaCO_3/CaSO_4$ weight ratio equal to 4 in Sahara dust, as given by Loye-Pilot et al. (1986) in their results from precipitation compositions in Corsica.

The diagram of Figure 12 takes into account this composition, but it must be kept in mind that an evolution in the calcite to gypse ratio may occur during the long range transport to the Alps, by a neutralisation of the calcite and its transformation into calcium sulphate. In this case the actual $CaCO_3/CaSO_4$ ratio could be lower. Finally the magnesium has been arbitrarily considered in the calculation under the chemical form of $MgSO_4$.

Cases B:

Cases A and C are extreme cases. Forty-five samples may be considered to be at an intermediate position since they are acidic but also contain relatively high amounts of impurities (mean ionic budget < 90 μeq l^{-1} - Figure 13, case B). Here, the insoluble fraction is only slightly higher than 1 mg l^{-1}, i.e. relatively low in comparison to case C. Ca^{++} still represents 50% of the cation amount, showing the important influence of crustal material.

This suggests that dust has been made soluble during its transport, according to a process already pointed out for the case C. Therefore, for these samples, the ratio R = (Ca, Mg)CO_3/(Ca, Mg) SO_4 is highly variable, depending on the neutralisation rate of the carbonates. The diagrams in Figure 14 represent three different neutralisation rates (from zero to ∞) for 3 identical measured meltwater chemical compositions. The anterior chemical history of the aerosol is unknown and cannot be determined from our glaciochemical measurements.

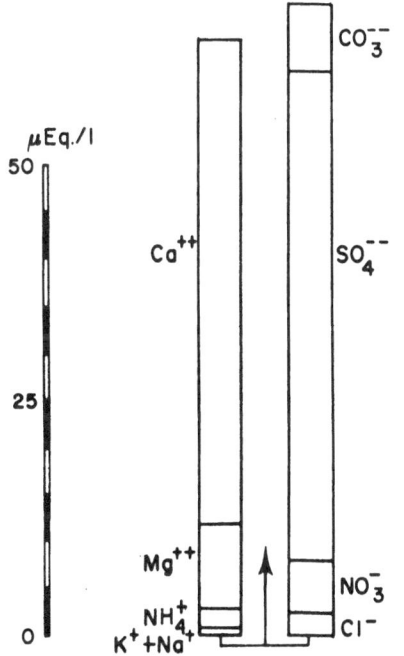

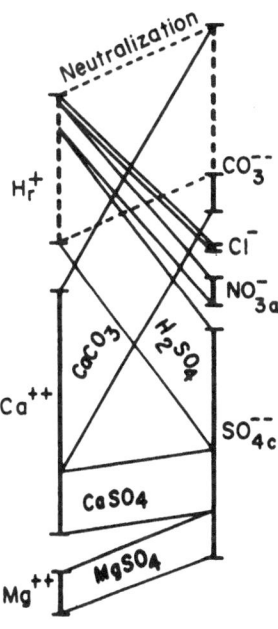

Figure 11. As in Figure 9, cases C.

Figure 12. As in Figure 10, cases C.

508

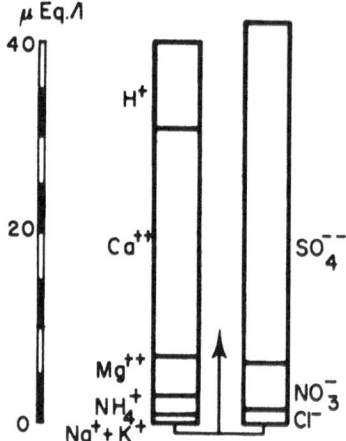

Figure 13   As in Figure 9, cases B.

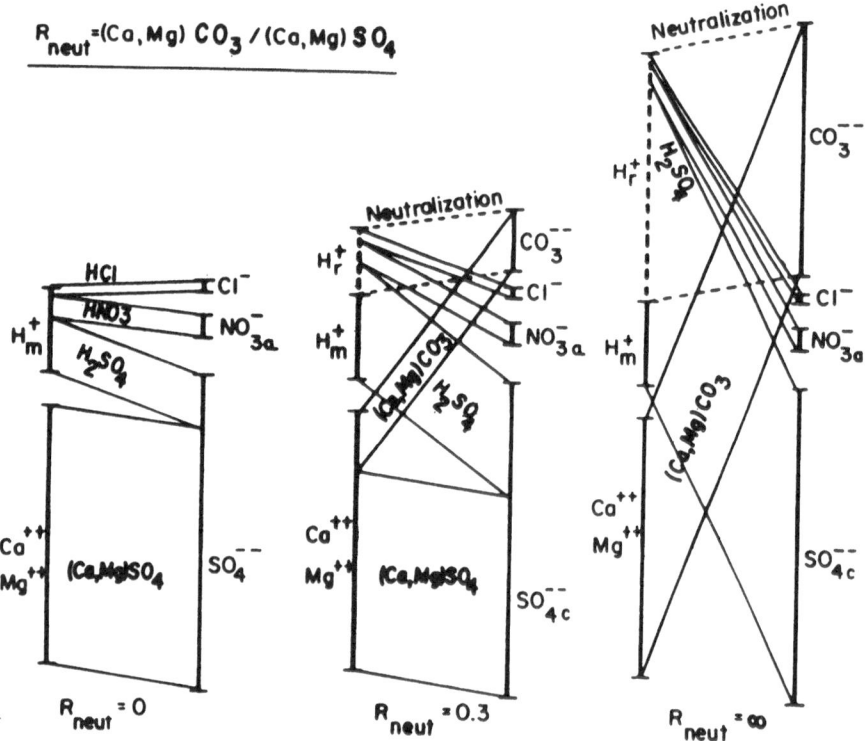

$$R_{neut} = (Ca, Mg) \, CO_3 \, / \, (Ca, Mg) \, SO_4$$

Figure 14   As in Figure 10, cases B. From the left to the right, the
diagrams correspond to a total, partial or negligible
neutralisation rate of the soil derived carbonate particles
(see text).

6. CONCLUSION

These preliminary chemical measurements carried out on a 10 m firn core drilled at a high elevation in the Mt Blanc area are promising. They demonstrate the importance of soil derived material in the chemical composition of the bulk aerosol deposition. In the Antarctic, the chemical impurities in snow are contributed both by sea-salt and gas-derived mineral acids (mostly H_2SO_4, HNO_3 and HCl). In the present case, the marine source is reimplaced by the crustal source, but anthropogenic acid pollutants (in particular H_2SO_4 and HNO_3) constitute an additional source of secondary aerosol. The role of NH_3 as a neutralising agent of background atmospheric acidity is not confirmed, the natural carbonates having a much more important influence on the acidity than expected.

It is hoped that the analysis of deeper firn cores, covering several decades, will allow an assessment of the influence of the anthropogenic pollution on the chemical composition of Alpine precipitation, a good indicator of the central European background atmosphere.

ACKNOWLEDGEMENTS

We are indebted to L Reynaud, M Pourchet and L Augustin for their help in the field. We acknowledge the financial support of the CEC, Brussels (Contract ENV-898-F).

REFERENCES

Asman, W.A.H. and Janssen, A.J. 1986. A long range transport model for ammonia and ammonium for Europe, and some model experiments. IMOU Report R-86-6, Rijksuniversiteit Utrecht (NL), pp. 79.

Batifol, F. and Boutron, C.F. 1984. Atmospheric heavy metals in high altitude surface snows from Mt Blanc, French Alps. Atmos. Environ. 18 (11), 2507-2515.

Boutron, C. 1979a. Alkali and alkaline earth enrichments in aerosols deposited in Antarctic snows. Atmos. Environ. 13 , 919-924.

Boutron, C. 1979b. Trace element content of Greenland snows along an east-west transect. Geochim. et Cosmochim. Acta 43 , 1253-1258.

Briat, M. 1978. Evaluation of levels of Pb, V, Cd, Zn and Cu in the snow of Mt Blanc during the last 25 years. In: Studies in Environmental Science, Vol 1., 225-228.

Davies, T., Abraham, P., Tranter, M., Blackwood, I., Brimblecombe, P. and Vincent, C. 1984. Black acidity snow in the remote Scottish Highlands. Nature 312 .

Delmas, R. and Aristarain, A. 1978. Recent evolution of strong acidity of snow at Mt Blanc. In: Studies in Environmental Science, Vol. 1, 233-237.

Delmas, R.J., Legrand, M. and Holdsworth, G. 1985. Snow chemistry on Mt Logan, Yukon Territory, Canada. Annals of Glaciology 7 , 213.

Fuhrer, J. 1984. Study of the chemical characteristics of wet and dry deposition in Switzerland. Proceedings of the 3rd European Symposium "Physico-chemical behaviour of atmospheric pollutants", Varese, Italy (20/9-1/10 1984), 423-432.

Haeberli, W., 1977. Sahara dust in the Alps - a short review. Z. Gletscherkunde und Glazialgeologie 13 , 206-208.

Haeberli, W., Schotterer, U., Wagenbach, D., Haeberli-Schwitter, H. and Bortenschlager, S. 1983. Accumulation characteristics on a cold, high alpine firn saddle from a snow pit study on Colle Gnifetti, Monte Rosa, Swiss Alps. J. Glaciol. 102 , 29, 260-271.

Hammer, C.U. 1977. Past volcanism revealed by Greenland ice sheet impurities. Nature 270 , 482-486.

Jouzel, J., Merlivat, L. and Pourchet, M. 1977. Tritium and activity in a snow core taken on the summit of Mont Blanc (French Alps). Determination of the accumulation rate. J. of Glaciology, Vol. 18 (80), 465-470.

Laird, L.B., Taylor, H.E., and Kennedy, V.C. 1986. Snow chemistry of the Cascade-Sierra Nevada Mountains. Environmental Science and Technology 20 , (3), 275-295.

Legrand, M.R., Aristarain, A.J. and Delmas, R.J. 1982. Acid titration of polar snow. Analyt. Chem. 54 (8), 1336-1339.

Legrand, M., De Angelis, M. and Delmas, R.J. 1984. Ion chromatographic determination of common ions at ultratrace levels in Antarctic snow and ice. Analyt. chim. Acta 156 , 181-192.

Legrand, M. and Delmas, R.J. 1984. The ionic balance of Antarctic snow: a 10-year detailed record. Atmos. Environ. 18 , 1867-1874.

Legrand, M. 1985. Chimie des neiges et glaces Antarctiques: un reflet de l'Environment. Thèse Université de Grenoble, pp. 439.

Legrand, M.R. and Delmas, R.J. (1987). Experimental protocol for the chemical analysis of snow, firn and ice cores. Seasonal snowcovers: Physics, Chemistry, Hydrology. H.G. Jones and W.J. Orville-Thomas (Eds.) NATO ASI Series, Vol. 211 (Series C), pp. 225-254. Reidel Publishing Co.

Likens, G.E., Wright, R.F., Galloway, J.N. and Butler, T.J. 1979. Acid rain. Scientific American 241 , (4), 39-47.

Lyons, W.B., Mayewski, P.A., Thompson, L.G. and Allen III B. 1985. The glaciochemistry of snow-pits from Quelccaya ice cap, Peru. Annals of Glaciology 7 , 84-88.

Loye-Pilot, M.D., Martin, J.M. and Morelli, J. 1986. Influence of Saharan dust on the rain acidity and atmospheric input to the Mediterranean. Nature 321 , 427-428.

Schotterer, U., Oeschger, H., Wagenbach, D. and Munnich, K.O. 1985. Z. Gletscherkunde und Glazialgeologie 21 , 379-388.

Stumm, W. and Morgan, J.J. 1970. Aquatic chemistry. Wiley Interscience, New York.

CHEMICAL COMPOSITION OF THE SEASONAL SNOWCOVER AT A SOUTHERN FRENCH
ALPS SITE: SOME PRELIMINARY RESULTS

V.M. Delmas, F. Ronseaux and R.J. Delmas
Laboratoire de Glaciologie et Geophysique de l'Environment
B.P. 96 38402 St Martin d'Hères Cedex
France

1. INTRODUCTION

In mountainous areas, snow is a major component of total deposition and
its relative importance increases with elevation. At 1500 to 2500 m
altitudes, snow accumulation periods range from 5 to 7 months depending
on many factors (exposure, latitude etc.). The main purpose of this
study is to physico-chemically characterise snowcover and snowmelt
water at an experimental site in the southern French Alps. Preliminary
sampling has been done during spring 1985 at 3 different altitudes in
the proximity of the site. The results are presented below.

2. STUDY AREA

The experimental area is located 20 km east of Briancon in the Parc
National des Ecrins (45°N and 4°05'E, Figure 1). Due to the
mediterranean influence up the Durance valley, the region is the French
Alps sunniest and driest area (Dollfus and Guet, 1981). The regional
geology is variable with a mixed acidic and basic substratum. The main
experimental site itself is situated on an openfield at an altitude of
1800 m and has an east exposure. Larch stands dominate the surrounding
vegetation. The bedrock is mainly composed of siliceous material.

3. ANALYTICAL PROCEDURES

All manipulations in the field and at the laboratory have been
conducted with great care to minimize snow or meltwater contamination
by gaseous components and dust (clean roon, sealed plastic bags, clean
sampling flasks) and by operators (clean clothes, plastic gloves,
masks) (Ronseaux, 1985).
    Three snowprofiles were sampled during early spring 1985: two of
them were done on March 7th (sampling site 1: "Le Casset", 1500 m;
sampling site 2: "Cret du cu", 1800 m) and the last one on April 27th

M. H. Unsworth and D. Fowler (eds.), Acid Deposition at High Elevation Sites, 511–516.
© *1988 by Kluwer Academic Publishers.*

512

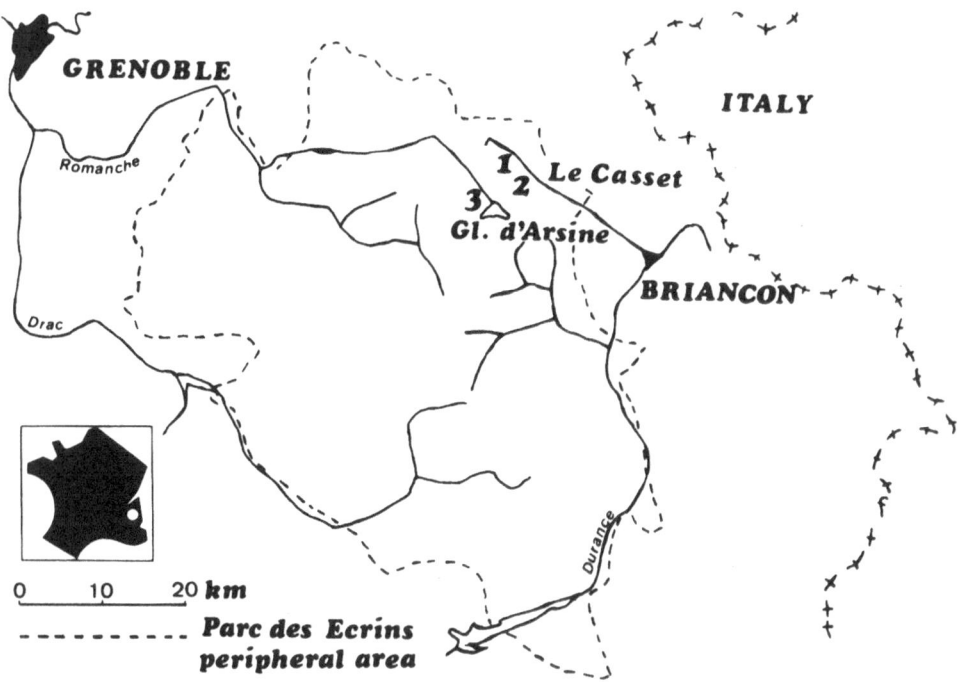

Figure 1. Sampling area and site, Parc des Ecrins, spring 1985.

(sampling site 3: "Glacier d'Arsine", 2500 m). Snowpits were dug and samples obtained by progressive sampling from the surface down to the bottom every 7 cm with the help of a stainless steel pallet and small laboratory plastic flasks in which they were kept frozen before melting for analysis.

In the laboratory, snow samples of the 3 snowprofiles have been melted in the capped plastic flasks at room temperature. Immediate analysis has been achieved for soluble NH_4^+, Na^+, K^+ cations and SO_4^{2-}, NO_3^-, Cl^- anions by ion chromatography (Dionex 2010) with an anaytical precision of 5-10% in most cases (Legrand et al., 1983). An acid titration method which eliminates atmospheric CO_2 influence, (Legrand et al., 1982) has been used for H^+ determination (precision $\pm$ 1 µeq 1^{-1}). Ca^{2+} has been measured by graphite furnace atomic absorption spectrophotometry (Perkin Elmer 2380 + HGA500).

4. RESULTS AND DISCUSSION

No significant melt occurred before sampling. It is assumed that snowprofiles give reasonable quantitative measures of total atmospheric deposition during winter. Sampled snowdepths were respectively of 110, 130 and 170 cm for sites 1 to 3. Mean ion concentrations are

represented for each site (Figure 2, dark bars). Acidity was not
measured for sampling site 2. The major anions observed in the
snowpack were sulphates and nitrates with mean SO_4/NO_3 equivalent
ratios of 4.2, 2.3 and 4.9 respectively, according to the site. Cl^-
contributed very few to the total anion equivalent (3%). At the
highest site (3) the H^+ concentration was very low (5.5 µeq 1^{-1}),
especially when compared with Ca^{2+} concentrations (15.8 µeq 1^{-1}).
This suggests that H^+ substantial neutralisation by alkaline
terrestrial dust probably occurred. At the lowest site, the snow was
found to be more acidic and H^+ concentration (24.5 µeq 1^{-1}) was

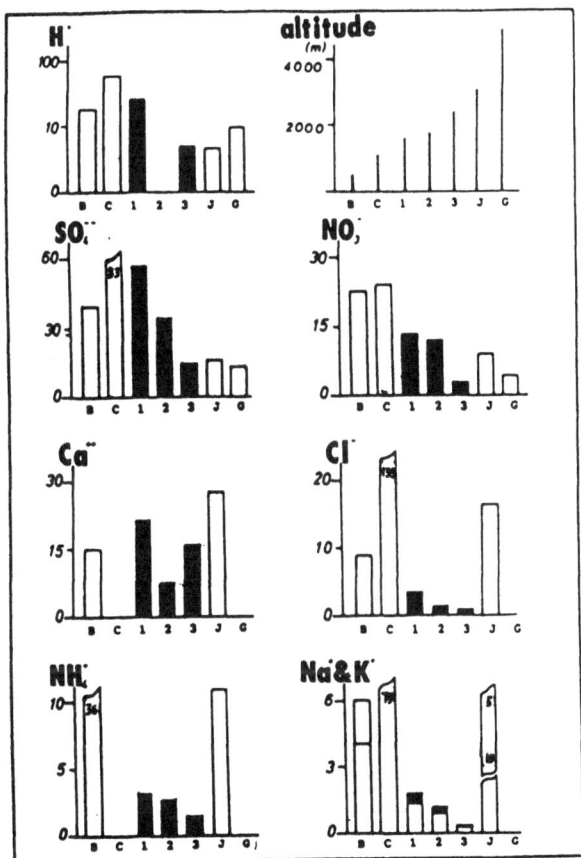

Figure 2.   Major ion concentrations (µeq 1^{-1}) at different sites
B:Berne, Switzerland (Fuhrer J., 1984)
C:Cairngorm, Scotland (Davies et al., 1984)
1, 2, 3:Parc des Ecrins, France (present study)
J. Jungfraujoch, Switzerland (Fuhrer J., 1984)
G:col Gnifetti, Switzerland (Schotterer et al., 1985)
At the upper right corner: altitude distribution of the
sites.

514

comparable with Ca^{2+} concentration (21.1 µeq l^{-1}). In the three
snowprofiles, NH_4^+ concentrations were relatively low (< 4 µeq l^{-1}).
Snow acidity neutralisation by ammonia was therefore considered to be
negligible.

For the four main ions (H^+, Ca^{2+}, SO_4^{2-}, NO_3^-), the results are in
good agreement with other values found in literature (Figure 2). At
the opposite, NH_4^+, Na^++K^+ and Cl^- concentations are much lower in this
study than in the others.

Bulk deposition loading of major ions at the different altitudes
have been estimated by taking into account the total quantity of water
sampled and the accumulation period duration for each site (Figure 3).

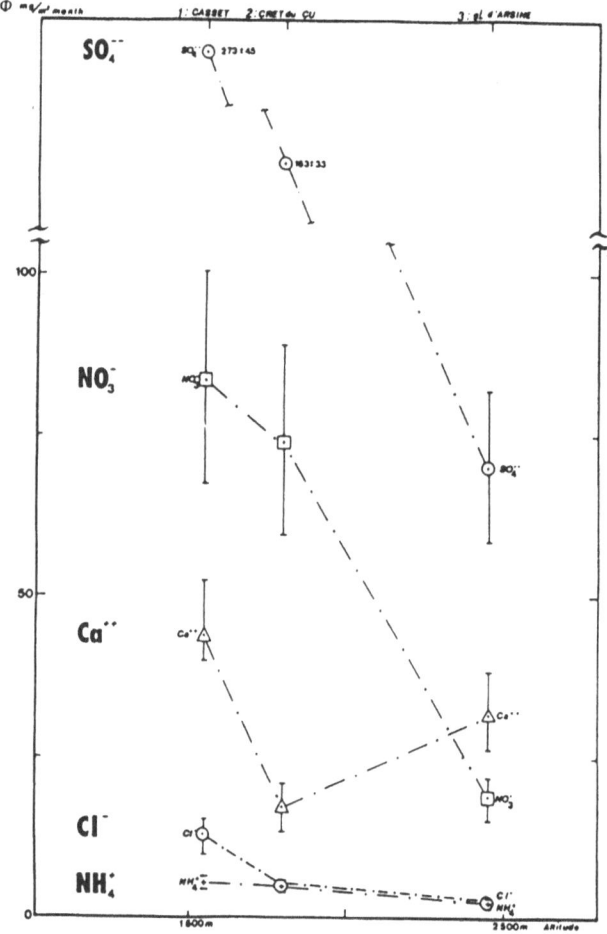

Figure 3.   Bulk deposition loading (mg m^{-2}month) of major ions at the
sampling sites, Parc des Ecrins, spring 1985.

The loading decline with altitude is clear for NO_3^- and SO_4^{2-}. This is not the case for Ca^{2+}, which comes from alkaline dust particles of different origins: local or regional contributions, Saharian dust events. These are often observed in Europe during spring when polar air masses penetrate into the northern Sahara for a short time period, generating strong low pressure systems (Schutz, 1980).

5. CONCLUSION

Preliminary sampling has permitted us to estimate atmospheric deposition loading during snow accumulation period at three different altitude sites. Acidity, sulphate and nitrate loading show a well marked decline at highest elevations. The comparison of the respective calcium and ammonium mean concentrations suggest that acidity is probably neutralized more efficiently by carbonates than by ammonia. To determine the evolution of both snowcover and snowmelt water quality, a more complete survey will be done during the snow season 1986-87. It will consist of snowcover and bulk deposition regular sampling and collection of snowmelt water flowing from 3 lysimeters (surface area: 1 m^2 each one). Also, the samples will be analysed more completely (aluminium, magnesium, organic acids). Attention will be focused on acid neutralisation mechanisms by alkaline dust particles within the snowpack, during the melt.

ACKNOWLEDGEMENTS

We acknowledge interest and assistance of the staff of Parc des Ecrins in our work, in particular J.P. Arnaud, R. Keck and J. Juge. This work was supported by the "Ministère de l'Environnement, Direction de la Protection de la Nature", the "Ministère de la Recherche et de l'Enseignement Superieur" and the "Ministere des Affaires Etrangères, collaboration France-Quebec".

REFERENCES

Davies, T.D., Abrahams, P.W., Tranter, M., Blackwood, I., Brimblecombe, P. and Vincent, C.E. 1984. Black acidic snow in the remote Scottish Highlands. Nature 312 , 58-61.
Dollfus, O. and Guet, J. 1981. Recherches en Brianconnais, report published by CEMAGREF, groupement de Grenoble 1 , 32pp.
Fuhrer, J. 1984. Study of the chemical characteristics of wet and dry deposition in Switzerland. In: Physico-chemical Behaviour of Atmospheric Pollutants. Proceedings of the third Europeans Symposium held in Varese, Italy (Eds. B. Versina and G. Angeletti). D. Reidel Publishing Company, 1 , 423-432.
Legrand, M., Aristarain, A.J. and Delmas, R.J. 1982. Acid titration of polar snow. Anal. Chem. 54 , 1336-1339.

516

Legrand, M., De Angelis, M. and Delmas, R.J. 1983. Ion chromatographic determination of common ions in Antarctic snow and ice. Anal. Chim. Acta **156** , 181-192.

Ronseaux, F., 1985. Chimie de la precipitation au Mont Blanc et dans les Ecrins. Internal report (DEA), laboratoire de glaciologie et de géophysique de l'environnement, 35pp.

Schotterer, U., Oeschger, H., Wagenbach, D., Munnich, K.O. 1985. Information on paleo-precipitation on a high-altitude glacier (Monte Rosa, Switzerland). In: Zeitschrift fur Gletscherkunde und Glazialgeoligie **21** , 379-388.

Schutz, L. 1980. Long range transport of desert dust with special emphasis on the Sahara. In: Aerosol: anthropogenic and natural sources and transport. (Eds. T.J. Kneip and P.J. Lioy). New York Academy of Sciences Pub. **338** , 515-532.

CHEMICAL COMPOSITION OF SNOW IN THE REMOTE SCOTTISH HIGHLANDS

T.D. Davies, P. Brimblecombe and I.L. Blackwood
School of Environmental Sciences
University of East Anglia
Norwich NR4 7TJ
UK

M. Tranter
Department of Oceanography
University of Southampton
Southampton
UK

P.W. Abrahams
Department of Geography
University College of Wales
Aberystwyth
Dyfed
UK

ABSTRACT. Altitudinal surveys suggest an increase in ionic
concentration in snow with increasing fall-path, although such studies
are complicated by the marked spatial variability in the chemical
composition of fresh snow-cover. This variability is characterised and
the implications for snowfall/snowcover sampling are addressed.
Nitrate concentrations in snow are relatively enhanced compared to
concentrations in rainfall. The NO_3^-/SO_4^{2-} ratios in snowfall appear
to be dependent on back-trajectory type and associated source region.
We consider the relevance of the observations for studies of the
chemical evolution of air masses. The Cairngorm range of mountains
experiences, not uncommonly, deeply-coloured snowfalls with a large
black carbon content which appears to be directly related to snowfall
acidity. The events can represent relatively large contributions to
the total impurity deposition in the mountains. The synoptic
conditions which lead to the black deposition events have been
identified.

1. INTRODUCTION

As part of a study of the removal of soluble ions from a melting
snowpack and interactions with soil chemistry, stream chemistry and

M. H. Unsworth and D. Fowler (eds.), Acid Deposition at High Elevation Sites, 517–539.

soil microbiology (e.g. Tranter et al., 1986; Thompson et al., 1987), attempts were made to collect representative samples of fresh snowfall. It proved impossible to sample every snowfall because of severe weather conditions regularly experienced in the study catchment (e.g. gusts up to 260 km hr^{-1}), but over three field seasons, we:

1) collected a number of fresh snowfalls at 1100 m,
2) undertook some altitudinal surveys, and
3) studied spatial variability in the chemical composition of freshly-fallen snowcover.

The chemical composition of snow is of obvious relevance to the "acidic flush", consequent upon snowmelt. This phenomenon may have ecological significance for the large parts of Europe and North America which experience seasonal snowcover (e.g. Leivestad and Muniz, 1976). Moreover, the nature of chemical deposition to an area is likely to be affected by the relative contributions of snow to the total precipitation. Dry deposition to a snow surface is different to that over the snow-free surface (e.g. Whelpdale and Shaw, 1974). In addition, there is accumulating evidence that sub-cloud scavenging is different for snow and rain. Snow is a more efficient scavenger of aerosols than is rainfall (Knutson et al., 1976; Raynor and Hayes, 1982) and it may remove relatively more nitrate (Raynor and Hayes, 1982; Sood and Jackson, 1970; Topol, 1986) although this might be related to the different precipitation amounts usually observed in snowfalls and rainfalls (Dasch, 1987). If there are differences in sub-cloud scavenging by snow and rain, then it is likely that altitudinal variations in precipitation composition will differ between the two forms of precipitation. The acidity of rainfall samples appears to increase with height in mountain areas (references in this volume), but altitudinal variations in snowfall composition have not received much attention. One reason for this is the enormous difficulty of collecting representative samples of the quantity and size-distribution of falling snowflakes (e.g. UKRGAR, 1987).

It is not always appropriate to directly compare observations from, and processes occurring in, different mountain areas. For example in the UK, detailed studies have been conducted on Great Dun Fell in northern England (references in this volume) and a less-intensive programme implemented on Cairngorm Mountain in Scotland. Cairngorm is higher, further north and more remote from the west coast; consequently it experiences relatively more snowfall. Elevated subsidence inversions, associated with anticyclones to the east, often occur at around the height of Cairngorm (~1250 m). Such synoptic conditions may lead of "acidic episodes" in northern Europe (e.g. Smith and Hunt, 1978) and can produce pronounced "black" acidic snowfalls in the Cairngorm range, where the presence of the elevated stable-layers (and interaction with the mountains) may play an important role (Davies et al., 1984). Great Dun Fell has a more "maritime" climate and is less likely to experience the conditions of pollutant deposition associated with long-range transport from East Europe/USSR sometimes found in the Cairngorms.

The "black" snowfalls in the Cairngorms, although frequently of small water-equivalent amounts, can make overwhelming contributions to the chemical composition of streamwater upon snowmelt (Tranter et al., 1987a). When the appropriate synoptic conditions prevail, tens of centimetres of snow may be highly-coloured (such an episode is discussed later). Grey snow episodes have occurred in other parts of the UK; large quantities of black particles have been observed in cloudwater collected on Great Dun Fell (M H Unsworth, S.A. Penkett, pers. comm.), and the southern Pennines in England (Gervat, 1985). A similar event to the black snow episode reported by Davies et al. (1984) appeared to have occurred on Whiteface Mountain, New York (Castillo, 1986) and Castillo et al. (in press) explore the possible relationship between elemental carbon and cloud-water acidity. Goldberg (1985) provides an extensive review of black carbon in the atmosphere. This review highlights its potential importance in atmospheric chemistry and many of the papers referred to by Goldberg indicate the role that black carbon may play in the production of acid in the atmosphere.

Black snow episodes in the Cairngorms appear to be relatively more frequent than elsewhere in the UK (Davies et al., 1984), probably because of the range's height and location, relative to prevailing synoptic patterns and the source regions of eastern Europe/USSR, although spectacular black showers were observed in NE Scotland in the late nineteenth century and impressed the recorder of the events sufficiently for him to claim that they would "in the end obtain world-wide fame" (Brimblecombe et al., 1986).

Implicit in many studies of snow and snowpack chemistry is the assumption that the spatial variability in the chemical composition of snowcover is small (Tranter et al., 1987b). In reality, this is not the case, especially in topographically complex areas. Unless future sampling strategies are explicitly designed to obviate the consequent problems, the general applicability of snow chemistry studies will continue to be restricted.

2. SITE, SAMPLING AND ANALYSIS

The main sampling site was Ciste Mhearad (57°06'N, 03°38'W), to the NE of Cairn Gorm Peak, at an altitude of 1080 m (Figure 1). The area is relatively remote with the major pollution sources around 150 km to the south (for further details see Tranter et al., 1986).

It was not possible to collect falling snows. Fresh snowfall was collected on the occasions when the mountain was accessible at, or within a few ($\sim$ 6) hours of snowfall, and when the observers were confident that there had been no wind redistribution of the consolidated layer, or melting subsequent to snowfall. The fresh snow-layer was immediately recognisable upon inspection, and the greater proportion of the new layer was sampled. Samples were collected by a PTFE-coated plastic scoop, and transferred to a field-laboratory before rapid melting in a water bath (Tranter et al., 1986; Abrahams et al., in press), whereupon they were suction-filtered through a 0.45 μm cellulose acetate filter. An aliquot was taken for

520

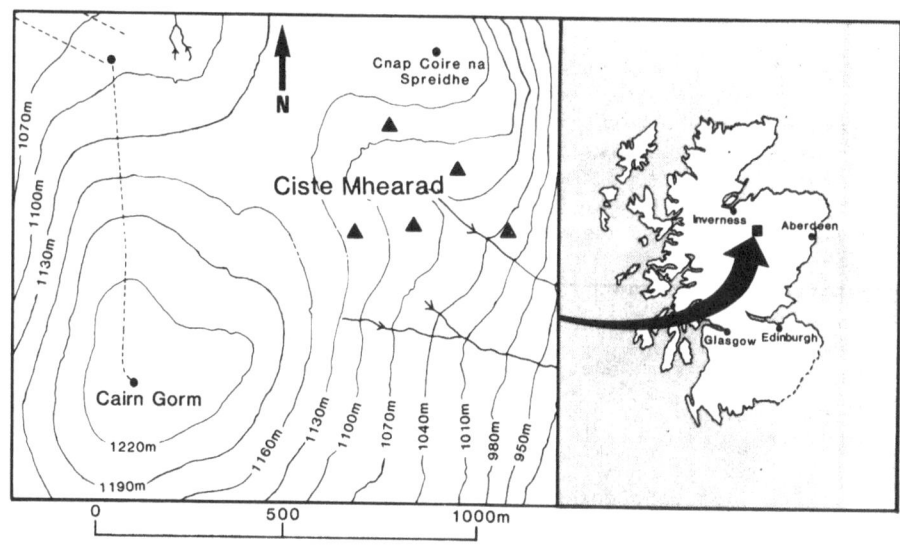

Figure 1   Location of the sampling site in Ciste Mhearad on Cairn Gorm
            Mountain in the Scottish Highlands.

immediate pH and conductance measurements. Samples for trace-element
analysis were drawn through filters which had been previously washed
with a solution of 3% (v/v) HNO$_3$. The pH measurements were made via a
technique modified from that of McQuaker et al. (1983). All samples
were then stored in pre-cleaned bottles at 4°C; the trace-element
samples being first acidified by the addition of HNO$_3$ (Abrahams et al.,
in press).

     Analysis for the major anions was performed by IC on a Dionex
Model 12; detection limits were 0.2 µeq l^{-1} and precision was better
than 5%. Major metal ion concentrations were measured by AAS on a Pye
Unicam SP9 Spectrophotometer. Precision was between ± 1.0 µeq l^{-1} and
± 2 µeq l^{-1} and detection limits were an order of magnitude less.
Trace elements were determined by graphite furnace AAS on the same
instrument. Detection limits of 4 µg Al l^{-1}, 0.1 µg Cd l^{-1}, 1.0 µg Cu
l^{-1}, 1.0 µg Fe l^{-1}, 0.3 µg Al l^{-1} and 0.35 µg Pb l^{-1} were attained.
The residual particles from some of the samples were subjected to EDAX
and INAA analysis (Davies et al., 1984; Abrahams et al., in press).

3. VARIATION OF SNOWFALL COMPOSITION WITH ALTITUDE

Seven individual altitudinal surveys of freshly-fallen snow were
conducted over the period February 1984 – January 1985. Between 4 and

6 samples were collected at different heights in each survey. Six of
the surveys included sampling points at 620 m, 750 m, 1085 m and 1130
m. One survey consisted of sample points at 280 m, 295 m, 320 m, 410
m, 500 m and 570 m. Figure 2 shows some of the results from the latter

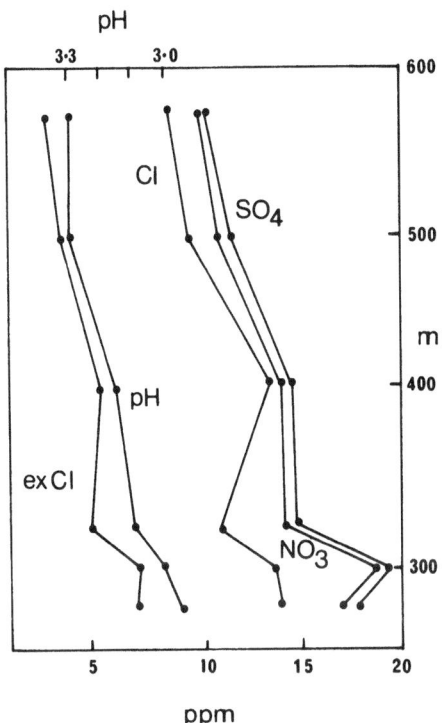

Figure 2   Variation of ionic concentrations in snowfall with elevation
            during one altitudinal survey.

survey. No attempt has been made to assess the statistical
significance of the observed gradient (the inverse relationship between
ionic concentration and altitude) because of the restraints placed upon
such analyses by the subsequent study of spatial variability (discussed
later), but Table 1 provides a summary of the results of linear
correlations of height against ion concentrations for all seven
surveys. The number of surveys is too small for a Chi-Squared, or
similar test, but it is of interest to note that the overall picture is
one of decreasing concentration with height. The trend, perhaps, is
stronger for H^+, NO_3^- and SO_4^{2-}, (considering both 'positive' and
'negative' count and 'probability' count). It might be expected that a

consequence of snowflakes' apparent relative efficiency at sub-cloud particle scavenging, and the fact that they remove relatively more nitrate, would be greater concentrations of pollutant ions with longer fall-paths through the atmosphere. However, there are important restrictions to be overcome in any similar study of fallen snow. The determination of snowfall amount and its variation with height is one, but the major constraint is the problem of collecting representative samples.

4. THE SPATIAL VARIABIITY IN THE CHEMICAL COMPOSITION OF FALLEN SNOW

Compositional variation in fallen snow, over altitude, is much smaller than the compositional variation between individual snowfalls (see, for example Davies et al., 1984, and later discussion). Consequently, spatial variability is less of a problem in comparative studies of a number of events than it is for studies of processes throughout individual events, or for assessing the total impurity loading in a catchment.

In order to quantify the possible extent of this constraint in our study area, we executed four surveys of freshly-fallen snow in the Ciste Mhearad catchment, sampling at 50 m intervals along a 500-800 m transect from the base to the head of the catchment. Between 13-15 samples were collected. Two additional, denser, surveys were conducted, at 1.0 m intervals at the same time as two of the '50 m surveys', by collecting a further 8-19 samples from a randomly-selected location within the catchment. Table 2a shows that the composition of the freshly-fallen snow is highly variable, with coefficients of variation (logged values) ranging from 1-144%. This degree of spatial variability in the chemical composition of fresh snowcover has important implications for the characterisation of snowfall (if freshly-fallen snow is collected) at similar upland catchments. Due regard should be given to the likely errors involved. For example, Table 2b shows the relationship between the number of samples, potential error of the mean (E_1) and coefficient of variation of log values. A full explanation is available in Tranter et al. (1987b). The data in Table 2b indicate that prohibitively large numbers of samples should be collected from log-normal distributions to derive a mean value with a small potential error.

Some of the observed variability may have resulted from sampling errors; slightly different depths of snow may have been sampled at each location. However, this would imply that even over scales of a few cm, freshly-fallen snow exhibits large variations in ionic concentrations with depth. Although it is frequently assumed that snowcover is chemically homogenous over a small scale in remote regions, some work (reviewed in Tranter et al., 1987b) has previously demonstrated that snowpack composition can be highly variable (e.g. Messer et al., 1981). At least some of the observed variability in snowpack composition is due to post-deposition wind redistribution and snowpack melting (Davies et al., 1987) but, in upland areas, it is also necessary to consider a variability impressed upon the snowcover during the deposition process. Atmospheric flow in the boundary-layer will be

highly perturbed in topographically-complicated terrain, and it is
likely that the deposition field will not be constant and some
redistribution will occur until the snowflakes become consolidated into
a layer. This process is important since the chemical composition of
precipitation is not constant throughout an event (e.g. Davies, 1985).
In addition, the mountain-induced atmospheric flow may lead to
mechanical sorting of the falling snowflakes by size and this could, in
turn, lead to spatial inhomogeneities in chemical composition.

As indicated earlier, the variability between individual events in
the Cairngorms is greater than the spatial variability of fresh
snowcover from one event. For example, the coefficient of variation
(logged values) for all the events collected varies from 16% (pH) to
9920% (NO_3^-). Consequently, we can examine the variation between
events, by collecting a relatively small number of samples for each
event, with some confidence. Inter- and intra-fall compositional
variability shall be discussed in detail elsewhere.

5. THE CHEMICAL COMPOSITION OF FRESH SNOWCOVER IN THE CAIRNGORMS

When discussing the composition of precipitation, it is important to
distinguish between the effects of overall concentration (ionic
strength) and the relative proportions of ions which comprise the
solute load (Brimblecombe et al., in press). Much of the variability
in the composition of precipitation is simply due to variability in the
overall concentration, and can be thought of as an effect of the
variable dilution of a given mass of solute by relatively pure water or
snow. Tranter et al. (1987b) have conducted a principal components
analysis on 173 surface snow samples from the Ciste Mhearad catchment,
including old, leached and wind-blown surfaces, and found that the
first three components described 96% of the total variance of H^+, Na^+,
Mg^{2+}, SO_4^{2-}, NO_3^- and Cl^- concentrations. The first component (71%
variance) weighted all ions equally and described the dominant effect
of dilution or overall concentration; the second component (a further
20% of the variance) differentiated the two main sources of solute to
snow, i.e. sea salt (Na^+, Mg^{2+}, Cl^-) and an acidic component (H^+, SO_4^{2-}
and NO_3^-); the third component described only a further 5% of the
variance and may have represented the process of differentiation
removal upon melt (Tranter et al., 1987b). Presenting the ion data on
triangular diagrams enables effects on composition other than dilution
to be examined. This expedient to interpretation is particularly
useful in this study, because the site does seem to offer the
opportunity to sample precipitation associated with trajectories having
distinctly different histories (Brimblecombe et al., in press). The
use of relative concentrations is an aid to interpretation because it
obviates the effects of correlation arising from dilution where, if the
variance in the dilution term is large enough, a correlation may exist
between two sets of precipitation concentration data even if the degree
of correlation between two sets of air concentration data approached
zero (details in Brimblecombe et al., in press). Correlations induced
by the dilution process are likely to be larger for instantaneous

precipitation samples than for bulked samples (where the variation in "dilution" would be smoothed out).

Figure 3 shows the proportional compositions of fresh snowcover in the $H^+-Mg^{2+}-Na^+$ and $SO_4^{2-}-NO_3^--Cl^-$ systems. Included in these diagrams

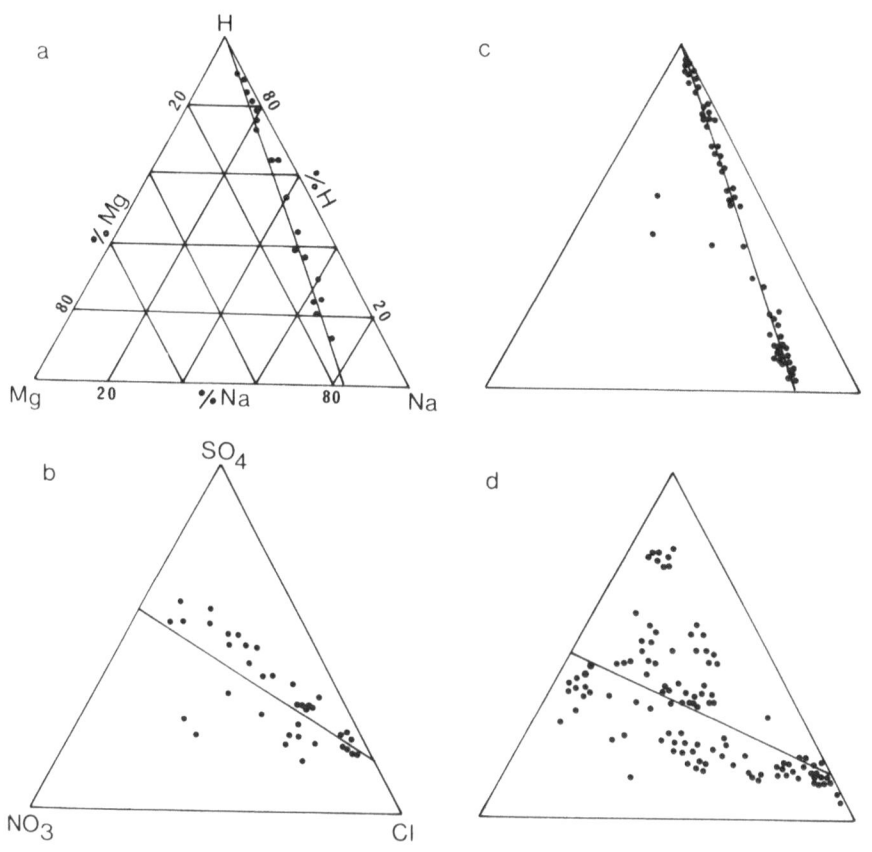

Figure 3   Triangle scattergrams which show the proportional contribution of individual ions in the $H^+-Mg^{2+}-Na^+$ and $SO_4^{2-}-NO_3^--Cl^-$ systems. Proximity to one apex indicates that ion represents close to 100% of the sum of the three ions (on an equivalent basis). A position in the centre of the triangle indicates that all ions contribute 33.3% to the total ion content in that system.

(a)   proportional cation triangle for rainfall samples with the best-fit line through the points
(b)   anion triangle for rainfall samples
(c)   cation triangle for snowfall samples
(d)   anion triangle for snowfall samples

are some replicate samples taken from around 30 separate events (but excluding the detailed spatial variability samples described earlier). These observations represent samples from 50-60% of the snowfall events throughout the sampling period. An average trend in the proportional ionic composition can be described by best fit lines through the data. These lines may be regarded as representing the mixing of sea salt and acidic components of the overall solute load. This is most easily seen on the cation triangle (Figure 3c). A number of rainfall events (between April - November) were also collected in the catchment (Tranter et al., 1987a) and the relative cation contributions in these events are represented in Figure 3a. The relative distributions of cations in snow and rainfall events (characterised by the best-fit lines) are very similar. This is unsurprising since seasalt has a fixed $Na^+:Mg^{2+}$ ratio and there will be no differential process operative upon scavenging. The distributions on the anion triangles for snow and rain (Figure 3d and 3b) however, are different. It is interesting to note that solute in snowfall is proportionally enriched in nitrate with respect to rain. The mean $NO_3^-/(NO_3^-+SO_4^{2-})$ ratio (equivalent basis) for all the snows collected is 0.36, for rainfall it is 0.32 (significantly different at the 0.1 level). This analysis includes the group of particularly sulphate-rich snows clustered near the SO_4^{2-} apex of Figure 3d); these are discussed in detail later, but they were the consequence of preceding meteorological conditions which were not associated with any of our collected rainfall events. There are a number of possible explanations for the apparent relative enrichment of nitrate in snow. It may reflect the more efficient scavenging of nitrate by snow, discussed earlier. Sulphate may be dry-deposited onto rain collectors (e.g. Fowler and Cape, 1984), and our rain collectors were not wet-only devices. The nitrate:sulphate ratio in UK precipitation is greater in the winter (UKRGAR, 1987) which may be a function of emissions, of photochemistry, or because more snowfall occurs in winter! (UKRGAR, 1987). A further contributory factor could be seasonal variations in back-trajectories with concomitant charges in air mass chemical evolution, and also the operation of varying source regions (with differing emission ratios) (see Brimblecombe et al., in press; UKRGAR, 1987).

Although our sampling strategy for snowfall and rainfall was loosely-constrained because of the operational restrictions discussed earlier, there is clear evidence here of distinct differences in proportional anionic composition of snowfall and rainfall incident upon the Cairngorm Mountains. In order to appreciate the processes which lead to the observed deposition in this location, we need to consider the composition of individual snowfall events each represented by a single point on a triangular diagram. Figure 4 is composed of points which reflect the median anionic concentrations of individual snowfalls when replicate samples were collected. Values of pH are also illustrated. The points are seen to cluster into three definable groups: (a) relatively high NO_3^-/SO_4^{2-} ratio, generally low pH, (b) relatively low NO_3^-/SO_4^{2-} ratio, generally low pH, and (c) high chloride with generally higher pH. Brimblecombe et al. (in press) demonstrated that these three groupings were quite closely associated

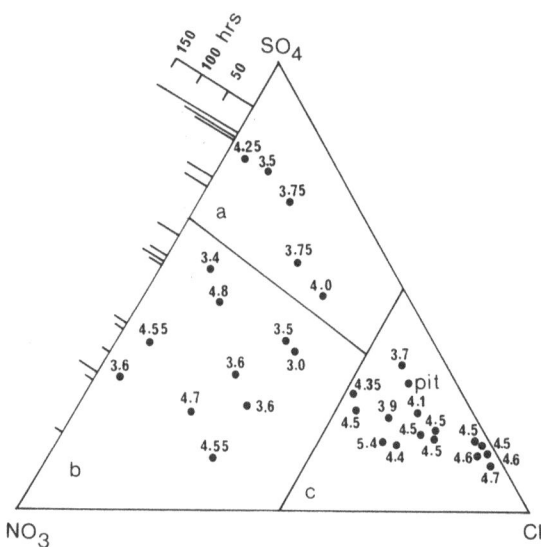

Figure 4   Relative anionic contributions in individual snowfalls. pH
value is illustrated. The bars on the SO_2^{2-}-NO_3^- axis
represent the estimated time of transit of the air mass from
the nearest identifiable large pollution source regions to
the point of deposition (on the basis of the trajectories
shown in Figure 5) for the snowfalls in categories (a) and
(b). The position on the axis is determined by a projection
from each data point along a constant NO_3^-/excess SO_4^{2-}
ratio line. The categories (a), (b), (c) are based on the
clustering (see Brimblecombe et al. (in press) for an
earlier version) of the points into three groups: (a) low
NO_3^-/excess SO_4^{2-} ratio, 0.5, with low pH (median 3.75);
(b) relatively high NO_3^-/excess SO_4^{2-} ratio, 0.5, with low
pH (median 3.6); (c) high Cl^- (50%), with generally high
pH (median 4.55).

with three particular classes of back-trajectories, although the
association was not completely exclusive (Figure 5). The snows in
category (a) were associated with sources of pollution at considerable
distance, in eastern Europe/USSR. The source regions were generally
2-7 days back along the trajectories. The three snowfalls with the
largest relative contributions of sulphate had the longest trajectories
(> 4 days). The remaining two snowfalls in this category had relative

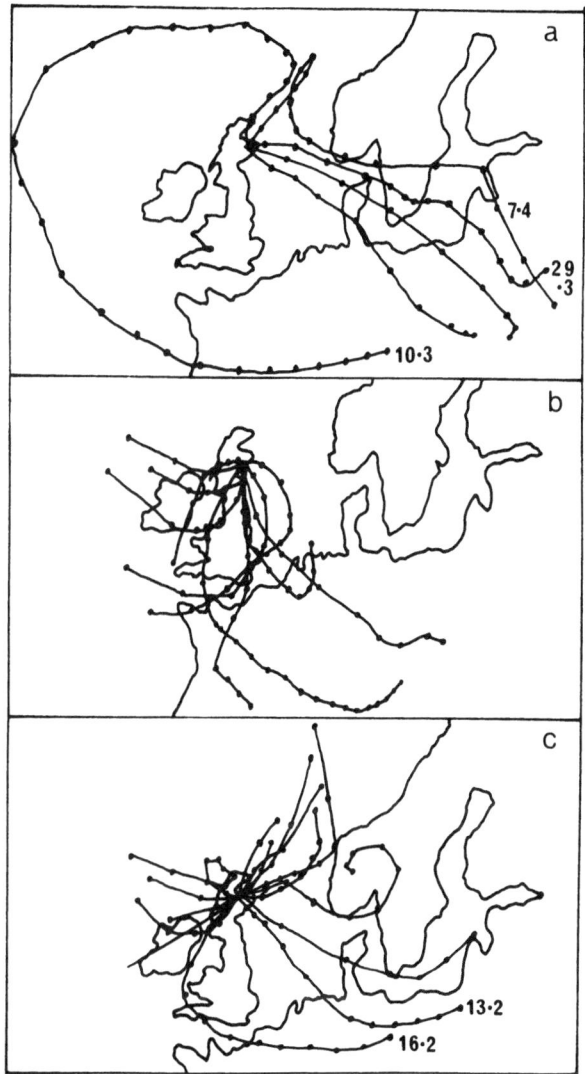

Figure 5   Back-trajectories associated with the Cairngorm snowfalls of
            compositional categories (a), (b), (c) (see Figure 4). The
            trajectories were determined by the UK Meteorological Office
            procedure. The dots represent six-hour positions. The
            classification was originally made, and justified, in
            Brimblecombe et al. (in press). The dates refer to
            trajectories which are discussed in the text.

sulphate contributions close to 50% and the associated trajectories
indicated that they were about 2 days from the likely pollution source
region.

Trajectories associated with snowfalls in category (b) (high
nitrate contributions) appeared to have passed over or near large
pollution sources in the UK or, on two occasions, the Low Countries or
the Ruhr. Four of the snowfalls in category (b) were associated with
Atlantic trajectories which subsequently passed over urban-industrial
areas of the UK or Ireland. Finally, snowfalls which were grouped into
category (c) were associated with trajectories that are broadly
maritime, although at least two passed over or near the
Glasgow/Edinburgh region. Some of the trajectories (Figure 5c) may
have been expected to be associated with snowfalls in category (a), but
explanations are suggested in terms of relatively high wind speeds over
the sea (thus collecting a relatively large marine aerosol component)
and relatively unstable atmospheric lapse rates (compared to those in
category (a)).

It has been noted that the NO_3^-/SO_4^{2-} ratio in precipitation
decreases with increasing travel time of polluted air (Skartveit,
1982). Nitrogen oxides are oxidised faster than sulphur dioxide, and
so the nitric acid becomes available for removal before sulphuric acid
(Rodhe et al., 1982). Brimblecombe et al. (in press) performed a model
calculation of the relative proportions of chloride, nitrate and
sulphate in precipitation associated with a polluted air mass that had
arisen from industrial sources and contained HCl, NO_x and SO_2. As the
air mass evolved, with advection, the concentrations of gas available
for scavenging changed in the fashion illustrated in Figure 6. Close

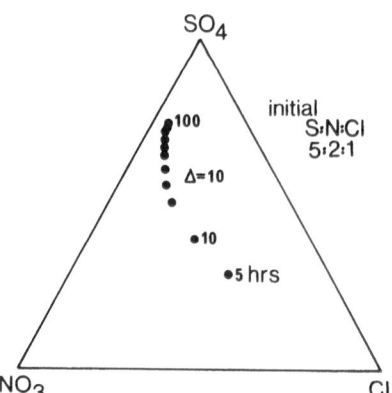

Figure 6   Evolution of a polluted air mass through oxidation of nitric
            oxides to nitric acid and the somewhat slower oxidation of
            sulphur dioxide to sulphur acid. In the model calculation,
            the relative proportions of Cl^-, NO_3^- and SO_4^{2-} in rainfall
            was calculated by assuming that only the acid species were
            removed by the precipitation. HCl was assumed to be present
            as the acid at the start of the calculation. For details of
            the model calculation, see Brimblecombe et al. (in press).

to the source HCl is the dominant species which is immediately available for efficient wet removal. Subsequently, nitric acid becomes available for removal before sulphuric acid.

Included in Figure 4 are bars which represent the estimated transit time of the trajectory from the last identifiable major pollution source region. The estimates are difficult to make in a wholly objective manner, but a general association between transit time and declining NO_3^-/SO_4^{2-} ratio is apparent. In fact, the estimated times for evolution of the air mass from pollution source are quite similar to the model calculations of Brimblecombe et al. (in press) in Figure 6. A complication is that "long" and "short" trajectories are generally associated with different source regions, so that some part of the difference in the NO_3^-/SO_4^{2-} ratio could be association with a different source composition.

There is also an association between the trace element concentrations of individual snowfalls and back-trajectory. For example, Mn and Pb concentrations of snow associated with trajectories which have passed over the UK were 5 times greater than those associated with snows with maritime trajectories; Al, Cd and Fe were approximately three times greater. The trace element concentrations associated with longer-range sources (e.g. eastern Europe or Scandinavia) were intermediate to those in the other two categories. A summary of the trace element concentrations is given in Table 3. A more detailed appraisal and a review of other studies of trace element deposition in snow is provided in Abrahams et al. (in press).

Of the individual snowfall events sampled, ten can be classified as being 'distincly' black or grey. Such episodes can make relatively large contributions to the total major ion and trace element loading in the mountain snowpack (Davies et al., 1984; Abrahams et al., in press). A simple measure of 'blackness' (i.e. total black carbon content) was made by determining the reflectance of the particulates on filter papers via an Eel reflectometer. For most samples, a 'standard' (500 ml) amount of melted snow had been filtered. For those samples where a different quantity had been filtered, the reflectance readings were corrected according to Lambert's Law. Figure 7a shows the relationship between reflectance and pH of the sample. The correlation coefficient is 0.91 (significant at the 0.01 level). During February 1986, the study catchment was covered by a ~60 cm layer of dark grey snow, associated with persistent winds from the east (see later). Although we were not able to sample freshly-fallen snow, on the 13 February a pit was dug and a vertical profile of snow samples extracted. Inspection of local meteorological records showed that no melting would have occurred subsequent to fall, and so a similar analysis was executed for these samples (Figure 7b). Here, a similar relationship is apparent, although it is weaker (r = 0.77, significant at 0.05 level).

The black snow episodes are plotted on a triangular anion diagram (Figure 8), which also indicates the pH value and the date of fall. The back-trajectories for each event are illustrated in Figure 9. The two most pronounced events, both in terms of black carbon content (Figure 7a) and acidity (20.2.84 and 13.3.86), have similar

530

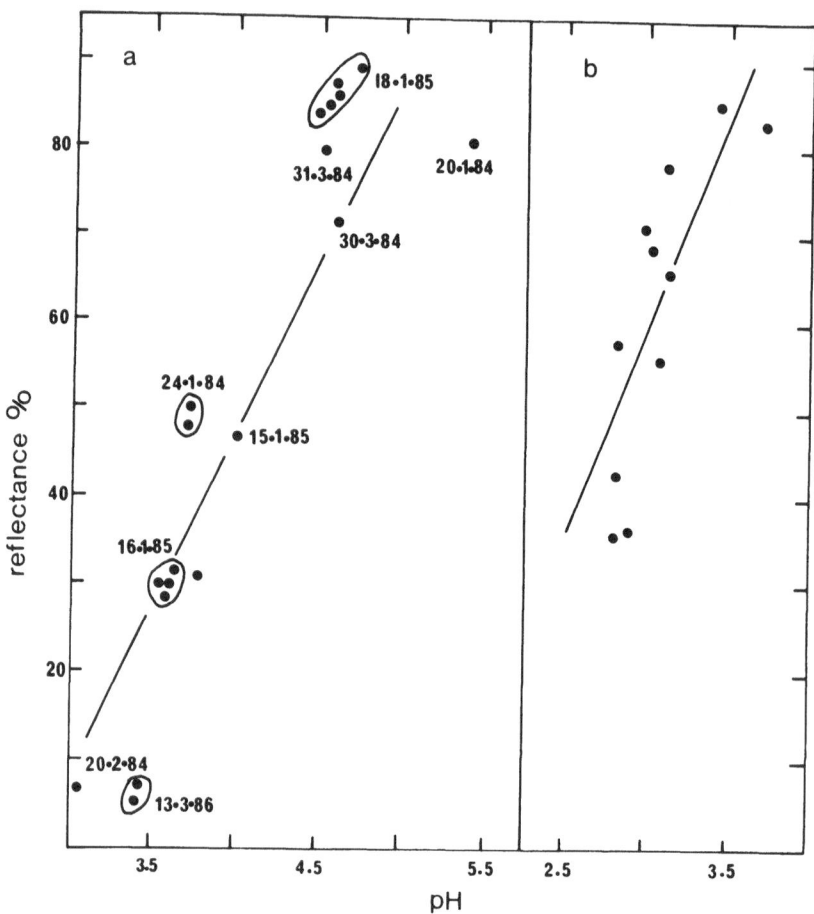

Figure 7a Relationship between filter reflectance and snowfall acidity
in 'black' snow episodes. Dates of fall are illustrated.
Where replicate samples of the same event were collected,
the data points are clustered. The best-fit line is shown
(r = 0.91, significant at the 0.01 level).

Figure 7b Similar relationship, but this time for snow collected in a
snowpit (before any melting occurred). The pH measurements
are in error (see Tranter et al., 1986, for a discussion of
the difficulties of field-laboratory measurements of the pH
of snow samples), but the data are shown since the readings
were repeatable and illustrate the general relationship
between variation in "blackness" and relative changes in pH
(r = 0.77, significant on at 0.05 level).

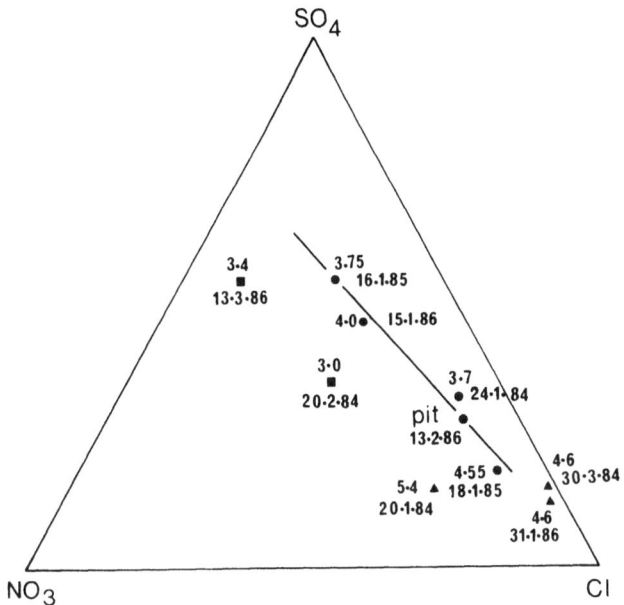

Figure 8   The black snow events shown on a $SO_4^{2-}-NO_3^--Cl^-$ triangular
            diagram. pH values and date of fall are included. The
            circles represent events associated with eastern Europe/USSR
            trajectories (Figure 9): the best-fit line is shown. The
            squares are associated with trajectories which passed over
            western Europe and the triangles are 'Arctic' black
            snow episodes.

trajectories. Both originiated in eastern Europe/USSR and passed over
western Europe, close to or over the UK. For some days before the
events, a strong anticyclone had been located to the east and, during
deposition, a frontal depression was approaching from the west. On
both occasions, the European Meteorological Bulletin reported
observations of smoke, over northern England on 19 and 20 February
1984, and in the Irish Sea region on the 12 and 13 March 1986. Both
trajectories were characterised by pronounced anticyclonic subsidence
inversions along their entire length from eastern Europe (as far as can
be determined from the daily upper air soundings reported in the EMB).
It is probable that these elevated stable layers play an important role
in inhibiting the dispersion and dilution of pollution plumes (Fisher,
1978; Smith et al., 1978; Davies et al., 1984). The surface synoptic
pressure pattern indicated a zone of horizontal convergence near the
UK, and it is possible that pollutant air concentrations could be
locally increased (Moore, D.J., personal communication).

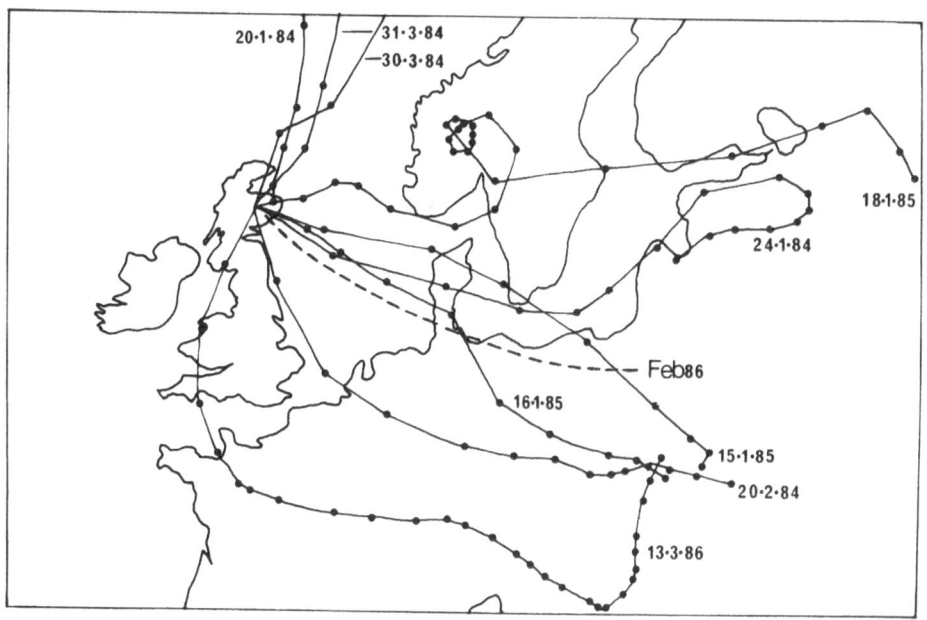

Figure 9   Back-trajectories for the black snow episodes.

Other events with an eastern Europe/USSR source are 24.1.84,
15.1.85, 16.1.85, and 18.1.85. These too, were associated with
anticyclones to the east or north-east of the UK, but the trajectories
did not subsequently pass over major sources more proximate to the UK.
These events were also characterised by elevated stable layers,
typically at 700-800 mb, along their path (EMB). The least direct,
slowest and convoluted trajectory (18.1.85) had a relatively high pH
value (~ 4.55) (see Figure 8). The snow which fell in February 1986,
and which was sampled from the snowpit, occurred during a blocking
pressure pattern, when the UK experienced persistent easterly winds.
Most of the snow fell after the first week of the month, and the falls
which occurred after the pit-sampling on the 13th were also grey in
colour but were excluded from this analysis because they were not
sampled before snowpack melting occurred. During the three-week
period, after the first week in February, subsidence inversions over
Scotland occurred on 19 out of 21 days (at ~ 700-800 mb). There were
many individual snowfalls during this period. The summary trajectory
for February 1986 was derived from the mean monthly sea-level pressure
chart published in the UK Monthly Weather Survey.
      The 'eastern Europe/USSR' black snow events all had similar

NO_3^-/SO_4^{2-} ratios, close to 0.41, compared with 0.85 for the trajectories which passed over western European sources.

The remaining three events were amongst the "least black" episodes (Figure 7a) and the back-trajectories had their origin in the Arctic. One to three days earlier the back trajectories were located between 76-79°N and 3°E-35°W. We assume that the snowfalls in the Cairngorms associated with these back-trajectories were a consequence of long-range transport of the "Arctic haze" (Goldberg, 1985), which is observed to be at its most intense in March and April. The pH values were relatively high. The NO_3^-/SO_4^{2-} ratios were similar to the eastern Europe/USSR trajectories for the 30.3.84 and 31.3.84 episodes, but for the 20.1.84 event the ratio was closer to that of the two blackest events associated with trajectory passage over western Europe (Figure 8). Inspection of Figure 9 indicates that the 20.1.84 trajectory could have passed over the vicinity of the city of Aberdeen. During the days preceding the "Arctic episodes", high pressure stretched from north of the UK to Greenland, with a stable layer apparent, close to the time of deposition, over northern Scotland (with the exception of the 30.3.84 event).

Figure 5 illustrates other occasions when, considering the trajectories from the east, black snowfalls may have been expected: 29.3.84 and 7.4.84 and possible 10.3.84 in category (a), and 16.2.84 and 13.2.85 in category (c). The latter two were originally placed in the 'maritime' category for reasons outlined in Brimblecombe et al. (in press). It may be instructive to speculate why black snowfalls did not result at the end-point of these trajectories. The 29.3.84, 7.4.84 and 10.3.84 snowfalls exhibit NO_3^-/SO_4^{2-} ratios similar to the eastern European/USSR trajectory black snows (compare Figure 4 with Figure 8), whereas the same ratio for the 16.2.84 and 13.2.85 is closer to the two nitrate-rich black snows associated with the trajectories which crossed western Europe. So, in terms of position on the triangular anion diagram, the snowfalls generally conform to the relationship with trajectory identified with respect to the black snows.

For the 7.4.84 snowfall, the associated air, during its passage over the North Sea was in close proximity to a frontal surface and a low pressure. Precipitation was observed in the North Sea area at this time, and no pronounced stable layer was evident over Scotland at the time of deposition. Similarly, for the 29.3.84 episode, a low pressure system and fronts were in the vicinity of the trajectory both on the day of, and the day before, deposition. Similarly, no clear anticyclonic subsidence inversion existed. The long trajectory for the event of the 10.3.84, which recurved over the Atlantic, was associated with a subsidence inversion over Scotland at the time of deposition but it was considerably higher than with the other events (~ 600 mb). At the trajectory's western-most position, there was interaction with a frontal depression, and it is possible that scavenging and deposition of particules in precipitation occurred at this location. On the 31.2.85, at the time of deposition, there existed a front and low pressure over Norway, although there was a quite strong inversion layer at ~ 860 mb over Scotland. During the 16.2.84 event, there did exist an inversion layer over Scotland, but it was lower than during the other events (at ~ 920 mb) and one day earlier, when the air parcel was over the southwest of England, the subsidence inversion was very low at 1000 mb (surface pressure was 1035 mb). Because of the low and strong stable layer, particles in the air may have been subjected to greater dry deposition than during other episodes, and so may not have been transported long distances.

So, qualitative analysis indicates that during those occasions when the back-trajectories might have been expected to lead to black snowfall in Scotland, such events were obviated by earlier precipitation scavenging, enhanced dispersion by low pressure systems, or restricted long-range transport through entrapment by a very low, intense, subsidence inversion.

Thus far, two of the black snow events have been studied in some detail (Davies et al., 1984; Abrahams et al., in press). Abrahams et al. (in press) examined the 16.1.85 event and pointed out that of the nine elements which exhibited enrichment factors greater than 10, five (As, Cu, In, Sb and Zn) are chalcophilic, and have a geochemical affinity for sulphides, which suggested a common source.

6. CONCLUSIONS

Although the observations are not statistically conclusive, there are indications from the limited number of altitudinal surveys conducted, that ionic concentrations in snowfall increase with greater fall-path. Consequently, observations of the relationship between precipitation composition and elevation from surveys in mountain areas should be considered in the light of the relative frequency of snowfall and rainfall.

The problems of such altitudinal studies are compounded, in the case of snowfall, by the difficulties of snowfall collection. Use of fresh snowcover as a surrogate warrants some caution because of the large degree of spatial variability in the chemical composition of the snowcover. Such variability seems to be pronounced, even in the absence of obvious post-fall wind redistribution. Topographically-complex terrain may produce local flows which sort snowflakes by size and which produce a non-uniform deposition field during the snowfall event. A large degree of spatial variability exacerbates the difficulties of determining deposition fluxes and calculating ionic loadings for snowmelt studies.

The issue of spatial variability is less important in the study of pollution transport and deposition processes from event-to-event because of the large inter-fall variability. There is evidence that snowfall removes relatively more nitrate than rainfall, although other processes may be making contributions (differing trajectories associated with varying source regions, etc.). The relative deposition of NO_3^- and SO_4^{2-} in snow in the Cairngorm mountains appear to exhibit an association with trajectory origin and the related atmospheric conditions. Some of the snowfall events are highly-coloured with large black carbon components. These episodes represent processes which can make a relatively large contribution to the total major ion and trace element loading to the mountain area (discussed elsewhwre, e.g. Davies

et al., 1984; Abrahams et al., in press). There is a strong
correlation between a simple index of carbonaceous content and snowfall
acidity. The major events are associated with transport from eastern
Europe/USSR, although they can be modified by air mass passage over
nearer sources. Less pronounced events originate in the Arctic.
During these black episodes, the long-range transport is characterised
by the existence of elevated anticyclonic subsidence stable layers,
which probably produce a pollution plume which has been relatively--
little dispersed, in the same elevation-range as the Cairngorm
summits. Snowflake scavenging, which removes the carbon particles
relatively efficiently, is the final stage in the deposition process.
The deposition of large quantities of particular material (e.g. 0.1
tonne km^{-2}; Davies et al., 1984), may be a consequence of large
quantities of polluted air cycling through large cumulus clouds,
although not all the events appear to be associated with convective
systems.

The coincidence of the geographical and meteorological/-
climatological attributes of the central Scottish Highlands may render
them unrepresentative of mountain areas elsewhere in the UK, but they
do seem to be a very suitable location for the closer examination of
air mass evolution and transport, and the process of the eventual
deposition of chemical deposition to the ground.

ACKNOWLEDGEMENTS

We are grateful to Dr J Crabtree (Meteorological Office) for kind
assistance with the back-trajectory analysis. Financial support was
provided by NERC and CEC. We are grateful to the Cairngorm Ski-lift
Company for logistical help.

REFERENCES

Abrahams, P.W., Davies, T.D., Tranter, M., Brimblecombe, P. and
    Blackwood, I. in press. Trace elements in snow in the remote
    Scottish Highlands. Water, Air and Soil Pollution.
Brimblecombe, P., Davies, T.D. and Tranter, M. 1986. Nineteenth
    century black Scottish showers. Atmos. Environ. 20(5),
    1053-1057.
Brimblecombe, P., Davies, T.D. and Tranter, M. in press. The
    composition of precipitation in the Scottish Highlands and the
    chemical evolution of air masses. In: Proceedings of the
    Symposium on Air Pollution Modelling, Leningrad, WNO, 1986.
Castillo, R. 1986. An analysis of black aerosol found in two winter
    Atlantic coastal storms at Whiteface Mountain, New York. Journal
    of Aerosol Science 17(4), 677-684.
Castillo, R. et al. in press. Organic and elemental carbon loading
    in interstitial aerosol and hydrometers at a rural northwest site,
    Whiteface Mountain, New York. Journal of Geophysical Research.

536

Dasch, J.M. 1987. On the difference between SO_4^{2-} and NO_3^- in winter-time precipitation. Atmos. Environ. **21(1)** , 137-141.

Davies, T.D. 1984. Rainborne SO_2, precipitation pH and airborne SO_2 over short sampling times throughout individual events. Atmos. Environ. **18(11)** , 2499-2502.

Davies, T.D., Abrahams, P.W., Tranter, M., Blackwood, I., Brimblecombe, P. and Vincent, C.E. 1984. Black acidic snow in the remote Scottish Highlands. Nature **312** , 58-61.

Davies, T.D. et al. 1987. The removal of soluble ions from melting snowpacks. In: The Chemical Dynamics of Seasonal Snowcover. (Eds. H.G. Jones and W.J. Orville-Thomas). Proceedings of the NATO ASI at Les Arcs, France, 1986. D. Reidel, 337-392.

Fisher, B.E.A. 1978. Long range transport and deposition of sulphur oxides. In: Sulphur in the Environment (Ed. Nriagu, J.O.), Wiley, New York, 243-296.

Fowler, D. and Cape, J.N. 1984. The contamination of rain samples by dry deposition on rain collectors. Atmos. Environ. **18** , 183-189.

Gervat, P. 1985. Clouds at ground level: samples from the southern Pennines. CERL TPRD/L/2700/N84. CERL, Leatherhead, UK.

Goldberg, E.D. 1985. Black carbon in the environment. Wiley, New York, 189 pp.

Knutson, E.O., Sood, S.K. and Stockham, J.d. 1976. Aerosol collection by snow and ice crystals. Atmos. Environ. **10** , 395-402.

McQuaker, N.R., Kluckner, P.D. and Sandberg, D.K. 1983. Environmetnal Science and Technology **17** , 431.

Leivestad, H. and Muniz, I.P. 1976. Fish kill at low pH in a Norwegian river. Nature **259** , 391-392.

Messer, J.J., Slezak, L. and Liff, C.E. 1982. Potential for acid snowmelt in the Wasatch Mountains. Utah Water Research Laboratory Water Quality Series UWRL, Q-82/06.

Raynor, G.S. and Hayes, J.V. 1982. Variation in chemical wet deposition with meteorological conditions. Atmos. Environ. **16** , 1647-1656.

Rodhe, H., Crutzen, P. and Vanderpol, A. 1981. Formation of sulphuric and nitric acids in the atmosphere during long-range transport. Tellus **33** , 132.

Skartveit, A. 1982. Wet scavenging of sea-salts and acid compounds in a rainy, coastal area. Atmos. Environ. **16** , 2715-2724.

Smith, T.B., Blumenenthal, D.L., Anderson, J.A. and Vanderpol, A.H. 1978. Transport of SO_2 in powerplant plumes. Atmos. Environ. **12** , 605-611.

Smith, F.B. and Hunt, R.D. 1978. Meteorological aspects of the transport of pollution over long distances. Atmos. Environ. **16** , 2715-2724.

Sood, S.K. and Jackson, M.R. 1970. Scavenging by snow and ice crystals. In: Precipitation Scavenging (Eds. Englemann, R.J. and Slinn, W.G.N.). Conf.-700601, National Tech. Inf. Service, Springfield, VA., 151-159.

Tranter, M., Brimblecombe, P., Davies, T.D, Vincent, C.E., Abrahams, P.W. and Blackwood, I. 1986. The composition of snowfall, snowpack and meltwater in the Scottish Highlands - evidence for preferential elution. Atmos. Environ. **20** , 517-525.

Tranter, M. et al. 1987a. Changes in streamwater chemistry during snowmelt. In: The Chemical Dynamics of Seasonal Snowcover. (Eds. H.G. Jones and W.J. Orville-Thomas). Proceedings of the NATO ASI at Les Arcs, France, 1986. D. Reidel, 575-597.

Tranter, M., Davies, T.D., Brimblecombe, P., Abrahams, P. and Blackwood, I. 1987b. Spatial variability of the chemical composition of snowcover in a small remote Scottish catchment. Atmos. Environ. 21 , 853-862.

Thompson, I.P., Blackwood, I. and Davies, T.D. 1987. The effect of polluted and leached snowmelt waters on the soil bacterial community - quantitative response. Environ. Poll. 43 , 143-154.

Topol, L.E. 1986. Differences in the ionic compositions and behaviour in winter rain and snow. Atmos. Environ. 20 , 347-360.

Whelpdale, D.M. and Shaw, R.W. 1974. Sulphur dioxide removed by turbulent transfer over grass, snow and water surfaces. Tellus 26 , 195-204.

United Kingdom Review Group on Acid Rain, 1987. Acid Deposition in the UK 1981-1985, Warren Spring Laboratory, Stevenage, UK, 104 pp.

Table 1   Summary of regression statistics for seven altitudinal surveys. 'Negative' denotes the number of analyses which demonstrated an inverse relationship between ionic concentration and height; 'positive' denotes a direct relationship. 'Significance' indicates whether or not the correlation was significant at the 0.2 level. None of the positive correlations were in this class. Cl^- data for one of the surveys were not available.

Correlation/Species	H^+	SO_4^{2-}	NO_3^-	Cl^-	Na^+	Mg^{2+}
Negative	6	6	5	5	4	5
Positive	1	1	2	1	3	2
Significance < 0.2	5	2	3	1	1	2

Table 2a Log mean (ppm) and coefficient of variation (%; standard deviation of logged values/log mean) of freshly-fallen surface snow (*, not available).

Survey		pH	Na^+	Mg^{2+}	SO_4^{2-}	NO_3^-	Cl^-	n
1	50 m	4.56(1)	-1.41(16)	-2.52(11)	-0.21(10)	0.03(93)	-1.16(24)	14
	1.0 m	4.72(3)	-0.97(34)	-1.90(19)	-0.14(58)	-0.03(72)	-0.66(55)	8
2	50 m	3.56(1)	-0.60(51)	-1.49(32)	-0.98(106)	0.20(144)	-0.32(79)	15
3	50 m	4.49(4)	-0.94(19)	-1.65(21)	-0.27(25)	-0.33(26)	-0.60(43)	14
	1.0 m	5.22(5)	-1.35(10)	-2.18(10)	-0.61(28)	-0.65(33)	-0.75(20)	19
4	50 m	*	-0.16(85)	-1.13(12)	-0.37(25)	-0.61(37)	-0.18(58)	13

Table 2b   Relationship between the number of samples, the potential error of the mean (E_1) and coefficient of variation of log values (confidence level = 95%).

		1%	5%	10%	20%	50%
E_1	2%	42	1030	4240	16960	106003
	5%	8	197	786	3145	19658
	10%	4	96	384	1537	9604
	20%	2	57	227	909	5683
	50%	1	33	133	532	3323
	100%	1	24	96	384	2401

Table 3   Median values of trace element concentrations ($\mu g\ 1^{-1}$) in snows associated with three different trajectory-types.

	Trajectory type		
	Maritime (n=4)	Long-range (n=7)	UK and near-W Europe (n=4)
Al	9.2	13.0	30.0
Cd	<0.1	0.2	0.32
Cu	<1.0	< 1.0	< 1.0
Fe	4.5	11.0	12.0
Mn	0.5	1.1	2.7
Pb	0.8	3.8	4.7

MEASURING AND MODELLING DRY DEPOSITION IN MOUNTAINOUS AREAS

B.B. Hicks and T.P. Meyers
NOAA/ARL Atmospheric Turbulence and Diffusion Division
P O Box 2456
Oak Ridge
TN 37831
USA

ABSTRACT. A trial programme has been initiated to test methods for estimating dry deposition rates from measurements of air concentration, using selected surface and atmospheric data to specify the appropriate deposition velocities. Results obtained in the testing programme are used here to compare a site on the slopes of Whiteface Mountain, New York, and a similarly forested site distant from mountains. Data are presented on atmospheric resistances to turbulent exchange, and on the surface resistance associated with the dry deposition of sulphur dioxide. All considerations involving the concept of a deposition velocity combining surface and vertical-diffusion components are limited in generality, because of the over-riding assumption that the aerodynamic transfer is dominated by vertical diffusion, therefore neglecting advective effects such as blowthrough. For the case of sulphur dioxide, the increase in computed dry deposition resulting when such advective effects are taken into account is found to be small. This result from the fact that SO_2 exchange is largely controlled by surface rather than atmospheric resistance; hence a similar result is expected for other trace gases most strongly influenced by surface resistance, such as ozone, and could be expected to extend to particle deposition in some size ranges. However, for nitric acid vapour (and presumably for all other trace gases having surface resistance small in comparison to atmospheric resistance) the consequences of surface heterogeneity and topographic complexity on deposition velocities could be very large.

1. INTRODUCTION

Although in some discussions there is a clear distinction between the dry deposition of airborne particles and the bi-directional exchange of gases between the atmosphere and the surface, in the present context "dry deposition" is assumed to involve the aerodynamic transfer of trace gases and aerosol particles, and the sedimentation of those particles large enough to settle under the influence of gravity.

M. H. Unsworth and D. Fowler (eds.), Acid Deposition at High Elevation Sites, 541–552.
© *1988 by Kluwer Academic Publishers.*

These processes result in a much smaller flux density than for wet deposition, with a dominant diurnal cycle replacing the quasi-random intermittency that is characteristic of precipitation. The chemical species of importance in the context of dry deposition are mainly trace gases, such as ozone, sulphur dioxide, nitrogen dioxide, and nitric acid vapour, although in some contexts particles can also be of importance. The concept of pH is rarely appropriate in discussion of dry deposition; the analogous consideration usually involves the effect of deposition of specific ions on ion balance at the surface.

In general, there is no particular method which is widely applicable to measure dry deposition, even in intensive research studies. Instead, considerable innovation is applied to extend methods of measurement developed in studies of geochemistry, hydrology, agricultural meteorology, forest meteorology, and particle physics in order to derive the desired deposition data. The practical approach which is usually used in routine measurement programmes relies on the measurement of appropriate atmospheric concentrations and the specification of a corresponding deposition velocity.

Three major areas of concern arise immediately: (i) the accuracy of the concentration data, (ii) the accuracy of the deposition velocity, and (iii) the applicability of the deposition velocity concept in the experimental circumstance. The first of these concerns is being addressed in a wide variety of air chemistry programmes. The second has been the subject of several recent papers (e.g. Baldocchi et al., 1987; Delany et al., 1986; Hicks and Matt, 1987; Meyers, 1987). Here, the last of these three concerns will be discussed. The present goal is to provide some initial guidance concerning the magnitude of errors likely to arise when terrain complexity is neglected in computing dry deposition from concentration data for high-altitude sites, drawing upon experience obtained in an ongoing evaluation of an inferential method for monitoring dry deposition.

In general, it is now apparent that the conceptual one-dimensionality of the dry deposition velocity (V_d) is potentially a severe limitation, of importance to both measurement and modelling programmes. The problems arising will be explored in the following section. Subsequent sections will present some suggestions concerning how to bound the estimates of dry deposition in locations where the standard V_d approach is suspected to break down. Measurements made at two locations in New York will be used to illustrate the differences between an area affected by complex topography and one that is not so greatly influenced.

2. LIMITATIONS OF V_d AND RESISTANCE ANALOGS

The direction of the average atmosphere-surface exchange of many pollutants is uncertain. Some gaseous nitrogen species, for example, can be emitted from the surface as a consequence of biological decay. Other species, however, can be assumed to be depositing at all times. Among these certainly-depositing species are highly reactive trace gases which attach to all available natural surfaces (e.g. nitric acid

vapour), materials emitted from distant sources and "imported" via
long-range transport (e.g. sulphur dioxide), chemicals constantly
produced by reactions in the atmosphere well above the surface (e.g.
ozone), and particles carrying the products of reactions of such gases
(e.g. sulphates). In such cases, surface deposition fluxes can be
inferred from atmospheric concentration data provided the uptake
characteristics of the situation in question can be formulated in terms
of properties that can be measured. This is the essence of the
so-called inferential method, in which a deposition velocity V_d is
computed from field observations and used to estimate the dry
deposition flux F_d (towards the surface, in this context) from air
concentration measurements C:

$$F_d = V_d.C.$$

We might note, in passing, that the philosophy of the inferential
approach is seen as a step ahead of the time-honoured "concentration
monitoring" methodologies, in which an air quality focus is paramount.
It should also be noted, however, that the credibility of any such
inferred dry deposition results relies on direct comparison against
independent measurements made wherever and whenever possible. This is
the over-riding philosophy of the nested network programme presently
underway in the United States, where measurements made at a small set
of Core Research Establishments (the CORE stations) provide
benchmarking for inferential techniques that are then deployed in an
array of satellite stations (see Hales et al., 1986). The locations
that will be discussed here, Whiteface Mountain and West Point, NY, are
two of these satellite stations.

Not all chemical species lend themselves to this approach. For
those that do, two independent pieces of information, C and V_d, are
necessary to derive the desired dry deposition rates F. Measurement of
atmospheric concentrations of relevant chemical species is a relatively
straightforward (but often demanding) task in experimental chemistry.
However, estimation of V_d involves a balanced consideration of
atmospheric physics, chemistry and biology. An analytical routine for
estimating V_d for sulphur dioxide, ozone, nitrogen dioxide, nitric acid
vapour, and submicron particles has recently been presented (Hicks et
al., 1985; Hicks and Matt, 1986) and is presently being used in the
CORE/satellite programme. In essence, any such deposition velocity
approach is one dimensional. It is derived from relationships
describing turbulent transport to surfaces that are effectively
horizontal and uniform. In the case of a nonuniform surface, such as a
checkerboard matrix of forest and cropland, edge effects will be of
considerable importance. The mechanisms involved are largely
advective, and hence once again the problem becomes three dimensional.
In this regard, it is evident that the one-dimensional deposition
velocity approach addresses but one part of the overall deposition
problem, and that another part remains to be addressed.

An over-riding problem is that a one-dimensional description of
deposition processes is used by almost all existing deposition models.
This framework is appropriate only in limited circumstances, which can

be fairly well defined as a result of micrometeorological guidelines arising from experience over the last half century, from modelling investigations conducted mainly in the 1960s, and recent micrometeorological field studies. In general, the one-dimensional approximation will break down whenever flow separation occurs, and whenever the constraint of horizontal homogeneity is violated. The homogeneity constraint concerns both changes in the nature of the surface and air quality variability. However, the most clearly evident problem lies in so-called edge effects; one example is the "blowthrough" penetration of air at edges of forests into the subcanopy air space.

Consider the multiple-resistance analog as it is used in the standard vertically-diffusive V_d approximation. In this analogy, we specify three major resistances:

- an aerodynamic resistance R_a that is a function only of atmospheric properties, such as turbulence and stratification;
- a near surface "quasi-laminar" resistance R_b that relates to the diffusion of pollutants across the near-surface layers where molecular and Brownian properties are important;
- a surface canopy resistance R_c that expresses the consequences of the chemical, morphological, and biological processes influencing pollutant adsorption or "capture" by the surface itself.

It is important to note that R_a is a dynamic property of the atmosphere, unaffected by pollutant properties, that R_b is a surface-related variable which depends on both pollutant and atmospheric factors, and that R_c is controlled by plant physiology (and by other surface properties), and is almost entirely independent of atmospheric flow and dynamic variables (although affected by insolation, temperature, and humidity, which all influence transpiration and photo-synthesis).

3. EFFECTS TERRAIN AND EDGES

Blowthrough at the edges of stands of tall plants constitutes a mechanism by which plants are exposed to levels of pollution without the full effect of the exchange constraint imposed by R_a. However, the plants at the edges of such stands will share much the same R_b and R_c as plants elsewhere in the same atmospheric and soil environment. Thus, as a first approximation, we can estimate the magnitude of the edge-effect problem by assessing the proportion of the surface that is affected, and by assuming that this area has depositional characteristics corresponding to $R_a = 0$, with R_b and R_c as specified by the standard methodologies.

It is acknowledged that processes considered here to be second-order may be important in some circumstances, such as when levels of oxidants in the blowthrough air are sufficient to influence R_c. Also, the assumption here that R_b is unaffected is acknowledged to be bold, and is doubtlessly a source of error. The present purpose is not to try to provide a complete and final answer, but to take a first step. In general, the intent is not to suggest that enough is known to quantify edge-affected deposition with confidence, but rather to propose that we can provide first-order BOUNDS on the problem. On the one hand we have the standard assessment which assumes no enhancement of deposition due to the edge-effect. On the other hand, an alternative estimate can be produced, based on the assumption that the edge-effect dominates vertical diffusion. The truth will lie somewhere between these two oversimplifications.

The case of complex topography can be considered along similar lines. The presence of an obstacle of any size can only increase area-wide atmospheric turbulence (although some local areas may be sheltered). Hence, areas of complex topography are characterized by enhanced atmospheric dispersion, relative to areas with simple topography. Moreover, in this context it is possible to specify "complex" topography in an objective manner. Classical meteorological guidelines indicate that the important feature is the local slope. A common criterion corresponds to a local change of slope of the order of 10%: any more, and a separation "bubble" or wake cavity region can develop. In unstable stratification (i.e. in convective conditions), such a cavity is likely to be small, but in stable flow (i.e. at night), such a coherent flow phenomenon provides a mechanism for transferring pollutants to the ground across the region of otherwise-limiting high stratification. It is for this reason that "complex terrain" is sometimes thought of as a nocturnal problem, at least in the context of dry deposition.

As a first-order estimator of the error imposed by neglecting complex topography when estimating area-wide dry deposition, an approach similar to that proposed above can be used - (i) determine which areas are "complex", (ii) assume $R_a = 0$ for such area, (iii) recompute V_d accordingly, and (iv) compare the resulting upper-limit dry deposition fluxes with the lower-limit values derived when terrain complexity is ignored.

4. WEST POINT WHITEFACE MOUNTAIN

The first-order considerations presented somewhat heuristically above imply that the importance of complex terrain in the context of dry deposition will vary for different chemical species. For example, "errors" resulting from neglecting effects of topography may not be as large for SO_2 as for HNO_3, since for SO_2 deposition R_a is not usually a limiting factor whereas for HNO_3 it is.

Data obtained as part of the CORE/satellite programme permit this matter to be explored further. Two sites of this trial network are located in similarly-vegetated regions of New York, one especially selected to be influenced by local slope (near the foot of Whiteface Mountain, see Figure 1), and the other in a flatter part of the state, near West Point. In both locations, the surface is forested with an

546

uneven-aged distribution of mixed deciduous species, predominantly oaks and maples. Neither site is satisfactory for micrometeorological flux studies, but both are considered adequate for the kinds of gross comparison presented here. At each site, routine measurements are made of atmospheric and surface properties that control R_a, R_b and R_c (see Hicks et al., 1985; 1987). The techniques for deriving estimates of these resistances are still evolving at this time, and hence the results presented here should not be taken to be accurate determinations of the various resistances. However, identical systems and analyses are employed at the two locations, and hence a comparison is considered quite valid.

ATDL-M86/646

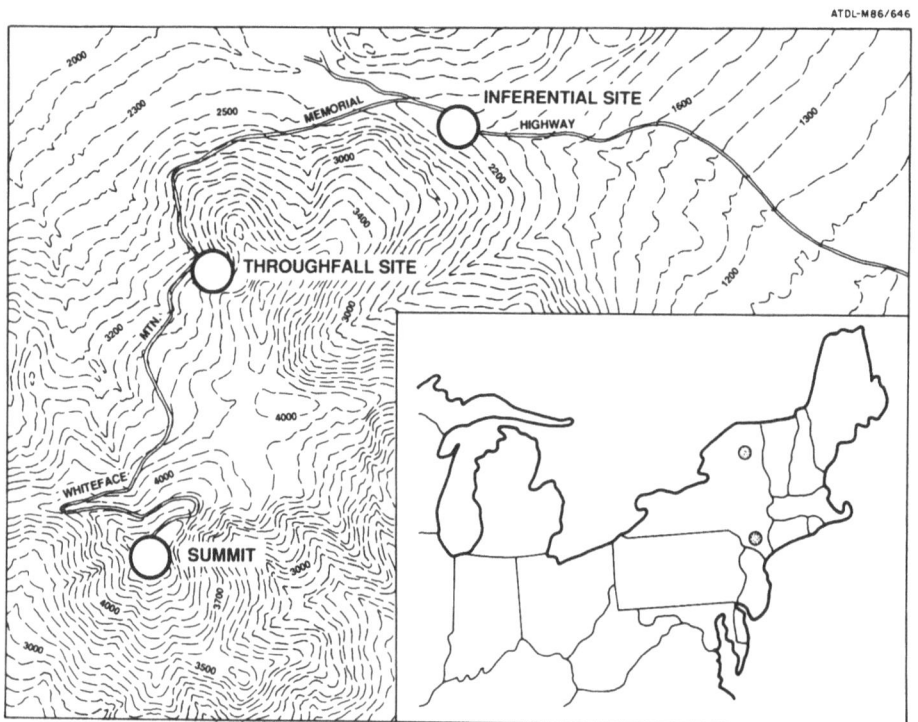

Figure 1. A topographic map of the surroundings of Whiteface Mountain, showing the locations of sites used for studies of cloud chemistry (at the summit: A), of dry deposition as reported here (near the foot of the mountain, B), and of throughfall, stemflow, and air chemistry (in a spruce/fir area, midway between A and B; C). An inset shows the locations of Whiteface Mountain and West Point, in New York.

ATDL-M86/643

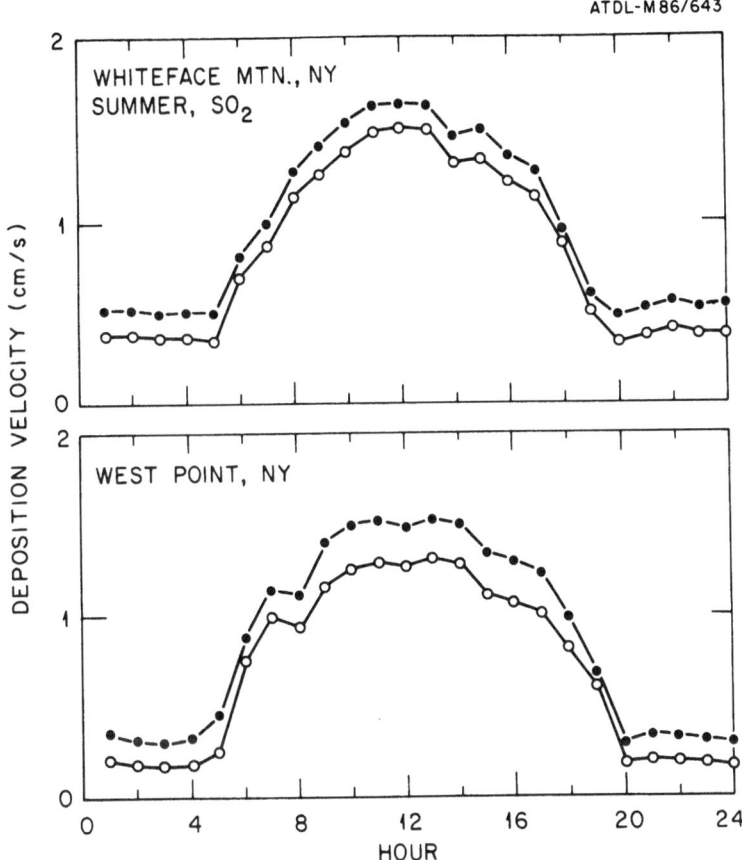

Figure 2. A comparison between average diurnal cycles of SO_2 deposition velocity at West Point (upper diagram) and at Whiteface Mountain (lower diagram). For each location, the two curves represent limits on the deposition velocity. The lower estimates correspond to the assumption of perfect-site vertically diffusive transfer through the lower atmosphere. The higher estimates represent the maximum deposition velocities expected at complex terrain sites with identical vegetation.

Figure 2 compares the average sulphur dioxide deposition velocity diurnal cycles, as derived from the observations at West Point and at Whiteface Mountain. Two curves are drawn for each location, representing two opposing interpretations of the observations: the squares correspond to the standard interpretation in which deposition velocities are based on the inclusion of an aerodynamic resistance

548

associated with vertical diffusion, and the crosses represent the
results obtained when R_a is assumed to be zero. Summer data have been
selected for presentation in the diagram. Results for spring are
similar. (For the other two seasons, differences in times of leaf
abscission and fall cause gross differences which are unrelated to the
subject here. These seasons are omitted rather than to risk further
confusion).

Figure 3 presents nitric acid deposition velocities, paralleling
the sulphur dioxide variability evident in Figure 2; each of R_a, R_b

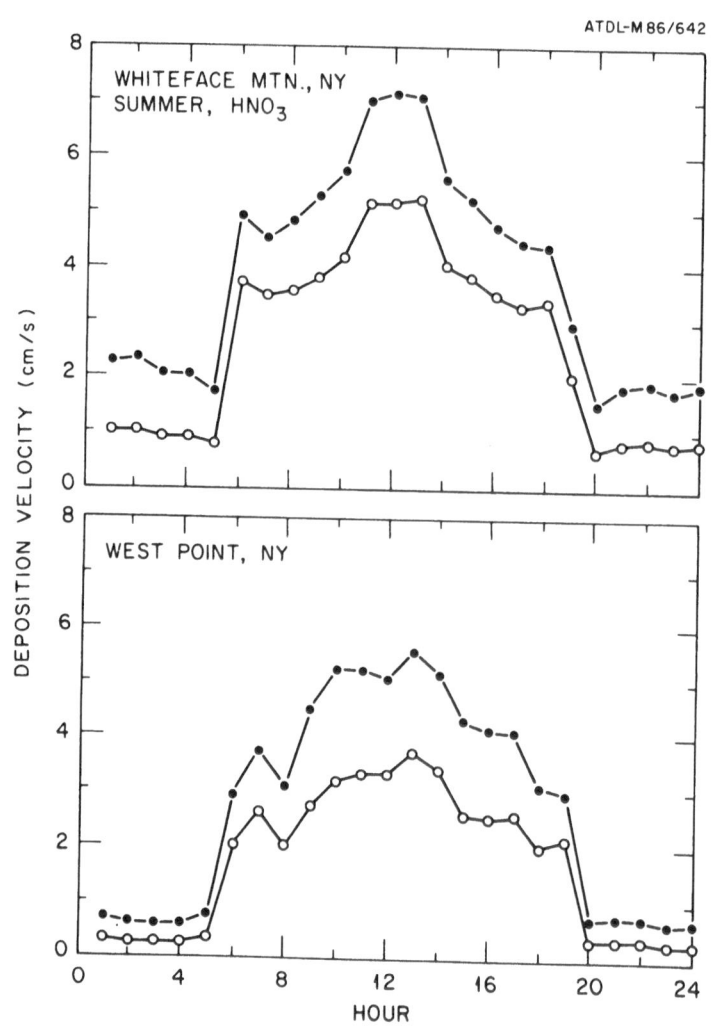

Figure 3. As in Figure 2, but for HNO3.

and R_c are affected by insolation in a manner that imposes an extremely strong diurnal cycle on the resulting deposition velocity.

In general, deposition velocities derived for the mountainside location exceed those obtained at the lower-level site, but the difference is not great. Closer scrutiny of the data reveals that the cause lies mainly in differences associated with R_a, but it is not yet clear whether the difference is a consequence of the different wind regimes at the two locations or due to imperfect siting of the instruments.

5. DISCUSSION

For the case of SO_2 deposition, the surface resistance is normally the rate-limiting factor. Errors due to overestimating the atmospheric resistance in the computation of dry deposition rates of gaseous sulphur should therefore be relatively small. Comparison of the two alternative deposition velocity cycles presented in Figure 2 indicates that the omission of consideration of terrain complexity in deriving SO_2 dry deposition fluxes could cause an error in the deposition velocity of about 0.15 cm s^{-1}, corresonding to a possible under-estimation of the surface flux rate by about 30% for the nightime periods, but only 15% for the daytime periods at the Whiteface Mountain low-level site, during the period of collection of the summer data presented here. This result was computed as the average of hourly evaluations of the ratio $F = (R_a + R_b + R_c)/(R_b + R_c)$. (For SO_2, the average value of F was 1.46). For HNO_3, the role of R_a is stronger. The possible error in V_d is about 1 cm s^{-1}, and a higher average value of F is found: 1.81 (indicating that nearly a factor of two under-estimation is possible). A greatly reduced effect is found for ozone; the maximum underestimation of the surface flux appears to be about 15% (F = 1.17).

In Figure 4, the importance of the considerations discussed here is illustrated by comparing time sequences of the different estimates of sulphur deposition at Whiteface Mountain with the smoother-terrain estimates at West Point. For this computation, weekly average deposition velocities have been used in conjunction with weekly average concentrations. This simple approach injects some error into the dry deposition computation, but the error is largely random and rarely exceeds 10% (Meyers and Yuen, 1987). Inspection of the diagram reveals that the uncertainty concerning the effects of terrain complexity obscure differences between locations. As yet, it is not possible to quantify the amount by which West Point receives more dry deposition of gaseous sulphur than the lower slopes of Whiteface Mountain.

The problem arising for the one-dimensionality of V_d is not limited to the interpretation of field measurements. All Eulerian models have difficulty with modelling dry deposition because atmosphere-surface exchange is largely a surface-controlled phenomenon. It is therefore necessary to take the details of surface features into account, but these features vary on scales far smaller

ATDL-M86/644

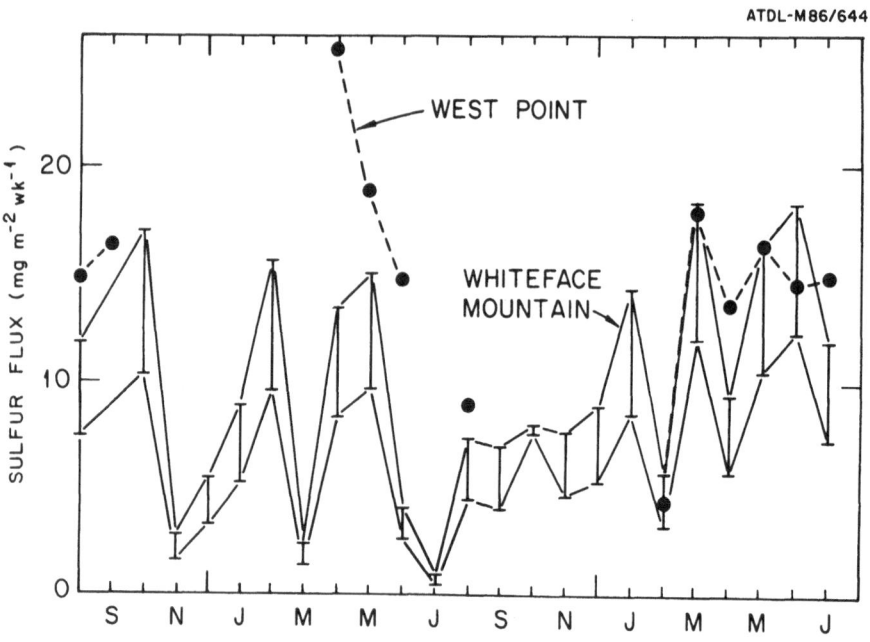

Figure 4. Comparisons of estimates of sulphur dry deposition as SO_2 derived at West Point (circles) with the range of similar values likely to be appropriate at the Whiteface Mountain location. Data are derived from the operation of a filterpack SO_2 monitoring system coupled with meteorological and surface-sampling apparatus designed to provide estimates of the deposition velocity. Note that the lower limits of the ranges shown for Whiteface Mountain correspond to the assumption that aerodynamic resistance is correctly computed using descriptions of vertical diffusion alone. The upper estimate follows from the assumption that atmospheric transfer is wholly by advective processes, and that the flux will be bounded by the assumption that $R_a = 0$. Only one value is given for West Point, which is not in such complex terrain.

than the grid areas of typical Eulerian models. Methods for deriving dry deposition fluxes from concentration data and from measurements of atmospheric and surface variables tailored to the needs of the computation are presently being tested (Hicks et al., 1985; 1987). Specific measurements are made to permit a more detailed and presumably better specification of the deposition velocity V_d than is possible using land-use categorization schemes presently used in models (e.g.

see Sheih et al., 1979). Thus the inferential methods used to estimate
V_d in field measurement programmes are not the same as the
parameterization routines used in numerical models, although the basic
multiple-resistance framework is used in each approach.

Note that the dry deposition estimates produced by any procedure
neglecting edge effects and complex terrain will be underestimated. For
want of something better, an interim procedure worthy of consideration
might be to base integrated assessments upon dry deposition estimates
intermediate between the simple-case (simple-world R_a) and first-order
correction ($R_a = 0$ when complex) estimates. A level of uncertainty
could then be specified objectively. For SO_2, this uncertainty is
likely to be less than 25%, even in the worst cases, since R_c will
normally dominate the overall exchange process regardless of whether R_a
is or is not relevant. For trace gases that are rapidly deposited upon
contact with the surface, the consequences of incorrectly assuming a
smooth-surface could be great, and existing field programmes and
numerical models could be providing large underestimates of dry
deposition.

The above discussion addresses the matter of trace gas exchange.
Particles impose a different set of considerations, that are not as
easily addressable.

6. CONCLUSIONS

Existing capabilities to model and to measure area-wide wet and dry
deposition are limited by difficulties associated with spatial
variability and representativeness. Nevertheless, routine measurements
of wet deposition are being made at selected locations, and attempts to
infer dry deposition rates are also commencing, on a trial basis. In
the case of dry deposition, most confidence results when sites are
selected so that the limited one-dimensional relationships presently
available for computing dry deposition from air concentration data are
appropriate. However, it is also necessary to consider less suitable
situations, such as mountaintops and areas of patchy forest, where the
assumption of simple one-dimensional vertically-diffusive transport
through the atmosphere is difficult to defend.

Available information does not yet warrant a detailed examination
of the errors that are likely to arise when computing dry deposition in
complicated situations. However, it appears possible to bound the
consequences of errors caused by incorrect specification of the
atmospheric transfer processes involved. For purposes of both routine
measurement and regional-scale numerical modelling, it should be
recognized that assessments made neglecting such factors as topographic
complexity and edge effects will normally result in underestimates of
area-averaged dry deposition. Field measurements made on the lower
slopes of Whiteface Mountain, New York, indicate that the errors are
likely to be no more than 30% for the case of SO_2 dry deposition, but
that much larger errors (perhaps as great as a factor of two) could
arise in computation of HNO_3 deposition. Comparisons between locations
(as when preparing isopleth presentations of deposition patterns)
should be correspondingly cautious.

552

ACKNOWLEDGEMENTS

This work was supported by the National Oceanic and Atmospheric
Administration as a contribution to the National Acid Precipitation
Assessment Programme. The contribution of the site operating teams at
West Point and at Whiteface Mountain is gratefully acknowledged. This
work is the product of continuing effort by a large team of people at
ATDD, among which special mention should be made of Detlef Matt, Dennis
Baldocchi, Ray Hosker, Jim Womack and Lynne Satterfield.

REFERENCES

Baldocchi, D.D., Hicks, B.B. and Camara, P. 1987. A Canopy Stomatal
    Resistance Model for Gaseous Deposition in Vegetated Surfaces.
    Atmos. Environ. **21**, 91–101.
Delany, A.C., Fitzjarrald, D.R., Lenschow, D.H., Pearson, R. Jr.,
    Wendel, G.J. and Woodruff, B. 1986. Direct Measurements of
    Nitrogen Oxides and Ozone Fluxes over Grasslands, submitted to J.
    Atmos. Chem.
Hales, J.M., Hicks, B.B. and Miller, J.M., 1986. The Role of Research
    Measurement-Networks as Contributors to Federal Assessments of
    Acid Deposition, Bull. Am. Meteorol. Soc. **68**, 216–225.
Hicks, B.B., Baldocchi, D.D., Hosker, R.P. Jr., Hutchison, B.A.,
    McMillen, R.T., Matt, D.R. and Satterfield, L.C. 1985. On the Use
    of Monitored Air Concentrations to Infer Dry Deposition (1985),
    NOAA Technical Memorandum ERL-141, 65 pp.
Hicks, B.B. and Matt, D.R. 1987. Combining Biology, Chemistry and
    Meteorology in Modelling and Measuring Dry Deposition, J. Atmos.
    Chem. (in press)
Meyers, T.P. 1987. The Sensitivity of Modelled SO_2 Fluxes and Profiles
    Within and Above a Deciduous Forest to Leaf Stomatal and Boundary
    Layer Resistances, Water, Air, and Soil Pollut. (in press)
Meyers, T.P. and Yuen, T.S. 1986. Assessment of Uncertainties
    Associated with Day/Night Sampling of Dry Deposition Fluxes of SO_2
    and O_3, J. Geophys. Res. **92 D6**, 6705–6712.
Sheih, C-M., Wesely, M.L. and Hicks, B.B. 1979. Estimated Dry
    Deposition Velocities of Sulphur over the Eastern United States
    and Surrounding Regions, Atmos. Environ. **13**, 1361–1368.

NUMERICAL SIMULATION OF SO_2 CONCENTRATION AND DRY DEPOSITION FIELDS IN THE TULLA EXPERIMENT

M. Baer and K. Nester
Institut für Meteorologie und Klimaforschung
Kernforschungszentrum Karlsruhe
Universitat Karlsruhe
Postfach 3640
D-7500 Karlsruhe 1
West Germany

ABSTRACT. The mesoscale field experiment TULLA was performed in Baden-Württemberg, Federal Republic of Germany, during March 1985. One of the aims of this experiment was the determination of a mass balance of SO_2 over Baden-Württemberg. For a selected day during the TULLA experiment, i.e. March 25, 1985, a mass balance was derived from experimentally determined SO_2 concentrations along the flight route and from simulated flow and dry deposition fields. The flow field and the dispersion coefficients were determined by a non-hydrostatic mesoscale model, which calculates these fields. The emissions of SO_2, available as hourly averages for a 1 km x 1 km grid and the SO_2 flux into Baden-Württemberg were taken into account as sources. The concentrations measured along the flight route were compared with the results of the DRAIS model. The agreement is quite satisfactory.

1. INTRODUCTION

There is at present a substantial lack of knowledge of the transport and chemical conversion of airborne pollutants in areas covered by a grid of several hundred square kilometers. In order to improve this knowledge the TULLA ('Transport and conversion of airborne pollutants in the state of Baden-Württemberg and from neighbouring states', in German: Transport und Umwandlung von Luftschadstoffen im Lande Baden-Württemberg und aus Anrainer Staaten) mesoscale field experiment was performed from March 18 until March 29 1985 (Fiedler, 1986). During intensive measuring phases in the experiment concentrations of pollutants such as SO_2, NO, NO_2 and O_3 and other species were determined and meteorological measurements were performed in the lower troposphere (Fiedler, 1986). One of the aims of the experiment was to establish a balance of the pollutants, i.e. to determine the ratio of foreign to local materials within the balancing area. The following influences must be taken into account when a balance is to be established:

553

M. H. Unsworth and D. Fowler (eds.), Acid Deposition at High Elevation Sites, 553–568.

- distribution of sources in space and time;
- knowledge of the atmospheric flow and turbulence taking into account the topography;
- chemical transformations of the pollutants dependant on the meteorological conditions;
- determination of depositions.

A preliminary estimate of the various contributions to the mass balance of SO_2 for March 25, 1985, shows that dry deposition plays an important part.

2. MODELLING DRY DEPOSITION

In most models the deposition rate is described by a single quantity, the pollutant deposition velocity v_D. the flux of material, F, directed towards the lower boundary surface, is defined by:

$$F = v_D \cdot c(z_r) \tag{1}$$

where $c(z_r)$ is the concentration of the material at a reference height z_r. The dry deposition velocity involves a complex linkage between turbulent diffusion in the surface boundary layer, molecular scale motion at the air-ground interface and interaction of the material with the surface.

In most efforts to model and parameterize v_D for the dry deposition of gases to vegetated surfaces a 'big-leaf' multiple resistance model is used (see Fig. 1). The most important individual resistances to pollutant transfer are the aerodynamic resistance (r_a) associated with atmospheric turbulence, a quasi-laminar boundary resistance (r_b), which takes into account the resistance to mass transfer through the quasi-laminar layer of air in contact with the surface elements and is influenced by the diffusivity of the material being transferred, and a net canopy resistance (r_c), which is dominated by biological surface factors and includes the stomatal, mesophyll and cuticular uptake resistances; however, the stomatal resistance tends to dominate for gases.

2.1. Determination of the Dry Deposition Velocity v_D

The aerodynamic, quasi-laminar boundary layer and canopy resistances are in series. Thus, v_D is given by:

$$v_D = \frac{1}{r_a + r_b + r_c} \tag{2}$$

As input data for the three dimensional diffusion model DRAIS (drei-dimensionales regionales Ausbreitungs – und Immissions-Simulationsmodell), a further development of a diffusion model (Fiedler et al., 1985) that was for the first time applied to the preliminary TULLA experiment, the dry deposition velocity v_D is calculated at each grid point. The modelling area is a complex

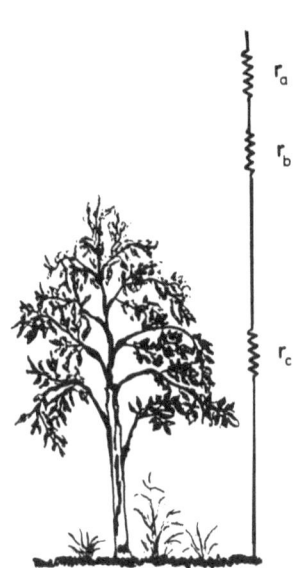

Figure 1. 'Big-leaf' multiple resistance model (r_a, r_b and r_c are explained in the text).

r_a: aerodynamic resistance
r_b: surface boundary layer resistance
r_c: canopy resistance

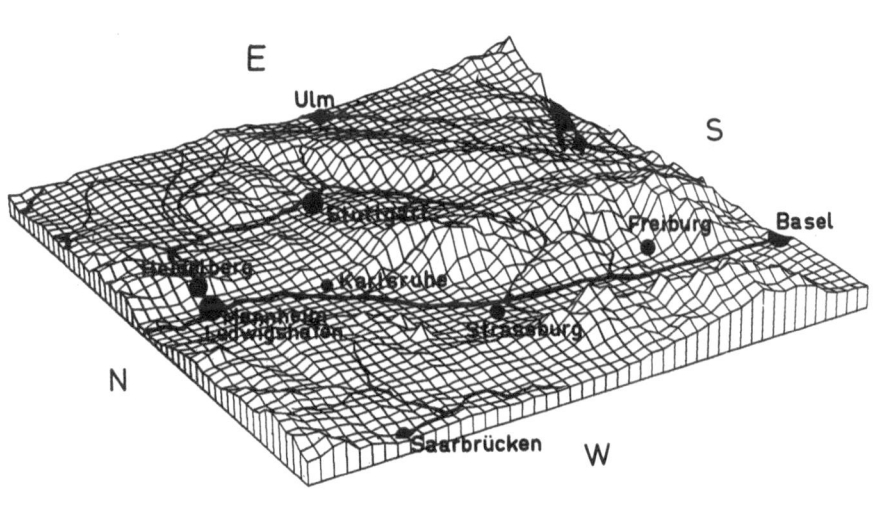

MODEL AREA

Figure 2. Modelling area for KAMM and DRAIS (53 x 49 grid points, 5 km x 5 km each).

irregular terrain with altitudes up to 1400 m (see Fig. 2). The highest elevations are to be found in the Black Forest. The modeling domain is subdivided into 2597 grid areas of 5 km x 5 km each.

2.1.1. Determination of the aerodynamic resistance. The aerodynamic resistance is a function of wind speed, surface properties and atmospheric stability. The parameterization used is given by Eq. 3.

$$
r_a(z_r) = \begin{cases}
\dfrac{1}{\kappa u_*} \left(0,74 \ln \left(\dfrac{z_r}{z_0}\right) + \dfrac{4,7}{L_*} (z_r - z_0)\right) & z_r/L_* > 0 \text{ stable} \\[3mm]
\dfrac{1}{\kappa u_*} \left(0,74 \ln \left(\dfrac{z_r}{z_0}\right)\right) & z_r/L_* = 0 \text{ neutral} \\[3mm]
\dfrac{1}{\kappa u_*} \cdot 0,74 \left[\ln \left(\dfrac{\sqrt{1-9\dfrac{z_r}{L_*}} - 1}{\sqrt{1-9\dfrac{z_r}{L_*}} + 1}\right) \right. \\[5mm]
\left. \qquad - \ln \left(\dfrac{\sqrt{1-9\dfrac{z_0}{L_*}} - 1}{\sqrt{1-9\dfrac{z_0}{L_*}} + 1}\right) \right] & z_r/L_* < 0 \text{ unstable}
\end{cases}
\tag{3}
$$

where u_* is the friction velocity, κ the von Karman constant ($\kappa = 0.4$), z the reference height, z_0 the roughness length, and L_* the Monin-Obukhov stability length.

Values for z_0 were parameterized by an interpolation of the z_0-values for the different known ground covers within one grid area of 5 km x 5 km size. The method has been described by van Dop et al, (1983). Only the weighting factors are not taken into account. The result is shown in Fig. 3. The mean z_0 for the whole modelling area was about 1m. The z_0-field does not differ very much from that obtained by a linear interpolation.

At every grid point values for u_* and L_* were calculated by KAMM (Karlsruher Atmosphärisches Mesoskaliges Modell) (Dorwarth, 1986). The KAMM model is three-dimensional, non-hydrostatic and instationary. A terrain following coordinate system is used. It calculates the flow, temperature and diffusion coefficient fields over an irregular terrain by solving

- the Navier Stokes equation of motion,
- the continuity equation, and
- the thermodynamic equation.

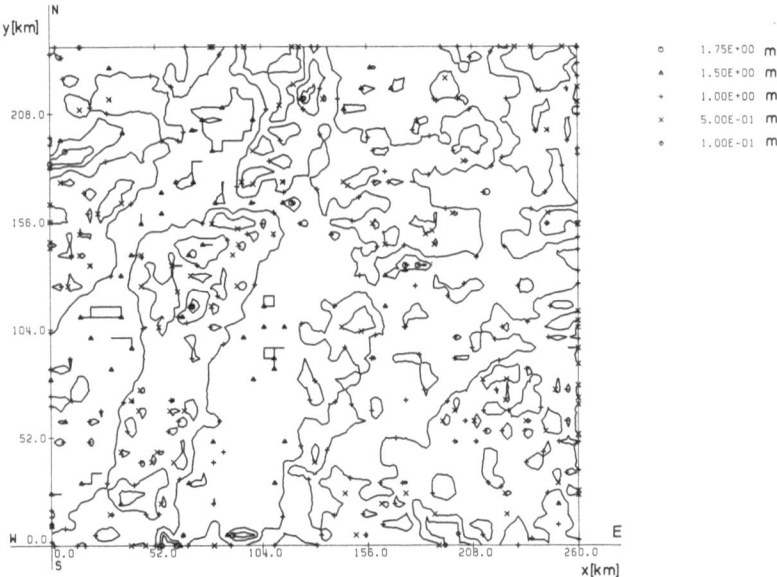

Figure 3.   Isolines of the roughness length z_0 for the modelling area.

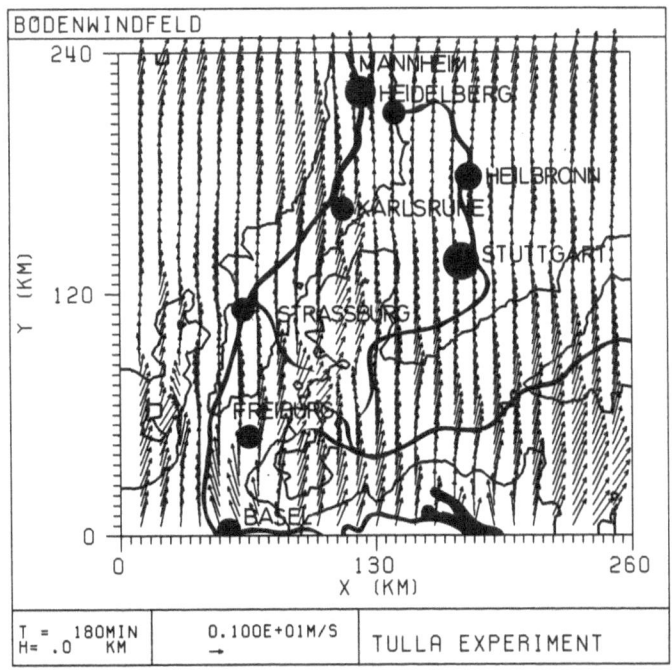

Figure 4.   Windfield at ground level (March 25, 1985).

558

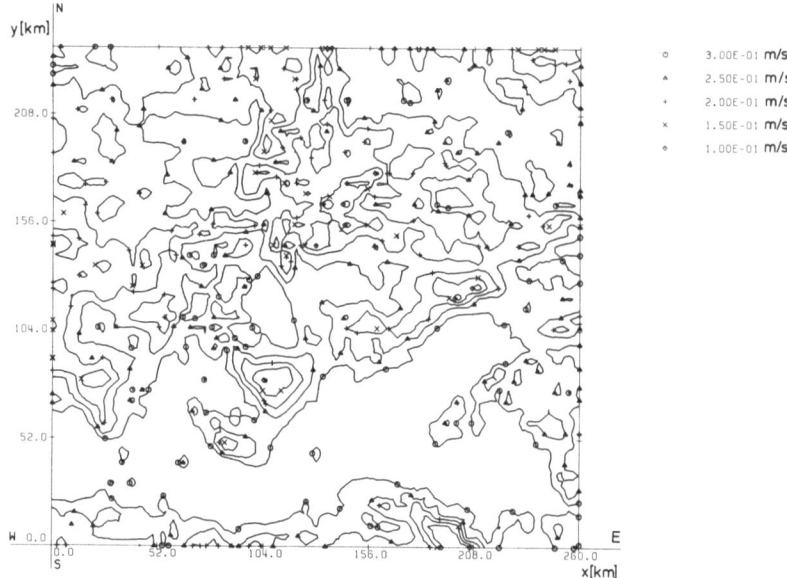

Figure 5.   Isolines of the friction velocity u* for the modelling
area (March 25, 1985).

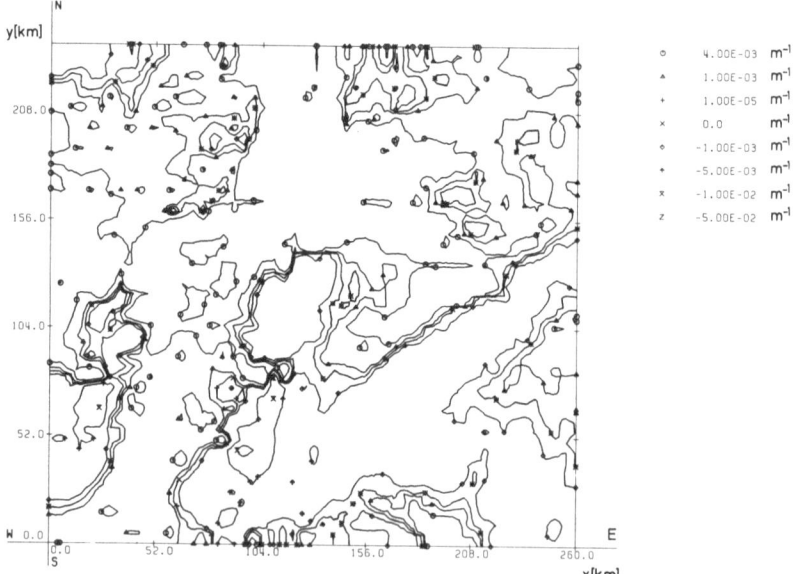

Figure 6.   Isolines of the inverse of the Monin-Obukhov stability
length (1/L*) for the modelling area (March 25, 1985).

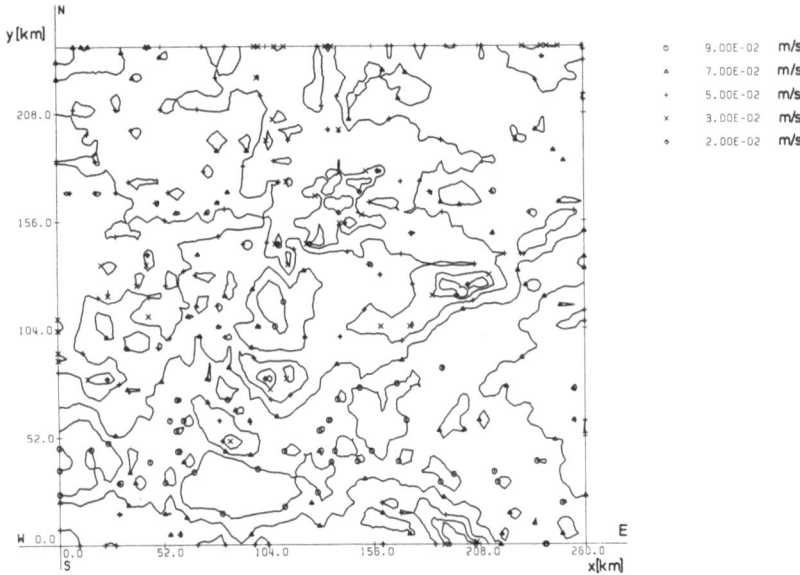

Figure 7. Isolines of the inverse of the aerodynamic resistance ($1/r_a$) for the modelling area (March 25, 1985).

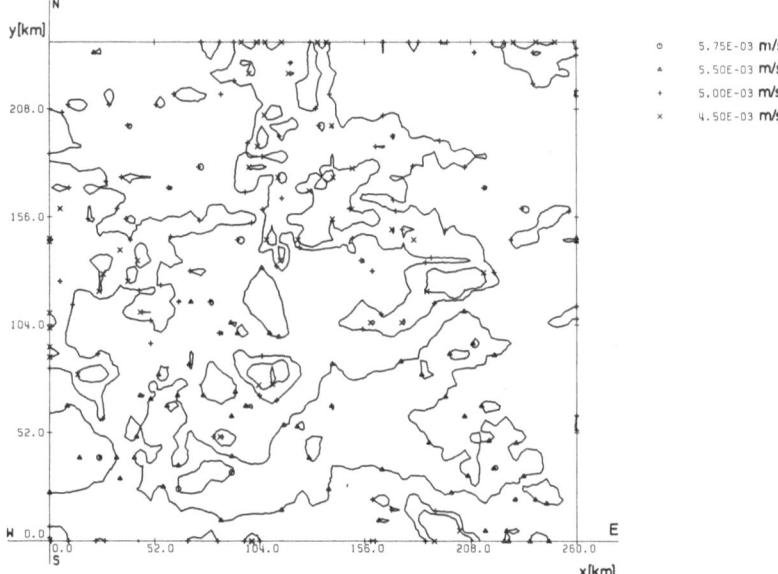

Figure 8. Isolines of the dry deposition velocity V_D for the modelling area (March 25, 1985).

As an example the data for March 25, 1985 were selected.

During the daytime of March 25, 1985 the atmosphere was almost neutrally stratified up to a height of 700 m. The wind at 200 m above ground level blew at a speed of about 6 m/s. The general wind direction was at 220 degrees.

The wind field over Baden-Württemberg calculated by KAMM, on the basis of the radiosonde measurements made in Stuttgart at noon is shown in Fig. 4. The corresponding values for u_* and $1/L_*$ are shown in Figs. 5 and 6. The maximum values of u_* were found at sites of maximum wind speeds (i.e. in the mountainous regions).

Maximum changes in $1/L_*$ occurred at the slopes of the mountains where the changes in stability are greatest.

The inverse of the aerodynamic resistance calculated with these values for the modelling area is displayed in Fig. 7.

Compared with Fig. 5 it becomes evident that $1/r_a$ is mostly influenced by u_*, where u_* is dominantly influenced by the topography.

2.1.2 <u>Determination of the quasi-laminar surface boundary layer resistance</u>. r_b is a function of wind speed and surface properties and is controlled by the molecular properties of the fluid. It can conveniently be written as:

$$r_b = \frac{1}{ku_*} \cdot kB^{-1} \qquad (4)$$

B^{-1} can be estimated in terms of the surface roughness Reynolds number and the Schmidt number.

For SO_2 we use

$$r_b = \frac{2.6}{ku_*} \qquad (5)$$

a value adopted from Wesely and Hicks (Wesely and Hicks, 1977).

For the canopy resistance r_c we have chosen a total value of r_c = 150 s/m for SO_2 at every grid point which is a mean value of canopy resistances observed for SO_2 with respect to grass (Fowler and Unsworth, 1979) and to a pine forest (Fowler and Cape, 1982) in early spring and autumn. The simulated dry deposition velocity field for March 25, 1985 according to Eq. 2 is shown in Fig. 8.

This is the field of dry deposition velocities we used as input data for DRAIS in order to simulate the concentration and dry deposition fields for March 25, 1985 (see Chapter 3). The input field is influenced mainly by topography which reflects dependence on u_* and L_*. The minimum and maximum values of v_D are 0.16 cm/s and 0.58 cm/s, respectively, and a mean value of v_D of 0.52 cm/s has been derived for the modelling area.

2.2 Dry Deposition as the Lower Boundary Condition in DRAIS

The conservation equation for gaseous or aerosol material in the atmosphere which emerges as a result of the averaging operations and closure approximations and which is the basis of most airshed models can be expressed in Cartesian coordinates by the following equation:

$$\frac{\partial \bar{c}_i}{\partial t} + \nabla(\overline{u}\overline{c}_i) = \nabla \cdot (K \cdot \nabla \bar{c}_i) + f_i(\bar{c}_i \ldots \bar{c}_p) \tag{6}$$

$\bar{c}_i$:   mean concentration of chemical species i

$\bar{u}$:   mean advective velocity field $\bar{u} = (\bar{u}, \bar{v}, \bar{w})$

K :   second order eddy diffusivity tensor

f_i: includes chemical formation (or depletion) rate of species i, sources and sinks of species i.

In a Cartesian coordinate system the corresponding lower boundary condition can be formulated as follows:

$$-K \ \overline{\nabla c}_i \cdot n_h = Q_i \ (x,y,h(x,y),t) \tag{7}$$

where K is the eddy diffusity tensor, n_h is the unit vector normal to the terrain directed into the atmosphere, Q_i is the mass flux of species i to the surface, and h(x,y) is the ground elevation above sea level at x,y (Reynolds et al., 1973).

When the conservation relations and the lower boundary conditions are applied it is not always desirable to use the Cartesian coordinate system. In the DRAIS diffusion model a terrain following coordinate system is used. The functional form of the coordinate transformation in terms of the original Cartesian system can be written as:

$$\begin{aligned}
\tilde{x}^1 &= x & x &= \tilde{x}^1 \\
\tilde{x}^2 &= y & y &= \tilde{x}^2 \\
\tilde{x}^3 &= \eta(x,y,z) & z &= \zeta(\tilde{x}^1, \tilde{x}^2, \tilde{x}^3)
\end{aligned} \tag{8}$$

The vertical coordinate in the terrain following coordinate system in DRAIS is given by:

$$a(1-\eta)^2 + b(1-\eta) = \frac{h(x,y)-z}{h(x,y)-H} \tag{9}$$

where h(x,y) is the ground elevation above sea level and H is the top of the modelled layer.

For z = H, η = 0 and for z = h(x,y), η = 1. In that η-coordinate system Eq. 6 takes the following form:

$$\frac{\partial \bar{c}_i}{\partial t} = \tilde{u}^1 \frac{\partial \bar{c}_i}{\partial x} + \tilde{u}^2 \frac{\partial \bar{c}_i}{\partial y} + \tilde{u}^3 \frac{\partial \bar{c}_i}{\partial \eta}$$

$$+ \frac{1}{\sqrt{\tilde{G}}} \frac{\partial}{\partial x} \left(\sqrt{\tilde{G}} \left\{ K^{11} \frac{\partial \bar{c}_i}{\partial x} + K^{11} \frac{\partial \eta}{\partial x} \frac{\partial \bar{c}_i}{\partial \eta} \right\} \right)$$

$$+ \frac{1}{\sqrt{\tilde{G}}} \frac{\partial}{\partial y} \left(\sqrt{\tilde{G}} \left\{ K^{22} \frac{\partial \bar{c}_i}{\partial y} + K^{22} \frac{\partial \eta}{\partial y} \frac{\partial \bar{c}_i}{\partial \eta} \right\} \right)$$

$$+ \frac{1}{\sqrt{\tilde{G}}} \frac{\partial}{\partial \eta} \left(\sqrt{\tilde{G}} \left\{ K^{11} \frac{\partial \eta}{\partial x} \frac{\partial \bar{c}_i}{\partial x} + K^{22} \frac{\partial \eta}{\partial y} \frac{\partial \bar{c}_i}{\partial y} \right. \right.$$

$$\left. \left. + \left[K^{11} \left(\frac{\partial \eta}{\partial x}\right)^2 + K^{22} \left(\frac{\partial \eta}{\partial y}\right)^2 + K^{33} \left(\frac{\partial \eta}{\partial z}\right)^2 \right] \frac{\partial \bar{c}_i}{\partial \eta} \right\} \right)$$

where $\quad \tilde{u}^1 = \bar{u}$

$$\tilde{u}^2 = \bar{v}$$

$$\tilde{u}^3 = \bar{u} \frac{\partial \eta}{\partial x} + \bar{v} \frac{\partial \eta}{\partial y} + \bar{w} \frac{\partial \eta}{\partial z}$$

K in Cartesian coordinates has the form:

$$K = \begin{pmatrix} K^{11} & 0 & 0 \\ 0 & K^{22} & 0 \\ 0 & 0 & K^{33} \end{pmatrix}$$

and $\tilde{G} = 1/_{(\partial \eta / \partial z)}$ is the determinant of the Jacobian of the three-dimensional transformations.

The lower boundary condition in the η-coordinate system reads (for $\eta = 1$):

$$K^{11} \frac{\partial \eta}{\partial} \frac{\partial \bar{c}_i}{\partial x} + K^{22} \frac{\partial \eta}{\partial y} \frac{\partial \bar{c}_i}{\partial y} + [\ K^{11} (\frac{\partial \eta}{\partial x})^2 + K^{22} (\frac{\partial \eta}{\partial y})^2 +$$

$$\tag{12}$$

$$K^{33} (\frac{\partial \eta}{\partial z})^2\ ]\ \frac{\partial \bar{c}_i}{\partial \eta} = v_{D,i}\ \bar{c}_{r,i}$$

where $v_{D,i}$ is the dry deposition velocity of chemical species i and $\bar{c}_{r,i}$ is the concentration at a reference height, where $v_{D,i}$ is determined conveniently.

3. SIMULATED CONCENTRATION AND DRY DEPOSITION FIELDS FOR MARCH 1985

3.1 SO_2 emissions and SO_2 measurements on March 25, 1985

Knowledge of the distribution of the main SO_2 sources is necessary for the calculation of the corresponding concentration fields. The SO_2 emissions for the state of Baden-Württemberg are available as hourly averages of a 1 km x 1 km grid for the whole period of the TULLA experiment ((Boysen et al., 1986). The relevant SO_2 emissions on March 25, 1985, from 9 a.m. until 3 p.m., are given in Fig. 9. The total SO_2 emission rate on that day was 30 t/h.

Since Fig. 9 contains also the topography, it can be seen that the sources are concentrated in the Upper Rhine and Neckar Valleys.

On March 25, 1985 four aircraft measured SO_2 concentration profiles on a flight pattern covering Baden-Württemberg. The lowest flight level was about 200 m above ground level, varying between 100 and 300 m. The other aircraft were flying at levels about 100 m distant from each other. The concentration distribution of SO_2 along the flight routes measured by the lowest aircraft together with the topography are plotted in Fig. 10. Due to the flow conditions on that day, the strength and locations of the sources, the highest concentrations were found in the northern part of Baden-Württemberg.

The SO_2 measurements of the highest aircraft show that on that day SO_2 was concentrated in the lowest part of the troposphere. On the flight route some higher concentrations were observed, by the highest flying aircraft (TULLA 4R) belonging to the plumes of elevated single sources. Only at one grid point the concentration measured by TULLA 4R was higher than the SO_2 concentration measured by the lowest flying aircraft. In this case we have assumed that the concentration drops to zero at a height of 100 m above the flight level of TULLA 4R. This height is derived from the neighbouring grid points. The influence on budgeting will be negligible.

3.2 Estimated Mass Balance of SO_2

The SO_2-fluxes around Baden-Württemberg were not experimentally determined during the TULLA experiment. Therefore these fluxes have to be determined from the aircraft measurements and the simulated wind field. From the aircraft measurements a SO_2 profile was derived at each grid point along the flight route. Using the simulated wind speeds at these grid points the SO_2 fluxes could be estimated.

If we assume stationary conditions the terms D and T sum up a value of 14.9 t/h. For a first estimate of the dry deposition over Baden-Württemberg, the ground level concentration of SO_2 along the flight route, derived from the SO_2 measurements of the aircraft, is used. At each grid point inside the flight pattern the SO_2 concentration is interpolated. The weighting function used is the inverse of the square distance to the grid points at the boundaries. By multiplying the concentration at each grid point by the corresponding dry deposition velocity (see Fig. 8) the dry deposition is determined (see Fig. 12). The main dry deposition can be observed in the northern part of Baden-Württemberg. The total dry deposition amounts to 10.8 t/h. This means that the chemical transformation of SO_2 contributes 4.1 t/h to the mass balance. On the average, a SO_2 containing air parcel takes 7.5 hours to pass the experimental site. Under these conditions a transformation rate of SO_2 of 1%/h is derived.

Wet deposition as a sink of SO_2 could be neglected since there was no precipitation during the time period of the experiment on March 25, 1985.

Although these are first estimates of the various contributions to the SO_2 mass balance they reflect the correct order of magnitude. The mass balance indicates that about 80% of the total emission of SO_2 are exported.

The difference between import and export is about half of the emissions. Therefore, it becomes evident that dry deposition and

	Southern boundary	Western boundary	Northern boundary	Eastern boundary	Balance
SO_2 flux t/h	6.1	3.2	−20.3	−4.1	−15.1

The mass balance can be written as:

$$\frac{d}{dt} m_{SO_2} = F + F_0 + D + T + E$$

F_I = input flux of SO_2 (9.3 t/h)
F_0 = output flux of SO_2 (24.4 t/h)
E = emission of SO_2 (30 t/h)
D = dry deposition of SO_2
T = chemical transformation of SO_2

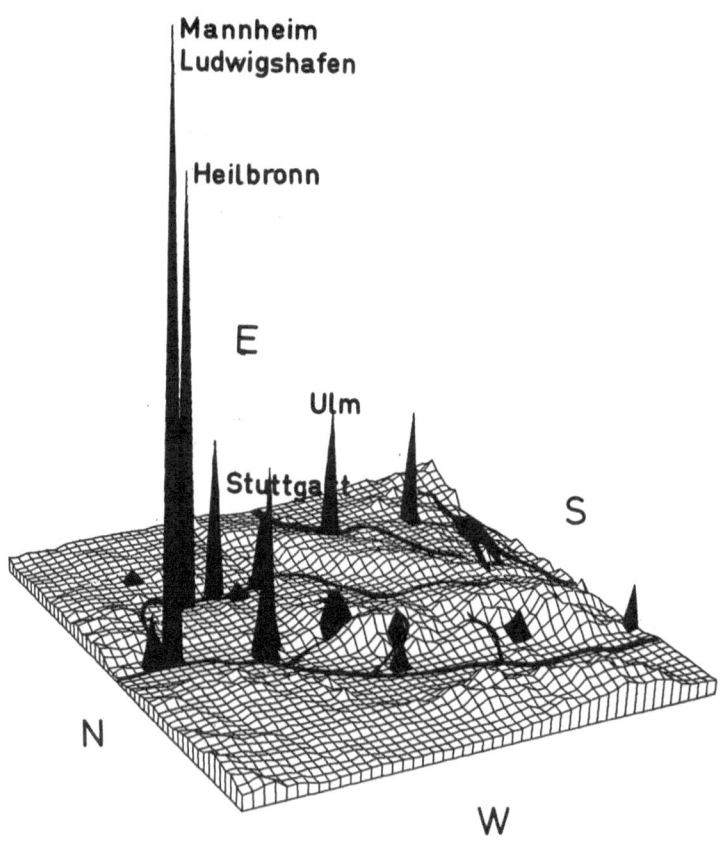

Figure 9.   SO$_2$-emissions from single sources (March 25, 1985).

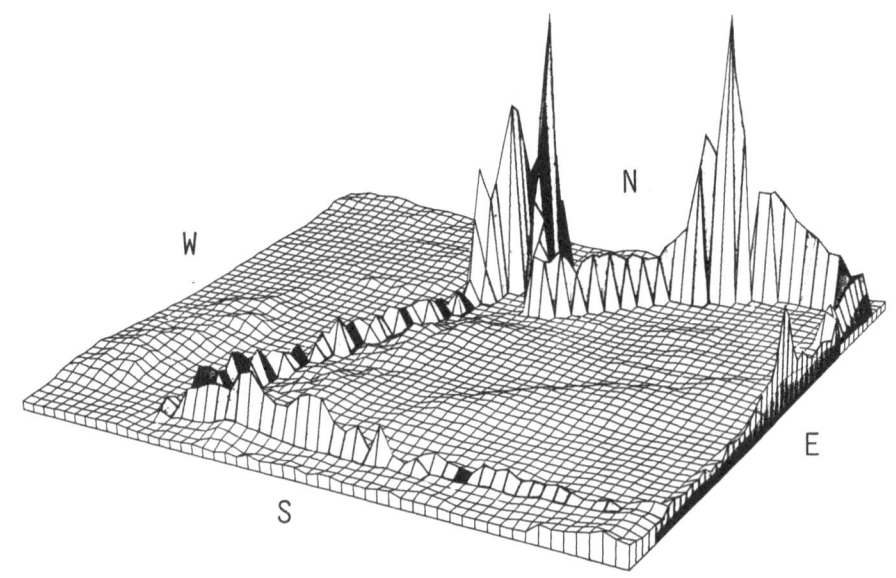

Figure 10. Measured SO_2-concentration along the flight route (lowest level, March 25, 1985).

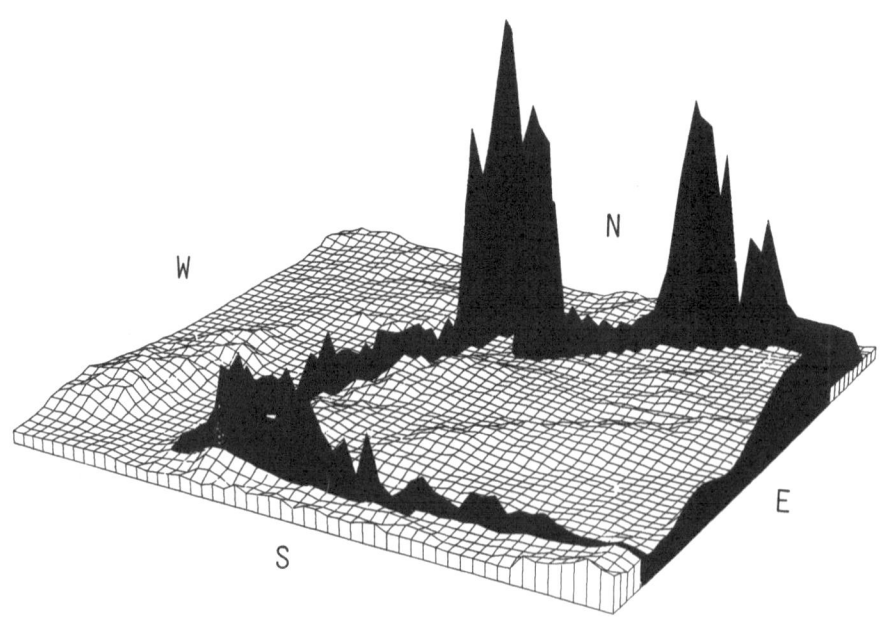

Figure 11. Calculated SO_2-concentration along the flight route as in Figure 10 (March 25, 1985).

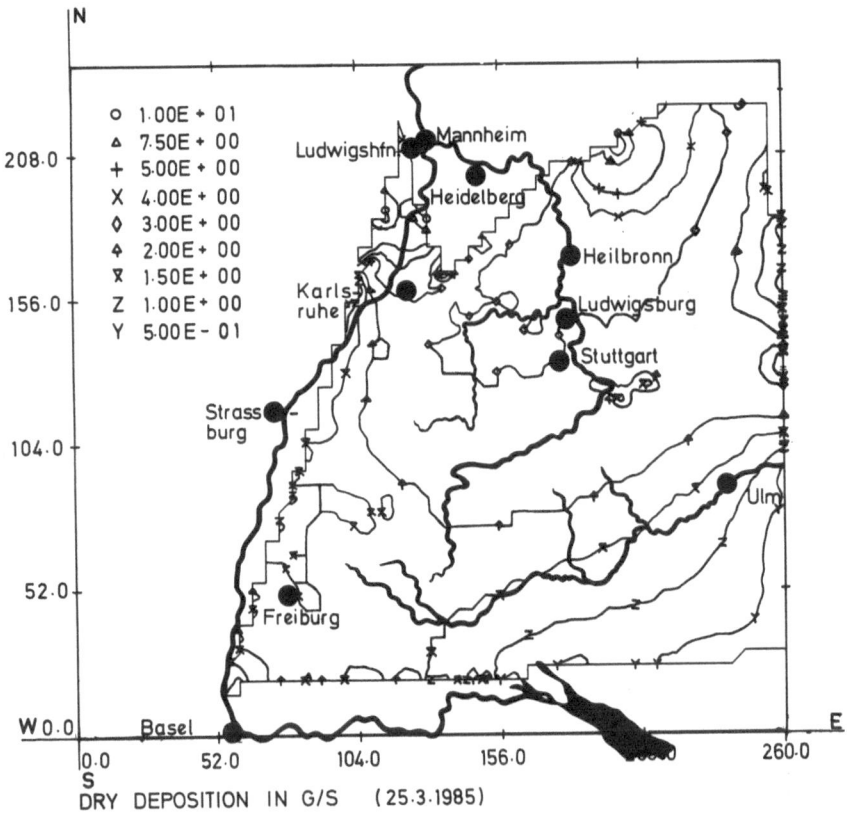

Figure 12. Dry deposition of SO$_2$ in g/s (March 25, 1985) (grid area 5 km x 5 km).

chemical transformations are important removal processes, with dry deposition playing a much more important part than chemical transformations. Although it is obvious, it should still be emphasized that the results are preliminary till all data will have been analysed.

To compare the SO_2 measurements at the outflow boundaries, a calculation with the DRAIS (Dreidimensionales regionales Ausbreitungs- und Immissions-Simulationsmodell) model was performed. The input flux and the sources inside of Baden-Württemberg were taken into account. Figure 11 shows the SO_2 concentration distribution around Baden-Württemberg which is comparable to that of Fig. 10. The agreement between both distributions is quite satisfactory. The maximum and minimum concentrations are located at almost the same positions. At the outflow boundaries the calculated SO_2 concentrations are lower than the measured ones. The mean relative error is 0.8.

When calculating the total dry deposition by DRAIS we obtain a value of about 10 t/h. This value agrees well with the experimentally derived one.

REFERENCES

Boysen, B., Friedrich, R., Müller, Th., Scheirle, N., and Voss, A. 1986. 'Feinmaschiges Kataster der SO_2 und NO_x Emissionen in Baden-Württemberg fur die Zeit der TULLA-Messkampagne'. 2. Statuskolloquium des PEF vom 4.-7.3.1986 im Kernforschungszentrum Karlsruhe, KfK-PEF 4, April 1986, Band **2** .

Dorwarth, G. 1986. 'Numerische Berechnung des Druckwiderstandes typischer Geländeformen. Wiss. Ber. Institut fur Meteorologie und Klimaforschung, Universität Karlsruhe, Nr **6** .

Fiedler, F. 1986. Ziele und Durchfuhrung des TULLA-Experiments' Statuskolloquium 4.-7.3 1986. Projekt Europäisches Forschungszentrum für Massnahmen zur Luftreinhaltung, Kernforschungszentrum Karlsruhe. KfK-PEF 4, April 1986, Band **2** .

Fiedler, F., Dorwarth, G., Laudenbach, I., Walk, O. 1985. 'Der Massenhaushalt von Schwefeldioxid im regionalen Bereich anhand des Beispiels von Baden-Württemberg (TULLA)', 1. Statuskolloquium 5.-7.3.1985, Projekt Europäisches Forschungszentrum für Massnamen zur Luftreinhaltung, im Kernforschungszentrum Karlsruhe. KfK-PEF 2, April 1985.

Fowler, D. and Unsworth, M.H. 1979. 'Turbulent transfer of sulphur dioxide to a wheat crop'. Quart J. R. Met. Soc. **105** , 767-783.

Fowler, D. and Cape, J.N. 1982. 'Dry deposition of SO_2 onto a Scots pine forest. In: Precipitation Scavenging, Dry Deposition and Resuspension, Vol. **2** (eds.) Pruppacher, H.R., Semonin, R.G. and Slinn, W.G.N., Elsevier Science Pub. N.Y., 763-773.

Reynolds, S.D., Roth, P.M. and Seinfeld, J.H. 1973. 'Mathematical Modelling of Photochemical Air Pollution - I Formulation of the Model'. Atmos. Environ **7** , 1033-1061.

Van Dop, H. 1983. 'Terrain Classification and derived meteorological parameters for interregional transport models'. Atmos. Environment **17** , 1099-1105.

Wesely, M.L. and Hicks, B.B. 1977. 'Some factors that effect deposition rates of SO_2 and similar gases on vegetation'. J. APCA **27** , 1100-1116.

A NEW INSTRUMENT FOR SO_2 EDDY FLUX MEASUREMENTS

M.G. Nestlen
Fraunhofer-Institut fur
Atmosphärisch Umweltforschung
Kreuzeckbahnstr. 19,
D-8100 Garmisch-Partenkirchen, F.R.G.

ABSTRACT. An instrument was designed to measure the vertical flux of SO_2 in the lower troposphere based on the eddy correlation technique. Differential absorption spectroscopy in the UV region (300nm) together with a White cell was used to measure SO_2 concentration fluctuations without interferences from other sulphur compounds. The concentration resolution achieved to date for SO_2, 2.8 $\mu g/m^3$, is mainly limited by the available optics. A better design should lower the concentration resolution to 0.1 $\mu g/m^3$. Preliminary measurements over patchy terrain covered with pine and beech trees and small buildings showed considerable variability in the apparent deposition velocity of SO_2 ranging from +2.5cm/s and -2.5cm/s, with a mean value of 0.36cm/s for a 24 hour interval.

1. INTRODUCTION

Dry deposition is an important sink for many trace gases. Among the various techniques available to measure vertical fluxes, the eddy correlation method yields the most direct and accurate results (Hicks 1978). Continuous fast response measurements of the vertical wind speed, w (m/s), and the concentration of the constituent, c ($\mu g/m^3$), at a fixed point above the surface have to be made in order to apply this method. The vertical flux of the constituent, j ($\mu g/(m^2 s)$), is given by the covariance $j=\overline{wc}$, where the overbar indicates the time average. If it is assumed that the mean averaged vertical wind velocity is zero, then $\overline{w}=0$ and $j=\overline{w'c'}$ where the prime indicates the instantaneous departure from the longterm mean. In reality, a better assumption is that $\overline{\rho w}=0$ (where ρ is the density of dry air); small but sometimes important errors arise if $\overline{w}=0$ is assumed (see Businger 1986, Webb et al. 1980).

The instruments used with this technique, however, must have time constants of less than one second and a resolution of better than 1% of the mean concentration (see Section 2). Only modified flame photometers, suffering from strong interferences due to other sulphur components (Fowler and Cape 1982, Galbally et al. 1979,

M. H. Unsworth and D. Fowler (eds.), Acid Deposition at High Elevation Sites, 569–582.
© *1988 by Kluwer Academic Publishers.*

Whelpdale and Shaw 1974, Neumann and den Hartog 1985) have been used to measure SO_2 fluxes. Therefore, a new instrument has been developed, which is fast and sensitive enough to be used for eddy correlation measurements and which is only sensitive to SO_2.

2. INSTRUMENT REQUIREMENTS

Trace gas analysers for eddy correlation measurements have to be designed to satisfy the specific requirements imposed by theoretical considerations (Hicks 1978). Three main points have to be considered:
- response time
- instrument size
- concentration resolution

2.1. Response Time

The eddy frequency n can be normalized according to

$$f = n(z/u(z)) \qquad (1)$$

where f is the normalised frequency and u(z) the mean horizontal wind velocity at height z (Lumley and Panofsky 1964). In order to evaluate the eddy fluxes properly, the total frequency range has to be covered. Literature data show that for usual atmospheric stability conditions, f is between 0.003 and 3 (Kaimal 1975, McBean and Miyake 1972, Panofsky and Mares 1968). At a measurement height of z=20m and a mean wind speed of u(z) = 5m/s, this yields a maximum response time of 0.7 seconds and a minimum integration time of 20 minutes.

2.2 Instrument Size

According to Taylor's hypothesis (Lumley and Panofsky 1964), an eddy size l is related to the eddy frequency:

$$l=u(z)/n \qquad (2)$$

$$\text{or from (1):} \quad l=z/f \qquad (3).$$

Assuming that f=3 and z=20m this gives l=6.6m. In order to detect these eddies, the instruments used must average over a distance less than half the size of the eddies, i.e. 3m (Nestlen 1984). The same holds true for the distance between the anemometer and the trace gas sensor when operated at a height of 20m.

2.3 Concentration Resolution

The mean vertical flux j of a substance with concentration c can be expressed as

$$j = r_{wc}*\delta_w*\delta_c \qquad (4)$$

where r_{wc} is the correlation coefficient of the vertical windspeed and the concentration (Hicks 1978); δ_w and δ_c are the variances of the vertical windspeed and the concentration respectively. Considering the definition of the deposition velocity (Roth 1975):

$$v_d = j/c(z) \qquad (5)$$

where $c(z)$ is the mean concentration at height z, the relative concentration standard deviation follows:

$$\frac{\delta_c}{c(z)} = \frac{v_d}{r_{wc} * \delta_w} \qquad (6)$$

Measurements have shown that r_{wc} is usually between 0.3 and 0.4 (Pasquill 1974) and that δ_w is in the range of 0.1m/s to 1m/s (Hicks 1978). Assuming a resolution of 10% of the mean concentration and a deposition velocity of 0.4cm/s, this yields a relative concentration resolution of 0.2% to 2%. Thus, when the mean SO_2 mixing ratio is 10ppb ($27\mu g/m^3$), the resolution of the instrument should be better than 20 to 200ppt ($0.05-0.5\mu g/m^3$) SO_2.

3. INSTRUMENTAL DETAILS

3.1. The SO_2 Sensor

The method for measuring the SO_2 concentration is based on differential optical absorption spectroscopy (DOAS) in the near ultraviolet region from 290nm to 310nm. SO_2 has several absorption bands in this wavelength region as shown in Fig. 1. Interference from other substances can be almost totally excluded; the only other atmospheric substance so far known absorbing in this wavelength region is ozone. The differential absorption coefficient of ozone, however, is about 100 times smaller than that of SO_2. Even in the presence of a high ozone concentration together with a low SO_2 concentration, the influence of ozone can be totally corrected: an O_3 reference spectrum is fitted to the measured spectrum and subsequently subtracted from the measured spectrum, which contained SO_2 and O_3. The result is a spectrum without any absorption bands from O_3.
    The concentration is then evaluated according to Beer's Law.

$$lg(Io/I) = O.D. = \varepsilon * l * c \qquad (7),$$

where O.D. is the differential optical density, ε the differential optical absorption coefficient, l the path length, c the concentration and Io and I are the intensities as shown in Fig. 1. The absorption coefficient for SO_2 has previously been determined as $2.7*10^{-7}$ $m^2/\mu g$ (Nestlen 1984, Kessler 1979).
    Differential absorption spectroscopy has been widely used to measure trace gas concentration along a linear light path of several hundred meters in the atmosphere (Platt et al. 1979, Perner 1979). To

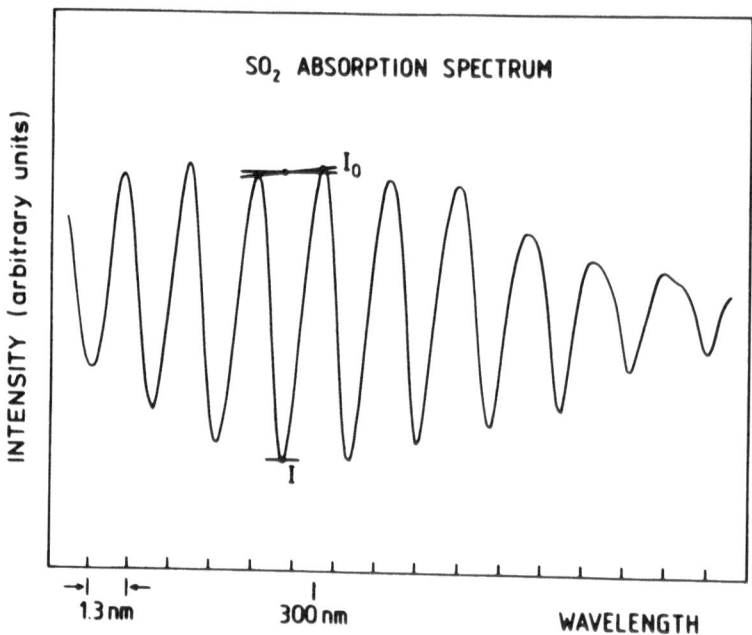

Figure 1. The absorption spectrum of SO_2 in the wavelength region of 300 nm.

use this technique for eddy correlation measurements, the light pathhas to be folded to reduce the instrument size. The instrumental set-up is illustrated in Fig. 2. The light source is a 75W high pressure Xe short-arc lamp of high brightness (400 Cd/mm^2). The light beam is focussed into the multiple reflection cell by a concave mirror (70mm diameter, focal length 380mm). This cell is similar to the one described by White (1942) and open to the air. The concave mirrors used have a focal length of 0.75m, a diameter of 0.2m (mirror A) and 0.25m (mirror B) and are placed at a distance of 1.5m. Their dielectric coating gives a reflectivity of 99.8% in the wavelength region of 300±15nm. Mirror B is cut along its diameter. The two parts can be adjusted separately in order to control the number of reflections. The light leaving the White cell is focussed on the entrance slit of an 0.25m spectrograph (Spex Minimate) equipped with a ruled grating (1200mm^{-1}).

A thin metal disc (200mm diameter) with 60 radial slits (0.1mm wide, 5mm apart, 20mm high) is placed in the outlet focal plane of the spectrograph. In this way, a section of 20nm width of the whole spectrum can be scanned repetitively. An infrared light barrier is used to trigger each scan and to define the spectral position of the slit. The disc is rotated at about 88rpm with an accuracy better than ±0.5%. The light intensity passing through the slits is monitored by a

573

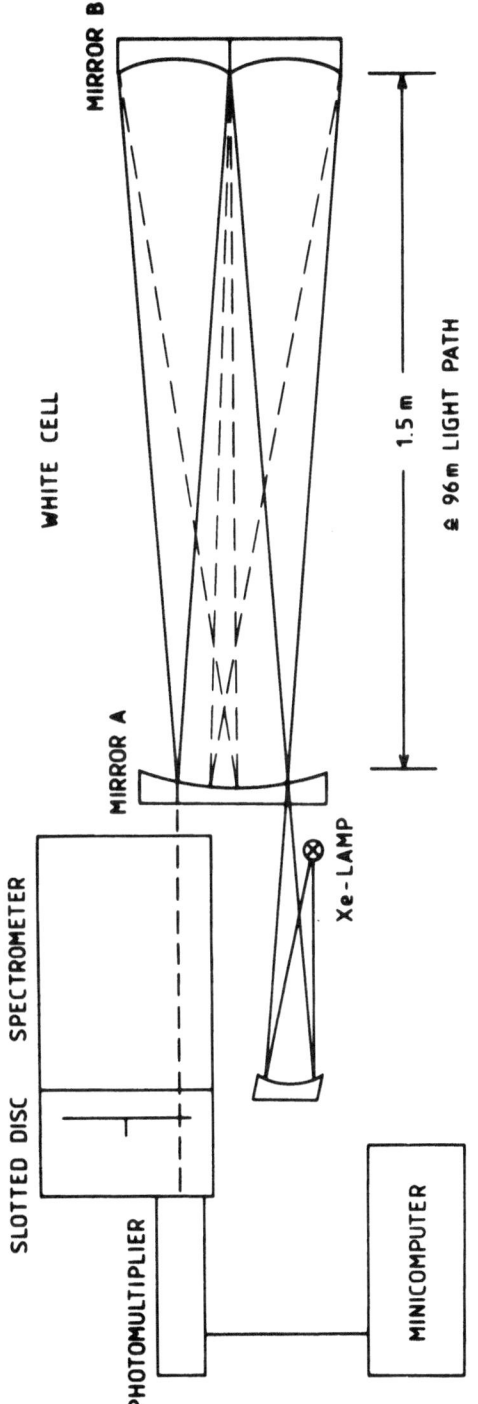

Figure 2. Optical setup of the SO$_2$ analyser.

photomultiplier (EMI 9750 QB). The signal from the photomultiplier is digitized and transmitted to a microcomputer (LSI 11/02). There, the scanned spectral section of 20nm is stored in 150 channels for further processing.

An individual scan requires about 10 ms. Thus, fluctuations of atmospheric density, movements of the lamp's arc and current fluctuations of the power supply are too slow to disturb the spectrum significantly during one scan.

The measured spectral noise was equal to an O.D. of about $8*10^{-5}$ (1h integration time, 34 wind intervals (see chapter 4), 104m light path), yielding a concentration resolution for SO_2 of 2.8 $\mu g/m^3$. This rather poor resolution is mainly due to light loss in the White cell due to astigmatism of the mirrors and to optical alignment problems. Fig. 3 shows the prototype of the instrument as mounted during tests at the meteorological tower of the Nuclear Research Center (KFA) in Jülich. The instrument's total weight was 98kg, its length 2.20m and its mean diameter about 0.5m. The frame of the open White cell consists of three stainless steel tubes of 50mm diameter. To avoid maladjustment of the mirror by thermal expansion of these tubes, the distance between the mirrors was fixed by quartz tubes (11mm diameter) inside the steel tubes.

3.2 Anemometer

A sonic anemometer built according to Campbell's and Unsworth's design was used to measure the fluctuations of the vertical wind component (Campbell and Unsworth 1979). This type uses the phase difference between signals sent and received to evaluate the fluctuations of the vertical wind speed. Since the phase shift caused by the crystal transducers used was strongly dependent on temperature, the zero point of the vertical windspeed could not be properly defined. Assuming that the mean vertical wind speed is identical to zero when properly aligned, the zero point could be evaluated as a median of the vertical windspeed distribution.

4. THE EDDY CORRELATION TECHNIQUE

The noise level of each scan depends mainly on the number of photons received by the photomultiplier. To minimize the error of the concentration measurements, several scans (about 1000) must be superimposed and processed before calculating the concentrations. Thus, a variation of the standard eddy correlation method was used.

The vertical wind speed was divided into 34 intervals, each interval representing about 4cm/s. For each velocity interval, the spectra which were measured together with this windspeed were superimposed. After an integration time of 30 minutes to 1 hour, enough spectra were superimposed to allow for a proper evaluation of the concentrations.

The net flux j was calculated according to

Figure 3. The prototype of the SO_2 sensor during the measuring campaign at the meteorological tower of the KFA Jülich.

$$j = \frac{1}{N} \sum_{i=1}^{B} (W_i * n_i * (C_i - C)) \qquad (8)$$

W_i : mean windspeed of interval i (m/s)
n_i : number of the spectra superimposed in interval i
B : number of intervals
C_i : mean concentration in interval i ($\mu g/m^3$)
C : mean concentration ($\mu g/m^3$)

$$\text{and } N = \sum_{i=1}^{B} n_i$$

5. ERROR ANALYSIS

The error of the deduced deposition velocities was calculated in a straightforward manner from equation (8). Several sensitivity analyses have been made to check the individual influence of the integration time, of the number of wind intervals, the wind offset and the concentration resolution. It was shown (Nestlen 1984) that the concentration resolution was the dominant factor limiting the precision of the deposition velocity to 30-40%. For SO_2 concentrations greater than $20\mu g/m^3$, doubling the number of wind intervals (i.e. increasing the resolution of the vertical velocity as used in the covariance calculation) yields an error reduction of not more than 20%, whereas doubling the concentration resolution halves the error. For SO_2 concentrations below $20\mu g/m^3$, the influence of the concentration resolution is even more severe.

6. RESULTS

Field experiments were carried out at the meteorological tower of the KFA Jülich in the summer of 1982. This tower is 120m high and is equipped with various meteorological instruments (anemometers, thermometers, humidity sensors) and has platforms at six different heights. Meteorological data were continuously recorded and evaluated.
The eddy flux instruments were mounted on a beam on the 50m platform. The tower was surrounded mainly by forest (pine and beech trees; canopy height about 19m), but with some nearby buildings of the KFA, roads and clearings. Only in the sector between 80° and 150° were the fetch and homogenity requirements (Hicks 1978) sufficiently fulfilled. In this sector, the trees were rather uniform up to a distance of 3 km, only occasionally interrupted by buildings (height similar to that of the surrounding trees) and few clearings. In this sector, no SO_2 sources were found within 10 km.
The vertical sensible heat flux was measured simultaneously with the vertical SO_2 flux, using the standard eddy correlation technique. The transport mechanisms for both the heat flux and the trace gas flux

are the same (Lumley and Panofsky 1964). Therefore, in order to verify the estimations of the maximum response time and the minimum integration time mentioned above (section 2), results from eddy heat flux measurements can be used. High- and lowpass filtering of the signals from the anemometer and the thermometer before calculating the correlation products showed that 90% of the heat flux was obtained in the frequency interval of n=0.33Hz to n=0.002Hz (Zenger 1983). The corresponding dimensionless frequencies were f=3 and f=0.02. These results showed that at a height z-d=35m (d : zeroplane displacement) and a mean horizontal windspeed of 5m/s a response time of 1s is sufficient to reduce the error of the measured eddy flux to 10%. Normally all the instruments were set to a response time of 1s. For SO_2 flux measurements, an integration time of 1 hour was used.

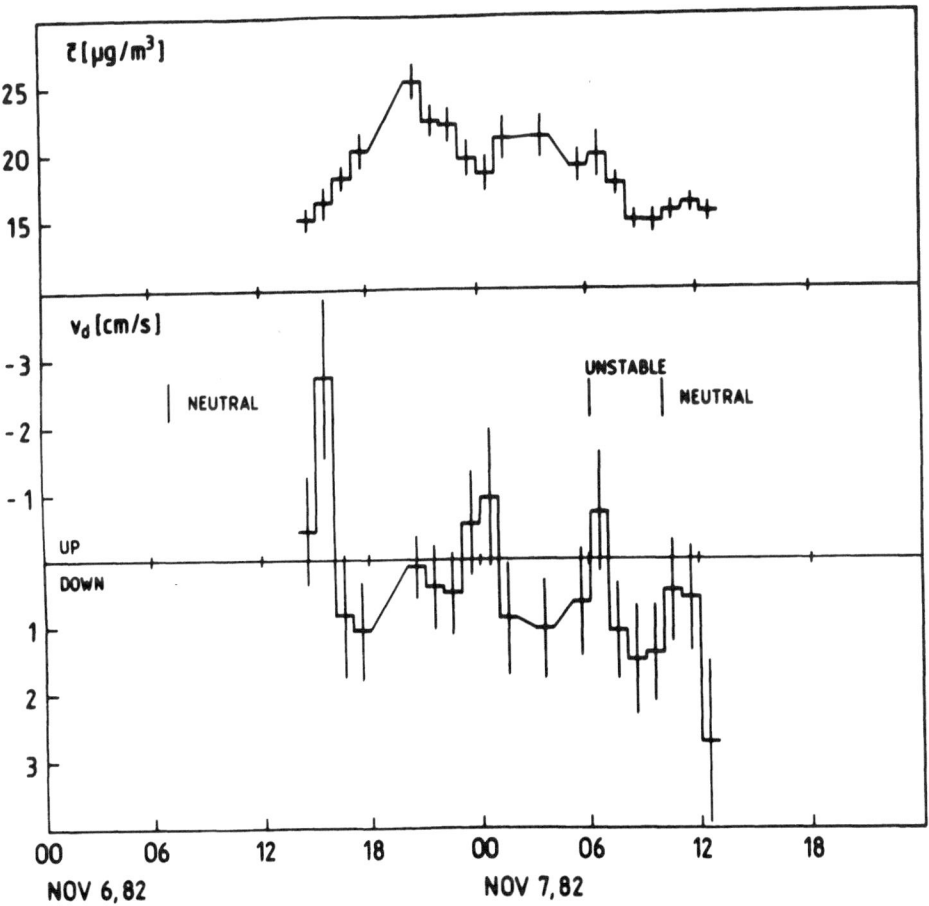

Figure 4. Mean SO_2 concentration and SO_2 deposition velocity. The vertical bars denote the measuring errors, the horizontal bars represent the integration time.

Technical problems with the SO_2 sensor did not allow continuous flux measurements on most days. Only on November 6th and 7th 1982 could a 24 hour interval be covered quasi-continuously. Fig. 4 shows the measured SO_2 fluxes. During the entire measuring period, the horizontal windspeed at 50m was about 7.5m/s from 110°. As mentioned above, the terrain in this sector was rather homogenous within a fetch of about 3km. The atmospheric stability was neutral, becoming slightly unstable after 11:00h on November 7. The atmospheric transfer resistance at z=50m was evaluated at

$$Ra(50m) = u(z)/u*^2 = 0.11s/cm,$$

the zeroplane displacement at d=15m, the roughness length at 1.2m.

The mean deposition velocity of SO_2 at z=50m was 0.36cm/s. This mean value, however, might be misleading since the deposition velocity and the vertical flux of SO_2 showed a great variability in the course of time as shown in Fig. 4 and Fig. 5. Even significant upward fluxes were measured. These upward fluxes might be correlated to changes of the mean SO_2 concentration with time due to advection and nonstationarity. Between 21:00h of Nov. 6 and 1:00h of Nov. 7, the mean SO_2 concentration decreased from $25\mu g/m^3$ to $18\mu g/m^3$ at a mean rate of $2\mu g/m^3$ per hour (see Fig. 4). Between 23:00h and 1.30h the mean upward SO_2 flux was $0.1 \pm 0.05\mu g/m^2s$.

The error j of a measured vertical flux due to nonstationarity can be deduced from mass balance considerations (Garland 1977, Wesely et al. 1985):

$$j^{noust.} - z(dc/dt) \qquad (9)$$

With the measured value of dc/dt ($-2 \ \mu gm^{-3} \ h^{-1}$), an error of the flux due to nonstationarity of approximately $+0.02 \ \mu g/(1m^2s)$ can be expected. This is only about 20% of the measured upward flux ($+0.1 \ \mu g/(m^2s)$). Even when the measuring error of $\pm0.05 \ \mu g/(m^2s)$ is considered, a residual upward flux of about $0.05 \ \mu g/(m^2s)$ remains to be explained. The cause for this might be the limited exchange between the air in the trunk space of the forest and the air above the canopy (Brutsaert 1979); effects of a concentration change caused by advection should be delayed. This means that due to storage of SO_2 below the canopy, the concentration gradient between the air masses above the measuring height and at canopy level could be negative, resulting in an upward flux much greater than assumed in the calculations above.

More detailed consideration of the fluxes and deposition velocities is not warranted, since many of the observed variations are not statistically significant. The upward flux at 6:00h could also be caused by short-time concentration variations, which are not resolved by the integration time. A hint might be the slight concentration increase between 2:00h and 4:00h. The deposition velocity and flux peak at 6:00h cannot be explained at present.

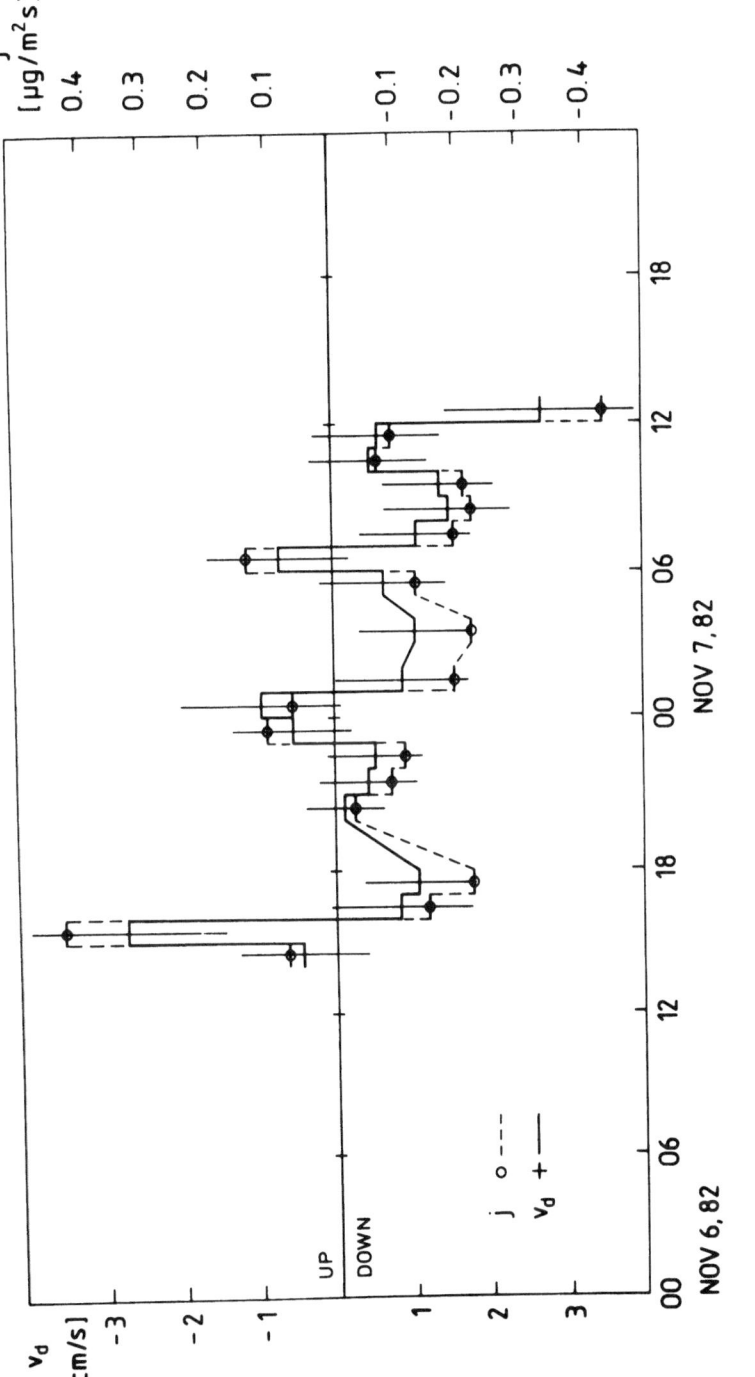

Figure 5. Deposition velocity and vertical flux of SO_2.

580

7. CONCLUSIONS

A newly developed SO_2 sensor for eddy flux measurements has been tested successfully under field conditions. Several technical problems, such as mechanical instability of the frame of the White cell and the mirror mounts, and image distortion due to astigmatism, severely limited the concentration resolution for SO_2. Thus, the error of the deposition velocity measurements was ±0.2 cm/s. Nevertheless, these test measurements have given valuable hints for the improvement of this prototype.

Primarily the White cell has to be redesigned. A different optical set-up (White 1976) will cancel out almost all vibration disturbance and will considerably reduce the astigmatism. A further benefit of this design will be a reduction of the White cell's length to about 1m.

Replacing the steel tubes by carbon fiber tubes and separating several components from the White cell will reduce the sensor's weight to approximately 40kg. The instrument could then be handled much more easily and could also be placed on smaller towers or masts.

Using a diode array "camera" instead of the slotted disc and the photomultiplier for scanning the spectrum will improve resolution by a factor of 3 since all the wavelength channels are exposed to light simultaneously. This provides a much better signal to noise ratio.

In the wavelength region between 195 nm and 210 nm, the differential absorption coefficient of SO_2 is about 30 times greater than in the region between 300 nm and 320 nm. Switching to the 200 nm region will improve concentration resolution by a factor of 4.

Taking into account all these improvements, a theoretical resolution for SO_2 better than $0.1 \mu g/m_3$ should be achieved.

ACKNOWLEDGEMENT

The author gratefully acknowledges the many helpful discussions with Dr U. Platt and Dr D. Flothmann and in particular the assistance of A. Zenger, who carried out the heat flux and wind speed fluctuation measurements. Many thanks are given to Dr G. Polster of the KFA Jülich for letting us use the meteorological tower and for providing meteorological data.

This work was performed at the Institute for Environmental Physics, Ruprecht Karls Universität Heidelberg, in cooperation with the Battelle Institute in Frankfurt. Financial support was provided by the Umweltbundesamt Berlin (UFOPLAN-Nr. 104 02 609).

REFERENCES

Brutsaert, W. 1979. Heat and mass transfer to and from surfaces with dense vegatation or similar permeable roughness. Boundary-Layer Meteorology **16** , 365-388.

Businger, J.A. 1986. Evaluation of the accuracy with which dry deposition can be measured with current micrometeorological techniques. Journal of Climate and Applied Meteorology 25 , 1100-1124.

Campbell, G.S. and Unsworth, M.H. 1979. An inexpensive sonic anemometer for eddy correlation. J. app. Met. 18 , 1072.

Fowler, D. and Cape, J.N. 1982. Dry deposition of SO$_2$ on to Scots pine forest. In: Precipitation Scavenging and Resuspension. Elsevier Science Publishing Co. Inc., New York, 763-772.

Galbally, I.E., Garland, J.A. and Wilson, M.J.G. 1979. Sulphur uptake from the atmosphere by forest and farmland. Nature 280 , 49-50.

Garland, J.A. 1977. The dry deposition of sulphur dioxide to land and water surfaces. Proc. R. Soc. Lond. A. 354 , 245-268.

Hicks, B.B. 1978. An examination of some micrometeorological methods for measuring dry deposition. EPA-Report, PB 285 765.

Kaimal, J.C. 1975. Sensors and techniques for direct measurements of turbulent fluxes and profiles in the atmosphere surface layer. Atmospheric Technology 7 , 7-14.

Kessler, C. 1979. UV-Spektroskopische Bestimmung der trockenen Deposition von SO$_2$ and NO$_2$ mittels Gradientenmethode. Diplomarbeit Institute für Umweltphysik, Ruprecht Karls Universität Heidelberg.

Lumley, J.L. and H.A. Panofsky. 1964. The structure of atmospheric turbulence. Interscience, John Wiley and Sons.

McBean, G.A. and M. Miyake. 1972. Turbulent transfer mechanism in the atmospheric surface layer. Quart. J.R. Met. Soc. 98, 383-398.

Nestlen, M.G. 1984. Ein neues Meßgerät zur Bestimmung der trockene Deposition von SO$_2$ mit Hilfe der Eddy-Korrelationsmethode. Dissertation, Institut für Umweltphysik der Ruprecht Karls Universität Heidelberg.

Neumann, H.H. and G. den Hartog. 1985. Eddy correlation measurements of atmospheric fluxes of ozone, sulphur, and particulates during the Champaign intercomparison study. J. of Geophys. Res. 90, No. D1, 2097-2110.

Panofsky, H.A. and E. Mares. 1968. Recent measurements of cospectra for heat-flux and stress. Quart. J.R. Met. Soc. 94, 2097-2110.

Pasquill, F. 1974. Atmospheric Diffusion. Halstead press, New York, 429pp.

Perner, D. 1979. Optischer Nachweis von Spurenstoffen in der Atmosphaure KFA-Mitteilungen, 1-2.

Platt, U., D. Perner and H.W. Pätz. 1979. Simultaneous measurement of atmospheric CH$_2$O, O$_3$ and NO$_2$ by differential optical absorption. J. Geophs. Res. 84, No. 10, 6329-6335.

Roth, R. 1975. Der vertikale Transport von Luftbeimengungen in der Prandtl-Schicht und die Deposition-Velocity. Meteorol. Rdsch. 28, 65-71.

Webb, E.K., G.I. Pearman and R. Leuning. 1980. Correction of flux measurements for density effects due to heat and water vapour transfer. Quart. J.R. Met. Soc. 106, 85-100.

Wesely, M.L., D.R. Cook and R.L. Hart. 1985. Measurement and parameterization of particulate sulphur dry deposition over grass. J. Geophys. Res. 90, No. D1, 2131-2143.

582

Whelpdale, D.M. and R.W. Shaw. 1974. Sulphur dioxide removal by
    turbulent transfer over grass, snow and water surface. <u>Tellus</u>
    16, 1-2.
White, J.U. 1942. Long paths of large aperture. <u>J. Opt. Soc. Am.</u> **32,**
    285-288.
White, J.U. 1976. Very long paths in air. <u>J. Opt. Soc. Am.</u> **66,** No. 5
    411-416.
Zenger, A. 1983. Die Eddy-Korrelation als Methode zur Besimmung von
    turbulenten Wärmeflüssen in der Atmosphäre. Diplomarbeit, Institut
    für Umweltphysik der Universität, Heidelberg.

GASEOUS DEPOSITION OF SO_2, NO_x AND O_3 TO A SPRUCE STAND IN THE NATIONAL PARK "BAYERISCHER WALD"

U. Teichmann, G. Enders
Lehrstuhl für Bioklimatologie und Angewandte Meteorologie
Universität München
Amalienstrasse 52
D-8000 München 40
FRG

ABSTRACT. Using the flux gradient approach dry deposition of SO_2, NO_x and O_3 to a 28 m tall spruce stand in the catchment "Grosse Ohe" in the National Park "Bayerischer Wald" is being studied. This catchment has a size of 19.1 km^2 and is situated about 200 km northeast of Munich. The elevation reaches from 770 m to 1452 m above sea level. The catchment was established in 1978 as a forest hydrological reference basin now serving as an experimental watershed for water quality studies too.

Before the main deposition began in July 1986, a pre-experiment was performed in the "Ebersberger Forst" near Munich to test the performance of the instrumentation. For a day of this period the turbulent diffusivities are calculated showing considerable discrepances between aerodynamic and energy balance approach. At the new site, for which first concentration profiles of SO_2 and NO_x are reported, a taller measuring tower than at the test site is operated providing meteorological and chemical measurements up to 51 m height.

1. INTRODUCTION

Different needs of different research interests require special experimental methods to address them (Hicks, 1985). This is especially true with respect to the actual forest decease research.

In 1983 the German Federal Ministry of Research and Technology began to support research work related to the "Forest decease/Air pollution" complex. As a contribution to the subprogramme area "Causes and Effects" the experimental project GASDEP (Gaseous Deposition of ...) has been established, which has to serve different requirements. One of the goals of GASDEP is to provide data helpful for effect studies. Therefore, besides meteorological site parameters concentrations of air pollutants and their vertical profiles within and above a spruce stand are measured. These measurements have to be performed continuously, but nevertheless with a maximum resolution in

M. H. Unsworth and D. Fowler (eds.), Acid Deposition at High Elevation Sites, 583–592.

space and time. The main objective, however, is to determine the dry
deposition of SO_2, NO_x and O_3 to this stand and to parameterise the
deposition velocities for the use in theoretical models.

In pursuance of both goals, to assist the effect people and to
determine deposition, the experimental set up has to be a compromise,
especially when running a real-world experiment, where the methods to
be used often are inferential substandard methods (Hicks, 1985).

2. MEASURING SITE AND METHOD

The outdoor part of the experiment began with a one year test phase in
the "Ebersberger Forst" near Munich. There a homogeneous terrain and a
uniform stand structure with good fetch provide quite optimal site
conditions for micrometeorological studies. During that time mainly
the performance of the complex measuring system was studied and
hardware problems were solved.

After disassembling the instrumentation was moved in April 1986 to
the catchment "Grosse Ohe" in the remote National Park "Bayerischer
Wald" (about 200 km northeast of Munich) and reinstalled. This
catchment has a size of 19.1 km^2, the elevation reaches from 770 m to
1452 m a.s.l. It was established in 1978 as a long-term reference
basin for water budget studies of a naturally forested area with 70%
spruce, and 28% beech and others. Besides routine and special
quantitative measurements of the separate components of the water
cycle, investigations are carried out to analyse pH and mineral loads
in fog, pure and bulk precipitation, stemflow and runoff (Gietl and
Rall, 1986). Measurements are taken in different altitudes and in
different stands. Beyond that work additional research programmes
focus on the National Park region, because the growth zone "Bavarian
Forest" belongs to the most damaged areas (inventory 1985: 63% of the
forested land) in Bavaria.

Site and stand conditions around the deposition station can be
characterised as follows: Elevation 807 m a.s.l.; mostly level,
uniform forest (86% spruce; 14% beech) up to a distance of 500 m;
complex, but also forested terrain beyond that area; average tree
height 28 m with needles down to 14 m; 712 trees/ha.

To measure the gaseous deposition the flux-gradient approach is
used in combination with a concentration accumulation technique. The
decision for that was principally made due to the lack of fast response
chemical sensors for long-term and continuous measurements at the time
the experiment was planned. The gradient-diffusion scheme, however,
works not very well or fails completely within and below the canopy
layer (Denmead and Bradley, 1985). Therefore, at present only fluxes
well above the canopy can be determined with the accuracy required.
The instrumental set up, however, will be completed in 1987 by an
improved eddy correlation system, which still needs intensive
attendance when permanently operated. Because the station is manned
day by day, this problem can be solved to intercompare the two
micrometeorological methods continuously and so to contribute to the
phenomena of counter-gradient fluxes.

3. INSTRUMENTATION, RESULTS AND DISCUSSION

3.1 Meteorology

In Figure 1, which gives the stand architecture in relation to the size
of the measuring tower, also the instrumental set up is shown.
Shortwave and longwave radiation fluxes (upward/downward) are measured
in two levels, soil heat fluxes at three different spots. Air
temperature, humidity and horizontal wind speed sensors, right now
operated at six levels, will be completed to a total of eleven by fall
'86. In addition wind direction, atmospheric pressure and rain events
are monitored.

From the readings of these sensors taken every 10 s the turbulent
diffusivities, which are substantial for the gradient-flux
relationship, have to be calculated. Depending on the meteorological
conditions either the aerodynamic approach or the energy balance method
can be used.

Meteorological measurements at the site "Bayerischer Wald" began
in late July '86, but at the very beginning not all readings were
reliable. Therefore, diffusivities from a data set gathered during the
test experiment are calculated to demonstrate the basic approach.
Aerodynamic stand parameters and stability correction functions used
had been evaluated in earlier measurements at Ebersberger Forst by
Baumgartner (1969) and Hager (1975).

a) Aerodynamic approach

$$K_0(z) = \frac{k^2 \cdot (z-d) \cdot u(z)}{\ln \frac{z-d}{z_0}} \text{ for } z > z_0 + d,$$

where $K_0(z)$ is diffusivity at reference level z under neutral
conditions, k is von Karman's constant, $u(z)$ is horizontal wind speed
at level z, d is zero plane displacement (= 20 m), and z_0 is roughness
length (= 3 m).

Stability is corrected by functions

$$K_A(z) = K_0(z) \cdot F$$

with $F = (1 + G \cdot Ri)^{-\frac{1}{2}}$ for stable conditions
        $F = (1 - F \cdot Ri)^{\frac{1}{2}}$ for unstable conditions,

where Ri is the Richardson number and G a constant (= 40).

b) Energy balance method

$$K_E(z) = \frac{-(Q + B + P)}{\rho \cdot (c_p \cdot \partial\Theta/\partial z + r \cdot \partial q/\partial z)}$$

586

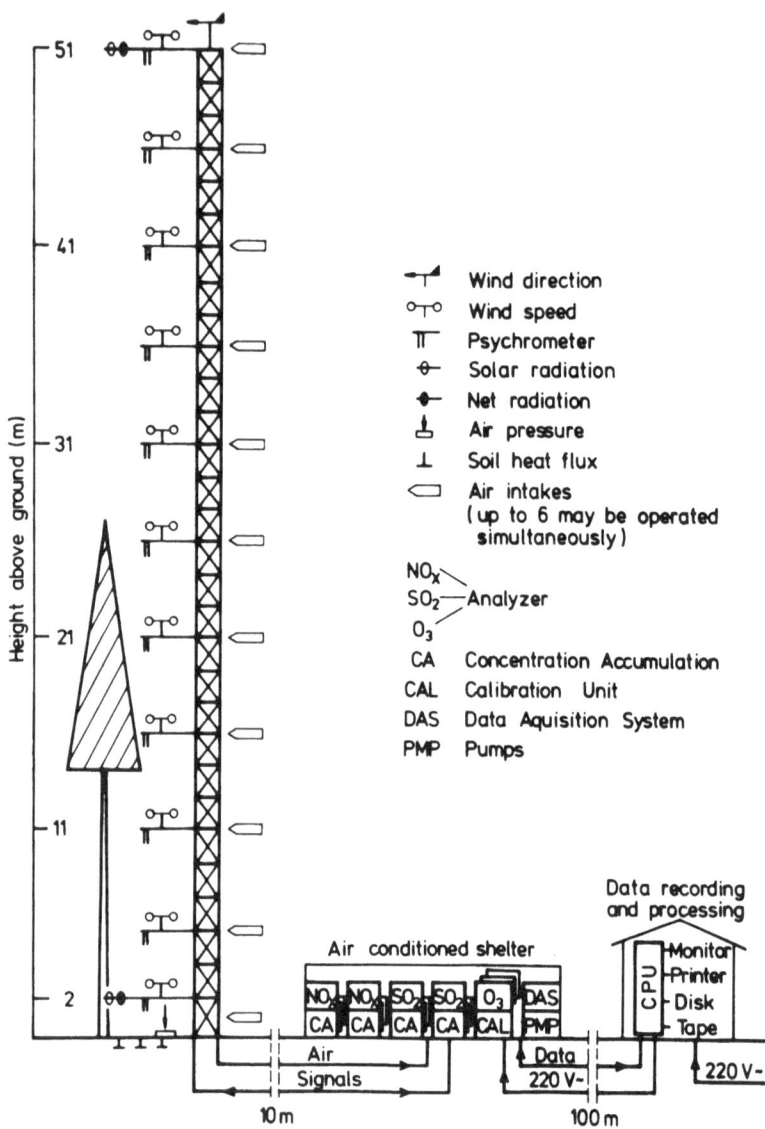

Figure 1   Instrumental set up at the site 'Bayerischer Wald'.

where $K_E(z)$ is diffusivity at reference height z, Q is net radiation, B is soil heat flux, P is plant heat flux, ρ is air density, c_p is specific heat of air at constant pressure, Θ is potential air temperature, r is latent heat of vaporisation, and q is specific humidity.

The plant heat flux P is calculated by

$$P = \rho_F \cdot c_F \cdot h \cdot \partial T / \partial t$$

where ρ_F is wood density, c_F is specific heat of wood depending on water content, h is equivalent thickness of biomass layer, and $\partial T / \partial t$ is the temperature change of the wood assumed to be equal the air temperature change (Strauss, 1971).

As one can see, this method fails completely during times of small net radiation and/or weak zero gradients of sensible and latent heat.

Data used were measured on September 22, 1985, a mostly clear day with air temperatures above the canopy between 15.4 °C and 25.0 °C, wind velocities from 1.0 to 2.8 m/s, and a net radiation up to 555 W/m^2 (all data given are hourly mean values). Air temperature and humidity gradients at reference level z = 36 m were calculated from measurements at 31 m and 41 m. The average tree height H was 32 m, the fetch during that day at least 20.H.

The results from both methods are given in Figure 2. During stable conditions the similarity seems quite acceptable, while unstable conditions result in differences up to a factor of 6. One explanation could be given by the use of 'old' z_0 and d values, and a reference height z only five roughness lengths above d.

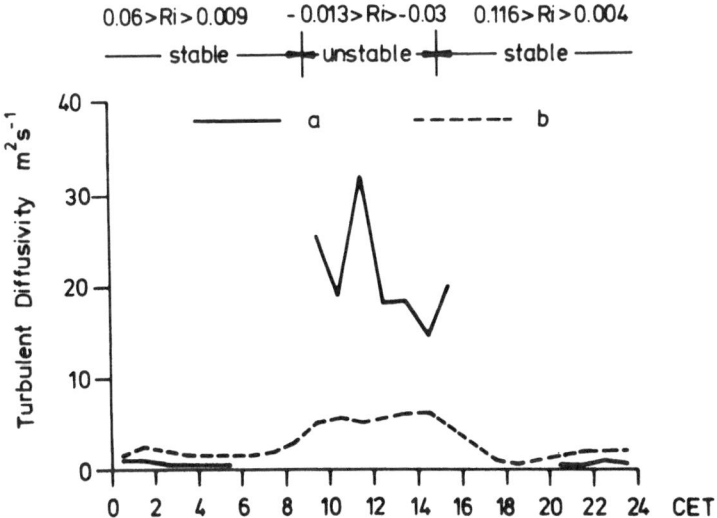

Figure 2. Mean hourly values of diffusivities K_A (aerodynamic approach; dashed line) and K_E (energy balance; solid line) for Ebersberger Forst on September 22, 1985.

Therefore, according to Tajchmann (1981) new calculations were made varying z_0 and d within $0.61 \leq d/H \leq 0.90$ and $0.02 \leq z_0/H \leq 0.26$. Even then the differences to the energy budget approach were not substantially reduced. If one believes in the aerodynamic approach, it is to assume that the plant heat flux is underestimated and, therefore, in reality not so much energy is available for exchange processes.

On the other hand, stability functions used seem rather to be comparable to 'smooth surface' values of F (Figure 3), especially when unstable conditions occur. Thom et al. (1975) and Raupach (1979) have shown on profile and eddy correlation data gathered over Thetford Forest, UK, that diffusitivities for heat and water vapour exceeded their values predicted from 'smooth surface functions' by a factor of 2 or more in unstable, near neutral and slightly stable conditions, even when the reference height was nine roughness lengths above the zero-plane displacement. In addition, in the years passed from Baumgartner's and Hager's measurements the stand architecture has changed due to the natural growth of the forest, a moderate thinning in the surroundings of the site, and a severe hail storm in July 1984. Therefore, for a better comparison roughness parameters and stability functions will be computed again from the complete data set gathered during the test study.

At the new site, the measuring tower is taller than at Ebersberger Forst, while the mean canopy height is less, so providing a better infrastructure for micrometeorological measurements.

3.2 Air sampling and analysing system

At rural sites like a National Park often the bare concentration values of SO_2 and NO_x may become close to the detection limit of standard analysers. Even when the concentrations are high, their vertical gradients can be very small. To overcome all the problems, which the use of different analysers in different measuring levels would have caused, a rather sophisticated air sampling and analysing system is being used (Figure 1). Air from individual measuring heights is pulled down the tower simultaneously and continuously through heated teflon tubes ($\emptyset$ 4.75 mm; air speed 25 m/s) to an air conditioned shelter. There the air streams are split up to reach concentration accumulation units, where $\underline{SO_2}$ and $\underline{NO_x}$ are trapped during $\overline{20}'$ and 8', respectively, at ambient temperatures. Three by three of those CAS and CAN units are connected to one specific analyser, and measured one after the other during desorption forced by strong heating. The ratio of desorption to absorption time allows three desorption times to fit into one absorption period, and decreases the detectable limit to the sub-ppb range. Each CAS and CAN is equipped with two accumulation cells to half the sampling time gap. Details of the set up are given by Enders and Teichmann (1986).

Three (six by fall '86) measuring heights are operated for each species; O_3 readings measured directly with separate analysers are taken from three heights too.

With this special arrangement some of the problems mentioned above were passed by, but others were traded in due to the complexity of the

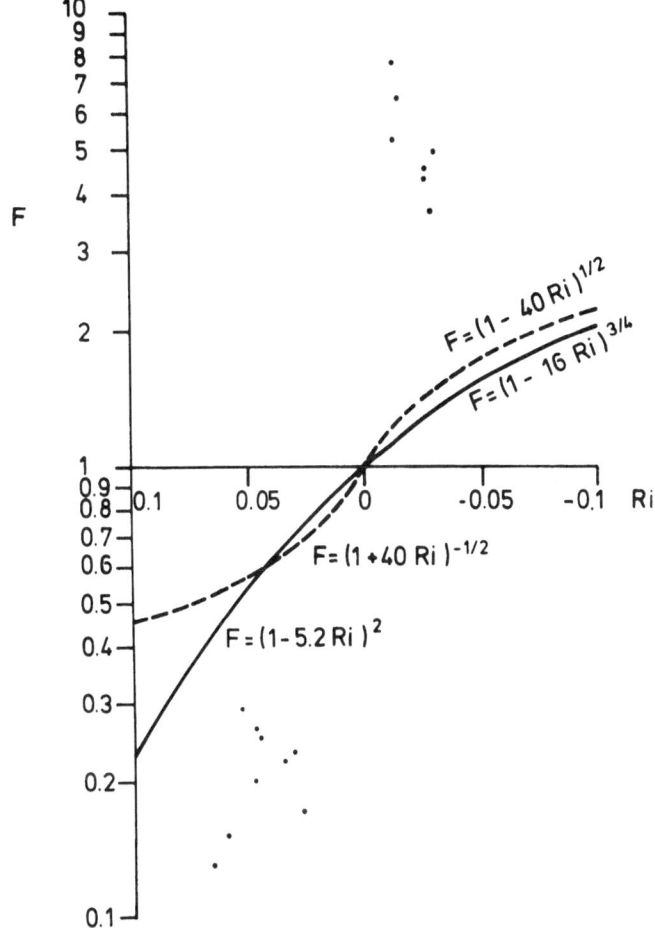

Figure 3   Stability factor F versus Richardson number Ri: F = (1-16
Ri)$^{3/4}$, F = (1-5.2 Ri)^2 for 'smooth' surfaces (Dyer and
Hicks, Webb: from Thom et al. 1975);   F = (1-40 Ri)$^{1/2}$, F =
(1+40 Ri)$^{-1/2}$ used for Ebersberger Forst.   Dots represent F =
K_E/K_o for Ebersberger Forst on September 22, 1985.

system and the use of this type of accumulation devices never used
before.   Some questions could be answered during the pre-experiment,
some still wait for their final solution:
-   Are there considerable wall effects or losses in the tubing
    system?
           Differences in O_3 measurements of about 3 ppb, which were
           obtained by two analysers connected to the same air intake
           (one of them very close, the other via the long tubing
           system) could be explained by different ambient conditions of
           the analysers.

- Do we really measure NO_x and not only a fraction of it?

  The absorption medium is specific only to NO_2. Therefore, NO is oxidised to NO_2 before entering the absorption cell. Because at desorption temperatures of 600 °C the NO_2 is partly reduced to NO again, the analyser is switched to the NO_x-channel for normal operation. For special investigations the oxidation cell may be passed by to allow for pure NO_2 measurements.

- How long can the absorption cells stand the permanent temperature changes during heating and cooling?

  By improvements of the resistance winding for heating the glass tubes the durability could be increased from a few days up to months.

- How differ calibration and operation conditions?

  At present an internal calibration system provides zero and span gases for all CAS, CAN and analysers under atmospheric pressure. In the operational mode however, individual pressure drops in the tubing system exist, causing variable massflows in the accumulation devices (depending on their measuring state) and, in addition, influencing directly the analyser readings. Therefore, from time to time massflow and pressure are measured very precisely to allow for a pressure correction on a calculation routine. For the future the development of an 'overall' calibration system is planned, which performs calibration under conditions equal to operating conditions.

Before a final check concentration data gathered during the test experiment have to be considered as raw data due to different malfunctions of the system, uncertainties in calibration and weak performance of the accumulation devices. For the National Park site on the other hand, diffusivities have not been calculated yet so far. Therefore, only concentration values form the new site are reported as an example.

Measurements shown in Figure 4 were made on August 22/23, 1986, two different days with respect to the radiation regime as characterised by the global radiation S+D. Wind speed at 44 m was most of the time more than 1 m/s. During night the system was shut down due to a power failure. Both NO_x and SO_2 show a strong gradient with decreasing height, which with the utmost caution may be interpreted to cause downward fluxes (note that the 31 m level is just above the canopy layer, while 16 m is below). On the clear day SO_2 in both measuring heights and NO_x in the upper level have remarkable amplitudes. The NO_x maximum occurs in the late afternoon, what often was observed to coincide with strong decreasing O_3, which unfortunately was not measured during that period.

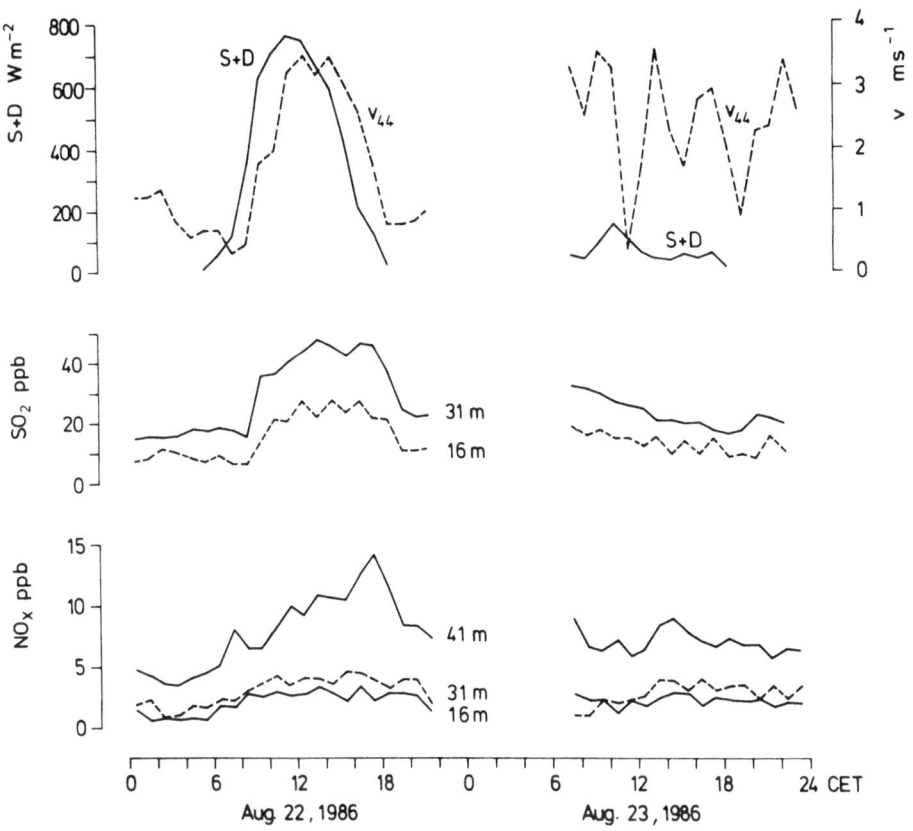

Figure 4. Mean hourly values of NO_x and SO_2 concentrations for the Bayerischer Wald site on August 22/23, 1986. Weather conditions are roughly characterised by global radiation S+D and wind speed v at 44 m.

ACKNOWLEDGEMENTS

This work is granted by the German Federal Ministry of Research and Technology under contract number 03-7339-0, the Bavarian State Ministry of Environmental Protection (8272-62-7923), and the Bavarian State Ministry of Food, Agriculture and Forestry (M 23). We also want to thank Mrs C. Mooslechner for drawing the figures and Mrs H. Dany for typing.

592

REFERENCES

Baumgartner, A. 1969. Meteorological Approach to the Exchange of CO_2 between the Atmosphere and Vegetation, particularly Forest Stands. <u>Photosynthetica</u> **3** , 127–149.

Denmead, O.T. and Bradley, E.F. 1985. Flux-Gradient Relationships in a Forest Canopy, pp. 421–442. In: <u>The Forest-Atmosphere Interactaction</u>', B.A. Hutchinson and B.B. Hicks (eds.). D. Reidel Publishing Company.

Enders, G. and Teichmann, U. 1986. GASDEP – Gaseous Deposition Measurements of SO_2, NO_x and O_3 to a Spruce Stand, pp. 13–24. In: <u>'Atmospheric Pollutants in Forest Areas'</u>, H.-W. Georgii (ed.). D. Reidel Publishing Company.

Gietl, G. and Rall, A. 1986. Bulk Deposition into the catchment "Grosse Ohe", pp. 79–88. In: <u>'Atmospheric Pollutants in Forest Areas'</u>. H.-W. Georgii (ed.). D. Reidel Publishing Company.

Hager, H. 1975. Kohlendioxydkonzentrationen, -Flüsse und Bilanzen in einem Fichtenhochwald. <u>Wiss. Mitt. Meteorol. Inst. Univ. Munchen</u> **26** , 182 pp.

Hicks, B.B. 1985. Comments on the "State of the science" of measurement and modelling atmosphere-canopy interaction. DOE Forest-Atmosphere Interaction Workshop, Lake Placid, NY, October 1–4, 1985.

Raupach, M.R. 1979. Anomalies in flux-gradient relationships over forest. <u>Boundary-Layer Meteorol.</u> **16** , 467–486.

Strauss, R. 1971. Energiebilanz und Verdunstung eines Fichtenwaldes im Jahre 1969. <u>Wiss. Mitt. Meteorol. Inst. Univ. Munchen</u> **22** , 66 pp.

Tajchman, S. 1967. Energie- und Wasserhaushalt verschiedener Pflanzenbestände bei München. <u>Wiss. Mitt. Meteorol. Inst. Univ. Munchen</u> **12** , 95 pp.

Tajchman, S. 1981. Comments of Measuring Turbulent Exchange Within and Above Forest Canopy. <u>Bull. Am. Met. Soc.</u> **62** , 1550–1559.

Thom, A.S., Stewart, J.B., Oliver, H.R. and Gash, J.H.C. 1975. Comparison of aerodynamic and energy budget estimates of fluxes over a pine forest. <u>Q.J.R. Meteorol. Soc.</u> **101** , 93–105.

DEPOSITION RATES OF AIRBORNE SUBSTANCES TO FOREST CANOPIES IN RELATION TO SURFACE STRUCTURE

J. Godt and R. Mayer
University of Kassel
Henschelstr 2.
P O Box 101380
3500 Kassel
West Germany

ABSTRACT. In this work the influence of interception deposition on input of air pollutants to forest ecosystems as a function of orographic and meteorological effects is discussed. Fluxes of airborne substances have been measured in a spruce stand (open field precipitation, canopy drip inside and at the edge of the stand) on the windward side of the Teutoburger Wald, Nordrhein-Westfalen (rural area). Flux balances indicate high total deposition rates especially for S. Interception deposition is increasing from inside to the edge of the stand.

Fractioning aerosols by aid of a low volume impactor shows presence of SO_4^{2-}, NO_3^-, Cl^- and heavy metals in fine fractions (Aitken- and accumulation mode) with low mass median diameter (MMD).

Spruce needles (biomonitoring system) and snow analyses for cadmium and lead indicate high deposition rates at the windward top of the Teutobuger Wald. As a consequence of orographic and meteorological effects one can expect high interception deposition rates of airborne pollutants in elevated parts of the landscape.

1. INTRODUCTION

Ulrich et al. (1979) developed flux balances for forest canopies, in which they distinguished 1. precipitation deposition and 2. interception deposition (Table 1) in order to point out the filtering effect of forest canopies (interception of gases and fine aerosols). Interception deposition is influenced by properties of the accepting surfaces (increasing with roughness of accepting surface, wetness of surfaces and wind speed) (Sehmel, 1980; Monteith, 1976; Hosker and Lindberg, 1982).

M. H. Unsworth and D. Fowler (eds.), Acid Deposition at High Elevation Sites, 593–606.
© 1988 by Kluwer Academic Publishers.

594

Table 1. Deposition processes (Ulrich et al., 1979, changed)

	DEPOSITION PROCESS	DOMINATING MECHANISM	TERM IN METEOROLOGY	TERM FROM ULRICH ET AL
particles	(a) precipitation of rain and snow	vertical transport due to gravity	wet deposition / sedimentation	precipitation deposition
particles	(b) precipitation of other particles Ø <2 μm	vertical transport due to gravity	wet deposition / sedimentation	precipitation deposition
particles	(c) impaction of particles Ø < 2 μm	turbulent transport and diffusion	dry deposition	interception deposition
gases	(d) dissolution / sorption, uptake of at. gases on surfaces	turbulent transport and diffusion	dry deposition	interception deposition

Flemming (1982) observed a higher degree of forest damages in elevated mountain areas and assumed a correlation of wind speed and higher deposition rates of air pollutants at these parts of a landscape. Hager and Kazda (1985) found higher concentrations of S in spruce needles which had been taken from sites with a loose canopy in comparison to a dense one. This result was explained by surface structure influencing interception deposition of S.

As physical properties of particles are influencing deposition processes, one can predict the importance of sedimentation by analysing the element-specific diameter of aerosols. For instance, Wiman et al. (1985) developed a model based on empirical data in which they found a depletion of aerosol concentration and change in particle diameter at the edge of a forest stand.

In general sedimentation is decreasing with particle diameter down to < 0.1 μm, where Brownian diffusion and impaction become relevant (Sehmel, 1980). Following Table 1 interception deposition compared to sedimentation plays a more important role for particles with a diameter < 2 μm.

2. HYPOTHESIS AND EXPERIMENTS

If interception deposition plays an important role as part of total input of airborne substances to forest canopies, high deposition rates should be found at the edge of a forest stand and at exposed parts of a landscape such as a windward top of a mountain area.

In order to study the influence of interception deposition, fluxes of several elements (bulk precipitation) have been measured in the open field, inside, and at the edge of a spruce stand in the Teutoburger Wald, Nordrhein-Westfalen, FRG. The stand is located at the windward top of the mountain area (400 m above sea level).

As interception deposition is furthermore influenced by diameter of aerosols, aerosols have been collected by a low volume impactor (Berner impactor) and have been analysed for elemental content.

If interception deposition is influenced by surface structure of a forest stand, differences should be found also in a larger scale, for instance when comparing deposition rates in lower/upper or windward/leeward sites of a mountain range. In order to study the effect of accepting surfaces on interception deposition in a larger scale, spruce needles, taken from 6 sites on a catena from the windward to the leeward side of the Teutoburger Wald, have been analysed for cadmium and lead content. These metals, especially lead, are discriminated at plant uptake via roots (Godt et al., 1986) and thereby can be used as indicators for surface contamination from atmospheric deposition. In addition to this, analyses of cadmium and lead in snow from sites on the same catena over the Teutoburger Wald have been used for analysing the pattern of precipitation deposition.

3. MATERIALS

3.1 Locations

The research area is located in the Teutoburger Wald, Nordrhein-Westfalen (Figure 1). The Teutoburger Wald is a mountain range stretching northwest/southeast with an altitude of max. 400 m above sea level and an average precipitation of 800-1400 mm per year. In higher altitudes foggy periods are quite frequent. The Teutoburger Wald is the first barrier for western and southwestern winds coming from the industrial agglomeration Rhein-Ruhr area (80-150 km distance).

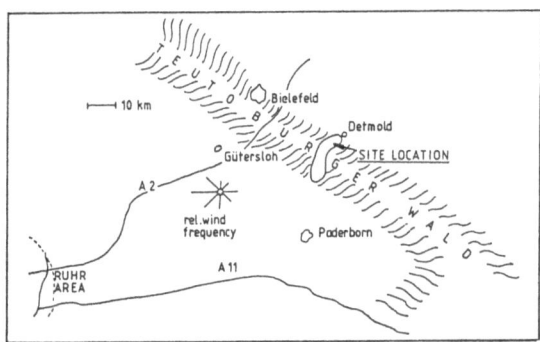

Figure 1 Site location (Nordrhein-Westfalen/FRG)

The location of the deposition measurement site (375 m above sea level, windward top of the Teutoburger Wald) and of other sites (spruce needles and snow sampling) is indicated in Figure 2. The 80 years old spruce stand, for which flux balances have been calculated, shows serious symptoms of tree damages, especially at the edge of the stand. The soil type is a shallow brown earth over limestone with very low pH-values in top mineral soil: 3.4 (demineralised water), 2.8 (KCl). The sites used for collecting snow are located near the spruce stands in the open field.

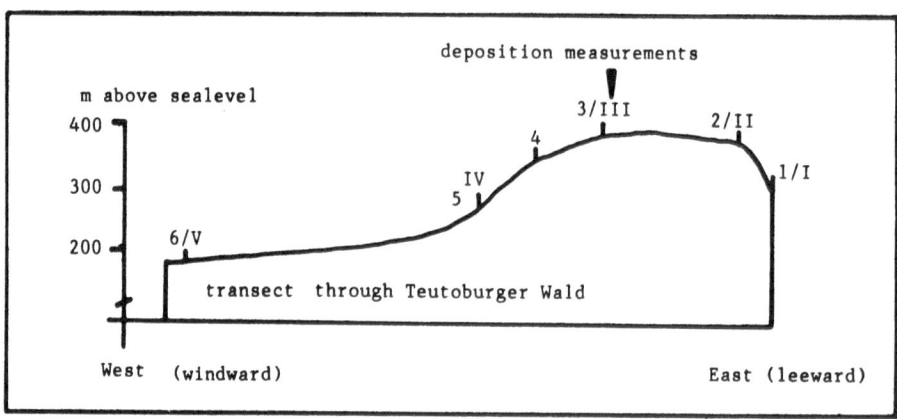

Figure 2 Experimental sites for deposition measurements, elemental stores in snow (I–V) and elemental concentration in spruce needles (1–6), Teutoburger Wald.

3.2 Methods

3.2.1 **Deposition measurements.** Bulk precipitation has been collected from 1.5.1982 to 31.4.1984 (open field precipitation, canopy drip (two years), and canopy drip at the edge of the stand (first year only)) in monthly intervals by using polethylene samplers. The precipitation has been analysed for pH, macroelements and heavy metals (reported in Godt et al., 1985 and Godt, 1986). Deposition rates have been calculated by multiplying concentration and precipitation amount. Flux balances for macroelements have been calculated by using equations given by Mayer (1985) and Meiwes (1984) (see also Bredemeier et al. in this volume).

3.2.2 **Spruce needles.** Spruce needles have been taken (3.4.1981, 30.3.1984) by using a standard method described by Knabe (1984) (seventh whorl, west exposed twigs). The needles have been ground and digested in teflon-lined pressure bombs.

3.2.3 **Aerosols.** Aerosols have been collected using a low volume impactor (Berner) with fractions going from 0.06–16.0 µm aerodynamic diameter. Anions have been eluted with demineralised water. Cations have been analysed after treatment with conc. HNO_3.

Table 2.   Fluxes of macroelements in a spruce stand, Teutoburger
Wald, 1982/1983 and 1983/1984 $kg.ha^{-1}.a^{-1}$

			Mn	Fe	Ca	Mg	K	Na	$SO_4{}^{2-}$-S
Edge of stand	**1982/1983**								
	(M) OD	deposition in open field	0.88	1.75	15.8	3.94	3.58	13.4	17.2
	(M) CD	canopy drip	6.10	2.45	30.7	8.54	14.5	29.2	131.4
	ID	interception deposition	1.05	2.08	18.8	4.69	4.26	15.8	114.2
	TD	total deposition	1.93	3.83	34.6	8.63	7.84	29.2	131.4
Inside stand	**1982/1983**								
	(M) OD	deposition in open field	0.88	1.75	15.8	3.94	3.58	13.4	17.2
	(M) CD	canopy drip	5.55	1.68	25.5	4.67	13.5	16.4	116.1
	ID	interception deposition	0.20	0.40	3.63	0.91	0.82	3.07	99.1
	TD	total deposition	1.08	2.15	19.4	4.85	4.40	16.4	116.1
	1983/1984								
	(M) OD	deposition in open field	0.44	0.80	10.5	2.74	2.70	12.2	11.8
	(M) CD	canopy drip	3.98	1.64	28.9	5.29	16.9	20.3	46.7
	ID	interception deposition	0.29	0.54	7.02	1.84	1.81	8.14	34.9
	TD	total deposition	0.73	1.34	17.5	4.58	4.51	20.3	46.7

(M) = measurements

598

3.2.4 <u>Snow</u>. Newly fallen snow has been collected (9 replicated/site) above a frozen compacted snow layer (14.12.1981). The volume (cm^3/m^2) and weight of snow have been measured in order to calculate deposition rates.

3.2.5 <u>Analytical methods</u>. Sulphate has been analysed by Ion Chromatography, metals and heavy metals by flame- and flameless Atomic Absorption Spectrophotometry (detailed description see Heinrichs <u>et</u> <u>al</u>., 1985).

4. RESULTS

In Table 2 the flux rates for macroelements, calculated according to the equations given by Mayer (1985) and Meiwes (1984) are shown. In Figure 3 the relation of canopy drip inside and at the edge of the stand is expressed in % of deposition rates in the open field. In addition to this, the relation of interception deposition to deposition in canopy drip (%) is illustrated.

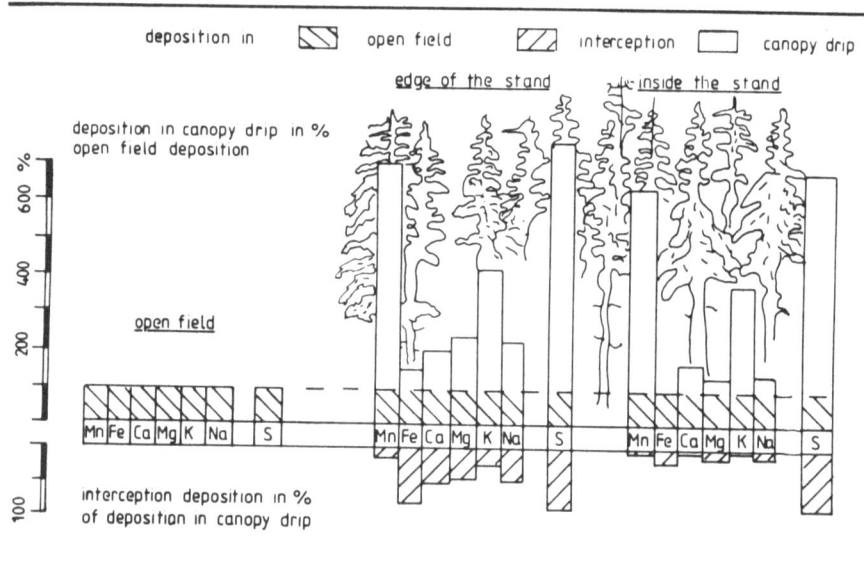

Figure 3   Fluxes in canopy drip in comparison to deposition in open field (100%) and the interception deposition in % of deposition in canopy drip, macroelements 1982/1983, spruce.

It is clearly visible that fluxes in canopy drip exceed fluxes in the open field, especially for Mn, K and S. Fe is obviously retained in tree canopy (canopy drip < open field deposition). Higher deposition rates in canopy drip can be observed at the forest edge for all elements considered. Interception deposition rates in % of canopy drip are increasing at the forest edge, especially for Fe and S. Comparing fluxes in 1982/1983 with those of 1983/1984 (Table 2), obviously higher deposition rates are found during the first year for Mn and S.

Table 3 and Figure 4 give meteorological data during the period when the impactor was used to collect aerosols (24.3.-1.04.1986). Some precipitation has been observed as well as high humidity during most of the days. Southwestern winds with low wind speed were predominating. Meteorological data and calculations of transport sectors (Deutscher Wetterdienst, 1984) showed that the air mass analysed during this period was deriving from the western part of the North Sea and was passing by the Rhein-Ruhr area (highly industrialised).

Table 3   Meteorological data (windward top of the Teutoburger Wald during impactor trial)

Day	Precipitation (mm)	Temperature (°C)	Rel. humidity (%)	Wind speed (m/sec.)
24.3.1984	2.2			
25.3.1984	0.1	1.5 - 9.5	50-95	1.7
26.3.1984	0	1.0 - 8.0	50-90	3.1
27.3.1984	4.9	1.5 - 10.5	60-100	2.9
28.3.1984	1.0	1.0 - 4.5	55-100	3.7
29.3.1984	0.3	-2.0 - 4.5	95-100	2.5
30.3.1984	0.2	-2.5 - 5.0	75-100	1.1
31.3.1984	5.4	-3.0 - 5.0	90-95	1.5
1.4.1984	6.1			

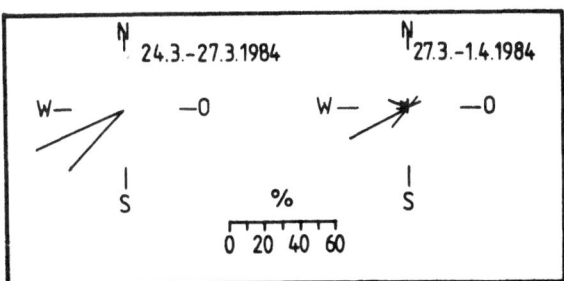

Figure 4   Relative frequency of wind directions during the two periods.

600

In Figure 5 the distribution of elements in aerosols (in % of total mass concentration) over aerodynamic diameter classes is given for the periods 24.3.-27.3. and 27.3.-1.04.1984. In addition to this, total concentration (ng/m^3 and $\mu g/m^3$) is listed up for the elements considered. It is obvious that Cd, Pb, Cu, Mn, Cl^-, SO_4^{2-} and NO_3^-

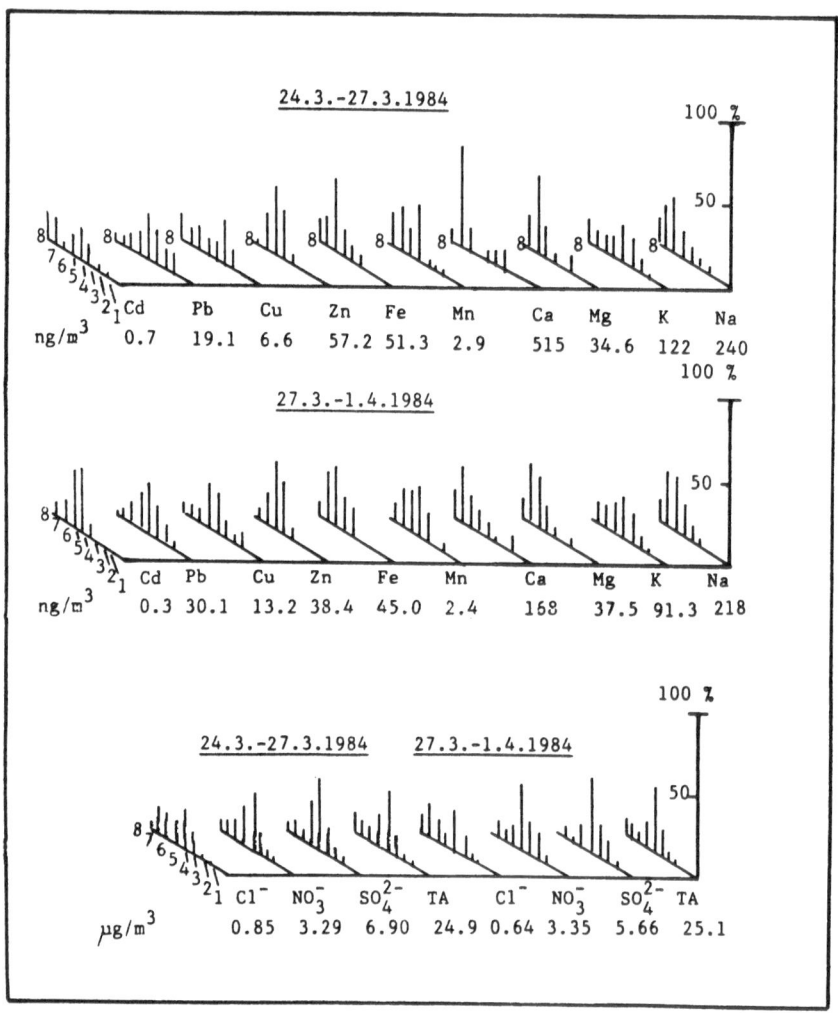

Figure 5   Aerosol content and relative distribution over aerodynamic mass diameter fractions (0.06-18 μm) in the Teutoburger Wald (24.3.-27.3. and 27.3.-1.4.1984) TA = Total aerosol

1 = 0.06-0.12 μm   5 = 1.00-2.00 μm
2 = 0.12-0.25 μm   6 = 2.00-4.00 μm
3 = 0.25-0.50 μm   7 = 4.00-8.00 μm
4 = 0.50-1.00 μm   8 = 8.00-16.0 μm

Table 4   Element-specific mass median diameter (µm) of
           aerosols, Teutoburger Wald, 24.3.-1.04.1984.

           1 = 24.3.-27.3.1984
           2 = 27.3.-1.04.1984

Element	1	2	$\bar{x}$
Cadmium	1.75	1.43	1.59
Lead	0.82	0.87	0.85
Copper	0.97	0.99	0.98
Zinc	1.52	1.49	1.51
Manganese	1.91	1.59	1.75
Iron	3.18	3.27	3.23
Calcium	4.46	3.98	4.22
Magnesium	2.87	3.68	3.28
Potassium	1.76	0.97	1.37
Sodium	3.43	3.58	3.51
Chloride	1.91	1.80	1.86
Nitrate	0.68	0.93	0.81
Sulphate	0.66	0.84	0.75

show relatively high quantities in fine fractions, whereas alkali- and
earthalkali elements are found in fractions with larger aerodynamic
diameter. Total concentrations of elements are changing from first to
second period. Sulphate and nitrate are the dominating anion species,
whereas Ca and Na are predominating among the cations considered.

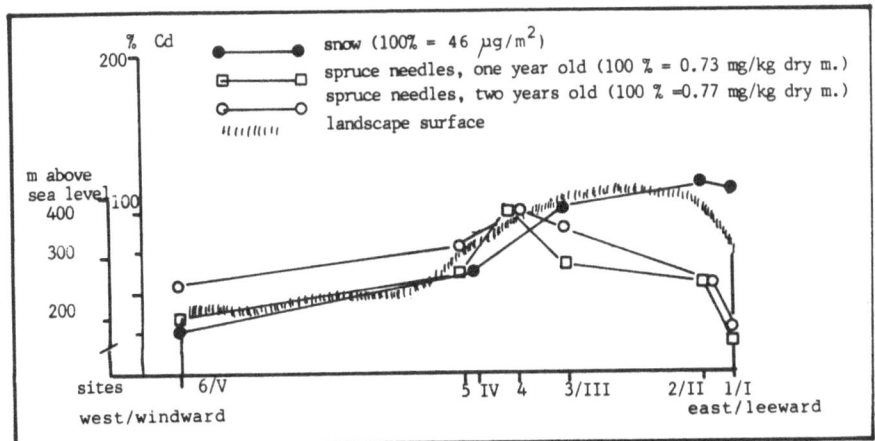

Figure 6   Relative distribution of cadmium in snow and spruce needles
            in the Teutoburger Wald.

In Table 4 the element-specific mass median diameters (MMD) for the investigated compounds are calculated.

In Figures 6 and 7 the relative concentration of cadmium and lead (site 4 = 100%) in spruce needles and deposition rates in snow (site III = 100%) are shown. Cadmium and lead in snow are increasing with elevation and show highest deposition on the leeward top of the Teutoburger Wald.

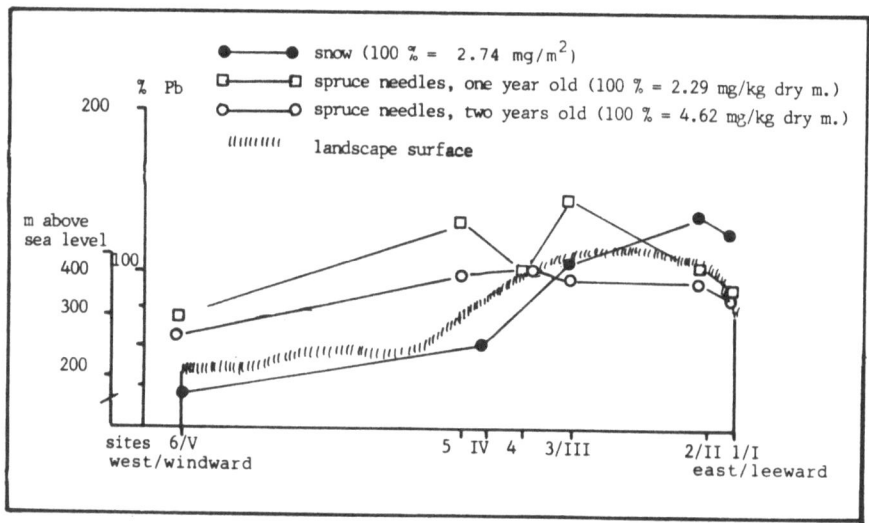

Figure 7   Relative distribution of lead in snow and spruce needles in the Teutoburger Wald.

5.   CONCLUSIONS

The basic hypothesis of this work is that due to interception deposition one should find higher dry deposition (and thereby total deposition) rates to elevated parts of the landscape, or to surface structures surmounting a given general level (as e.q. edge of a forest stand, solitary tree, windexposed top of a mountain range).

Our results are in good agreement with the basic hypothesis when comparing atmospheric input inside and at the edge of the examined spruce stand. By these findings one can show that interception deposition plays an important role especially for S, which is partially deposited as SO_2.

The concentrations of cadmium and lead in spruce needles indicate high deposition rates at the windward exposed part of the Teutoburger Wald and thereby support the hypothesis as well (see also Korff et al., 1980 and Glavac, 1986). In addition to interception deposition, high precipitation deposition (sedimentation) can be expected in higher mountain ranges as a function of higher precipitation amount (rain, snow etc., see also Figs. 7 and 8 (snow)).

The deposition measurements in a spruce stand which is located in
the windexposed highest altitudes of the Teutoburger Wald showed high
total deposition rates of macroelements (esp. S) and also heavy metals
(detailed discussion see Godt, 1986). The very high input of S is
derived from the industrial agglomeration Rhein-Ruhr area, as could be
shown by aid of meteorological analysis (Godt, 1986). Similar results
have been found by Pfeffer (1985) for immision measurements of SO_2 in
the Eggegebirge in the neighbourhood of the Teutoburger Wald. Schwela
and Rademacher (1985) examined immission rates (using IRMA method) on
the top of the Teutoburger Wald (25 m above ground). Their experiments
were conducted in a 3 km distance of the spruce stand of this project.
Schwela and Rademacher found in the Teutoburger Wald even higher
S-immission rates (IRMA) than in parallel measurements in the centre of
the highly industrialised area Rhein-Ruhr (Duisburg).

Although the time for the reported impactor experiments was
restricted to one week, the findings are in good agreement with those
of Höfken and Gravenhorst (1986) and Schmidt et al. (1985), who
conducted their experiments in a rural forest site (Solling area) as
well. Other authors (Lindberg and Harris, 1981; Dlugi et al., 1981;
Laskus et al., 1979; Dannecker et al., 1981; Müller, 1984 and Rohbock,
1985) conducted impactor experiments in more or less industrialised
areas (detailed discussion see Godt, 1986). Nevertheless these authors
found higher concentrations of aerosols and elements in aerosols even
in rural areas compared to the results from the Teutoburger Wald.
Condensation of water vapour on particulate material and subsequent
gravitational deposition during the period of the experiment might have
led to low aerosol concentration in the Teutoburger Wald experiment
(see also Fowler, 1978).

In comparison to industrial areas, one can assume for rural areas
like Teutoburger Wald a relative increase of aerosols in the Aitken-
and accumulation mode in comparison to aerosols with a diameter > 2 µm
due to gravitational deposition of larger particles close to source.
Thereby interception deposition compared to precipitation deposition
should become more important in rural areas. When considering dry
deposition processes, interception deposition will be of importance for
heavy metals and anions under investigation, whereas sedimentation
might play a role for Ca, Mg, K and Na (Table 4).

Due to the correlation of interception deposition and orography in
combination with meteorological effects, deposition rates in a highly
structured landscape will be very variable. The high interception
deposition rates at exposed parts of the landscape can lead to
accumulation of air pollutants on plant parts (Godt, 1986) and also
lead to accumulation in other parts of the ecosystem (for instance in
soil). Considering this, it is important to mention that tree damages
are primarily observed in exposed parts of a landscape. Keeping this
in mind the future development of the whole site (at continuing input
of air pollutants) might be indicated at exposed parts of the
ecosystem.

604

ACKNOWLEDGEMENTS

We want to thank the Deutsche Forschungsgemeinschaft for financial
support. Also we thank Dr Georgi/Hannover for helping us in carrying
out the impactor experiments.

REFERENCES

Berner, A. 1984. 'Design principles of the AERAS low pressure
    impactor'. Aerosols, Elsevier Science Publication New York.
Dannecker, W., Naumann, K. and Herbst, R. 1981. 'Zeitgleiche
    Schwebstaubprobenahme in verschiedenen Höhen im Ballungsraum
    Hamburg und die nachfolgende chemische Analyse'.
    Staub-Reinhaltung der Luft 41, 7, 254 ff.
Deutscher Wetterdienst 1984. Zentralamt Offenbach.
Dlugi, R., Jordan, S. and Lindemann, E. 1981. 'Physikalisch-chemische
    Eigenschaften von Rauchgasaerosolen'. In: Aerosols in Science,
    Medicine and Technology, S. 163-168, 9th Conference, 23-25 Sept.
    1981, Duisburg, FRG, Hrsg.: Gesellschaft für Aerosolforschung.
Flemming, G. 1982. 'Die Windgeschwindigkeit als Verstärkungsfaktor
    für Rauchschäden im Wald in Abhängigkeit vom Waldaufbau und
    Relief'. Z. Meteor. 31, 1, 14-22.
Fowler, D. 1978. 'Wet and dry deposition of sulphur and nitrogen
    compounds from the atmosphere'. In: T.C. Hutchinson and M.
    Havas: 'Effects of acid precipitation on terrestrial
    ecosystems'. NATO conference series I, Ecology 4 ,9-27.
Glavac, V. 1986. 'Die Abhängigkeit der Schwermetalldeposition in
    Waldbeständen von der Höhenlage'. Natur und Landschaft 61 (1986)
    2 ,43-47.
Godt, J. 1986. 'Untersuchung von Prozessen im Kronenraum von
    Waldökosystemen und deren Berücksichtigung bei der Erfassung von
    Schadstoffeinträgen unter desonderer Beachtung der
    Schwermetalle'. Diss. Gesamthochschule Kassel, FB Stadt- und
    Landschaftsplanung, Berichte des Forschungszentrums
    Waldökosysteme/Waldsterben Göttingen 19, 265, S.
Godt J., Mayer, R. and Georgi, B. 1985. 'Die Interceptionsdeposition
    als wichtiger Faktor der Schwermetallbelastung von
    Waldökosystemen'. VDI-Berichte 560 ,333-356.
Godt, J., Schmidt, M. and Mayer, R. 1986. 'Processes in the canopy
    of trees: Internal and external turnover of elements'. In:
    H.-W. Georgii (ed.): Atmospheric pollutants in forest areas. D.
    Reidel Publishing Company, Dordrecht.
Hager, H. and Kazda, M. 1985. 'The influence of stand density and
    canopy position on sulphur content in needles or Norway spruce
    (Picea abies L. Karst.). Water, Air and Soil Pollution 25,
    321-329.
Heinrichs, N., König, N. and Schultz, R. 1985. 'Atomabsorption-
    und-Emissionsspektroskopische Bestimmungsmethoden für Haupt- und
    Spurenelemente in Probelösungen aus Waldökosystem-
    untersuchungen'. Berichte des Forschungszentrums Waldökosysteme/
    Waldsterben Göttingen 8, 92 S.

Höfken, K.D. and Gravenhorst, G. 1980. 'Concentration and size distribution of trace substances in aerosols above and beneath a forest canopy'. Tagung Gesellschaft für Aerosolforschung Schmallenberg, Okt. 1980.

Hosker, R.P. and Lindberg, S.E. 1982. 'Review: Atmospheric deposition and plant assimilation of gases and particles'. Atmos. Environ. 16, No. 5, 889-910.

Knabe, W. 1984. 'Merkblatt zur Entnahme von Blatt- und Nadelproben für chemische Analysen'. Allgemeine Forst Zeitschrift 33/34, 847-848.

Korff, H.C. et al. 1980. Einfluss der Orographie auf die räumliche Verteilung von Schadstoffen im Steigerwald und im Fichtelgebirge'. Bayreuther Geowissensch. Arbeiten 1, Bayreuth, 1980.

Laskus, L., Bake, D., Kura, J. and Möller, M. 1979. 'Konzentration und Korngrössenverteilung von Inhaltsstoffen im Luftstaub deutscher Städte. WaBoLu-Berichte 2, Dietrich Reimer Verlag, Berlin.

Lindberg, S. and Harris, 1981. 'The role of atmospheric deposition in an Eastern US deciduous forest'. Water, Air and Soil Pollution 16 , 13-31.

Mayer, R. 1985. 'Verfahren zur Erfassung der Schadstoffzufuhr in Waldökosystemen'. Staub Reinhaltung der Luft 6, 267-268.

Meiwes, K.J., Hauhs, M., Gerke, H., Asche, N. und Lamersdorf, N. 1984. 'Die Erfassung des Stoffkreislaufes in Waldökosystemen - Konzept und Methodik -'. Berichte des Forschungszentrums Waldökosysteme/ Waldsterben Göttingen 7, 69-142.

Monteith, J.L. (ed.) 1976. 'Vegetation and the atmosphere'. Vol. 2, case studies. Academic Press London.

Müller, J. 1984. 'Measurement of heavy metals in airborne particulate matter, dry and wet deposition'. EMEP workshop on heavy metals Lillestrom, 27.-29. Aug. 1984, Proc.: c/o NILU, Lillestrom-Norway.

Pfeffer, H.-U. 1985. 'Immissionserhebungen in quellfernen Gebieten Nordrhein-Westfalens'. Staub-Reinhaltung der Luft 45, 6, 287-293.

Rohbock, E. 1985. 'Trockene und feuchte Deposition von Schwermetallen' Bielefelder Ökologische Beiträge 1, 65-82.

Schmidt, M., Mayer, R. and Georgi, B. 1985. 'Beeinflussung der Interceptionsdeposition (trockene Deposition) durch die Aerosolkonzentration und -grössenverteilung in einem Buchenbestand (Solling)'. VDI-Berichte 560, 423-438.

Schwela, D. und Rademacher, R. 1985. 'Untersuchungen zur Belastung durch Luftverunreinigungen in quellfernen Gebieten mittels Bioindikatoren, IRMA- und Staubniederschlagsmessungen'. Staub-Reinhaltung der Luft 45, 6, 284-287.

Sehmel, G.A. 1980. 'Particle and gas dry deposition: a review'. Atmos. Environ. 14 , 983-1011.

606

Ulrich, B., Mayer, R. and Khanna, P.K. 1979. 'Deposition von Luftverunreinigungen und ihre Auswirkungen in Waldökosystemen des Solling'. Schriften der Forstlichen Fakultät Universität Göttingen und der Niedersächsischen Forstlichen Versuchsanstalt 58, 291 S.

Wiman, B.L.B. and Lannefors, H.O. 1985. 'Aerosol depletion and deposition in forests - a model analysis'. Atmos. Environ. 19, No. 2, 335-347.

A SIMPLE AND APPROPRIATE METHOD FOR THE ASSESSMENT OF TOTAL ATMOSPHERIC
DEPOSITION IN FOREST ECOSYSTEM MONITORING

M. Bredemeier, E. Matzner and B. Ulrich
Forschungszentrum Waldökosysteme/Waldsterben
University of Goettingen
Büsgenweg 2
D-3400 Goettingen
W. Germany

ABSTRACT. A method to assess long-term rates of total atmospheric
deposition to forest canopies is proposed for discussion. Rates of
total deposition are estimated from measurements of above- and
below-canopy element fluxes (precipitation and throughfall),
assumptions being made about binding forms and phases, in which special
chemical constituents are deposited. The approach considers canopy
interactions such as proton buffering and cation exchange and leaching
from the leaves. Its application delivers plausible total deposition
data and plausible differentiations between sites. It should be tested
further under various environmental conditions and in connection with
other methods.

1. INTRODUCTION

There is some difficulty in determining total deposition rates from the
atmosphere to terrestrial ecosystems. Precipitation input cannot be
regarded as total deposition in most cases, because interception
deposition of gases, particles and cloud droplets is not taken into
account. The flux of chemical elements in forest canopy throughfall,
however, is a result of both deposition from the atmosphere (external
input) and canopy processes (internal cycling). Canopy processes act
as sink or source functions for special chemical constituents. Some
constituents may, however, pass the canopy inertly.
    Element flux in throughfall, therefore, does not necessarily
correspond to the rate of deposition onto the canopy. Furthermore, dry
deposition cannot be determined accurately by only measuring deposition
rates to artificial inert collector surfaces. The canopy differs from
artificial collectors in both surface structure and adsorption
characteristics. Additionally, determination of long-term deposition
rates with artificial collector installations is relatively expensive
and requires extensive field and laboratory work (Lindberg and Lovett,
1985). Investigations have not yet been pursued for time periods
longer than weeks to months.

M. H. Unsworth and D. Fowler (eds.), Acid Deposition at High Elevation Sites, 607–614.

608

Clearly, then, a need exists for a reliable method to quantify long-term total deposition rates in order to construct element budgets and balances for forest ecosystems.

2. QUANTIFICATION APPROACH

The approach, developed by Ulrich and co-workers (Mayer and Ulrich, 1974; Ulrich et al., 1979; Ulrich, 1983), is designed to assess long-term (e.g. yearly) total deposition rates in forests. It is based upon measurements of major element fluxes in forest canopy input and output. It considers binding forms and phases, in which elements exist in the atmosphere and are deposited, and plant-physiological

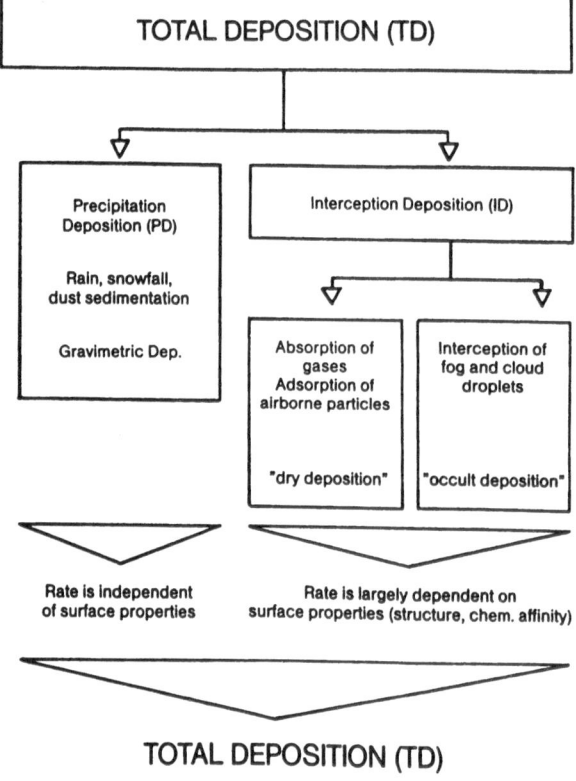

Figure 1.   Partition of total deposition.

implications. Compared to micro-meteorological methods of deposition
measurement (e.g. Gravenhorst and Höfken, 1982; Lindberg and Lovett,
1985), which can deliver data with high time resolution, its
experimental installations in the field are simple and easy to handle
over long periods. Principally, only open collectors for precipitation
(above canopy or in adjacent open plots) and throughfall, with a
sufficient number of replicates, are used. Experimental design and
calculations are aimed at long-term monitoring, i.e. over years or
longer periods. Inferences about air pollutant concentrations and
temporal variability, and about short-term deposition rates and peaks,
cannot be made.

Figure 1 shows the different fractions of total deposition
considered in our calculations. Precipitation deposition (PD), is
defined here as the flux to a bulk collector in the open land (see
Fig. 1). Interception deposition (ID) is the element amount which is
additionally deposited upon the forest surface. The sum of both yields
total deposition (TD) for each chemical constituent:

$$(1) \quad TD_x = PD_x + ID_x; \quad x = \text{chem. Element}$$

An important presumption is that for the constituents considered the
canopy has only one output, the throughfall solution: particulate or
gaseous releases from the leaves into the atmosphere are considered
negligible. We know of no experimental evidence to contradict this
assumption for forest stands. Gaseous exchange of H_2O, CO_2, O_2 and
N-gases is not considered in the balance. Release of S-containing
gases (H_2S, SO_2) from leaves or phyllospheric epiflora is negligibly
low compared to deposition of S-compounds (SO_2, SO_4^{2-}), at least under
Central European conditions.

If throughfall (TF) is the only output flux from the canopy, the
following equation is valid:

$$(1a) \quad TF_x = PD_x + ID_x \pm S_x$$

where S indicates the canopy sink/source function with respect to the
chemical element x (i.e. in-canopy assimilation or canopy leaching).

Calculating interception (ID) and total deposition (TD) rates for
all major elements starts with those elements for which the canopy can
be assumed to act as an "inert sampler", i.e., for which S_x is ca. 0.
These elements are Na, Cl and S:

$$(2) \quad ID_x = TF_x - ND_x; \quad x = Na, Cl, S$$
$$(\text{"Canopy-difference"})$$

The assumption of inert flow of deposited Na through the canopy layer
is theoretically based upon three considerations:

1. Sodium is not a physiologically important plant nutrient (compared
    to K, Ca, Mg).
2. Its concentrations in leaves/needles are accordingly low.

3. The Na^+-flux in throughfall does not show a "physiological seasonality", that means it does not peak at times of increased physiological activity or physiological changes, such as start of growth and leaf development in spring or the beginning of leaf fall in autumn. K^+, which is leached from the canopy in large amounts, shows a clear seasonality in throughfall flux (Matzner, 1984).

Experimental evidence for this asumption also exists. A greenhouse leaching experiment with young spruce in an aerosol-filtered atmosphere (Fassbender, 1977), showed needle release of K^+, Mn^{++} and P, but neither Na^+- nor SO_4^{2-}-leaching. These results are confirmed by Hoefken (1981). He additionally showed negligibly low rates of Cl^--leaching.

Sodium in the atmosphere is found only in aerosols, predominantly from sea spray. From the ID of Na^+, the particulate ID of other elements can be calculated as follows:

$$(2) \quad (ID/PD)_{Ne} = (ID/PD)_{Xpart};$$
$$X = H, K, Ca, Mg, Al, Fe, Mn, S$$

Equation (2) implies that the fraction ID/PD is constant for particles with different mass median diameter (MMD), because the elements given in Eq. 2 occur in different MMD-classes of aerosols (Höfken and Gravenhorst, 1982).

We think this assumption is acceptable, because precipitation deposition (washout) and interception deposition of particles are, at least to some extent, due to the same physical processes. These are diffusion for MMD < 0.1 μ and impaction for MMD > 1.0 μ. Fowler (1984) notes: "The problem is similar whether particle transfer is to the ground or to cloud droplets". The assumption is further supported by measurements of aerosol concentrations above and below canopy. Gravenhorst and Hoefken (1982) found atmospheric concentrations of Cd, Pb, Mn, Fe, SO_4^{2-}, Cl^-, NH_4^+ and NO_3^- below canopy to be consistently 63-75% of those above canopy, though the mass median diameters of the respective aerosols are different. This similar range of concentration gradients indicates similar deposition velocities for different particle sizes.

Rearranging equation (2), one can calculate particulate ID rates:

$$(2a) \quad ID_{Xpart} = (ID/PD)_{Na} \cdot PD_{Xpart}$$
$$Xpart = H, K, Ca, Mg, Al, Fe, Mn, S$$

In addition to particulate interception deposition, some elements are also dry deposited as gases:

$$(3) \quad ID_X = ID_{xpart} + ID_{xgas}; \quad x = S, (N)$$

There is a variety of possible transformations of nitrogen compounds, comprising both sinks and sources in the canopy. An estimate of N-deposition from the canopy flux balance would, therefore, be very approximate. For Cl, a gaseous deposition fraction is neglected, because it should be only important in close vicinity to HCl sources. However, sulphur which is dry deposited from the gas phase (SO_2) is an important fraction of total deposition. This rate can be calculated as the difference between total interception of S (Eq. 2) an ID_s part (Eq. 3).

$$(3a) \quad ID_{sgas} = (TF - PD)_s - ID_{spart}$$

The above equation implies, that there is no sulphur leaching from the canopy. Ulrich (1983) reports a mean value of ca. 1.5 (0. - 5.) $kg.ha^{-1}.a^{-1}$ for S leaching from senescent leaves of a beech stand in northwestern Germany (Solling). The corresponding mean S flux in throughfall is 52. kg. $ha^{-1}.a^{-1}$. Leaching should not exceed 5% of total sulphur in throughfall, due to high deposition rates in central Europe. Therefore, neglecting sulphate leaching from the canopy is acceptable for our calculations.

Within the pH-range found in precipitation samples and with ambient concentrations of oxidants, sulphur-dioxide will be oxidised, according to:

$$SO_2 + H_2O + \tfrac{1}{2} O_2 \longrightarrow SO_4^{2-} + 2 H^+$$

Thus, the calculated S_{gas}-deposition rate means an equivalent deposition of protons:

$$(4) \quad ID_{Sgas} = ID_{Hgas} \ [eq]$$

Total deposition of H^+ is therefore

$$(5) \quad TD_H^+ = PD_H^+ + ID_{H^+part} + ID_{H^+gas}$$

Protons (H_3O^+-ions) deposited in the canopy can be buffered by base cation (esp. Ca^{2+}, Mg^{2+}) exchange from leaf tissue (Ulrich et al., 1973; Matzner and Ulrich, 1984). In this case, deposited protons do not appear as measurable acidity in throughfall. The leaching rates of base cations from the canopy, however, are increased in equivalent amounts. H^+-buffering rate in the canopy is:

$$(6) \quad H^+\text{-buffering} = TD_H^+ - TF_H^+$$

Canopy leaching rates of leachable cations, which are released in physiological processes or exchanged in H^+-buffering reactions, are calculated as follows:

$$(6) \quad CL_x = TF_x - TD_x; \quad x = Ca^{2+}, Mg^{2+}, K^+, Mn^{2+}$$
$$CL = \text{"canopy leaching"} \quad \text{(quantitatively relevant)}$$

The set of equations to calculate the canopy ion flux balance is now complete.

3. EXAMPLE RESULTS

An example of long-term total deposition rates, calculated in the above described way is given in Table 1. Data stem from four experimental forest ecosystems which have been monitored in long-term studies.

The Solling sites, situated at elevations of ca. 500 m above sea level in a mountaineous region with a moist, cool climate in northwestern Germany (Ellenberg, 1971) are more impacted by air pollution than the Lüneburger Heide sites, which are located at about 80 m elevation in the Northern German flat plains in a large, relatively sheltered forest area. Soils are Dystric cambisols (podzolic acid brown earths) at all sites, derived from loess over sandstone in the Solling and from glacial sands in the Lüneburger Heide. In both regions, a deciduous and coniferous forest ecosystem were monitored. The stands in the Solling are a beech (<u>Fagus sylvatica</u> L.), 135 year old, and a spruce (<u>Picea abies</u> Karst.), 100 year old stand. The sites in the Lüneburger Heide are stocked with 105 year old oak (<u>Quercus robur</u> L.) and 100 year old pine (<u>Pinus sylvestris</u> L.), respectively.

4. DISCUSSION

For long-term ecosystem monitoring, a simple, inexpensive yet reliable method of assessing total deposition rates is needed. We are proposing a quantification mode for critical discussion.

Table 1   Long-term mean rates of bulk precipitation deposition (PD), throughfall (TF) and calculated total deposition (TD) in four forested ecosystems in northwestern Germany ($keq.ha^{-1}.a^{-1}$).

| | Solling (mean 1969–83) | | | | | Lüneburger Heide (mean 1980–84) | | | | |
| | beech | | | spruce | | | oak | | pine | |
	PD	TF	TD	TF	TD	PD	TF	TD	TF	TD
H_2O (mm)	1017	864	–	747	–	830	688	–	639	–
H^+	.79	1.34	1.92	3.11	3.81	.43	.51	.97	.99	1.23
Na^+	.34	.62	.62	.76	.76	.67	.83	.83	1.09	1.09
K^+	.09	.72	.16	.73	.20	.11	.62	.14	.33	.18
Ca^{2+}	.50	1.20	.91	1.62	1.12	.26	.66	.32	.95	.42
Mg^{2+}	.15	.33	.27	.40	.34	.19	.32	.24	.36	.31
Fe^{3+}	.04	.06	.07	.11	.09	.01	.02	.01	.04	.02
Mn^{2+}	.01	.14	.02	.20	.02	.01	.10	.01	.03	.02
Al^{3+}	.12	.18	.22	.33	.27	.09	.07	.11	.14	.15
SO_4^{2-}	1.46	3.14	3.14	5.31	5.31	1.12	1.82	1.82	2.35	2.35
$H_2PO_4^-$	.01	.02	–	.02	–	.02	.04	–	.00	–
Cl^-	.47	.92	.92	1.11	1.11	.79	1.06	1.06	1.31	1.31

In spite of some uncertainties, partly resulting from experimental design (wet + dry samplers) and partly from assumptions that must be made, we believe that the proposed calculation mode is an adequate means for the assessment of long-term deposition rates. It delivers plausible data and plausible differentiations of data over a range of different forest ecosystems. H^+ and $SO_2{}^{2-}$-S deposition rates, for instance, come out consistently higher for coniferous stands compared to hardwood stands at the same location (due to higher leaf surface area and/or exposure throughout the year). Exposed sites (e.g. Solling, Table 1) show significantly higher rates than relatively sheltered sites (e.g. Lüneburger Heide, Table 1).

The approach allows calculations of canopy interactions such as proton buffering in the canopy and cation leaching. The plausibility and consistency of data calculated with this approach should be tested under various environmental conditions and in connection with micro-meteorological methods.

REFERENCES

Ellenberg, H. (ed.) 1972. Integrated experimental ecology, methods and results of ecosystem research in the German-Solling-Project. Ecological Studies 2 , Springer Verlag.
Fassbender, H.W. 1977. Modellversuche mit jungen Fichten zur Erfassung des internen Nährstoffumsatzes. Ecologia plantarum 12 , 263-272.
Fowler, D. 1984. Transfer to terrestrial surfaces. In: The ecological effects of deposited sulphur and nitrogen compounds. Phil. Trans. R. Soc. Lond. 305 , 259-279.
Gravenhorst, G. and Höfken, K.D. 1982. In: H.W. Georgii and J. Pankrath: Deposition of atmospheric Pollutants. D. Reidel Publ. Comp., Dordrecht.
Höfken, K.D. 1981. Untersuchungen über die Deposition atmosphärischer Spurenstoffe an Buchen- und Fichtenwald. Dissertation, Institute für Meteorologie und Geophysik der J.W. Geothe-Univ. Frankfurt.
Höfken, K.D. and Gravenhorst, G. 1982. Deposition of atmospheric aerosol particles to beech and spruce forest. In: H.W. Georgii and J. Pankrath: Deposition of atmospheric pollutants. D. Reidel Publ. Comp., Dordrecht.
Lindberg, S.E. and Lovett, G.M. 1985. Field measurements of particle dry deposition rates to foliage and inert surfaces in a forest canopy. Environ. Sci. Technol. 19(3) .
Matzner, E. 1984. Deposition und Umsatz chemischer Elemente im Kronenraum von Waldbeständen. Ber. d. Fz. Waldökosys./Waldsterben, Univ. Göttingen 2, 61-87.
Matzner, E. and Ulrich, B. 1984. Raten der Deposition, der internen Produktion und des Umsatzes von Protonen in zwei Waldökosystemen. Z. Pflanzenernähr. Bodenkde 147 , 290-308.
Mayer, R. and Ulrich, B. 1974. Conclusions on the filtering action of forests from ecosystem analysis. Oecol. Plant 9(2) , 157-168.

614

Ulrich, B., Steinhardt, U. and Müller-Suhr, A. 1973. Untersuchungen über den Bioelementgehalt in der Kronentraufe. Göttinger Bodenkdl. Ber. **29** , 133–192.

Ulrich, B., Mayer, R. and Khanna, P.K. 1979. Die Deposition von Luftverunreinigungen und ihre Auswirkungen in Waldökosystemen im Solling. Schriften aus der Forstl. Fak. d. Univ. Göttingen, Bd. 58, Sauerländer-Verlag.

Ulrich, B. 1983. Interaction of forest canopies with atmospheric constituents: SO_2, alkali and earth alkali cations and chloride. In: B. Ulrich and J. Pankrath (eds.): Effects of accumulation of air pollutants in forest ecosystems. D. Reidel Publ. Comp., 1983, 33–45.

OBSERVATIONS ON WET AND DRY DEPOSITION TO FOLIAGE AT A HIGH ELEVATION SITE

P.H. Schuepp[1], D.N. McGerrigle[1], H.G. Leighton[2],
G. Paquette[1], R.S. Schemenauer[3] and S. Kermasha[1]

McGill University, Departments of Renewable Resources (1) and
Meteorology (2), Ste. Anne-de-Bellevue, Quebec H9X 1C0, and
Atmospheric Environment of Canada (3).

ABSTRACT. A series of experiments to measure acidic deposition due
to various meteorological processes (rain, snow, fog and dry weather)
were carried out between August 1985 and April 1986 at the south summit
(Pic White, 860 m) of Mt. Tremblant, Quebec, Canada. They involved
natural, varathane-coated and artificial shoots of fir trees. Anion
and (for some samples) cation content of deposition was measured and
the 850 mb trajectories for selected events of high deposition
examined.
  Results show deposition of anions during periods of dry weather in
the fall season, with buildup of sulphate greater than that of nitrates
and chlorides, and effective rinse-off of deposition by precipitation
events. Cumulative winter deposition was strongly dependent on
orientation of sampling branches. The use of standardized artificial
branches in idealized exposure conditions gave the most consistent
results and enabled a comparison to be made between the various
deposition processes.

1. INTRODUCTION

Little is known about the relative magnitude of wet vs. dry
deposition at elevated sites, where clouds (fog) are a significant
deposition input, nor about the significance of chemical interaction
between deposited ions and foliage. The problem is of particular
interest in the Province of Quebec in Canada which contains large areas
of elevated (500 - 900 m), mountainous terrain downwind of major
pollution source areas in the United States and Canada. Most tree
cover above 800 m is coniferous, retaining its high collection
efficiency year-round. The frequency of fog events at elevated Quebec
sites has been discussed elsewhere (Schemenauer et al., 1986); it is of
the order of 40% on a probability-per-event-per-day basis and monitored
as part of the Atmospheric Environment Service's Chemistry of High
Elevation Fog (CHEF) programme.

M. H. Unsworth and D. Fowler (eds.), Acid Deposition at High Elevation Sites, 615–637.
© *1988 by Kluwer Academic Publishers.*

The higher ion concentrations and lower pH values of fog relative to rain has been documented (e.g. Houghton, 1955; Mrose, 1966; Jacob et al., 1983; Waldman et al., 1985; Schemenauer, 1986). Of particular interest, in addition to the Canadian studies, are those carried out at Whiteface Mountain in New York State and Camel's Hump Mountain in Vermont, because of their relative proximity to the Quebec sites. Falconer and Falconer (1980) suggested that acidic input from clouds at the Whiteface site was significantly more important than that of other precipitation while Vogelmann et al. (1968) demonstrated the high collection efficiency for fog water at the Camel's Hump site (artificial foliage in sampling buckets collected 1.7 times the amount of open buckets). Schlesinger and Reiners (1974) made similar observations in the White Mountains of New Hampshire, with ratios of foliar collectors to open buckets averaging 4.5, probably due to stronger winds at the higher elevation (1370 vs. 1097 m).

The relative importance of fog and rain as deposition pathways in Eastern North America has been discussed by Barrie and Schemenauer (1986). Galloway and Whelpdale (1980) estimated, also for Eastern North America, the ratios of dry to wet deposition for sulfur compounds to be of the order of 0.3 to 1.1, increasing with increasing proximity to pollution sources, with absorption by foliage the most prominent sink. Some correlation of dry deposition with stomatal opening is indicated by Fowler and Cape (1983) and Johansson et al. (1983) for dry canopies but diurnal variation of deposition did not follow a simple cycle of stomatal movements. The latter authors also found a strong seasonal variation of deposition, with little deposition in winter. Grennfelt et al. (1983) found similar correlation between dry deposition and stomatal action (or transpiration). For wet canopies the chemistry of the surface film appears to have a significant effect (Fowler and Cape, 1983).

Damage to forests has been hypothesized to be linked to the leaching of nutrients from the soil (e.g. Ulrich et al., 1980) and possible to direct physiochemical interaction between deposition and foliage. Sulphuric acid may be formed directly on and in the leaf through oxidation of dry-deposited SO_2 (e.g. Tomlinson, 1984), and acid occurring in fog water and deposited to foliage may be much more damaging than that which occurs in rain and which is largely shed by the waxy surface-coating of the leaf.

The effect of wet acidic deposition on the leaching of nutrients has been shown for deciduous and coniferous folige, involving e.g. calcium on tobacco (Fairfax and Lepp, 1975), calcium, magnesium and potassium on beans and maple (Wood and Bormann, 1975) and amino acids, potassium and calcium on birch and spruce (Scherbatskoy and Klein, 1983). Also, throughfall observations by Hoffman et al. (1980) suggested cation exchange of H^+ with foliar Ca^{++}.

The above cursory literature review shows the existence of studies on selected aspects of dry and wet deposition and its possible effects on foliage. However, to our knowledge, no observations have been reported on the various deposition processes (wet deposition by rain, snow and fog as well as dry deposition) in their natural sequence on natural or realistic artificial foliage at an elevated site. The

studies reported here are a first attempt to assess realistically, in different seasons, the relative magnitudes of acidic deposition – and by implication its effects – on the ecosystem at such sites. Specifically, the goals of the study were:
- to compare the relative magnitude of deposition of chloride, nitrate and sulphate from rain, snow, fog and dry deposition to natural and artificial coniferous foliage which had been exposed for varying lengths of time;
- to provide first-order trajectory analysis for air mass movements associated with high deposition of the above-mentioned anions;
- to compare cation concentrations on artificial and natural folige elements as evidence for possible significant leaching of nutrients from foliar surfaces.

2. EXPERIMENTAL APPROACH

Six series of experiments were carried out between August 1985 and April 1986, most of them at the south summit of Mt. Tremblant, Quebec (46.21 N, 74.55 W; elevation 860 m). Experimental goals were as follows: **Experiments 1 to 3** measured ion deposition to shoots of balsam fir with and without a chemically inert coating, to assess the magnitude of deposition to real foliage under realistic conditions at a mountain site and to look for evidence of ion exchange processes between deposition and foliage. **Experiments 4 and 5** attempted to monitor cumulative wet and dry deposition during winter conditions. **Experiment 6** was an event-by-event collection of deposition to artificial foliage during spring to assess relative contribution towards acid loading from the different deposition processes. Methodologies for the various experiments are summarized below.

Experiment no. 1 (August 19 – September 18, 1985):

One hundred and twelve fir shoots (three-pronged with approximate overall length of 11 cm) were selected at random from among trees at the mountain site. Half of them were coated with a 'varathane' (polyurethane) film; all were cleaned (rinsed) prior to exposure. Deposition was monitored by rinsing of shoots on designated sampling days (days 1 through 10 and days 15, 19, 25 and 30 of the test period). Each sampling day, two natural and two coated shoots were sampled after exposure from the last sampling day (e.g. 24 hours for days 1 to 10 to monitor daily accumulation rates) and a similar set of pairs after exposure from the start of the experiment, to test for cumulative deposition.

Rinsing was done by spray bottle, involving 30 – 40 ml of distilled, deionized water for cleaning and 5 ml for sampling after exposure. Separate tests on sampling efficiency of 5 ml rinsings of shoots showed approximately 60 – 70% removal of deposition by a first rinsing, a factor to be considered in interpretation of test results as to total deposition.

Experiment no. 2 (October 1 to 31, 1985):

Methodology was similar to experiment 1 but only cumulative deposition to days 1 through 6 and days 10, 15, 20 and 30 of the test period were monitored.

Experiment no. 2-A (October 1 to 31, 1985):

Identical in scope and methodology to experiment 2 but running concurrently at a near-urban low-level site (Macdonald Campus of McGill University at Ste. Anne-de-Bellevue; elevation 40 m). Sampling shoots were balsam fir.

Experiment no. 3 (October 17 to November 2, 1986):

Methodology similar to that of experiments 1 and 2 but only coated shoots were used, with two samples for daily accumulation and two cumulative samples for sampling days 1 through 14 and 16.

Experiment no. 4 (January 22 to March 13, 1986):

Three series of tests, overlapping in time, consisting of clean, inert, simulated balsam fir branches (artificial Christmas trees; branch lengths 48 and 38 cm) attached to real balsam fir at the Mt. Tremblant site. They were subsequently detached in duplicate after exposure varying from 5 to 50 days and deposits melted and rinsed off, to evaluate cumulative deposition from winter events.

Experiment no. 5 (March 17 to April 23, 1986):

Methodology similar to that of experiment 4, involving 4 plastic branches attached to each of three balsam fir trees. At 10 day intervals one branch was sampled from each tree, by melting and rinsing off deposit.

Experiment no. 6 (April 13 to 23, 1986):

Artificial balsam fir branches (such as used in experiments 4 and 5) were exposed in triplicate in free-air flow at the top of the Atmospheric Environment Service station at the Mt. Tremblant site. Each branch had an overall length of 25 cm and consisted of 12 individual shoots. They were removed and replaced after each 'event' (clear weather, snowfall, rain, fog etc.) and deposition sampled as above. Three branches were exposed for the duration of the tests to check for cumulative deposits.

Preliminary analysis of air-mass trajectories, based on the 850 mb pressure field (as outlined by Kurtz et al., 1984) was performed for events that showed high anion content in an attempt to establish tentative connections between high deposition events and possible source regions.

Ion concentrations in deposits (or rinses of deposits) were
determined by ion chromatography. Obviously, the concentrations thus
determined do not represent the total amount of ions deposited during
the period of exposure, partly because of incomplete rinse-off
efficiency as discussed above and partly because of post-deposition
transfer and exchange processes, such as foliar absorption of gases and
washing off of deposits by precipitation. The term 'deposition', as
used below, refers to net deposition to foliage as observed in
rinse-off at the time of sampling.

3. OBSERVATIONS

Given the different methodologies and timing of the three main groups
of tests (experiments 1 - 3, experiments 4 -5 and experiment 6), the
observations obtained from each will be discussed separately.

3.1 Experiments 1 to 3

Observations from experiment 1 will be discussed in detail. Results
from experiments 2 and 3 will be mentioned only insofar as they

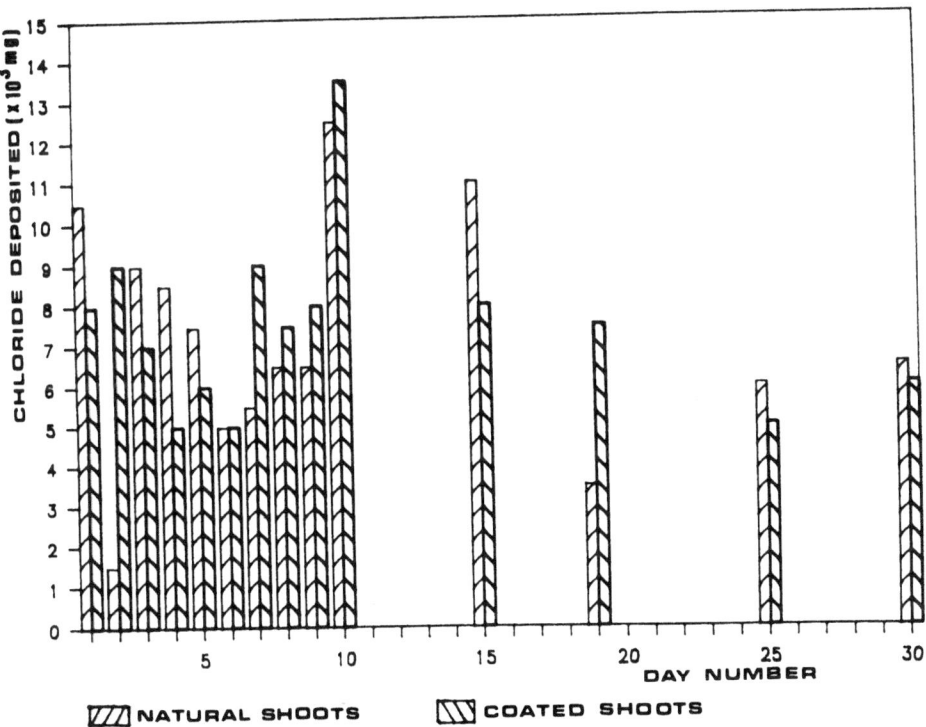

Figure 1. Experiment 1: Daily chloride deposition (days 1 to 10) and 4
to 5 depositions from days 11 to 30.

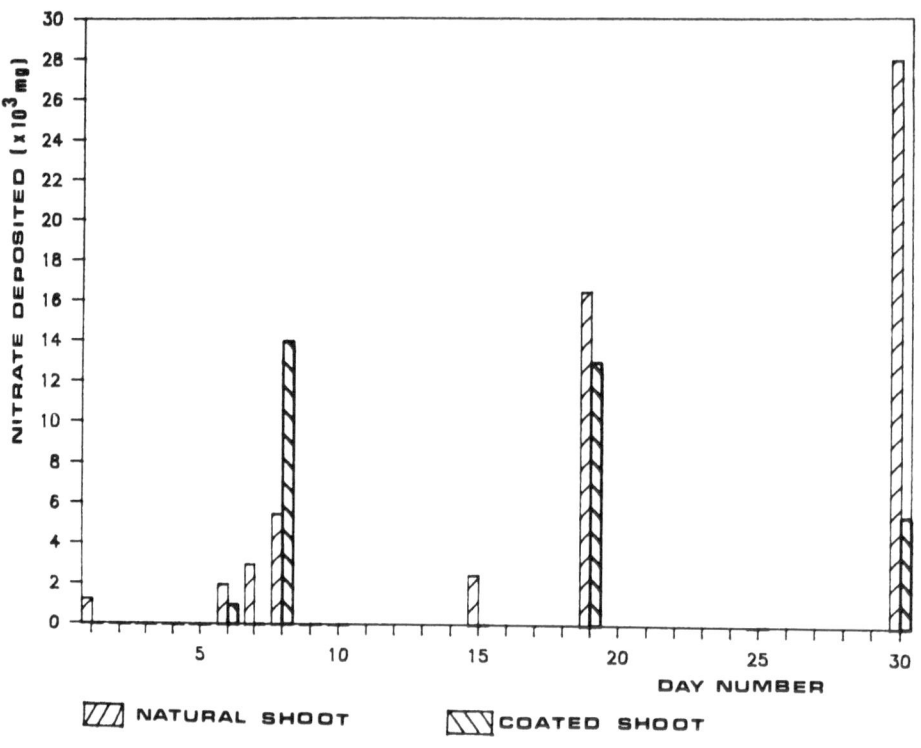

Figure 2. Experiment 1:   Daily nitrate deposition (days 1 to 10) and 4
to 5 day depositions from days 11 to 30.

confirmed or put into question specific observations of experiment 1.

Figures 1 to 3 show deposition (in μg rinsed off per shoot) from
natural and coated balsam fir for chloride, nitrate and sulphate
respectively. Values up to day 10 represent daily accumulation, values
after day 10 accumulation between the day of observation and the
previous sampling day (e.g. between days 10 and 15, etc.). Some
indication of meteorological conditions during the test period can be
obtained from Figure 5.

Deposition of chloride is more or less constant between sampling
days while that of nitrate and sulphate is highly variable (episodic),
with nitrate deposition often falling below the detection limit. Apart
from day 30, which may have shown high readings due to steam fog
occurring during the morning of days 27 to 30, the significant
depositions of nitrate and sulphate appear on the two fog days 8 and
19.

These days were examined in terms of the 850 mb airmass
trajectories. In the first case an upper level Low over southern Lake
Michigan on days 5 and 6 brought airflow from the SW, i.e. from the
direction of primary source regions in central and eastern North
America, towards the sampling site. Moderate SW flow of 10 m s^{-1}
persisted through day 7 and by day 8 winds had shifted to W-NW, with

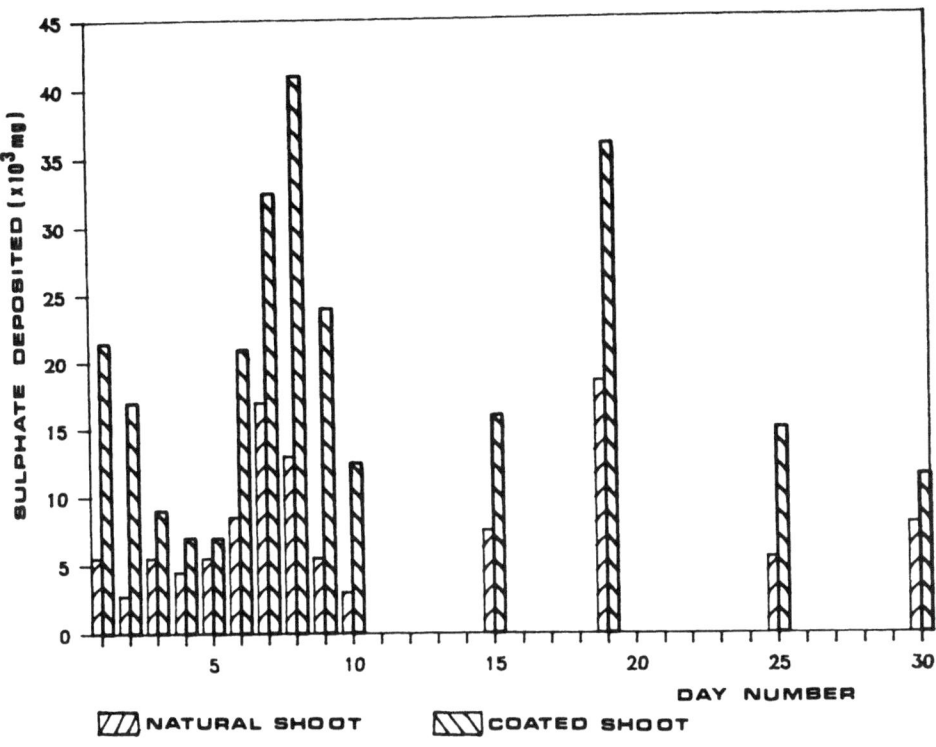

Figure 3. Experiment 1: Daily sulphate deposition (days 1 to 10) and 4 to 5 day depositions from days 11 to 30.

airflow coming from NW Ontario. The second fog day (day 19) was preceeded on days 15 to 18 by a stationary 850 mb High over Cape Hatteras, bringing a steady flow of air into the sampling region from the Ohio Valley. Upper level winds over Lake Erie and Lake Ontario were steady at 15 - 20 m s^{-1}. The High had moved NE by day 19 and,combined with a new Low over Hudson Bay, caused a W-NW flow. However, such perfunctory analysis must be used with great reservations: the closest upper air sampling station to Mt. Tremblant is Maniwaki, approximately 100 km W of the sampling site. It recorded SE flow on day 6; only examination of the larger circulation picture showed the link with the cyclonic center to the SW.

Figures 4 and 5 show cumulative deposition (as determined by the rinsing procedure) for anions on natural and inert (varathane-coated) shoots. Error bars indicate the spread of results obtained from duplicate samples, most likely attributable to differences in exposure of the randomly-selected shoots on different parts of the trees. Long-term accumulation of deposits, under the given meteorological conditions, does not exceed that reached after 5 to 10 days, largely due to wash-off from precipitation events (indicated in Figure 5). Approximate dry deposition rates estimated from daily data (e.g. a maximum of 17 to 22 μg of sulphate per shoot per day on days 1 and 2)

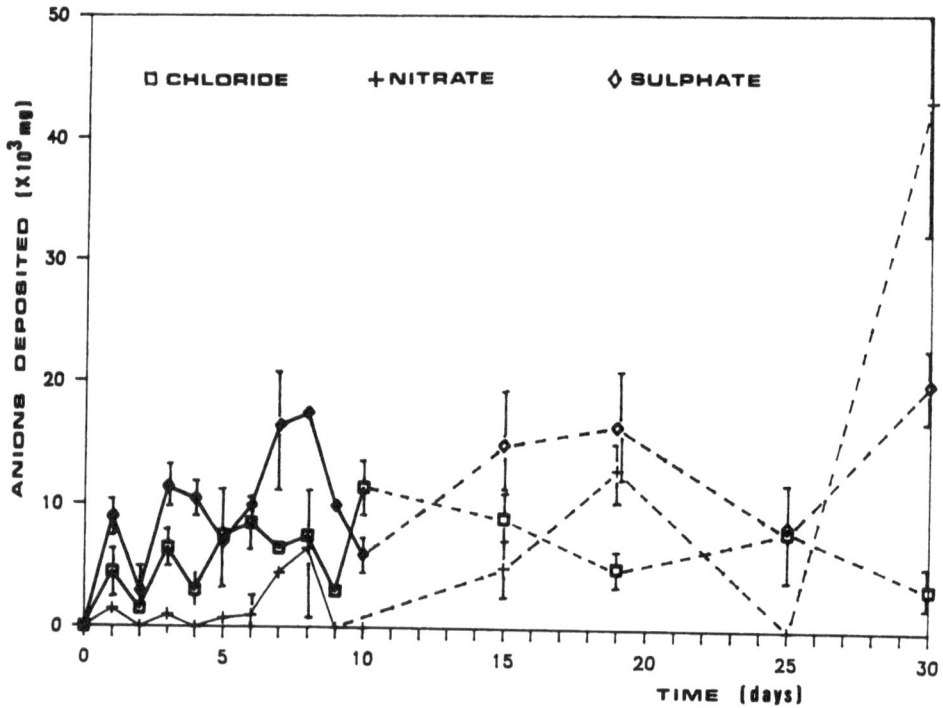

Figure 4. Experiment 1: Cumulative deposition of anions to natural shoots. Vertical bars indicate variability of individual samples.

or from cumulative deposits of up to 40 µg during the first three, precipitation-free days on coated shoots will be given in section 4.

Interestingly, deposits of sulphate on inert foliage are considerably higher than those on natural foliage (Figures 3, 4 and 5). Further studies will have to show if this could be due to differences in surface characteristics (sticking coefficients, solubility effects on SO_2 resulting from adsorbed ammonium, etc.) between natural and coated surfaces or, perhaps, by release of SO_4 from decaying biomass through cracks in the varathane coating. If real, the observation challenges the assumption that dry deposition of sulfur particles is not an important process, with the main portion of deposition in the form of SO_2, absorbed by foliage through stomata and subsequently leached to plant surfaces either as a sulphate salt or as sulfuric acid (e.g. Ulrich, 1983). Nitrate/sulphate ratios (on an equivalent weight basis) for the fog days 8 and 19 vary between 0.33 and 0.69 for the natural shoots and between 0.26 and 0.28 for the coated shoots. Given the high variability of such ratios for single events, and the fact that the data cited by Calvert et al. (1985) lie between our data for natural and coated shoots, these observations do not provide evidence for or against the validity of deposition values observed on coated shoots.

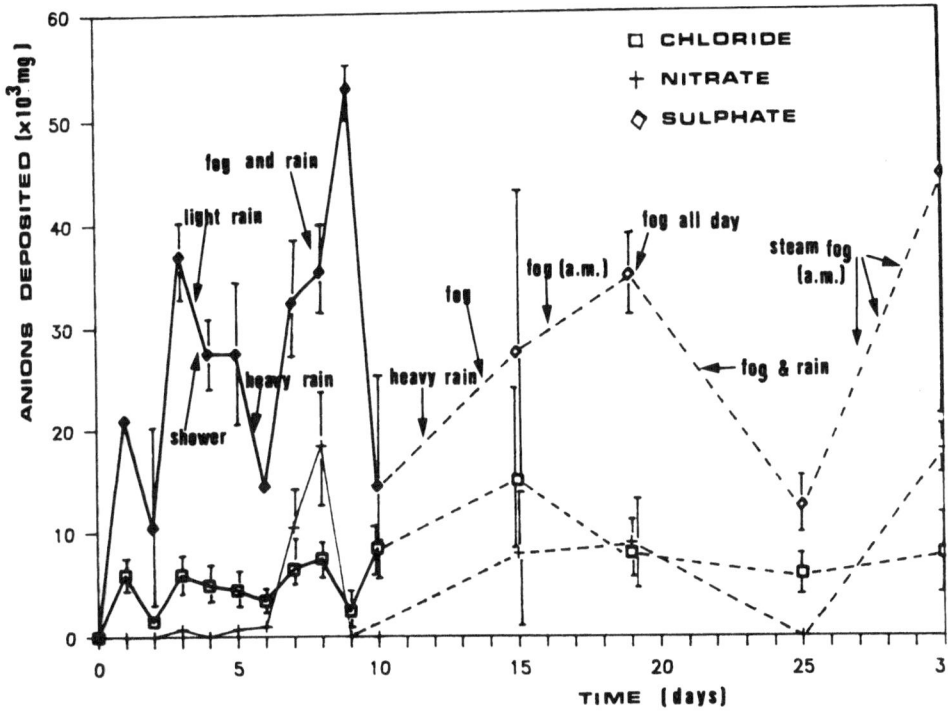

Figure 5. Experiment 1: Cumulative deposition of anions to varathane-coated shoots and indication of precipitation events.

Experiment 2 essentially confirmed the observations of experiment 1, with peak deposit of sulphate on coated shoots exceeding that of natural shoots by an average factor of > 2. Relative proportions of sulphate-, nitrate- and chloride deposition were similar to those of experiment 1, with absolute values lower by approximately 30 to 50%. Equivalent nitrate-sulphate ratios averaged 0.52 on natural shoots during the buildup of cumulative deposition.

Experiment 2A differed from the observations of experiments 1 and 2 insofar as chloride deposition to coated shoots exceeded that to natural shoots by an average factor between 1.5 and 2, perhaps indicating the presence of hydrochloric acid at the near-urban site. Sulphate deposition again exceeded that of other anions, but nitrate and chloride contents were about twice those of experiment 1 and the concurrent experiment 2 at the Mt. Tremblant mountain site. Particularly noticeable was an increase in nitrate, raising the average equivalent nitrate/sulphate ratio to between 0.6 and 1.1, possibly a function of the relative proximity of the Arboretum parking lot and Trans Canada Highway.

Experiment 3 used only coated shoots and the results of daily sampling are given in Figures 6 to 8. Nitrate deposition was again episodic relative to that of the other anions, but absolute amounts were very low compared to previous experiments. Concentrations of

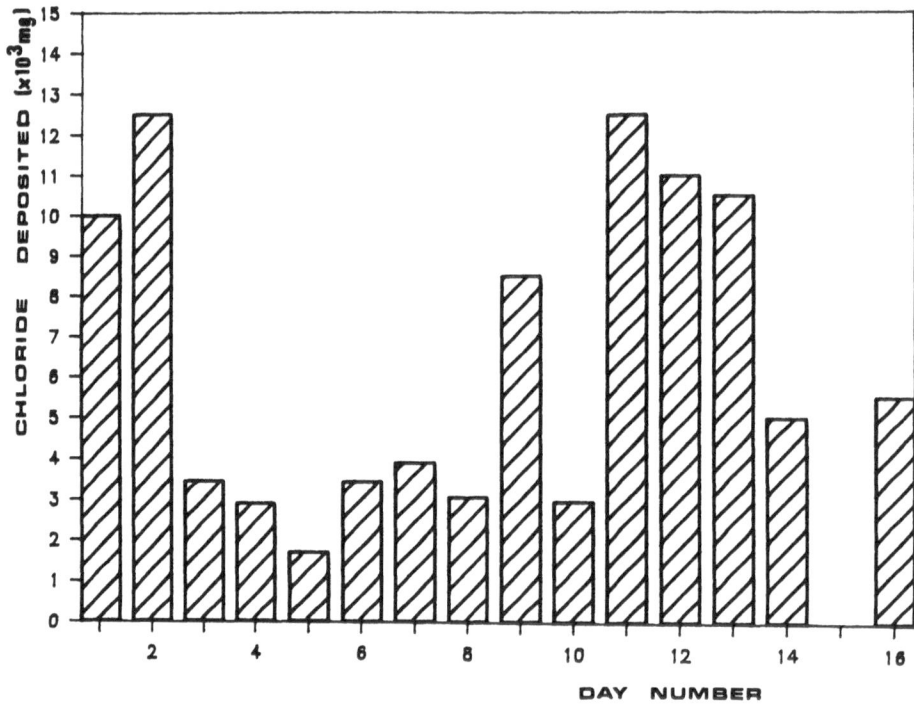

Figure 6. Experiment 3: Daily chloride deposition.

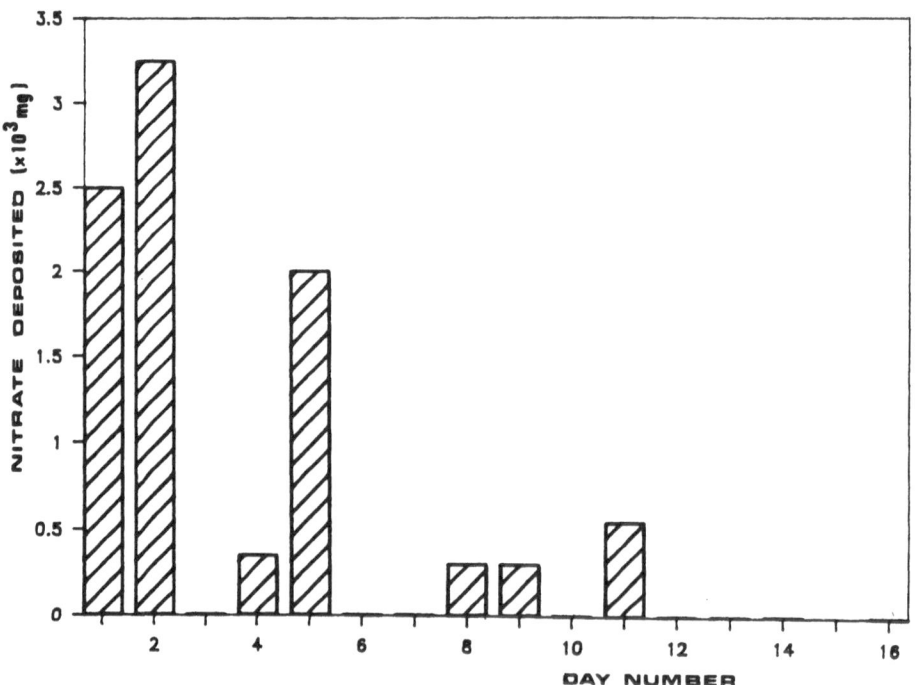

Figure 7. Experiment 3: Daily nitrate deposition.

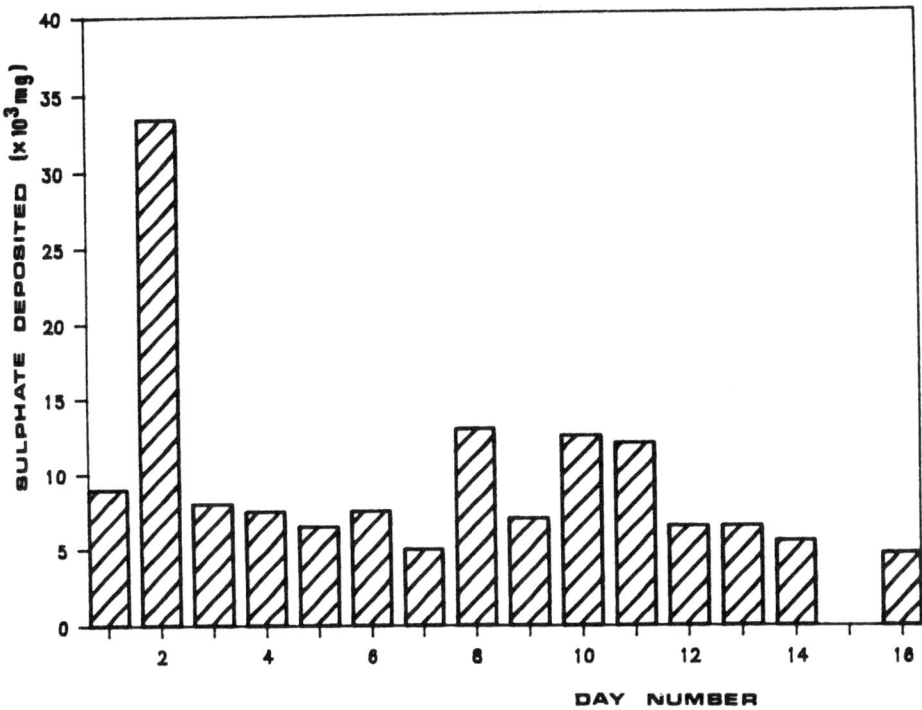

Figure 8. Experiment 3: Daily sulphate deposition.

chloride were more variable than in experiment 1. Significant
deposition for all three anions occurred on day 2. The 850 mb flow for
two days prior to day 0 had W winds with speeds of 12 to 15 m s^{-1} at
Maniwaki. On day 0 flow shifted to SW at 10 m s^{-1}, increasing to 20 m
s^{-1} on day 1. By the evening of day 2 winds had shifted to NW at 10 m
s^{-1}. The weather was clear throughout the first 6 days of this series.

Cumulative deposition, shown in Figure 9, is most evident in the
case of sulphate. The dip on day 5 is most likely due to randomness of
sampling, but light rain on day 7 and heavy thundershowers with hail on
day 8 effectively rinsed off most deposition. Drizzle occurred on day
9 – 10.

Cation analysis was only done for selected samples due to limited
equipment availability. Analyses were performed for days 4, 7 and 30
of experiment 1 and days 1 and 6 of experiment 3. Average results from
duplicate samples are summarized in Table I.

Data presented in Table I do not indicate characteristic
differences between coated and natural shoots, i.e. evidence for
leaching of cations from natural foliage. With the possible exception
of calcium, they also show no evidence for significant cumulative
buildup of cations although the values on days 4 and 7 (experiment 1)
may be artificially depressed because of overnight fog and showers
preceeding day 4 and heavy rain on day 6 which might have washed off
deposits. The data given for experiment 3, with six days of dry

weather, may be taken to show the accumulation of cations on an inert surface in the absence of leaching.

Table I. Cation concentrations (ppm) on samples from experiments 1 and 3.   Data from natural and coated (C) shoots.   (nd) indicates concentration < 0.1 ppm.

Expt.-day	Na	NH$_4$	K	Mg	Ca
1-day 4	0.80	0.22	0.68	(nd)	0.12
1-day 7	0.70	0.48	0.62	(nd)	0.20
1-day 30	0.78	0.25	1.05	0.48	4.18
1-day 4 (C)	0.58	0.10	0.35	(nd)	0.12
1-day 7 (C)	0.65	0.72	0.82	(nd)	0.35
1-day 30 (C)	1.70	0.38	1.38	0.38	3.20
3-day 1 (C)	1.40	0.45	0.55	(nd)	0.10
3-day 6 (C)	1.05	1.42	0.80	(nd)	1.12

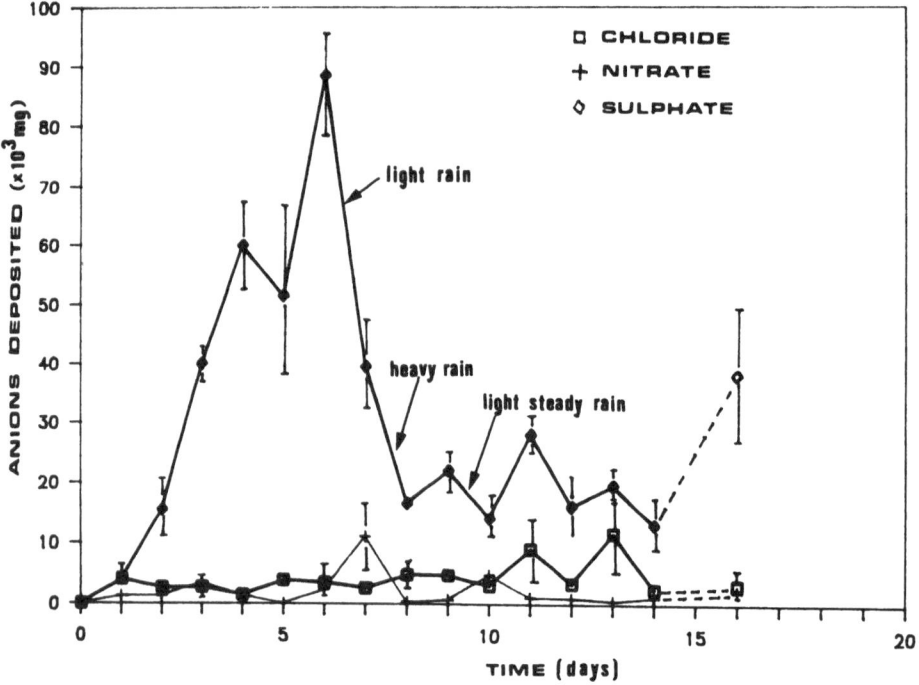

Figure 9. Experiment 3: Cumulative deposition of anions.

3.2 Experiments 4 and 5

These tests, intended to study accretion of supercooled fog droplets
and snow on foliage under winter conditions, were interrupted by
uncharacteristic warm spells with periods of melting. Deposition was
also lost through blow-off or wind abrasion and interpretation of
results was complicated by local variability of turbulence
characteristics around trees with corresponding differences in snow
deposition.

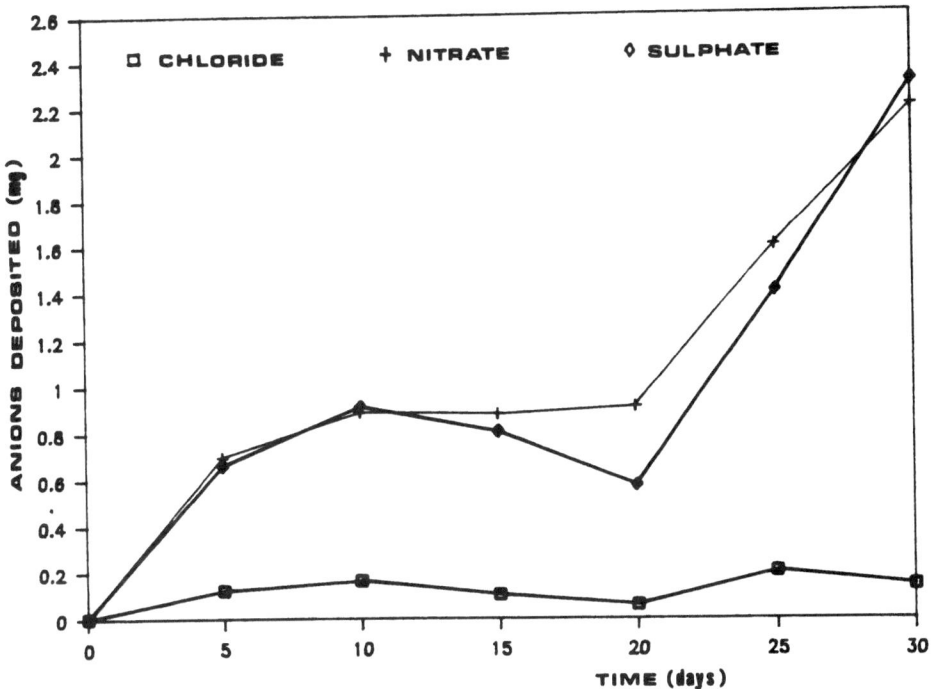

Figure 10. Experiment 4: Cumulative deposition of anions.

Figure 10 shows results of the first of the three test series
(days 1 to 30 of experiment 4) and illustrates the main observations.
The curves in Figure 10 do not represent running graphs of deposits but
a sequence of point observations in time interspersed by events that
might reduce deposition. Snow initially deposited on the test branches
had melted off by day 17. A snowfall of 7 - 10 cm occurred on day 19,
followed by steady 1 - 3 cm/day accumulation from day 21 - 25. Strong
winds on day 26 blew off much of the deposited snow but 10 - 15 cm of
wet snow on the eve of day 30 led to another significant deposition for
the observations of day 30. Cumulative deposits after day 30 dropped
again to about 1 mg (or sulphate or nitrate) by days 40 - 50 for
reasons stated above. Typical for these winter conditions is the
approximately equal deposition of sulphate and nitrate (on a per-mass
basis), with an equivalent nitrate/sulphate ratio between 0.5 and 1.2.

628

The relatively stronger nitrate deposition in winter confirms observations of seasonal variation of nitrate/sulphate ratios, as reviewed e.g. by Calvert et al. (1985).

Experiment 5, at a later time of season (March–April), showed more frequent periods of melt-off but observations very similar to those of experiment 4. Nitrate and sulphate deposition was again approximately equal in magnitude, with mean equivalent ratios 0.77, and maximum sampled depositions between 1 and 2 mg per branch.

3.3 Experiment 6

Sampling branches, exposed in free airflow during experiment 6, were sampled after each meteorological 'event', with events characterized in Table II. Three branches were exposed for the total duration of the experiment of 243 hours.

Anion concentrations in melted deposits or rinses from sampling branches for the events given in Table II are plotted in Figure 11. As observed in previous experiments, chloride deposition was far exceeded by that of nitrate or sulphate, except for the snowfall event 12, occurring after a heavy fog. For nitrate, snow events (1, 2 and 12) did not contain large amounts; dry deposition events (4 – 6) contributed significantly, however, and even higher deposition occurred during fog events (9 – 11) with magnitudes of deposition comparable to that of sulphate. The largest desposition of nitrate by a single event was due to event 7, the light rainfall after five days of dry weather.

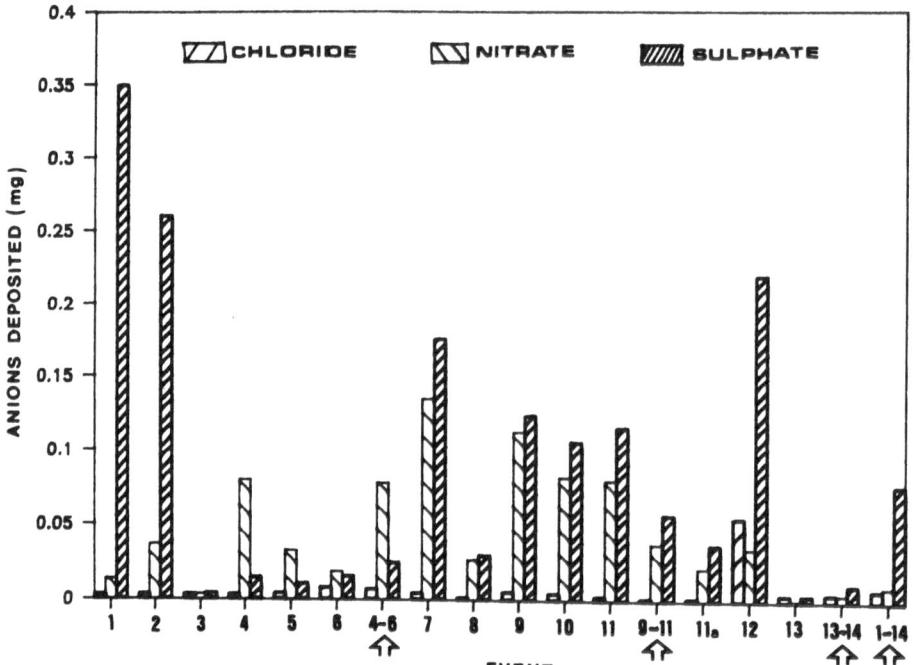

Figure 11. Experiment 6: Event deposition of anions. Cumulative samples (events 4-6, 9-11, 13-14, and overall accumulation 1-14) are indicated by arrows.

Table II. Meteorological events, with their duration and timing, during experiment 6.

Date	Event No.	Duration (hrs)	Meteorological conditions at the sampling site
April 13	1	5	Stratcumul. 10/10, Wind W, gusting to 8 m/s. Visibil. 100-200 m. Temp. -1 C. Snow flurries (fine snow - ice pellents).
13-14	2	18.5	Overnight snow sample. By morning strato-cumul. 5/10. Wind W 3 m/s. Visibil. > 50 km. Temp. 0 C. No precipitation.
April 14	3	6	No cloud. Wind W 1-1.5 m/s. Visibil. > 50 km. Temp. 7 C. No precipitation.
14-15	4	24	Cirrus < 1/10. Wind E 5-6 m/s (upper air winds from SW). Visibil. > 25 km. Temp. 8 C. No precipitation.
15-17	5	47.5	Same as above.
17-19	6	44.5	Same as above.
19-20	7	22.5	Nimbostrat. 7/10. Wind SW 10 m/s. Visibil. 10 km. Temp. 11 C. Pressure falling. Light rain 2 hrs before sample was taken.
20-21	8	22.5	Overnight sample, preceeded (Apr. 20) by conditions similar to those of event 7. Heavy rain overnight. By morning stratus 10/10. Wind SW 12-15 m/s. Visibl. 10 m. Temp. 6.5 C. Rain and fog.
April 21	9	2	Same as above but much lighter rain. Almost pure fog.
April 21	10	2	Same as above. Fog sample on foliage.
April 21	11	2.5	Slight change to stratus, stratocumul. 10/10. Wind SW 11-14 m/s. Visibil. 10 m. Temp. 4 C. Almost pure fog.
21-22	12	16.5	Overnight snow sample. Morning: cirrus 3/10. Wind W 0.5-1 m/s. Visibil. 10 km. Temp. -7 C. No precip. but snow collected on branches.
April 22	13	6	Same meteorol. conditions as above; dry deposition only on branches.
22-23	14	24	Dry deposition during previous day and overnight. By morning no cloud cover. Wind E-NE 4-5 m/s. Visibil. > 20 km. Temp. 0 C. No precipitation.

Most likely, this indicates accumulation of gaseous NO_2, HNO_3 and NH_4NO_3 effectively scavenged from the previously dry atmosphere by the nimbostratus precipitation event. Very little nitrate remained in the cumulative samples (1 - 14). In the case of sulphate, snow, the light rainfall of event 7 and fog contributed significantly to measured net deposition. Dry deposition, while still measurable, was much smaller than wet deposition. Table V (presented in section 4) contains some concurrent anion measurements obtained in precipitation from the Chemistry of High Elevation Fog (CHEF) programme's bulk samplers. It shows generally high nitrate and sulphate readings for events 7 to 11, with approximately equal concentration in terms of mass, and low chloride, in basic agreement with Figure 11. However, event 12 exhibits a noticeable difference in relative distribution of nitrate and sulphate between the CHEF data and those of experiment 6 for which no obvious explanation is as yet available.

Analysis of the 850 mb flow showed N-NW winds at 10 m s^{-1} for the three days preceeding event 1 and persisting through the snowfall events 1 and 2. SW upper air flow of 10 m s^{-1} developed during events 3 and 4 followed by advection of maritime air from the NE at 5 to 10 m s^{-1} for event 5. During event 6, upper level winds were calm and variable in direction. By event 7 a relatively deep Low over northern Ontario brought SW winds of 12 - 15 m s^{-1} into the region. These winds persisted through event 8 and by next morning (April 21), the centre of the Low had moved to a location approximately 150 km N of the sampling site. During the strong advected fog events (9 - 11) the upper level winds at Maniwaki were SW at 5 m s^{-1}, but winds of 12 - 15 m s^{-1} were recorded at the summit of Mt. Tremblant, perhaps a localized orographic effect. For event 12 upper level winds changed to N at 5 to 7 m s^{-1} and remained there for the remainder of the experiment, perhaps explaining the low deposition observed during events 13 and 14.

Quantitative interpretation of these observations must consider the loss of deposited fog and rain through dripping or blowoff, as discussed by Merriam (1973) and illustrated by fog events 9 - 11: two branches exposed throughout events 9 - 11 showed less deposit than any of the 2 hour samplings. A separate half-hour sample taken during fog event 11 (event 11a in Figure 11) showed approximately 30% of the 2 hour accumulation, indicating the expected less severe loss due to dripping during the shorter period of exposure.

Cation concentrations in selected samples of experiment 6 are listed in Table III. Where a second sample was taken, its results are given in brackets under those of the first one. Also included are two samples taken from natural balsam fir foliage during events 11 (fog-drip) and 12 (melting of snow collected on branches).

Results given in Table III show significant dry-deposition of calcium during the 5 days of dry weather (events 4 - 6), for a total of 0.098 mg on the sampling branch. This is comparable to the 0.114 mg deposited by the overnight snowfall of event 2. Dry-deposition was also high for potassium, relative to wet deposition by snow of fog. To what extent this might be due to interception of dust or soil particles is a matter of speculation at this time. The fog event 9 was characterised by elevated ammonium concentration; it also had high

Table III. Cation concenteration (ppm) on samples from experiment 6
and simultaneous samplings from natural (nat) foliage.
Absence of symbol means absence of measurement; (nd)
indicates concentration below the detection limit of 0.1
ppm.

vent	Na	NH$_4$	K	Mg	Ca
1				n.d.	2.22
2	0.30	0.10	0.10	n.d.	1.20
4				0.15	10.0
5				0.20	5.8
4-6 cumul.	1.05	0.65	0.65	0.15	8.70
(4-6 cumul.)				(0.20)	(13.0)
9	0.70	6.30	n.d.	0.20	2.60
10				0.25	2.40
9-11 cumul.	0.25	2.50	n.d.	0.20	1.20
12	0.45	0.20	0.10	n.d.	n.d.
(12)				(n.d.)	(n.d.)
11 nat.	3.50	0.15	12.20	0.90	2.45
12 nat.	0.45	n.d.	2.85	0.20	0.55

nitrate concentration (of 15.8 ppm), as shown in Figure 11, consistent
with the expectation of significant concentrations of NH$_4$NO$_3$ as
condensation nucleus (e.g. Robinson and Robbins, 1970). It is
interesting to note the low cation concentration of event 12, a
snowfall after a day-long fog event. Perhaps the fog was an efficient
scavenger for atmospheric cations.

The similarity of calcium concentrations on natural and inert
branches during fog events 9 to 11 gives no edvidence for proton
exchange with calcium, but interaction of fog with natural foliage
appears to enhance the concentrations of NA$^+$, Mg^{++} and especially K$^+$.
K$^+$ concentrations on the natural foliage are much higher than those
obtained from wet or dry deposition processes as monitored on
artificial foliage. Even the melt-off of the deposit on balsam fir
(event 12) indicates leaching through relatively elevated K$^+$
concentration.

4. DISCUSSION AND QUANTITATIVE DEPOSITION ESTIMATES

Observations from experiments 1 to 3 (summer-fall) show, not
unexpectedly, that sulphate deposits in larger quantities on coniferous
foliage than chloride and nitrate, with nitrate deposition more
sporadic than the other two and most prevalent during fog conditions.
Precipitation, even in small amounts has been shown to be effective in
rinsing off deposited anions from shoots. The high variability of
results based on the sampling of individual or duplicate shoots
underlines the need for sampling based on multiple (5 to 10) samples

randomly distributed over trees under consideration. Larger volumes of rinse water (15 to 20 ml per shoot) might also assure more complete removal of deposition from sampled foliage.

Analyses presented in section 3 strongly suggest a link between high daily deposition and upper-air flow from major pollution sources located to the W and SW of the sampling site. However, since upper air flow analysis was not done for low-depostion events a clear correlation between wind direction and daily deposition has not been established; also, the limited reliability of this cursory type of trajectory analysis must be kept in mind.

The observed excess concentration of sulphate in deposits on varathane-coated shoots compared to natural shoots, at variance with some accepted hypotheses on the role of stomatal exchange in uptake (absorption) of SO_2 for oxidation into leachable sulphate, has already been discussed in section 3. It will be followed up by more detailed studies on surface properties of natural and coated shoots so that potential implications of this finding - if real - for our understanding of sulfur deposition may be fully utilized.

Finally, the analysis of cations has not shown, at the given site and time, any indication of significant leaching except for potassium.

Quantitative interpretation of results are bound to be speculative, given inadequate data on forest geometry and foliage density. Shoot geometry for experiments 1 - 3 was estimated at 500 needles of average length 2 cm and radius 0.06 cm, giving about 160 shoots/m^2 if an average cumulative leaf area index of 6 m^2/m^3 is assumed. The latter estimate is based on sparse areas of forest at the sampling site. Observed dry deposition rates of 17 to 22 µg of SO_4 per shoot per day (days 1 and 2 of experiment 1) or cumulative deposition up to 40 µg per coated shoot during the first three precipitation-free days of experiment 1 would correspond to 3 to 4 kg/ha/yr of sulfur which is low compared to figures given by Mayer and Ulrich (1978) or Lindberg et al. (1986) for European and North American forests, with total depositions of up to 90 kg/ha/yr, or dry-depositions of the order of 40 kg/ha/yr by the latter authors. However, their figures relate to forest systems much closer to major pollution sources. Apart from the fact of increased distance from most pollution sources, our estimates may be low because of selection of sampling branches from the lower, less exposed, part of trees, underestimation of foliage density, inadequate rinse-off efficiency etc., pointing out the desirability of using these data as relative rather than absolute measures of deposition.

Even though the winter experiments 4 to 5 were inadequate in quantifying absolute amounts of cumulative wet and dry deposition for prolonged periods of time (due to unforeseen periods of melt-off), they give quantitative information about deposition from winter events. Expectedly, the exposure of branches on the tree is a dominant factor for wet deposition and the fact that measurable amounts of sulphate and nitrate remained on foliage after melting of snow and ice may be significant in the interpretation of potential hazards to trees.

The standard size and exposure of artificial branches in experiment 6 gave the most consistent results (usually within 25% for duplicate samples) although at the cost of a certain degree of realism. Considering shoot geometry it is possible, for example, to make an estimate of net sulfur deposition to foliage for the main meteorological events listed in Table II, keeping in mind that post-deposition losses of precipitation from foliage are excluded. Data based on an estimated branch area of 0.08 m^2, or a total of 80 similar small branches per square meter of ground, are given in Table IV; they have been corrected for the average 65% rinse-off efficiency of dry deposition.

Table IV. Net sulfur deposition to foliage, estimated from the sampling procedure used in experiment 6.

Event - description - duration	estim. S deposition per hectare
1 - snowfall - 5 hrs	93 g
2 - overnight snowfall - 18 hrs total	70 g
3, 4, 5, 6 - clear weather - 122 hrs	27 g
7 - rain following dry period - 2 hrs	45 g
8 - heavy rain following event 7 and fog - 23 hrs	20 g
9, 10, 11 - fog - 7 hrs	101 g
12 - overnight snow - 17 hrs	59 g
13, 14 - clear weather - 24 hrs	22 g

The figures given in Table IV are consistent with expected low dry deposition under winter conditions (Johansson et al., 1982). Total sampled deposition (extrapolated to the per-hectare basis) for the 220 hours listed above is 437 g of sulfur, equivalent to a per-year net deposition to foliage of about 17 kg/ha. Order-of-magnitude of total deposition for precipitation events may be estimated from bulk sampling (bucket collectors and modified ASRC string fog collectors) carried out simultaneously by the Chemistry of High Elevation Fog (CHEF) programme, considering geometry and boundary-layer characteristics of samplers.

Table V summarizes CHEF observations on wet precipitation according to time of observation, corresponding event number (Table II), sample volume and composition, concentrations of major anions (in ppm) and estimated total wet deposition (for sulfur only). Sampling sequences were not synchronized with the event-sampling of experiment 6 but the approximate daily sampling of the CHEF programme caught the major precipitation events of that period. Any sample including fog was recorded from the ASRC string collector, non-fog samples from the bucket samplers. Deposition estimates for the bucket samplers were made on the basis of the area covered by the 0.5 m diameter bucket opening. Estimates based on the string collector were corrected for differences in collecting area (0.25 m^2 per sampler vs. a hypothesized 6 m^2/m^2 of coniferous foliage) and in deposition velocity v_g (with v_g of foliage estimated 0.1 of v_g of the sampler because of the smaller diameter of the sampling element and higher degree of exposure).

A number of observations are of interest: estimated deposition rates for events that were expected to have been sampled adequately by

Table V. CHEF sampling data - and associated deposition estimtes -
for major anions in wet precipitation at Mt. Tremblant
during experiment 6. S = snow; IP = ice pellets; R = rain;
F = fog. Each sample is the mean of a simultaneous pair.

Date	Event no.	Sample Comp.	Volume	Concentration (in ppm) Cl^-	NO_3^-	SO_4^{--}	Est. deposition (sulfur)
Apr 12-13	1	S,IP	1467	0.20	0.26	1.59	119 g/ha
Apr 13-14	2	S	658	0.18	0.21	2.73	91 g/ha
Apr 13-14	2	F,S,IP	325	0.37	0.36	3.58	28 g/ha
Apr 19-20	7	F,R	35	0.81	8.45	9.63	8 g/ha
Apr 20-21	8	F,R	788	1.11	62.0	35.0	662 g/ha
Apr 20-21	8	R	1385	0.28	3.45	5.61	396 g/ha
Apr 21	9-11	F,R	135	0.83	15.07	18.68	61 g/ha
Apr 21-22	12	F,R,S	533	0.33	2.53	4.35	56 g/ha
Apr 21-22	12	R,S	310	0.30	2.10	3.78	60 g/ha

the artificial foliage of experiment 6, such as snow, are of similar
magnitude than estimates presented in Table IV. Differences, as
expected, are found for precipitation events (mainly rain) where net
deposition to foliage is small compared to total deposition. Combining
the information presented in Tables IV and V it is possible to make a
tentative sulfur balance, including wet and dry deposition: using mean
values for estimates in Table V where two independent samples were used
(events 8 and 12), total deposition based on Table V was 894 g/ha.
Adding to this figure the estimated 49 g from dry deposition (Table IV
- events 3 - 6 and 13 - 14), the total deposition for the period of
April 12 to 23 is estimated to have been of the order of 950 g/ha,
corresponding to an annual rate of deposition of sulfur of about 32
kg/ha. While it may be premature to assess the reliability of such
estimates, they indicate surprisingly small overall contributions from
dry deposition at the given site and time of year, a factor which
should be further investigated.

Comparison of cation ratios with literature data may help to give
some idea about realism of collected data. Schlesinger and Reiners
(1974) reported weighted averages of 0.58 for K/Na, 1.11 for Ca/Na and
0.31 for Mg/Na for the New England Mountains, far above the lower limit
of 0.036, 0.21 and 0.04 respectively for seawater. Observed cation
ratios for available samples are listed in Table VI for comparison.

The data given in Table VI agree fairly well with literature data
cited, particularly for experiment 1. Notable differences, such as the
high K/Na ratio in melt off from natural foliage can probably be
explained in terms of leaching and the high Ca/Na ratio in experiment 6
has been biased by the elevated Ca concentrations associated with the
dry deposition events 4 to 6.

Table VI. Average weighted cation ratios available for experiments 1, 3 and 6. Asterisk indicates highly variable data.

Experiment	K/Na	Ca/Na	Mg/Na
expt. 1 - natural shoots	0.58	1.14	0.58
expt. 2 - coated shoots	0.52	0.73	0.21
expt. 3 - (all shoots coated)	0.30	0.29	not avail.
expt. 6 - where data available	0.28	3.01*)	0.39*)
expt. 6 - drip/melt off from natural foliage	2.90	0.56	0.33

5. CONCLUSION

The difficulty encountered with analysis of observations presented in the previous secions stems largely from the fact that any study of depositon processes under realistic field conditions is affected by a large number of site-specific (soil type, vegetation) and time-specific (meteorological) parameters. However, it is precisely because extrapolation of observations made in different ecosystems, at different geographic locations and elevations, are difficult to extrapolate to the mountaneous areas under consideration by the CHEF programme, and because of the seasonal variability of any such observations, that experiments such as these appear desirable and (within limits) necessary.

Of particular interest appears an extension of event-type sampling (experiment 6), at different times of season, in conjunction with the routine sampling of precipitation (snow, rain and fog) carried out at the CHEF site. The general agreement between the tentative deposition estimtes made from bulk sampling with those based on foliage rinse-off, for those deposition processes that are effectively accumulated on the foliage, is encouraging and more analysis about the degree of reliability of deposition estimates to foliage, based on the routine sampling of the CHEF programme, will follow. Quantitative estimtes on dry vs. wet deposition, as presented in the preceeding section, could then be refined and repeated for different times of season, to provide a year-round picture of the contribution towards total deposition at the given site from the various deposition pathways.

ACKNOWLEDGEMENTS

The support of this study by the Atmospheric Environment Service of Environment Canada is gratefully acknowledged. The authors would also like to thank Dr G. Tomlinson (Domtar Inc.) for many helpful discussions and Mr K. Parker for his unstinting help with field work.

636

REFERENCES

Barrie, L.A. and Schemenauer, R.S. 1986. 'Pollution Wet Deposition
    Mechanisms in Precipitation and Fog Water'. Accepted: Water Air
    and Soil Poll.
Calvert, J., Lazrus, A., Kok, G.L., Keikes, B.G., Walega, J.G., Lind,
    J. and Cantrell, C.A. 1985. 'Chemical Mechanisms of Acid
    Generation in the Troposphere'. Nature 317 , 27-35.
Fairfax, J.A.W. and Lepp, N.W. 1975. 'Effect of Simulated Acid Rain on
    Cation Loss from Leaves'. Nature 255 , 324-325.
Falconer, R.E. and Falconer, P.D. 1980. 'Determination of Cloud Water
    Acidity at a Mountain Observatory in the Adirondack Mountains of
    New York State'. J. Geophys. Res. 85 , 7465-7470.
Fowler, D. and Cape, J.N. 1983. 'Dry Deposition of SO_2 onto a Scots
    Pine Forest'. In: Precipitation Scavenging, Dry Deposition, and
    Resuspension, Volume I: Precipitation Scavenging. Elsevier, New
    York, 763-773.
Galloway, J.N. and Whelpdale, D.M. 1980. 'An Atmospheric Sulfur Budget
    for Eastern North America'. Environment 14 , 409-417.
Grennfelt, P., Bengtson, C. and Skarby, L. 1983. 'Dry Deposition of
    Nitrogen Dioxide to Scots Pine Needles'. In: Precipitation
    Scavenging, Dry Deposition, and Resuspension, Volume I:
    Precipitation Scavenging. Elsevier, New York, 753-762.
Hoffman, W.A. Jr., Lindberg, S.E. and Turner, R.R. 1980.
    'Precipitation Acidity: The Role of the Forest Canopy in Acid
    Exchange'. J. Environ. Qual. 9 , 95-100.
Houghton, H. 1955. 'On the Chemical Composition of Fog and Cloud
    Water'. J. Meteorol. 12 , 355-357.
Jacob, D.J., Flagan, R.C., Waldman, J.M. and Hoffman, M.R. 1983.
    'Design and Calibration of a Rotating Arm Collector for Ambient
    Fog Sampling'. In: Precipitation Scavenging, Dry Deposition, and
    Resuspension, Volume I: Precipitation Scavenging. Elsevier, New
    York, 125-136.
Johansson, C., Richter, A. and Granat, L. 1983. 'Dry Deposition on
    Coniferous Forest of SO_2 at ppb Levels'. In: Precipitation
    Scavenging, Dry Deposition, and Resuspension, Volume II: Dry
    Deposition and Resuspension. Elsevier, New York, 775-783.
Kurtz, J., Tang, A.J.S., Kirk, R.W. and Chan, W.H. 1984. 'Analysis of
    an Acidic Deposition Episode at Dorset, Ontario'. Atmosph.
    Environment 18 , 387-394.
Lindberg, S.E., Lovett, G.M., Richter, D.D. and Johnson, D.W. 1986.
    'Atmospheric Deposition and Canopy Interactions of Major Ions in a
    Forest'. Science 231 , 141-145.
Mayer, R. and Ulrich, B. 1978. 'Input of Atmospheric Sulfur by Dry and
    Wet Deposition to Two Central European Forest Ecosystems'.
    Atmosph. Environment 12 , 375-377.
Merriam, R.A. 1973. 'Fog Drip from Artificial Leaves in a Fog Wind
    Tunnel'. Water Res. Res. 9 (6), 1591-1598.
Mrose, H. 1966. 'Measurements of pH, and Chemical Analyses of Rain-
    Snow-, and Fog-Water'. Tellus 18 , 266-270.

Robinson, E. and Robbins, R.C. 1970. 'Gaseous Nitrogen Compound Pollutants from Urban and Natural Sources'. A.P.C.A. Journal 20 (5), 643-659.

Schemenauer, R.S. 1986. 'Acidic Deposition to Forests: the 1985 Chemistry of High Elevation Fog (CHEF) Project'. Accepted: Atmosphere-Ocean.

Schemenauer, R.S., Schuepp, P.H. and Kermasha, S. 1986. 'Measurements of the Properties of High Elevation Fog in Quebec, Canada'. Proceedings Nato Advanced Research Workshop on Acid Deposition to High Elevation Sites, Edinburgh, 8-12 September, 1986.

Scherbatskoy, T. and Klein, R.M. 1983. 'Response of Spruce and Birch Foliage to Leaching by Acidic Mists'. J. Environ. Qual. 12 (2), 189-195.

Schlesinger, W.H. and Reiners, W.A. 1974. 'Deposition of Water and Cations on Artificial Foliar Collectors in Fir Krummholz of New England Mountains'. Ecology 55 , 378-386.

Tomlinson, G.H. 1984. 'Calcium and Magnesium Deficiencies in Conifers and their effect on Foliage'. Project No. 84-8031-01, Domtar Research Centre, Senneville, Quebec H9X 3L7.

Ulrich, B., Mayer, R. and Khanna, P.K. 1980. 'Chemical Changes due to Acid Precipitation in a Loss-Derived Soil in Central Europe'. Soil Science 130 , 193-199.

Vogelmann, H.W., Siccama, T., Leedy, D. and Ovitt, D.C. 1968. 'Precipitation from Fog Moisture in the Green Mountains of Vermont'. Ecology 49 (6), 1205-1207.

Waldman, J.M. Munger, J.W., Jacob, D.J. and Hoffmann, M.R. 1985. 'Chemical Characterization of Stratus Cloudwater and its Role as a Vector for Pollutant Deposition in a Los Angeles Pine Forest'. Tellus 37B , 91-108.

Wood, T. and Bormann, F.H. 1975. 'Increases in Foliar Leaching Caused by Acidification of an Artificial Mist'. Ambio 4 (1), 169-171.

CHEMICAL INTERACTIONS BETWEEN CLOUD DROPLETS AND TREES

J.N. Cape
Institute of Terrestrial Ecology
Bush Estate
Penicuik
Midlothian EH26 0QB
U.K.

ABSTRACT. The chemical composition of both liquid droplet and leaf surface determines the behaviour of a cloud droplet on interception by vegetation. In some circumstances, particularly in wind-driven cloud, droplets may not adhere to foliage. Once on the leaf surface, the contact angle between the droplet and surface will determine the amount of water held by the canopy. Factors influencing this contact angle are discussed in the light of field measurements from Scots pine. Chemical modification of canopy water in terms of dissolution and ion exchange is briefly considered, including oxidation of SO_2. Evaporation of water from the canopy leads to changes in solute concentrations, which influence fluxes of volatile solutes and transfer of involatile solutes between a leaf and surface water. In all these processes the physical and chemical nature of the leaf surface is of the greatest importance, and is expected to depend on both the environment in which the tree grows and the tree species.

The chemical interaction of a cloud or rain droplet with a leaf surface begins as soon as it is intercepted by the leaf. Transfer of small droplets typical of cloud (5–50 μm diameter) is effected by the turbulence of the atmosphere, and gravitational settling is not significant when compared with potential deposition rates to forests at windspeeds typical of high elevations. At high windspeeds, mean droplet velocity (and hence kinetic energy) is large, and this must be dissipated on collision with a leaf. If the leaf is already covered by a layer of water, this energy may be expended as heat by elastic deformation of the water layer. However, not all leaf surfaces are wetted efficiently, and the droplet may have sufficient energy to be reflected, or "bounce" off the leaf, while losing some of its momentum. The factors affecting droplet reflection have been studied in relation to pesticide formulation and application, where droplets (usually much larger, 100–250 μm) fall under gravity at terminal velocities in air. However, the formalism developed (Hartley & Brunskill, 1958) may be used also for smaller droplets in wind-driven cloud.

639

M. H. Unsworth and D. Fowler (eds.), Acid Deposition at High Elevation Sites, 639–649.
© *1988 by Kluwer Academic Publishers.*

640

A droplet radius r has a surface energy (Eo) of $4\pi r^2 \cdot \gamma_{AL}$ where γ_{AL} is the interfacial tension of the liquid (in this case water) in air. When the drop is at equilibrium on a surface, the surface energy is given by:

$$E_s = A_1 \gamma_{AL} + A_2 \cdot \gamma_{LS} - A_2 \cdot \gamma_{AS} \tag{1}$$

where A_1 is the area of contact between drop and air, and A_2 is the area of contact between drop and surface.

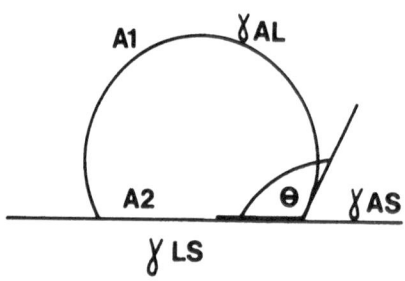

Figure 1. Schematic representation of a water droplet on a leaf surface
$\gamma_{AS} = \gamma_{LS} + \gamma_{AL} \cdot \cos\theta$ contact angle

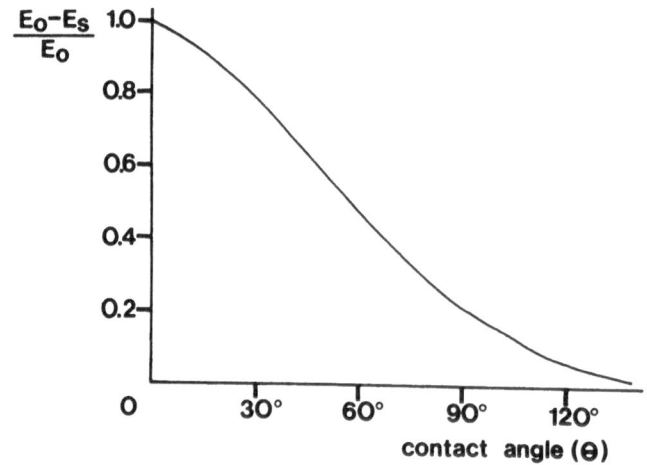

Figure 2. Ratio $\dfrac{E_0 - E_s}{E_0}$ is the minimum energy barrier between attached and free drop, relative to E_0, and is a function only of θ.

The air-surface interfacial tension, γ_{AS}, is given by:

$$\gamma_{AS} = \gamma_{LS} + \gamma_{AL} \cdot \cos \theta \qquad (2)$$

where θ is the contact angle of the droplet on the surface (Fig. 1).
The trigonometry of the system permits the calculation of Es as:

$$E_s = \pi r^2 \cdot \gamma_{AL} \cdot (4/3)^{2/3} \cdot [2(1-\cos\theta)-\sin^2\theta\cos\theta]/[1-\cos\theta+(\cos^3\theta-1)/3]^{2/3} \qquad (3)$$

A droplet will only reflect from a surface if its kinetic energy is sufficient to exceed the difference between the equilibrium energy of the attached droplet and the surface energy of the free drop. The ratio (Eo-Es)/Eo represents the minimum energy barrier relative to the surface energy of the free drop, and is a function of θ alone (independent of r) (Fig. 2). A droplet will only be reflected when its kinetic energy (Ek/Eo) exceeds this ratio. A water droplet diameter 100 μm falling at terminal velocity in air (0.25 m s^{-1}) has Ek/Eo $\cong$ 0.014 and will not be reflected. In contrast, a much larger rain drop will have a kinetic energy much greater than the surface energy and will tend to fragment on impact.

Wind-driven cloud droplets will have much greater kinetic energies, and the critical velocities are shown in Figure 3 for a number of droplet sizes as a function of contact angle. For contact

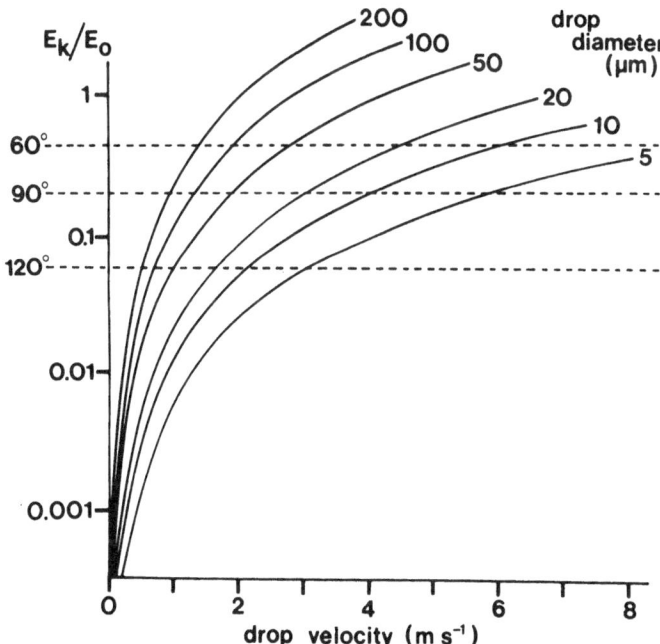

Figure 3. For 'bounce' to occur k.e. (E$_k$) > E$_0$-E$_s$
and k.e. (E$_k$) $\leqslant$ E$_0$.

angles of 90°, a 10 μm droplet would need a velocity greater than 4 m s^{-1} for its kinetic energy to exceed the energy barrier for reflection. In practice, the contact angle of a water drop on a leaf surface may exhibit hysteresis, with a smaller value as the drop recedes from the surface. This suggests that the critical velocity for droplet reflection given in Figure 3 is a lower limit if based on the equilibrium contact angle.

Equation (2) above shows the dependence of the contact angle, θ, on the interfacial tensions γ_{AL}, γ_{LS} and γ_{AS}. The surface tension of water droplets in air (γ_{AL}) depends upon the chemical composition of the droplet, particularly if surfactants are present. The ionic composition of cloud droplets is unlikely to alter the surface tension significantly at typical concentrations ($<10^{-2}$M), but there is evidence that droplets may contain surface active agents which could significantly reduce the contact angle by reducing the surface tension of droplets in air (γ_{AL}). For example, salt-spray damage to coastal trees has been attributed to the incorporation of detergents from sewage effluent in sea-spray droplets, which enhanced the ability of such drops to wet foliage and cause salt damage (Moodie et al., 1986). The sea surface is also the source of natural organic materials, but even in continental areas rain and cloud may contain large quantities of organic material, particularly from combustion sources. The widespread occurrence of black particles in cloud has been reported by several authors in this volume, and it is likely that surface active materials are also present. Direct measurements of the surface tension of cloud droplets would clearly be difficult as the surface properties of bulk (i.e. collected) cloudwater would be different from that of individual drops. Measurements on rainwater have shown evidence of organic 'coatings' on cloud drops from a number of studies (see e.g. Weschler & Graedel, 1982). If such surface active materials are long-chain organic acids or bases it is likely that the surface tension will depend to some extent on droplet pH. The behaviour of droplets with a small surface tension is relatively well understood, as agricultural sprays are often formulated to include wetting agents (Ford et al., 1965).

The interfacial tension between water and the leaf surface (γ_{LS}) depends on both the chemical and physical characteristics of the surface. Leaf epicuticular waxes may be hydrophobic or hydrophilic (Holloway, 1969), and contact angles are also dependent on surface roughness. This may be on a microscopic scale (1 μm, wax structure) or on a macroscopic scale (1 mm, leaf structure) (Warburton, 1963). Chemical and physical properties may both be altered by abrasion, or by the presence of particles deposited on the surface from the air, cloud or rain. Indeed, it has been suggested (Leece, 1976) that insoluble particles should be added to spray formulations to enhance penetration of surface waxes. The presence of large surface concentrations of particles on needles of Scots pine trees growing in polluted air has been associated with measured decreases in contact angles (Cape, 1983). Even in clean air, the surface properties of leaves will change with the age of the leaf and the stage of development of the plant. The production and form of epicuticular wax is known to be strongly

dependent upon several environmental factors, including air pollution
(e.g. Hull et al., 1975; Bystrom et al., 1968).

For coniferous trees, it is generally assumed that once needle
expansion is complete, no further wax formation takes place, and that
subsequent changes in chemistry or form result from external
influences. This is not the case for other plants, where rapid
exchange of internal and external lipids has been shown to occur
(Cassagne & Lessire, 1975). Examples of typical contact angles for
Scots pine trees are shown in Figure 4. These data cover a wide range
of environments and genotypes and show the general decrease with needle
age. They also illustrate the change in wettability from young needles
($\theta > 90°$) to older needles ($\theta < 90°$), which affects the water-holding
capacity of the forest canopy. These measurements were made on the
middle part of the abaxial surface of pine needles, but there is
considerable variation with position on the needles (Figure 5). It has
been suggested that the small contact angles in the sheath of pine
needles is an adaptation which allows pine trees to use water by
interception of cloud droplets (Leyton & Juniper, 1963). As indicated

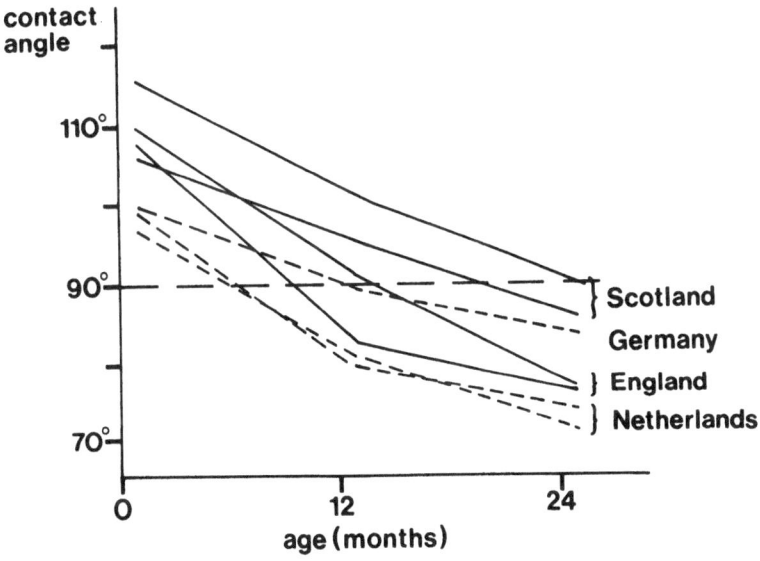

Figure 4. Variation of contact angle with pine needle age (1μl drop
deionised water: means of 25-80 measurements).

above, the value of the contact angle controls both leaf wettability
and canopy capacity for water with increased retention of water
droplets by rough surfaces with larger contact angles (Merrall, 1981).
In practice this will affect processes such as ion exchange, leaching
and dissolution of surface material, by changing the average contact
time and area of water (from rain or cloud) with a leaf surface. It

644

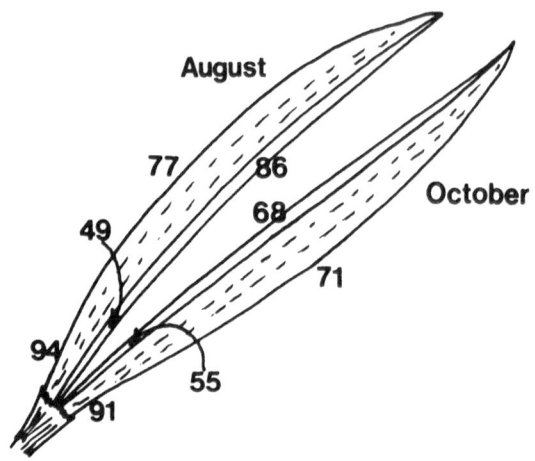

Figure 5. Variation of contact angle with position on leaf (Leyton and
Juniper, 1963).

will also determine whether surface water exists as discrete droplets
or as a draining surface film; evaporation rates of surface water will
be different in each case, as will the possibility of droplet
resuspension by wind (either directly or by mechanical agitation of
branches).

Before considering the chemical changes which occur as water
evaporates from leaf surfaces, it will be useful to review briefly the
processes which operate when leaf surfaces are wetted. In general,
exchange of material between a layer of water and a leaf surface must
occur through the cuticle, rather than stomata, as aqueous solutions
will not penetrate stomatal apertures unless contact angles are very
small (Schönherr & Bukovac, 1972).

The exchange of cations between surface water and leaf surfaces is
well established, although it is difficult to separate the exchange
processes from cation leaching. Leaves are sources of K^+, Mg^{2+}, Ca^{2+}
and Mn^{2+}, and may be a net sink of H^+ in rain, although the acidity of
throughfall may be greater than incident rainfall. Field measurements,
however, do not easily distinguish the mechanisms involved. In
addition to net ion exchange and leaching, water may dissolve surface
deposits of soluble material. These surface exchange processes depend
on the nutrient status of the plant, and its water status, and recent
experiments suggest that this largely passive transport of cations
(Mecklenburg et al., 1966) may be moderated by the anions present in
precipitation (Skiba et al., 1986; McFarlane & Berry, 1974). It is not
only cations which may enter surface water; a wide range of organic
substances may also be leached, from carbohydrates to proteins (Tukey,
1970; Scherbatskoy & Klein, 1983), and many of these may be surface

active, thereby affecting water behaviour on the leaf surface. They may also serve to buffer the acidity of surface water. Leaching and ion exchange may not be restricted to leaf surfaces; measurements of throughfall under deciduous trees show marked changes in chemical composition (relative to rainfall) during leafless periods (Madgwick & Ovington, 1959). Surfactants may also have an active role in ion exchange by increasing the effective permeability of the leaf surface (Cantliffe & Wilcox, 1972).

The apparent ubiquity of black particles associated with cloudwater and leaf surfaces suggests that leaf surfaces may act as very efficient sites for oxidation reactions, for example, for SO_2 to sulphate, given large concentrations of carbon (Rogowski et al., 1982) and the presence of ions such as Mn^{2+} from leaching. In such circumstances, oxidation rates of SO_2 by molecular oxygen (which is normally a slow process in cloud or rain) may approach those of the more usual oxidation reactions with H_2O_2 or O_3. This may be particularly important if in situ buffering by leaf leachates and/or ion exchange prevents the increase in acidity which would otherwise inhibit solution of SO_2.

If we now consider the dynamics of the interaction of cloud or rain with leaf surfaces, three phases can be identified; initial wetting, 'steady state' (with deposition, evaporation and removal processes in balance) and final drying. Initial wettings will dissolve surface deposited materials remaining from evaporation of earlier precipitation, accumulated as dry deposition or through guttation. Large ionic concentrations will arise if the rate of wetting (amount of water) is small. In this context, it is of interest to note that many experiments which ostensibly are designed to show effects of low pH solutions on plants may in practice be showing effects of ionic strength, as 'controls' use distilled water or simulated rain with much smaller ionic strength. As cations other than H^+ are known to have an effect on ion leaching (Tukey, 1970), the interpretation of such experiments, while likely to be essentially correct, may be confounded by a need to separate effects of H^+ from those of high ionic strength. Experiments by Adams & Hutchinson (1984) suggest that increased ionic strength caused by leaching of cations to neutralise acidity was correlated with increased resistance to acid damage and would point to H^+ as the damaging factor, but this may not be generally true. The 'steady state' condition has been considered (for involatile solutes) by Unsworth (1984), who showed that under conditions where the rate of evaporation exceeded the rate of deposition, very large concentrations of material in water on leaf surfaces could be expected. In practice, leaf exchange processes (uptake/leaching) will alter ion concentrations, depending upon the relative time scales of exchange and water movement through the canopy. Recent experiments using artificially applied salts to leaf surfaces, however, suggest that the transition from initial wetting (and removal of surface deposits) to a 'steady state' may take a considerable time, and may be more complex than the above suggests (McCune & Lauver, 1986). The potential for continued exchange over a long period was illustrated by field measurements during rain which showed that neutralisation of rainfall acidity still occurred after 15 h continuous rain (Adams & Hutchinson, 1984).

On final drying, large concentrations of salts (and other material) will be produced. This process has been studied for agricultural crops in relation to residues left after pesticidal spraying, and formulations are designed to minimise the formation of 'islands' of wetness and to encourage uniform deposition of active components (Ford et al., 1965). The rate of evaporation is strongly dependent on the drop size distribution (Brain & Butler, 1985) which in turn will depend on contact angle and surface tension. In making the contact angle measurements described above (Cape, 1983), it was observed that distilled water droplets evaporating from a leaf surface tended to maintain a fixed area of contact rather than a spherical shape. If this occurs in the field, any dissolved material would be concentrated over a larger area than if the contact angle had been maintained during evaporation.

For volatile solutes, however, evaporation of water on leaf surfaces will lead to displacement of gas-solution equilibria, and release of dissolved gases from solution. If, in addition, gas solubility depends on solution acidity (as for SO_2) then the loss to the gas phase may exceed the rate of water loss. As a numerical example, consider a water droplet with strong acidity equivalent to 31.6 μM H^+ (pH 4.5) in equilibrium with SO_2 gas at 30 μg m^{-3} (10 ppbv) at 10°C. The resultant aqueous concentrations are:

$$[SO_2] \text{ aq} = 2.2 \times 10^{-8}M$$
$$[HSO_3^-] = 9.93 \times 10^{-6}M$$
$$[H^+] = 41.5 \times 10^{-6}M \text{ (pH 4.38)}$$

If the canopy storage of water is equivalent to a precipitation depth of 2 mm (2 $1/m^2$) then the net release of SO_2 to the gas phase may be calculated:

(i)   Water evaporation rate 0.5 g $m^{-2}s^{-1}$ typical of mid-day conditions.
        The SO_2 release rate is 5 $\times$ 10^{-7} g $m^{-2}s^{-1}$, or for comparison with atmospheric removal rates equivalent (but opposite in direction) to a deposition velocity of 18 mm/s. The loss of S^{IV} present in the evaporated water amounts to only 60% of this figure, the remainder being the result of increasing solution acidity.

(ii)   Evaporation rate 0.05 g $m^{-2}s^{-1}$ (night time) gives an SO_2 release rate of 5.4 $\times$ 10^{-8} g $m^{-2}s^{-1}$ equivalent to a (negative) deposition velocity of 1.9 mm s^{-1}.

Evidence that fluxes of this magnitude can occur is shown in the results of Fowler and Cape (1983), where an eddy correlation technique was used to measure SO_2 fluxes above a Scots pine canopy. After a period of rain during the night a consistent upward flux of SO_2 was observed over one hour as water evaporated from the canopy. In the absence of oxidation reactions or uptake, wetted forest canopies may act as short-term stores of volatile gases during rainfall or cloud interception.

In summary, interactions of leaf surfaces with rain or cloud droplets involve a number of well-known processes, but there are four areas where quantitative knowledge is lacking:

(i)   The effect of droplet surface tension on the retention of impacting water droplets, and extent of subsequent wetting of leaf surface.

(ii) The importance of leaf surface heterogeneity (physical and chemical) in relation to wetting.

(iii) The rates of chemical reaction, particularly oxidation of SO_2 and other gases, as a heterogeneous process on leaf surfaces in the presence of particulate carbon and/or metal ion catalysts leached from the leaf or surface deposits.

(iv) The magnitude and importance of the fluxes of volatile gases in the absence of absorption or reaction when surface water evaporates from a canopy.

In addition, the possible effects of increased rain or cloud acidity on leaf surfaces should be separated from the effects of increased ionic strength by the design of experiments with appropriate control treatments.

REFERENCES

Adams, C.M. and Hutchinson, T.C. 1984. A comparison of the ability of leaf surfaces of three species to neutralise acidic rain drops. New Phytologist **97** , 463-478.

Brain, P. and Butler, D.R. 1985. A model of drop size distribution for a system with evaporation. Plant, Cell & Environment **8** , 247-252.

Bystrom, B.G., Glater, R.B., Scott, F.M. and Bowler, E.S.C. 1968. Leaf surface of Beta vulgaris - electron microscope study. Botanical Gazette **129** , 133-138.

Cantliffe, D.J. and Wilcox, G.E. 1972. Effect of Surfactant on Ion Penetration through Leaf Wax and a Wax Model. J. Amer. Soc. Hort. Sci. **97** , 360-363.

Cape, J.N. 1983. Contact angles of water droplets on needles of Scots pine (Pinus sylvestris) growing in polluted atmospheres. New Phytologist **93** , 293-299.

Cassagne, C. and Lessire, R. 1975. Studies on alkane biosynthesis in epidermis of Allium porrum L. leaves, IV. Wax movement into and out of epidermal cells. Plant Science Letters **5** , 261-268.

Ford, R.E., Furmidge, C.G.L. and Montagne, J.T.W. 1965. The role of surface-active agents in the performance of foliar sprays. In: Surface Activity and the Microbial Cell, pp. 214-243, SCI Monograph No. 19, Society of Chemical Industry, London.

648

Fowler, D. and Cape, J.N. 1983. Dry deposition of SO_2 on to a Scots pine forest. In: Precipitation Scavenging, Dry Deposition, and Resuspension. H.R. Pruppacher, R.G. Sermonin & W.G.N. Slinn (Eds.), pp. 763-774. Elsevier, Amsterdam.

Hartley, G.S. and Brunskill, R.T. 1958. Reflection of water drops from surfaces. In: Surface Phenomena in Chemistry and Biology. J.F. Danielli, K.G.A. Pankhurst and A.C. Riddiford (eds.) pp. 214-223 Pergamon Press, London.

Holloway, P.J. 1969. Effects of superficial wax on leaf wettability. Annals of Applied Biology 63 , 145-153.

Hull, H.M., Morton, H.L. and Wharrie, J.R. 1975. Environmental influences on cuticle development and resultant foliar penetration. Botanical Review 41 , 421-452.

Leece, D.R. 1976. Composition and ultrastructure of leaf cuticles from fruit trees, relative to differential foliar absorption. Aust. J. Plant Physiol. 3 , 833-847.

Leyton, L. and Juniper, B.E. 1963. Cuticle structure and water relations of pine needles. Nature 198 , 770-771.

McCune, D.C. and Lauver, T.L. 1986. Experimental modelling of the interaction of wet and dry deposition on conifers. In: Aerosols, S.D. Ure, T. Schneider, L.D. Grant and P.J. Verkerk (eds.) pp. 1715-1722, Lewis Publishers, Chelsea, Michigan.

McFarlane, J.C. and Berry, W.L. 1974. Cation penetration through isolated leaf cuticles. Plant Physiology 53 , 723-727.

Madgwick, H.A.I. and Ovington, J.D. 1959. The chemical composition of precipitation in adjacent forest and open plots. Forestry 32 , 14-22.

Mecklenburg, R.A., Tukey, H.B. and Morgan, J.V. 1966. A mechanism for the leaching of calcium from foliage. Plant Physiology 41 , 610-613.

Merrall, G.T. 1981. Physical factors that influence the behaviour of chemicals on leaf surfaces. In: Microbial Ecology of the Phylloplane. J.P. Blakeman (ed.) pp 265-281. Academic Press, London.

Moodie, E.G., Stewart, R.T. and Bowen, S.E. 1986. The impact of surfactants on Norfolk Island Pines along Sydney coastal beaches since 1973. Environmental Pollution A 41 , 153-164.

Rogowski, R.S., Schryer, D.R., Cofer, W.R. and Edahl, R.A. 1982. Oxidation of SO_2 by NO_2 and air in an aqueous suspension of carbon. In: Heterogeneous Atmospheric Chemistry. D.R. Schryer (ed.) pp 174-177. American Geophysical Union, Monograph 26, Washington D.C.

Scherbatskoy, T. and Klein, R.M. 1983. Response of Spruce and Birch Foliage to leaching by Acidic Mists. J. Environ. Qual. 12 , 189-195.

Schönherr, J. and Bukovac, M.J. 1972. Penetration of stomata by liquids. Plant Physiology 49 , 813-819.

Skiba, U., Peirson-Smith, T.J. and Cresser, M.S. 1986. Effects of simulated precipitation acidified with sulphuric and/or nitric acid on the throughfall chemistry of Sitka spruce Picea sitchensis and heather Calluna vulgaris. Environ. Pollution B. 11, 255-270.

Tukey, H.B. 1970. The leaching of substances from plants. Ann. Review Plant Physiology 21, 305-324.

Unsworth, M.H. 1984. Evaporation from forests in cloud enhances the effects of acid deposition. Nature 312, 262-264.

Warburton, F.L. 1963. The effect of structure on waterproofing. In: Waterproofing and Water-Repellency, J.L. Moillet (ed.) pp 24-51. Elsevier Scientific, Amsterdam.

Weschler, C.J. and Graedel, T.E. 1982. Theoretical limitations on heterogeneous catalysis by transition metals in aqueous atmospheric aerosols. In: Heterogeneous Atmospheric Chemistry, D.R. Schryer (ed.) pp 196-202. American Geophysical Union, Monograph 26, Washington D.C.

CHEMICAL AND PHYSICAL PROCESSES IN ACID RAIN DROPS ON LEAF SURFACES

B Richter Larsen
RISO National Laboratory
Department of Energy Technology
4000 Roskilde
Denmark

ABSTRACT. The results of research on factors directly influencing the impact of acid rain on plants are reviewed. Foliage wettability determines the retained amount and the contact area of rain on leaf surfaces. Contact angles between drops and surfaces have been used as a measure of wettability, but recent studies show that physico-chemical techniques for analysis of surface tensions may provide more detailed information on surface forces complementing microscopy and chemical analysis of leaf cuticles. Results of surface tension studies on agricultural crops related to impacts of acid rain are described.

Leaf surfaces may neutralise acid rain drops. The processes involved have been investigated by many workers and include cation exchange, leaching, dissolution of surface deposits and evaporation of rain water. The state of the art is reviewed.

A modern approach to the acid rain problem is to consider acid rain as one among many plant stresses. The results of recent experiments with quantification of plant stress are summarized, together with the possible role of stress in changing leaf surface characteristics.

1. INTRODUCTION

Any input from the atmosphere to a terrestrial ecosystem may reach a plant by two routes: directly through the shoots or indirectly via the soil through the roots. In forest ecosystems, the impact of acidic precipitation is believed to occur mainly as soil acidification, leaching of nutrients from the forest floor, and mobilisation of toxic compounds (Ulrich and Pankrath, 1983). In agricultural systems, with cultivated soils, the influence of acid rain on plants occurs via the leaf surface. Simulated rain has been shown to cause leaf damage at a pH below 3.5. However, only a few examples have been given of leaf damage from ambient rain, although pH occasionally may be lower than 3 (Evans, 1984). Accelerated leaching of the foliar cations Ca^{++}, K^+ and Mg^{++} after exposure to acid rain has been demonstrated (Wood and Bormann, 1975; Fairfax and Lepp, 1975; Evans et al., 1981; Scherbatskoy

651

M. H. Unsworth and D. Fowler (eds.), Acid Deposition at High Elevation Sites, 651–666.
© 1988 by Kluwer Academic Publishers.

and Klein, 1983; Adams and Hutchinson, 1984), but other chemical substances may also be leached by acid rain, e.g. Na^+, Mn^{++}, trace metals, sulphate, ammonia, nitrate, organic acids, radionuclides (Myttenaere et al., 1980; Hoffmann et al., 1980; Cronan and Reiners, 1983; Lovett and Lindberg, 1984; Evans et al., 1985). It is conceivable that the leaching of substances from plant leaves may cause reductions in productivity, especially in connection with specific nutrient deficiency (Amthor, 1986). However, only a few investigations report effects of acid rain on tissue concentrations of nutrients (Proctor, 1983; Larsen, 1986a; 1986d; 1986e).

The past one or two decades of acid rain research have been carried out with emphasis on growth and reproduction. In recent years attention has been drawn to interaction of acid rain drops with leaf surfaces. The ability of foliage to neutralise acidity has been investigated in vitro (Scholz and Reck, 1977; Czuchajowska and Przybylski, 1978; Sidhu and Zakrevsky, 1982; Polypec and Redmann, 1984; Craker and Bernstein, 1984) or in vivo (Lepp and Dickenson, 1976; Adams and Hutchinson, 1984; Evans et al., 1985; Hutchinson et al., 1986; Larsen, 1986b).

In theory, the extent of interaction between plant foliage and acid rain depends upon the area of leaves in contact with rain water. Many surface characteristics are important in determining the wettability of leaves, e.g. chemical groups exposed at the surface of leaf wax and the surface roughness, governed by the chemical composition and the morphology of the epicuticular wax (Holloway, 1969; 1970). The contact angle of the water drops at leaf surfaces, and the water holding capacity, have been used as measure of wettability in pesticide technology (Cutler et al., 1982), and, recently in acid rain research have been shown between correlations of foliar injury from acid rain and leaf wettability (Keever and Jacobson, 1983; Haines et al., 1985). Research is in progress on the possible changes in surface characteristics of crop and tree species upon acid rain exposure (Baker and Hunt, pers. comm.), and on the comparative abilities of leaf surfaces to neutralise acid rain drops as influenced by wettability, leaf age, rain duration, surface deposits, and calcium nutrition (Adams and Hutchinson, 1986; Hutchinson and Adams, 1986).

Modern physico-chemical methods have been developed to determine surface tensions arising from polar and non-polar forces in a surface (Chattoraj and Birdi, 1984). These methods have been applied in a study of neutralisation of simulated acid rain by leaf surfaces (Larsen, 1986b), and in studies of the change of leaf surfaces following acid rain exposure (Birdi et al., 1986; Larsen, 1986e).

The present paper reviews the recent advances in the research on chemical and physical processes in rain drops on leaf surfaces.

2. CHEMICAL AND PHYSICAL PROPERTIES OF RAIN DROPS

As a consequence of varying weather conditions, the physical properties of rain drops are subject to great changes, even in localised geographical areas. Terminal drop velocity and drop size are important

factors in determining the amount of water retained on leaves
(Jorgensen, 1984). In order to be retained at a surface the kinetic
energy of any rain drop must be converted to various other forms of
energy. Some is absorbed by the plant as kinetic energy and
subsequently converted to heat, some is converted to heat as a result
of the inelastic collision, and some is lost to the process of dividing
into smaller droplets. In principle, if the drop still possesses
enough energy to overcome the work of adhesion it will bounce off the
surface. For the topic of the present paper, the interesting point is
the role of adhesion, which is a factor likely to be affected by air
pollution.

It has recently been shown that the retention of drops may be
predicted from knowledge of surface tension, size and terminal velocity
of the drops (Jorgensen, 1984). The ratio between the work of
adhesion, W_{adh}, and the kinetic energy, E_K, can be written as

$$\frac{W_{adh}}{E_K} = \frac{\gamma_L(1+COS\Theta)}{2/3.r.v^2.\rho} \left(\frac{4sin^4\Theta}{2-3COS\Theta+COS^3\Theta}\right)2/3 \qquad (1)$$

where γ_L denotes the surface tension of the liquid, Θ the advancing
contact angle between leaf and drop, r the drop radius before contact
with the leaf, v the terminal velocity, and ρ the density of the
liquid. It has been shown that, when this ratio exceeds 0.5, drops are
most likely to be retained, whereas almost no drops are retained when
the ratio is below 0.1 (Jorgensen, 1984).

This information clearly shows that, in the selection of
experimental parameters for simulated acid rain exposure experiments,
great attention should be paid to the rain drop velocity, which
influences the retention quadratically, while the drop size only
influences linearly. The contact angle is determined solely by surface
forces in the liquid and in the solid, which will be discussed in
section 3.

Although the chemical composition of ambient rain may vary
considerably with the degree of air pollution, none of the typical
chemical components have significant surfactant properties at ambient
concentration levels. Thus, the surface tension, γ_L, of acid rain is
practically constant. Recent investigations by the author
(unpublished) have confirmed this. Haines et al. (1985) have observed
that the wettability of leaves was independent of the rain pH in the
range of 2 to 5. These observations agree with a constant γ_L of acid
rain.

A recent interesting study, although not directly relevant to the
acid rain problem, has shown that surfactants coming from coastal
sewage discharges near Sydney, Australia may mediate foliar uptake of
Na^+ and Cl^-. This has resulted in forest decline due to salt-burning
from wind-borne seaspray (Moodie et al., 1986). Surfactants reduce the
surface tension of sea water. This may reduce the drop size due to the
presence of lower surface forces to keep the droplet water together.
Thus the terminal velocity may be affected. As will be shown in
section 3 a reduced γ_L means an increased wettability of the surface.

In order to be able to evaluate the relative importance of changes in the surface tension of the liquid (water) and the solid (leaf) the analytical expression of the work of adhesion is derived in section 3.

3. PHYSICAL PROPERTIES OF THE LEAF SURFACE

The surface tension of a liquid, γ_L, and of a solid, γ_S, arise from polar forces (P) (hydrogen bonds, dipole-dipole interactions, dipole-induced interactions, π-bonds, donor-acceptor bonds, and electrostatic interactions) and disperson forces (D). Thus,

$$\gamma_L = \gamma_L^D + \gamma_L^P$$

and

$$\gamma_S = \gamma_S^D + \gamma_S^P \tag{3}$$

$$(2)$$

The Young-Fowkes equation provides the relation between the surface angle, Θ, and the various surface tensions (Chattoraj and Birdi, 1984):

$$\cos\Theta = -1 + \frac{2(\gamma_L^D \cdot \gamma_S^D)^{1/2}}{\gamma_L} + \frac{2(\gamma_L^P \cdot \gamma_S^P)^{1/2}}{\gamma_L} \tag{4}$$

For the sake of simplicity we disregard the polar forces for a moment*. Hence,

$$\cos\Theta = -1 + 2 \cdot \left(\frac{\gamma_S}{\gamma_L}\right)^{1/2} \tag{5}$$

The work adhesion can according to Chattoraj and Birdi (1984) be described by:

$$W_{adh} = 2(\gamma_L \cdot \gamma_S)^{1/2} \cdot A_r \tag{6}$$

where A_r is the contact area between drop and leaf.

Assuming that a rain drop at a leaf surface has the shape of a spherical cap with the radius, r, the contact area, A_r, can be expressed as,

$$A_r = \pi \cdot \left(\frac{1}{4r^3}(2 - 3\cos\Theta + \cos^3\Theta)\right)^{-2/3} \cdot (1 - \cos^2\Theta) \tag{7}$$

* It can be shown that qualitatively the same results are obtained when both polar and dispersion forces are considered. The proof goes beyond the scope of this paper.

The author has applied this technique on the study of surface tensions of leaves. The results are summarised in Table I. In general γ_S^D was found to be rather constant while γ_S^P was highly variable among species, in good agreement with previously reported variations in chemical compositions of cuticular wax (Cutler et al., 1982).

Table I. Surface tensions* of adaxial leaf surfaces of unexposed plants (in mN m^{-1}).

Surface		γ_S^D	γ_S^P	γ_S
Teflon	(control)	17–19	0	17–19
Cress,	3 weeks	25–26	9–10	35–36
Lettuce,	3–6 weeks	30–34	3– 5	34–40
Mustard,	3–6 weeks	19–32	9–31	33–51
Bean,	3–6 weeks	20–34	17–28	46–57

* Values are the range of observations in published and unpublished work by the author (Larsen, 1986b; 1986e; Birdi et al., 1986).

The ability of plant leaves to neutralise acid rain drops has been demonstrated to be favoured by high γ_S^P (Larsen, 1986b). Whether this is caused by increased contents of neutralising polar chemical components in the surface or merely by the increased contact area with increasing γ_S^P and thereby γ_S remains unexamined.

Plant leaves consist of living tissue; thus, surface characteristics may change during a life cycle. The cuticular wax chemistry changes as leaves develop. This is reflected by the wettability of the leaves, which typically is reduced during leaf expansion and increased again at the senescing stage (Cutler et al., 1982; Bukovac et al., 1979; Keever and Jacobson, 1983). Great differences are also found between true leaves and cotyledons, the latter being much more wettable. This has been shown to be reflected by the greater interaction of cotyledons with acid rain compared to true leaves (Jacobson et al., 1985; Adams and Hutchinson, 1986; Caporn and Hutchison, 1986). The author has examined the surface tensions of mustard leaves at different physiological ages. The results are shown in Table II.

The data in Table II clearly show that γ_S^D is rather constant, whereas γ_S^P varies. The pattern of variation is consistent with the above- mentioned observations on wettability and interaction with acid rain. Values of γ_S^P and thereby γ_S, decrease as the leaves mature and markedly increase when senescence occurs.

The surface tension of leaves may also change as the plants are exposed to air pollution or acid rain. Recent information shows that needles of trees growing in polluted atmospheres become more wettable with smaller contact angles than needles from unpolluted sites. These

656

Combining equations (5), (6), and (7) one gets,

$$W_{adh} = 8\pi r^2 \cdot \gamma_S \left(1 - \left(\frac{\gamma_S}{\gamma_L}\right)^{1/2}\right) \cdot \left(2\left(\frac{\gamma_S}{\gamma_L}\right)^{3/2} - 3 \cdot \frac{\gamma_S}{\gamma_L} + 1\right)^{-2/3} \tag{8}$$

For an arbitrary value of r and varying values of γ_S and γ_L the work of adhesion is plotted in Figure 1. It is seen that this work increases markedly as γ_S approached γ_L, which means that the wettability increases. Most leaves have γ_S in the range of 30 to 50 mN m^{-1}. Thus it appears that changes in γ_L from 72 to 60 mN m^{-1}, simulating increased pollution of rain water, have only small effects on the wettability. However, leaves with higher γ_S, such as cotyledons or senescing leaves, are more affected by decreased γ_L of the rain.

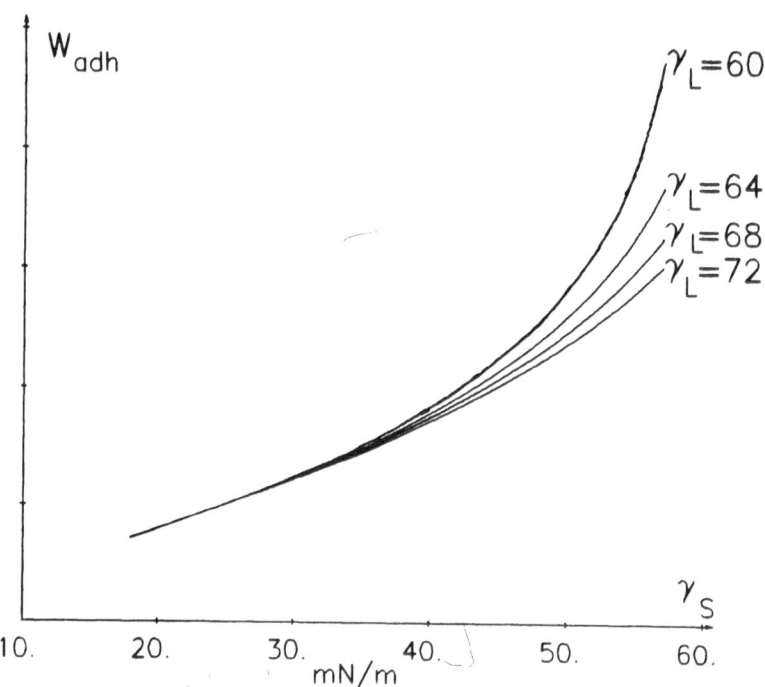

Figure 1. The dependence of the work of adhesion (arbitrary units) on γ_L and γ_S in Equation 8.

Recent work by Chattoraj and Birdi (1984) has demonstrated the use of Young-Fowkes equation (4) for the determination of surface tensions of a solid. The method is briefly described here.
    Several techniques exist for the measurements of γ_L, γ_L^D and subsequently γ_L^P of liquids. Data for mixtures of water and alcohols, for example, have been tabulated (Chattoraj and Birdi, 1984). Droplets of such mixtures can be placed on the surface under investigation and contact angles measured. For surfaces with non-polar forces, plots of cos vs. $(\gamma_L^D)^{1/2} \cdot \gamma_L^{-1}$ for a series of mixtures produce straight lines

with origin at $\cos\Theta = -1$ and slopes of $2.(\gamma_S^D)^{1/2}$. However, for surfaces with polar forces, the lines intercept with the ordinate at $-1 < \cos\Theta < 1$. In these cases the value of γ_S^D can be derived from the intercept with $\cos\Theta = 1$, where $\gamma_S^D = ((\gamma_L^D)^{1/2}.\gamma_L^1)^{-2}$ (see Figure 2). After calculating γ_S^D the second term of the Young-Fowkes equation (4), $2.(\gamma_S^P.\gamma_L^P)^{1/2}$ can be estimated. These data can be replotted as $2(\gamma_S^P.\gamma_L^P)^{1/2}$ vs. $(\gamma_L^P)^{1/2}$ and from the slope of the lines γ_S^P can be determined (see Figure 3).

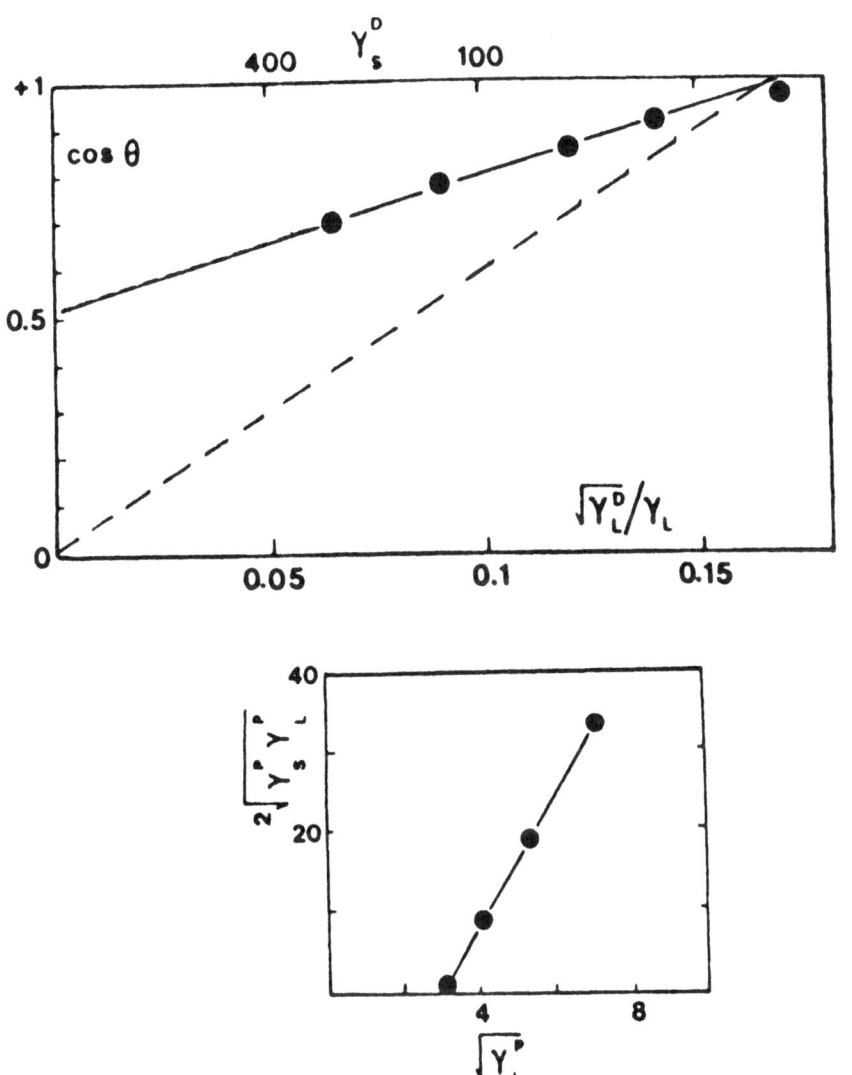

Figures 2 and 3. Examples of plots from which γ_S^D (Fig. 2, left) and (Fig. 3, right) can be determined. Reconstructed from Chattoraj and Birdi (1984).

Table II. Surface tensions* of leaves at different physiological age (in mN m^{-1}).

Surface	γ_S^D	γ_S^P	γ_S
2 week old mustard			
Leaf No. 1, expanding	34	11	45
Cotyledon, ceased	37	9	46
3 week old mustard			
Leaf No. 1, ceased	31	9	40
Cotyledon, senescent	32	38	70
6 week old mustard			
Leaf No. 6, just unfolding	27	24	51
Leaf No. 5, expanding	30	19	49
Leaf No. 4, just expanded	29	16	45
Leaf No. 2, ceased	30	6	36
Leaf No. 1, senescent	28	19	47

* Each value represents the mean of 4–8 contact angle measurements on a series of four water-propanol mixtures. Standard deveations of the angle data were 4–6% of the means.

changes are associated with different chemical composition of epicuticular wax (Cape, 1983; 1986). Baker and Hunt have demonstrated similar effects on leaves of a variety of crop and tree species after simulated acid rain exposure. They concluded that both direct erosion of wax (Baker and Hunt, 1986) and indirect pH-induced stress mechanisms altering the wax synthesis (pers. comm.) may be responsible for the changes in the surfaces. These findings agree with observations by Birdi et al. (1986) and Larsen (1986e), who demonstrated significant increases in γ_S^P of various crop species after two weeks exposure to simulated acid rain. The results are summarised together with new data in Figure 4.

The effect of acid rain on γ_S^P seems to vary among species. The most affected species, bean and mustard, have been shown to have a high buffering ability towards acid rain, whereas the least affected species, lettuce and cress, have a low buffering ability (Larsen, 1986b). These observations, together with studies of susceptibility to acid rain of different species, point to the following hypothesis (Larsen, 1985; 1986a; 1986d; 1986e): Plants are protected from acid rain damage in two different ways. One category of species has a hydrophobic, water repellent surface with low water-holding capacity.

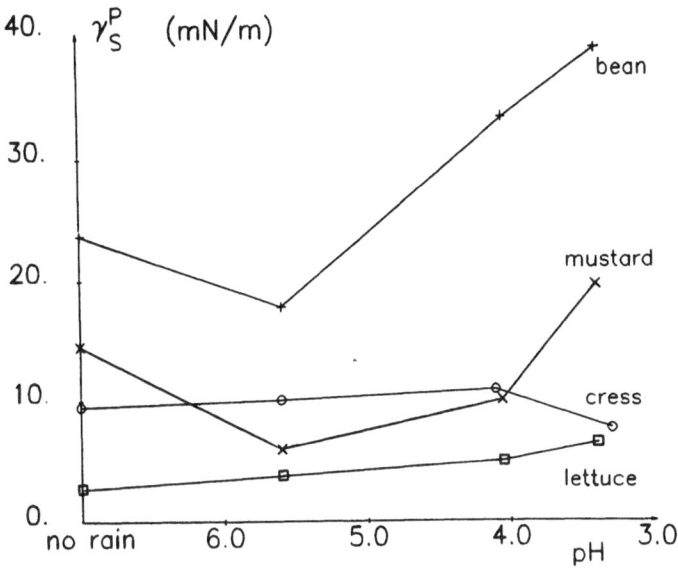

Figure 4. Changes in γ_S^P of leaf surfaces of various plant species
exposed for two weeks to acid rain. The plotted points
represent means of published and unpublished data (Larsen,
1986e; Birdi et al., 1986).

Rain drops bounce off such plants easily due to their low surface
tension, γ_S. The high contact angles ensure a small area of leaves
in contact with the rain, whereby the interaction with the rain is
minimised. However, the neutralisation of acid is also minimised,
which may lead to extensive tissue damage in the small spots of
contact. The typical damage picture for such species is 'pitted'
necrosis, but low overall effects on the growth. Leaves from the other
category of species are easily wetted, and a high degree of interaction
with acid rain is possible. From these plants chemical substances may
leach to rain drops, which, on the one hand, neutralise the acid, but,
on the other hand use important nutrients and resources from the plant
(Amthor, 1986). This type of species typically develops non-localised
visible damage, and is more susceptible to acid rain regarding growth
effects.

An interesting observation in Figure 4 is that some plants become
more hydrophobic from exposure to acid rain of moderate pH compared to
no rain exposure. Baker and Hunt have observed the same phenomenon at
even lower pH (pers. comm.). From this, it may be concluded that
irrigation of plants with rain of low or moderate acidity stimulates
the growth and development of cuticular wax. Stimulative responses to
acid rain down to pH 4.1 have also been reported by the author (Larsen,
1986d; 1986e). This phenomenon may also be explained by a rinsing

effect of rain. Hutchinson and Adams (1986) have recently demonstrated that deposited particles are very important for processes in the leaf surface. Removal of these particles may very well reduce the wettability of the leaves.

The species in Figure 4 are glabrous, with smooth cuticles apart from occasional trichomes. Many species have pubescent or glossy leaves with cuticles covered by epicuticular wax crystals of a considerable size. These surfaces are highly hydrophobic and less subject to damage from acid rain (Haines et al., 1985). The author has investigated the effect of acid rain on surface tensions of rape plants, which have glabrous leaves with a glossy surface due to a dense epicuticular wax covering. Under magnification, the surface of rape leaves appear to have a micro-roughness. The roughness of a surface, r, is defined in equation 9 from the apparent contact angle with a water drop, θ_{app} and the theoretical contact angle on a smooth surface of the same material, θ_{smooth} (Holloway, 1970).

$$r = \frac{\cos\theta_{app}}{\cos\theta_{smooth}} \qquad (9)$$

From the knowledge of the chemical composition of the epicuticular wax of rape plants, and from the information published by Holloway (1969) θ_{smooth} was estimated for rape leaves and the roughness calculated. The fraction of the water actually in contact with the rough surface, $f_{L/S}$ and in contact with only air, $f_{A/L}$ was estimated from equation 10 (Holloway, 1970),

$$\cos\theta_{app} = f_{L/S} \cdot \cos\theta_{smooth} - f_{A/L} \qquad (10)$$

The results are shown in Table III.

Table III Micro-roughness* of adaxial lead surfaces of rape plants exposed to acid rain.

Rain pH	θ_{app}	r	$f_{L/S}$	$f_{A/L}$
No rain	146°	3.69	0.21	0.79
pH 5.6	138°	3.30	0.32	0.68
pH 4.1	135°	3.14	0.36	0.64
pH 3.3	132°	2.97	0.40	0.60

* Each value of θ_{app} represents the mean of 4-8 contact angle measurements with distilled water.

It can be seen that the micro-roughness of rape leaf surfaces is reduced by acid rain exposure. This increases the area of leaf in

actual contact with rain and consequently increases the susceptibility
to continued acid rain exposure. The observations are in agreement
with recent observations for the same species by Baker and Hunt (pers.
comm.).

Summarising, it has been shown that the wettability, governed
mainly by polar forces and epicuticular wax morphology, is a decisive
factor in determining the potential interaction of plant leaves with
acid rain. The wettability varies between and within species and is
subject to changes during growth and pollution exposure.

4. CHEMICAL INTERACTION OF DROPS AND LEAVES

When an acid rain drop contacts a leaf chemical processes occur. The
acidity of rain is normally too small to induce direct chemical changes
of the cuticle through, for example, hydrolysis. However, the activity
gradient across the cuticle may initiate passive diffusion of H^+ into
the internal tissue, and diffusion of cations and anions into the rain
water. Adams and Hutchinson (1984) suggested that acid droplet
neutralisation involves exchange reactions on the surface, in which
cations on exchange sites of the cuticle and cell walls are exchanged
for H^+ from the rain. The exchange sites may be the free carboxyl
groups of pectic cell wall matrix substances, or polygalacturonic acids
(Lauchli, 1976). The exchanged sites may be replenished again after
adaxial cations are transported with the evaporation stream. Recent
work has demonstrated the prevalence of trans-cuticular micro-pores of
more rapid ion-transport in several taxa (Miller, 1985). It has been
suggested that the abundance of such micro-channels may control the
different acid rain buffering abilities among species (Larsen, 1986b).
Consistent with this, Brassica is one among many taxa which does not
contain cuticular micro-pores, and which has a low buffering ability
(Larsen, 1986b; Hutchinson et al., 1986; Adams and Hutchinson, 1986).
Together with the leaching (exchanged) cations, organic or inorganic
substances may be transported as counter-ions (malate, citrate,
oxalate, bicarbonate, sulphate, nitrate, or chloride are possible
substances). This would contribute to the neutralisation of rain
drops. Recent investigations of leaf extracts have indicated
reductions in internal tissue pH after acid mist exposure at very low
pH (Bytnerowicz et al., 1986). This observation may place some support
to the suggested mechanisms. It has been speculated that the calcium
status of leaves may determine the different neutralising abilities
among species (Adams and Hutchinson, 1984; Hutchinson et al., 1986).
In a study with plants grown in filtered air, it was clearly
demonstrated that the dissolution of surface particles is of major
importance for the buffering of acid rain, possibly even more important
than exchange of H^+ and leaching of basic substances (Hutchinson and
Adams, 1986).

In the last three years, different authors have performed studies
on the time dependence of pH changes in rain drops on leaves (Adams and
Hutchinson, 1984; 1986; Hutchinson and Adams, 1986; Hutchinson et al.,
1986; Evans et al., 1985; Larsen, 1986b). The results are all in good

agreement. At rain drop pH > 3.0 all plant species were able to increase droplet pH before drying. The effect appeared to be two-phased, with a rapid initial neutralisation probably due to dissolution of surface particles and leaf exudates, or due to collapse of epidermal cells, followed by a slower subsequent phase, probably due to exchange and leaching processes. There seemed to be a positive relation of leaf surface wettability and buffering ability. At lower rain drop pH, most plants were unable to neutralise the acid, and droplet pH decreased as the drops dried out. This evaporation effect has previously been pointed out by other investigators (Unsworth, 1984; Frevert and Klemm, 1984). In species with very hydrophobic surfaces and thus very low buffering abilities (e.g. coniferous trees) the evaporation effect is dominant, and pH may become very low as the drops dry. However in most species it appears that neutralisation can more than balance evaporation.

5. THE ROLE OF PLANT STRESS

Exposure to acid rain may stress plants. Evidence of acid rain-induced plant stress has been given, not only indirectly in the form of growth effects at pH levels where visible injury does not occur (Evans, 1984), but also directly by production of the stress hormone, ethylene (Arny and Pell, 1986; Larsen, 1986a; 1986c). The author has speculated that leaching of K^+ and Ca^{++} from guard cells, causing inexpedient stomatal actions, may lead to disturbances in the evapotranspiration control system, thereby stressing the plants. This hypothesis is based on observations by Evans et al. (1981), demonstrating significant changes in leaf conductivity after acid rain exposure, and on my own observations with water stressed plants exposed to acid rain (Larsen, 1986f).

Several investigators have demonstrated the effect of environmental conditions on surface characteristics of plant leaves (Martin and Juniper, 1970; Whitecross and Armstrong, 1972; Cutler et al., 1982). The interaction of environmental stress and acid rain was first studied by Keever and Jacobson (1983), and later by the author (Larsen, 1986b; 1986c; 1986d; 1986e). The results seem to confirm that plants under environmental stress are more sensitive to acid rain exposure, and that acid rain may increase the symptoms of environmental stress. Both types of stress affect the evapotranspiration control system, both induce stress-ethylene evolution, and both affect the epicuticular wax ultra-structure so that plants become more wettable and thus more susceptible to trace organic chemical pollution (Larsen, 1986c; 1986d; 1986e).

Summarising, recent experiments have drawn attention to aspects of acid rain effects on plants other than visible damage and growth reduction. It has been emphasised that interactions with environmental conditions and other stresses occur. Many of these interactions seem to be mediated by chemical and physical processes at the leaf-drop interface. The state of the art in this field is still not adequate. Hence, future research on acid rain should be carried out in this area.

ACKNOWLEDGEMENTS

The author wishes to thank Mr Einar Danielsen for valuable computer
assistance and Dr Peter Fynbo for mathematical support.

REFERENCES

Adams, C.M. and Hutchinson, T.C. 1984. A comparison of the ability
    of leaf surfaces of three species to neutralise acidic rain
    drops. New Phytologist 97, 463-478.
Adams, C.M. and Hutchinson, T.C. 1986. Comparative abilities of leaf
    surfaces to neutralise acidic raindrops. II. The influence of
    leaf wettability, leaf age and rain duration on changes in droplet
    pH and chemistry on leaf surfaces. New Phytologist (in press).
Amthor, J.S. 1986. An estimate of the 'cost' of nutrient leaching from
    forest canopies by rain. New Phytologist 102, 359-364.
Arny, C.J. and Pell, E.J. 1986. Ethylene production by potato, radish
    and soybean leaf tissue treated with simulated acid rain.
    Environmental and Experimental Botany 26, 9-15.
Baker, E.A. and Hunt, G.M. 1986. Erosion of waxes from leaf surfaces
    by simulated rain. New Phytologist 102, 161-173.
Birdi, K.S., Larsen, B.R. and Sanches, R. 1986. Effects of simulated
    acid rain on the surface tension of selected leaves. Langmuir
    (submitted, January 1986).
Bukovac, M.J., Flore, J.A. and Baker, E.A. 1979. Peach leaf surfaces:
    Changes in wettability, retention, cuticular permeability and
    epicuticular wax chemistry during expansion with special reference
    to spray application. Journal of the American Society for
    Horticultural Science 104, 611-617.
Bythnerowicz, A., Temple, P.J. and Taylor (1986). Effects of
    simulated acid fog on leaf acidification and injury development of
    into beans. Canadian Journal of Botany 64, 918-922.
Cape, J.N. 1983. Contact angles of water droplets on needles of Scots
    pine (Pinus sylvestris) growing in polluted atmospheres. New
    Phytologist 93, 293-299.
Cape, J.N. 1986. Effects of air pollution on the chemistry of surface
    waxes of Scots pine. Water, Air and Soil Pollution 31, 393-399.
Caporn, S. and Hutchinson, T.C. 1986. The contrast in response to
    simulated acid rain of leaves and cotyledons of cabbage (Brassica
    oleracea L.). Water, Air and Soil Pollution (in press).
Chattoraj, D.K. and Birdi, K.S. 1984. Adsorption and the Gibbs Surface
    Excess. Plenum Press, New York.
Craker, L.E. and Bernstein, D. 1984. Buffering of acid rain by leaf
    tissue of selected crop plants. Environmental Pollution (A) 36,
    375-381.
Cronan, C.S. and Reiners, W.A. 1983. Canopy processing of acidic
    precipitation by coniferous and hardwood forests in England.
    Oecologia 59, 216-223.
Cutler, D.F., Alvin, K.F. and Price, C.E. (eds.) 1982. The Plant
    Cuticle. Academic Press, London.

Czuchajowska, Z. and Przybylski, T. 1978. The seasonal changes of acidity and buffer capacity of aqueous homogenates of Pinus silvestris needles and the influence of zinc-plant immissions. Bull. Acad. Pol. Sci. **26** , 361-368.

Evans, L.S. 1984. Acidic precipitation effects on terrestrial vegetation. Annual Review of Phytopathlogy 22 , 397-420.

Evans, L.S., Curry, T.M. and Lewin, K.F. 1981. Responses of Phaseolus vulgaris to simulated acidic rain. New Phytologist **88** , 403-420.

Evans, L.S., Santucci, K.A. and Patti, M.J. 1985. Interactions of simulated rain solutions and leaves of Phaseolus vulgaris. Environmental and Experimental Botany **25** , 31-40.

Fairfax, J.A.W. and Lepp, N.W. 1975. Effects of simulated acid rain on cation loss from leaves. Nature, London **255** , 324-325.

Frevert, T. and Klemm, O. 1984. Wie anderen sich pH-Werte im Regen und Nebelwasser beim Abtrocknen auf Pflanzenoberflachen? Archives for Meteorology, Geophysics and Bioclimatology., Series B **34** , 75-81.

Haines, B.L., Jernstedt, J.A. and Neufeld, H.S. 1985. Direct foliar effects of simulated acid rain. II. Leaf surface charateristics. New Phytologist **99** , 407-416.

Hoffman, W.A., Lindberg, S.W. and Turner, R.R. 1980. Precipitation acidity: the role of the forest canopy in acid exchange. Journal of Environmental Quality **9** , 95-100.

Holloway, P.J. 1969. Chemistry of leaf waxes in relation to wetting. Journal of the Science of Food and Agriculture **20** , 124-128.

Holloway, P.J. 1970. Surface factors affecting the wetting of leaves. Pesticide Science **1** , 156-163.

Hutchinson, T.C. and Adams, C.M. 1986. Comparative abililties of leaf surfaces to neutralise acidic raindrops. I. The influence of calcium nutrition and filtered air on leaf response. New Phytologist (in press).

Hutchinson, T.C., Adams, C.M. and Gaber, B. 1986. Neutralisation of acidic rain drops on leaves of agricultural crop and boral forest species. Water, Air and Soil Pollution (in press).

Jacobson, J.S., Troiand, J. and Heller, L. 1985. Stage of development response and recovery of radish plants from episodic exposure to simulated acidic rain. Journal of Experimental Botany **36** , 159-167.

Jorgensen, L. 1984. Retention in relation to drop-velocity, drop-size and surface tension (Danish). 1. Danske plantevaernskonference. Ukrudt. Statens Planteavlsforsog, Plantevaerncentret. Institut for Ukrudtsbekaempelse, pp. 196-214.

Keever, G.J. and Jacobson, J.S. 1983. Response of Glycine max L. cv Merrill to simulated acid rain. I. Environmental and morphological influences on the foliar leaching of [86]Rb. Field Crop Research **6** , 241-250.

Larsen, B.R. 1985. Effects of simulated acid rain and (+)-2-(2, 4-dichlorophenoxy) propanoic acid on selected crops. Ecotoxicology and Environmental Safety **10** , 228-238.

Larsen, B.R. 1986a. Synergistic effects of acid rain and (+)-2-(2, 4-dichlorophenoxy) propanoic acid (2, 4-DP) on growth, mineral content and stress-induced ethylene in lettuce. <u>Environmental Pollution (A)</u> **41** , 179-196.

Larsen, B.R. 1986b. <u>In vivo</u> buffering and concentration of acid rain drops on leaves of selected crop plants. <u>Water, Air and Soil Pollution</u> (in press).

Larsen, B.R. 1986c. Hydrocarbon gases produced in plants indicate interaction of acid rain and phytotoxic trace compounds. <u>Ecotoxicology and Environmental Safety</u> (in press).

Larsen, B.R. 1986d. Combined effects of simulated acid rain and Chlorsulfuron or pentachlorophenol in <u>Sinapis abla</u> L. (in prep.).

Larsen, B.R. 1986e. Simulated acid rain increases surface tension of leaves and effects of Chlorsulfuron in <u>Phaseolus vulgaris</u> L. <u>Environmental and Experimental Botany</u> (submitted).

Larsen, B.R. 1986f. Effects of acid rain and organic chemical pollution on crops. Ph.D. Thesis. The Technical University of Denmark.

Lauchli, A. 1987. Apoplasmic transport in tissues. In: <u>Transport in Plants II.</u> part B. <u>Encyclopedia of Plant Physiology 2.</u> (Ed. Luttge, U. and Pitman, M.G.), pp. 3-29. Springer-Verlag, New York.

Lepp, N.W. and Dickenson, N.M. 1976. <u>The pH of leaf surfaces and its modification by atmospheric pollution.</u> Proc. Kupio Meeting on plant damages caused by air pollution. Kupio.

Lovett, G.M. and Lindberg, S.E. 1984. Dry deposition and canopy exchange in a mixed oak forest as determined by analysis of throughfall. <u>Journal of Applied Ecology</u> **21** , 1013-1027.

Martin, J.T. and Juniper, B.E. 1971. <u>The Cuticles of Plants.</u> Edward Arnold Publishers Ltd., London.

Miller, R.M. 1985. The prevalance of pores and chanals in leaf cuticular membranes. <u>Annals of Botany</u> **55** , 459-471.

Moodie, E.G., Stewart, R.T. and Bowen, S.E. 1986. The impact of surfactants on Norfolk Island Pines along Sydney coastal beaches since 1973. <u>Environmental Pollution (A)</u> **41** , 153-164.

Myttenaere, L, Daoust, C. and Roucoux, P. 1980. Leaching of Technicum from foliage by simulated rain. <u>Environmental and Experimental Botany</u> **20** , 415-419.

Proctor, J.T.A. 1983. Effects of simulated sulphuric acid rain on apple tree foliage nutrient content, yield and fruit quality. <u>Environmental and Experimental Botany</u> **23** , 167-174.

Polypec, B. and Redmann, R.E. 1984. Acid-buffering capacity of foliage from boral forest species. <u>Canadian Journal of Botany</u> **62** , 2650-2653.

Scherbatskoy, T. and Klein, R.M. 1983. Response of spruce and birch foliage to leaching by acidic mists. <u>Journal of Environmental Quality</u> **12** , 263-270.

Scholz, F. and Reck, P. 1977. Effects of acids on forest trees as measured by titration <u>in vitro</u>, inheritance of buffering capacity in <u>Picea abies</u>. <u>Water, Air and Soil Pollution</u> **8** , 41-45.

666

Sidhu, S.S. and Zakrevsky, G.J. 1982. A standard method for determining buffering capacity of plant foliage. Plant and Soil 66 , 173-179.

Ulrich, B. and Pankrath, J. 1983. Effects of accumulation of air pollutants in forest ecosystems. Reidel, London.

Unsworth, M.H. 1984. Evaporation from forests in cloud enhances the effect of acid deposition. Nature 312 , 262-264.

Whitecross, M.I. and Armstrong, D.J. 1972. Environmental effects on epicuticular waxes of Brassica napus L. Australian Journal of Botany 20 , 87-95.

Wood, T. and Bormann, F.H. 1975. Increases in foliar leaching caused by acidification of an artificial mist. Ambio 4 , 169-171.